Lecture Notes in Computer Science 4948

Commenced Publication in 1973
Founding and Former Series Editors:
Gerhard Goos, Juris Hartmanis, and Jan van Leeuwen

Editorial Board

David Hutchison
  Lancaster University, UK
Takeo Kanade
  Carnegie Mellon University, Pittsburgh, PA, USA
Josef Kittler
  University of Surrey, Guildford, UK
Jon M. Kleinberg
  Cornell University, Ithaca, NY, USA
Alfred Kobsa
  University of California, Irvine, CA, USA
Friedemann Mattern
  ETH Zurich, Switzerland
John C. Mitchell
  Stanford University, CA, USA
Moni Naor
  Weizmann Institute of Science, Rehovot, Israel
Oscar Nierstrasz
  University of Bern, Switzerland
C. Pandu Rangan
  Indian Institute of Technology, Madras, India
Bernhard Steffen
  University of Dortmund, Germany
Madhu Sudan
  Massachusetts Institute of Technology, MA, USA
Demetri Terzopoulos
  University of California, Los Angeles, CA, USA
Doug Tygar
  University of California, Berkeley, CA, USA
Gerhard Weikum
  Max-Planck Institute of Computer Science, Saarbruecken, Germany

Ran Canetti (Ed.)

Theory of Cryptography

Fifth Theory of Cryptography Conference, TCC 2008
New York, USA, March 19-21, 2008
Proceedings

 Springer

Volume Editor

Ran Canetti
IBM T.J. Watson Research Center
NY, USA
E-mail: canetti@us.ibm.com

Library of Congress Control Number: 2008921565

CR Subject Classification (1998): E.3, F.2.1-2, C.2.0, G, D.4.6, K.4.1, K.4.3, K.6.5

LNCS Sublibrary: SL 4 – Security and Cryptology

ISSN	0302-9743
ISBN-10	3-540-78523-X Springer Berlin Heidelberg New York
ISBN-13	978-3-540-78523-1 Springer Berlin Heidelberg New York

This work is subject to copyright. All rights are reserved, whether the whole or part of the material is concerned, specifically the rights of translation, reprinting, re-use of illustrations, recitation, broadcasting, reproduction on microfilms or in any other way, and storage in data banks. Duplication of this publication or parts thereof is permitted only under the provisions of the German Copyright Law of September 9, 1965, in its current version, and permission for use must always be obtained from Springer. Violations are liable to prosecution under the German Copyright Law.

Springer is a part of Springer Science+Business Media

springer.com

©International Association for Cryptologic Research 2008

Typesetting: Camera-ready by author, data conversion by Scientific Publishing Services, Chennai, India
Printed on acid-free paper   SPIN: 12237708   06/3180   5 4 3 2 1 0

Preface

TCC 2008, the 5th Theory of Cryptography Conference, was held in New York, New York, March 19–21, 2008, at New York University. TCC 2008 was sponsored by the International Association for Cryptologic Research (IACR) and was organized in cooperation with the Department of Computer Science at New York University and the Courant Institute for Mathematical Sciences. The General Chairs of the conference were Yevgeniy Dodis and Victor Shoup.

The conference received 81 submissions, of which the Program Committee selected 34 for presentation at the conference. The authors of two papers then decided to merge their papers, resulting in a total of 33 presented papers. The Best Student Paper Award was given to Paul Valiant for his paper "Incrementally Verifiable Computation or Knowledge Implies Time/Space Efficiency." These proceedings consist of revised versions of the presented papers. The revisions were not reviewed. The authors bear full responsibility for the contents of their papers.

The conference program also included four special events: an invited talk entitled "Randomness Extractors and Their Cryptographic Applications" by Salil Vadhan; a tutorial entitled "Bridging Cryptography and Game Theory: Recent Results and Future Directions," given by Jonathan Katz (with an accompanying tutorial in the proceedings); a panel discussion on "Game Theory and Cryptography: Towards a Joint Point of View?" with Tal Rabin as moderator and Jonathan Katz, Silvio Micali, and Moni Naor as panelists; and a Rump Session chaired by Anna Lysyanskaya.

In spite of the relatively small number of submissions, many of them were of high quality. Consequently, the selection process was challenging and very competitive. Indeed, a number of good papers were not accepted due to lack of space in the program. The main considerations in selecting the program were conceptual and technical innovation, quality of presentation, and relevance to the theory of cryptography. An attempt was made to maintain the unique character of the conference as a stage for presenting innovative work on the foundations of cryptography.

I would like to thank the TCC Steering Committee for entrusting me with the role of Program Committee Chair. In the few years since its inception, TCC has been tremendously successful in attracting high-quality papers and in providing a home and identity for the theory of cryptography community. I am honored to have had the opportunity to contribute to the continuation of this success.

Special thanks are due to the Program Committee members, who have dedicated so much time and effort to provide a thorough and in-depth review of the submissions, with high standards of professional integrity. I also thank the many external reviewers who assisted the Program Committee in its work. Most importantly, I thank the authors of submitted papers for their contributions; these papers are, after all, the only reason for TCC to exist.

I am grateful to the General Chairs, Yevgeniy Dodis and Victor Shoup, and their assistant Anna Mackay for their invaluable work in making the conference happen. Another special thanks is due to Shai Halevi for writing the software that greatly facilitated the committee work, and for his responsiveness in attending to our whims.

I thank our corporate sponsors, the D. E. Shaw group, IBM, and Microsoft for their generous sponsorship of the conference, and Cynthia Dwork, Rosario Gennaro, Jonah Kolb, Christine Mathias, and Tal Rabin for their help in obtaining the sponsorships.

Finally, I appreciate the assistance provided by the Springer LNCS editorial staff, including Alfred Hofmann, Frank Holzwarth, and Anna Kramer, in assembling these proceedings.

January 2008   Ran Canetti
                                              TCC' 2008 Program Chair

TCC 2008

5th Theory of Cryptography Conference

New York University, New York, New York, USA
March 18–21, 2008

General Chairs

Yevgeniy Dodis and Victor Shoup, New York University

Sponsoring Institutions

The D. E. Shaw group
IBM Inc.
Microsoft Inc.

Program Committee

Boaz Barak	Princeton University
Ran Canetti	IBM Research (Chair)
Yevgeniy Dodis	New York University
Marc Fischlin	Darmstadt University of Technology
Jens Groth	University College, London
Dennis Hofheinz	Centrum voor Wiskunde en Informatica (CWI)
Susan Hohenberger	Johns Hopkins University
Russell Impagliazzo	University of California San Diego and the Institute of Advanced Study
Eyal Kushilevitz	Technion
Yehuda Lindell	Bar Ilan University
Ueli Maurer	ETH Zürich
Moni Naor	Weizmann Institute
Tatsuaki Okamoto	NTT Laboratories
Adriana Palacio	Bowdoin College
Christopher J. Peikert	Stanford Research Institute
Leonid Reyzin	Boston University
abhi shelat	University of Virginia
Hoeteck Wee	University of California, Berkeley and Columbia University

TCC Steering Committee

Mihir Bellare	University of California, San Diego
Ivan Damgård	University of Aarhus
Oded Goldreich (Chair)	Weizmann Institute of Science
Shafi Goldwasser	MIT and Weizmann Institute of Science
Johan Håstad	Royal Institute of Technology
Russell Impagliazzo	University of California, San Diego
Ueli Maurer	ETH Zürich
Silvio Micali	Massachusetts Institute of Technology
Moni Naor	Weizmann Institute of Science
Tatsuaki Okamoto	NTT Laboratories

External Reviewers

Michel Abdalla	Chiu Yuen Koo	Amit Sahai
Joel Alwen	Hugo Krawczyk	Louis Salvail
Benny Applebaum	Anja Lehmann	Palash Sarkar
Giuseppe Ateniese	Richard Lindner	Dominique Schroeder
Zuzana Beerliova	Moses Liskov	Gil Segev
Amos Beimel	Vadim Lyubashevsky	Sarah Shoup
Andrej Bogdanov	Joshua Mason	Thomas Shrimpton
Seung Geol Choi	Alexander May	Adam Smith
Sherman Chow	Anton Mityagin	Stefano Tessaro
Nenad Dedić	Kirill Morozov	Luca Trevisan
Serge Fehr	Joern Mueller-Quade	Dominique Unruh
Matthias Fitzi	Gregory Neven	Salil Vadhan
Eiichiro Fujisaki	Ryo Nishimaki	Vinod Vaikuntanathan
Juan Garay	Carles Padr	Bogdan Warinschi
Sharon Goldberg	Omkant Pandey	Brent Waters
Robbert de Haan	Saurabh Panjwani	John Watrous
Iftach Haitner	Rafael Pass	Stephen A. Weis
Kristiyan Haralambiev	Krzysztof Pietrzak	David Woodruff
Danny Harnik	Manoj Prabhakaran	Juerg Wullschleger
Yuval Ishai	Bartosz Przydatek	David Xiao
Bhavana Kanukurthi	Omer Reingold	Scott Yilek
Jonathan Katz	Renato Renner	Shengyu Zhang
Robert Koenig	Thomas Ristenpart	Yunlei Zhao
Gillat Kol	Guy Rothblum	

Table of Contents

Technical Session 9

Technical Session 10

Technical Session 11

Technical Session 12

Technical Session 13

Incrementally Verifiable Computation
or
Proofs of Knowledge Imply Time/Space Efficiency

Paul Valiant

Massachusetts Institute of Technology
pvaliant@mit.edu

Abstract. A probabilistically checkable proof (PCP) system enables proofs to be verified in time polylogarithmic in the length of a classical proof. Computationally sound (CS) proofs improve upon PCPs by additionally shortening the length of the transmitted proof to be polylogarithmic in the length of the classical proof.

In this paper we explore the ultimate limits of non-interactive proof systems with respect to time and space efficiency. We present a proof system where the prover uses space polynomial in the space of a classical prover and time essentially linear in the time of a classical prover, while the verifier uses time and space that are essentially constant. Further, this proof system is *composable*: there is an algorithm for merging two proofs of length k into a proof of the conjunction of the original two theorems in time polynomial in k, yielding a proof of length *exactly* k.

We deduce the existence of our proposed proof system by way of a natural new assumption about proofs of knowledge. In fact, a main contribution of our result is showing that knowledge can be "traded" for time and space efficiency in noninteractive proof systems. We motivate this result with an explicit construction of noninteractive CS proofs of knowledge in the random oracle model.

1 Introduction

Perhaps the simplest way to introduce the computational problem we address is by means of the following.

Human motivation. Suppose humanity needs to conduct a very long computation which will span super-polynomially many generations. Each generation runs the computation until their deaths when they pass on the computational configuration to the next generation. This computation is so important that they also pass on a proof that the current configuration is correct, for fear that the following generations, without such a guarantee, might abandon the project. Can this be done?

Computational setting. In a more computational context, this problem becomes: How can we compile a machine M into a new machine M' that frequentlyoutputs

R. Canetti (Ed.): TCC 2008, LNCS 4948, pp. 1–18, 2008.
© International Association for Cryptologic Research 2008

pairs (c_i, π_i) where the ith output consists of the ith memory state c_i of machine M, and a proof π_i of its correctness, while keeping the resources of M intact?[1]

1.1 A New Problem

We motivate our problem by way of a few examples of how current techniques fail to achieve our goal. Suppose we are given a computation M that takes time t and space $k \ll t$.

A natural approach is have the compiled machine M' keep a complete record of all the memory states of M it has simulated so far; every time it simulates a new state of M, it uses this record to output a proof that its simulation of M is thus far correct. However, this approach has the clear drawbacks that the compiled machine M' uses space tk to store the records, and the proofs it outputs consist simply of this record of size tk; this requires the verifier of the proofs to also use time tk and space tk to verify *each* proof. If t is polynomial in k, then all these parameters are polynomial in k and this simple system is in fact "optimal up to polynomial factors in k." We concern ourselves here with the much more interesting case where the running time t is much larger than k —exponentially larger, even— in which case this naive system is not at all efficient. What we need is a more efficient proof system.

We note that the problems of improving the efficiency of the construction, transmission, and verification of proofs have been important themes in our field, and have fueled a long line of research. One major milestone on this path was the discovery of *probabilistically checkable proofs* (PCPs) (see [1,2,5,10] and the references therein). Under a PCP proof system statements with classical proofs of exponential length could now be verified in polynomial time, via randomized sampling of an encoded version of the classical proof. A PCP system still uses exponential resources to construct and transmit the proof, but verification is now polynomial time.

The second milestone we note is the theory of *computationally sound* (CS) proofs as formalized by Kilian and Micali [12,13]. This notion improves on the PCP system by keeping verification polynomial time while shortening the length of the transmitted proof from exponential to polynomial in k. If we instruct the compiled machine M' to output (noninteractive) CS proofs, then the length of the transmitted proofs, and the time and space required by the verifier are now polynomial in k, but the compiler still requires memory at least t, and a time interval of at least t between consecutive proofs.[2]

---

[1] More generally one might consider a machine that, instead of outputting proofs π_i, engages in some interactive proof protocol.

[2] A third major approach for improving the efficiency of proofs, arguably the most historically successful, is that of adding interaction between the prover and verifier[11,15,3]. Unfortunately, this approach does not help us here: our prover has only k memory so he may transfer his entire knowledge to the verifier at the start of their interaction; any further correspondence between the prover and verifier may be simulated by the verifier with no loss of efficiency.

1.2 Intuitive Idea of Our Solution

The ideal way to achieve incrementally verifiable computation consists of *efficiently merging two CS proofs of equal length into a single CS proof* which is as short and easy to verify as each of the original ones. Letting $c_0, c_1, \ldots$ be the sequence of configurations of machine M, and for $i < j$, intuitively denote by $(M : c_i \xrightarrow{t} c_j)$ the assertion that configuration c_j is correctly obtained from configuration c_i by running M for t steps. After running M for 1 step from the initial configuration c_0 so as to reach configuration c_1 one could easily produce a CS proof of $(M : c_0 \xrightarrow{1} c_1)$. Running M for another step from configuration c_1, one can easily produce a CS proof that $(M : c_1 \xrightarrow{1} c_2)$. At this point, if CS proofs can be easily merged as hypothesized above, one could obtain a CS proof that $(M : c_0 \xrightarrow{2} c_2)$. And so on, until a final configuration c_f is obtained, together with a CS proof that $(M : c_0 \xrightarrow{t} c_f)$

Unfortunately, we have no idea of how to achieve such efficient and length preserving merging of CS proofs. However, if a variant of CS proofs —which we call CS *proofs of knowledge*— exist, we show a sufficient approximation of this ideal strategy. The main idea is to construct recursively embedded CS proofs: to merge proofs π_1 and π_2 I prove that "I have seen convincing π_1 and π_2." In a nutshell, the CS proof methodology enables us to work with very short proofs, and proofs of knowledge enable the soundness of the proof system to persist across many levels of recursion.

1.3 A New Role for a New Type of Proof of Knowledge

Proofs of knowledge may be seen as a restricted form of classical proofs. While classically, proofs of a statement "There exists w such that $R(x, w) = 1$"[3] can take a wide variety of non-constructive forms, the proof of knowledge form asserts essentially "I have seen a w such that $R(x, w) = 1$." We note here that the inapplicability of classical proofs to our setting results from the combination of two circumstances: we require our proofs to be embeddable in other proofs, and we must work in merely computationally sound proof systems where deceptive proofs— while almost impossible to find— exist in abundance. We see the problem, intuitively, if we try to embed two computationally sound proof systems. The result would be a (computationally sound) proof that "There exists a computationally sound proof π of x." The problem is that *of course* there exists a computationally sound proof of x, *even when x is false*. So a proof that there exists a computationally sound proof of x implies nothing about the truth or falsehood of x.

Significantly, however, proofs of knowledge *can* be combined in this way: the result is a (computationally sound) proof that "Prover A has seen a computationally

---

[3] We remind the reader that since classical proofs are verifiable in polynomial time, we may consider any classical theorem as being a statement of membership in an NP-language of the form "There exists a proof w such that the verifier R accepts the pair consisting of the theorem x and proof w."

sound proof that Prover B has seen a witness w of x." Intuitively, this is the difference between saying "A is convinced that B is convinced of x" and saying "A is convinced that B *could be* convinced of x" —the first statement is reasonable evidence of x when both A and B are reasonable, but the second statement holds no weight since even a reasonable person *could be* mislead. In essence, the proof of knowledge property lets "reasonableness" be transferred down a sequence of provers. The formal statement of this assertion is that by sufficient repeated application of the *knowledge extractor E* associated with the proof system one can extract a valid witness w from any procedure that returns embedded proofs.

Remark 1: This simple intuition unfortunately translates into neither simple definitions nor simple proofs. Because this work seeks to optimize both prover and verifier time and space as well as the overall soundness of the proofs, we need to keep track everywhere not only of who is proving who's knowledge of what to whom, but also the time and space bounds of all involved parties, along with the security parameters. Nevertheless, it is our hope that the simple intuition underlying the constructions here will make the technical details less opaque.

Remark 2: We note that embedding proof systems deprives us of another principal tool: the use of random oracles. Specifically, suppose we have an oracle-based prover-verifier system (P^O, V^O) that can prove statements about the results of computation like "Machine M accepts the following string within t time steps. . . ." When we try to recursively embed this system the recursion breaks down because, even at the first level of recursion, we are no longer trying to prove statements about classical computation but rather statements of the form "M with oracle access to O accepts the following string...." Thus standard applications of random oracles do not appear to help. It remains an interesting question whether the goals of this paper may be attained in some other way using random oracles.

The Noninteractive CS Knowledge Assumption. Random oracles are intricately tied to CS proofs, in that the only known constructions of noninteractive CS proofs make use of random oracles (see [13]). Nevertheless, as with most random oracle constructions, the hope is that in practice the random oracle may be replaced by a suitably strong hash function plus access to a common random string.

In Section 4 we extend Micali's construction of CS proofs to a construction of CS proofs of knowledge: there exists an efficient extractor E that, given a statement X, a CS proof π, and access to the CS prover that produced π, outputs in quasilinear time a (classical) proof Π of X. We highlight this construction as a motivation for our assumption that oracle-less CS proofs of knowledge exist.

In essence, our assumption states that, in a specific construction of noninteractive CS proofs (Constructions 4 and 5), it is possible to replace the random oracle with a random string and still preserve the strength of the proofs. (That is, we do not invoke the random-oracle hypothesis in its general form. As shown by Canetti, Goldreich, and Halevi [8] and others in different contexts, we expect that there may be other non-interactive CS proof constructions for which no way to replace the oracle exists.)

We note that, while the Fiat-Shamir heuristic of replacing random oracle calls with a deterministic hash function yields feasible proposals for how to remove the oracle calls from the prover and verifier, it says nothing about how to translate the knowledge extractor into this new setting. For this reason we cannot explicitly conjecture a noninteractive CS proof of knowledge. However, in the context of this paper, the knowledge extractor component of the CS proof system serves only as a technique to argue security and is not invoked in our construction of incrementally verifiable computation. Thus we may propose the following much more explicit conjecture: our construction of incrementally verifiable computation (Theorem 1) works when using the prover-verifier pair (P, U) from Construction 4, modified by replacing the random oracle with a suitably strong hash function plus access to a common random string.

Knowledge $\Rightarrow$ Time/Space Efficiency. In this work we start with an unusual and very strong assumption about (proofs of) knowledge and conclude with a proof system of unprecedented time and space efficiency. In this paragraph we wish to draw the reader's attention not to the assumption or the conclusion, but to the nature of the relationship between them. On the left we make an assumption about *knowledge* in CS proofs: we take a restricted system that only deals with witnesses of length $3k$ and compresses them to proofs of length k, the security parameter, and assume that there is a linear-time *knowledge extractor* that can extract the witness given access to the prover. On the right we conclude with a proof system that compresses any proof to length $poly(k)$, uses space polynomial in the space needed to classically accept the language, and is time-efficient in the tightest possible sense, using only $poly(k)$ time to process each step of the classical acceptance algorithm. We note that current constructions of non-interactive CS proofs based on random oracles need time *polynomial* in the time to classically accept, and space of the same order as their *time*[13]. Our results constitute a new technique to leverage *knowledge* to gain time and space efficiency, and is in a sense a completeness result for CS proof systems.

2   Definitions

2.1   Noninteractive Proofs and the Common Random String Model

It is a well-known aphorism in cryptography that "security requires randomness". In many standard settings, a participant in a protocol injects randomness into his responses to protect him from some pre-prepared deviousness on the part of the other participant.

In the noninteractive proof setting such an approach is inadequate: the verifier is unable to protect himself with randomized messages to the prover, since he cannot even *communicate* with the prover. To address these issues, the *common random string* (CRS) model was introduced [7,6].

The CRS model —sometimes called the common *reference* string model— assumes that all parties have access to the same random string, and further that each can be confident that this string is truly random and not under the influence of the other parties. Potential examples of such a string are measurements of cosmic background radiation or, for a string that will appear in the future, tomorrow's weather.

In the analysis of the security of a CRS protocol leeway must be given for "unlucky" choices of strings, since if every choice of string worked in the protocol we would not need a random one. Thus even if a CRS protocol has a chance of failing, we still consider it secure if this chance is negligible as a function of the size of the random string.

2.2 Incremental Computation

Basic notation. We denote a Turing machine M with no inputs by $M()$, a Turing machine with one input by $M(\cdot)$, a Turing machine with two inputs by $M(\cdot, \cdot)$, etc. We assume a standard encoding, and denote by $|M|$ the length of the description of M. For a Turing machine M running on input s, we denote by $time_M(s)$ the time M takes on input s, and by $space_M(s)$ the space M takes on input s; we denote the empty input by ϵ, so that $space_M(\epsilon)$ is the space of Turing machine M when run on no input.

Incremental outputs. Commonly, Turing machines make an output only once, and making this output ends the computation. Instead, we interpret Turing machines as being able to output their current memory state at certain times in their operation: explicitly, consider a Turing machine with a special state "Output" where whenever the machine is in state "Output" the entire contents of its tape are outputted. [4] This captures our intuitive notion of an "incremental computation," namely one divided into "generations" where at the end of each generation the entire memory configuration is output so that the next generation may resume the computation from the current configuration.

2.3 Incrementally Verifiable Computation

We formally define incrementally verifiable computation here. We consider a Turing machine $M()$ that we wish to simulate for t time steps using k memory, where $k \geq \log t$. We consider a fixed *compiler* $C(\cdot, \cdot)$ that produces from (M, k) an *incrementally verifiable* version of M, namely a machine $C(M, k) = T(\cdot)$ that takes as input the common random string, runs in time $t \cdot k^{O(1)}$, uses memory $k^{O(1)}$, and every $k^{O(1)}$ time steps outputs its memory configuration. The jth

---

[4] We note that this is a slightly unusual model of output, as the machine would be unable to output a string such as "Hello World" without first deleting all other memory locations on the tape. In the context of this paper, we expect machines to not delete this other information: since we consider only $poly(k)$-space machines, it imposes no undue burden on the prover to output this information, and no undue burden on the verifier to ignore it.

memory configuration output should be interpreted as a pair consisting of a claim about the memory configuration of M at time j, and a CS proof of its correctness. There is a fixed machine V, the verifier, that will accept all pairs of configurations and proofs generated in this way, and will reject other pairs, subject to the usual condition of the CRS model that the verifier may be fooled with negligible probability, and the computational soundness caveat that an adversary with enormous computational resources may also fool the verifier.

Definition 1. *An increasing sequence of integers $\{t_j\}$ is an α-incremental time-line if for any j, $t_j - t_{j-1} \leq \alpha$.*

Definition 2. *A Turing machine that makes outputs at every time on an α- incremental timeline is called an α-incremental output Turing machine.*

Definition 3 (Feasible Compiler). *Let $C(\cdot, \cdot)$ be a polynomial time Turing machine. We say that C is a* feasible compiler *if there exists a constant c such that for all $k > 0$ and all $M()$ such that $|M| \leq k$, $C(M, k)$ is a Turing machine $T(\cdot)$ taking as input the common random string, such that*

1. *T is a k^c-incremental output Turing machine.*
2. *$space_T(r) = k^c$ for all inputs r.*

In other words, properties 1 and 2 guarantee that each compiled machine T outputs its internal configuration "efficiently often" while working in "efficient space."

Definition 4 (Incrementally Verifiable Computation). *The pair (C, V) is an* incrementally verifiable computation scheme *(in the CRS model) with security K if C is a feasible compiler, V is a polynomial-time Turing machine ("the verifier") and $K(k) : \mathbb{Z}^+ \to \mathbb{Z}^+$, such that the following properties hold: For any Turing machine M with $|M| \leq k$ let the jth output of the compiled machine $C(M, k)$ be parsed as an ordered pair (m_j, π_j^r), representing a claim about the jth memory configuration of M, and its proof; and let r denote the common random string of length k^2. We require:*

1. *(Correctness) The compiled machine accurately simulates M, in that m_j is indeed the jth memory configuration of $M(\epsilon)$ for all j, independent of r.*
2. *(Completeness) The verifier V accepts the proofs π_j^r: $\forall r, V(M, j, m_j, \pi_j^r, r) = 1$.*
3. *(Computational soundness) For any constant c and for any machine P' that for any length k^2 input r outputs a triple $(j, m_j'^r, \pi_j'^r)$ in time K, we have for large enough k that*

$$Prob_r[m_j'^r \neq m_j \wedge V(M, j, m_j'^r, \pi_j'^r, r) = 1] < k^{-c}.$$

We note that for the incrementally verifiable computation scheme to be secure against polynomial-time adversaries we must have K super-polynomial.

2.4 Noninteractive CS Proofs of Knowledge

We now specify the assumption we make: the existence of noninteractive CS proofs of knowledge.

We note that proofs of knowledge are typically studied in the form of *zero knowledge proofs of knowledge*. In this setting, one party wants to convince another party that he possesses certain knowledge without revealing this knowledge. The reason why he does not simply transmit all his evidence to the other party is that he wishes to maintain his privacy.

In our setting the reason one generation does not just transmit all its evidence to the next generation is not a privacy concern, but rather the concern that the following generation will not have the time to listen to all this evidence.

In both settings, the "knowledge" that must be proven may be considered to be a witness for a member of an NP-complete language: one party proves to the other that he knows, for example, a three-coloring of a certain graph.

In the zero-knowledge setting, our prover does not wish for the verifier to learn a three-coloring of the graph. In the incremental computation setting, our prover is worried that the verifier may not want to spare the resources to learn a three-coloring of the graph.

Related issues were considered in a paper of Barak and Goldreich where they investigated efficient (interactive) ways of providing proofs and proofs of knowledge [4]. Our definition of a noninteractive CS proof of knowledge contains elements from their definition of a *universal argument*.

For the sake of concreteness, we work with a specific NP-complete language, which has the property that for any k the strings in the language of length $4k$ have witnesses of length $3k$. We will require of our CS proof system that instead of returning proofs of length $3k$ (for example, the witnesses) the proofs are shortened to be of length k.

Definition 5. *Let c be a constant. The language $\mathcal{L}_c$ consists of the ordered pairs (M, x) where M is a Turing machine and x is a string such that, letting $k = |M|$ we have:*

1. *$|x| = 3k$.*
2. *There exists a string w of length $3k$ such that M when run on the concatenation (x, w) accepts within time k^c.*

We note that the string w may be thought of as the *NP witness* for (M, x)'s membership in the language. Further, since M may express any polynomial-time computation (for large enough k), the language $\mathcal{L}_c$ is NP complete.[5]

Definition 6 (Noninteractive CS proof of knowledge). *The pair (P, U) is a noninteractive CS proof of knowledge (in the CRS model) with parameters $K'(k) : \mathbb{Z}^+ \to \mathbb{Z}^+, c, c_1, c_2$ if P and U are Turing machines such that for all machines M, defining $k = |M|$, and all strings x of length $3k$ the following properties hold:*

---

[5] One can easily manipulate any NP language into one whose members and witnesses have lengths in the 4:3 ratio by appropriate padding.

1. *(Efficient prover) For any (CRS) string r of length k, $time_P(M, x, w, r) = k^{O(1)}$*
2. *(Length shrinking) For any (CRS) string r of length k, $|P(M, x, w, r)| = k$.*
3. *(Efficient verification) For any (CRS) string r of length k, $time_U(P(M, x, w, r), M, x, r) \le k^{c-1}$*
4. *(Completeness) For any (CRS) string r of length k, $U(P(M, x, w, r), M, x, r) = 1$*
5. *(Knowledge extraction) There exists a constant c_2 such that for any Turing machine P' there exists a randomized Turing machine $E_{P'}$, the extractor, such that for any input (M, x) of length $4k$ such that for all r of length k, $time_{P'}(M, x, r) \le K'(k)$ and $Pr_r[U(P'(M, x, r), M, x, r) = 1] = \alpha > 1/K'$ we have*

$$Prob[w \leftarrow E_{P'}(M, x) : M(x, w) = 1] > 1/2$$

and the running time of $E_{P'}(M, x)$ is at most k^{c_2}/α times the expected running time (over choices of r) of $P'(M, x, r)$.

3 Constructing Incrementally Verifiable Computation

3.1 Merging Proofs

We aim here to reexpress claims of the form $M : s_1 \xrightarrow{t} s_2$ as claims of membership in the language $\mathcal{L}_c$. The equivalence will not be exact but instead, in light of the goals of this paper, computationally sound. We define this relation inductively, for t that are powers of 2. The base case, when $t = 1$, is an exact relation.

Construction 1 (Base Case). *Let T_0 be the machine that interprets its input as a pair of length $3k$ strings (x, w) where x is interpreted as a triple of length k strings $x = (M, s_1, s_2)$, and checks that M when simulated for one step on configuration s_1 ends up in configuration s_2, ignoring the auxiliary input w.*

We note that for strings M, s_1, s_2 of length k, the pair $(T_0, (M, s_1, s_2))$ is in $\mathcal{L}_c$ if and only if $M : s_1 \xrightarrow{1} s_2$. The language $\mathcal{L}_c$ is crucial here, because this is the language which (by assumption) we may find CS proofs for.

We extend this construction, defining machines T_i such that $(T_i, (M, s_1, s_2)) \in \mathcal{L}_c$ is equivalent in a computationally sound sense to $M : s_1 \xrightarrow{2^i} s_2$. In particular, T_i is such that, given CS proofs of the claims $(T_i, (M, s_1, s_2)) \in \mathcal{L}_c$ and $(T_i, (M, s_2, s_3)) \in \mathcal{L}_c$ we can construct a CS proof of the claim $(T_{i+1}, (M, s_1, s_3)) \in \mathcal{L}_c$. Reexpressing these three statements, we see that given a CS proof that "$(M : s_1 \xrightarrow{2^i} s_2)$" and a CS proof that "$(M : s_2 \xrightarrow{2^i} s_3)$" we may construct a CS proof that "$(M : s_1 \xrightarrow{2^{i+1}} s_3)$." Since the lengths of each of these CS proofs is (by definition) k, this is our desired notion of *merging proofs*.

Construction 2. *Define T_{i+1} as a machine that interprets its input as the pair (x, w) where x is interpreted as (M, s_1, s_3) and w is interpreted as (p_1, p_2, s_2), and does the following:*

Check if p_1, p_2 are CS proofs of knowledge respectively that $(T_i, (M, s_1, s_2)) \in \mathcal{L}_c$ and $(T_i, (M, s_2, s_3)) \in \mathcal{L}_c$.

Given x, w, i such that w witnesses the fact that $(T_{i+1}, x) \in \mathcal{L}_c$, we can efficiently construct a CS proof of this fact as $P(T_{i+1}, x, w, r_{i+1})$ by assumption. (We note that we take the common random string r_{i+1} to be dependent on i.) We prove that this construction is computationally sound. In the following, we call a pair $(x = (M, s_1, s_2), p)$ *deceptive* if p proves to the verifier that $(T_i, x) \in \mathcal{L}_c$ but it is not the case that running M for 2^i steps from memory state s_1 reaches memory state s_2. The proof is by induction; the base case of T_0, as observed above, is trivial.

Lemma 1. *For $\alpha \in (\frac{1}{K'}, 1)$ and $b \in (2(2^i + k), K')$, if T^i has the property that no machine running in time b, outputs a deceptive pair $((M, s_1, s_2), p)$ with probability $\frac{1}{2}$ over the random strings $r_0, \ldots, r_i$, then no machine running in time $\frac{\alpha}{2} b / k^{c_2}$ outputs a deceptive pair for the machine T_{i+1} with probability α, over the random strings $r_0, \ldots, r_{i+1}$.*

Proof. This result is a straightforward consequence of the *knowledge extraction* property of the proofs in Definition 6. Assume we have a machine P' that outputs deceptive pairs $(x = (M, s_1, s_3), p')$ for T_{i+1} with probability α (over r) in time $\frac{\alpha}{2} b / k^{c_2}$. We apply the *extractor* $E_{P'}$, and have by definition that $E_{P'}(T_{i+1}, x)$ returns a classical witness w (relative to r_{i+1}) with probability at least $1/2$ in time at most $b/2$. The witness w is a classical witness for (T_{i+1}, x) in the language L, and thus (by the definition of T_{i+1}) w may be interpreted as $w = (p_1, p_2, s_2)$. Further, since w is a classical witness, both the proofs p_1 and p_2 are accepted by the verifier. However, since p' is deceptive, at least one of p_1, p_2 must be deceptive (with respect to T, r_i). In time $2^i + k \le b/2$ we can classically check which one of p_1, p_2 is deceptive, by simply simulating M for 2^i steps on s_1 comparing the current state against s_2, and reporting "p_1" if they agree, "p_2" if they do not. Thus using $b/2 + b/2 = b$ time we have recovered a deceptive pair for T_i with probability at least $1/2$, contradicting our assumption. □

Applying Lemma 1 inductively starting from $b = K'$, letting $\alpha = \frac{1}{2}$ for the first $i - 1$ iterations and $\alpha = \epsilon$ for the last yields:

Lemma 2. *No machine running in time $2\epsilon K'/(4k^{c_2})^i$ outputs a deceptive pair for the machine T_i with probability ϵ, over the random strings $r_0, \ldots, r_i$.*

3.2   The Main Result

Theorem 1. *Given a noninteractive CS proof of knowledge (P, U, K', c, c_1, c_2), there exists an incrementally verifiable computation scheme (C, V, K) provided $K k^{2 \log k + c_2 \log K} \le K'$.*

Proof. Making use of the CS proof of knowledge, Construction 2 describes a recursive procedure for generating a proof for 2^i steps of the computation using i levels of a binary recursion. Consider the tree that such a recursion would induce. The leaves of the recursive tree are the memory configurations of M, and the internal nodes j levels above the leaves are proofs of knowledge of recursive depth j (by way of machine T_j) asserting the results of simulating M for 2^j steps. Each node is computable in time polynomial in k from its two children, as this requires just one application of the polynomial-time prover P.

Let $C(M)$ be a machine that performs a depth-first traversal of the binary tree, starting at the leaf corresponding to time 0, visiting each leaf in order, and computing the value of every node it visits. At any moment in such a traversal the "stack" consists of the values of nodes on a path from a leaf to the root. Every time a leaf is visited, let $C(M)$ output the values of all the nodes along this path as a *proof of incremental correctness*. We note that processing any node takes time polynomial in k, and the depth of the recursion is less than k, and so a leaf is visited every $k^{O(1)}$ time. Thus this procedure uses the desired time and space.

We now show that these "stack dumps" in fact constitute computationally-sound proofs.

Consider a subtree whose leaves consist of a range $[t_1, t_2]$. (If the subtree has depth j then t_1 and t_2 will be consecutive multiples of 2^j.) When the recursion finishes processing this subtree, it will store in the parent node parameters $x = (M, s_1, s_2)$ and a proof of knowledge that M when starting in configuration s_1 reaches configuration s_2 in time $t_2 - t_1$.

We note that when the recursion processes leaf t' it must have finished processing all the leaves before t', and thus the leaves spanned by those subtrees in the "stack" must constitute all the leaves before t'. Thus these proofs of knowledge, when considered together, assert the complete result of simulating M from time 0 to time t'.

To check such a sequence of proofs, V verifies their individual correctness, and checks that the start and end memory states for each of the corresponding "theorems" match up.

We note, as above, that if such a sequence of proofs is *deceptive*, then we can (classically) isolate the deceptive proof using $O(t)$ additional time by simulating M. From Lemma 2 with $\epsilon = k^{-\log k}$, the probability that this incrementally verifiable computation scheme fools the verifier is negligible in k provided the time to execute of $C(M)$ plus the additional $O(t)$ classical verification time is at most $2k^{-\log k}K'/(4k^{c_2})^{\log t}$. We note that $C(M)$ consists essentially of constructing t CS proofs, each of which takes time $k^{O(1)} < k^{\log k}$. Thus (C, V, K) is an incrementally verifiable computation scheme for computations of length $t \le K$ provided $Kk^{\log k} \le k^{-(\log k)-c_2 \log K}K'$. Rearranging terms yields the desired result. □

4   CS Proofs of Knowledge in the Random Oracle Model

To explicitly introduce CS proofs of knowledge, and support our hypothesis that there exist noninteractive CS proofs of knowledge in the common reference string model we provide details of such proofs in the random oracle model.

Specifically, our construction will satisfy Definition 6 modified by replacing the string r everywhere with access to an oracle $\mathcal{R}$.

The construction of the proofs is based closely on the constructions of Kilian and Micali[12,13]. The construction of the witness extractor is inspired by a construction of Pass[14].

4.1   Witness-Extractable PCPs

One of the Principal tools in the construction of CS proofs is the probabilistically checkable proof (PCP)[1,2]. The PCP theorem states that any witness w for a string x in a language in NP can be encoded into a probabilistically checkable witness, specifically, a witness of length n can be encoded into a PCP of length $n \cdot (\log n)^{O(1)}$ with an induced probabilistic scheme (based on x) for testing $O(1)$ bits of the encoding such that:

- For any proof generated from a valid witness the test succeeds.
- For any x for which no witness exists the test fails with probability at least $\frac{2}{3}$.

In practice, the test is run repeatedly to reduce the error probability from $\frac{1}{3}$ to something negligible in n. In addition to the above properties of PCPs, we require one additional property that is part of the folklore of PCPs but rarely appears explicitly:

Definition 7 (Witness Extracting PCP). *A PCP is* witness-extracting with radius γ *if there exists a polynomial time algorithm W that, given any string s on which the PCP test succeeds with probability at least $1 - \gamma$, extracts an NP witness w for x.*

We sketch briefly how this additional property can be attained. Consider the related notion of a *PCP of proximity* (PCPP)[5]:

Definition 8 (Probabilistically checkable proof of proximity). *A pair of machines (P, V) are a PCPP for the NP relation $L = \{(x, w)\}$ with proximity parameter ϵ if*

- *When $(x, w) \in L$ the verifier accepts the proof output by the prover:*

$$Prob[V(P(x, w), (x, w)) = 1] = 1.$$

- *If for some x, w is ϵ-far from any w' such that $(x, w') \in L$, then the verifier will reject any proof π with high probability:*

$$Prob[V(\pi, (x, w)) = 1] < \frac{1}{3}.$$

We note that this property is stronger than the standard PCP property since in addition to rejecting if no witness exists, the verifier also rejects if the prover tries to significantly deceive him about the witness. Ben-Sasson et al. showed the existence of PCPPs with $O(1)$ queries and length $n \cdot (\log n)^{O(1)}$[5]. We use these PCPPs to construct witness-extractable PCPs:

Construction 3. *Let R be an error-correcting code of constant rate that can correct ϵ fraction of errors, with ϵ the PCPP parameter as above. Let $L = \{(x, w)\}$ be the NP relation for which we wish to find a witness-extractable PCP. Modify L using the code R to obtain a relation*

$$L' = \{(x, R(w)) : (x, w) \in L\}.$$

Let P be a PCPP prover for this relation. The verifier for this proof system is just the PCPP verifier for L', which expects inputs of the form $(P(x, R(w)), (x, R(w)))$. Let the witness extractor W for the proof system run the decoding algorithm on the portion of its input corresponding to $R(w)$ and report the result.

Claim. Construction 3 is a witness-extractable PCP with quasilinear expansion, where the verifier reads only a constant number of bits from the proof.

Proof. We note that since R is a constant-rate code and P expands input lengths quasilinearly, this scheme also has quasilinear expansion. Since the PCPP system reads only $O(1)$ bits of the proof, this new system does too.

For any pair $(x, w) \in L$ the proof generated will be accepted by the verifier, so this scheme satisfies the first property of PCPs. If x is such that no valid w exists for the L relation, then no valid $R(w)$ exists under the L' relation and the verifier will fail with probability at least $\frac{2}{3}$, as required by the second property of PCPs.

Finally, to show the witness extractability property we note that by definition of a PCPP, if the verifier succeeds with probability greater than $\frac{1}{3}$ on $(\pi, (x, s))$ then s is within relative distance ϵ from the encoding of a valid witness $R(w)$. Since the code R can correct ϵ fraction errors, we apply the decoding algorithm to s to recover a fully correct witness w. We have thus constructed a witness-extractable PCP for $\gamma = \frac{2}{3}$. □

4.2   CS Proof Construction

We now outline the construction of noninteractive CS proofs of knowledge, which is essentially the CS proof construction of Kilian and Micali[12,13]. We present the knowledge extraction construction in the next section.

The main idea of this CS proof construction is for the prover to construct a (witness-extractable) PCP, choose random queries, simulate the verifier on this PCP and queries, and send only the results of these queries to the real verifier, along with convincing evidence that the queries were chosen randomly and independent of the chosen PCP. For security parameter k' (we differentiate from the parameter k used in the non-oracle-based definitions.) the prover sends only data related to k' runs of the PCP verifier, and thus the length of the proof essentially depends only on the security parameter k'.

The technical challenge in the construction is to convince the verifier that the queries to the PCP are independent of the PCP. To accomplish this we use a *random oracle*. Let $\Re$ denote the set of functions

$$\mathcal{R} : \{0,1\}^{2k'} \to \{0,1\}^{k'}.$$

By a *random oracle* we mean a function $\mathcal{R}$ drawn uniformly at random from the set $\Re$. The machines in our construction will have oracle access to such an $\mathcal{R}$.

We start by defining a *Merkle hash*:

Definition 9 (Merkle hash). *Given a string s and a function $\mathcal{R} : \{0,1\}^{2k'} \to \{0,1\}^{k'}$, do the following:*

- *Partition s into chunks of length k', padding out the last chunk with zeros.*
- *Let each chunk be a leaf of a full binary tree of minimum depth.*
- *Filling up from the leaves, for each pair of siblings s_0, s_1, assign to their parent the string $\mathcal{R}(s_0, s_1)$.*

To aid in the notation we define a *verification path* in a tree:

Definition 10 (Verification path). *For any leaf in a full binary tree, its verification path consists of all the nodes on the path from this node to the root, along with each such node's sibling.*

The construction of CS proofs is as follows:

Construction 4. *Given a security parameter k', a polynomial-time relation $L = \{(x,w)\}$ with $|w| < 2^{k'}$ and a corresponding witness-extractable PCP with prover and verifier PP, PV respectively, we construct a CS prover P and verifier U.*

P on input (x,w) and a function $\mathcal{R} : \{0,1\}^{2k'} \to \{0,1\}^{k'}$ does the following:

1. *Run the PCP prover to produce $s = PP(x,w)$.*
2. *Compute the Merkle hash tree of s, letting s_r denote the root.*
3. *Using $\mathcal{R}$ and s_r as a seed, compute enough random bits to run the PCP verifier PV k' times.*
4. *Run PV k' times with these random strings; let the CS proof $P^{\mathcal{R}}(x,w)$ consist of the $k' \cdot O(1)$ leaves accessed here, along with their complete verification pathways.*

U on input x, a purported proof π and a function $\mathcal{R}$ does the following:

1. *Check for consistency of the verification pathways, i.e. for each pair of claimed children (s_0, s_1) verify that $\mathcal{R}(s_0, s_1)$ equals the claimed parent.*
2. *From the claimed root s_r run the procedure in steps 3 and 4 of the construction of P, failing if the procedure asks for a leaf from the tree that does not have a verification pathway.*
3. *Accept if both steps succeed, otherwise reject.*

These are essentially the CS proofs of Killian and Micali. In the next section we exhibit the knowledge extraction property of these proofs, and thereby infer their soundness; further properties and applications may be found in the original papers.

4.3   Knowledge Extraction

We now turn to new part of this construction, the *knowledge extractor* from part 5 of Definition 6. We construct a *black-box extractor*, that is, a fixed E that takes a description of the machine P' as an input argument, instead of depending arbitrarily on P'.

Recall that we want to construct a machine E that when given a (possibly deceptive) prover P' will efficiently extract a witness w for any x on which

$$Pr[U^{\mathcal{R}}(P'^{\mathcal{R}}(x), x) = 1] > 1/K'.$$

In other words, if P' reliably constructs a proof for a given x, then there is a witness "hidden" inside P', and E can extract one. The general idea of our construction is to simulate $P'^{\mathcal{R}}(x)$ while noting each oracle call and response, construct all possible Merkle trees that P' could have "in mind", figure out based on the output of P' which Merkle tree it finally chose, read off the PCP at the leaves of the tree, and use the PCP's witness extraction property to reveal a witness.

We note that this extractor is slightly unusual in that it does not "rewind" the computation at any stage, but merely examines the oracle calls P' makes; such extractors have been recently brought to light in other contexts under the names *straight-line extractors*[14] or *online extractors*[9]. The principal reason we need such an extractor is that we require the extractor to run in time *linear* in the time of P', up to multiplicative constant k^{c_2}, and we cannot afford the time needed to match up data from multiple runs.

We show that the following extractor fails with negligible probability on the set of $\mathcal{R}$ where $P'^{\mathcal{R}}(x)$ is accepted by the verifier; to obtain an extractor that never fails, we re-run the extractor until it succeeds.

Construction 5 (CS extractor). *Simulate $P'^{\mathcal{R}}(x)$, and let $q_1, ..., q_t$ be the queries P' makes to $\mathcal{R}$, in the order in which they are made, duplicates omitted. Assemble $\{q_i\}$ and separately $\{\mathcal{R}(q_i)\}$ into data structures that can be queried in time logarithmic in their sizes, $\log t$ in this case. If for some $i \neq j$ $\mathcal{R}(q_i) = \mathcal{R}(q_j)$, or if for some $i \leq j$ $q_i = \mathcal{R}(q_j)$, then abort.*

Consider $\{q_i\}$ as the nodes of a graph, initially with no edges. For any q_i whose first k' bits equal some $\mathcal{R}(q_j)$ and whose second k' bits equal some $\mathcal{R}(q_l)$, draw the directed edges from q_i to both q_j and q_l.

In the proof output by $P'^{\mathcal{R}}(x)$ find the string at the root, s_r. If s_r does not equal $\mathcal{R}(q_r)$ for some r, then abort. If the verification paths from the proof are not embedded in the tree rooted at q_r, abort.

Compute from x the depth of the Merkle tree one would obtain from a PCP derived from a witness for x. (Recall that for the language $\mathcal{L}_c$ in Definition 5, witnesses have length identical to that of x; in general we could pad witnesses to a prescribed length.) Read off from the tree rooted at q_r all strings of this depth from the root; where strings are missing fill in $0^{2k'}$ instead. Denote this string by pcp.

Apply the PCP witness extractor to pcp, and output the result.

Lemma 3. *Construction 5 when given (P', x) such that $P'^{\mathcal{R}}(x)$ always runs in time at most $2^{k'/4}$ and that convinces the verifier with probability $Pr_{\mathcal{R}}[U^{\mathcal{R}}(P'^{\mathcal{R}}(x), x) = 1] = \alpha > 2^{-k'/8}$, will return a witness w for x on all but a negligible fraction of those $\mathcal{R}$ on which P' convinces the verifier in time $O(k/\alpha)$ times the expected running time of P'.*

Proof. We show that this construction fails with negligible probability. We begin by showing that the probability of aborting is negligible.

Suppose P' has already made $i - 1$ queries to the oracle, and is just about to query $\mathcal{R}(q_i)$. This value is uniformly random and independent of the view of P' at this point, so thus the probability that $\mathcal{R}(q_i)$ equals any of q_j or $\mathcal{R}(q_j)$ for $j < i$ is at most $2i \cdot 2^{-k'}$. The probability that this occurs for any $i \leq t$ is thus at most $t^2 2^{-k'}$, which bounds the probability that the extractor aborts in the first half of the extractor.

We note that since no two q_i's hash to the same value, the trees will be constructed without collisions, and since $q_i \neq \mathcal{R}(q_j)$ for $i \leq j$, the graph will be acyclic and thus a valid binary tree. We may now bound the probability that some node on a verification path (including possibly the root) does not lie in the graph we have constructed. Let s_0, s_1 be a pair of siblings on a verification pathway for which the concatenation $(\mathcal{R}(s_0), \mathcal{R}(s_1))$ is not in the graph. Thus P' does not ever query $\mathcal{R}(\mathcal{R}(s_0), \mathcal{R}(s_1))$. Since the proof P' generates is accepted by the verifier, the value of $\mathcal{R}(\mathcal{R}(s_0), \mathcal{R}(s_1))$ must be on the verification path output by P'. Thus P' must have *guessed* this value without evaluating it, and further, the guess must have been right. This occurs with probability at most $2^{-k'}$. Thus the total probability of aborting is at most $(t^2 + 1)2^{-k'}$.

We now show that if the extractor does not abort, it extracts a valid witness on all but a negligible fraction of $\mathcal{R}$'s. Recall that the CS verifier makes k' calls to the PCP verifier, each of which, if seeded randomly, fails with probability $\frac{2}{3}$ whenever the string pcp does not encode a valid witness w.

Consider for some non-aborting $\mathcal{R}$ and some $i \leq t$ the distribution ρ on $\mathcal{R}$ obtained by fixing those values of $\mathcal{R}$ that $P'^{\mathcal{R}}(x)$ learns in its first i oracle calls, and letting the values of $\mathcal{R}$ on the remaining inputs be distributed independently at random. Consider an $\mathcal{R}$ drawn from the distribution ρ. Construct a Merkle tree from the values $\{(q_j, \mathcal{R}(q_j)) : j \leq i\}$ rooted at q_i, i.e., pretending that P', when it finishes, will output $\mathcal{R}(q_i)$ as the root, and let pcp be the string read off from the leaves, as in the construction of the extractor. Compute from $\mathcal{R}$ and $\mathcal{R}(q_i)$ as in step 3 of the construction of the CS prover P the k' sets of queries to the PCP verifier. Unless the oracle calls generated here collide with the i previous calls, the PCP queries will be independent and uniformly generated; if witness extraction fails on pcp then by definition, these PCP tests will succeed with probability at most $\frac{1}{3}^{k'}$. Adding in the at most $t^2 2^{-k'}$ chance that, under this distribution, one of the new oracle calls will collide with one of the old calls, the total probability that pcp is not witness-extractable, yet the tests succeed, is at most $(t^2 + 1)2^{-k'}$.

Consider all distributions ρ with i fixed values as above. We note that the distributions have disjoint support, since no fixed $\mathcal{R}$ could give rise to two different initial sequences of oracle calls. We note also that any $\mathcal{R}$ either aborts or induces such a distribution ρ with i fixed values. We now vary i from 1 to t. Consider the set of non-aborting $\mathcal{R}$ for which there is some i such that the string $pcp_i^{\mathcal{R}}$ is not witness-extractable yet the PCP tests generated by $\mathcal{R}$ all succeed. By the above arguments and the union bound this set has density at most

$$t(t^2 + 1)2^{-k'}.$$

By assumption the set of $\mathcal{R}$ for which the verifier accepts $P'^{\mathcal{R}}(x)$ has density at least $2^{-k'/8}$. Thus for all but a negligible fraction of these $\mathcal{R}$, the string pcp is witness-extractable, and we may recover a witness w as desired. $\square$

We note that our extractor runs logarithmic factor slower than P'. Since the running time of P' is subexponential in k, the extractor takes time $o(k)$ more than P'. As noted above, if P' returns an acceptable proof with probability α we may have to run the extractor $1/\alpha$ times (in expectation) before it returns a witness. Since by the above construction $\alpha \sim 1$, our extractor runs k times slower than P' and always returns acceptable proofs, as desired.

Acknowledgements

I am indebted to Silvio Micali for introducing me to the area of CS proofs and for many invaluable discussions along the way to writing this paper. I am grateful to Ran Canetti, Brendan Juba, and Rafael Pass for their many helpful comments on earlier versions of this paper. I would also like to thank the anonymous referees for their suggestions; those mistakes that remain are my own, and likely due to lack of understanding of what the referees were recommending.

References

1. Arora, S., Lund, C., Motwani, R., Sudan, M., Szegedy, M.: Proof verification and the hardness of approximation problems. Journal of the ACM 45(3), 501–555 (1998)
2. Arora, S., Safra, S.: Probabilistic checking of proofs: A new characterization of NP. Journal of the ACM 45(1), 70–122 (1998)
3. Babai, L., Fortnow, L., Lund, C.: Non-deterministic exponential time has two-prover interactive protocols. Computational Complexity 1, 3–40 (1991)
4. Barak, B., Goldreich, O.: Universal Arguments. In: Proc. Complexity (CCC) (2002)
5. Ben-Sasson, E., Goldreich, O., Harsha, P., Sudan, M., Vadhan, S.: Robust PCPs of proximity, shorter PCPs and applications to coding. In: STOC 2004, pp. 1–10 (2004)
6. Blum, M., De Santis, A., Micali, S., Persiano, G.: Noninteractive Zero-Knowledge. SIAM J. Comput. 20(6), 1084–1118 (1991)
7. Blum, M., Feldman, P., Micali, S.: Non-Interactive Zero-Knowledge and Its Applications (Extended Abstract). In: STOC 1988, pp. 103–112 (1988)
8. Canetti, R., Goldreich, O., Halevi, S.: The Random Oracle Methodology, Revisited. In: STOC 1998, pp. 209–218 (1998)

9. Fischlin, M.: Communication-efficient non-interactive proofs of knowledge with on-line extractors. Advances in Cryptology (2005)
10. Goldreich, O., Sudan, M.: Locally testable codes and PCPs of almost-linear length. In: FOCS 2002 (2002)
11. Goldwasser, S., Micali, S., Rackoff, C.: The Knowledge Complexity of Interactive Proof Systems. SIAM J. on Computing 18(1), 186–208 (1989)
12. Kilian, J.: A note on efficient zero-knowledge proofs and arguments. In: STOC, pp. 723–732 (1992)
13. Micali, S.: Computationally Sound Proofs. SIAM J. Computing 30(4), 1253–1298 (2000)
14. Pass, R.: On deniability in the common reference string and random oracle model. In: Advances in Cryptology, pp. 316–337 (2003)
15. Shamir, A.: IP = PSPACE. Journal of the ACM 39(4), 869–877 (1992)

On Seed-Incompressible Functions

Shai Halevi[1], Steven Myers[2], and Charles Rackoff[3]

[1] IBM Research
shaih@alum.mit.edu
[2] Indiana University
samyers@indiana.edu
[3] University of Toronto
rackoff@cs.toronto.edu

Abstract. We investigate a new notion of security for "cryptographic functions" that we term *seed incompressibility* (SI). We argue that this notion captures some of the intuition for the alleged security of constructions in the random-oracle model, and indeed we show that seed incompressibility suffices for some applications of the random oracle methodology. Very roughly, a function family $f_s(\cdot)$ with $|s| = n$ is seed incompressible if given (say) $n/2$ bits of advice (that can depend on the seed s) and an oracle access to $f_s(\cdot)$, an adversary cannot "break $f_s(\cdot)$" any better than given only oracle access to $f_s(\cdot)$ and no advice.

The strength of this notion depends on what we mean by "breaking $f_s(\cdot)$". We first show that for any family f_s there exists an adversary that can distinguish $f_s(\cdot)$ from a random function using $n/2$ bits of advice, so seed incompressible pseudo-random functions do not exist. Then we consider the weaker notion of seed-incompressible correlation intractability. We show that although the negative results can be partially extended also to this weaker notion, they cannot rule it out altogether. More importantly, the settings that we cannot rule out still suffice for many applications. In particular, we show that they suffice for constructing collision-resistant hash functions and for removing interaction from Σ-protocols (3-round honest verifier zero-knowledge protocols).

1 Introduction

Identifying useful security notions of "cryptographic functions" was proposed ten years ago by Canetti [4], as a plausible way of putting random-oracle-based constructions on a firmer theoretical footing. The challenge is to find specific "random-oracle-like" properties, such that functions with these properties (a) can be realized in the standard model and (b) can be securely used in some cryptographic applications in lieu of access to a truly random function. However, very little progress along this line has been made since then, in fact the only non-obvious notion along this line that we know of is the "perfect one-way hashing" notion of Canetti [4,7].

In this work we study a very different security notion that we term *seed incompressibility*. On a very high level, this notion is meant to capture the intuition

R. Canetti (Ed.): TCC 2008, LNCS 4948, pp. 19–36, 2008.
© International Association for Cryptologic Research 2008

that a random function has no structure. At a first glance, it seems hopeless to define an efficiently computable function that has no structure, since the fact that the function is computed by a small circuit is itself some structure. However, we may still hope that this small circuit is the only interesting "small property" of the function. That is, no adversary can find a *significantly smaller property* that differentiates it from your average random function. Roughly, if you do not get enough bits to describe the entire function, then you get nothing.

Toward formalizing this intuition, let $F = \{f_s\}_s$ be a family of functions with n-bit seeds, and consider an adversary that works in two phases: In the first phase the adversary gets the n-bit seed s, compresses it to (say) an $n/2$-bit string σ, and keeps only σ in memory. Then the adversary gets an oracle access to the function $f_s(\cdot)$, and it tries to "break it" (according to some notion of security). We call this the *seed compression* attack model. We say that the family F satisfies the underlying notion of security under seed-compression attack (or that it is *seed incompressible* with respect to the underlying notion of security), if breaking the function knowing σ is not any easier than breaking it without knowing σ.[1]

The choice of $n/2$ as the compression threshold is quite arbitrary. The results that we present in this paper remain unchanged whenever the threshold for the length of the compressed seed is anywhere from n^ϵ to $n - n^\epsilon$ for any fixed $0 < \epsilon < 1$. Below we stick to the $n/2$ threshold for convenience.

Following the intuition from above, we would have liked a construction where it is not possible to distinguish f_s from a random function, even given σ. Some care must be taken when defining this notion to avoid obvious pitfalls (such as σ being the first $n/2$ bits of $f_s(0)$), but this can be handled using ideas similar to the ones of Coron et al. [9] (see details in Section 3). Unfortunately, even with these ideas we show that the resulting notion cannot be realized, namely no function family can be pseudo-random under seed-compression attacks. Roughly, the reason is that the adversary can encode in σ a CS-proof [21] for the statement that f_s is computed by a small circuit. This impossibility result is somewhat disheartening, as it does not really show the existence of some property of the function that is smaller than its description; rather, it is simply a fact that convincing someone that a function has a small circuit takes much fewer bits than actually telling them what the circuit is. Further, from a security perspective the fact that the function is being computed by a small circuit is clearly information that the adversary knows.

Faced with this negative result, we investigate weaker notions of security. One direction that seems promising is the notion of *correlation-intractability* under seed-compression attacks. The notion of correlation-intractability was defined by Canetti et al. [6] as the inability of the attacker to find any input x such that the pair $(x, f_s(x))$ satisfies any "non-trivial relation" (cf. Section 4). Canetti et al. proved that correlation-intractability is not realizable when the adversary sees the entire seed s, but we point out that it may be realizable when the adversary is only given the "compressed seed" σ. We note that the negative

---

[1] Of course, this is only meaningful if "breaking f_s" is hard without any knowledge of the seed s.

results from [6] do not seem to extend to this model. On the other hand, our negative results for PRFs can be partially extended also to this weaker notion. However, it seems that these negative results hit an inherent limitation for some parameter settings, and we show in Section 4 that the remaining parameter settings are still useful. For example, we show that we can use them to construct collision-resistant hash functions, and perhaps more interestingly that we can use them to remove interaction from three-move public-coin honest-verifier zero-knowledge proofs.

Briefly, the primitives that can be constructed from seed-incompressible functions are those for which a "break" can be encoded with only a few bits. (For example, one can encode a collision in a hash function using only two inputs to the function.) When constructing such primitives from seed-incompressible functions, we let the seed of the function be sufficiently longer than the number of bits that are needed to encode a break, and then any adversary that breaks the resulting primitive can be converted into a "compressor" (that given the seed outputs a break), thus violating the seed-incompressibility of the underlying function.

We unfortunately were not able to find a construction that provably achieves seed-incompressible correlation intractability under a better-known computational hardness assumption. Still, one can conjecture that ad-hoc constructions such as AES or HMAC-SHA1 have this property. Such a conjecture is theoretically more appealing than using the random-oracle model since, at the very least, we do not have a proof that it is false, while still providing a conjecture open to disproof. We explain below our intuition for why one might conjecture that AES and SHA type constructions satisfy our definitions.

1.1   Seed-Incompressibility and Contemporary Block-Ciphers

Here is one way to use the intuition behind DES and AES like block-cipher constructions to possibly construct seed-incompressible functions. We will use AES to denote any similar composition-of-round based block-cipher construction. Each AES function is expressed as the composition of r rounds of permutations $p_{k_1} \circ \ldots \circ p_{k_r}$, where each n-bit k_i is determined by the key. For this theoretical presentation, instead, we assume that all the "round keys" k_i are chosen randomly and independently. Consider now using not r rounds but nr (independent) rounds. It seems unlikely that a key of this new construction can be "compressed" to only (say) $n/2$ bits. The intuition (at least for the case where the compressed seed consists of actual key bits), is that if the compressed seed is so short then there must be r consecutive rounds for which the key is completely undetermined, which in some intuitive sense is as strong as a standard r-round construction of an AES like construction. One issue for block-cipher like constructions is their invertibility, but by choosing a compressed seed of only $n/2$ bits, then it is not enough to help inverting the AES function. For example giving $n/2$ bits from the pre-image of zero does not appear to help when the AES permutation is mapping over an n-bit domain.

1.2 Related Work

The Random-Oracle Model. Following the Fiat-Shamir heuristic for transform-ing public-coin identification protocols into signature schemes [14] and several other uses of random-oracles in the literature, although sometimes used in dif-ferent contexts (e.g., [18]), Bellare and Rogaway formalized the random-oracle heuristic as a "general-purpose" design methodology for cryptographic schemes [2] and emphasized the need to develop formal proofs of security within it. The methodology requires that one first design an ideal system in which all parties (including the adversary) have oracle access to a truly random function and prove the security of this ideal system. (The proof is called "a security proof in the random-oracle model.") Next, one replaces the random oracle by a "crypto-graphic hash function" (such as SHA), where all parties (including the adversary) have a succinct description of this function. Thus, one obtains an implementation of the ideal system in a "real-world" where random oracles do not exist.

The random-oracle methodology has been used quite extensively since then, often resulting in very efficient and seemingly secure schemes. A drawback of this methodology, however, is that it is not at all clear what security properties are needed from the cryptographic hash function in order for a specific scheme to be secure. In fact, Canetti et al. demonstrated that this methodology is not sound in general, in that there exist secure "ideal schemes" that have no secure implementation in the "real world" [6]. A similar negative result was later proved by Goldwasser and Kalai also for the original Fiat-Shamir heuristic [16].

Still, there are many cryptographic schemes whose only known security proof is in the random-oracle model, some of which withstood substantial cryptanalysis and are widely implemented and deployed. Seeking to provide some theoretical footing to the security of such schemes, we would like to be able to describe "random-oracle like" properties that are (a) well-defined, (b) realizable, and (c) sufficient for the security of some instances of the random-oracle methodology. As we mentioned above, a first step in this direction was taken a decade ago by Canetti et al. with the notion of perfect one-way hashing [4,7].

With respect to the realizability of such notion, one thing that we could have hoped for is to prove its existence based on a more standard cryptographic assumption (e.g., the hardness of factoring). We point out, however, that at least as important is that there will be some hope (or intuition) that typical cryptographic functions such as SHA or AES actually fulfill this notion (as it is these functions that are used in actual implementations of protocols proven correct in the Random Oracle model).

Conditional Entropy Hash Functions. Barak et al. [1] conjectured the existence of families of hash functions h_s for which no attacker can generate an input that has a predictable output. Specifically, a keyed function $h_s(x)$ is said to "ensure conditional entropy e" if for every attacker A that takes as input the key s and produces as output an input x to h, it holds that the conditional entropy $H(h_s(A(s))|A(s)) \geq e$. This notion appears to be very close to the notion of correlation intractable functions of [6] (since if $h_s(x)$ has "high entropy" then it is unlikely to hit any evasive relation).

Barak et al. show that such functions are sufficient to implement the Fiat-Shamir heuristic to Σ-protocols, as is done herein with seed incompressible functions. (This result from [1] casts doubt on the informal claim made in [6] that correlation intractability was insufficient for such constructions.) Yet, other than via the connection to correlation-intractability, their notion seems unrelated to seed-incompressibility. For example, we do not see a obvious way in which conditional entropy hashes imply collision-resistance, as we show seed-incompressibility does in Section 5.1.

Exposure-Resilient Functions. The notion that we investigate in this work can be seen as an enhancement of exposure-resilient functions (ERFs) as defined and constructed by Canetti et al. [5]. Recall that an ERF is a function whose output looks random even when some of the input bits are known. It is easy to see that if we restrict the seed compression attack to only output some of the bits of the seed, then a "seed incompressible PRF" can be constructed from an ERF and a standard PRF (by first applying the exposure-resilient function to the seed).

The Bounded-Retrieval Model. Our seed-compression attack model can also be seen as an instance of the "bounded-retrieval model" that was introduced by Dziembowski and by Di Crescenzo et al. [13,11]. In this model, an adversary installs a virus on a target machine; the virus can observe all the secrets on the target machine, but only has a limited available bandwidth with which to communicate these secrets back to its "home base". The works of Dziembowski, Di Crescenzo et al. and Cash et al. [13,11,8] investigate obtaining secure key-exchange and authentication protocols in this model.

The current work can be thought of as trying to obtain primitives similar to pseudo-randomness in the same setting. (However, our focus is quite different, we view this model merely as a tool in order to establish primitives that can be used in other more standard models.)

Compressibility of NP Languages. A different notion of "compressibility" with applications to cryptography was recently proposed by Harnik and Naor [17]. In their notion, we are given an NP language and a word that is potentially in that language, and we try to produce a shorter word that is in the language if and only if the original is. For example, we are given a CNF formula ϕ and we try to compress it to a shorter ϕ' such that ϕ' is satisfiable if and only if ϕ is.[2] Harnik and Naor proved that if SAT is compressible then collision-resistant hashing can be constructed from one-way functions.

Our notion of compression seems quite different from the one of Harnik and Naor: roughly the difference is that they consider compressing the instance, whereas we are interested in compression of the witness (i.e., the secret seed of the function in our case).

---

[2] The length of ϕ' should be poly-logarithmic in the length of ϕ, but can be polynomial in the number of variables of ϕ.

2   Notations and CS Proofs

Notations. We define some notation used throughout the paper. Given a bit b, we use b^ℓ to denote the bit-string of ℓ bits b. Concatenation of bit strings is denoted with $||$. Given an ℓ bit-string $s = s_1, ..., s_\ell$ and $c \leq \ell$ we denote its first c significant bits $s_1, ..., s_c$ by $\lfloor s \rfloor_c$. We use $a \in_{\mathcal{R}} S$ to denote choosing uniformly at random an element a from a set S. We use $\mathsf{negl}(n)$ to denote some function f, such that for all c and sufficiently large n $f(n) \leq 1/n^c$ and $\mathsf{poly}(n)$ denotes some polynomial function $p \in \mathcal{O}(n^d)$ for some constant d.

CS Proofs. Our negative results use CS-proofs as constructed by Micali [20] (using techniques from Kilian [19]), as well as a variant of them due to Naor and Nissim [22]. Below, we briefly recall the definition. For our purposes, we view a CS-proof system as consisting of a prover, PRV, that wants to convince a verifier, VER, of the validity of an assertion $x \in L$ where L is some NP-language and PRV is in possession of a witness w for x.[3] In our context, we use non-interactive CS-proofs that work in the Random Oracle Model; that is, both the prover and verifier have access to a common random oracle. The prover generates an alleged proof that is examined by the verifier.

Definition 1 (Non-interactive CS proofs in the Random Oracle Model).
A CS-proof system for a language $L \in NP$ (with relation R_L), consists of two deterministic polynomial-time oracle machines, a prover PRV and a verifier VER, operating as follows:

- *On input $(1^k, x, w)$ such that $(x, w) \in R_L$ and access to an oracle $\mathcal{O}$, the prover computes a proof $\pi = \mathrm{PRV}^{\mathcal{O}}(1^k, x, w)$ such that $|\pi| \leq \mathsf{poly}(k, \log(|x| + |w|))$.*
- *On input $(1^k, x, \pi)$ and access to $\mathcal{O}$, the verifier decides whether to accept or reject the proof π (i.e., $\mathrm{VER}^{\mathcal{O}}(1^k, x, \pi) \in \{\mathsf{accept}, \mathsf{reject}\}$).*

The proof system satisfies the following conditions, where the probabilities are taken over the random choice of the oracle $\mathcal{O}$:

Perfect completeness: *For any $(x, w) \in R_L$ and for any k,*

$$\Pr_{\mathcal{O}} \left[\pi \leftarrow \mathrm{PRV}^{\mathcal{O}}(1^k, x, w), \ \mathrm{VER}^{\mathcal{O}}(1^k, x, \pi) = \mathsf{accept} \right] = 1.$$

Computational soundness: *For any polynomial time oracle machine BAD and any input $x \notin L$ it holds that*

$$\Pr_{\mathcal{O}} \left[\pi \leftarrow \mathrm{BAD}^{\mathcal{O}}(1^k, x), \ \mathrm{VER}^{\mathcal{O}}(1^k, x, \pi) = \mathsf{accept} \right] \leq \mathsf{negl}(k).$$

We sometimes also require a stronger soundness condition by replacing the negligible function $\mathsf{negl}(k)$ with an exponentially small function $\frac{poly(k+|x|)}{2^k}$. (This stronger condition can still be proven in the random-oracle model.)

---

[3] Micali defined CS-proofs more generally, but we do not need this extra generality for our purposes.

3   Seed-Incompressible Pseudo-random Functions

Following the intuition as presented in the introduction, we would have liked to have a construction $F = \{f_s\}$ such that f_s looks random even when given a "compressed version of s." Formalizing this takes some care, since this "compressed version of s" could be, for example, the first $|s|/2$ bits of $f_s(0)$ (which would make it easy to distinguish f_s from an unrelated random function). This technicality can be solved by borrowing the technique used by Coron et al. (in the context of domain extenders for random oracles) [9]. Namely, the second phase of the adversary gets either the compressed seed σ and access to $f_s(\cdot)$, or access to a random function $f(\cdot)$ and a "simulated compressed seed" that was generated by a simulator S^f (where S has access to the same random f).

In the formal definition below, we fix some polynomially-bounded length functions ℓ_1, ℓ_2 and consider function families from $\ell_1(n)$ bits to $\ell_2(n)$ bits with n-bit seeds. We denote by $\mathcal{F}_n$ the set of all functions $f : \{0,1\}^{\ell_1(n)} \to \{0,1\}^{\ell_2(n)}$.

Definition 2 (Seed-Incompressible PRFs). *Let $\{F_n\}_{n \in \mathbb{N}}$ be a family of functions such that $F_n : \{0,1\}^n \times \{0,1\}^{\ell_1(n)} \to \{0,1\}^{\ell_2(n)}$ can be efficiently computed, and denote $f_s(\cdot) \equiv F(s, \cdot)$.*

The family $\{F_n\}$ is pseudo-random under seed-compression attacks if for every two-phase efficient adversary $\mathtt{Adv} = (A, B)$ there exists an efficient simulator S and a negligible function negl such that

$$\left| \begin{array}{l} \Pr\left[s \in_{\mathcal{R}} \{0,1\}^n, \ \sigma \leftarrow A(s) \ : \ |\sigma| \leq n/2 \text{ and } B^{f_s}(\sigma) = 1 \right] \\ - \Pr\left[f \in_{\mathcal{R}} \mathcal{F}_n, \ \sigma \leftarrow S^f(1^n) \ : \ |\sigma| \leq n/2 \text{ and } B^f(\sigma) = 1 \right] \end{array} \right| \leq \mathsf{negl}(n)$$

Seed-Incompressible PRFs would have been very useful, but unfortunately they do not exist, as will be shown below. While we show that SI-PRFs do not exist, a related concept will be introduced later, and therefore the discussion is useful for this later topic.

Theorem 1. *Seed-Incompressible PRFs as defined in Definition 2 do not exist.*

Proof. Let $\{F_n\}_{n \in \mathbb{N}}$ be a family of functions as in Definition 2. We show a two-phase adversary $\mathtt{Adv} = (A, B)$ for which no simulator exists. Fix some n and let $\ell_1 = \ell_1(n)$ and $\ell_2 = \ell_2(n)$. Let $j = \lceil 2n/\ell_2 \rceil$ (i.e., the output of f_s on $1, 2, \ldots, j$ contains at least $2n$ bits).

The first phase of the adversary, A, gets as input a seed $s \in \{0,1\}^n$. It computes $y_i = f_s(0\|i)$ for $i = 1, 2, \ldots, j$, and then prepares a CS-proof π for the true NP statement

"there exists a seed s' such that $y_i = f_{s'}(0\|i)$ for $i = 1, 2, \ldots, j$"   ($\star$)

The proof is prepared relative to the oracle $\mathcal{O}(\cdot) = f_s(1\|\cdot)$ and security parameter $k = \sqrt{n}$. Then A outputs the proof π as the "compressed seed", to be used by the second phase B. (Notice that the length of this proof is $k \cdot \mathrm{polylog}(n) < n/2$.)

The second phase B, on input π, first uses its oracle f to compute $y_i = f(0\|i)$ for $i = 1, 2, \ldots, j$, thereby recovering the statement that π is supposed to be a

CS-proof for. Then B attempts to verify the proof π relative to $f(1||\cdot)$ (where f is the provided oracle) and security parameter $k = \sqrt{n}$. It accepts if the proof is valid and rejects otherwise. By the perfect completeness of CS-proofs, B accepts with probability one when given the proof that A generated and access to the same f_s for which that proof was generated.

On the other hand, the soundness of CS-proofs implies that no simulator can make B accept with non-negligible probability. Indeed, when f is a random function in $\mathcal{F}_n$ then $y_1, \ldots, y_j$ consist of at least $2n$ random bits, hence the probability that the statement $(\star)$ from above is true is at most 2^{-n}. And if the statement is not true, then no efficient simulator with access to a random f can generate a valid proof for it with probability better than $\mathrm{poly}(n) \cdot 2^{-\Theta(k)} = \mathsf{negl}(n)$.

4   Seed-Incompressible Correlation Intractability

Canetti et al. [6] introduced the concept of Correlation Intractability to capture the intuition that the adversary cannot "hit" any rare input-output relation. Roughly, an *evasive relation* R is one where it is hard to find an input x such that $(x, f(x)) \in R$ for a random function f, and a function family F is *correlation intractable* if for any evasive relation R it is hard to find $(x, f(x)) \in R$ for a random member $f \in F$. These notions can be extended to $2p$-ary relations in the obvious way (see below).

Canetti et al. proved that correlation-intractable function families do not exist, in that an adversary that knows the short description of $f \in F$ can always find some $(x, f(x)) \in R_F$ for a particular relation R_F that depends on F. In our case, however, we are interested in an adversary that does not see the entire description of $f \in F$ but only gets a "compressed description". We provide the formal definitions below, and then discuss the extent to which the negative results from [6] and from Section 3 do or do not extend to this new notion. Below we again fix some polynomially bounded length functions ℓ_1, ℓ_2, and denote by $\mathcal{F}_n$ the set of all functions $f : \{0,1\}^{\ell_1(n)} \to \{0,1\}^{\ell_2(n)}$.

Definition 3 (Evasive Relations). *A $2p$-ary relation R is evasive if for any efficient adversary A, there is a negligible function negl such that for all sufficiently large n*

$$\Pr_{f \in \mathcal{F}_n} [\langle x_1, \ldots, x_p \rangle \leftarrow A^f(1^n) : \langle x_1, \ldots, x_p, f(x_1), \ldots, f(x_p) \rangle \in R] \leq \mathsf{negl}(n).$$

Sometimes we are interested only in *efficient* relations, namely relations R for which the membership problem $\langle x_1, \ldots, x_p, y_1, \ldots, y_p \rangle \overset{?}{\in} R$ can be efficiently decided (i.e., in polynomial time).

Definition 4 (Seed-Incompressible Correlation Intractability)
Let $\{F_n\}_{n \in \mathbb{N}}$ be a family of functions where $F_n : \{0,1\}^n \times \{0,1\}^{\ell_1(n)} \to \{0,1\}^{\ell_2(n)}$ can be efficiently computed, and denote $f_s(\cdot) \equiv F(s, \cdot)$.

For some polynomial $p = p(n)$, we say that the family $\{F_n\}$ is correlation intractable under seed-compression attacks with respect to $2p$-ary relations if for every $2p$-ary evasive relation R, and for every two-phase efficient adversary $\mathtt{Adv} = (A, B)$, *there is a negligible function* negl *such that for all sufficiently large* n

$$\Pr \begin{bmatrix} s \in_{\mathcal{R}} \{0,1\}^n, \ \sigma \leftarrow A(s), \ \langle x_1, \ldots, x_p \rangle \leftarrow B^{f_s}(\sigma) \ : \\ |\sigma| \leq n/2 \ \text{and} \ \langle x_1, \ldots, x_p, f_s(x_1), \ldots, f_s(x_p) \rangle \in R \end{bmatrix} \leq \mathsf{negl}(n).$$

We also call such function families seed-incompressible correlation-intractable (*with respect to $2p$-ary relations*), or SI-CorInt($2p$), *for short.*

In the case that we restrict the above quantification on all evasive relations to only efficient evasive relations, we say that the family $\{F_n\}$ is weakly seed-incompressible correlation-intractable with respect to $2p$-ary relations (*wSI-CorInt($2p$)*).

4.1 Do SI Correlation Intractable Functions Exist?

The first question to answer with respect to the seed-incompressible correlation intractability as defined above is whether we can extend the impossibility result from Theorem 1 (or from [6]) to show that it too cannot be realized.

One first observes that for some setting of parameters, an attacker in the seed-compression model is just as powerful as an attacker that has the full uncompressed seed. Specifically, if the seed is more than $2p$ times the length of the input to h, then the first phase of an attacker in the seed-compression model can output the vector that breaks the correlation intractability as the "compressed seed". However, the impossibility results from [6] do not extend to very long seeds, so this simple observation does not appear to shed new light on the existence of SI-CorInt functions. Below we show, however, that the technique from Theorem 1 can be extended for some settings of parameters:

As opposed to the case of Theorem 1, here the adversary needs not only to distinguish f_s from random (which can be done with CS-proofs), but also to compute some "unpredictable relation". The idea that we exploit here is that the CS-proof itself can be thought of as an "unpredictable relation." Roughly, we have a relation of the form

$$\left\{ \begin{array}{c} \langle (1, \ldots, t, v_1, \ldots, v_m), (x_1, \ldots, x_t, y_1, \ldots, y_m) \rangle \ : \\ \text{The CS-proof } (v_1, \ldots, v_m) \text{ is valid for the instance } (x_1 = f(1), \\ \ldots, x_t = f(t)) \text{ w.r.t. } V \text{ receiving oracle answers } (y_1, \ldots, y_m). \end{array} \right\}$$

Tracing through the various parameters we see that to use Micali's construction for CS-proofs with a relation such as above we need $t = (n + \omega(\log n))/\ell_2(n)$ and $m = polylog(n)$. Hence we get an impossibility result for $2p$-ary relations where $p = n/\ell_2(n) + polylog(n)$. Moreover, if we assume the existence of collision-resistant hashing then we can use the variant of CS-proofs with few oracle calls due to Naor and Nissim [22], and then we can get by with a relation that only depends on $m = O(n/\ell_1(n))$ of the v_i's. Hence for function families with n-bit seeds and input/output bit lengths of size $\ell_1(n), \ell_2(n) = \Omega(n)$ we also obtain an impossibility result for $2p$-ary relations where $p = O(1)$ (we can get as low as $p = 3$ when $\ell_1(n), \ell_2(n) > n$.)

Lemma 1. *An efficiently computable functions family* $\{F_n : \{0,1\}^n \times \{0,1\}^{\ell_1(n)} \to \{0,1\}^{\ell_2(n)}\}$ *cannot be weakly correlation intractable under seed-compression attacks with respect to $2p$-ary relations, for any $p \geq \lceil (n + \omega(\log n))/\ell_2(n) \rceil + \lceil polylog(n)/\ell_2(n) \rceil$.*

Moreover, if collision resistant hash function family $\{H_n : \{0,1\}^n \times \{0,1\}^{\ell_4(n)} \to \{0,1\}^{\ell_5(n)}\}$ *exist, for polynomials $\ell_4(n) > \ell_5(n)$, then no family as above can be correlation intractable under seed-compression attacks with respect to $2p$-ary relations where $p \geq \lceil (n + \omega(\log n))/\ell_2(n) \rceil + \lceil n^\epsilon/\ell_2(n) \rceil + \lceil n^\epsilon/\ell_1(n) \rceil$ (for any $\epsilon > 0$).*

A proof of this lemma will be in the full version of this paper.

Smaller relations. We speculate that current techniques cannot be used to rule out relations with arity less than 6. This is because with the current technique of using CS-proofs, you would need at least one oracle call to specify the function instance, at least one oracle call for the CS-proof, and you would have to use at least one more v_i to describe the CS-proof itself. Thus it is still plausible that seed-incompressible correlation intractable function families exist with respect to such low-arity relations.

5   Implications of Seed-Incompressible Correlation Intractability

We demonstrate the usefulness of seed-incompressible functions by showing how they can be used to easily construct two primitives: specifically collision-resistant hash functions and (single-theorem) NIZK systems via the Fiat-Shamir methodology. More generally, the primitives that can be constructed from seed-incompressible functions are those for which a "break" can be encoded with only a few bits. (For example, one can encode a collision in a hash function using only two inputs to the function. Similarly, for a NIZK that was derived from a 3-move Σ-protocol, one can encode a false proof using only the two messages that the prover sends.) When constructing such primitives from seed-incompressible functions, we let the seed of the function be sufficiently longer than the number of bits that are needed to encode a break, and then any adversary that breaks the resulting primitive can be converted into a "compressor" (that given the seed outputs a break), thus violating the seed-incompressibility of the underlying function.

5.1   Collision Resistant Hashing

We show that seed-incompressible correlation-intractable function with respect to quaternary relations must be "essentially collision-resistant." We view this feature as a minimal requirement for any primitive that one hopes to use in lieu of a random-oracle, since heuristic implementations of random-oracle-based constructions always use collision-resistant hash functions such as SHA to replace the oracle. Note that it is not true that any seed-incompressible function is also collision-resistant (for example, seed-incompressible functions need not be length decreasing). Rather, we show below that any seed-incompressible function must have "embedded in it" a collision-resistant function.

Specifically, given a seed-incompressible correlation-intractable function $f_s(\cdot)$, we consider shortening the inputs and outputs of f_s so that the inputs are shorter than one quarter of the seed and the outputs are shorter than the inputs. We then observe that an algorithm that finds collisions in the resulting (length-decreasing) function can be (trivially) converted to a "compressor" that breaks the seed-incompressibility of f_s: the "compressor" only needs to output the collision.

We formally state the definition of collision-resistance for completeness and then state the theorem with proof.

Definition 5 (Collision Resistant Hash Functions (CRHF))
Fix polynomially-bounded length functions $\ell_2(n) < \ell_1(n)$. A function generator $\{H_n : \{0,1\}^n \times \{0,1\}^{\ell_1(n)} \to \{0,1\}^{\ell_2(n)}\}_{n \in \mathbb{N}}$ is collision resistant if for every probabilistic polynomial time adversary A there is a negligible functions negl such that for all sufficiently large n:

$$\Pr[s \leftarrow \{0,1\}^n, (x_1, x_2) \leftarrow A(s) \mid x_1 \neq x_2 \wedge h_s(x_1) = h_s(x_2)] \leq negl(n).$$

Theorem 2. *If there exists a function family $\{F_n : \{0,1\}^n \times \{0,1\}^{\ell_1(n)} \to \{0,1\}^{\ell_2(n)}\}$ with super-logarithmic length functions $\ell_1, \ell_2 = \omega(\log n)$, which is correlation intractable under seed-compression attacks with respect to quaternary relations, then collision resistant hash functions exist.*

Proof. Let ℓ_1, ℓ_2 be super-logarithmic length functions, $\ell_1, \ell_2 = \omega(\log n)$, and assume that a family $F = \{F_n : \{0,1\}^n \times \{0,1\}^{\ell_1(n)} \to \{0,1\}^{\ell_2(n)}\}_{n \in \mathbb{N}}$ is correlation intractable under seed-compression attacks with respect to quaternary relations.

Consider a modification of the family F to operate on potentially shorter inputs and outputs. Namely, let $\ell_1' = \min(\ell_1, n/4)$ and $\ell_2' = \min(\ell_2, \ell_1' - 1)$, and consider the family $H = \{H_n : \{0,1\}^n \times \{0,1\}^{\ell_1'(n)} \to \{0,1\}^{\ell_2'(n)}\}_{n \in \mathbb{N}}$ which if defined as follows: on seed s of length n and input x' of length ℓ_1', first append zeros to x' up to length of ℓ_1 bits, then apply $F(s, \cdot)$ to the result, and finally take only the first ℓ_2' bits of the outcome.

$$H(s, x') \stackrel{\text{def}}{=} \lfloor F(s, x) \rfloor_{\ell_2'}, \text{ where } x = x' \| 0^{\ell_1(|s|) - \ell_1'(|s|)}.$$

We prove that if F is SI-CorInt with respect to quaternary relations then H is collision-resistant. In particular, consider the relation $R \subset \{0,1\}^{\ell_1} \times \{0,1\}^{\ell_1} \times \{0,1\}^{\ell_2} \times \{0,1\}^{\ell_2}$:

$$R \stackrel{\text{def}}{=} \{(x_1, x_2, y_1, y_2) \mid x_1 \neq x_2 \text{ and } \lfloor y_1 \rfloor_{\ell_2'} = \lfloor y_2 \rfloor_{\ell_2'}\}.$$

This relation is clearly evasive, as every polynomial-tie adversary has probability at most $poly(n) \cdot 2^{-\ell_2'} = \mathsf{negl}(n)$ of outputting x_1, x_2 for which $(x_1, x_2, f(x_1), f(x_2)) \in R$ where f is a random function (since ℓ_2' is super-logarithmic).

Assume for contradiction that the family H from above is not collision resistant, and let C be a collision finding adversary that given a random $s \in \{0, 1\}^n$ outputs a pair of strings $x_1, x_2 \in \{0, 1\}^{\ell_1}$ such that $H(s, x_1) = H(s, x_2)$ with probability at least $1/n^c$. Then, define the adversary $\mathsf{Adv} = (A, B)$ for the underlying SI-CorInt family as follows: The "compressor" $A(s)$ simply outputs the collision $(x_1, x_2) = C(s)$, and note that $|A(s)| = n/2$. The second phase of the attack is just translate H-inputs into F-inputs by appending zeros, namely $B(x_1, x_2)$ outputs (x_1', x_2') where $x_i' = x_i || 0^{\ell_1 - \ell_1'}$. By definition if $H(s, x_1) = H(s, x_2)$ (which happens with non-negligible probability) then $\lfloor F(s, x_1') \rfloor_{\ell_2'} = \lfloor F(s, x_2') \rfloor_{\ell_2'}$, and therefore $(x_1', x_2', F(s, x_1'), F(s, x_2')) \in R$, contradicting the security of F.

5.2   From Σ-Protocols to NIZK Arguments

One of the "signature uses" of the random-oracle heuristic is to remove interaction from zero-knowledge protocols using the Fiat-Shamir heuristic. Specifically, given a public-coin honest verifier zero-knowledge proof system (known as Σ-protocols in the case where the number of rounds is three), it is possible to transform it into a non-interactive protocol by replacing the verifier's messages with the output of a "cryptographic hash function", applied to the transcript up to that point.

It is well known that if the original protocol has negligible soundness error, then the resulting non-interactive protocol can be proven in the random-oracle model to be a non-interactive zero-knowledge argument system, and can also be used as a secure signature scheme. On the other hand, Goldwasser and Kalai proved in [16] that there exist interactive Σ-protocols with negligible soundness for which their resulting protocols are not secure signature schemes in the standard model, no matter what function family is used to replace their interaction.

However, that negative result still leaves the possibility of a function family that will convert the interactive protocol into a NIZK argument system for a single theorem. Indeed, below we show that the latter is possible if seed-incompressible correlation-intractable functions exist.

We begin by recalling the definitions of Σ-protocols and NIZK argument systems. Below let L be an NP-language and let R_L be a binary relation that defines L, namely $L = \{x : \exists w \text{ s.t. } (x, w) \in R_L\}$ (where the witness w has length polynomial in $|x|$).

Σ-Protocols. For a pair (P, V) of interacting protocols, we denote by (P, V) (x, w) a run in which P has input $(x, w) \in R_L$, V has input x, and P attempts to convince V of the validity of the assertion $x \in L$. For a three-move protocols as above (with P going first), denote by α, β, and γ the three messages that are exchanged in the protocol, and let P_1, and P_2 be the randomized functions that the prescribed prover uses to compute its two messages, namely we have

$$(\alpha, \text{state}) \leftarrow P_1(x, w), \text{ and } \gamma \leftarrow P_2(x, w, \text{state}, \alpha, \beta).$$

We also denote by V^* the function that the verifier employs to decide whether to accept or reject the proof,

$$V^*(x, \alpha, \beta, \gamma) \in \{\text{accept,reject}\}.$$

Typically, the first flow α is called a *commitment*, the second flow β is called a *challenge*, and the third flow γ is called a *response*.

Definition 6. *A 3-move protocol (P, V) as above is a Σ protocol for a language L if it satisfies the following properties:*

Public-coin verifier. *The message β sent by V is always a sequence of $t(|x|)$ uniformly chosen random bits (for some length function t).*
Perfect completeness. *For every $(x, w) \in R$:*

$$\Pr\left[\begin{array}{c} (\alpha, \text{state}) \leftarrow P_1(x, w); \ \beta \in_{\mathcal{R}} \{0,1\}^{t(|x|)}; \ \gamma \leftarrow P_2(x, w, \text{state}, \alpha, \beta) \ : \\ V^*(x, \alpha, \beta, \gamma) = \text{accept} \end{array} \right] = 1,$$

where the probability is over the random choices of P_1, P_2 and β.
Soundness. *There is a negligible function negl such that for every $x \notin L_R$, for every pair of adversarial (computationally unlimited) prover circuits P_1^*, P_2^*:*

$$\Pr\left[\alpha \leftarrow P_1^*(x); \ \beta \in_{\mathcal{R}} \{0,1\}^{t(|x|)}; \ \gamma \leftarrow P_2^*(x, \beta) : V^*(x, \alpha, \beta, \gamma) = acpt\right] = \frac{1}{2^{t(|x|)}},$$

where the probability is over the random choice of β. [4] We note a property of Σ-protocols of interest to our later arguments: For every Σ-protocol, simple parallel repetition and padding arguments allow the length $t(|x|)$ of the verifier's challenge to be set to an arbitrary positive integer. [5] For the remainder of the paper we therefore assume that $t \in \omega(\log n)$, and that the adversaries probability of success, as stated above, is negligible in $|x|$.
Zero-knowledge. *There exists a polynomial-time simulator S that on input x and $\beta' \in \{0,1\}^{t(|x|)}$ outputs $(\alpha', \beta', \gamma')$ (we have S outputting its input β' to simplify the notations somewhat).*
We require that for every $(x, w) \in R_L$ and every $\beta' \in \{0,1\}^{t(|x|)}$, the distribution on $(\alpha', \beta', \gamma') = S(x, \beta')$ is identical to the conditional distribution on the transcript $(\alpha, \beta, \gamma) = (P, V)(x, w)$, when conditioned on $\beta = \beta'$.

---

[4] Traditionally Σ-protocols are defined with a stronger soundness condition called extractability that clearly implies the current soundness definition.
[5] We refer the reader to [10] for a complete discussion on this and other properties of Σ-protocols.

NIZK Arguments. We remind the reader that the common reference string (CRS) model is one in which all participants and the adversary have access to a polynomial sized common reference string chosen by a trusted third party from a pre-specified distribution (which we denote $\mathcal{D}$).

Definition 7. *A pair of efficient probabilistic algorithms (P, V) are a single-theorem NIZK argument system for a language L (specified by a binary relation R_L) in the CRS model, if it satisfies:*

Completeness. $\forall (x, w) \in R,\ \Pr_{crs, P, V}[\sigma \leftarrow P(crs, x, w) : V(crs, x, \sigma) = 1] = 1.$

Computational Soundness. For every (possibly cheating) efficient probabilistic prover P^* there is a negligible function negl such that for all $x \notin L$:

$$\Pr_{crs, P^*, V}[\sigma \leftarrow P^*(crs, x) : V(crs, x, \sigma) = 1] \leq negl(|x|).$$

Zero-Knowledge. There exists an efficient simulator S such that for every $(x, w) \in R$, the output of the following two experiments are (computationally, statistically, perfectly)-indistinguishable.

$\mathsf{Exp}_1(x, w)$	$\mathsf{Exp}_2(x)$
$crs \leftarrow \mathcal{D}$	$(crs', \sigma') \leftarrow S(x)$
$\sigma \leftarrow P(crs, x, w)$	
Output (crs, σ)	Output (crs', σ')

The Fiat-Shamir Transformation. Fiat-and Shamir described in [14] a transformations that turns Σ-protocols into non-interactive argument systems. Specifically, instead of having the verifier choose a random challenge β, one computes β by applying a hash function to the input x and the commitment α, setting $\beta = f(x, \alpha)$. The non-interactive proof σ then consists of the elements α, γ of the Σ-protocol (i.e., the commitment and the response). Given the input x and (α, γ), the verifier computes $\beta = f(x, \alpha)$ and checks that $V^*(x, \alpha, \beta, \gamma)$ =accept. It is easy to show that when the hash function f is modeled as a random oracle then the resulting protocol is still computationally sound (since the challenge β is still a string of random bits that the adversary cannot control, other than attempting to select a polynomial number of them). Moreover, if the simulator can program the random oracle then this protocol also remains zero-knowledge.

We next show that using SI-CorInt families (with respect to quaternary relations), we can construct function families for which the Fiat-Shamir transformation yields a single-theorem NIZK argument system in the CRS model.

- Let (P, V) be a Σ-protocol in which the commitment, challenge, and response are of lengths $|\alpha| = t_1(|x|)$, $|\beta| = t_2(|x|)$, and $|\gamma| = t_3(|x|)$, respectively.
- Also, let $\{F : \{0,1\}^n \times \{0,1\}^{\ell_1(n)} \rightarrow \{0,1\}^{\ell_2(n)}\}_{n \in \mathbb{N}}$ be a function family, where ℓ_1, ℓ_2 are polynomially bounded from above and below. (That is $n^{1/c} \leq \ell_1(n), \ell_2(n) \leq n^c$ for some constant $c > 1$ and every sufficiently large n).

For inputs of length $|x| = m$, we choose the security parameter n (which defined the seed-length for F) as $n = \max\{ 2(m + t_1(m) + t_3(m)), \ell_1^{-1}(m + t_1(m)), \ell_2^{-1}(t_2(m)) \}$. Namely, n is chosen large enough so that $m + t_1(m) + t_3(m) \leq n/2$ and also $\ell_1(n) \geq m + t_1(m)$ and $\ell_2(n) \geq t_2(m)$. Note that since ℓ_1, ℓ_2 and the t_i's are polynomially-bounded, then n is polynomial in m. Below we view n, ℓ_1, ℓ_2 as functions of the input length m. We then reset the input and output length to be exactly $\ell_1' = m + t_1(m)$ and $\ell_2' = t_2(m)$ by setting

$$F'(s, x') \stackrel{\text{def}}{=} \lfloor F(s, x) \rfloor_{\ell_2'}, \text{ where } x = x' || 0^{\ell_1 - \ell_1'}$$

Finally we define $H : \{0,1\}^{n+\ell_2'} \times \{0,1\}^{\ell_1'} \to \{0,1\}^{\ell_2'}$ as

$$H(\langle s, z \rangle, x') \stackrel{\text{def}}{=} z \oplus F'(s, x') = z \oplus \left\lfloor F\left(s, x' || 0^{\ell_1 - \ell_1'}\right) \right\rfloor_{\ell_2'} \quad (1)$$

We are now ready to describe the NIZK argument system. The CRS consists of a pair (s, z) where $s \in \{0,1\}^n$ is a seed for the underlying function F and z is a random string of length $\ell_2' = t_2(m)$ (so together $\langle s, z \rangle$ are a seed for the function H from Eq. (1)). The NIZK argument system is obtained by applying the Fiat-Shamir transformation to the original Σ-protocol (P, V) using the function $H_{s,z}$.

Namely, on input $(x, w) \in R_L$ with $|x| = m$ and $\mathsf{crs} = (s, z)$, the prover sets $(\alpha, \text{state}) = P_1(x, w)$, $\beta = H_{s,z}(x, \alpha)$ and $\gamma = P_2(x, w, \text{state}, \alpha, \beta)$. The proof is the string $\sigma = (\alpha, \gamma)$. Given x, $\mathsf{crs} = (s, z)$, and the proof $\sigma = (\alpha, \gamma)$, the verifier computes $\beta = H_{s,z}(x, \alpha)$ and checks that $V^*(x, \alpha, \beta, \gamma) = $accept.

Theorem 3. *Let (P, V) be a three round Σ-protocol for the language L, defined by the NP relation R_L, and let $F = \{F : \{0,1\}^n \times \{0,1\}^{\ell_1(n)} \to \{0,1\}^{\ell_2(n)}\}_{n \in \mathbb{N}}$ be a function family with polynomially-bounded length functions.*

If (P, V) has a negligible soundness error and F is correlation intractable under seed-compression attacks with respect to quaternary relations, then applying the Fiat-Shamir transformation to (P, V) using the function family H from Eq. (1) yields a single theorem NIZK argument system for L in the CRS model.

Proof. The perfect completeness of the resulting NIZK system follows immediately from the perfect completeness of the Σ-protocol.

For the zero-knowledge property, the simulator S^* for the NIZK system uses the simulator S given by the Σ-protocol. It first chooses a random value $\beta' \in \{0,1\}^{t_2(|x|)}$ and uses S to compute $S(z) = (\alpha', \beta', \gamma')$. It then chooses a random seed $s \in \{0,1\}^n$ for the function F and computes

$$z = \beta' \oplus F'(s, \langle x, \alpha' \rangle) = \beta' \oplus \left\lfloor F\left(s, \langle x, \alpha' \rangle || 0^{\ell_1 - \ell_1'}\right) \right\rfloor_{\ell_2'}$$

Note that by definition, we have $H(\langle s, z \rangle, \langle x, \alpha' \rangle) = \beta'$. The simulator S^* outputs the CRS $\langle s, z \rangle$ and proof $\langle \alpha', \gamma' \rangle$. Clearly, the distribution on the output of S^* is identical to the distribution on the real pairs of CRS and proof.

It is left to prove computational soundness. Suppose for contradiction that there existed an efficient cheating prover $\widehat{P}'$ such that for a constant $c > 0$ and infinitely many $x \notin L$ it holds that:

$$\Pr_{s,z}\left[\sigma \leftarrow \widehat{P}'((s,z),x) : \widehat{V}((s,z),x,\sigma) = \text{accept}\right] \geq |x|^{-c}. \qquad (2)$$

We then show a *non-uniform* seed-compression adversary $\text{Adv} = (A,B)$ that breaks the correlation-intractability of the underlying family F.

For each $x \notin L$, denote by $z(x)$ the auxiliary string z that maximizes the success probability of $\widehat{P}'$. That is,

$$z(x) = \operatorname{argmax}_z\left\{\Pr_s\left[\sigma \leftarrow \widehat{P}'((s,z),x) : \widehat{V}((s,z),x,\sigma) = \text{accept}\right]\right\}.$$

An easy averaging argument implies that whenever Eq. (2) holds:

$$\Pr_s\left[\sigma \leftarrow \widehat{P}'((s,z(x)),x) : \widehat{V}((s,z(x)),x,\sigma) = \text{accept}\right] \geq |x|^{-c}.$$

On the other hand, for any $x \notin L$ the relation

$$\widehat{R}_x = \left\{(x_1,x_2,y_1,y_2) \;\middle|\; \begin{array}{l} x_1 = \left(\langle x,\alpha\rangle \,\|0^{\ell_1 - \ell_1'}\right) \text{ and} \\ V^*\left(x,\alpha,\left(z(x) \oplus \lfloor y_1 \rfloor_{\ell_2'}\right),x_2\right) = \text{accept} \end{array}\right\}$$

is evasive by the soundness of the original Σ-protocol. (Note, that $z(x)$ is a constant in this relation, so XOR-ing it to $\lfloor y_1 \rfloor_{\ell_2'} = \lfloor f(x_1)\rfloor_{\ell_2'}$ has no effect on soundness when f is a random function.) It follows that for our evasive relations $\widehat{R}_x$ $\widehat{R} = \bigcup_{x \notin L} \widehat{R}_x$ is also an evasive relation.

Since our choices of parameters imply that $|x| + |\alpha| + |\gamma| \leq n/2$, then we can use $\widehat{P}'$ to construct a seed-compression attacker $\text{Adv} = (A,B)$. The first part A gets as advice string the values $x, z(x)$ for some $x \notin L$ for which Eq. (2) holds, as well as the seed s for F. It uses $\widehat{P}'$ to compute α, γ, and outputs (x,α,γ) as the "compressed seed", and indeed the length of this "compressed seed" is at most $n/2$. The second part B outputs $\langle x,\alpha\rangle$ and γ, and indeed we have by definition $(\langle x,\alpha\rangle, \gamma, F(s,\langle x,\alpha\rangle), F(s,\gamma)) \in \widehat{R}$.

5.3 Non-uniformity in the Proof of the Fiat-Shamir Transform

We comment that the proof of Theorem 3 seems inherently non-uniform. We use non-uniformity in two places: one is to select $x \notin L$ for which $\widehat{P}'$ has good success probability, and the other to select the auxiliary $z(x)$. The first use can be eliminated by switching to a uniform soundness condition on the underlying Σ-protocol (i.e., when a uniform cheating prover needs to output some $x \notin L$ together with a convincing proof for it). The latter use of non-uniformity seems harder to eliminate, however. Maybe this can be done by switching to 6-ary relations and setting $z = F(s,0)$, so we get

$$\widehat{R}' = \left\{(x_1,x_2,x_3,y_1,y_2,y_3) \;\middle|\; \begin{array}{l} x_1 = 0, \; x_2 = \langle x,\alpha\rangle, \; x_3 = \gamma, \\ x \notin L \text{ and } V^*\left(x, \; \alpha, \; \lfloor y_1 \oplus y_2\rfloor_{\ell_2'}, \; \gamma\right) = \text{accept} \end{array}\right\}$$

It is not hard to see that this $\hat{R}'$ is evasive, but it is not clear how to translate the success of $\hat{P}'$ in breaking the soundness (when z is chosen at random) to success against this $\hat{R}'$ (when z is set as $z = F(s, 0)$ for a random s).

6 Future Work and Open Problems

The most intriguing open question that results from this work is whether seed-incompressible correlation-intractable functions can be constructed under more traditional computational assumptions. Given that their existence implies the existence of NIZK protocols without the aid of any apparent trapdoor feature, such a construction will likely need substantial insight. Alternately, and just as interesting, would be an argument showing that knowledge of small numbers of key-bits really does allow one to say something meaningful about composition based block-cipher and hash-function designs.

In a slightly orthogonal direction, the impossibility results presented in this paper are derived through proving exactly the one property of the function generators that we know the adversary has direct knowledge of: the function is computed by a small circuit. Any definition along the lines of seed-incompressibility that also managed to circumvent this problem would be interesting.

Finally, it would be nice if one could show that the OAEP scheme proposed by Bellare and Rogaway[3] (or a close relative of it) could be proven secure under such an assumption, as it is this random-oracle protocol that is probably used in practice on the most frequent basis (due to its inclusion in the TLS protocol [12] for secure web transactions), and thus further evidence of its security would be heartening.

References

1. Barak, B., Lindell, Y., Vadhan, S.: Lower Bounds for Non-Black-Box Zero-Knowledge. The Journal of Computer and System Sciences 72(2), 321–391 (2006) (JCSS FOCS 2003 Special Issue)
2. Bellare, M., Rogaway, P.: Random oracles are practical: a paradigm for designing efficient protocols. In: 1st Conference on Computer and Communications Security, pp. 62–73. ACM, New York (1993)
3. Bellare, M., Rogaway, P.: Optimal asymmetric encryption. In: De Santis, A. (ed.) EUROCRYPT 1994. LNCS, vol. 950, pp. 92–111. Springer, Heidelberg (1995)
4. Canetti, R.: Towards realizing random oracles: Hash functions that hide all partial information. In: Kaliski Jr., B.S. (ed.) CRYPTO 1997. LNCS, vol. 1294, pp. 455–469. Springer, Heidelberg (1997)
5. Canetti, R., Dodis, Y., Halevi, S., Kushilevitz, E., Sahai, A.: Exposure-Resilient Functions and All-or-Nothing Transforms. In: Preneel, B. (ed.) EUROCRYPT 2000. LNCS, vol. 1807, pp. 453–469. Springer, Heidelberg (2000)
6. Canetti, R., Goldreich, O., Halevi, S.: The random oracle methodology, revisited. Journal of the ACM 51(4), 209–218 (2004) Preliminary version in STOC 1998, pp. 209–218.

7. Canetti, R., Micciancio, D., Reingold, O.: Perfectly one-way probabilistic hashing. In: Proceedings of the 30th Annual ACM Symposium on the Theory of Computing, Dallas, TX, May 1998, pp. 131–140. ACM Press, New York (1998)
8. Cash, D., Ding, Y.Z., Dodis, Y., Lee, W., Lipton, R.J., Walfish, S.: Intrusion-Resilient Key Exchange in the Bounded Retrieval Model. In: Vadhan, S.P. (ed.) TCC 2007. LNCS, vol. 4392, pp. 479–498. Springer, Heidelberg (2007)
9. Coron, J., Dodis, Y., Malinaud, C., Puniya, P.: Merkle-Damgrd Revisited: How to Construct a Hash Function. In: Shoup, V. (ed.) CRYPTO 2005. LNCS, vol. 3621, pp. 430–448. Springer, Heidelberg (2005)
10. Damgard, I.: On Σ-Protocols. Lecture notes for Cryptologic Protocol Theory course, Aarhus University (2005), http://www.daimi.au.dk/%7Eivan/Sigma.pdf
11. DiCrescenzo, G., Lipton, R.J., Walfish, S.: Perfectly Secure Password Protocols in the Bounded Retrieval Model. In: Halevi, S., Rabin, T. (eds.) TCC 2006. LNCS, vol. 3876, pp. 225–244. Springer, Heidelberg (2006)
12. Dierks, T., Allen, C.: RFC2246:The TLS Protocol. RFC 2246, The Internet Society, Network Working Group (1999)
13. Dziembowski, S.: Intrusion-Resilience Via the Bounded-Storage Model. In: Halevi, S., Rabin, T. (eds.) TCC 2006. LNCS, vol. 3876, pp. 207–224. Springer, Heidelberg (2006)
14. Fiat, A., Shamir, A.: How to prove yourself. practical solutions to identification and signature problems. In: Odlyzko, A.M. (ed.) CRYPTO 1986. LNCS, vol. 263, pp. 186–189. Springer, Heidelberg (1987)
15. Goldreich, O.: Modern Cryptography, Probabilistic Proofs and Pseudorandomness. Algorithms and Combinatorics, vol. 17. Springer, Heidelberg (1998)
16. Goldwasser, S., Kalai, Y.T.: On the (In)security of the Fiat-Shamir Paradigm. In: 44th Symposium on Foundations of Computer Science (FOCS 2003), pp. 102–115. IEEE Computer Society Press, Los Alamitos (2003)
17. Harnik, D., Naor, M.: On the Compressibility of NP Instances and Cryptographic Applications. In: 47th Annual IEEE Symposium on Foundations of Computer Science (FOCS 2006), pp. 719–728. IEEE Computer Society Press, Los Alamitos (2006)
18. Impagliazzo, R., Rudich, S.: Limits on the provable consequences of one-way permutations. In: Proceedings of the 21st Annual ACM Symposium on Theory of Computing, pp. 44–61. ACM Press, New York (1989)
19. Kilian, J.: A note on efficient zero-knowledge proofs and arguments. In: Proceedings of the 24th Annual ACM Symposium on the Theory of Computing, May 1992, pp. 723–732. ACM Press, New York (1992)
20. Micali, S.: CS proofs. In: 35th Annual Symposium on Foundations of Computer Science (FOCS 1994), pp. 436–453. IEEE Computer Society Press, Los Alamitos (1994)
21. Micali, S.: Computationally Sound Proofs. SIAM Journal on Computing 30(4), 1253–1298 (2000)
22. Naor, M., Nissim, K.: Computationally sound proofs: Reducing the number of random oracle calls (manuscript, 1999)

Asymptotically Efficient Lattice-Based Digital Signatures*

Vadim Lyubashevsky and Daniele Micciancio

University of California, San Diego
La Jolla, CA 92093-0404, USA
{vlyubash,daniele}@cs.ucsd.edu

Abstract. We give a direct construction of digital signatures based on the complexity of approximating the shortest vector in ideal (e.g., cyclic) lattices. The construction is provably secure based on the worst-case hardness of approximating the shortest vector in such lattices within a polynomial factor, and it is also asymptotically efficient: the time complexity of the signing and verification algorithms, as well as key and signature size is almost linear (up to poly-logarithmic factors) in the dimension n of the underlying lattice. Since no sub-exponential (in n) time algorithm is known to solve lattice problems in the worst case, even when restricted to cyclic lattices, our construction gives a digital signature scheme with an essentially optimal performance/security trade-off.

1 Introduction

Digital signature schemes, initially proposed in Diffie and Hellman's seminal paper [9] and later formalized by Goldwasser, Micali and Rivest, [15], are among the most important and widely used cryptographic primitives. Still, our understanding of these intriguing objects is somehow limited.

The definition of digital signatures clearly fits within the public key cryptography framework. However, efficiency considerations aside, the existence of secure digital signatures schemes can be shown to be equivalent to the existence of conventional (symmetric) cryptographic primitives like pseudorandom generators, one-way hash functions, private key encryption, or even just one-way functions [23,27]. There is a big gap, both theoretical and practical, between the efficiency of known constructions implementing public-key and private-key cryptography. In the symmetric setting, functions are often expected to run in time which is linear or almost linear in the security parameter k. However, essentially all known public key encryption schemes with a supporting proof of security are based on algebraic functions that take at least $\Omega(k^2)$ time to compute, where 2^k is the conjectured hardness of the underlying problem. For example, all factoring based schemes must use keys of size approximately $O(k^3)$ to achieve k bits of

---

* Research supported in part by NSF grant CCF-0634909. Any opinions, findings, and conclusions or recommendations expressed in this material are those of the author(s) and do not necessarily reflect the views of the National Science Foundation.

© International Association for Cryptologic Research 2008

security to counter the best known sub-exponential time factoring algorithms, and modular exponentiation raises the time complexity to over $\omega(k^4)$ even when restricted to small k-bit exponents and implemented with an asymptotically fast integer multiplication algorithm.

When efficiency is taken into account, digital signatures seem much closer to public key encryption schemes than to symmetric encryption primitives. Most signature schemes known to date employ the same set of number theoretic techniques commonly used in the construction of public key encryption schemes, and result in similar complexity. Digital signatures based on arbitrary one-way hash functions have also been considered, due to the much higher speed of conjectured one-way functions (e.g., instantiated with common block ciphers as obtained from ad-hoc constructions) compared to the cost of modular squaring or exponentiation operations typical of number theoretic schemes. Still, the performance advantage of one-way function is often lost in the process of transforming them into digital signature schemes: constructions of signature schemes from non-algebraic one-way functions almost invariably rely on Lamport and Diffie's [9] one-time signature scheme (and variants thereof) which requires a number of one-way function applications essentially proportional to the security parameter. So, even if the one-way function can be computed in linear time $O(k)$, the complexity of the resulting signature scheme is again at least quadratic $\Omega(k^2)$.

Therefore, a question of great theoretical and practical interest, is whether digital signature schemes can be realized at essentially the same cost as symmetric key cryptographic primitives. While a generic construction that transforms any one-way function into a signature scheme with similar efficiency seems unlikely, one may wonder if there are specific complexity assumptions that allow to build more efficient digital signature schemes than currently known. Ideally, are there digital signature schemes with $O(k)$ complexity, which can be proved as hard to break as solving a computational problem which is believed to require $2^{\Omega(k)}$ time?

1.1   Results and Techniques

The main result in this paper is a construction of a provably secure digital signature scheme with key size and computation time almost linear (up to polylogarithmic factors) in the security parameter. In other words, we give a new digital signature scheme with complexity $O(k \log^c k)$ which can be proved to be as hard to break as a problem which is conjectured to require $2^{\Omega(k)}$ time to solve.

The problem underlying our signature scheme is that of approximating the shortest vector in a lattice with "cyclic" or "ideal" structure, as already used in [22] for the construction of efficient lattice based one-way functions, and subsequently extended to collision resistant functions in [25,18]. As in most previous work on lattices, our scheme can be proven secure based on the *worst case* complexity of the underlying lattice problems.

Since one-way functions are known to imply the existence of many other cryptographic primitives (e.g., pseudorandom generators, digital signatures, private key encryption, etc.), the efficient lattice based one-way functions of [22] immediately yield corresponding cryptographic primitives based on the complexity of cyclic lattices. However, the known generic constructions of cryptographic primitives from one-way functions are usually very inefficient. So, it was left as an open problem in [22] to find *direct* constructions of other cryptographic primitives from lattice problems with performance and security guarantees similar to those of [22]. For the case of collision resistant hash functions, the problem was resolved in [25,18], which showed that various variants of the one-way function proposed in [22] are indeed collision resistant. In this paper we build on the results of [22,25,18] to build an asymptotically efficient lattice-based digital signature scheme.

Theorem 1. *There exists a signature scheme such that the signature of an n-bit message is of length $\tilde{O}(k)$ and both the signing and verification algorithms take time $\tilde{O}(n) + \tilde{O}(k)$. The scheme is strongly unforgeable in the chosen message attack model, assuming the hardness of approximating the shortest vector problem in all ideal lattices of dimension k to within a factor $\tilde{O}(k^2)$.*

Our lattice based signature scheme is based on a standard transformation from one-time signatures (i.e., signatures that allow to securely sign a single message) to general signature schemes, together with a novel construction of a lattice based one-time signature. We remark that the same transformation from one-time signatures to unrestricted signature schemes was also employed by virtually all previous constructions of digital signatures from arbitrary one-way functions (e.g., [21,23,27]). This transformation, which combines one-time signatures together with a tree structure, is relatively efficient and allows one to sign messages with only a logarithmic number of applications of a hash function and a one-time signature scheme [28]. The bottleneck in one-way function based signature schemes is the construction of one-time signatures from one-way functions. The reason for the slowdown is that the one-way function is typically used to sign a k-bit message one bit at a time, so that the entire signature requires k evaluations of the one-way function. In this paper we give a direct construction of one-time signatures, where each signature just requires two applications of the lattice based one-way function of [22,25,18]. The same lattice based hash function can then be used to efficiently transform the one-time signature into an unrestricted signature scheme with only a logarithmic loss in performance.

The high level structure of our lattice based one-time signature scheme is easily explained. The construction is based on the generalized compact knapsack functions of [22,25,18]. These are keyed functions (indexed by a key $(a_1, \ldots, a_k)$) of the form

$$h(x_1, \ldots, x_m) = \sum_i a_i \cdot x_i,$$

where $a_1, \ldots, a_m, x_1, \ldots, x_m$ are elements of some large ring R, and the result of the function is also in R. The domain of the function is restricted to $x_i \in D$, where D is a subset of R of small elements. For example, if R is the ring of integers, and $D = \{0, 1\}$, then h is just the subset-sum function. Notice that if

D is not restricted, then h is certainly not a one-way function: the function can be easily inverted over the integers using the extended Euclid algorithm for greatest common divisor computation. For efficiency reasons, here (as in [22,25,18]) we use a different ring R and a much larger subset $D \subset R$, so that a single element of D can be used to encode a k-bit message (see section 2.3). We now give very high level overviews of our one-time signature and the proof of its security.

One-time signature. When the user wants to generate a key for the one-time signature scheme, he simply picks two "random" inputs $x, y \in D^m$, and computes their images under the hash function $(h(x), h(y))$. (The key $(a_1, \ldots, a_m)$ to the hash function h can also be individually chosen by the user, or shared among all the users of the signature scheme.) The secret key is the pair (x, y) while the public key is given by their hashes $(h(x), h(y))$. Then, the signature of a message z is simply obtained as a "linear combination" $x \cdot z + y$ of the two secret vectors, with coefficient being the message z to be signed. (The multiplication $x \cdot z$ is defined as the ring multiplication of each coordinate of x by z.) Signatures can be easily verified using the homomorphic properties of the lattice based hash function $h(x \cdot z + y) = h(x) \cdot z + h(y)$.

Security proof. If the domain D^m were closed under the ring addition and multiplication operations, then one could show that the public key $(h(x), h(y))$ and signature $x \cdot z + y$ do not reveal enough information to obtain the signer's secret key (x, y), and a forgery relative to a *different* secret key will yield a collision to the hash function. But because the domain is restricted, there is a possibility that the signer's secret key was the only one that could have produced $h(x), h(y)$ and signature $x \cdot z + y$, and so an adversary who sees these values might be able to deduce the secret key. This turns out to be the main difficulty in carrying out our proof. We overcome this technical problem by choosing the secret key elements x, y according to a carefully crafted (non-uniform) probability distribution, which can be intuitively thought as a "fuzzy" subset of the full domain R^m. It turns out that if the appropriate distribution is used, then we can have the domain D^m be closed under the ring operations in an approximate probabilistic sense, and still have h be a function that's hard to invert.

1.2 Related Work

Lamport showed the first construction of a one-time signature based on the existence of one-way functions. In that scheme, the public key consists of the values $f(x_0), f(x_1)$, where f is a one-way function and x_0, x_1 are randomly chosen elements in its domain. The elements x_0 and x_1 are kept secret, and in order to sign a bit i, the signer reveals x_i. This construction requires one application of the one-way function for every bit in the message. Since then, more efficient constructions have been proposed in (e.g. [20,7,6,11,4,5,16]), but there was always an inherent limitation in the number of bits that could be signed efficiently with one application of the one-way function [12].

Provably secure cryptography based on lattice problems was pioneered by Ajtai in [2], and attracted considerable attention within the complexity theory community because of a remarkable worst-case/average-case connection: it is possible to show that breaking the cryptographic function on the average is at least as hard as solving the lattice problem in the worst-case. Unfortunately, functions related to k-dimensional lattices typically involve an k-dimensional matrix/vector multiplication, and therefore require k^2 time to compute (as well as k^2 storage for keys). A fundamental step towards making lattice based cryptography more attractive in practice, was taken by Micciancio [22] who proposed a variant of Ajtai's function which is much more efficient to compute (thanks to the use of certain lattices with a special cyclic structure) and still admits a worst-case/average-case proof of security. The performance improvement in [22] (as well as in subsequent work [25,18],) comes at a cost: the resulting function is as hard to break as solving the shortest vector problem in the worst case over lattices with a cyclic structure. Still, since the best known algorithms do not perform any better on these lattices than on general ones, it seems reasonable to conjecture that the shortest vector problem is still exponentially hard. It was later shown in [25,18] that, while the function constructed in [22] was only one-way, it is possible to construct efficient collision-resistant hash functions based on the hardness of problems in lattices with a similar algebraic structure.

1.3   Open Problems

Our work raises many interesting open problems. One such problem is constructing a one-time signature with similar efficiency, but based on a weaker hardness assumption. For instance, it would be great to provide a one-time signature with security based on the hardness of approximating the shortest vector problem (in ideal lattices) to within a factor of $\tilde{O}(n)$. Also, with the recent results of Peikert and Rosen [26], showing a possible way to build cryptographic functions whose security is based on approximating the shortest vector in special lattices to within a factor $O(\sqrt{\log n})$, we believe that it is worthwhile exploring whether one-time signatures can be built based on similar assumptions.

Another direction to try to build efficient signature schemes based directly on the hardness of lattice problems without going through one-time signatures and an authentication tree. The main advantage of such a scheme would be that the signer would not have to "keep a state" and remember which verification keys have already been used. Such constructions have been achieved based on problems from number theory [13,8] but they are not as efficient, in an asymptotic sense, as the signature scheme presented here.

While the scheme presented here has almost optimal asymptotic efficiency, it is not yet ready to be used for practical applications (see Section 4). The main issue is that lattice reduction algorithms perform much better in practice than in theory, and thus our signature scheme may be insecure for parameters appropriate for practical schemes. Nevertheless, the recent advances in lattice-based cryptography are a very encouraging sign that with some novel ideas, our construction can be modified into a serviceable signature scheme.

2   Preliminaries

2.1   Signatures

We recall the definitions of signature schemes and what it means for a signature scheme to be secure.

Definition 2. *A signature scheme consists of a triplet of polynomial-time (possibly probabilistic) algorithms (G, S, V) such that for every pair of outputs (s, v) of $G(1^n)$ and any n-bit message m,*

$$Pr[V(v, m, S(s, m)) = 1] = 1$$

where the probability is taken over the randomness of algorithms S and V.

In the above definition, G is called the key-generation algorithm, S is the signing algorithm, V is the verification algorithm, and s and v are, respectively, the signing and verification keys.

A signature scheme is said to be secure if there is only a negligible probability that any adversary, after seeing signatures of messages of his choosing, can sign a message whose signature he has not already seen [15]. One-time security means that an adversary, after seeing a signature of a single message of his choosing, cannot produce a valid signature of a different message.

Definition 3. *A signature scheme (G, S, V) is said to be one-time secure if for every polynomial-time (possibly randomized) adversary $\mathcal{A}$, the probability that after seeing $(m, S(s, m))$ for any message m of its choosing, $\mathcal{A}$ can produce $(m' \neq m, \sigma')$ such that $V(v, m', \sigma') = 1$, is negligibly small. The probability is taken over the randomness of G, S, V, and $\mathcal{A}$.*

In the standard security definition of a signature scheme, the adversary should not be able to produce a signature of a message he hasn't already seen. A stronger notion of security, called *strong unforgeability* requires that in addition to the above, an adversary shouldn't even be able to come up with a different signature for a message whose signature he has already seen. The scheme presented in this paper satisfies this stronger notion of unforgeability.

Another feature of signatures that is sometimes desirable is the ability of the legitimate signer to prove that a message was not actually signed by her. Of course, it should be impossible for the signer to repudiate a message that she actually signed. Signatures schemes that have this feature are called Fail-Stop [24]. Our scheme has this property as well.

2.2   Notation

Let $R = \mathbb{Z}_p[x]/\langle f \rangle$ be a ring where f is an irreducible monic polynomial of degree n over $\mathbb{Z}[x]$ and p is some small prime. For the rest of the paper, the variables n, p, and f will always be associated with the ring R. We will denote elements in R by bold letters and elements of R^m, for some positive integer m, by a bold letter with a hat. That is, $\hat{a} = (a_1, \ldots, a_m) \in R^m$ when all the a_i's are in R. For an element $\hat{a} = (a_1, \ldots, a_m) \in R^m$ and an element $z \in R$, we define $\hat{a}z = (a_1 z, \ldots, a_m z)$. For two elements $\hat{a}, \hat{b} \in R^m$, addition is defined as $\hat{a} + \hat{b} = (a_1 + b_1, \ldots, a_m + b_m)$ and the dot product as $\hat{a} \odot \hat{b} = a_1 b_1 + \ldots + a_m b_m$.

Notice that with the operations that we defined, the set R^m is an R-module. That is, R^m is an abelian additive group such that for all $\hat{a}, \hat{b} \in R^m$ and $r, s \in R$, we have

1. $(\hat{a} + \hat{b})r = \hat{a}r + \hat{b}r$
2. $(\hat{a}r)s = \hat{a}(rs)$
3. $\hat{a}(r + s) = \hat{a}r + \hat{a}s$

We will now give a definition for the "length" of elements in R. To do so, we will first need to specify their representations in the ring. For our application, we will represent elements in R by a polynomial of degree $n - 1$ having coefficients in the range $[-\frac{p-1}{2}, \frac{p-1}{2}]$, and so when we talk about reduction modulo p, we mean finding an equivalent element modulo p in the aforementioned range. For an element $a = a_0 + a_1 x + \ldots + a_{n-1} x^{n-1} \in R$, we define $\|a\|_\infty = max_i(|a_i|)$. Similarly, for elements $\hat{a} = (a_1, \ldots, a_m) \in R^m$, we define $\|\hat{a}\|_\infty = \max_i (\|a_i\|_\infty)$. Notice that $\| \cdot \|_\infty$ is not exactly a norm because $\|\alpha a\|_\infty \neq \alpha \|a\|_\infty$ for all integers α (because of the reduction modulo p), but it still holds true that $\|a + b\|_\infty \leq \|a\|_\infty + \|b\|_\infty$ and $\|\alpha a\|_\infty \leq \alpha \|a\|_\infty$.

While putting an upper-bound on $\|a + b\|_\infty$ is straight-forward, it turns out that upper-bounding $\|ab\|_\infty$ is somewhat more involved. Suppose that we are trying to determine the upper bound on $\|ab\|_\infty$. For a moment, let's pretend that a and b are polynomials in $\mathbb{Z}[x]$. Then, the product ab will have degree at most $2n - 2$ and the absolute value of the maximum coefficient of ab will be at most $n\|a\|_\infty \|b\|_\infty$. Reducing ab modulo p will not increase the absolute value of the maximum coefficient, but reducing modulo the polynomial f can (and usually does). So if we want to upper bound $\|ab\|_\infty$, we need to account for the increase in the coefficient size when we reduce a polynomial in $\mathbb{Z}[x]$ of degree $2n - 2$ modulo f.

For any ring R, we define a constant $\phi(R)$ as,

$$\phi(R) = \min \{j : \forall a, b \in R, \ \|ab\|_\infty \leq jn\|a\|_\infty \|b\|_\infty\}.$$

The constant $\phi(R)$ is intimately tied to the concept of "expansion factor" introduced in [18]. It is also somewhat related to the root discriminant of a number field as described in [26]. We will not go into many details here, other than to mention that there are many polynomials f which result in $\phi(R)$ being small and it is not too hard to upper bound the value of $\phi(R)$. For example for $f = x^n + 1$, $\phi(R) = 1$ and for $f = x^n + x^{n-1} + \ldots + 1$, $\phi(R) \leq 2$. In the rest of the paper, we will omit the parameter R, and just write ϕ.

2.3 A Hash Function Family

We now define a function family $\mathcal{H}_{R,m}$ that maps R^m to R. The functions $h \in \mathcal{H}_{R,m}$ are indexed by elements $\hat{a} \in R^m$. The input to the function is an element $\widehat{z} \in R^m$, and the output is $\widehat{a} \odot \widehat{z}$. So the functions in $\mathcal{H}_{R,m}$ map elements from R^m to R. To summarize,

$$\mathcal{H}_{R,m} = \{h_{\widehat{a}} : \widehat{a} \in R^m\}, \text{ where } h_{\widehat{a}}(\widehat{z}) = \widehat{a} \odot \widehat{z}$$

Throughout the paper, we will write h rather than $h_{\widehat{a}}$ with the understanding that there is an $\widehat{a}$ associated with the function h. Notice that we can efficiently generate random functions from the function family $\mathcal{H}_{R,m}$ by simply generating a random $\widehat{a} \in R^m$.

It was shown in [18] that finding two "small" elements $\widehat{s}, \widehat{s}' \in R^m$ such that $h(\widehat{s}) = h(\widehat{s}')$ for randomly chosen $h \in \mathcal{H}_{R,m}$ is at least as hard as solving the approximate shortest vector problem for *all* lattices of a certain type (a problem which is believed to be hard). The security of our signature scheme will be based on the hardness of this collision problem. We now define the problem formally.

Definition 4. *The collision problem,* $Col_{d,\mathcal{H}_{R,m}}(h)$ *takes as input a random function* $h \in \mathcal{H}_{R,m}$, *and asks to find two distinct elements* $\widehat{s}, \widehat{s}' \in R^m$ *with* $\|\widehat{s}\|_\infty, \|\widehat{s}'\|_\infty \leq d$ *such that* $h(\widehat{s}) = h(\widehat{s}')$.

We now make some useful observations about the function family $\mathcal{H}_{R,m}$. The first observation is that the functions in $\mathcal{H}_{R,m}$ are module homomorphisms.

Claim. $\mathcal{H}_{R,m}$ is a set of module homomorphisms. That is, for every $\widehat{k}, \widehat{l} \in R^m$, $z \in R$, and $h \in \mathcal{H}_{R,m}$, the following two conditions are satisfied:

1. $h(\widehat{k} + \widehat{l}) = h(\widehat{k}) + h(\widehat{l})$
2. $h(\widehat{k}z) = h(\widehat{k})z$

Proof. By the definition of the hash function h, we have

1. $h(\widehat{k} + \widehat{l}) = \widehat{a} \odot (\widehat{k} + \widehat{l}) = \widehat{a} \odot \widehat{k} + \widehat{a} \odot \widehat{l} = h(\widehat{k}) + h(\widehat{l})$
2. $h(\widehat{k}z) = \widehat{a} \odot (k_1 z, \ldots, k_m z) = a_1 k_1 z + \ldots + a_m k_m z$
   $= (a_1 k_1 + \ldots + a_m k_m)z = (\widehat{a} \odot \widehat{k})z = h(\widehat{k})z$

$\square$

The next observation is that the kernel of every $h \in \mathcal{H}_{R,m}$ contains an exponential number of "small" elements.

Lemma 5. *For every* $h \in \mathcal{H}_{R,m}$, *there exist at least* 5^{mn} *elements* $\widehat{y} \in R^m$ *such that* $\|\widehat{y}\|_\infty \leq 5p^{1/m}$ *and* $h(\widehat{y}) = 0$.

Proof. Let S be the set containing all elements in R^m with coefficients between 0 and $5p^{1/m}$. Since $|S| = (5p^{1/m} + 1)^{mn} > 5^{mn}p^n$ and $|R| = p^n$, by the pigeonhole principle, there exists a $t \in R$ and a subset $S' \subseteq S$ such that $|S'| \geq 5^{mn}$ and for all $\widehat{s}' \in S'$, $h(\widehat{s}') = t$. If $t = 0$, then we're done, otherwise let $S' = \{\widehat{s}'_1, \widehat{s}'_2, \ldots, \widehat{s}'_k\}$ and consider the set $Y = \{\widehat{s}'_1 - \widehat{s}'_1, \widehat{s}'_1 - \widehat{s}'_2, \ldots, \widehat{s}'_1 - \widehat{s}'_k\}$ of size $|S'|$. Note that for each $\widehat{y} \in Y$, $\|\widehat{y}\|_\infty \leq 5p^{1/m}$ and $h(\widehat{y}) = 0$ because of the homomorphic property of h. $\square$

2.4   Lattices

In this subsection, we explain the relationship between the collision problem from Definition 4 and finding shortest vectors in certain types of lattices.

An n-dimensional integer lattice $\mathcal{L}$ is a subgroup of $\mathbb{Z}^n$. A lattice $\mathcal{L}$ can be represented by a set of linearly independent generating vectors, called a basis. For any lattice vector $y \in \mathcal{L}$, the infinity norm of y, $\|y\|_\infty$, is the absolute value of the largest coefficient of y. The *minimum distance* (in the infinity norm[1]) of a lattice $\mathcal{L}$, denoted by $\lambda_1(\mathcal{L})$, is defined as:

$$\lambda_1(\mathcal{L}) = \min_{y \in \mathcal{L} \setminus \{0\}} \{\|y\|_\infty\}$$

Computing the $\lambda_1(\mathcal{L})$ of a lattice was first shown to be NP-hard by van Emde Boas [29], and it was shown hard to approximate to within a factor of $n^{1/\log\log n}$ by Dinur [10]. It is conjectured that approximating $\lambda_1(\mathcal{L})$ to within any polynomial factor is a hard problem (though not NP-hard [14,1]) since the fastest known algorithm takes time $2^{O(n)}$ to accomplish this [17,3].

Micciancio [22] defined a cyclic lattice to be a lattice $\mathcal{L}$ such that if the vector $(a_1, \ldots, a_{n-1}, a_n) \in \mathcal{L}$, then the vector $(a_n, a_1, \ldots, a_{n-1})$ is also in the lattice $\mathcal{L}$. Such lattices correspond to ideals in $\mathbb{Z}[x]/\langle x^n - 1 \rangle$. In [22], Micciancio gave a construction of an efficient family of one-way functions with security based on the worst case hardness of approximating $\lambda_1(\mathcal{L})$ in cyclic lattices. Subsequently, it was shown in [25,18] how to modify Micciancio's function in order to make it collision resistant. In addition, it was shown in [18] how to create efficient collision resistant hash functions with security based on approximating $\lambda_1(\mathcal{L})$ in lattices that correspond to ideals in rings $\mathbb{Z}[x]/\langle f \rangle$ for general f. A lattice corresponding to an ideal means that the vector $(a_0, \ldots, a_{n-1})$ is in the lattice, if and only if the polynomial $a_0 + a_1 x + \ldots + a_{n-1} x^{n-1}$ is in the ideal. Despite the added structure of these algebraic lattices, the best algorithms to solve the shortest vector problem are the same ones as for arbitrary lattices.

The following theorem is a weaker special case of the main result of [18] that is most pertinent to this work:

Theorem 6. *Let f be an irreducible polynomial in $\mathbb{Z}[x]$ of degree n and define integers $p = (\phi n)^3$ and $m = \lceil \log n \rceil$. If there exists a polynomial-time algorithm that solves $\mathrm{Col}_{d,\mathcal{H}_{R,m}}(h)$ for $R = \mathbb{Z}_p[x]/\langle f \rangle$ and $d = 10\phi p^{1/m} n \log^2 n$, then there exists a polynomial-time algorithm that approximates $\lambda_1(\mathcal{L})$ to within a factor of $\tilde{O}(\phi^5 n^2)$ for every lattice $\mathcal{L}$ corresponding to an ideal in the ring $\mathbb{Z}[x]/\langle f \rangle$.*

We point out that in [18, Theorem 2] (which is the main result of [18]), it is shown that solving the $\mathrm{Col}_{d,\mathcal{H}_{R,m}}(h)$ problem for certain parameters $p, d,$ and m implies approximating the shortest vector to within a factor of $\tilde{O}(n)$. Unfortunately, in the current paper we cannot show that breaking the one-time signature implies

---

[1] All the results in this paper can be adapted to any ℓ_p norm. For simplicity, we concentrate on the ℓ_∞ case, since it is the most convenient one in cryptographic applications.

solving the $Col_{d,\mathcal{H}_{R,m}}(h)$ problem for such optimal parameters (mainly, we cannot get the parameter d to be too small). And so Theorem 6 is a weaker version of [18, Theorem 2] where the parameters p, m, and d are set in a way such that breaking the one-time signature implies solving $Col_{d,\mathcal{H}_{R,m}}(h)$.

We also notice in the above theorem that the approximation factor heavily depends on ϕ. Thus it's prudent to choose a polynomial f that results in a small ϕ. Choosing irreducible polynomials of the form $x^n + 1$ or $x^n + x^{n-1} + \ldots + 1$ makes ϕ a small constant (1 and 2 respectively). We also point out that the integer p needs not be a prime for the proof of security to hold, but there are some practical advantages to setting it to a prime when implementing functions that involve multiplications of elements in $\mathbb{Z}_p[x]/\langle f \rangle$ [19].

3   The One-Time Signature Scheme

In this section we present our one-time signature scheme. The security of the scheme will be ultimately based on the worst-case hardness of approximating the shortest vector in all lattices corresponding to ideals in the ring $\mathbb{Z}[x]/\langle f \rangle$ for any irreducible polynomial f. The approximation factor is determined by the polynomial f as in Theorem 6. The key-generation algorithm for the signature scheme allows us to specify the polynomial f that we want to use for the hardness assumption.

Key-Generation Algorithm:
Input: 1^n, irreducible polynomial $f \in \mathbb{Z}[x]$ of degree n.
1: Set $p \leftarrow (\phi n)^3$, $m \leftarrow \lceil \log n \rceil$, $R \leftarrow \mathbb{Z}_p[x]/\langle f \rangle$
2: For all positive i, let the sets DK_i and DL_i be defined as:

$$DK_i = \{\widehat{\boldsymbol{y}} \in R^m \text{ such that } \|\widehat{\boldsymbol{y}}\|_\infty \le 5ip^{1/m}\}$$

$$DL_i = \{\widehat{\boldsymbol{y}} \in R^m \text{ such that } \|\widehat{\boldsymbol{y}}\|_\infty \le 5in\phi p^{1/m}\}$$

3: Choose uniformly random $h \in \mathcal{H}_{R,m}$
4: Pick a uniformly random string $r \in \{0,1\}^{\lfloor \log^2 n \rfloor}$
5: **if** $r = 0^{\lfloor \log^2 n \rfloor}$ **then**
6:     set $j = \lfloor \log^2 n \rfloor$
7: **else**
8:     set j to the position of the first 1 in the string r
9: **end if**
10: Pick $\widehat{\boldsymbol{k}}, \widehat{\boldsymbol{l}}$ independently and uniformly at random from DK_j and DL_j respectively
11: Signing Key: $(\widehat{\boldsymbol{k}}, \widehat{\boldsymbol{l}})$. Verification Key: $(h, h(\widehat{\boldsymbol{k}}), h(\widehat{\boldsymbol{l}}))$

Signing Algorithm:
Input: Message $z \in R$ such that $\|z\|_\infty \le 1$; signing key $(\widehat{\boldsymbol{k}}, \widehat{\boldsymbol{l}})$
Output: $\widehat{\boldsymbol{s}} \leftarrow \widehat{\boldsymbol{k}}z + \widehat{\boldsymbol{l}}$

Verification Algorithm:
Input: Message z; signature $\widehat{s}$; verification key $(h, h(\widehat{k}), h(\widehat{l}))$
Output: "ACCEPT", if $\|\widehat{s}\|_\infty \le 10\phi p^{1/m} n \log^2 n$ and $h(\widehat{s}) = h(\widehat{k})z + h(\widehat{l})$
"REJECT", otherwise.

At this point we would like to draw the reader's attention to the particulars of how the key-generation algorithm generates the secret signing key $(\widehat{k}, \widehat{l})$. Because of the way that the integer j is generated, the secret key $\widehat{k}$ (resp. $\widehat{l}$) gets chosen uniformly at random from the set DK_j (resp. DL_j) with probability 2^{-j} for $1 \le j < \lfloor \log^2 n \rfloor$ and with probability 2^{-j+1} for $j = \lfloor \log^2 n \rfloor$. Since $DK_1 \subset DK_2 \subset \ldots \subset DK_{\lfloor \log^2 n \rfloor}$ and $DL_1 \subset DL_2 \subset \ldots \subset DL_{\lfloor \log^2 n \rfloor}$, the keys $\widehat{k}$ and $\widehat{l}$ end up being chosen from the sets $DK_{\lfloor \log^2 n \rfloor}$ and $DL_{\lfloor \log^2 n \rfloor}$, but *not* uniformly at random. Notice that keys with smaller coefficients are more likely to be chosen, and it's also extremely unlikely that we will ever end up with keys that are not in $DK_{\lfloor \log^2 n \rfloor - 1}$ and $DL_{\lfloor \log^2 n \rfloor - 1}$. So with probability negligibly close to 1, there will always be valid secret keys that are "larger" than the ones generated by the key-generation algorithm. This will be crucial to the proof of security.

We will first show that the verification algorithm will always accept the signature generated by the signing algorithm of any message $z \in R$. Note that the signing keys $\widehat{k}, \widehat{l}$ are contained in sets $DK_{\log^2 n}$ and $DL_{\log^2 n}$ respectively. Thus $\|\widehat{k}\|_\infty \le 5p^{1/m} \log^2 n$ and $\|\widehat{l}\|_\infty \le 5\phi p^{1/m} n \log^2 n$. Therefore,

$$\|\widehat{s}\|_\infty = \|\widehat{k}z + \widehat{l}\|_\infty \le \|\widehat{k}z\|_\infty + \|\widehat{l}\|_\infty \le \phi n \|\widehat{k}\|_\infty \|z\|_\infty + \|\widehat{l}\|_\infty \le 10\phi p^{1/m} n \log^2 n$$

Also, by the homomorphic property of functions $h \in \mathcal{H}_{R,m}$,

$$h(\widehat{s}) = h(\widehat{k}z + \widehat{l}) = h(\widehat{k})z + h(\widehat{l}).$$

We next show that the above signature scheme is secure against forgery. More precisely, we show that forging a signature implies being able to solve the $Col_{d,\mathcal{H}_{R,m}}(h)$ problem for the parameters in Theorem 6, which in turn implies being able to approximate $\lambda_1(\mathcal{L})$ for any lattice $\mathcal{L}$ that corresponds to an ideal in the ring $\mathbb{Z}[x]/\langle f \rangle$.

Theorem 7. *If there exists a polynomial-time adversary that, after seeing a signature $\widehat{s} = \widehat{k}z + \widehat{l}$ of a message z, can output a valid signature of another message z' with probability $1/poly(n)$, then there exists a polynomial time algorithm that can solve the $Col_{d,\mathcal{H}_{R,m}}(h)$ problem for $d = 10\phi p^{1/m} n \log^2 n$.*

Proof. Let $\mathcal{A}$ be an adversary who can break the one-time signature scheme. This means that after seeing a signature for any message of his choice, $\mathcal{A}$ can then successfully sign a different message of his choice.

Before proceeding any further, we point out that an adversary who succeeds in forging a signature with non-negligible probability must succeed with non-negligible probability in the case that $j < \lfloor \log^2 n \rfloor$ in the key-generation step. This is because j equals $\lfloor \log^2 n \rfloor$ with probability only $2^{-\lfloor \log^2 n \rfloor + 1}$, and so an adversary must also be able to forge signatures for other values of j if he is to have a non-negligible success probability. In the remainder of the proof, we will be assuming that the j generated in the key-generation step was less than $\lfloor \log^2 n \rfloor$. In other words, we'll be assuming that $\widehat{k} \in DK_{\lfloor \log^2 n-1 \rfloor}$ and $\widehat{l} \in DL_{\lfloor \log^2 n-1 \rfloor}$.

The algorithm below uses the message-forging adversary $\mathcal{A}$ to solve the $Col_{d,\mathcal{H}_{R,m}}(h)$ problem for the parameters specified in Theorem 6.

$Col_{d,\mathcal{H}_{R,m}}(h)$

1: Run the Key-Generation algorithm (but use the given h instead of generating a random one).
2: Receive message z from $\mathcal{A}$.
3: Send $\widehat{k}z + \widehat{l}$ to $\mathcal{A}$.
4: Receive message z' and its signature $\widehat{s}'$ from $\mathcal{A}$
5: Output $\widehat{s}'$ and $\widehat{k}z' + \widehat{l}$

We now need to show that the outputs of the above algorithm are a collision for the function h with non-negligible probability. If $\mathcal{A}$ succeeds in forging a signature $\widehat{s}'$ for z' (which happens with non-negligible probability), then $\|\widehat{s}'\|_\infty \leq 10\phi p^{1/m} n \log^2 n$ and $h(\widehat{s}') = h(\widehat{k})z' + h(\widehat{l}) = h(\widehat{k}z' + \widehat{l})$. So if $\widehat{s}' \neq \widehat{k}z' + \widehat{l}$, then our algorithm outputted two distinct elements that form a collision for the function h.

On the other hand, if $\widehat{s}' = \widehat{k}z' + \widehat{l}$, then we do not get a collision. To complete the proof of Theorem 7, we will show that it's extremely unlikely that an adversary (even one with unlimited computational power) can produce an $\widehat{s}'$ and a z' such that $\widehat{s}' = \widehat{k}z' + \widehat{l}$. This will be done in two steps. In the first step, we show that being able to produce such an $\widehat{s}'$ and z' implies uniquely determining the signing key $(\widehat{k}, \widehat{l})$. Then in the second step we show that given the public key $(h, h(\widehat{k}), h(\widehat{l}))$ and a signature $\widehat{k}z + \widehat{l}$ of message z, it is *information theoretically* impossible to determine the signing key $(\widehat{k}, \widehat{l})$. This means that if $\mathcal{A}$ is able to forge a signature $\widehat{s}'$ for some message z', then almost certainly $\widehat{s}' \neq \widehat{k}z' + \widehat{l}$.

We now show that obtaining an $\widehat{s}'$ and a z' such that $\widehat{s}' = \widehat{k}z' + \widehat{l}$ uniquely determines $\widehat{k}, \widehat{l}$. Since we know that $\widehat{s} = \widehat{k}z + \widehat{l}$ and $\widehat{s}' = \widehat{k}z' + \widehat{l}$, it follows that $\widehat{s} - \widehat{s}' = \widehat{k}(z - z')$. Since $\|\widehat{k}\|_\infty \leq 5p^{1/m} \log^2 n$ and $\|z - z'\|_\infty \leq 2$, multiplying $\widehat{k}$ by $z - z'$ in the ring $\mathbb{Z}_p[x]/\langle f \rangle$ is the same as multiplying them in the ring $\mathbb{Z}[x]/\langle f \rangle$ because the coefficients never get big enough to get reduced modulo p. This is because

$$\|\widehat{k}(z - z')\|_\infty \leq 10\phi p^{1/m} n \log^2 n = 80\phi^{1+\frac{3}{\log n}} n \log^2 n = \phi^{1+o(1)} \cdot o(n^2),$$

but in order to get reduced modulo p, the absolute value of the coefficients would have to be at least $p/2 = \Theta(\phi^3 n^3)$, which is a much larger quantity. Now, since the ring $\mathbb{Z}[x]/\langle f \rangle$ is an integral domain and $z - z' \neq 0$, there cannot exist another key $\widehat{k}' \neq \widehat{k}$ such that $\widehat{k}'(z - z') = \widehat{k}(z - z')$. And so the key $\widehat{k}$ is uniquely determined (and is equal to $\frac{\widehat{s} - \widehat{s}'}{z - z'}$), and similarly the key $\widehat{l} = \widehat{s} - \widehat{k}z$ is also unique.

Now we move on to showing that by knowing only $h, h(\widehat{k}), h(\widehat{l}), z$, and $\widehat{k}z+\widehat{l}$, it is information theoretically impossible to determine the signing key $(\widehat{k}, \widehat{l})$ (and thus, information theoretically impossible to come up with $\widehat{s}', z'$ such that $\widehat{s}' = \widehat{k}z' + \widehat{l}$). The idea is to show that for every $h, h(\widehat{k}), h(\widehat{l}), z, \widehat{k}z + \widehat{l}$ there is an exponential number of signing keys $(\widehat{k}', \widehat{l}')$, other than $(\widehat{k}, \widehat{l})$, that satisfy $h(\widehat{k}) = h(\widehat{k}')$, $h(\widehat{l}) = h(\widehat{l}')$, and $\widehat{k}z + \widehat{l} = \widehat{k}'z + \widehat{l}'$. And in addition, the total probability that one of these other keys was chosen in the key-generation step (conditioned on $h, h(\widehat{k}), h(\widehat{l}), z, \widehat{k}z + \widehat{l}$) is almost one.

We point out that we are not proving *witness-indistinguishability*. It's actually quite possible that for every other key $(\widehat{k}', \widehat{l}')$, the probability that it was the key that was used to sign the message is exponentially smaller than the probability that $(\widehat{k}, \widehat{l})$ was the key. What we will be showing is that the sum of probabilities of all other possible keys *combined* being the secret key is exponentially *larger* than the probability that $(\widehat{k}, \widehat{l})$ was the key.

Lemma 8. *Let (h, K, L) be the verification key of the signature scheme and $\widehat{s}$ is the signature of some message z. Then for any signing key $(\widehat{k}, \widehat{l})$ such that $\widehat{k} \in DK_{\lfloor \log^2 n - 1 \rfloor}, \widehat{l} \in DL_{\lfloor \log^2 n - 1 \rfloor}$, $h(\widehat{k}) = K$, $h(\widehat{l}) = L$ and $\widehat{s} = \widehat{k}z + \widehat{l}$, the probability that this was the actual signing key generated by the key-generation algorithm is negligibly small.*

Proof. We define the set Y to be the elements of the kernel of h that have "small lengths". In particular,

$$Y = \{\widehat{y} \in R^m \text{ such that } \|\widehat{y}\|_\infty \leq 5p^{1/m} \text{ and } h(\widehat{y}) = 0\}.$$

For every $\widehat{y} \in Y$, consider the elements $\widehat{k}' = \widehat{k} - \widehat{y}$ and $\widehat{l}' = \widehat{l} + \widehat{y}z$. Notice that

$$h(\widehat{k}') = h(\widehat{k} - \widehat{y}) = h(\widehat{k}) - h(\widehat{y}) = K - 0 = K,$$

$$h(\widehat{l}') = h(\widehat{l} + \widehat{y}z) = h(\widehat{l}) + h(\widehat{y})z = L + 0 = L,$$

$$\widehat{k}'z + \widehat{l}' = (\widehat{k} - \widehat{y})z + \widehat{l} + \widehat{y}z = \widehat{k}z + \widehat{l} = \widehat{s}.$$

Thus, for every $\widehat{y} \in Y$, if $\widehat{k}'$ happens to be in $DK_{\lfloor \log^2 n \rfloor}$ and $\widehat{l}'$ happens to be in $DL_{\lfloor \log^2 n \rfloor}$, then $(\widehat{k}', \widehat{l}')$ is another valid signing key that could have been used to sign the message z. Since $\|\widehat{y}\|_\infty \leq 5p^{1/m}$ and $\|\widehat{y}z\|_\infty \leq 5n\phi p^{1/m}$, we get the following bounds on the norms of $\widehat{k}'$ and $\widehat{l}'$:

$$\|\widehat{k}'\|_\infty \leq \|\widehat{k}\|_\infty + \|\widehat{y}\|_\infty \leq \|\widehat{k}\|_\infty + 5p^{1/m},$$

$$\|\widehat{l}'\|_\infty \leq \|\widehat{l}\|_\infty + \|\widehat{y}z\|_\infty \leq \|\widehat{l}\|_\infty + 5n\phi p^{1/m}.$$

For the remainder of the proof, let i be the smallest integer such that $\widehat{k}$ and $\widehat{l}$ are contained in DK_i and DL_i respectively. Then $\widehat{k}'$ and $\widehat{l}'$ are definitely contained in DK_{i+1} and DL_{i+1} for every $\widehat{y} \in Y$. And since we assumed that $\widehat{k} \in DK_{\lfloor \log^2 n - 1 \rfloor}$ and $\widehat{l} \in DL_{\lfloor \log^2 n - 1 \rfloor}$, it turns out that $(\widehat{k}', \widehat{l}')$ is a perfectly valid signing key. To prove the lemma, we will need to upper-bound the probability that the generated secret keys were $\widehat{k}, \widehat{l}$ given that the public keys are $K = h(\widehat{k})$ and $L = h(\widehat{l})$ and the signature of z is $\widehat{s} = \widehat{k}z + \widehat{l}$. Let E be the event that the verification key are K and L and the signature of z is $\widehat{s}$.

$$Pr[\text{signing key} = (\widehat{k}, \widehat{l}) | E] = \frac{Pr[\text{key} = (\widehat{k}, \widehat{l}) \& E]}{Pr[E]} = \frac{Pr[\text{key} = (\widehat{k}, \widehat{l})]}{Pr[E]}$$

We now calculate the probability that the keys were $\widehat{k}, \widehat{l}$. This is computed by noting that $\widehat{k}, \widehat{l}$ were generated by selecting $j \geq i$ with probability 2^{-j} and then selecting $\widehat{k}, \widehat{l}$ from DK_j and DL_j. Since $\widehat{k}$ and $\widehat{l}$ are chosen uniformly and independently at random from DK_j and DL_j, the probability that they are both chosen is $\frac{1}{|DK_j| \cdot |DL_j|}$. So,

$$Pr[\text{signing key} = (\widehat{k}, \widehat{l})] = \frac{1}{2^i |DK_i||DL_i|} + \frac{1}{2^{i+1}|DK_{i+1}||DL_{i+1}|} + \dots \quad (1)$$

To calculate the probability of event E, we need to figure out the probability that the keys chosen will result in public keys K and L and when given the message z, the signature will be $\widehat{s}$. We have shown above that for every $\widehat{y} \in Y$, choosing the keys $\widehat{k} - \widehat{y}, \widehat{l} + \widehat{y}z$ will produce public keys K, L and signature $\widehat{s}$. Since we know that $\widehat{k} - \widehat{y}$ and $\widehat{l} + \widehat{y}z$ are contained in DK_{i+1} and DL_{i+1} respectively, we get

$$Pr[E] > \frac{|Y|}{2^{i+1}|DK_{i+1}||DL_{i+1}|} + \frac{|Y|}{2^{i+2}|DK_{i+2}||DL_{i+2}|} + \dots \quad (2)$$

If we let $q = Pr[\text{signing key} = (\widehat{k}, \widehat{l})]$, then combining (1) and (2) we get

$$Pr[E] > |Y| \left(q - \frac{1}{2^i |DK_i||DL_i|} \right)$$

and so,

$$\frac{Pr[\text{signing key} = (\widehat{k}, \widehat{l})]}{Pr[E]} < \frac{q}{|Y| \left(q - \frac{1}{2^i |DK_i||DL_i|} \right)} = \frac{q 2^i |DK_i||DL_i|}{|Y|(q 2^i |DK_i||DL_i| - 1)}$$

$$= \frac{1}{|Y|} \left(1 + \frac{1}{q 2^i |DK_i||DL_i| - 1} \right)$$

Before proceeding, we will state the following inequality that will be used later,

$$\frac{|DK_{i+1}||DL_{i+1}|}{|DK_i||DL_i|} = \frac{(2 \cdot 5(i+1)p^{1/m})^{mn}(2 \cdot 5(i+1)n\phi p^{1/m})^{mn}}{(2 \cdot 5ip^{1/m})^{mn}(2 \cdot 5in\phi p^{1/m})^{mn}}$$

$$= \left(1 + \frac{1}{i} \right)^{2mn} \leq 2^{2mn} = 4^{mn}$$

Now we use the above inequality to lower bound the quantity $q2^i|DK_i||DL_i|$. Recall that q was defined to be the probability that the signing key is $(\widehat{k}, \widehat{l})$, and so from Equation (1), we obtain

$$q2^i|DK_i||DL_i| = 2^i|DK_i||DL_i| \left(\frac{1}{2^i|DK_i||DL_i|} + \frac{1}{2^{i+1}|DK_{i+1}||DL_{i+1}|} + \cdots \right)$$

$$> 2^i|DK_i||DL_i| \left(\frac{1}{2^i|DK_i||DL_i|} + \frac{1}{2^{i+1}|DK_{i+1}||DL_{i+1}|} \right)$$

$$= 1 + \frac{|DK_i||DL_i|}{2|DK_{i+1}||DL_{i+1}|} \geq 1 + \frac{1}{2 \cdot 4^{mn}}$$

Using the above inequality, we obtain

$$\frac{Pr[\text{signing key} = (\widehat{k}, \widehat{l})]}{Pr[E]} < \frac{1}{|Y|} \left(1 + \frac{1}{q2^i|DK_i||DL_i| - 1} \right) \leq \frac{1}{|Y|}(1 + 2 \cdot 4^{mn})$$

and since by Lemma 5 we know that $|Y| \geq 5^{mn}$, we are done.   □

This concludes the proof of the theorem.   □

3.1   Strong Unforgeability

We now show that our one-time signature scheme also satisfies a stronger notion of security, called *strong unforgeability*. In the previous section we showed that if an adversary can produce a signature for an unseen message, then $Col_{d, \mathcal{H}_{R,m}}(h)$ can be solved in polynomial time. Now we point out that $Col_{d, \mathcal{H}_{R,m}}(h)$ can be solved in polynomial time even if the adversary is able to produce a different signature of a message whose signature he has seen. Suppose that after seeing the signature $\widehat{s} = \widehat{k}z + \widehat{l}$ of a message z, the adversary $\mathcal{A}$ sends back another valid signature $\widehat{s}' \neq \widehat{s}$ of z. Then $\widehat{s}$ and $\widehat{s}'$ form a collision for h. This is because

$$h(\widehat{s}') = h(\widehat{k})z + h(\widehat{l}) = h(\widehat{k}z + \widehat{l}) = h(\widehat{s}).$$

4   Practical Attacks

While our scheme is efficient and secure in an asymptotic sense, it is not yet secure for parameters that one would want to use in practical applications. In this section we demonstrate an attack against our one-time signature scheme by showing how an adversary would go about forging a signature for the message $z = 0$. We demonstrate the attack for this message because it is the simplest to explain, but the attack can be easily adapted to any other message.

Knowing the public keys $h(\widehat{k})$ and $h(\widehat{l})$, we can forge a signature for the message $z = 0$ by finding an element $\widehat{l}'$ of length less than $10\phi p^{1/m} n \log^2 n$ such that $h(\widehat{l}') = h(\widehat{l})$ and outputting it as the signature $\widehat{s}$. Note that $h(\widehat{s}) = h(\widehat{k})0 + h(\widehat{l}) = h(\widehat{l}) = h(\widehat{l}')$ and also $\|\widehat{s}\|_\infty = \|\widehat{l}'\|_\infty$. So $\widehat{s}$ will be a valid signature of $\mathbf{0}$. The hard part is of course finding an $\widehat{l}'$ such that $h(\widehat{l}') = h(\widehat{l})$. But while this problem is believed to be exponentially hard in n (the degree of the polynomial $h(\widehat{l})$),for small values of n, this problem is heuristically solvable. We will now give an overview of how one would go about finding an $\widehat{l}'$ with small coefficients when given $h(\widehat{l}')$.

The idea is to use lattice reduction and so first we will need to view multiplication in the ring $\mathbb{Z}_p[x]/\langle f \rangle$ as matrix-vector multiplication. Every polynomial in $\mathbb{Z}_p[x]/\langle f \rangle$ can be associated with an n-dimensional vector in $\mathbb{Z}_p$ in the obvious way. Also, for any element $\boldsymbol{a} \in \mathbb{Z}_p[x]/\langle f \rangle$, define $M(\boldsymbol{a})$ to be an $n \times n$ matrix where the ith column (for $0 \leq i \leq n - 1$) corresponds to the vector representation of the polynomial $\boldsymbol{a}x^i$. Now we can see that the multiplication of two polynomials $\boldsymbol{a}, \boldsymbol{b} \in \mathbb{Z}_p[x]/\langle f \rangle$ can be written as the multiplication modulo p of the matrix $M(\boldsymbol{a})$ by the vector representation of $\boldsymbol{b}$.

By the above observation, the evaluation of the function $h_{\widehat{a}}(\widehat{l}')$ can be interpreted as as a multiplication of an $n \times nm$ matrix $\boldsymbol{A} = (M(\boldsymbol{a}_1)| \ldots |M(\boldsymbol{a}_m))$ by the vector representation of $\widehat{l}'$ modulo p. And so when we're given the public key $h(\widehat{l}')$, we can interpret it as a vector (call it $\boldsymbol{y}$), and then try to find a vector $\boldsymbol{b}$ with coefficients at most $10\phi p^{1/m}n \log^2 n$ such that $\boldsymbol{A}\boldsymbol{b} = \boldsymbol{y}(\bmod p)$. We will now explain how to use lattice reduction to find such a vector $\boldsymbol{b}$.

We first define a matrix $\boldsymbol{A}' = (\boldsymbol{A}|\boldsymbol{y})$, and then try to find a vector $\boldsymbol{b}'$ such that $\boldsymbol{A}'\boldsymbol{b}' = \mathbf{0}(\bmod p)$ where the last coordinate of $\boldsymbol{b}'$ is -1. Notice that this problem is equivalent to the previous one. We now observe that all the vectors $\boldsymbol{b}' \in \mathbb{Z}^{mn+1}$ that satisfy $\boldsymbol{A}'\boldsymbol{b}' = \mathbf{0}(\bmod p)$ form an additive group, and thus an integer lattice of dimension $mn+1$. And since we are trying to find a $\boldsymbol{b}'$ with small coordinates, this is akin to finding a short vector in the aforementioned lattice. The basis of this lattice can be constructed in polynomial time (by viewing $\boldsymbol{A}'$ as a linear transformation mapping $\mathbb{Z}^{mn+1}$ to $\mathbb{Z}_p^n$, and computing the kernel of this transformation). And now all we need to do is find a vector in this $mn + 1$ dimensional lattice such that all its coordinates are less than $10\phi p^{1/m}n \log^2 n$, and the last coordinate is -1.

Suppose that n is around 512, then $p = n^3 = 2^{27}$, $m = \log n = 9$, and suppose that $\phi = 1$. Thus we need to find a vector whose coordinates are less that $80n \log^2 n \approx 2^{21}$ in a lattice of dimension $512 * 9 + 1 = 4609$. It's important to notice that this lattice has a vector all of whose coefficients have absolute value at most 1, and all we need is a vector whose coefficients are less than 2^{21}. Such a large vector (relative to the shortest vector) can easily be found by using standard lattice reduction algorithms that find an approximate shortest vector of the lattice. And heuristically, the algorithm can find such a short vector with the added requirement that the last coordinate is -1.

At this point it is unclear exactly how large we would have to set n in order to avoid the above attack, but it is certainly above any parameter that could be useful in practical applications. Nevertheless, we believe that by using the general structure of the scheme presented in this paper as a starting point, it may be possible to construct a practical and secure signature scheme, and this could prove to be a fruitful direction for further research.

Acknowledgements

We would like to thank the anonymous referees for their comments which helped improve the presentation of this paper.

References

1. Aharonov, D., Regev, O.: Lattice problems in NP ∩ coNP. Journal of the ACM 52(5), 749–765 (2005)
2. Ajtai, M.: Generating hard instances of lattice problems. Complexity of Computations and Proofs, Quaderni di Matematica 13, 1–32 (2004) (Preliminary version in STOC 1996)
3. Ajtai, M., Kumar, R., Sivakumar, D.: A sieve algorithm for the shortest lattice vector problem. In: STOC, pp. 601–610 (2001)
4. Bleichenbacher, D., Maurer, U.: On the efficiency of one-time digital signatures. In: Kim, K.-c., Matsumoto, T. (eds.) ASIACRYPT 1996. LNCS, vol. 1163, pp. 145–158. Springer, Heidelberg (1996)
5. Bleichenbacher, D., Maurer, U.: Optimal tree-based one-time digital signature schemes. In: Puech, C., Reischuk, R. (eds.) STACS 1996. LNCS, vol. 1046, pp. 363–374. Springer, Heidelberg (1996)
6. Blum, M., Micali, S.: How to generate cryptographically strong sequences of pseudo-random bits. SIAM J. Comput. 13(4), 850–864 (1984)
7. Bos, J., Chaum, D.: Provably unforgeable signatures. In: McCurley, K.S., Ziegler, C.D. (eds.) Advances in Cryptology 1981 - 1997. LNCS, vol. 1440, pp. 1–14. Springer, Heidelberg (1999)
8. Cramer, R., Shoup, V.: Signature schemes based on the strong RSA assumption. ACM Trans. Inf. Syst. Secur. 3(3), 161–185 (2000)
9. Diffie, W., Hellman, M.: New directions in cryptography. IEEE Transactions on Information Theory IT-22(6), 644–654 (1976)
10. Dinur, I.: Approximating SVP_∞ to within almost-polynomial factors is NP-hard. Theor. Comput. Sci. 285(1), 55–71 (2002)
11. Even, S., Goldreich, O., Micali, S.: On-line/off-line digital signatures. J. Cryptology 9(1), 35–67 (1996)
12. Gennaro, R., Gertner, Y., Katz, J., Trevisan, L.: Bounds on the efficiency of generic cryptographic constructions. SIAM Journal on Computing 35(1), 217–246 (2005)
13. Gennaro, R., Halevi, S., Rabin, T.: Secure hash-and-sign signatures without the random oracle. In: Stern, J. (ed.) EUROCRYPT 1999. LNCS, vol. 1592, pp. 123–139. Springer, Heidelberg (1999)
14. Goldreich, O., Goldwasser, S.: On the limits of nonapproximability of lattice problems. J. Comput. Syst. Sci. 60(3) (2000)
15. Goldwasser, S., Micali, S., Rivest, R.: A digital signature scheme secure against adaptive chosen-message attacks. SIAM J. Comput. 17(2), 281–308 (1988)
16. Hevia, A., Micciancio, D.: The provable security of graph-based one-time signatures and extensions to algebraic signature schemes. In: Zheng, Y. (ed.) ASIACRYPT 2002. LNCS, vol. 2501, pp. 379–396. Springer, Heidelberg (2002)

17. Kumar, R., Sivakumar, D.: On polynomial-factor approximations to the shortest lattice vector length. SIAM J. Discrete Math. 16(3), 422–425 (2003)
18. Lyubashevsky, V., Micciancio, D.: Generalized compact knapsacks are collision resistant. In: Bugliesi, M., Preneel, B., Sassone, V., Wegener, I. (eds.) ICALP 2006. LNCS, vol. 4052, pp. 144–155. Springer, Heidelberg (2006)
19. Lyubashevsky, V., Micciancio, D., Peikert, C., Rosen, R.: Provably secure FFT hashing. Technical report, 2nd NIST Cryptographic Hash Function Workshop (2006)
20. Merkle, R.: A digital signature based on a conventional encryption function. In: McCurley, K.S., Ziegler, C.D. (eds.) Advances in Cryptology 1981 - 1997. LNCS, vol. 1440, pp. 369–378. Springer, Heidelberg (1999)
21. Merkle, R.: A certified digital signature. In: McCurley, K.S., Ziegler, C.D. (eds.) Advances in Cryptology 1981 - 1997. LNCS, vol. 1440, pp. 218–238. Springer, Heidelberg (1999)
22. Micciancio, D.: Generalized compact knapsacks, cyclic lattices, and efficient one-way functions. Computational Complexity (2007) (Special issue on worst-case versus average-case complexity, in print. Available on-line as doi:10.1007/s00037-007-0234-9. Preliminary version in FOCS 2002)
23. Naor, M., Yung, M.: Universal one-way hash functions and their cryptographic applications. In: STOC, pp. 33–43 (1989)
24. Pedersen, T., Pfitzmann, B.: Fail-stop signatures. SIAM J. Comput. 26(2), 291–330 (1997)
25. Peikert, C., Rosen, A.: Efficient collision-resistant hashing from worst-case assumptions on cyclic lattices. In: Halevi, S., Rabin, T. (eds.) TCC 2006. LNCS, vol. 3876, Springer, Heidelberg (2006)
26. Peikert, C., Rosen, A.: Lattices that admit logarithmic worst-case to average-case connection factors. In: STOC (2007)
27. Rompel, J.: One-way functions are necessary and sufficient for secure signatures. In: STOC, pp. 387–394 (1990)
28. Szydlo, M.: Merkle tree traversal in log space and time. In: Cachin, C., Camenisch, J.L. (eds.) EUROCRYPT 2004. LNCS, vol. 3027, pp. 541–554. Springer, Heidelberg (2004)
29. van Emde Boas, P.: Another NP-complete problem and the complexity of computing short vectors in a lattice. Technical Report 81-04, University of Amsterdam (1981), http://turing.wins.uva.nl/~peter/

Basing Weak Public-Key Cryptography on Strong One-Way Functions

Eli Biham[1], Yaron J. Goren[2,*], and Yuval Ishai[3,**]

[1] Technion, Israel
biham@cs.technion.ac.il
[2] Rosetta Genomic, Israel
yaron.goren@gmail.com
[3] Technion, Israel and UCLA, USA
yuvali@cs.technion.ac.il

Abstract. In one of the pioneering papers on public-key cryptography, Ralph Merkle suggested a heuristic protocol for exchanging a secret key over an insecure channel by using an idealized private-key encryption scheme. Merkle's protocol is presumed to remain secure as long as the gap between the running time of the adversary and that of the honest parties is at most *quadratic* (rather than super-polynomial). In this work, we initiate an effort to base similar forms of public-key cryptography on well-founded assumptions.

We suggest a variant of Merkle's protocol whose security can be based on the *one-wayness* of the underlying primitive. Specifically, using a one-way function of exponential strength, we obtain a key agreement protocol resisting adversaries whose running time is nearly quadratic in the running time of the honest parties. This protocol gives the adversary a small (but non-negligible) advantage in guessing the key. We show that the security of the protocol can be amplified by using a one-way function with a strong form of a hard-core predicate, whose existence follows from a conjectured "dream version" of Yao's XOR lemma. On the other hand, we show that this type of hard-core predicate cannot be based on (even exponentially strong) one-wayness by using a black-box construction.

In establishing the above results, we reveal interesting connections between the problem under consideration and problems from other domains. In particular, we suggest a paradigm for converting (unconditionally) secure protocols in Maurer's *bounded storage model* into (computationally) secure protocols in the random oracle model, translating storage advantage into computational advantage. Our main protocol can be viewed as an instance of this paradigm. Finally, we observe that a *quantum* adversary can completely break the security of our protocol (as well as Merkle's heuristic protocol) by using the quadratic speedup of Grover's quantum search algorithm. This raises a speculation that there might be a closer relation between (classical) public-key cryptography and quantum computing than is commonly believed.

Keywords: Public-key cryptography, one-way functions, Merkle's puzzles, bounded storage model, quantum computing.

* Most of this work was done while at the Department of Computer Science, Technion.
** Supported by ISF grant 1310/06, BSF grant 2004361, and NSF grants 0205594, 0430254, 0456717, 0627781, 0716835, 0716389.

R. Canetti (Ed.): TCC 2008, LNCS 4948, pp. 55–72, 2008.
© International Association for Cryptologic Research 2008

1   Introduction

The fundamental cryptographic primitives and protocols can be roughly divided into two categories: "private cryptography" which includes private key encryption, pseudo-random generators, pseudo-random functions, bit commitment and digital signatures, and "public cryptography" which includes public key encryption, key agreement, oblivious transfer, and secure function evaluation. Since the existence of these primitives implies that P≠NP, given the current state of complexity theory we need to base it on unproven computational assumptions. These assumptions may turn out to be false; thus, basing primitives on the minimal possible assumptions has been put forward as one of the most important goals in cryptography.

The weakest assumption that is commonly used in cryptography is the existence of *one-way functions*; it is weakest in the sense that if such functions do not exist, none of the above primitives exist [15]. The existence of one-way functions implies the existence of all of the private cryptography primitives (cf. [16,6] and references therein). In contrast, it is not known how to base public cryptography primitives on one-way functions. Obtaining such a construction is arguably one of the most intriguing open questions in cryptography. In addition to its fundamental importance, this question is also motivated by the big efficiency gap between the best current implementations of public primitives and the (much more efficient) implementations of private primitives. This gap is mostly due to the fact that current approaches for obtaining public primitives rely on *algebraic* intractability assumptions. Since the underlying algebraic objects are highly structured, there are sophisticated attacks that exploit this structure. Thus, the underlying objects must be very large in order to defeat known attacks. An additional, more recent, motivation for basing public cryptography on one-way functions is the advent of efficient *quantum* algorithms that break most (but not all) of the concrete algebraic intractability assumptions that currently underly public cryptography [22].

In light of the above, basing public-key cryptography on one-way functions can be viewed as a "holy grail" both from a theoretical and from a practical point of view. In fact, from the latter point of view even a heuristic construction based on a random oracle might be considered satisfactory (as the random oracle can often be replaced in practice by a sufficiently "structureless" function). However, a seminal result of Impagliazzo and Rudich [16] suggests that standard methods cannot be used to realize such constructions. Specifically, this result rules out the possibility of a *black-box* construction based on a one-way permutation (see also [20]). Furthermore, the result of [16] shows that a provable construction of a public-key primitive based on a random permutation oracle is unlikely to be found, as it would imply a proof that P≠NP.

Weak public-key cryptography. An implicit assumption in the last statements is that the gap between the resources of the honest parties and those of the adversary must be super-polynomial. It is natural to relax this assumption and consider a weaker variant of public-key cryptography, where the resource gap between the adversary and the honest parties is bounded by some *fixed* polynomial. Such a weaker form of public-key cryptography has a similar qualitative flavor as standard public-key cryptography, and might be relevant to practice. Indeed, even with a quadratic resource gap, the *ratio* between the amount of time required by the adversary and that required by honest parties grows linearly with the computing power. Thus, security gets better with technology. In this work, we study the possibility of basing such weak public-key cryptography on one-way functions and related primitives.

Merkle's puzzles. Our point of departure is the pioneering work of Merkle [19], who proposed the following protocol for secret key agreement over public channels. Merkle's protocol involves two honest parties, Alice and Bob. It relies on the ability to efficiently create "puzzles" which encapsulate a value chosen by the puzzle creator and require a "moderate" amount of time T to be solved by another party. The protocol proceeds by letting Alice pick a large number S of random pairs (k_i, id_i) and send to Bob S puzzles encapsulating these pairs. Bob picks a random puzzle r and, after spending time T solving it, obtains a pair (k_r, id_r). It then sends id_r to Alice. Now both parties have a common key k_r. The time spent by Alice in this protocol is roughly S (assuming that a puzzle can be generated at a unit cost), and the time spent by Bob is roughly T. However, from the point of view of an external eavesdropper Eve, r remains secret. Thus, the intuition is that Eve will need to solve $S/2$ puzzles on average, spending $\Omega(ST)$ time, before she can learn k_r. Setting $S = T$, both Alice and Bob have a quadratic advantage over Eve. Merkle suggested a heuristic implementation of the puzzles using a weakened version of a private-key encryption scheme, where solving the puzzle amounts to exhaustively searching over a (moderately sized) key space.

Trying to instantiate the puzzles in the above protocol using a standard (semantically secure) private-key encryption scheme is problematic for several reasons. First, an implicit assumption that underlies the security of the protocol is that there is a sharp bound T between the maximal time required by honest parties to *completely* solve a puzzle and the minimal time required by an adversary to gain *some* information about the solution. One might try to achieve this goal by requiring the encryption to have "exponential strength" in its key size. However, it is not clear how to realize such a strong primitive based on (even strong versions of) low-level primitives such as a one-way function. A second problem is that the security of the resulting protocol seems to rely on the assumption that the adversary has no better strategy for recovering the key k_r than by trying to solve the puzzles one by one until finding the one that contains s_r. Again, this is an unsubstantiated assumption in a complexity-based cryptography.

1.1 Our Contribution

The question whether *weak* public key cryptography can be based on one-way functions, or some variation of them, is largely unexplored. Our goal is to understand what kinds of weak public key cryptography are possible and under what assumptions.

We start by suggesting a variant of Merkle's protocol which admits a simple proof of security in the random oracle model. In this protocol, each party (independently) evaluates a random *permutation* on a random set of inputs whose size is roughly the square root of the domain size, and the parties communicate the set of the outputs of these evaluations. By the birthday paradox, the two sets of outputs intersect with high probability, and the preimage of this intersection can be used to extract a common key. (The above protocol can be viewed as based on a similar protocol of Cachin and Maurer [2] in the *bounded storage model* – see below.)

We then show that the random permutation oracle in this protocol can be instantiated with an exponentially strong one-way permutation (OWP), or even an exponentially strong 1-1 OWF, yielding a key agreement protocol with a polynomial gap between the bounded parties and the adversary. Specifically, if the OWP is secure against adversaries that run in time $2^{(1-\delta)n}$, the protocol is secure as long as the running time of the adversary is less than the running time of the honest parties to the power of $2 - 2\delta$. Thus, we approach quadratic security as δ tends to 0. Towards obtaining a similar result under any one-way function, we show a way for transforming an exponentially strong OWF into a *family* of exponentially strong OWFs that are "almost" 1-1. (We stress that this transformation inherently relies on the exponential strength of the underlying OWF; its analysis gives a general method for redistributing the hardness of OWFs which may be of independent interest.) Using this transformation we obtain a similar key agreement protocol based on an exponentially strong OWF.

On the existence of strong one-way functions. Our protocols rely on one-way functions whose strength goes beyond the birthday paradox bound of $2^{n/2}$. The existence of such OWFs can be regarded as a very mild assumption from a cryptanalytic point of view. For instance, an explicit attack against AES that runs in time $2^{0.9n}$, where n is the key size, would be considered as indicating a major vulnerability. Exponentially strong OWFs were recently exploited in several cryptographic contexts. In the context of program obfuscation, Wee [23] uses a OWF with a form of exponential strength which is even stronger than ours in terms of the ratio between the adversary's time bound and its success probability. OWFs with milder forms of exponential strength were recently employed for constructing pseudorandom generators [14,11,4]. It should be noted that given generic time-space tradeoffs for inverting functions [12,5], one cannot expect a *fixed* function (rather than a collection of functions) to be *non-uniformly* one-way with a very good exponential strength (say, better than $2^{2n/3}$). Thus, in the context of this work one should either restrict adversaries to be *uniform*, or alternatively rely on a *collection* of strong one-way functions.

On reducing the adversary's advantage. The key agreement protocols described above allow the adversary to gain an inverse polynomial advantage in guessing the secret key. This type of insecurity may be viewed as reasonable in the context of weak public-key cryptography, but it is still desirable to obtain the standard notion of security with negligible advantage with respect to a weaker class of adversaries. Unfortunately, known techniques for converting weak key agreement to strong one (e.g., those of Holenstein [13]) do not seem sufficient for this purpose. We show that the security of the protocol can be boosted to allow only a negligible advantage if one assumes the underlying primitive to have a strong form of a hard-core predicate [8] which we call a *multi-source hardcore predicate* (MSHCP). Roughly speaking, an MSHCP applies a predicate to several inputs, such that an (exponentially strong) adversary can only predict the value of the predicate on *independently chosen* inputs with an advantage that is negligible *in the size of the input domain*. (This should be contrasted with a standard hard-core predicate in which the predicate is applied to a single input, and it is only guaranteed that the advantage is negligible in the *bit-length of the input*.) We show that the existence of an MSHCP follows from a conjectured "dream version" of Yao's XOR lemma (a close variant of a conjecture appearing in [9]). On the other hand, we show that in contrast to standard hard-core predicates, the existence of an MSHCP cannot be based on (even exponentially strong) one-wayness by using a black-box construction.

Our results reveal some interesting and perhaps unexpected connections between the problem under consideration and problems from other domains.

Relation with the Bounded Storage Model. In Maurer's *bounded storage model* (BSM) [18], it is assumed that a large random source is transmitted, out of which the adversary can only store a limited amount of information. Viewing the random source as an oracle, the transmission of the random source can be replaced by local computation. In terms of security, the resulting model is incomparable to the original BSM: the adversary here is weaker in that it can only access "physical" bits of the source by querying the oracle (rather than store an arbitrary function of the source), but it is stronger in the sense that it is allowed access to the source even *after* the execution of the protocol. To get around the latter problem, a natural approach is to code the source in the *image* of the oracle. That is, the evaluation of the oracle f at point x gives a pair (i, b) indicating that the ith bit of the source is b. When the honest parties in the BSM protocol only need to access the source at *random* locations, the protocol can still be efficiently implemented using the random oracle. Our main protocol can be viewed as applying this conversion paradigm to the BSM protocol from [2]. A similar transformation can be applied to the oblivious transfer protocol from [3] to yield an oblivious transfer protocol (with quadratic security) in the random oracle model.

Relation with quantum computing. Finally, we observe that a *quantum* adversary can completely break the security of our protocol (as well as that of Merkle's heuristic protocol) by using the quadratic speedup of Grover's quantum search algorithm [10]. Thus, the two most prominent examples for speedup by quantum algorithms – the strong speedup of Shor's algorithm and the weaker speedup of Grover's algorithm – seem to be "tailored" to break the two main types of public-key cryptosystems – strong ones based on number-theoretic assumptions[1] and weak ones based on Merkle's technique. While this can be dismissed as a pure coincidence, it also raises the interesting speculation that there might be a closer relation between (classical) public-key cryptography and quantum computing than is commonly believed. This speculation may be supported by the relative scarcity of useful algorithms in the two domains.

It is important to stress that the quadratic speedup that can be achieved using Grover's algorithm is by no means universal, and applies only in scenarios that involve *parallel* search. An interesting problem left open by our work is that of obtaining a weak key agreement protocol, even in the random oracle model, that resists this kind of quantum attack. A natural approach for achieving this is by obtaining efficient implementations of puzzles that resist parallel search attacks. A similar problem was considered by Boneh and Naor [1] in the context of timed commitments. However, the only known implementations of this primitive rely on number-theoretic assumptions that do not resist a quantum attack. The possibility of implementing such "non-parallelizable" puzzles using a one-way function, or even a random function, remains open.

---

[1] One should note in this context that we *do* have candidates for strong public-key cryptosystems that resist quantum attacks, mostly ones based on lattice problems and error-correcting codes. However, because of the strong algebraic structure of the underlying computational problems, the existence of efficient quantum algorithms for these problems does not seem unlikely.

Organization. The remainder of this paper is organized as follows. Following some preliminaries, in Section 3 we describe a key agreement protocol based on a random function. The protocol resists adversaries whose running time is nearly quadratic in the running time of the honest parties. In Section 4 we replace the random function with an exponentially strong one-way permutation, and in Section 5 we show how to base a variant of this protocol on an exponentially strong one-way function. Some details and proofs that were omitted from this version can be found in the full version.

2   Definitions

In contrast to conventional cryptography, in this work we assume the resource gap between the honest parties and the adversary to be bounded by a fixed polynomial. This requires us to introduce an "exact" variant for some common definitions and to set a concrete model of computation which is sensitive to such gaps. We use a RAM model (e.g., a "log-cost" RAM) as our default model of computation, for both honest parties and adversaries. A $T(n)$-bounded algorithm is an algorithm whose running time on input of length n is bounded by $T(n)$.

Our results are stated for uniform adversaries, but are valid for non-uniform adversaries as well. In this case the bound on the running time serves also as a bound on the size of the advice string given to the adversary. Specific differences between the results for uniform and non-uniform adversaries will be discussed when relevant.

Notation. We write $f(n) = \tilde{O}(g(n))$ if there exists some constant c such that $f(n) = O(g(n) \log^c(g(n)))$. We say that a function $\epsilon(\cdot)$ is negligible and denote such a function by $neg(\cdot)$ if for any constant c, $\epsilon(n) < \frac{1}{n^c}$ for sufficiently large n. We say that $\epsilon(\cdot)$ is bounded away from c if $\epsilon(n) \leq c - 1/p(n)$ for some polynomial p and all sufficiently large n. We denote by U_n the uniform distribution over $\{0,1\}^n$. By IP we denote the modular inner product function defined by $IP(x,r) = \sum_{i=1}^{n}(x_i \cdot r_i) \bmod 2$.

2.1   Key Agreement Protocols

An $l(\cdot)$-bit key agreement protocol is an interactive protocol in which Alice and Bob receive a security parameter k, exchange messages over a public channel and each output a key in $\{0,1\}^{l(k)}$. Throughout the paper we deal only with 1-bit key agreement as defined below. Our protocols can be extended to $l(k)$-bit key agreement for any $l(k) \leq \text{polylog}(k)$ with similar asymptotic parameters by independent repetition. (A longer key will reduce the polynomial advantage of the honest parties.) Such an l-bit key can then be used to encrypt longer messages using a (conventional) symmetric encryption scheme. We note that the key agreement protocols presented in this paper are limited to two rounds. Hence what we achieve can be viewed as a (weak) *public key encryption* scheme, where the first message serves as the public key.

Definition 1 (Key agreement). *A protocol (Alice, Bob) is a (d, ϵ)-secure key agreement protocol if the following conditions hold:*

– Correctness: *Alice and Bob are $\tilde{O}(k)$-bounded and they output the same bit except for a failure probability $\delta(k) = neg(k)$.*

- Security: *For any constant $d' < d$ and any $O(k^{d'})$-bounded adversary, for suffi-ciently large k the probability that the adversary guesses Bob's output on a random transcript of the protocol is bounded by $\frac{1}{2} + \epsilon(k)$.*

We say that the protocol has quadratic security *if it is (d, ϵ)-secure for $d = 2$ and some negligible $\epsilon(\cdot)$.*

2.2   Strong One-Way Functions and Hard-Core Predicates

Definition 2 (One-way function). *An efficiently computable function $f : \{0, 1\}^* \to \{0, 1\}^*$ is a (T, ϵ) one-way function if for any $T(n)$-bounded adversary A and for all sufficiently large n, $Pr_{x \in U_n}[f(A(1^n, f(x))) = f(x)] < \epsilon(n)$. If in addition f is a permutation we say that it is a (T, ϵ) one-way permutation. If f is (T, ϵ) one-way with $\epsilon(n) \le \frac{1}{16}$, we say that it is a $T(n)$ one-way function.*

We note that a standard one-way function is $(n^c, \frac{1}{n^c})$ one-way for every constant $c > 1$.

Definition 3 (Hard-core predicate). *An efficiently computable function $h : \{0, 1\}^* \to \{0, 1\}$ is a (T, ϵ) (randomized) hard-core predicate for f if for any $T(n)$-bounded adversary A, for sufficiently large n,*

$$Pr_{x \in U_n, r \in U_n}[A(f(x), r) = h(x, r)] < 1/2 + \epsilon(n).$$

The following definition generalizes the concept of hard-core predicates to allow the predicate to depend on several pre-images.

Definition 4 (Multi-source hard-core predicate). *A polynomial time computable function $H : \{0, 1\}^* \to \{0, 1\}$ is a (T, ϵ) multi-source (randomized) hard-core pred-icate (MSHCP) for f if there exist two polynomials $t(\cdot)$ and $s(\cdot)$ such that for any $T(n)$-bounded adversary A and all sufficiently large n,*

$$Pr_{x_1 \ldots x_{t(n)} \in U_n, r \in U_{s(n)}}[A(1^n, f(x_1) \ldots f(x_{t(n)}), r) = H(x_1 \ldots x_{t(n)}, r)] < 1/2 + \epsilon(n).$$

If H is a (T, ϵ) MSHCP with $\epsilon(n) = neg(2^n)$ we say that it is a strong MSHCP.

Note that a strong MSHCP can be guessed only with an advantage that is negligible in the size of the input domain. This is not possible with standard hard-core predicates since a single invocation of f is enough for finding a pre-image with probability which is the inverse of the input domain size. We also note that it is easy to show that a *random* function has a strong MSHCP; however, its security does not seem to follow from one-wayness alone. A relevant black-box separation is given in the full version.

3   Key Agreement in the Random Oracle Model

We describe a variant of Merkle's key agreement protocol in which all parties have access to a random function oracle, and show that adversaries whose running time is nearly quadratic in the running time of the honest parties can only have a negligible advantage in guessing the agreed key. For simplicity of presentation we assume that the function which the oracle computes is chosen uniformly from some set of functions *after* the adversary has been set, and in this scenario we bound the adversary's advantage in guessing the key. This result can be extended, using a standard argument [16], to show that the protocol is secure against any uniform adversary when the oracle is set to a specific function from the set, with probability 1 over the choice of functions. Alternatively, the protocol is secure against non-uniform adversaries when the function is chosen after the adversary is set (i.e. no single non-uniform adversary can break the protocol for a significant fraction of the functions).

We start with a protocol which uses an oracle to a random permutation and a random predicate and then extend the result for the case of a random function.

The ROM protocol: For a security parameter k we use an oracle to a random permutation $f : [k^2] \to [k^2]$ and a random predicate $h : [k^2] \to \{0, 1\}$. We set a minimal intersection size parameter $l(k)$ to be $l(k) = \frac{1}{2} \log^2(k)$.

- Alice chooses a random set $\mathcal{A} \subset [k^2]$ of size $k \cdot \log k$, queries the oracle on these inputs and sends $f(\mathcal{A}) = \{f(a) | a \in \mathcal{A}\}$ to Bob.
- Bob chooses a random set $\mathcal{B} \subset [k^2]$ of size $k \cdot \log k$ and queries the oracle on these inputs. If $|f(\mathcal{A}) \cap f(\mathcal{B})| < l(k)$ Bob aborts and both parties output random values. Otherwise, Bob randomly chooses $l(k)$ common outputs $c_1, \ldots, c_{l(k)} \in f(\mathcal{A}) \cap f(\mathcal{B})$ and sends them to Alice.
- Alice and Bob find the common inputs $s_1, \ldots, s_{l(k)} \in \mathcal{A} \cap \mathcal{B}$ such that $f(s_i) = c_i$ and output $\bigoplus_{i=1}^{l(k)} h(s_i)$.

In the full version we prove that the above protocol has quadratic security. We also show a similar protocol which uses a random function instead of a permutations and a predicate and prove the following theorem.

Theorem 1. *Given an oracle to a random function $f : \{0, 1\}^* \to \{0, 1\}$, there exists a key agreement protocol with quadratic security.*

4   Key Agreement from One-Way Permutations

In order to construct key agreement from one-way permutations we replace the random permutation and random predicate used in the ROM protocol with an exponentially strong one-way permutation and a hard-core predicate. The analysis is divided into two parts: first we show how to construct a key agreement protocol from a one-way permutation with an MSHCP and then we show how to construct an MSHCP for any one-way function. The maximal possible advantage in guessing the MSHCP determines the advantage the protocol allows adversaries in guessing the key. Using a conjectured dream XOR lemma we construct a strong MSHCP (in which the maximal advantage is negligible in the size of the input domain) and hence key agreement protocols in which the adversary's advantage is negligible. Without this conjecture we do not know how to construct strong MSHCPs, but can get a weaker MSHCP that suffices for limiting the adversary's advantage to $1/\text{poly}(k)$. In the full version we show that the limitation on the strength of the MSHCP is inherent to black-box constructions.

The results in this section hold in both uniform and non-uniform settings, depending on the setting in which the permutations are assumed to be one-way. However, it follows from generic time-space tradeoffs for inverting functions [12,5] that one cannot expect a *fixed* function (rather than a collection of functions) to be non-uniformly one-way with a very good exponential strength. In order to achieve meaningful results in the non-uniform case, one may use a collection of one-way functions in the protocol described in Section 5.4. Finally, we note that the results of this section can be generalized to any 1-1 one-way function.

4.1   Key Agreement from a One-Way Permutation with an MSHCP

We use a variant of the ROM protocol in which the random permutation is replaced by a one-way permutation and the random predicate is replaced by an MSHCP.

The OWP protocol: For a security parameter k we set $n = 2 \cdot \log k$ (i.e. $k = 2^{n/2}$) and use a one-way permutation $f : \{0,1\}^n \to \{0,1\}^n$ for which H is a (T, ϵ) MSHCP with $T = 2^{n(1-\delta)}$. We set the minimal size of the intersection to be $l(k) = t(n)$ where $t(\cdot)$ is the number of inputs for H as in Definition 4.

- Alice chooses a random set $\mathcal{A} \subset \{0,1\}^n$ of size $k\sqrt{2l(k)}$, applies f to these inputs and sends $f(\mathcal{A}) = \{f(a) | a \in \mathcal{A}\}$ to Bob.
- Alice also sends Bob a random string $r \in \{0,1\}^{s(n)}$, where $s(\cdot)$ is the size of the random input for H as in Definition 4.
- Bob chooses a random set $\mathcal{B} \subset \{0,1\}^n$ of size $k\sqrt{2l(k)}$ and applies f to these inputs. If $|f(\mathcal{A}) \cap f(\mathcal{B})| < l(k)$ Bob aborts and both parties output random values. Otherwise, Bob randomly chooses $l(k)$ common outputs $c_1, \ldots, c_{l(k)} \in f(\mathcal{A}) \cap f(\mathcal{B})$ and sends them to Alice.
- Alice and Bob find the common inputs $s_1, \ldots, s_{l(k)} \in \mathcal{A} \cap \mathcal{B}$ such that $f(s_i) = c_i$ and output $H(s_1, \ldots, s_{l(k)}, r)$.

Theorem 2 (Key agreement from a one-way permutation with an MSHCP). *For any constant $\delta < \frac{1}{2}$, if there exists a one-way permutation with a (T, ϵ) MSHCP such that $T = 2^{n(1-\delta)}$ then there exists a (d, ϵ)-secure key agreement protocol with $d = 2 - 2\delta$.*

Proof. The proof of correctness is the same as in the ROM protocol. The proof of security is by contradiction. Suppose that for some constant $d' < 2 - 2\delta$, an $O(k^{d'})$-bounded adversary A guesses the agreed bit with probability at least $\frac{1}{2} + \epsilon$ when given a random transcript of the protocol. We show how to use A to guess $H(x_1, \ldots, x_{l(k)}, r)$, given $f(x_1), \ldots, f(x_{l(k)})$ and r on random $x_1, \ldots, x_{l(k)} \in \{0,1\}^n$ and $r \in \{0,1\}^{s(n)}$. We create a random transcript of the protocol using the following procedure:

- Randomly choose a set $\mathcal{A} \subset \{0,1\}^n$ of size $k\sqrt{2l(k)} - l(k)$, and apply f to these inputs. Randomly interleave $f(x_1), \ldots, f(x_{l(k)})$ within the set $f(\mathcal{A})$ and use the result as the first part of Alice's message to Bob.
- Use the random string r as the second part of Alice's message to Bob.
- Use $f(x_1), \ldots, f(x_{l(k)})$ as Bob's message to Alice.

It is easy to verify that the result is indeed distributed identically to a random transcript created by Alice and Bob. We then apply A to the transcript and output the same bit as

A does. By our assumption A's output is equal to $H(x_1, \ldots, x_{l(k)}, r)$ with probability at least $\frac{1}{2} + \epsilon$. The transcript can be created in time $\tilde{O}(k)$ and since $\delta < \frac{1}{2}$, for large enough n the total running time is bounded by $k^{2-2\delta} = 2^{n(1-\delta)}$, contradicting the hardness of H. □

4.2 Construction of an MSHCP for any One-Way Function

We first construct a (single-source) randomized hard-core predicate defined by $h(x, r) = IP(x, r)$ for a random r and prove its security using an exact version of the Goldreich-Levin lemma. Then we use h to construct an MSHCP defined by $H((x_1, \ldots, x_t), (r_1, \ldots, r_t)) = \bigoplus_{i=1}^{t} h(x_i, r_i)$ and prove its security using an exact version of Yao's XOR lemma.

Lemma 1 (Goldreich-Levin). *If f is a (T, ϵ) one-way function then $h(x, r) = IP(x, r)$ (where IP denotes inner product modulo 2) is a (T', ϵ') randomized hard-core predicate for f with $T'(n) = T(n) \cdot \frac{\epsilon^4}{n^3}$ and $\epsilon'(n) = 4\epsilon$.*

The lemma follows from the alternative version of Proposition 2.5.3 in [7]. A consequence of this lemma is that every $T(n)$ one-way function has a (T', ϵ') (randomized) hard-core predicate with $T' = T(n)/\text{poly}(n)$ and $\epsilon' = 1/4$.

Definition 5 (Hard predicate). *We say that $P : \{0, 1\}^* \to \{0, 1\}$ is a (T, ϵ) hard predicate if for any $T(n)$-bounded adversary A and all sufficiently large n, $Pr_{x \in U_n}[A(x) = P(x)] < 1/2 + \epsilon(n)$.*

Lemma 2 (Yao's XOR lemma). *If P is a (T, ϵ) hard predicate and it is possible to efficiently sample from the distribution $(U_n, P(U_n))$, then for any $\mu(n)$ and $t = \text{poly}(n)$, $P^{(t)}(x_1, \ldots, x_t) = \bigoplus_{i=1}^{t} P(x_i)$ is a (T', ϵ') hard predicate for $T' = T \cdot \frac{\mu^2}{\text{poly}(n)}$ and $\epsilon' = (2\epsilon)^t + \mu$.*

The lemma can be derived by a careful analysis of Levin's proof for Yao's XOR lemma given in [17,9]; see full version. Combining Lemma 1 with Lemma 2 allows us to construct an MSHCP but there is an inherent limitation to the strength of the MSHCP which we may construct in this way. A (T', ϵ') hard predicate constructed using Lemma 2 has the property that $\epsilon' > \mu$ and $T' < T \cdot \mu^2$. For any (T, ϵ) hard-core predicate for an *efficiently computable* one-way function, $T = \tilde{O}(2^n)$ since it is possible to invert the one-way function by an exhaustive search. For our key agreement protocol we need a (T', ϵ') MSHCP in which $T' = 2^{(1-\delta)n}$ for some $\delta < \frac{1}{2}$. Any (T', ϵ') MSHCP constructed under the above restrictions will have $\epsilon' > \mu > \sqrt{T'/T} > 2^{-n/4}$. The following conjecture allows us to construct a (T', ϵ') hard predicate with $\epsilon' = neg(2^n)$ while T' remains close to T, and thus can be used to construct a strong MSHCP.

Conjecture 1 (Dream XOR lemma). *If P is a (T, ϵ) hard predicate for some ϵ that is bounded away from $\frac{1}{2}$ and it is possible to efficiently sample from the distribution $(U_n, P(U_n))$, then there exists a constant $c < 1$, a negligible $\mu(\cdot)$ and some $\eta(\cdot)$ which is bounded away from 1 such that for any $t = \text{poly}(n)$, $P^{(t)}(x_1, \ldots, x_t) = \bigoplus_{i=1}^{t} P(x_i)$ is a (T', ϵ') hard predicate for $T' = T \cdot 2^{-o(n)}$ and $\epsilon' = 2^{cn} \cdot \eta^t + \mu(2^n)$.*

A similar "dream version" of Yao's XOR lemma was conjectured in [9] and it was observed that it can not be proved using a black-box analysis. (The variant appearing in [9] requires ϵ to be smaller but allows T to be smaller as well.) Theorem 3 states the parameters of MSHCP that can be obtained with and without the dream XOR lemma conjecture, while in the full version we show that a strong MSHCP cannot be obtained from a OWF using a black-box construction.

We apply Lemma 2 and Conjecture 1 with $t = \mathrm{poly}(n)$ to the predicate defined in Lemma 1 to get the following result (smaller values of t suffice for the first two cases, but this does not improve the asymptotic result).

Theorem 3. *For any $\delta < 1$, every $2^{n(1-\delta)}$ one-way function has a (T, ϵ) MSHCP with the following T and ϵ:*

- $T = 2^{n(1-\delta)}/\mathrm{poly}(n)$ $\epsilon = \frac{1}{\mathrm{poly}(n)}$ *using $\mu = O(\epsilon)$*
- $T = 2^{n(1-\delta-\tau)}/\mathrm{poly}(n)$ $\epsilon = 2^{-\tau n/2}$ *using $\mu = O(2^{-\tau n/2})$*
- $T = 2^{n(1-\delta)}/2^{o(n)}$ $\epsilon = neg(2^n)$ *assuming the dream XOR lemma*

It is easy to verify that Theorem 2 holds also when $T = 2^{n(1-\delta)}$ is replaced with $2^{n(1-\delta)}/2^{o(n)}$. Combining this with Theorem 3 we get the main result regarding the construction of key agreement from one-way permutations. It relates the strength of the underlying one-way permutation to the security of the key agreement protocol that can be constructed from it.

Corollary 1. *For any constant $\delta < \frac{1}{2}$ if there exists a $2^{n(1-\delta)}$ one-way permutation then there exists a (d, ϵ) secure key agreement protocol for the following d and ϵ:*

- $d = 2 - 2\delta$ $\epsilon = 1/\log^c k$ *for any constant c*
- $d = 2 - 2\delta - 2\tau$ $\epsilon = k^{-\tau}$ *for any $\tau < 1 - \delta$*
- $d = 2 - 2\delta$ $\epsilon = neg(k)$ *assuming the dream XOR lemma*

5 Key Agreement from One-Way Functions

We extend the result from the previous section to obtain weak key agreement from exponentially strong one-way functions. The main technical result in this section is a construction of a collection of one-way functions which is almost 1-1 from an exponentially strong one-way function which is not necessarily 1-1. The construction applies a restriction to the domain of the one-way function such that the restricted function usually remains one-way and is almost always 1-1. The resulting collection of one-way functions is then used to construct a key agreement protocol. The results in this section hold in both uniform and non-uniform settings, depending on the setting in which the functions are assumed to be one-way. We refer to the uniform setting by default.

5.1 Definitions

A collection of one-way functions is defined by a pair of functions G and F such that $G(1^n)$ generates a key i of length $l(n)$ which defines a function $f_i : \{0,1\}^n \to \{0,1\}^*$ and $F(i, x)$ computes $f_i(x)$.

Definition 6 (Collection of one-way functions). *Let $\mathcal{F} = \bigcup F_n$ be a collection of functions where $F_n = \{f_i : \{0,1\}^n \to \{0,1\}^* | i \in I_n\}$. We say that $\mathcal{F}$ is (T, ϵ) one-way if there exist two PPT algorithms G and F such that the following holds:*

1. Easy to compute: *There exists a polynomially bounded function $l(\cdot)$ such that the output of G on input 1^n is in the set $I_n \subseteq \{0,1\}^{l(n)}$. On input $i \in I_n$ and $x \in \{0,1\}^n$, $F(i,x) = f_i(x)$.*
2. Hard to invert: *For every $T(n)$-bounded adversary A, for all sufficiently large n's,*

$$Pr_{i \in G(1^n), x \in U_n}[f_i(A(1^n, i, f_i(x))) = f_i(x)] < \epsilon(n).$$

We say that $\mathcal{F}$ is $T(n)$ one-way if it is (T, ϵ) one-way with $\epsilon \leq 1/32$. We say that $\mathcal{F}$ is almost 1-1 if the probability that $f_i \in F_n$ is not 1-1 is bounded by 2^{-n}.

We will be using a family of injective *length-increasing*, pairwise independent hash functions in order to restrict the domain of a one-way function and increase the probability that it is 1-1. The function $m(\cdot)$ in the definition below determines the length of the output of the hash functions relative to the input length.

Definition 7 ($m(\cdot)$ **pairwise independent family of hash functions**). *Let $\mathcal{H} = \bigcup H_n$ be a collection of functions where $H_n = \{h_i : \{0,1\}^n \to \{0,1\}^{m(n)} | i \in I_n\}$. We say that $\mathcal{H}$ is a $m(\cdot)$ pairwise independent family of hash functions if there exist two PPT algorithms G and H such that the following holds:*

1. Easy to compute: *There exists a polynomially bounded function $l(\cdot)$ such that the output of G on input 1^n is in the set $I_n \subseteq \{0,1\}^{l(n)}$. On input $i \in I_n$ and $x \in \{0,1\}^n$, $H(i,x) = h_i(x)$.*
2. Pairwise independent: *For every $x_1 \neq x_2 \in \{0,1\}^n$ and $y_1 \neq y_2 \in \{0,1\}^m$, $Pr_{h \in H_n}[h(x_1) = y_1] = \frac{1}{2^m}$ and $Pr_{h \in H_n}[h(x_2) = y_2 \mid h(x_1) = y_1] = \frac{1}{2^m - 1}$.*

Definition 7 can be instantiated with the collection of functions of the form $h_{a,b}(x) = ax + b$ where $a, b \in GF(2^m)$, $a \neq 0$, both addition and multiplication are over $GF(2^m)$, and every x is interpreted as a distinct element of the subfield $GF(2^n)$. Definition 7 implies that each function from the collection must injective. Moreover, it also implies the following *balance* property that will be useful in our analysis: For every n and every $y \in \{0,1\}^{m(n)}$, $Pr_{h \in H_n}[\exists x \text{ such that } y = h(x)] = \frac{2^n}{2^m}$.

5.2 Restricted Exponentially Strong One-Way Functions are 1-1

We show that strong one-way functions do not have many collisions to begin with (Lemma 3), and that by restricting the domain of such a function we get a function which is 1-1 with high probability (Theorem 4).

Definition 8 (**Collision group of y relative to f**). $[y]_f = \{y' | f(y') = f(y)\}$

We denote the input length of the one-way function in the following lemma by m, as we will use n for the input length of the family of one-way functions that we construct.

Lemma 3 (**Exponentially strong one-way functions have few collisions**). *If f is a (T, ϵ) one-way function then there exists a polynomial $t(\cdot)$ such that for sufficiently large m and for every $y \in \{0,1\}^m$, $|[y]_f| \leq 2^m \cdot \max\{2\epsilon(m), t(m)/T(m)\}$.*

Proof. We give the proof for the uniform setting, for the non-uniform case a stronger version of the lemma can be proved using the fact that a pre-image for y^* defined below can be given to the algorithm as advice. Let $t(\cdot)$ be the time required for evaluation of f. Assume for contradiction that the statement does not hold for y^*. We first show a $T(m)$-bounded algorithm that finds a pre-image for $f(y^*)$ with probability $\frac{1}{2}$. Randomly choose $y_1, \ldots, y_{T/t} \in \{0,1\}^m$ and apply f to them. By the assumption $|[y^*]_f| > 2^m \cdot t/T$, hence for each i, $y_i \in [y^*]_f$ with probability at least t/T. Therefore the probability that there is no i for which $x_i \in [y^*]_f$ is at most $(1 - t/T)^{T/t} < \frac{1}{2}$.

By the assumption $|[y^*]_f| > 2^m \cdot 2\epsilon$, therefore $y \in [y^*]_f$ with probability at least 2ϵ. If indeed $y \in [y^*]_f$, finding a pre-image for $f(y)$ is the same as finding a pre-image for $f(y^*)$ and the algorithm described above will find a pre-image with probability at least $\frac{1}{2}$. The conclusion is that there exists a $T(m)$-bounded adversary which finds a pre-image for $f(x)$ on a random x with probability ϵ, contradicting the fact that f is a (T, ϵ) one-way function. $\square$

Notation. From here on we let $f_h(\cdot) \stackrel{\text{def}}{=} f(h(\cdot))$.

Theorem 4 (Restricted exponentially strong one-way functions are 1-1). *If* $\mathcal{H} = \bigcup H_n$ *is an* $m(\cdot)$ *pairwise independent family of functions and* f *is a* (T, ϵ) *one-way function with* $T(m) \geq 2^{\mu \cdot m}$ *and* $\epsilon(m) \leq 2^{-\mu \cdot m}$ *for some* $\mu > 0$, *then the probability over* $h \in H_n$ *that* f_h *is not 1-1 is bounded by* $\text{poly}(m) \cdot 2^{2n - \mu m}$.

Proof. Fix $x_1 \neq x_2 \in \{0,1\}^n$. We bound the probability that f_h maps both inputs to the same image for a random $h \in H_n$.
$Pr_{h \in H_n}[f_h(x_1) = f_h(x_2)] = Pr_{h \in H_n}[h(x_2) \in [h(x_1)]_f]$

$$= \sum_{y_1 \in \{0,1\}^m} Pr_{h \in H_n}[h(x_1) = y_1] \cdot Pr_{h \in H_n}[h(x_2) \in [y_1]_f \mid h(x_1) = y_1]$$

$$= \sum_{y_1 \in \{0,1\}^m} Pr_{h \in H_n}[h(x_1) = y_1] \cdot \sum_{y_2 \in [y_1]_f} Pr_{h \in H_n}[h(x_2) = y_2 \mid h(x_1) = y_1]$$

$$\leq 2^m \cdot 2^{-m} \cdot 2^m \cdot \max\{2\epsilon, \text{poly}(m)/T\} \cdot \frac{1}{2^m - 1} \tag{1}$$

$$\leq 2^{-\mu \cdot m} \cdot \text{poly}(m) \tag{2}$$

where (1) follows from the pairwise independence of $\mathcal{H}$ and from Lemma 3, and (2) follows from the hypothesis of the theorem about ϵ and T. As there are 2^{2n} pairs of inputs in $\{0,1\}^n$, by a union bound the probability that h_f maps *any* two inputs of length n to the same output is bounded by $\text{poly}(m) \cdot 2^{2n} \cdot 2^{-\mu \cdot m}$. $\square$

5.3   Restricted Exponentially Strong One-Way Functions are One-Way

We show that if f is a strong one-way function and $\mathcal{H}$ is a pairwise independent family of hash functions, then $\mathcal{F} = \{f_h \mid h \in \mathcal{H}\}$ is a collection of strong one-way functions. The idea is that if we have an algorithm that finds a pre-image for $z = f_h(x)$ given z and h for a random h and x, we can use it to find a pre-image for $z = f(y)$ for a random

y in the following way. We randomly select sufficiently many functions $h \in \mathcal{H}$ so that one of them will have y in its range, and apply the inversion algorithm to z and each such h. If the algorithm succeeds, we get some x such that $f_h(x) = f(h(x)) = z$ and $h(x)$ is a pre-image of z under f. This approach gives the following theorem.

Theorem 5. *Suppose f is (T, ϵ) one-way and $\mathcal{H}$ is a $m(\cdot)$ pairwise independent family of functions. Then $\mathcal{F} = \{f_h \mid h \in \mathcal{H}\}$ is a (T', ϵ') one-way collection of functions for any T', ϵ' such that for all sufficiently large n the following conditions hold with $m = m(n)$:*

1. $T'(n) \leq \frac{1}{2} \cdot T(m)$
2. $\epsilon'(n) \geq 2 \cdot \epsilon(m)$
3. $\frac{T'(n)}{\epsilon'(n)} \leq \frac{1}{4} \cdot \frac{2^n}{2^m} \cdot \frac{T(m)}{\epsilon(m)}$
4. $T'(n) > m(n)^c$ *for every constant c*

Proof. Throughout the proof we view T' and ϵ' as functions of n and T and ϵ as functions of m (where m is itself a function of n). We omit m and n in order to simplify notation. Assume for contradiction that f is a (T, ϵ) one-way function, but $\mathcal{F}$ is not a (T', ϵ') one-way collection for $T, T', \epsilon, \epsilon'$ as stated in the theorem. By Definition 6 there exists a T'-bounded adversary A' such that for infinitely many n's $Pr_{h \in H_n, x \in U_n}[f_h(A'(1^n, f_h(x), h)) = f_h(x)] \geq \epsilon'$. We construct a T-bounded algorithm A such that for infinitely many m's $Pr_{y \in U_m}[f(A(1^m, f(y))) = f(y)] \geq \epsilon$, in contradiction with the assumption that f is (T, ϵ) one-way.

The algorithm A, described below, finds a pre-image for $z = f(y)$. On input $(1^m, z)$, A repeats the following steps $t = \frac{T}{2T'}$ times:

1. Choose a random $h \in H_n$.
2. Compute $x = A'(1^n, z, h)$
3. If $f_h(x) = z$ stop and output $h(x)$

Assuming $z = f(y)$, we define the random variables $A_i(y)$ and $B_i(y)$ for $i = 1, \ldots, t$ as follows: $A_i(y) = 1$ if in the i'th iteration there exists a pre-image for y under h, and $B_i(y) = 1$ if in the i'th iteration $f_h(x) = f(y)$. (Probabilities in these variables are taken over the random coins of A.) The following two claims allow us to complete the proof of Theorem 5.

Claim 1. $Pr_{y \in U_m}[\exists i : A_i(y) = 1] \geq \frac{\epsilon}{\epsilon'}$

Claim 2. *For every i, $Pr_{y \in U_m}[B_i(y) = 1 \mid A_i(y) = 1] \geq \epsilon'$*

By the above claims and the definitions of A_i and B_i, we can lower bound the probability that A finds a pre-image for $f(y)$.

$$
\begin{aligned}
Pr_{y \in U_m}[f(A(1^m, f(y))) = f(y)] &= Pr_{y \in U_m}[\exists i : B_i(y) = 1] \\
&\geq Pr[\exists i : A_i(y) = 1] \cdot Pr[B_i(y) = 1 \mid A_i(y) = 1] \\
&\geq \frac{\epsilon}{\epsilon'} \cdot \epsilon' \\
&= \epsilon
\end{aligned}
$$

A's running time in every iteration is bounded by $T' + m^c$ for some constant c. As there are $t = \frac{T}{2T'}$ iterations and $T' > m^c$ the total running time is bounded by $\frac{T}{2T'} \cdot (T' + m^c) < T$, contradicting the assumption that f is (T, ϵ) one-way.   □

Proof of Claim 1. We show that for any fixed $y \in \{0,1\}^m$, $Pr[\exists i : A_i(y) = 1] \geq \frac{\epsilon}{\epsilon'}$ and the claim follows. By the balance property we have $Pr[A_i(y) = 1] = Pr_{h \in H_n}[\exists x$ such that $y = h(x)] = \frac{2^n}{2^m}$ for every $i = 1, \ldots, t$. We view $A_1, \ldots, A_t$ as $t = \frac{T}{2T'}$ independent experiments, each with success probability $\delta = \frac{2^n}{2^m}$. The probability that none of the experiments succeed can be bounded by $(1-\delta)^t < \frac{1}{t\delta+1}$. Therefore $Pr[\exists i : A_i(y) = 1] \geq 1 - \frac{1}{t\delta+1} = \frac{t\delta}{t\delta+1}$. If $t\delta > 1$, $Pr[\exists i : A_i(y) = 1] \geq \frac{1}{2} \geq \frac{\epsilon}{\epsilon'}$, otherwise, $Pr[\exists i : A_i(y) = 1] \geq \frac{t\cdot\delta}{2} = \frac{1}{4} \cdot \frac{2^n}{2^m} \cdot \frac{T}{T'} \geq \frac{\epsilon}{\epsilon'}$. In both cases the last inequality follows from the assumptions in the hypothesis of the theorem. $\square$

Proof of Claim 2. We use the following notation:

$p_1 \triangleq Pr_{y \in U_m}[B_i(y) = 1 \mid A_i(y) = 1]$

$p_2 \triangleq Pr_{h \in H_n, y \in U_m}[f_h(A'(1^n, f(y), h)) = f(y) \mid \exists x : y = h(x)]$

$p_3 \triangleq Pr_{h \in H_n, x \in U_n}[f_h(A'(1^n, f_h(x), h)) = f_h(x)]$

By the definition of A_i and B_i we have $p_1 = p_2$ for every i, and by our assumption that $\mathcal{F}$ is not (T', ϵ') one-way, we have $p_3 \geq \epsilon'$. Since Claim 2 is that $p_1 \geq \epsilon'$ for every i, it remains to show that $p_2 = p_3$. We show this by proving that the pair (y, h) under the conditions in p_2 is distributed identically to the pair $(h(x), h)$ under the conditions in p_3. Fix some (y, h) such that y is in the range of h. We calculate the probability of getting this pair under both distributions. Under the conditions in p_2, y is chosen uniformly from a set of size 2^m and h is chosen uniformly from a set of size $\frac{2^n}{2^m}|H_n|$, altogether the probability is $\frac{1}{|H_n|2^n}$. Under the conditions in p_3, h is chosen uniformly from a set of size $|H_n|$ and x is chosen from a set of size 2^n. Since h is 1-1, the probability for getting $y = h(x)$ is 2^{-n}, thus the overall probability is again $\frac{1}{|H_n|2^n}$. $\square$

The above construction allows us to 'redistribute' the hardness of a one-way function. For example a (T, ϵ) one-way function which is strongly secure (small ϵ) against weak adversaries (small T) can be used to construct a (T', ϵ') one-way function family $\mathcal{F}$ that is weakly secure against strong adversaries. The conditions in the theorem give us the boundaries of the possible redistribution. Condition 1 limits the maximal gap between T' and T. Condition 2 does the same for ϵ. Both conditions are easy to satisfy when $m(\cdot)$ is large. Condition 3 defines the loss in the time over success ratio caused by the transformation. The loss is bigger when $m(\cdot)$ is large. For $m = c \cdot n$, if $\frac{T(n)}{\epsilon(n)} = 2^{n(1-\delta)}$ we will get $\frac{T'(n)}{\epsilon'(n)} < 2^{n(1-c\delta)}$.

Corollary 2. *If there exists a (T, ϵ) one-way function with $T, \epsilon^{-1} \geq 2^{m/3}$ and $T/\epsilon \geq 2^{m(1-\delta)}$ then there exists a collection of $2^{n(1-10\delta)}$ one-way functions which is almost 1-1.*

Proof. By applying Theorem 4 to the family $\mathcal{H}$ of $m(\cdot)$ pairwise independent family of hash functions with $m = 9n$ and a one-way function f, we get a family $\mathcal{F} = \{f_h | h \in \mathcal{H}\}$ which is 1-1, except for probability at most 2^{-n}. $\mathcal{F}$ is $2^{n(1-10\delta)}$ one-way if it is (T', ϵ') one-way with $T' = 2^{n(1-10\delta)}$ and $\epsilon' = 1/32$ and it remains to verify that for sufficiently large n, the conditions of Theorem 5 are fulfilled.

1. $T'(n) = 2^{n(1-10\delta)} < \frac{1}{2} \cdot 2^{3n} = \frac{1}{2} \cdot 2^{m/3} \leq \frac{1}{2} \cdot T(m)$

2. $\epsilon'(n) = \frac{1}{32} > 2 \cdot 2^{-m/3} \geq 2 \cdot \epsilon(m)$

3. $\frac{T'(n)}{\epsilon'(n)} = 32 \cdot 2^{n(1-10\delta)} < \frac{1}{4} \cdot 2^{n(1-9\delta)} = \frac{1}{4} \cdot 2^{n-\delta m} = \frac{1}{4} \cdot \frac{2^n}{2^m} \cdot 2^{m(1-\delta)} \leq \frac{1}{4} \cdot \frac{2^n}{2^m} \cdot \frac{T(m)}{\epsilon(m)}$

4. $T'(n) = 2^{n(1-10\delta)} > (9n)^c = m^c$ $\square$

We note that in the non-uniform setting the parameters in the corollary can be improved by using a stronger version of Lemma 3.

5.4   Key Agreement from a Collection of 1-1 One-Way Functions

We show a key agreement protocol which is based on a collection of one-way functions $\mathcal{F}$ which is almost 1-1. The protocol is similar to the one described for constructing key agreement from a one-way permutation. However, there are two obstacles which prevent us from directly applying the previous protocol to a random $f_i \in \mathcal{F}$. First, unlike one-way permutations, $f_i \in \mathcal{F}$ is not always 1-1 and hence Alice and Bob may have different outputs. Second, since $\mathcal{F}$ is hard to invert only on average, it is possible that a specific function $f_i \in \mathcal{F}$ is easy to invert. We use standard techniques for amplifying both correctness and security. Specifically, we begin by describing a basic protocol which in itself lacks in both security and in correctness. By combining several copies of the basic protocol we create an intermediate protocol which is secure but has a big error probability. By combining several copies of the intermediate protocol we get the final protocol which is both secure and correct. We note that the copies of the basic protocol can be run concurrently and hence the final protocol remains a two message protocol.

The basic protocol. Let $\mathcal{F} = \bigcup F_n$, be a collection of $2^{n(1-\delta)}$ one-way functions which is almost 1-1. For a security parameter 1^k, we set $n = 2 \cdot \log k$. In each copy of the protocol, Alice chooses a random $r \in \{0,1\}^n$, a random index $i \in I_n$ which defines a function $f_i \in F_n$ and a random set $\mathcal{A} \subset \{0,1\}^n$ of size $k \cdot \log k$. She applies f_i to the inputs in $\mathcal{A}$, and sends the outputs i and r to Bob. Bob randomly chooses a similar set of inputs $\mathcal{B}$, and applies f_i to them. If $f(\mathcal{A}) \cap f(\mathcal{B}) = \emptyset$, he aborts and both parties output random values; otherwise, he randomly chooses a common output $c \in f(\mathcal{A}) \cap f(\mathcal{B})$ and sends c to Alice. Alice and Bob each identify a source for the common input, $x_A \in \mathcal{A}$ and $x_B \in \mathcal{B}$, so that $f_i(x_A) = f_i(x_B) = c$. Their outputs are $s_A = IP(x_A, r)$ and $s_B = IP(x_B, r)$ respectively.

The intermediate protocol. We denote Alice and Bob's outputs in the i'th copy of the basic protocol by s_A^i and s_B^i. The intermediate protocol consists of $l = \text{polylog}(k)$ copies of the basic protocol, where Alice and Bob's outputs are $S_A = \bigoplus_{i=1}^{l} s_A^i$ and $S_B = \bigoplus_{i=1}^{l} s_B^i$ respectively.

The final protocol. We denote Alice and Bob's outputs in the i'th copy of the intermediate protocol by S_A^i and S_B^i. The final protocol consists of $l = \text{polylog}(k)$ copies of the intermediate protocol with the following addition to Bob's messages. Bob chooses a random $S \in \{0,1\}$ and for each copy of the intermediate protocol sends $S \oplus S_B^i$ to Alice. Bob's output is S and Alice's output is $MAJ\{S \oplus S_A^i \oplus S_B^i\}_{i=1}^{l}$.

A straightforward analysis of the final protocol (appearing in the full version) gives the following theorem:

Theorem 6 (Key agreement from a collection of one-way functions which is almost 1-1). *For any constant* $\delta < 1/2$, *if there exists a collection of* $2^{n(1-\delta)}$ *one-way functions which is almost 1-1, then there exists a* (d, ϵ) *secure key agreement protocol for the following d and* ϵ:

- $d = 2 - 2\delta$	$\epsilon = 1/\log^c k$	*for any constant c*
- $d = 2 - 2\delta - 2\tau$	$\epsilon = k^{-\tau}$	*for any* $\tau < 1 - \delta$
- $d = 2 - 2\delta$	$\epsilon = neg(k)$	*assuming the dream XOR lemma*

Combining Theorem 6 with Corollary 2, we get our main result on weak public-key cryptography from strong one-way functions.

Corollary 3 (Key agreement from one-way functions). *For any constant* $\delta < 1/10$, *if there exists a* (T, ϵ) *one-way function with* $T, \epsilon^{-1} \geq 2^{m/3}$ *and* $T/\epsilon \geq 2^{m(1-\delta)}$ *then there exists a* (d, ϵ) *secure key agreement protocol for the following d and* ϵ:

- $d = 2 - 20\delta$	$\epsilon = 1/\log^c k$	*for any constant c*
- $d = 2 - 20\delta - 2\tau$	$\epsilon = k^{-\tau}$	*for any* $\tau < 1 - 10\delta$
- $d = 2 - 20\delta$	$\epsilon = neg(k)$	*assuming the dream XOR lemma*

6 Conclusions and Open Problems

We established the feasibility of basing weak public-key cryptography on strong, but arguably reasonable, forms of one-way functions. We leave open the possibility of basing weak public-key cryptography on standard (polynomially strong) one-way functions, as well as the possibility of amplifying the security of our protocols without relying on a conjectured dream version of Yao's XOR Lemma. Finally, an interesting open question that was already discussed in the Introduction is the possibility of resisting quantum attacks in our setting. The discussion in Section 1.1 referred to the case where the honest parties are classical and the adversary is quantum. If the honest parties are quantum (and can therefore also exploit the quadratic speedup of Grover's algorithm), it seems possible to retain some of the efficiency gap between the honest parties and the adversary. Setting $T = S^2$ in the description of Merkle's protocol from Section 1, honest quantum parties can run in time $O(T)$ whereas a quantum adversary needs to run in time $\Omega(T^{3/2})$. The optimality of this gap, as well as the possibility of basing it on (quantum) one-way functions, remain to be further studied.

Acknowledgement. We thank the anonymous referees for helpful suggestions and comments.

References

1. Boneh, D., Naor, M.: Timed Commitments. In: Bellare, M. (ed.) CRYPTO 2000. LNCS, vol. 1880, pp. 236–254. Springer, Heidelberg (2000)
2. Cachin, C., Maurer, U.: Unconditional Security Against Memory-Bounded Adversaries. In: Kaliski Jr., B.S. (ed.) CRYPTO 1997. LNCS, vol. 1294, pp. 292–306. Springer, Heidelberg (1997)

3. Ding, Y., Harnik, D., Shaltiel, R., Rosen, A.: Constant-Round Oblivious Transfer in the Bounded Storage Model. In: Naor, M. (ed.) TCC 2004. LNCS, vol. 2951, Springer, Heidelberg (2004)
4. Dubrov, B., Ishai, Y.: On the randomness complexity of efficient sampling. In: Proceedings of STOC 2006, pp. 711–720 (2006)
5. Fiat, A., Naor, M.: Rigorous Time/Space Trade-offs for Inverting Functions. SIAM J. Comput. 29(3), 790–803 (1999)
6. Gertner, Y., Kannan, S., Malkin, T., Reingold, O., Viswanathan, M.: The Relationship between Public Key Encryption and Oblivious Transfer. In: Proc. of the 41st Annual Symposium on Foundations of Computer Science (FOCS) (2000)
7. Goldreich, O.: Foundations of Cryptography Basic Tools. Cambridge University Press, Cambridge (2001)
8. Goldreich, O., Levin, L.: A Hard-Core Predicate for all One-Way Functions. In: STOC 1989, pp. 25–32 (1989)
9. Goldreich, O., Nisan, N., Wigderson, A.: On Yao's XOR lemma. Technical Report TR95-50, Electronic Colloquium on Computational Complexity (1995)
10. Grover, L.: A Fast Quantum Mechanical Algorithm for Database Search. In: Proceedings of the 28th Annual ACM Symposium on the Theory of Computing, May 1996, p. 212 (1996)
11. Haitner, I., Harnik, D., Reingold, O.: Efficient Pseudorandom Generators from Exponentially Hard One-Way Functions. In: Bugliesi, M., Preneel, B., Sassone, V., Wegener, I. (eds.) ICALP 2006. LNCS, vol. 4052, pp. 228–239. Springer, Heidelberg (2006)
12. Hellman, M.E.: A Cryptanalytic Time-Memory Trade-Off. IEEE Transactions on Information Theory IT-26(4), 401–406 (1980)
13. Holenstein, T.: Key Agreement from Weak Bit Agreement. In: Proceedings of STOC 2005, pp. 664–673 (2005)
14. Holenstein, T.: Pseudorandom Generators from One-Way Functions: A Simple Construction for Any Hardness. In: Halevi, S., Rabin, T. (eds.) TCC 2006. LNCS, vol. 3876, pp. 443–461. Springer, Heidelberg (2006)
15. Impagliazzo, R., Luby, M.: One-Way Functions are Essential for Complexity-Based Cryptography. In: 30th IEEE Symposium on Foundations of Computer Science (FOCS), pp. 230–235. IEEE, Los Alamitos (1989)
16. Impagliazzo, R., Rudich, S.: Limits on the Provable Consequences of One-way Permutations. In: Proceedings of the ACM Symposium on Theory of Computing, pp. 44–61 (1989)
17. Levin, L.A.: One-Way Functions and Pseudorandom Generators. Combinatorica 7(4), 357–363 (1987) Earlier version in STOC 1985
18. Maurer, U.: Conditionally-Perfect Secrecy and a Provably Secure Randomized Cipher. Journal of Cryptology 5(1), 53–66 (1992)
19. Merkle, R.: Secure communications over insecure channels. CACM, 294–299 (April 1978)
20. Reingold, O., Trevisan, L., Vadhan, S.: Notions of Reducibility between Cryptgraphic Primitives. In: Naor, M. (ed.) TCC 2004. LNCS, vol. 2951, pp. 1–20. Springer, Heidelberg (2004)
21. Rudich, S.: Limits on the Provable Consequences of One-way Functions. Ph.D. thesis
22. Shor, P.W.: Polynomial-Time Algorithms for Prime Factorization and Discrete Logarithms on a Quantom Computer. SIAM J. Comp. 26(5), 1484–1509 (1997)
23. Wee, H.: On obfuscating point functions. In: STOC 2005, pp. 523–532 (2005)

Which Languages Have 4-Round Zero-Knowledge Proofs?

Jonathan Katz[*]

Department of Computer Science
University of Maryland
jkatz@cs.umd.edu

Abstract. We show that if a language L has a 4-round, black-box, computational zero-knowledge proof system with negligible soundness error, then $\bar{L} \in \mathrm{MA}$. Assuming the polynomial hierarchy does not collapse, this means in particular that NP-complete languages do not have 4-round zero-knowledge proofs (at least with respect to black-box simulation).

1 Introduction

A zero-knowledge proof system [23] for a language L is a protocol that enables a prover $\mathcal{P}$ to convince a polynomial-time verifier $\mathcal{V}$ that a given instance x is indeed a member of L. Roughly speaking, the guarantees provided are:

Completeness: If $x \in L$ then the honest prover $\mathcal{P}$ will convince the honest verifier $\mathcal{V}$ to accept, except possibly with some small probability. If $\mathcal{P}$ always convinces $\mathcal{V}$ to accept when $x \in L$ then we say the proof system has *perfect completeness*.

Soundness: If $x \notin L$ a cheating prover $\mathcal{P}^*$ will be unable to falsely convince the honest verifier that x is in L, except with some small probability known as the *soundness error*.

Zero knowledge: When $x \in L$ and the prover is honest, even a malicious verifier $\mathcal{V}^*$ "learns nothing" beyond the fact that $x \in L$.

There are various ways of formalizing the above properties. In this paper, we are interested in the case when the soundness property holds against all-powerful provers — i.e., we focus on *proofs* rather than *arguments* [13] — and we are interested in proof systems with negligible soundness error. For the proof system to be non-trivial, the completeness error should not be too large; we will consider both the case of perfect completeness as well as the case when, for $x \in L$, the honest verifier accepts with some noticeable (i.e., inverse polynomial) probability. Finally, we focus on the case of *computational* zero knowledge (CZK) where, informally, the requirement is that a non-uniform *polynomial-time* cheating verifier learns nothing from the interaction. (Formal definitions are provided

---

[*] This work was supported by NSF CAREER award #0447075 and US-Israel Binational Science Foundation grant #2004240.

© International Association for Cryptologic Research 2008

in Section 2.) We let CZK denote the class of languages that admit a computational zero-knowledge proof system.

In this paper we study the round complexity of CZK proof systems, where a round consists of a message sent from one party to the other and we assume that the prover and the verifier speak in alternating rounds. We briefly survey what is known in this regard:

Unconditional constructions. The only languages currently known to be in CZK *unconditionally* are those that admit *statistical* zero-knowledge (SZK) proofs [23] where, informally, even an all-powerful cheating verifier learns nothing from its interaction with the prover; we denote the class of languages admitting statistical zero-knowledge proofs by SZK. While it is not known[1] whether all languages in SZK have constant-round statistical zero-knowledge proof systems, such proof systems are known for specific languages. In particular, graph non-isomorphism [21] (cf. [21, Remark 12]) as well as languages related to various number-theoretic problems [23,30,34,15,31,14] have 4-round SZK proof systems, and graph isomorphism [7] has a 5-round SZK proof system.

Constructions based on one-way functions/permutations. Assuming the existence of one-way functions, every language in NP has an $\omega(1)$-round CZK proof system where the honest prover runs in polynomial time given an NP-witness for the statement being proved [21]. (Actually, this result holds for MA as well.[2]) If no computational restrictions are placed on the honest prover, then any language in AM has an $\omega(1)$-round CZK proof system under the same assumption, and any language in IP = PSPACE has a CZK proof system with polynomially-many rounds [29,10].

Assuming the existence of one-way permutations, Feige and Shamir [17] show a 4-round computational zero-knowledge *argument* for any language in NP. Their techniques yield a 5-round CZK argument based on one-way functions, and this was later improved to 4 rounds by Bellare et al. [6].

Constructions based on stronger assumptions. Assuming the existence of a two-round statistically-hiding commitment scheme, there exists a 5-round CZK proof system for any language in NP [19] (or even AM if the honest prover can be unbounded. (More generally, given a constant-round statistically-hiding commitment scheme, there exists a constant-round CZK proof system for any language in AM.(Two-round statistically-hiding commitment schemes, in turn, can be constructed based on a variety of number-theoretic assumptions [12,13,24] or the existence of collision-resistant hash functions [16,28].

Although statistically-hiding commitment schemes can be constructed from any one-way function [27], constructions of *constant-round* statistically-hiding commitment schemes from one-way functions are unlikely to exist [26].

---

[1] The constant-round proofs in [8], based on specific number-theoretic assumptions, consider a weaker variant of SZK where the verifier is assumed to run in polynomial time during its interaction with the prover. See also [35].

[2] The class MA is defined in Section 2. AM denotes the class of languages having *constant-round* Arthur-Merlin proofs.

Lower bounds. Goldreich and Oren [22] show that 2-round CZK proofs exist only for languages in BPP. (Their result applies to *auxiliary-input* zero knowledge proofs, the type we will be concerned with here as well.) Extending this result, Goldreich and Krawczyk [20] show that 3-round *black-box* CZK proofs exist only for languages in BPP. (A definition of black-box CZK is given in Section 2.) Both these results hold for arguments as well as proofs.

Pass [33] gives evidence of the difficulty of showing a black-box *construction* of a constant-round CZK proof for NP based on any one-way function (even if non black-box *simulation* is allowed). We refer to his paper for a precise statement of this result.

1.1 Our Result

We show that 4-round black-box CZK proofs, even with imperfect completeness, exist only for languages whose complement is in MA. This result is unconditional, and holds independent of any cryptographic assumptions one might make. Other than the fact that the bound holds only with respect to black-box simulation, this result is essentially the best one could hope for:

– Under widely-believed number-theoretic assumptions, there exist 5-round CZK proofs for all of NP [19]. Assuming the polynomial hierarchy does not collapse [11], our result indicates that the round complexity in this case is optimal.
– Our result applies only to proofs, but not arguments. Indeed, as noted earlier, there exist 4-round CZK *arguments* for all of NP under relatively weak assumptions [17,6].
– There exist unconditional constructions of 4-round CZK proofs for languages believed to be outside of BPP, such as graph non-isomorphism [21].

We remark also that for the case of *uniform* zero-knowledge (i.e., protocols which are zero knowledge for *uniform* polynomial-time verifiers), a 4-round protocol for all of NP is possible [19] assuming the existence of 1-round statistically-hiding commitment schemes (that are computationally binding for uniform adversaries).

Besides shedding further light on the finer structure of the class CZK, our result indicates that (black-box) 4-round CZK proofs for all of NP are impossible and so the round complexity of the Goldreich-Kahan protocol [19] is optimal. Our result also gives an "explanation" as to why the known SZK proof for graph isomorphism requires five rounds [7] even though graph *non*-isomorphism has a 4-round SZK proof [21].

Limitations of black-box impossibility results. We prove our result only for the case of *black-box* zero-knowledge protocols. The work of Barak [3], however, shows that black-box impossibility results and lower bounds need not carry over to the general case.[3] Nevertheless, black-box bounds are useful insofar as they

---

[3] Barak's work gives a constant-round, public-coin, CZK argument for all of NP, something that was ruled out with respect to black-box simulation by Goldreich and Krawczyk [20].

rule out a particular approach for solving a problem. We remark further that many of the known (natural) zero-knowledge *proofs* are in fact black-box zero knowledge; in particular, the protocols of Barak [3] as well as those based on "knowledge of exponent" assumptions [25,9] are zero-knowledge *arguments*. On the other hand, non black-box zero-knowledge *proofs* using four or fewer rounds are known to exist based on various non-standard assumptions [5,32].

Our current ability to prove *general* (as opposed to black-box) lower bounds for zero-knowledge protocols is, unfortunately, relatively limited [22,5].

High-level overview of our technique. Our lower bound for 4-round protocols is proved by extending the Goldreich-Krawczyk lower bound [20] for 3-round protocols. (We assume familiarity with their proof in what follows.) To prove their result, Goldreich and Krawczyk consider a cheating verifier V^* who generates its message, in the second round of the protocol, using fresh random coins that are determined as a function of the prover's first message. On an intuitive level this means that rewinding is useless because every time V^* is rewound, and a different first message is sent by the simulator, it is as if the protocol execution is being started again from scratch.

We use the same basic idea, now applied to the verifier's message sent in the *third* round of the protocol. A problem is that the verifier's first-round message may "commit" the verifier, in a computational sense, to only one possible third-round message. (Roughly speaking, the verifier cannot be committed in an information-theoretic sense because then an all-powerful prover could guess the third-round message *in advance* based on the first-round message alone. This is one reason why our result applies only to proofs, and not to arguments.) For this reason, we use some "all-powerful" entity to provide the verifier with *collisions*, i.e., multiple third-round messages consistent with the same first-round message. This idea was directly inspired by the recent work of Haitner et al. [26], who use such collisions to prove lower bounds on the round complexity of black-box constructions of interactive protocols in other settings. In their work, an oracle provides collisions. Here, we do not have an oracle; instead, we have an all-powerful prover that provides collisions as part of an interactive MA-proof for some language. See Section 3 for further intuition, as well as the details of the proof.

1.2 Outline of the Paper

Standard definitions, as well as some terminology specific to this paper, are provided in Section 2. In Section 3 we prove our result for the case of CZK proof systems with perfect completeness. Technical modifications necessary to deal with the case of imperfect completeness are deferred to Section 4. We conclude with some open questions in Section 5.

2   Definitions

Given interactive algorithms $\mathcal{P}$ and $\mathcal{V}$, we let $\langle \mathcal{P}(x), \mathcal{V}(y) \rangle$ denote the interaction of $\mathcal{P}$, holding input x, with $\mathcal{V}$, holding input y. We let $\langle \mathcal{P}(x), \mathcal{V}(y) \rangle = 1$ denote the event that $\mathcal{V}$ outputs 1 in the indicated interaction, where an output of "1" is interpreted as "accept" and an output of "0" is interpreted as "reject". We now give the standard definition of an interactive proof system [23] for a language L.

Definition 1. *Interactive algorithms* $\mathcal{P}, \mathcal{V}$ *form an* interactive proof system *for a language L if $\mathcal{V}$ runs in probabilistic polynomial time and there exist non-negative functions c, s such that:*

- *For all $x \in L$, it holds that* $\Pr[\langle \mathcal{P}(x), \mathcal{V}(x) \rangle = 1] \geq c(|x|)$. *(Note that we do not require $\mathcal{P}$ to run in polynomial time.)*
- *For all $x \notin L$ and any $\mathcal{P}^*$ we have* $\Pr[\langle \mathcal{P}^*, \mathcal{V}(x) \rangle = 1] \leq s(|x|)$.
- *There exists a polynomial p such that* $c(|x|) \geq s(|x|) + 1/p(|x|)$.

We call c the acceptance probability, *and s the* soundness error. *If $c(|x|) = 1$ for all x, we say the proof system has* perfect completeness. *If s is negligible, we say the proof system has* negligible soundness error. $\diamond$

We will only consider zero-knowledge proof systems having negligible soundness error.

A *round* of an interactive proof system consists of a message sent from one party to the other, and we assume that the prover and the verifier speak in alternating rounds. Following [2], we let MA denote the class of languages having a 1-round proof system and in this case refer to the prover as *Merlin* and the verifier as *Arthur*; that is:

Definition 2. $L \in$ MA *if there exists a probabilistic polynomial-time verifier $\mathcal{V}$, a non-negative function s, and a polynomial p such that the following hold for all sufficiently-long x:*

- *If $x \in L$ then there exists a string w (that can be sent by Merlin) such that*

$$\Pr[\mathcal{V}(x, w) = 1] \geq s(|x|) + 1/p(|x|).$$

- *If $x \notin L$ then for all w (sent by a cheating Merlin) it holds that*

$$\Pr[\mathcal{V}(x, w) = 1] \leq s(|x|).$$

$\diamond$

In fact, it is known that an equivalent definition is obtained even if we require perfect completeness and negligible soundness error.

2.1 Zero Knowledge Proof Systems

A *distribution ensemble* $\{X(a)\}_{a \in \{0,1\}^*}$ is an infinite sequence of probability distributions, where a distribution $X(a)$ is associated with each value of a. Two distribution ensembles X and Y are *computationally indistinguishable* if for all polynomial-time algorithms D, there exists a negligible function μ such that for every a we have

$$|\Pr[D(X(a), a) = 1] - \Pr[D(Y(a), a) = 1]| \leq \mu(|a|).$$

(We do not need to consider non-uniform distinguishers here since non-uniformity can be incorporated via the auxiliary input that we will provide to the cheating verifier, below.)

Given interactive algorithms $\mathcal{P}, \mathcal{V}^*$, we let $\mathsf{trans}_{\mathcal{V}^*}\langle \mathcal{P}(x), \mathcal{V}^*(y)\rangle$ denote the transcript of the indicated interaction; for convenience, this includes both messages of the prover as well as those of the verifier. (We remark that we do not need to consider the entire *view* of $\mathcal{V}^*$ since we will restrict to deterministic verifiers, as justified below, and the input y of $\mathcal{V}^*$ will be provided to the distinguisher as per our definition of computational indistinguishability, above.) We now review the standard definitions for computational zero-knowledge proofs.

Definition 3. *An interactive proof system* $\mathcal{P}, \mathcal{V}$ *for a language* L *is said to be a* computational zero-knowledge *proof system if for any probabilistic polynomial-time algorithm* $\mathcal{V}^*$, *there exists an expected polynomial-time simulator* $\mathcal{S}$ *such that the following distribution ensembles are computationally indistinguishable:*

$$\{\mathsf{trans}_{\mathcal{V}^*}\langle \mathcal{P}(x), \mathcal{V}^*(x,z)\rangle\}_{x \in L, z \in \{0,1\}^*} \quad and \quad \{\mathcal{S}(x,z)\}_{x \in L, z \in \{0,1\}^*} .$$

$\Diamond$

The above definition incorporates an auxiliary input z provided to $\mathcal{V}^*$, and we may therefore restrict our consideration to verifiers $\mathcal{V}^*$ that are deterministic. Note also that we allow simulation in *expected* polynomial time; this makes our results stronger. (Also, constant-round, black-box CZK proofs with strict polynomial-time simulation are already ruled out by Barak and Lindell [4].)

A computational zero-knowledge proof system $(\mathcal{P}, \mathcal{V})$ is *black-box* zero knowledge if there exists a "universal" simulator that takes oracle access to the cheating verifier $\mathcal{V}^*$. That is:

Definition 4. *A computational zero-knowledge proof system* $\mathcal{P}, \mathcal{V}$ *is* black-box zero-knowledge *if there exists an expected polynomial-time oracle machine* Sim *(the* black-box simulator*) such that for any probabilistic polynomial-time algorithm* $\mathcal{V}^*$ *the following distribution ensembles are computationally indistinguishable:*

$$\{\mathsf{trans}_{\mathcal{V}^*}\langle \mathcal{P}(x), \mathcal{V}^*(x,z)\rangle\}_{x \in L, z \in \{0,1\}^*} \quad and \quad \left\{\mathsf{Sim}^{\mathcal{V}^*(x,z)}(x)\right\}_{x \in L, z \in \{0,1\}^*} .$$

$\Diamond$

We denote by $^{bb}\mathsf{CZK}(r)$ the class of languages that have r-round, *black-box*, computational zero-knowledge proof systems with negligible soundness error.

Terminology and simplifying assumptions. We will be concerned with 4-round CZK proof systems, where (without loss of generality) the verifier sends the first message and the prover sends the final message. We use $\alpha, \beta, \gamma, \delta$ to denote the first, second, third, and fourth messages, respectively. We let $\mathcal{P}_x$ (resp., $\mathcal{V}_x$) denote the honest prover (resp., honest verifier) algorithm when the common input is x.

We let $\alpha = \mathcal{V}_x(r)$ denote the first message sent by $\mathcal{V}_x$ when its random coins are fixed to r, and let $\gamma = \mathcal{V}_x(\alpha, \beta; r)$ denote the third message sent by $\mathcal{V}_x$ in this case. Finally, $\mathcal{V}_x(\alpha, \beta, \gamma, \delta; r)$ is a bit denoting whether the verifier accepts (i.e., outputs 1) or rejects. We say that $(\alpha, \beta, \gamma, \delta, r)$ is an *accepting transcript* for a given x if $\mathcal{V}_x(\alpha, \beta, \gamma, \delta; r) = 1$. Note that we do not require the verifier's decision to depend on the actual transcript alone, but allow its decision to also possibly depend on its random coins.

Without loss of generality, we make a number of simplifying assumptions about the behavior of black-box simulator Sim. The first query of Sim to $\mathcal{V}^*$ will simply be a "prompt" query to which $\mathcal{V}^*$ responds with α. Subsequent queries by Sim are all of the form (α, β) (for some β of Sim's choice), to which $\mathcal{V}^*$ will respond with some γ. (We can assume Sim makes no queries of the form $(\alpha, \beta, \gamma, \delta)$ since $\mathcal{V}^*$ can simply refuse to respond to such queries.) We assume Sim makes a given query only once. Finally, if the simulator outputs the transcript $(\alpha', \beta, \gamma, \delta)$ we assume that $\alpha' = \alpha$, and that the simulator previously queried (α, β) to $\mathcal{V}^*$ and received response γ.

3 CZK Proof Systems with Perfect Completeness

We now state our main result:

Theorem 1. $^{bb}\mathsf{CZK}(4) \subseteq \mathsf{coMA}$.

In this section we prove this result in the easier case when the proof system in question has perfect completeness; we handle the case of imperfect completeness in the following section.

As intuition for the proof, consider first the case of a malicious verifier $\hat{\mathcal{V}}$ who acts in the following way: it sends an initial message α, and then in response to the prover's second message β it chooses a random message γ consistent with α. (For now, we do not worry about the fact that this does not necessarily represent a feasible polynomial-time strategy.) Formally, if we let R_α denote the set of random coins consistent with α (i.e., $r \in R_\alpha$ implies $\mathcal{V}_x(r) = \alpha$), then in response to β the malicious verifier chooses a random $r \in R_\alpha$ and computes $\gamma = \mathcal{V}_x(\alpha, \beta; r)$. Intuitively, it will be difficult to simulate an accepting transcript for such a verifier since each time the simulator "rewinds" $\hat{\mathcal{V}}$ it will be given a message γ consistent with a *different* set of random coins. In fact, we can prove that if $x \notin L$ then the simulator will *not* be able to simulate an accepting transcript for such a verifier, since the ability to do so with non-negligible probability could be translated into the ability to violate the soundness condition of the proof system with non-negligible probability. (A proof of this fact goes along similar lines as the proof in [20].)

On the other hand, consider the case when $x \in L$. From the perspective of the honest prover, the behavior of $\hat{\mathcal{V}}$ is identical to that of the honest verifier, and so the honest prover's interaction with $\hat{\mathcal{V}}$ leads to an accepting transcript with probability 1. We would like to claim that the zero-knowledge condition implies that Sim simulates an accepting transcript for such a verifier with high probability. Unfortunately, $\hat{\mathcal{V}}$ as described above may not run in polynomial time, whereas simulation is only guaranteed for polynomial-time verifiers.

It is possible, however, to obtain a polynomial-time cheating verifier with the desired behavior by providing the verifier as *auxiliary input* a sequence of sufficiently-many coins $r_1, \ldots, r_s$ that are all consistent with the same first message α. Specifically, consider the verifier $\mathcal{V}^*$ defined as follows: given auxiliary input $r_1, \ldots, r_s$ (all consistent with the same first message α) and a poly-wise independent hash function h, send α as the first message. In response to the prover's second message β, compute $i = h(\beta)$ and use r_i to compute the next message $\gamma = \mathcal{V}_x(\alpha, \beta; r_i)$. Note that if $r_1, \ldots, r_s$ are chosen at random (subject to the constraint that they are all mutually consistent) then the behavior of $\mathcal{V}^*$ is identical to the behavior of $\hat{\mathcal{V}}$ as far as the honest prover is concerned. Since $\mathcal{V}^*$ runs in polynomial time, we are now able to argue that Sim simulates an accepting transcript for $\mathcal{V}^*$ with high probability when $x \in L$. Furthermore, it is still possible to show (using a slightly more complicated argument) that, with overwhelming probability, Sim *fails* to simulate an accepting transcript for this verifier whenever $x \notin L$.

Based on the above, we obtain an MA proof system for $\bar{L}$: Merlin sends Arthur a sequence $r_1, \ldots, r_s$ of random coins that are all consistent with the same first message α, and Arthur simulates an execution of $\mathsf{Sim}^{\mathcal{V}^*}(x)$. If this does *not* result in an accepting transcript then Arthur accepts, while if it does lead to an accepting transcript then Arthur rejects.

We now formalize the above intuition and show how to handle various technicalities that arise. Fix $L \in {}^{bb}\mathsf{CZK}(4)$. This means that, for this language, there exists a prover $\mathcal{P}$, a verifier $\mathcal{V}$, and a black-box simulator Sim satisfying Definitions 1–4 (except that, in this section, we are assuming perfect completeness). Assume without loss of generality that the second message of the protocol always has length $m(\cdot)$, and let $\ell(\cdot)$ denote the number of random coins used by $\mathcal{V}$. Let $T(\cdot)$ denote an upper-bound on the expected running time of Sim.

Consider the following MA proof system for the language $\bar{L}$, where Merlin (i.e., the prover) and Arthur (i.e., the verifier) share in advance an input x of length n:

Notation: Let $\ell = \ell(n)$, $m = m(n)$, and $T = T(n)$. Set $s = 50 \cdot T^2$; note that s is polynomial in n.

Merlin's message: Merlin sends a sequence of s coins $r_1, \ldots, r_s \in \{0, 1\}^\ell$. (For the honest Merlin, these are all consistent with the same first message α.)

Arthur's actions: Arthur proceeds as follows:

1. Set $\alpha = \mathcal{V}_x(r_1)$. Check that $\alpha = \mathcal{V}_x(r_i)$ for all $1 < i \leq s$, i.e., that all the random coins are consistent with the same first message α. If not, reject; otherwise, go to the next step.
2. Choose a random $5T$-wise independent hash function $h \colon \{0, 1\}^m \to \{1, \ldots, s\}$. Construct the following deterministic verifier $\mathcal{V}^*$:

   (a) Send first message α to the prover.

(b) Upon receiving message β from the prover, compute $i = h(\beta)$ and send the message $\gamma = \mathcal{V}_x(\alpha, \beta; r_i)$ to the prover.

3. Run $\mathrm{Sim}^{\mathcal{V}^*}(x)$ for at most $5T$ steps using uniformly-chosen random coins for Sim. If Sim does *not* output an accepting transcript within this time bound, output "accept". Otherwise, output "reject". (Formally, output "reject" iff Sim outputs $(\alpha, \beta, \gamma, \delta)$, within the allotted time bound, such that $\mathcal{V}_x(\alpha, \beta, \gamma, \delta; r_{h(\beta)}) = 1$.)

The following claims show that the above is a valid MA-protocol for $\bar{L}$, thus proving Theorem 1 for the case of protocols having perfect completeness.

Claim 1. *For any $x \notin \bar{L}$ sufficiently long and for any message $r_1, \ldots, r_s$ sent by Merlin, the probability that Arthur accepts is at most $2/5$.*

Proof. Fix some $r_1, \ldots, r_s$ sent by Merlin. Assume $\mathcal{V}_x(r_i) = \mathcal{V}_x(r_j)$ for all $1 \leq i, j \leq s$ since, if not, Arthur rejects immediately. When $x \notin \bar{L}$ we have $x \in L$ and, by perfect completeness, the interaction of the honest prover $\mathcal{P}_x$ with $\mathcal{V}^*$ would result in an accepting transcript with probability 1. (To see this, note that an execution of $\mathcal{V}^*$ is equivalent to an execution of the honest verifier $\mathcal{V}_x$ using random coins $r_{h(\beta)}$.) The zero-knowledge condition thus implies that, for x sufficiently long, $\mathrm{Sim}^{\mathcal{V}^*}(x)$ outputs an accepting conversation with probability at least $4/5$. It follows that even the truncated version of Sim, where its execution is halted after $5T$ steps, outputs an accepting conversation with probability at least $3/5$. Arthur thus accepts with probability at most $2/5$, as claimed.

Claim 2. *For any $x \in \bar{L}$ sufficiently long, there exists a message $r_1, \ldots, r_s$ such that Arthur will accept with probability at least $1/2$.*

Proof. Fix $x \in \bar{L}$. We show a randomized strategy that allows Merlin to convince Arthur with probability at least $1/2$; this implies the claim.

Merlin proceeds as follows: choose random $r_1 \in \{0,1\}^\ell$ and compute $\alpha = \mathcal{V}_x(r_1)$. Define $R_\alpha \stackrel{\mathrm{def}}{=} \{r \mid \mathcal{V}_x(r) = \alpha\}$; i.e., R_α is the set of coins for the honest verifier consistent with the first message α. Then choose $r_2, \ldots, r_s$ uniformly from R_α. (These need not be distinct.) Send $r_1, \ldots, r_s$ to Arthur. Let p^* denote the probability that Arthur rejects. Note that this is exactly the probability that $\mathrm{Sim}^{\mathcal{V}^*}(x)$ outputs an accepting transcript within the allotted time bound.

We upper-bound p^* by considering a slightly different experiment involving an all-powerful cheating prover $\mathcal{P}^*$ attempting to falsely convince the honest verifier $\mathcal{V}_x$ that $x \in L$. The strategy of $\mathcal{P}^*$ is defined as follows:

1. Receive message α from the verifier. Let $R_\alpha \stackrel{\mathrm{def}}{=} \{r \mid \mathcal{V}_x(r) = \alpha\}$.
2. Run Sim using uniformly-chosen random coins, for at most $5T$ steps. Sim expects to be given oracle access to a (cheating) verifier, and $\mathcal{P}^*$ simulates the actions of such a verifier as follows:
   (a) Choose a random index $q \leftarrow \{1, \ldots, 5T\}$.
   (b) Send α as the verifier's first message.

(c) In response to the ith simulator message (α, β_i) for $i \neq q$, choose a random $r_i \leftarrow R_\alpha$, compute $\gamma_i = \mathcal{V}_x(\alpha, \beta_i; r_i)$, and give γ_i to Sim. (Recall we assume that Sim never makes the same query twice.)

(d) In response to the qth simulator message (α, β_q), send β_q to the (external) honest verifier, and receive in return a message γ_q. Give γ_q to Sim.

3. If Sim outputs a conversation $(\alpha, \beta, \gamma, \delta)$ with $\beta = \beta_q$ within the allotted time bound, then send δ to the (external) honest verifier.

In the above experiment, each "query" β_i of Sim is answered by using a random element $r_i \leftarrow R_\alpha$ to compute the response $\gamma_i = \mathcal{V}_x(\alpha, \beta_i; r_i)$. This is immediate for $i \neq q$, but is true also for $i = q$ since, from the perspective of $\mathcal{P}^*$ and Sim, the coins being used by the external, honest verifier are uniformly-distributed in R_α. Let $\hat{p}$ denote the probability that Sim outputs an accepting transcript in this case, within the allotted time bound. Since Sim makes at most $5T$ queries to its oracle in the above experiment, $\mathcal{P}^*$ convinces the honest verifier to accept with probability $\hat{p}/5T$. Since the proof system has negligible soundness error we have that, for x sufficiently long, $\hat{p} \leq 1/4$.

We return now to consideration of p^*. When Arthur runs $\mathsf{Sim}^{\mathcal{V}^*}(x)$, he does so by first choosing a random h and then answering the simulator's i^{th} query (α, β_i) by using $r_{h(\beta_i)}$ to compute the response $\gamma_i = \mathcal{V}_x(\alpha, \beta_i; r_{h(\beta_i)})$. Since Merlin chooses each of the r_i uniformly from R_α, these responses are distributed identically to the above experiment unless there is a collision in h; that is, unless there exist some $\beta_i \neq \beta_j$ with $h(\beta_i) = h(\beta_j)$. Because h is chosen in a $5T$-wise independent fashion and Sim is restricted to making only $5T$ queries, a standard birthday bound shows that the probability of such a collision is at most $(5T)^2/2s = 1/4$. Conditioned on a collision not occurring, the probability that $\mathsf{Sim}^{\mathcal{V}^*}(x)$ outputs an accepting conversation is exactly $\hat{p} \leq 1/4$. We conclude that $p^* \leq 1/4 + 1/4 = 1/2$, and so Arthur rejects with probability at most $1/2$ (and accepts with probability at least $1/2$).

4   Handling Imperfect Completeness

In the previous section we assumed perfect completeness, and in fact this is essential for the MA proof system given there. To see the problem, assume $\mathcal{P}, \mathcal{V}$ is such that the honest verifier immediately rejects whenever its random coins are all 0. Then a cheating Merlin can send $r_1 = \cdots = r_s = 0^\ell$ and this will cause Arthur to accept with probability 1 even when $x \notin \bar{L}$.

In the modified proof system, we have Arthur "verify" that Merlin sends "representative" random coins $r_1, \ldots, r_s$ by checking that $\mathsf{Sim}^{\mathcal{V}_x(r_i)}(x)$, for a random element r_i in the set sent by Merlin, outputs an accepting transcript with "high" probability. Then Arthur checks whether $\mathsf{Sim}^{\mathcal{V}^*(x; r_1, \ldots, r_s, h)}(x)$ *fails* to output an accepting transcript, as in the previous section. Unfortunately, this may make the honest Merlin's job harder when $x \in \bar{L}$ since in this case $\mathsf{Sim}^{\mathcal{V}_x(r_i)}(x)$ might (legitimately) *never* output an accepting transcript. But Arthur can easily check for this by running $\mathsf{Sim}^{\mathcal{V}_x(r)}(x)$ using random coins r that it chooses itself.

We remark that if we were content to show inclusion in AM (rather than MA), the proof could be simplified somewhat.

Before presenting the modified proof system, we introduce some notation. For a given randomized experiment Expt that can be run in polynomial time, we let $\mathsf{estimate}_\varepsilon(\Pr_r[\mathsf{Expt}])$ denote a procedure that outputs an estimate to the given probability (taken over randomness r) to within an additive factor of ε, except with probability at most ε. That is:

$$\Pr\left[\left|\mathsf{estimate}_\varepsilon(\Pr_r[\mathsf{Expt}=1]) - \Pr_r[\mathsf{Expt}=1]\right| \geq \varepsilon\right] \leq \varepsilon.$$

This can be done in the standard way using $\Theta(\varepsilon^{-2}\log\frac{1}{\varepsilon})$ independent executions of Expt. The important thing to note is that when ε is noticeable, this estimation can be done in polynomial time. In the experiments we will be considering, some variables will be fixed as part of the experiment and others will be chosen at random; we will always subscript those variables being chosen at random (as done above with the subscripted r).

Below, we let $\mathcal{V}^*$ denote the same malicious verifier as in the previous section. Specifically, on input x and auxiliary input $z = r_1, \ldots, r_s, h$, where each r_i represents coins for the honest verifier and h is a hash function, $\mathcal{V}^*$ acts as follows:

1. Send first message $\alpha = \mathcal{V}_x(r_1)$ to the prover.
2. Upon receiving message β from the prover, compute $i = h(\beta)$ and send the message $\gamma = \mathcal{V}_x(\alpha, \beta; r_i)$ to the prover.
3. Receive final message δ from the prover.

We say an interaction of $\mathcal{P}_x$ with $\mathcal{V}^*(x, z)$ *results in an accepting transcript* if $(\alpha, \beta, \gamma, \delta, r_i)$ is an accepting transcript.

Let $L \in {}^{bb}\mathsf{CZK}(4)$, and assume L has a 4-round CZK proof system $\mathcal{P}, \mathcal{V}$ with acceptance probability $c(\cdot)$ where c is noticeable (i.e., $c = \Omega(1/p)$ for some polynomial p). Let ℓ, m, and T be as in the previous section. Once again, Merlin and Arthur share in advance an input x of length n. The MA proof system for the language $\bar{L}$ follows:

Notation: Let $c = c(n)$, $\ell = \ell(n)$, $m = m(n)$, and $T = T(n)$. Assume n is large enough so that $c > 0$. Set $\varepsilon = c/20$, and $s = 4T^2\varepsilon^{-3}$. (Note that ε is noticeable, and s is polynomial.) Let $\widetilde{\mathsf{Sim}}$ denote an execution of Sim for at most $2T/\varepsilon$ steps.

Merlin's message: Merlin sends a sequence of s coins $r_1, \ldots, r_s \in \{0,1\}^\ell$.

Arthur's actions: Arthur proceeds as follows:

1. Compute
$$p_1 = \mathsf{estimate}_\varepsilon\left(\Pr_{r',r}\left[\widetilde{\mathsf{Sim}}^{\mathcal{V}_x(r')}(x;r) \text{ outputs an accepting transcript}\right]\right).$$

   If $p_1 < c - 2\varepsilon$ then accept; otherwise, continue to the next step.
2. Set $\alpha = \mathcal{V}_x(r_1)$. Check that $\alpha = \mathcal{V}_x(r_i)$ for all $1 < i \leq s$. If not, reject; otherwise, continue to the next step.

3. Choose $i \leftarrow \{1, \ldots, s\}$ and coins r and run $\widetilde{\mathsf{Sim}}^{\mathcal{V}_x(r_i)}(x; r)$. If this does not result in an accepting transcript, reject; otherwise, continue to the next step.

4. Let H denote a family of $2T/\varepsilon$-wise independent hash functions $h\colon \{0,1\}^m \to \{1, \ldots, s\}$. Compute

$$p_2 =$$

$$\mathsf{estimate}_\varepsilon\left(\Pr_{h \leftarrow H, r}\left[\widetilde{\mathsf{Sim}}^{\mathcal{V}^*(x; r, h)}(x; r) \text{ outputs an accepting transcript}\right]\right),$$

where $r = (r_1, \ldots, r_s)$. If $p_2 < c - 10\varepsilon$ accept; else reject.

(It should be clear that we have not attempted to optimize any of the parameters of the above proof system.) We now prove claims analogous to those in the previous section.

Claim 3. *For any $x \notin \bar{L}$ sufficiently long and for any message $r_1, \ldots, r_s$ sent by Merlin, the probability that Arthur accepts is at most $c - 6\varepsilon$.*

Proof. If $x \notin \bar{L}$ then $x \in L$ and so the interaction of $\mathcal{P}_x$ with $\mathcal{V}_x$ results in an accepting transcript with probability at least c. The zero-knowledge condition implies that, for x sufficiently long,

$$\Pr_{r', r}[\widetilde{\mathsf{Sim}}^{\mathcal{V}_x(r')}(x; r) \text{ outputs an accepting transcript}] \geq c - \varepsilon.$$

This means that, except with probability at most ε, the value p_1 computed by Arthur satisfies $p_1 \geq c - 2\varepsilon$; thus, Arthur accepts in the first step with probability at most ε.

Fix some $r = (r_1, \ldots, r_s)$ sent by Merlin. We may assume $\mathcal{V}_x(r_i) = \mathcal{V}_x(r_j)$ for all $1 \leq i, j \leq s$ since, if not, Arthur rejects in the second step. Define

$$\hat{p} = \Pr_{i \leftarrow \{1, \ldots, s\}, r}\left[\widetilde{\mathsf{Sim}}^{\mathcal{V}_x(r_i)}(x; r) \text{ outputs an accepting transcript}\right].$$

There are two cases to consider:

Case 1: If $\hat{p} < c - 7\varepsilon$, then the probability that Arthur does not reject in step 3 is at most $c - 7\varepsilon$.

Case 2: On the other hand, if $\hat{p} \geq c - 7\varepsilon$ then (again using the zero-knowledge property)

$$\Pr_{i \leftarrow \{1, \ldots, s\}, r}[\langle \mathcal{P}_x(r), \mathcal{V}_x(r_i)\rangle = 1] \geq c - 8\varepsilon.$$

By definition of $\mathcal{V}^*$ it holds that

$$\Pr_{h \leftarrow H, r}[\langle \mathcal{P}_x(r), \mathcal{V}^*(x, r, h)\rangle \text{ results in an accepting transcript}]$$
$$= \Pr_{i \leftarrow \{1, \ldots, s\}, r}[\langle \mathcal{P}_x(r), \mathcal{V}_x(r_i)\rangle = 1].$$

Thus, relying on the zero-knowledge property once again,

$$\Pr_{h \leftarrow H, r}\left[\widetilde{\mathsf{Sim}}^{\mathcal{V}^*(x; r_1, \ldots, r_s, h)}(x; r) \text{ outputs an accepting transcript}\right] \geq c - 9\varepsilon.$$

So, except with probability at most ε, the value p_2 computed by Arthur satisfies $p_2 \geq c - 10\varepsilon$; thus, Arthur accepts in the last step with probability at most ε.

Combining the above, we see that Arthur accepts with probability at most $\varepsilon + \max\{c - 7\varepsilon, \varepsilon\}$, which is at most $c - 6\varepsilon$.

Claim 4. *For any $x \in \bar{L}$ sufficiently long, there exists a message $r_1, \ldots, r_s$ such that Arthur will accept with probability at least $c - 5\varepsilon$.*

Proof. Fix $x \in \bar{L}$. Define

$$\hat{p} = \Pr_{r', r}\left[\widetilde{\mathsf{Sim}}^{\mathcal{V}_x(r')}(x; r) \text{ outputs an accepting transcript}\right].$$

There are two cases to consider:

Case 1: If $\hat{p} < c - 3\varepsilon$ then, except with probability at most ε, the value p_1 computed by Arthur satisfies $p_1 < c - 2\varepsilon$; thus, Arthur accepts in the first step with probability at least $1 - \varepsilon \geq c - 5\varepsilon$.

Case 2: On the other hand, say $\hat{p} \geq c - 3\varepsilon$. As in the proof of Claim 2, Merlin proceeds as follows: choose random $r_1 \in \{0,1\}^\ell$ and compute $\alpha = \mathcal{V}_x(r_1)$. Let $R_\alpha \stackrel{\text{def}}{=} \{r \mid \mathcal{V}_x(r) = \alpha\}$, and choose $r_2, \ldots, r_s$ uniformly from R_α. Send $r = (r_1, \ldots, r_s)$ to Arthur. We show that Arthur will accept with high probability.

Arthur can reject in either step 3 or step 4. We upper-bound the probability that Arthur rejects in either of these steps individually, and then apply a union bound to upper-bound the total probability that Arthur rejects.

Each r_i, taken individually, is uniformly distributed in $\{0,1\}^\ell$. Thus, in step 3, choosing a random $i \in \{1, \ldots, s\}$ and using coins r_i is equivalent to choosing uniformly-random coins for $\mathcal{V}_x$. It follows that the probability that Arthur rejects in step 3 is exactly equal to $1 - \hat{p} \leq 1 - c + 3\varepsilon$.

We proceed to analyze step 4. As in the proof of Claim 2, say a *collision* occurs in an execution of $\widetilde{\mathsf{Sim}}^{\mathcal{V}^*(x; r_1, \ldots, r_s, h)}(x; r)$ if the simulator makes two distinct queries (α, β_i) and (α, β_j) for which $h(\beta_i) = h(\beta_j)$. Let coll denote such an event. As before, we have

$$\Pr_{r,h,r}\left[\widetilde{\mathsf{Sim}}^{\mathcal{V}^*(x; r, h)}(x; r) \text{ outputs an accepting transcript}\right] \leq \qquad (1)$$

$$\Pr_{r,h,r}[\text{coll}] + \Pr_{r,h,r}\left[\widetilde{\mathsf{Sim}}^{\mathcal{V}^*(x; r, h)}(x; r) \text{ outputs an accepting transcript} \mid \overline{\text{coll}}\right],$$

where $r = (r_1, \ldots, r_s)$ are chosen by Merlin as described above (and not uniformly and independently at random). The probability of a collision is independent of $r_1, \ldots, r_s$, and is upper-bounded by $\Pr[\text{coll}] \leq \frac{(2T/\varepsilon)^2}{2s} = \frac{\varepsilon}{2}$. As in the proof of Claim 2, for sufficiently-long x it holds that

$$\Pr_{r,h,r}\left[\widetilde{\mathsf{Sim}}^{\mathcal{V}^*(x; r, h)}(x; r) \text{ outputs an accepting transcript} \mid \overline{\text{coll}}\right] \leq \varepsilon^2/2;$$

this means that, except with probability at most ε, the $r_1, \ldots, r_s$ chosen by Merlin satisfy

$$\mathrm{Pr}_{h,r}\left[\widetilde{\mathsf{Sim}}^{\mathcal{V}^*(x;r_1,\ldots,r_s,h)}(x;r) \text{ outputs an accepting transcript} \mid \overline{\mathsf{coll}}\right] \leq \varepsilon/2.$$

Using Equation (1), we see that except with probability at most ε, the $r_1, \ldots, r_s$ chosen by Merlin satisfy

$$\mathrm{Pr}_{h,r}\left[\widetilde{\mathsf{Sim}}^{\mathcal{V}^*(x;r_1,\ldots,r_s,h)}(x;r) \text{ outputs an accepting transcript}\right] \leq \varepsilon < c - 11\varepsilon.$$

Assuming the above to be the case, Arthur will reject in step 4 with probability at most ε. Taken together, this means that Arthur rejects in step 4 with probability at most 2ε.

Summing the probabilities of rejection in steps 3 and 4, we see that, overall, Arthur rejects with probability at most $1 - c + 5\varepsilon$, or accepts with probability at least $c - 5\varepsilon$.

5   Future Directions

Coupled with the obvious fact that $^{bb}\mathsf{CZK}(4) \subseteq \mathsf{AM}$, this work shows that $^{bb}\mathsf{CZK}(4) \subseteq \mathsf{AM} \cap \mathsf{coMA}$. Due to the similarity with the fact that $\mathsf{SZK} \subseteq \mathsf{AM} \cap \mathsf{coAM}$ [18,1], as well as the fact that the only languages known to be in $^4\mathsf{CZK}$ (under any assumption) are also in SZK, it is natural to conjecture that $^{bb}\mathsf{CZK}(4) \subseteq \mathsf{SZK}$.

Another interesting direction would be to show any broad positive results for $^4\mathsf{CZK}$: say, along the lines of proving that $\mathsf{NP} \cap \mathsf{coNP} \subseteq {}^4\mathsf{CZK}$.

In a slightly different direction, suggested by Hoeteck Wee: can a tighter bound be shown for languages L having 4-round zero-knowledge proofs *of knowledge* (beyond the fact that $L \in \mathsf{NP}$)?

Finally, is it possible to apply the techniques from [26] to show that there are no black-box constructions of *constant-round* (black-box) zero-knowledge proofs for NP?

Acknowledgments

I would like to thank Hoeteck Wee for many illuminating discussions regarding the round complexity of zero knowledge in general, and for helpful remarks on a previous draft of this paper. Thanks also to Dov Gordon and Arkady Yerukhimovich for reading and commenting on a preliminary version of this manuscript. The suggestions of an anonymous member of the ECCC scientific board and the TCC referees also helped to clarify the presentation.

References

1. Aiello, W., Håstad, J.: Statistical zero-knowledge languages can be recognized in two rounds. J. Computer and System Sciences 42(3), 327–345 (1991)
2. Babai, L., Moran, S.: Arthur-Merlin games: A randomized proof system and a hierarchy of complexity classes. J. Computer and System Sciences 36(2), 254–276 (1988)
3. Barak, B.: How to go beyond the black-box simulation barrier. In: Proc. 42nd Annual Symposium on Foundations of Computer Science (FOCS), pp. 106–115. IEEE Computer Society Press, Los Alamitos (2001)
4. Barak, B., Lindell, Y.: Strict polynomial-time in simulation and extraction. SIAM J. Computing 33(4), 738–818 (2004)
5. Barak, B., Lindell, Y., Vadhan, S.: Lower bounds for non-black-box zero knowledge. J. Computer and System Sciences 72(2), 321–391 (2006)
6. Bellare, M., Jakobsson, M., Yung, M.: Round-optimal zero-knowledge arguments based on any one-way function. In: Fumy, W. (ed.) EUROCRYPT 1997. LNCS, vol. 1233, pp. 280–305. Springer, Heidelberg (1997)
7. Bellare, M., Micali, S., Ostrovsky, R.: Perfect zero knowledge in constant rounds. In: Proc. 22nd Annual ACM Symposium on Theory of Computing (STOC), pp. 482–493. ACM, New York (1990)
8. Bellare, M., Micali, S., Ostrovsky, R.: The (true) complexity of statistical zero knowledge. In: Proc. 22nd Annual ACM Symposium on Theory of Computing (STOC), pp. 494–502. ACM, New York (1990)
9. Bellare, M., Palacio, A.: The knowledge-of-exponent assumptions and 3-round zero-knowledge protocols. In: Franklin, M. (ed.) CRYPTO 2004. LNCS, vol. 3152, pp. 273–289. Springer, Heidelberg (2004)
10. Ben-Or, M., Goldreich, O., Goldwasser, S., Håstad, J., Kilian, J., Micali, S., Rogaway, P.: Everyting provable is provable in zero knowledge. In: Goldwasser, S. (ed.) CRYPTO 1988. LNCS, vol. 403, pp. 37–56. Springer, Heidelberg (1990)
11. Boppana, R., Håstad, J., Zachos, S.: Does coNP have short interactive proofs? Information Proc. Letters 25(2), 127–132 (1987)
12. Boyar, J., Kurtz, S., Krentel, M.: Discrete logarithm implementation of perfect zero-knowledge blobs. J. Cryptology 2(2), 63–76 (1990)
13. Brassard, G., Chaum, D., Crépeau, C.: Minimum disclosure proofs of knowledge. J. Computer and Systems Sciences 37(2), 156–189 (1988)
14. Cramer, R., Damgård, I., MacKenzie, P.: Efficient zero-knowledge proofs of knowledge without intractability assumptions. In: Imai, H., Zheng, Y. (eds.) PKC 2000. LNCS, vol. 1751, pp. 354–372. Springer, Heidelberg (2000)
15. Di Crescenzo, G., Persiano, G.: Round-optimal perfect zero-knowledge proofs. Information Proc. Letters 50(2), 93–99 (1994)
16. Damgård, I., Pedersen, M., Pfitzmann, B.: On the existence of statistically-hiding bit commitment schemes and fail-stop signatures. J. Cryptology 10(3), 163–194 (1997)
17. Feige, U., Shamir, A.: Zero knowledge proofs of knowledge in two rounds. In: Brassard, G. (ed.) CRYPTO 1989. LNCS, vol. 435, pp. 526–544. Springer, Heidelberg (1990)
18. Fortnow, L.: The complexity of perfect zero knowledge. In: Micali, S. (ed.) Advances in Computing Research, vol. 5, pp. 327–343. JAC Press, Inc. (1989)
19. Goldreich, O., Kahan, A.: How to construct constant-round zero-knowledge proof systems for NP. J. Cryptology 9(3), 167–190 (1996)

20. Goldreich, O., Krawczyk, H.: On the composition of zero-knowledge proof systems. SIAM J. Computing 25(1), 169–192 (1996)

21. Goldreich, O., Micali, S., Wigderson, A.: Proofs that yield nothing but their validity, or all languages in NP have zero-knowledge proof systems. J. ACM 38(3), 691–729 (1991)

22. Goldreich, O., Oren, Y.: Definitions and properties of zero-knowledge proof systems. J. Cryptology 7(1), 1–32 (1994)

23. Goldwasser, S., Micali, S., Rackoff, C.: The knowledge complexity of interactive proof systems. SIAM J. Computing 18(1), 186–208 (1989)

24. Goldwasser, S., Micali, S., Rivest, R.: A digital signature scheme secure against adaptive chosen-message attacks. SIAM J. Computing 17(2), 281–308 (1988)

25. Hada, S., Tanaka, T.: On the existence of 3-round zero-knowledge protocols. In: Krawczyk, H. (ed.) CRYPTO 1998. LNCS, vol. 1462, pp. 408–423. Springer, Heidelberg (1998) (See also http://eprint.iacr.org/1999/009)

26. Haitner, I., Hoch, J.J., Reingold, O., Segev, G.: Finding collisions in interactive protocols — a tight bound on the round complexity of statistically-hiding commitments. In: Proc. 48th Annual Symposium on Foundations of Computer Science (FOCS), pp. 669–679. IEEE, Los Alamitos (2007), http://eprint.iacr.org/2007/145

27. Haitner, I., Reingold, O.: Statistically-hiding commitment from any one-way function. In: Proc. 39th Annual ACM Symposium on Theory of Computing (STOC), pp. 1–10. ACM Press, New York (2007)

28. Halevi, S., Micali, S.: Practical and provably-secure commitment schemes from collision-free hashing. In: Koblitz, N. (ed.) CRYPTO 1996. LNCS, vol. 1109, pp. 201–215. Springer, Heidelberg (1996)

29. Impagliazzo, R., Yung, M.: Direct minimum-knowledge computations (extended abstract). In: Pomerance, C. (ed.) CRYPTO 1987. LNCS, vol. 293, pp. 40–51. Springer, Heidelberg (1988)

30. Itoh, T., Sakurai, K.: On the complexity of constant round ZKIP of possession of knowledge. In: Matsumoto, T., Imai, H., Rivest, R.L. (eds.) ASIACRYPT 1991. LNCS, vol. 739, pp. 331–345. Springer, Heidelberg (1993)

31. Kurosawa, K., Ogata, W., Tsujii, S.: 4-move perfect ZKIP for some promise problems. IEICE Trans. on Fundamentals of Electronics, Communications, and Computer Sciences E78-A(1), 34–41 (1995)

32. Lepinski, M.: On the existence of 3-round zero-knowledge proofs. Master's thesis, MIT (2002), Available at http://theory.lcs.mit.edu/Simcis/cis-theses.html

33. Pass, R.: On Arthur-Merlin games and the possibility of basing cryptography on NP-hardness. In: 21st Annual IEEE Conference on Computational Complexity, pp. 88–95. IEEE Computer Society Press, Los Alamitos (2006)

34. Saito, T., Kurosawa, K., Sakurai, K.: 4-move perfect SKIP of knowledge with no assumption. In: Matsumoto, T., Imai, H., Rivest, R.L. (eds.) ASIACRYPT 1991. LNCS, vol. 739, pp. 320–331. Springer, Heidelberg (1993)

35. Vadhan, S.: A Study of Statistical Zero-Knowledge Proofs. PhD thesis, MIT (1999)

How to Achieve Perfect Simulation
and
A Complete Problem for Non-interactive Perfect Zero-Knowledge

Lior Malka

Department of Computer Science
University of Victoria, BC, Canada
liorma@cs.uvic.ca

Abstract. We study perfect zero-knowledge proofs (**PZK**). Unlike statistical zero-knowledge, where many fundamental questions have been answered, virtually nothing is known about these proofs.

We consider reductions that yield hard and complete problems in the statistical setting. The issue with these reductions is that they introduce errors into the simulation, and therefore they do not yield analogous problems in the perfect setting. We overcome this issue using an *error shifting technique*. This technique allows us to remove the error from the simulation. Consequently, we obtain the first complete problem for the class of problems possessing non-interactive perfect zero-knowledge proofs (**NIPZK**), and the first hard problem for the class of problems possessing public-coin **PZK** proofs.

We get the following applications. Using the error shifting technique, we show that the notion of zero-knowledge where the simulator is allowed to fail is equivalent to the one where it is not allowed to fail. Using our complete problem, we show that under certain restrictions **NIPZK** is closed under the OR operator. Using our hard problem, we show how a *constant-round, perfectly hiding* instance-dependent commitment may be obtained (this would collapse the round complexity of public-coin **PZK** proofs to a constant).

Keywords: cryptography, non-interactive, perfect zero-knowledge, perfect simulation, error shifting, complete problems.

1 Introduction

Zero-knowledge protocols allow one party (the prover) to prove an assertion to another party (the verifier), yet without revealing anything beyond the validity of the assertion [18,5]. These protocols protect the privacy of the prover, which makes them very useful to cryptography. Zero-knowledge protocols can guarantee three levels of privacy: perfect, statistical, and computational. This is formulated using the notion of simulation. When the simulation error is zero, the protocol is *perfect zero-knowledge*. This means that the verifier learns absolutely nothing from the prover. When the simulation error is negligible, the protocol is either *statistical zero-knowledge* or *computational*

R. Canetti (Ed.): TCC 2008, LNCS 4948, pp. 89–106, 2008.
© International Association for Cryptologic Research 2008

zero-knowledge. This means that the prover leaks a little amount of information to the verifier.

In this paper we focus on perfect zero-knowledge protocols. These protocols are interesting from a cryptographic perspective because, unlike statistical or computational zero-knowledge protocols, they provide the highest level of privacy to the prover. Such protocols exist for a variety of well known languages, such as GRAPH-ISOMORPHISM, DISCRETE-LOG, variants of QUADRATIC-RESIDUOUSITY, and more ([31,13,6,26,21]). The fact that the complexity of these languages is an open question also makes perfect zero-knowledge protocols interesting from a complexity theoretic perspective.

Unfortunately, working with perfect zero-knowledge protocols is difficult. This is so because they do not allow any error in the simulation. In contrast, statistical zero-knowledge protocols allow a small error in the simulation. This means that in the statistical setting we can use a variety of techniques, even if they introduce a small error into the simulation. Indeed, such techniques were used to prove many fundamental results about statistical zero-knowledge proofs (**SZK**). These results include complete problems, equivalence between private-coin and public-coin, equivalence between honest and malicious verifier, and much more ([24,26,14,16,32,23]). These results do not apply to the perfect setting because they use techniques that introduce error into the simulation, and such techniques cannot be used in the perfect setting. Consequently, virtually nothing is known about perfect zero-knowledge proofs (**PZK**).

1.1 Our Results

In this paper we consider reductions that yield hard and complete problems in the statistical setting. The issue with these reductions is that they introduce errors into the simulation, and therefore they do not yield analogous problems in the perfect setting.

Our goal is to overcome this issue. This is important because if we understand why techniques from the statistical setting introduce error into the simulation, then we might be able to fix these techniques, and then apply them to the perfect setting. This will enable us to translate the results from statistical zero-knowledge to perfect zero-knowledge. In addition, results from the statistical setting are proved using the tool of complete and hard problems. Thus, to be able to prove these results in the perfect setting it is important that we obtain complete and hard problems for the perfect setting.

We remark that we are not the first to observe the fact that reductions from the statistical setting do not apply to the perfect setting. Specifically, in the case of hard problems, [26] showed that the reduction could eliminate the error using approximation techniques. However, this solution does not yield a hard problem in the perfect setting because it only applies in certain cases (for example, when the underlying problem has perfect completeness).

In this paper we modify the reductions from the statistical setting so that they yield hard and complete problems in the perfect setting. To do this we use what we call *the error shifting technique*. What is new about this technique is that instead of dealing with the error in the reduction itself, we shift it forward to the protocol. Intuitively, our reduction isolates the error that the underlying problem incurs, and shifts it forward to the protocol, where it is no longer a simulation error. Consequently, we obtain complete and hard problems for the perfect setting. We remark that the error shifting technique is also useful in other contexts (e.g., in Section 4 we use it to show that the notion of zero-knowledge where the simulator is allowed to fail is equivalent to the one where it is not allowed to fail).

The error shifting technique applies to reductions in both the interactive and the non-interactive models. In the non-interactive model we apply it to the reduction of [30,15], thus obtaining the first complete problem for the class of problems possessing non-interactive perfect zero-knowledge proofs (**NIPZK**).

Theorem 1. The problem UNIFORM (UN) is complete for **NIPZK**.

Informally, instances of UNIFORM are circuits that have an additional output bit. Ignoring this bit, we can think of YES instances of UN as circuits that represent the uniform distribution, whereas NO instance are circuits that hit only a small fraction of their range. This problem is identical to the **NISZK**-complete problem STATISTICAL DISTANCE FROM UNIFORM (SDU) [15], except that YES instances of UNIFORM represent the uniform distribution, whereas YES instances of SDU represent a distribution that is only "close" to uniform. This difference is natural because it reflects the difference between perfect and statistical simulation.

In the interactive model we obtain a similar result. That is, we apply the error shifting technique to the reduction of [26], thus obtaining a hard problem for the class of problems possessing public-coin **HVPZK** proofs. Instances of our hard problem are triplets of circuits. Again, ignoring one of these circuits, our problem is a variant of STATISTICAL-DISTANCE (SD) [26]. That is, we can think of YES instances of our problem as pairs of circuits representing the same distribution, whereas instances of the reduction of [26] are circuits representing "close" distributions.

Theorem 2 (informal). Essentially, $\overline{SD}^{1/2,0}$ is hard for **public-coin-HVPZK**.

To demonstrate the usefulness of our **NIPZK**-complete problem we prove that under certain restrictions **NIPZK** is closed under the OR operator. What is special about this result is that even in statistical setting, where we have more techniques to work with, it is not clear how to prove (or disprove) it.[1] Also, we show how our hard problem may lead to a *constant-round, perfectly hiding* instance-dependent commitment-scheme. Notice that except for [21], who used the techniques of [8] to construct such a scheme for V-bit protocols, all the known instance-dependent commitment-schemes are only statistically hiding [33,23,22]. Thus, using our hard problem it might be possible to collapse the round complexity of public-coin **PZK** proofs to a constant. These applications can be found in Section 4.

1.2 Related Work

As we mentioned, virtually nothing is known about perfect zero-knowledge proofs. The only exception is the result of [9], who showed a transformation from *constant-round, public-coin* **HVPZK** proofs to ones that are **PZK**. Also, a **HVPZK**-complete problem was given by [26], but it is unnatural, and defined in terms of the class itself. We remark

---

[1] [30] claimed that **NISZK** is closed under the OR operator, but this claim has been retracted.

that perfect zero-knowledge *arguments* for **NP** languages have been constructed under various unproven assumptions (e.g., [19,7]), but we are interested in the unconditional study of perfect zero-knowledge proofs.

Variants of QUADRATIC-RESIDUOUSITY and QUADRATIC-NONRESIDUOUSITY were shown to to be in **NIPZK** by [6,27]. Bellare and Rogaway [3] showed that a variant of GRAPH-ISOMORPHISM is in **NIPZK**. They also showed basic results about **NIPZK**, but their notion of zero-knowledge allows simulation in expected (as opposed to strict) polynomial-time. This notion is disadvantageous, especially when non-interactive protocols are executed as sub-protocols. Other aspects of **NIPZK** were studied in [27,28,29], but they apply to problems with special properties.

1.3 Organization

We use standard definitions, to be found in Appendix A. In Section 2 we present the error shifting technique, and use it to obtain a **NIPZK**-complete problem. In Section 3 we apply this technique to the interactive setting, where we obtain a hard problem. In Section 4 we show some applications of these results.

2 A Complete Problem for NIPZK

In this section we introduce the *error shifting technique*. Using this technique we modify the reduction of [15], hence obtaining a **NIPZK**-complete problem.

Starting with some background, we give the definition of STATISTICAL DISTANCE FROM UNIFORM (SDU), the **NISZK**-complete problem of [15]. Instances of this problem are circuits. These circuits are treated as distributions, under the convention that the input to the circuit is uniformly distributed. Specifically, YES instances are circuits representing a distribution that is close to uniform, and NO instances are circuits representing a distribution that is far from uniform.

Definition 2.1. *Define* SDU $\overset{def}{=} \langle \text{SDU}_Y, \text{SDU}_N \rangle$ *as*

$$\text{SDU}_Y = \{X | \Delta(X, U_n) < 1/n\}, \text{ and}$$
$$\text{SDU}_N = \{X | \Delta(X, U_n) > 1 - 1/n\},$$

where X is a circuit with n output bits, and U_n is the uniform distribution on $\{0, 1\}^n$.

We informally describe the reduction of [15] to SDU. This reduction originated from the work of [30]. Given a **NISZK** problem Π, this reduction maps instances x of Π to circuits X of SDU. The circuit uses the simulator S from the proof of Π. Specifically, X executes $S(x)$, and obtains a transcript. This transcript contains a simulated message of the prover, and a simulated reference string. If the verifier accepts in this transcript, then X outputs the simulated reference string. Otherwise, X outputs the all-zero string. Intuitively, this reduction works because if x is a YES instance, then the simulated reference string is almost uniformly distributed, and thus X is a YES instance of SDU. Conversely, if x is a NO instance, then the verifier rejects on most reference strings, and thus X is a NO instance of SDU.

The issue with the reduction of [15]. When we apply the above reduction to **NIPZK** problems, it is natural that we should get a **NIPZK**-complete problem whose instances are circuits that represents the uniform distribution. This is so because the circuit X outputs the simulated reference string, and when the simulation is perfect, this string is uniformly distributed. Indeed, if we apply the above reduction to **NIPZK** problems that have perfect completeness, then the verifier will accept, and thus we will get a circuit X that represents the uniform distribution. However, if the underlying problem does not have perfect completeness, then the distribution represented by X will be skewed. This will cause problems later, when we try to construct a proof system and a simulator for our complete problem. Hence, this reduction does not apply to **NIPZK**.

To overcome the above issue, instead of working only with the reduction to SDU, our idea is to modify both the reduction and the proof system for SDU at the same time.

The Error Shifting Technique. In its most general form, *the error shifting technique shifts into the protocol errors that would otherwise become simulation errors.* This description is a very loose, but we chose it because our technique can be applied in various different contexts, and in each of these contexts it takes a different form. However, the following application will clarify our technique.

▶ **The first step of the error shifting technique** is to identify where the simulation error comes from, and then isolate it. In our case, the error comes from the reduction: if the verifier rejects, then the circuit X does not represent the uniform distribution. Thus, the error comes from the completeness error of the underlying problem. To separate this error, we add an extra output bit to the circuit X. That is, X executes the simulator, and it outputs the simulated reference string followed by an extra bit. This bit takes the value 1 if the verifier accepts, and 0 if the verifier rejects.

▶ **The second step of the error shifting technique** is to shift the error forward, to the completeness or the soundness error of the protocol. In our case, from the circuit X to the protocol of our complete problem. This step is not trivial because we cannot just use the protocol of [15] for SDU. Specifically, in this protocol the prover sends a string r, and the verifier accepts if $X(r)$ equals the reference string. If we use this idea in our case, then we will get a simulation error. Thus, we modify this protocol by starting with the simulator, and constructing the prover based on the simulator. Informally, the simulator samples the circuit X, and the verifier accepts if the extra bit in this sample is 1. The prover simply mimics the simulator. This shows that the error was shifted from X to the completeness error (of a new protocol).

The above reduction yields our **NIPZK**-complete problem UNIFORM. A formal description of the above reduction and our proof system is given in the next section.

2.1 A Complete Problem for NIPZK

In this section we formalize the intuition given in the previous section, thus proving our first result.

Theorem 2.1. UNIFORM *is* **NIPZK**-complete.

We start with the definition of UNIFORM (UN). Recall that when we applied the error shifting technique we got circuits X with an extra output bit. We use the convention that $n + 1$ denotes the number of output bits of X. We need the following notation.

- T_X is the set of outputs of X that end with a 1. Formally, $T_X \overset{\text{def}}{=} \{x | \exists r \ X(r) = x,$ and the suffix of x is $1\}$. As we shall see, the soundness and completeness properties will imply that the size of T_X is large for YES instances of UN, and small for NO instances of UN.
- X' is the distribution on the first n bits that X outputs. That is, X' is obtained from X by taking a random sample of X, and then outputting the first n bits. As we shall see, the zero-knowledge property will imply that if X is a YES instance of UN, then X' is the uniform distribution on $\{0, 1\}^n$.

Now, letting X be a circuit with $n + 1$ output bit, we say that X is β-*negative* if $|T_X| \leq \beta \cdot 2^n$. That is, T_X is small, and contains at most $\beta \cdot 2^n$ strings. We say that X is α-*positive* if X' is the uniform distribution on $\{0, 1\}^n$ and $\Pr_{x \leftarrow X}[x \in T_X] \geq \alpha$. This implies that T_X is large, and contains at least $\alpha \cdot 2^n$ strings.

Definition 2.2. *The problem* UNIFORM *is defined as* UN $\overset{\text{def}}{=} \langle \text{UN}_Y, \text{UN}_N \rangle$, *where*

$$\text{UN}_Y = \{X | X \text{ is } 2/3 - positive\}, \text{ and}$$
$$\text{UN}_N = \{X | X \text{ is } 1/3 - negative\}.$$

To prove that UN is **NIPZK**-complete we first show that the reduction from the previous section reduces every **NIPZK** problem to UN.

Lemma 2.1. UN *is* **NIPZK**-hard.

Proof. Let $\Pi = \langle \Pi_Y, \Pi_N \rangle$ be a **NIPZK** problem. Fix a non-interactive protocol $\langle P, V \rangle$ for Π with completeness and soundness errors $1/3$. Let r_I denote the common reference string in $\langle P, V \rangle$, and fix i such that $|r_I| = |x|^i$ for any $x \in \Pi_Y \cup \Pi_N$. Fix a simulator S for $\langle P, V \rangle$. Since S is efficient, we can fix an efficient transformation t and an integer ℓ such that on input $x \in \Pi_Y \cup \Pi_N$ the output of $t(x)$ is a circuit S' that executes S on inputs x and randomness r_S of length $|x|^\ell$. That is, $t(x) = S'$, and on input a string r_S of length $|x|^\ell$ the output of $S'(r_S)$ is the output of $S(x; r_S)$.

We show that Π Karp reduces to UN. That is, we define a polynomial-time Turing machine that on input $x \in \Pi_Y \cup \Pi_N$ outputs a circuit X such that if $x \in \Pi_Y$, then $X \in \text{UN}_Y$, and if $x \in \Pi_N$, then $X \in \text{UN}_N$. The circuit $X : \{0, 1\}^{|x|^\ell} \to \{0, 1\}^{|x|^i + 1}$ carries out the following computation.

- Let r_S be the $|x|^\ell$-bit input to X, and let $S' = t(x)$. Execute $S'(r_S)$, and obtain $S(x; r_S) = \langle x, r'_I, m' \rangle$.
- If $V(x, r'_I, m') = \texttt{accept}$, then output the string $r'_I 1$ (i.e., the concatenation of r'_I and 1). Otherwise, output $r'_I 0$.

Now we analyze our reduction. Let $x \in \Pi_Y$, and let X be the output of the above reduction on x. We show that X is $2/3$-positive. Consider the distribution on the output $\langle x, r'_I, m' \rangle$ of $S(x)$. Since $S(x)$ and $\langle P, V \rangle(x)$ are identically distributed, r'_I is uniformly distributed. Thus, X' (i.e., the distribution on the first $|x|^i$ output bits of X) is uniformly distributed. It remains to show that $\Pr[X \in T_X] \geq 2/3$. This immediately follows from the perfect zero-knowledge and completeness properties of $\langle P, V \rangle$. That is, the output of S is identically distributed to $\langle P, V \rangle(x)$, and V accepts in $\langle P, V \rangle$ with probability at least $2/3$.

Let $x \in \Pi_N$, and let X be the output of the above reduction on x. We show that X is $1/3$-negative. Assume towards contradiction that X is β-negative for some $\beta > 1/3$. We define a prover P^* that behaves as follows on CRS r_I. If $r_I 1 \in T_X$, then there is an input r_S to X such that $X(r_S) = r_I 1$. By the construction of X, there is randomness r_S for the simulator such that $S(x; r_S) = \langle x, r_I, m' \rangle$, and $V(x, r_I, m') = 1$. In this case P^* sends r_S to V. If $r_I 1 \notin T_X$, then P^* fails. Notice that P^* makes V accept on any r_I such that $r_I 1 \in T_X$. Since $|T_X| > 2^{|x|^i}/3$, and since r_I is uniformly chosen in $\langle P^*, V \rangle$, the probability that $r_I 1 \in T_X$ is strictly greater than $1/3$. Thus, V accepts in $\langle P^*, V \rangle(x)$ with probability strictly greater than $1/3$, and contradiction to the soundness error of $\langle P, V \rangle$. Hence, X is $1/3$-negative.

To prove Theorem 2.1 it remains to give a **NIPZK** proof for UN.

Lemma 2.2. UN *has a* **NIPZK** *proof with a deterministic verifier.*

Proof. We start with our non-interactive proof for UN. This proof is based on our simulator, which we describe later. On input $X : \{0, 1\}^\ell \rightarrow \{0, 1\}^{n+1}$ and common reference string $r_I \in \{0, 1\}^n$ the prover P picks z according to the distribution X such that the n-bit prefix of z equals r_I. Such a z exists because X' (i.e., the distribution on the first n bits of X) is the uniform distribution when $X \in UN_Y$. The prover uniformly picks $r \in X^{-1}(z)$, and sends r to the verifier V. The deterministic verifier accepts if $X(r) = r_I 1$, and rejects otherwise. Our prover is based on the following simulator. Let S be a probabilistic, polynomial-time Turing machine that on input X uniformly picks $r' \in \{0, 1\}^\ell$, and computes $z' = X(r')$. The simulator assigns the n bit prefix of z' to r_I' (i.e., the simulated reference string), and outputs $\langle X, r_I', r' \rangle$.

Let $X \in \Pi_Y$. We show that S perfectly simulates $\langle P, V \rangle$. Consider the distribution $S(X)$ on simulated transcripts $\langle X, r_I', r' \rangle$, and the distribution $\langle P, V \rangle(X)$ on the view $\langle X, r_I, r \rangle$ of V. Since X' is uniformly distributed over $\{0, 1\}^n$, the string r_I' obtained by the simulator is uniformly distributed over $\{0, 1\}^n$. Since r_I is uniformly distributed, r_I' and r_I are identically distributed. It remains to show that r and r' are identically distributed conditioned on $r_I = r_I'$. For each $y \in \{0, 1\}^n$, we define B_y to be the set of all strings $\hat{r}$ for which the prefix of $X(\hat{r})$ is y. Now, for any simulated reference string r_I', the randomness r' chosen by the simulator is uniformly distributed in $B_{r_I'}$. Similarly, for any reference string r_I the message of the prover is a string r chosen uniformly from B_{r_I}. Hence, conditioned on $r_I = r_I'$, the strings r and r' are identically distributed. We conclude that $S(X)$ and $\langle P, V \rangle(X)$ are identically distributed for any $X \in \Pi_Y$.

Turning our attention to the completeness property, we show that V accepts X with probability at least $2/3$. By the zero-knowledge property, the output $\langle X, r_I', r' \rangle$ of $S(X)$ is identically distributed to the view $\langle X, r_I, r \rangle$ of V on X. Thus, it is enough to show that when choosing a transcript $\langle X, r_I', r' \rangle$ according to $S(x)$ the probability that $V(X, r_I', r') = 1$ is at least $2/3$. Since S uniformly chooses r', and since X is $2/3$-positive, the probability that $X(r) \in T_X$ is at least $2/3$. Thus, the probability that the suffix of $X(r)$ is 1 is at least $2/3$. Hence, V accepts X with probability at least $2/3$.

The soundness property follows easily. Let $X \in \mathrm{UN_N}$. Since X is $1/3$-negative, $|T_X| \leq 1/3 \cdot 2^n$. Since r_I is uniformly distributed, the probability that $r_I 1 \in T_X$ is at most $1/3$. Hence, if $X \in \mathrm{UN_N}$, then V accepts X with probability at most $1/3$.

3   A Hard Problem for Public-Coin PZK Proofs

In this section we use the error shifting technique to modify the reduction of [26] for public-coin **HVSZK** proofs. Hence, we obtain a hard problem for the class of problems possessing public-coin **HVPZK** proofs (**AM ∩ HVPZK**). We start with motivation.

The reduction of [26] originated from the works of [11,1]. Informally, given a problem Π that has a public-coin **HVSZK** proof, this reduction maps instances x of Π to pairs of circuits $\langle X_0, X_1 \rangle$. The circuits X_0 and X_1 are statistically close when x is a YES instance of Π, and statistically far when x is a NO instance of Π.

The issue with this reduction is that it does not apply to the perfect setting. Specifically, when we apply it to YES instances of a problem that has a public-coin **HVPZK** proof, we get a pair of circuits $\langle X_0, X_1 \rangle$ that are only statistically close, but not identically distributed. This is unnatural because the closeness between X_0 and X_1 reflects the closeness of the simulation. Thus, in the perfect setting we expect X_0 and X_1 to be *identically distributed*, as in the complement of $\mathrm{SD}^{1/2,0}$.

Definition 3.1. *The problem* $\overline{\mathrm{SD}^{1/2,0}}$ [26] *is the pair* $\langle \overline{\mathrm{SD}^{1/2,0}}_Y, \overline{\mathrm{SD}^{1/2,0}}_N \rangle$, *where*

$$\overline{\mathrm{SD}^{1/2,0}}_Y = \{\langle X_0, X_1 \rangle \mid \Delta(\mathrm{X_0}, \mathrm{X_1}) = 0\}, \text{ and}$$
$$\overline{\mathrm{SD}^{1/2,0}}_N = \{\langle X_0, X_1 \rangle \mid \Delta(\mathrm{X_0}, \mathrm{X_1}) \geq 1/2\}.$$

Sahai and Vadhan [26] were aware of this issue, and they addressed it by directly calculating the errors of the underlying problem. However, their technique applies only in certain cases (for example, when the underlying problem has a proof with perfect completeness). In the next section we will show how to overcome this issue by using the error shifting technique. Essentially, we obtain a hard problem where YES instances are pairs of circuits representing identical distributions, and NO instances are circuits representing statistically far distributions. Formally, our hard problem is as follows.

Definition 3.2. *The problem* IDENTICAL DISTRIBUTIONS *is* $\mathrm{ID} \stackrel{def}{=} \langle \mathrm{ID_Y}, \mathrm{ID_N} \rangle$, *where*

$$\mathrm{ID_Y} = \{\langle X_0, X_1, Z \rangle \mid \Delta(\mathrm{X_0}, \mathrm{X_1}) = 0 \text{ and } \Pr[Z = 1] \geq 2/3\}, \text{ and}$$
$$\mathrm{ID_N} = \{\langle X_0, X_1, Z \rangle \mid \Delta(\mathrm{X_0}, \mathrm{X_1}) \geq 1/2 \text{ or } \Pr[Z = 1] \leq 1/3\}.$$

3.1   Modifying the Reductions for Public-Coin HVSZK Proofs

In this section we show that ID is hard for **AM ∩ HVPZK**, and then we conclude that, essentially, $\overline{\mathrm{SD}^{1/2,0}}$ is also hard for **AM ∩ HVPZK**. Starting with some background, we describe the reduction of [26].

Notation. Let $\langle P, V \rangle$ be a public-coin **HVPZK** proof for a problem Π with a simulator S. Given a string x we use $v \stackrel{\text{def}}{=} v(|x|)$ to denote the number of rounds in the interaction between P and V on input x. That is, in round i the prover P sends m_i and V replies with a random string r_i, until P sends its last message m_v, and V accepts or rejects. We denote the output of $S(x)$ by $\langle x, m_1, r_1, \ldots, m_v \rangle$.

The reduction of [26] maps instances x of Π to pairs of circuits $\langle X_0', X_1' \rangle$. These circuits are constructed from the circuits X_i and Y_i, defined as follows. The circuit X_i chooses randomness, executes $S(x)$ using this randomness, and outputs the simulated transcript, truncated at the i-th round. That is, X_i obtains $\langle x, m_1, r_1, \ldots, m_v \rangle$, and outputs $\langle m_1, r_1, \ldots, m_i, r_i \rangle$. The circuit Y_i is defined exactly the same, except that it replaces r_i with a truly random string r_i'.

- $X_i(r)$: execute $S(x; r)$ to obtain $\langle x, m_1, r_1, \ldots, m_v \rangle$. Output $\langle m_1, r_1, \ldots, m_i, r_i \rangle$.
- $Y_i(r, r_i')$: execute $S(x; r)$ to obtain $\langle x, m_1, r_1, \ldots, m_v \rangle$. Output $\langle m_1, r_1, \ldots, m_i, r_i' \rangle$.

Notice that X_i and Y_i represent the same distribution when x is a YES instance. This is so because $S(x)$ perfectly simulates the view of the verifier, and therefore r_i is uniformly distributed, just like r_i'. We define $X = X_1 \otimes \cdots \otimes X_v$. That is, X executes all the circuits X_i and outputs the concatenation of their outputs. Similarly, we define $Y = Y_1 \otimes \cdots \otimes Y_v$. Again, X and Y are identically distributed when x is a YES instance. Now, the pair $\langle X_0', X_1' \rangle$ is defined from $\langle X, Y \rangle$ as follows. The circuit X_1' outputs 1 followed by the output of Y. The circuit X_0' outputs the output of Z followed by the output of X, where Z is the circuit that outputs 1 if with high probability $S(x)$ outputs accepting transcripts, and 0 otherwise.

The issue with the reduction of [26]. The above reduction does not apply to the perfect setting (except for the case where $\langle P, V \rangle$ have perfect completeness). This is so because there is a non-zero probability that Z will output 0, in which case X_0' and X_1' will not represent the same distribution. To overcome this issue we use the error shifting technique in two steps, just like we did in the previous section. Our goal is to show that, essentially, $\overline{\text{SD}}^{1/2,0}$ is hard for **AM** ∩ **HVPZK**

Our first step is to separate the error that the circuit Z incurs. Thus, instead of including Z in the circuits X_0' and X_1', our reduction simply maps an instance x of Π to the triplet $\langle X, Y, Z \rangle$. By the analysis from [26], if x is a YES instance, then X and Y are identically distributed, and Z outputs 1 with high probability. Such a triplet is a YES instance of or hard problem. Similarly, if x is a NO instance, then either X and Y are statistically far, or Z outputs 0 with a high probability. Such a triplet is a NO instance of our hard problem. The following lemma shows that IDENTICAL DISTRIBUTIONS (ID) is hard for **AM** ∩ **HVPZK**.

Lemma 3.1 (Implicit in [26]). *For any problem* $\Pi = \langle \Pi_Y, \Pi_N \rangle$ *possessing a public-coin* **HVPZK** *proof there is a Karp reduction mapping strings* x *to circuits* $\langle X, Y, Z \rangle$ *with the following properties.*

- *If* $x \in \Pi_Y$, *then* $\Delta(X, Y) = 0$ *and* $\Pr[Z = 1] \geq 2/3$.
- *If* $x \in \Pi_N$, *then* $\Delta(X, Y) \geq 1/2$ *or* $\Pr[Z = 1] \leq 1/3$.

Indeed, instances of ID are triplets of circuits $\langle X, Y, Z \rangle$, as opposed to pairs $\langle X, Y \rangle$. Thus, we are not done yet. We need to show that ID and $\overline{\mathrm{SD}}^{1/2,0}$ are essentially the same. Hence, we continue to our second step.

Recall that the second step of the error shifting technique is to shift the error forward to the protocol. However, $\overline{\mathrm{SD}}^{1/2,0}$ is not known to have a **PZK** proof. Thus, our second step is to modify any **PZK** protocol $\langle P, V \rangle$ for this problem into a **PZK** protocol $\langle P, V' \rangle$ for ID. That is, we take an arbitrary protocol $\langle P, V \rangle$ for $\overline{\mathrm{SD}}^{1/2,0}$, and then we show that even if the input to this protocol is an instance $\langle X, Y, Z \rangle$ of ID (instead of a pair $\langle X, Y \rangle$), then the behavior of P and the modified verifier V' on input $\langle X, Y, Z \rangle$ is identical to the behavior of $\langle P, V \rangle$ on input $\langle X, Y \rangle$. This will show that the two problems are essentially the same, and therefore we will be done.

Our modification is as follows. On input $\langle X, Y, Z \rangle$ the first step of the modified verifier V' is to estimate the value of $\Pr[Z = 1]$, and reject if this value is at most $1/3$. If V' did not reject, then P and V' execute $\langle P, V \rangle$ on input $\langle X, Y \rangle$. This modification is a part of the error shifting technique because we shift the error from the circuit Z into an arbitrary protocol $\langle P, V \rangle$ for $\overline{\mathrm{SD}}^{1/2,0}$.

We analyze the modified protocol $\langle P, V' \rangle$ for our hard problem. We observe that V' is very unlikely to reject if $\Pr[Z = 1] \geq 2/3$. We also observe that if the protocol continues, then either $\langle X, Y, Z \rangle$ is a YES instance of our hard problem and $\Delta(\mathrm{X}, \mathrm{Y}) = 0$, or $\langle X, Y, Z \rangle$ is a NO instance of our hard problem and $\Delta(\mathrm{X}, \mathrm{Y}) \geq 1/2$. Thus, in this case the behavior of P and V' on instances of our hard problem is identical to the behavior of P and V on instances of $\overline{\mathrm{SD}}^{1/2,0}$.

Our modification shows that although we did not prove that $\overline{\mathrm{SD}}^{1/2,0}$ is hard for **AM ∩ HVPZK**, it can be treated as such (because any protocol that we design for this problem can be immediately modified to a protocol with the same properties for ID).

4   Applications

We show an application of the error shifting technique. We also show how our complete and hard problems can facilitate the study of zero-knowledge in the perfect setting.

4.1   Obtaining Simulators That do Not Fail

We use the error shifting technique to show that the notion of zero-knowledge where the simulator is allowed to fail is equivalent to the one where it is not allowed to fail. This holds in both the interactive and the non-interactive models, and regardless of whether the simulator runs in strict or expected polynomial-time.

Starting with background, we recall that the notion of perfect zero-knowledge requires that the view of the verifier be identically distributed to the output of the simulator [18]. Later, this notion was relaxed by allowing the simulator to output fail with probability at most $1/2$, and requiring that, conditioned on the output of the simulator not being fail, it be identically distributed to the view of the verifier [9].

A known trick to remove the fail output is to execute the simulator for $|x|$ times (where x is the input to the simulator), and output the first transcript, or fail if the simulator failed in all $|x|$ executions [12]. This works for statistical and computational zero-knowledge, but not for perfect zero-knowledge. Notice that in all of these cases we actually introduce an extra error into the simulation, and we do not understand why. Furthermore, despite the fact that important problems have **PZK** proofs (e.g., GRAPH-ISOMORPHISM, QUADRATIC-RESIDUOUSITY [18,13,31]), all of these proofs have a simulator that outputs `fail` with probability $1/2$. Now we fix this issue.

The transformation. Let $\langle P, V \rangle$ be a **PZK** proof for a problem Π, and let S be a simulator for $\langle P, V \rangle$. Notice that S may fail with some probability. We use the error shifting technique to obtain a simulator S' that does not fail.

Recall that the error shifting technique is applied in two steps: we need to find where the error is coming from, and then we shift it forward. For the first step, we observe that when S outputs `fail`, the verifier V actually learns that S failed. This is something that V does not learn from the prover P (because transcripts between P and V are never of the form `fail`). Thus, the error comes from the fact that P is not teaching V that $S(x)$ may output fail with some probability. We are done with the first step. In the second step we shift this error forward by letting P teach V that $S(x)$ may output `fail`. That is, on input x, the new prover P' executes $S(x)$ for $|x|$ times, and if $S(x) = $ `fail` in all of these executions, then P' outputs `fail`. Otherwise, P' behaves like P. In other words, we shifted the error from the simulation to the protocol.

The new simulator S' simply executes S, and if all executions failed, then it behaves just like P'. Namely, it outputs the transcript $\langle x, $ `fail`$; r_V \rangle$, where r_V is the randomness of V. Otherwise, S' outputs a simulated transcript of S. Notice that we increased the completeness error by $1/2^n$, but by executing $S(x)$ polynomially many times, the probability that P' will fail can be made extremely small. We conclude that $\langle P', V \rangle$ is a **PZK** proof for Π with a simulator S' that never fails.

4.2 Under Certain Restrictions NIPZK Is Closed Under the OR Operator

We use our **NIPZK**-complete problem to show that under certain restrictions **NIPZK** is closed under the OR operator. We remark that these restrictions are severe, but our goal is to show the usefulness of our complete problem, rather than proving a closure result (in fact, even with these restrictions it is hard to see how to prove this result).

Motivation. We want to construct a **NIPZK** proof where the prover and the verifier are given two instances x and y of some problem $\Pi \in$ **NIPZK**, and the verifier accepts only if either x or y are YES instances of Π. Since we now have a **NIPZK**-complete problem, we can construct a protocol where the prover and the verifier reduce x and y to circuits X and Y, respectively, and then work with these circuits.

A natural approach to design our protocol is to ask what is the difference between YES and NO instances of UN, and then, based on this difference, to design a protocol and a simulator. As we saw, instances of UN differ in their number of output strings that end with a 1. That is, $|T_X| + |T_Y|$ is large if either X or Y is a YES instance, and small if both X and Y are NO instances. Thus, it seems that we should use lower bound protocols [17]. However, we avoid using these protocols because they incur error into the simulation, and we do not know how to remedy this problem.

Thus, we take a different approach. Instead of focusing on the difference between YES and NO instances, we focus on the simulation. That is, instead of starting with the protocol, taking care of completeness and soundness, we start with the simulator, taking care of perfect zero-knowledge. Indeed, this approach is implicit in Section 2, where we first modified the simulator, and then modified the prover to mimic the simulator. This approach has the advantage that we retain perfect simulation, but on the other hand we are forced to make restrictions in order to guarantee completeness and soundness.

The protocol. Recall that the prover and the verifier are given instances X and Y of UN, and the verifier should accept if $X \in \mathrm{UN_Y}$ or $Y \in \mathrm{UN_Y}$. As usual, we use $n+1$ to denote the number of output bits of X and Y. Since the main obstacle is how to achieve perfect simulation, we start with the zero-knowledge property. That is, we start with the simulator, and then we design the protocol based on the simulator.

Consider a simulator that uniformly picks r_X and r_Y, and computes $z = X(r_X) \oplus Y(r_Y)$. The simulator may not know which of X or Y is a YES instance of UN. However, the n-bit prefix of z is uniformly distributed because either X' or Y' represent the uniform distribution. This observation allows us to use the n-bit prefix of z as the simulated reference string.

Our simulator informs the following protocol: on reference string r_I the prover sends r_X and r_Y to the verifier such that the n-bit prefix $X(r_X) \oplus Y(r_Y)$ equals r_I. The issue with this protocol is that we need to make two restrictions in order to prove completeness and soundness.

Achieving completeness. Suppose that the verifier accepts only if the last bit of both $X(r_X)$ and $Y(r_Y)$ is 1. This works when both circuits X and Y are YES instances of UN. However, if one of the circuits is a NO instance of UN, then it is possible that all the strings outputted by this circuit end with a 0 (e.g., for any r_X the suffix of $X(r_X)$ is 0), and this will make V reject.

Since we do not know how to overcome this issue without introducing error into the simulation, we add the restriction that instances of $\mathrm{PU_Y}$ be 1-positive. That is, for any circuit $Z \in \mathrm{PU_Y}$, all the strings that Z outputs have 1 as the rightmost bit. Intuitively, this restriction helps the simulator in identifying NO instances. For example, if a sample of X ends with a 0, then X must be a NO instance. However, notice that X could be a NO instance and still have outputs that end with a 1. Thus, this help is limited.

We redefine the simulator based on the above restriction. As before, the simulator uniformly picks r_X and r_Y, computes $z = X(r_X) \oplus Y(r_Y)$, and if both $X(r_X)$ and $Y(r_Y)$ end with a 1, then the simulator uses the n-bit prefix of z to simulate the reference string. Otherwise, one of the samples ends with a 0. For example, suppose that $X(r_X)$ ends with a 0. This implies that Y is a YES instance. Hence, the simulator uses the n bit prefix of $Y(r_x)$ to simulate the reference string. Similarly, we redefine the verifier. That is, when the verifier receives $\langle r_X, r_Y \rangle$ from the prover, it only checks that the n-bit prefix of $Y(r_Y)$ equals to the reference string, and that $Y(r_Y)$ ends with a 1.

Achieving soundness. Notice that even when both X and Y are NO instances, there could be many combinations for $X(r_X) \oplus X(r_Y)$. That is, for most reference strings

r_I a cheating prover may find r_X and r_Y such that the n bit prefix of $X(r_X) \oplus Y(r_Y)$ equals r_I, and both $X(r_X)$ and $Y(r_Y)$ end with a 1. This compromises the soundness property. Since we do not know how to overcome this issue without introducing error into the simulation, we restrict the number of such pairs.

Discussion. We used our **NIPZK**-complete problem to show that under certain restrictions **NIPZK** is closed under the OR operator (See Appendix B for the proof). Indeed, we added severe restrictions to retain perfect simulation, but without our complete problem it is not clear how to prove this result (even with these restrictions). Thus, we interpret these restrictions as evidence that in the perfect setting there are few techniques to work with. Recall that even in the statistical setting, where we have more techniques to work with, such closure result is not known.

4.3   Applications of the AM ∩ HVPZK-Hard Problem

In this paper we showed that IDENTICAL DISTRIBUTIONS (ID) is hard for the class of problems admitting public-coin **HVPZK** proofs, and that we can treat it as $\overline{\mathrm{SD}}^{1/2,0}$. Unfortunately, our result is restricted to public-coin. In contrast, the reduction of [26] for **HVSZK** (which follows from the works of [11,1,25]) is not restricted to public-coin, but it manipulates distributions in a way that skews the distributions, and we do not know how to apply it to **HVPZK**.

However, what is special about ID, and what makes it different from SD, is that its YES instances are pairs of circuits $\langle X_0, X_1 \rangle$ representing identical (as opposed to statistically close) distributions. Thus, they can be used to obtain perfectly (as opposed to statistically) hiding instant-dependent commitment-schemes. Using the observation of [20], such schemes could then be plugged into the protocols for **NP** [4,13], thus yielding a **HVPZK** proof for ID. We mention that, except for [21], who used the techniques of [8] to construct a perfectly hiding scheme for V-bit protocols, all the known instance-dependent commitment-schemes are only statistically hiding [33,23,22].

Notice that if we can use instances $\langle X_0, X_1 \rangle$ of ID to construct a *constant-round*, perfectly hiding, instance-dependent commitment-scheme, then we would collapse the round complexity of public-coin **HVPZK** proofs. One idea for a commitment is to take a sample of X_b. That is, given common input $\langle X_0, X_1 \rangle$, a commitment to a bit b is computed by uniformly choosing r and outputting $X_b(r)$. Thus, on YES instances the scheme is perfectly hiding. However, the scheme may not be binding on NO instances because there could be r and r' for which $X_0(r) = X_1(r')$. Thus, other techniques are needed to make sure that the binding property is achieved.

5   Conclusion

We explained why reductions that apply to the statistical setting do not apply to the perfect setting. Using the error shifting technique we modified these reductions. Thus, we obtained complete and hard problems, and interesting applications. We believe that insight provided here will be useful in the study of perfect zero-knowledge proofs.

References

1. Aiello, W., Håstad, J.: Statistical zero-knowledge languages can be recognized in two rounds. J. of Computer and System Sciences 42(3), 327–345 (1991)
2. Babai, L., Moran, S.: Arthur-merlin games: A randomized proof system and a hierarchy of complexity classes. J. of Computer and System Sciences 36, 254–276 (1988)
3. Bellare, M., Rogaway, P.: Noninteractive perfect zero-knowledge. Unpublished manuscript (June 1990)
4. Blum, M.: How to prove a theorem so no one else can claim it. In: Proceedings of the ICM, pp. 1444–1451 (1986)
5. Blum, M., Feldman, P., Micali, S.: Non-interactive zero-knowledge proofs and their applications. In: Proceedings of the 20th STOC, ACM, New York (1988)
6. Blum, M., Santis, A.D., Micali, S., Persiano, G.: Noninteractive zero-knowledge. SIAM J. Comput. 20(6), 1084–1118 (1991)
7. Brassard, G., Crépeau, C., Yung, M.: Everything in NP can be argued in perfect zero-knowledge in a bounded number of rounds (extended abstract). In: Quisquater, J.-J., Vandewalle, J. (eds.) EUROCRYPT 1989. LNCS, vol. 434, pp. 192–195. Springer, Heidelberg (1990)
8. Damgård, I.: Interactive hashing can simplify zero-knowledge protocol design without computational assumptions (extended abstract). In: McCurley, K.S., Ziegler, C.D. (eds.) Advances in Cryptology 1981 - 1997. LNCS, vol. 1440, pp. 100–109. Springer, Heidelberg (1999)
9. Damgård, I., Wigderson, O.G.A.: Hashing functions can simplify zeroknowledge protocol design (too). Technical Report RS-94-39, BRICS (November 1994)
10. Even, S., Selman, A.L., Yacobi, Y.: The complexity of promise problems with applications to public-key cryptography. Information and Control 61(2), 159–173 (1984)
11. Fortnow, L.: The complexity of perfect zero-knowledge. In: Micali, S. (ed.) Advances in Computing Research, vol. 5, pp. 327–343. JAC Press, Inc (1989)
12. Goldreich, O.: Foundations of Cryptography, vol. 1. Cambridge University Press, Cambridge (2001)
13. Goldreich, O., Micali, S., Wigderson, A.: Proofs that yield nothing but their validity or all languages in NP have zero-knowledge proof systems. J. ACM 38(3), 691–729 (1991)
14. Goldreich, O., Sahai, A., Vadhan, S.P.: Honest-verifier statistical zero-knowledge equals general statistical zero-knowledge. In: STOC, pp. 399–408 (1998)
15. Goldreich, O., Sahai, A., Vadhan, S.P.: Can statistical zero knowledge be made noninteractive? or on the relationship of SZK and NISZK. In: Wiener, M.J. (ed.) CRYPTO 1999. LNCS, vol. 1666, Springer, Heidelberg (1999)
16. Goldreich, O., Vadhan, S.P.: Comparing entropies in statistical zero-knowledge with applications to the structure of SZK. In: IEEE Conference on Computational Complexity, pp. 54–73 (1999)
17. Goldwasser, S., Sipser, M.: Private-coins versus public-coins in interactive proof systems. In: Micali, S. (ed.) Advances in Computing Research, vol. 5, pp. 73–90. JAC Press, Inc (1989)
18. Goldwasser, S., Micali, S., Rackoff, C.: The knowledge complexity of interactive proof systems. SIAM J. Comput. 18(1), 186–208 (1989)
19. Groth, J., Ostrovsky, R., Sahai, A.: Perfect non-interactive zero knowledge for NP. In: Vaudenay, S. (ed.) EUROCRYPT 2006. LNCS, vol. 4004, pp. 339–358. Springer, Heidelberg (2006)
20. Itoh, T., Ohta, Y., Shizuya, H.: A language-dependent cryptographic primitive. J. Cryptology 10(1), 37–50 (1997)

21. Kapron, B., Malka, L., Srinivasan, V.: Characterizing non-interactive instance-dependent commitment-schemes (NIC). In: Arge, L., Cachin, C., Jurdziński, T., Tarlecki, A. (eds.) ICALP 2007. LNCS, vol. 4596, pp. 328–339. Springer, Heidelberg (2007)
22. Nguyen, M.-H., Ong, S.J., Vadhan, S.: Statistical zero-knowledge arguments for NP from any one-way function. In: Proceedings of the 47th Annual IEEE Symposium on Foundations of Computer Science (FOCS 2006), Berkeley, CA, October 2006, pp. 3–14 (2006)
23. Nguyen, M.-H., Vadhan, S.: Zero knowledge with efficient provers. In: STOC 2006: Proceedings of the thirty-eighth annual ACM symposium on Theory of computing, pp. 287–295. ACM Press, New York, USA (2006)
24. Okamoto, T.: On relationships between statistical zero-knowledge proofs. J. Comput. Syst. Sci. 60(1), 47–108 (2000)
25. Petrank, E., Tardos, G.: On the knowledge complexity of NP. In: FOCS, pp. 494–503 (1996)
26. Sahai, A., Vadhan, S.P.: A complete problem for statistical zero-knowledge. J. ACM 50(2), 196–249 (2003)
27. De Santis, A., Di Crescenzo, G., Persiano, G.: The knowledge complexity of quadratic residuosity languages. Theor. Comput. Sci. 132(1-2), 291–317 (1994)
28. De Santis, A., Di Crescenzo, G., Persiano, G.: Randomness-efficient non-interactive zero-knowledge (extended abstract). In: Automata, Languages and Programming, pp. 716–726 (1997)
29. De Santis, A., Di Crescenzo, G., Persiano, G.: On NC^1 boolean circuitcomposition of non-interactive perfect zero-knowledge. In: Fiala, J., Koubek, V., Kratochvíl, J. (eds.) MFCS 2004. LNCS, vol. 3153, pp. 356–367. Springer, Heidelberg (2004)
30. De Santis, A., Di Crescenzo, G., Persiano, G., Yung, M.: Image densityis complete for non-interactive-SZK (extended abstract). In: Larsen, K.G., Skyum, S., Winskel, G. (eds.) ICALP 1998. LNCS, vol. 1443, pp. 784–795. Springer, Heidelberg (1998)
31. Tompa, M., Woll, H.: Random self-reducibility and zero-knowledge interactive proofs of possession of information. In: 28th FOCS, pp. 472–482 (1987)
32. Vadhan, S.P.: A study of statistical zero-knowledge proofs. PhD thesis, MIT (1999)
33. Vadhan, S.P.: An unconditional study of computational zero knowledge. In: FOCS, pp. 176–185 (2004)

A   Preliminaries

We use standard definitions [12]. We study *promise-problems* [10], which are a generalization of languages. Formally, $\Pi \stackrel{\text{def}}{=} \langle \Pi_Y, \Pi_N \rangle$ is a problem if $\Pi_Y \cap \Pi_N = \emptyset$. The set Π_Y contains the YES instances of Π, and the set Π_N contains the NO instances of Π. We define $\overline{\Pi} \stackrel{\text{def}}{=} \langle \Pi_N, \Pi_Y \rangle$.

Let $X : \{0,1\}^m \to \{0,1\}^n$ be a circuit. We treat X both as a circuit and as a distribution (under the convention that the input to the circuit is uniformly distributed). For example, given a set T, the probability $\Pr_{x \leftarrow X}[x \in T]$ equals $\Pr_{r \leftarrow U_m}[X(r) \in T]$, where U_m is the uniform distribution on $\{0,1\}^m$, and $d \leftarrow D$ denotes choosing an element d according to the distribution D. The *statistical distance* between two discrete distributions X and Y is $\Delta(X,Y) \stackrel{\text{def}}{=} \sum_\alpha |\Pr[X = \alpha] - \Pr[Y = \alpha]|$. We define non-interactive protocols.

Definition A.1 (Non-interactive protocols). *A non-interactive protocol $\langle c, P, V \rangle$ is a triplet (or simply a pair $\langle P, V \rangle$, making c implicit), where P and V are functions, and $c \in \mathbb{N}$. We denote by r_P the random inputs to P. The interaction between P and V on common input x is the following random process.*

1. *Uniformly choose r_P, and choose a* common random string $r_I \in \{0,1\}^{|x|^c}$.
2. *Let* $\pi = P(x, r_I; r_P)$, *and let* $m = V(x, r_I, \pi)$.
3. *Output* $\langle x, r_I, \pi, m \rangle$.

We call $\langle P, V \rangle(x) \stackrel{def}{=} \langle x, r_I, \pi \rangle$ *the view of* V *on* x. *We say that* V *accepts* x *(respectively, rejects* x) *if* $m = \texttt{accept}$ *(respectively,* $m = \texttt{reject}$).

Definition A.1 considers a deterministic V, and is equivalent to a the definition that considers a probabilistic V [2]. We define non-interactive proofs.

Definition A.2 (Non-interactive proofs). *A non-interactive protocol* $\langle c, P, V \rangle$ *is a non-interactive proof for a problem* Π *if there is* $a \in \mathbb{N}$ *and* $c(n), s(n) : \mathbb{N} \to [0,1]$ *such that* $1 - c(n) \geq s(n) + 1/n^a$ *for any* n, *and the following conditions hold.*

- *Efficiency: V runs in time polynomial in $|x|$.*
- *Completeness: V accepts all $x \in \Pi_Y$ with probability at least $1 - c(|x|)$ over r_I and r_P.*
- *Soundness:* $\Pr_{r_I}[V(x, r_I, P^*(x, r_I)) = \texttt{accept}] \leq s(|x|)$ *for any function P^* and any $x \in \Pi_N$.*

The function c *is called the* completeness error, *and the function* s *is called the* soundness error. *We say that* $\langle P, V \rangle$ *has* perfect completeness *if* $c \equiv 0$.

We proceed to zero-knowledge. Our definition considers simulators that do not fail, which is justified by our result from Section 4.

Definition A.3 (Non-interactive, zero-knowledge protocols). *A non-interactive protocol* $\langle P, V \rangle$ *is perfect zero-knowledge (**NIPZK**) for a problem* $\Pi = \langle \Pi_Y, \Pi_N \rangle$ *if there is a probabilistic, polynomial-time Turing machine* S, *called the* simulator, *such that the ensembles*

$$\{\langle P, V \rangle(x)\}_{x \in \Pi_Y} \quad \text{and} \quad \{S(x)\}_{x \in \Pi_Y}$$

are statistically identical.

If these ensembles are statistically indistinguishable, then $\langle P, V \rangle$ *is a non-interactive statistical zero-knowledge (**NISZK**) protocol for* Π. *Similarly, if the ensembles are computationally indistinguishable, then* $\langle P, V \rangle$ *is non-interactive computational zero-knowledge (**NICZK**) protocol for* Π.

*The class of problems possessing **NIPZK** (respectively, **NISZK**, **NICZK**) protocols is also denoted **NIPZK** (respectively, **NISZK**, **NICZK**).*

B Under Certain Restrictions NIPZK Is Closed Under OR

Lemma B.1. *Let Π be a **NIPZK** problem with a proof $\langle P', V' \rangle$, and let $c \in \mathbb{N}$ such that on input of length n the reference string is of length n^c. If $\langle c, P', V' \rangle$ has perfect completeness and soundness error $2^{1-n^c/2}$, then $\Pi \vee \Pi$ has a **NIPZK** proof with perfect completeness, and soundness error $1/3$.*

Proof. Let $\langle x_0, x_1 \rangle$ such that $x_i \in \Pi_Y \cup \Pi_N$ for each $i \in \{0,1\}$, and let $n = |x_0|$. We start with the case where $|x_0| = |x_1|$ because when we reduce x_0 and x_1 to UN we get circuits whose output length is equal. As we will see, the general case follows easily using the same proof.

We construct a **NIPZK** protocol $\langle P, V \rangle$ for $\Pi \vee \Pi$. Initially, P sets $i = 0$ if both x_0 and x_1 are in Π_Y. Otherwise, there is a unique i such that $x_i \in \Pi_N$, and P fixes this i. In addition, for each $i \in \{0,1\}$ both P and V reduce x_i to an instance X_i of UN.

Recall that $\langle c, P', V' \rangle$ is a **NIPZK** proof for Π such that on input of length n the reference string is of length n^c. By the properties of the reduction to UN, for each $i \in \{0,1\}$ the circuit X_i has $n^c + 1$ output gates and the following properties hold. If $x_i \in \Pi_Y$, then X_i' is the uniform distribution on $\{0,1\}^{n^c}$, and samples of X_i end with a 1. If $x_i \in \Pi_N$, then $|T_{X_i}| \leq 2^{-(n^c/2+1)} \cdot 2^{n^c} = 2^{n^c/2-1}$.

The protocol proceeds as follows. Recall that P initially computes i. Thus, the first step of P is to uniformly choose a string r_i, and assign y the output of $X_i(r_i)$, excluding the rightmost bit. On reference string r_I, if $X_i(r_i) = y0$, then P uniformly chooses $r_{\bar{i}} \in X_{\bar{i}}^{-1}(r_I 1)$, and sends $\langle r_0, r_1 \rangle$ to V. Otherwise, $X_i(r_i) = y1$, in which case P uniformly chooses $r_{\bar{i}} \in X_{\bar{i}}^{-1}(y1 \oplus r_I 0)$, and sends $\langle r_0, r_1 \rangle$ to V. The verifier accepts if $\langle r_0, r_1 \rangle$ are correctly computed. Namely, V computes $X_0(r_0)$ and $X_1(r_1)$, and if there is $i \in \{0,1\}$ such that $X_i(r_i)$ ends with a 0 and $X_{\bar{i}}(r_{\bar{i}}) = r_I 1$, then V accepts. Otherwise, if $X_0(r_0) \oplus X_1(r_1) = r_I 0$ (that is, both $X_0(r_0)$ and $X_1(r_1)$ end with a 1), then V accepts. Otherwise, V rejects.

The completeness property of $\langle P, V \rangle$ follows from its zero-knowledge property. Thus, we start the simulator S for $\langle P, V \rangle$. As in $\langle P, V \rangle$, the simulator reduces $\langle x_0, x_1 \rangle$ to $\langle X_0, X_1 \rangle$. The simulator uniformly chooses r_0 and r_1, and computes $X_0(r_0)$ and $X_1(r_1)$. If there is $i \in \{0,1\}$ such that $X_i(r_i)$ ends with a 0 (i.e., $X_i \in \text{PU}_N$), then S outputs $\langle \langle x_0, x_1 \rangle, r_I', \langle r_0, r_1 \rangle \rangle$, where r_I' equals the n^c-bit prefix of $X_{\bar{i}}(r_{\bar{i}})$. Otherwise, S outputs $\langle \langle x_0, x_1 \rangle, r_I', \langle r_0, r_1 \rangle \rangle$, where r_I' equals the n^c-bit prefix of $X_0(r_0) \oplus X_1(r_1)$. In both cases r_I' is uniformly distributed, and $\langle r_0, r_1 \rangle$ are distributed as in $\langle P, V \rangle$. Thus, S perfectly simulates $\langle P, V \rangle$. Since S always outputs accepting transcripts, $\langle P, V \rangle$ has perfect completeness.

We turn our attention to the soundness property. Let $x_0, x_1 \in \Pi_N$, and let $\langle r_0, r_1 \rangle$ be the message received by V. We consider two cases in which V accepts. In the first case there is $i \in \{0,1\}$ such that $X_i(r_i)$ ends with a 0, and $X_{\bar{i}}(r_{\bar{i}}) = r_I 1$. Since $|T_{X_{\bar{i}}}| \leq 2^{n^c/2-1}$, and r_I is uniformly distributed, it follows that in the first case V accepts with probability at most $2 \cdot \Pr_{r_I}[X_{\bar{i}}(r_{\bar{i}}) = r_I 1] \leq 2 \cdot 2^{-(n^c/2+1)}$. The reason we multiplied the probability by 2 is because a cheating P^* may use either X_0 or X_1. In the second case the suffix of both $X_0(r_0)$ and $X_1(r_1)$ is 1, and $X_0(r_0) \oplus X_1(r_1) = r_I 0$. In this case the probability over r_I that $X_0(r_0) \oplus X_1(r_1) = r_I 0$ is at most $1/4$ because $|T_{X_0}| \cdot |T_{X_1}| \leq 2^{n^c/2-1} \cdot 2^{n^c/2-1} = 2^{n^c}/4$, and r_I is uniformly distributed. We conclude that in total V accepts with probability at most $1/4 + 2 \cdot 2^{-(n^c/2+1)}$, which is $1/3$ for sufficiently large inputs.

Recall that in the beginning of this proof we considered the case where $|x_0| = |x_1|$. In this case the length of the output of X_0 equals that of X_1. The general case can be treated exactly the same, except that X_0 and X_1 are modified before the protocol begins. For example, if $|x_0| = n$ and $|x_1| = n + a$ (for some $a \in \mathbb{N}$), then we simply add $(n+a)^c - n^c$ input gates to X_0. These gates are outputted as the prefix of X_0. Call this new circuit X_0'. Now both X_0' and X_1 have $(n+a)^c + 1$ output bits, and X_0' inherits the properties of X_0 (that is, for any α and β, if X_0 is α-positive, then X_0' is α-positive, and if X_0 is β-negative, then X_0' is β-negative). Thus, we can apply the proof as above. The lemma follows.

General Properties of
Quantum Zero-Knowledge Proofs

Hirotada Kobayashi

Principles of Informatics Research Division, National Institute of Informatics
2-1-2 Hitotsubashi, Chiyoda-ku, Tokyo 101-8430, Japan
hirotada@nii.ac.jp

Abstract. This paper studies general properties of quantum zero-knowledge proof systems. Among others, the following properties are proved on quantum computational zero-knowledge proofs:

- *Honest-verifier* quantum zero-knowledge equals general quantum zero-knowledge.
- *Public-coin* quantum zero-knowledge equals general quantum zero-knowledge.
- Quantum zero-knowledge with *perfect completeness* equals general quantum zero-knowledge with imperfect completeness.
- Any quantum zero-knowledge proof system can be transformed into a *three-message public-coin* quantum zero-knowledge proof system of perfect completeness with polynomially small error in soundness (hence with arbitrarily small constant error in soundness).

All the results proved in this paper are unconditional, i.e., they do not rely any computational assumptions. The proofs for all the statements are direct and do not use complete promise problems, and thus, essentially the same method works well even for quantum statistical and perfect zero-knowledge proofs. In particular, all the four properties above hold also for the statistical zero-knowledge case (the first two were shown previously by Watrous), and the first two properties hold even for the perfect zero-knowledge case. It is also proved that allowing a simulator to output "FAIL" does not change the power of quantum perfect zero-knowledge proofs. The corresponding properties are not known to hold in the classical perfect zero-knowledge case.

1 Introduction

Background. Zero-knowledge proof systems were introduced by Goldwasser, Micali, and Rackoff [13], and have played a central role in modern cryptography since then. Intuitively, an interactive proof system is zero-knowledge if *any* verifier who communicates with the *honest* prover learns nothing except for the validity of the statement being proved in that system. By "learns nothing" we mean that there exists a polynomial-time *simulator* whose output is indistinguishable from the output of the verifier after communicating with the honest prover. Depending on the strength of this indistinguishability, several variants of

R. Canetti (Ed.): TCC 2008, LNCS 4948, pp. 107–124, 2008.
© International Association for Cryptologic Research 2008

zero-knowledge proofs have been investigated: *perfect* zero-knowledge in which the output of the simulator is identical to that of the verifier, *statistical* zero-knowledge in which the output of the simulator is statistically close to that of the verifier, and *computational* zero-knowledge in which the output of the simulator is indistinguishable from that of the verifier in polynomial time. The most striking result on zero-knowledge proofs would be that every problem in NP has a computational zero-knowledge proof system under certain intractability assumptions [10]. It is also known that some problems have perfect or statistical zero-knowledge proof systems. Among others, the GRAPH ISOMORPHISM problem has a perfect zero-knowledge proof system [10], and some lattice problems have statistical zero-knowledge proof systems [9].

Another direction of studies on zero-knowledge proofs has been to prove their general properties. Sahai and Vadhan [22] were the first who took an approach of characterizing zero-knowledge proofs by complete promise problems. They showed that the STATISTICAL DIFFERENCE problem is complete for the class HVSZK of problems having *honest-verifier* statistical zero-knowledge proof systems. Here, the honest-verifier zero-knowledge is a weaker notion of zero-knowledge in which now zero-knowledge property holds only against the *honest* verifier who follows the specified protocol. Using this complete promise problem, they proved a number of general properties of HVSZK and simplified the proofs of several previously known results, including that HVSZK is in AM [6,2], that HVSZK is closed under complement [21], and that any problem in HVSZK has a public-coin honest-verifier statistical zero-knowledge proof system [21]. Goldreich and Vadhan [12] presented another complete promise problem for HVSZK, called the ENTROPY DIFFERENCE problem, and obtained further properties of HVSZK. Since Goldreich, Sahai, and Vadhan [11] proved that HVSZK = SZK, where SZK denotes the class of problems having statistical zero-knowledge proof systems, all the properties proved for HVSZK are inherited to SZK (except for those related to round complexity). More recently, Vadhan [24] gave two characterizations, the INDISTINGUISHABILITY characterization and the CONDITIONAL PSEUDO-ENTROPY characterization, for the class ZK of problems having computational zero-knowledge proof systems. These are not complete promise problems, but more or less analogous to complete promise problems and play essentially same roles as complete promise problems in his proofs. Using these characterizations, he proved a number of general properties of ZK unconditionally (i.e., not assuming any intractability assumptions), such as that honest-verifier computational zero-knowledge equals general computational zero-knowledge, that public-coin computational zero-knowledge equals general computational zero-knowledge, and that computational zero-knowledge with perfect completeness equals that with imperfect completeness.

Quantum zero-knowledge proofs were first studied by Watrous [25] in a restricted situation of *honest-verifier* quantum statistical zero-knowledge proofs. He gave an analogous characterization to the classical case due to Sahai and Vadhan [22] by showing that the QUANTUM STATE DISTINGUISHABILITY problem is complete for the class HVQSZK of problems having honest-verifier quantum statistical zero-knowledge proof systems. Using this, he proved a number of general properties of HVQSZK, such as that HVQSZK is closed under complement, that any problem in HVQSZK has a public-coin honest-verifier quantum statistical zero-knowledge proof system, and that HVQSZK is in PSPACE. Very recently, Ben-Aroya and Ta-Shma [3] presented another complete promise problem for HVQSZK, called the QUANTUM ENTROPY DIFFERENCE problem, which is a quantum analogue of the result by Goldreich and Vadhan [12]. It has been a wide open problem if there are nontrivial problems that have quantum zero-knowledge proofs secure even against any dishonest quantum verifiers, because of the difficulties arising from the "rewinding" technique [14], which is commonly used in classical zero-knowledge proofs. Damgård, Fehr, and Salvail [4] studied zero-knowledge proofs against dishonest quantum verifiers, but they assumed the restricted setting of the common-reference-string model to avoid this rewinding problem. Very recently, Watrous [27] settled this affirmatively. He established a quantum "rewinding" technique by using a method that was originally developed in Ref. [19] for the purpose of amplifying the success probability of QMA, a quantum version of NP, without increasing quantum witness sizes. With this quantum rewinding technique, he proved that the classical proof system for the GRAPH ISOMORPHISM problem in Ref. [10] has a perfect zero-knowledge property even against any dishonest *quantum* verifiers, and under some reasonable intractability assumption, the classical proof system for NP in Ref. [10] has a computational zero-knowledge property even against any dishonest *quantum* verifiers. He also proved that HVQSZK $-$ QSZK, where QSZK denotes the class of problems having quantum statistical zero-knowledge proof systems. Together with his proof construction, this implies that all the properties proved for HVQSZK in Ref. [25] are inherited to QSZK (except for those related to round complexity), in particular, that any problem in QSZK has a public-coin quantum statistical zero-knowledge proof system.

Our contribution. This paper proves a number of general properties on quantum zero-knowledge proofs, not restricted to the statistical zero-knowledge case. Specifically, for quantum computational zero-knowledge proofs, letting QZK and HVQZK denote the classes of problems having quantum computational zero-knowledge proof systems and *honest-verifier* quantum computational zero-knowledge proof systems, respectively, the following are proved among others:

Theorem 1. HVQZK = QZK.

Theorem 2. *Any problem in* QZK *has a public-coin quantum computational zero-knowledge proof system.*

Theorem 3. *Any problem in* QZK *has a quantum computational zero-knowledge proof system of perfect completeness.*

Theorem 4. *Any problem in* QZK *has a three-message public-coin quantum computational zero-knowledge proof system of perfect completeness with soundness error at most* $\frac{1}{p}$ *for any polynomially bounded function* $p\colon \mathbb{Z}^+ \to \mathbb{N}$.

All the properties proved in this paper on quantum computational zero-knowledge proofs hold unconditionally, meaning that they hold without any computational assumptions such as the existence of quantum one-way functions or permutations. Some of these properties may be regarded as quantum versions of the results by Vadhan [24]. It is stressed, however, that our approach to prove these properties is completely different from those the existing studies took to prove general properties of classical or quantum zero-knowledge proofs. No complete promise problems nor characterizations are used in our proofs. Instead, we *directly* prove these properties.

The idea is remarkably simple. We start from any proof system of *honest-verifier* quantum zero-knowledge, and apply several transformations so that we finally obtain another proof system of honest-verifier quantum zero-knowledge that possesses a number of desirable properties. For instance, to prove that HVQZK = QZK, we show that any proof system of honest-verifier quantum computational zero-knowledge can be transformed into another proof system of honest-verifier quantum computational zero-knowledge (with some smaller gap between completeness and soundness accepting probabilities) such that (i) the proof system consists of three messages and (ii) the proof system is public-coin in which the message from the honest verifier consists of a single bit that is an outcome of a classical fair coin-flipping. This can be done by first achieving negligible completeness error by sequential repetition, then applying the parallelization method for usual quantum interactive proofs due to Kitaev and Watrous [16] to obtain a *three-message* honest-verifier quantum zero-knowledge proof system, and finally applying the Marriott-Watrous construction for usual quantum interactive proofs [19] to obtain a three-message *public-coin* honest-verifier quantum zero-knowledge proof system. It is proved that the Kitaev-Watrous parallelization method preserves the honest-verifier zero-knowledge property if completeness error is negligible, and that the Marriott-Watrous construction also preserves the honest-verifier zero-knowledge property. Now, by applying the quantum rewinding technique due to Watrous [27], this three-message public-coin proof system is proved to be zero-knowledge even against any *dishonest* quantum verifiers. The final piece is the sequential repetition, which makes completeness and soundness errors arbitrarily small. This simultaneously shows the equivalence of public-coin quantum computational zero-knowledge and general quantum computational zero-knowledge. To show that any quantum computational zero-knowledge proofs can be made perfectly complete, now we have only to show that any *honest-verifier* quantum computational zero-knowledge proofs can be made perfectly complete. Again we can use another construction for usual quantum interactive proofs due to Kitaev and Watrous [16], but now we need to carefully and explicitly design a protocol for the honest prover in their construction so that the honest-verifier zero-knowledge property is preserved. Using this construction as a preprocessing, the previous argument shows the equivalence of quantum computational zero-knowledge with perfect completeness and that with imperfect completeness. Combining all the desirable properties of honest-verifier quantum computational zero-knowledge proofs shown in this paper with a careful application of the quantum rewinding technique, we can show that any problem in QZK has a three-message public-coin quantum computational zero-knowledge proof system of perfect completeness with soundness error at most $\frac{1}{p}$ for any polynomially bounded function p.

In fact, our approach above is very general and basically works well even for quantum statistical and perfect zero-knowledge proofs. In the quantum statistical zero-knowledge case, all the properties shown for the quantum computational zero-knowledge case also hold. This gives alternative proofs of the facts that HVQSZK = QSZK and that public-coin quantum statistical zero-knowledge equals general quantum statistical zero-knowledge, which were originally shown by Watrous [27] using his previous results [25], and also shows the following new properties of quantum statistical zero-knowledge proofs:

Theorem 5. *Any problem in QSZK has a quantum statistical zero-knowledge proof system of perfect completeness.*

Theorem 6. *Any problem in QSZK has a three-message public-coin quantum statistical zero-knowledge proof system of perfect completeness with soundness error at most $\frac{1}{p}$ for any polynomially bounded function $p\colon \mathbb{Z}^+ \to \mathbb{N}$.*

In the quantum perfect zero-knowledge case, however, not all the properties above can be shown to hold, because very subtle points easily lose the *perfect* zero-knowledge property. In particular, our method of making proof systems perfectly complete no longer works well for quantum perfect zero-knowledge case. Also, we need a careful modification of the protocol when parallelizing to three messages. Still, we can show the following properties for the classes QPZK and HVQPZK of problems having quantum perfect zero-knowledge proof systems and *honest-verifier* quantum perfect zero-knowledge proof systems, respectively:

Theorem 7. HVQPZK = QPZK.

Theorem 8. *Any problem in QPZK has a public-coin quantum perfect zero-knowledge proof system.*

Note that no such general properties are known for the classical perfect zero-knowledge case. As a bonus property, it is also proved that quantum perfect zero-knowledge with a worst-case polynomial-time simulator that is not allowed to output "FAIL" is equivalent to the one in which a simulator is allowed to output "FAIL" with small probability. Again, such equivalence is not known in the classical case.

Due to space limitations, most of the technical proofs are relegated to the full version of this paper [18].

2   Preliminaries

We assume the reader is familiar with classical zero-knowledge proof systems and quantum interactive proof systems. Detailed discussions of classical zero-knowledge proof systems can be found in Refs. [7,8], for instance, while quantum interactive proof systems are discussed in Refs. [26,16,19]. We also assume familiarity with the quantum formalism, including the quantum circuit model and definitions of mixed quantum states, admissible transformations (completely-positive trace-preserving mappings), trace norm, diamond norm, and fidelity (all of which are discussed in detail in Refs. [20,15], for instance). Some of the notions and notations that are used in this paper are summarized in this section.

Throughout this paper, let $\mathbb{N}$ and $\mathbb{Z}^+$ denote the sets of positive and nonnegative integers, respectively. Let poly denote the set of all functions $p\colon \mathbb{Z}^+ \to \mathbb{N}$ such that there exists a polynomial-time deterministic Turing machine that outputs $1^{p(n)}$ on input 1^n. For every $d \in \mathbb{N}$, let I_d denote the identity operator of dimension d. Also, for any Hilbert space $\mathcal{H}$, let $I_\mathcal{H}$ denote the identity operator over $\mathcal{H}$. In this paper, all Hilbert spaces are of dimension power of two.

For any Hilbert space $\mathcal{H}$, let $|0_\mathcal{H}\rangle$ denote the quantum state in $\mathcal{H}$ of which all the qubits are in state $|0\rangle$, and let $\mathbf{D}(\mathcal{H})$ and $\mathbf{U}(\mathcal{H})$ denote the sets of density and unitary operators over $\mathcal{H}$, respectively. For any Hilbert spaces $\mathcal{H}$ and $\mathcal{K}$, let $\mathbf{T}(\mathcal{H}, \mathcal{K})$ be the set of admissible transformations from $\mathbf{D}(\mathcal{H})$ to $\mathbf{D}(\mathcal{K})$. An admissible transformation $\Phi \in \mathbf{T}(\mathcal{H}, \mathcal{K})$ is q_{in}-*in* q_{out}-*out* if $\mathcal{H}$ and $\mathcal{K}$ consist of q_{in} and q_{out} qubits, respectively. Let $\mathcal{N}$, $\mathcal{X}$, and $\mathcal{Y}$ be Hilbert spaces such that $\mathcal{H} \otimes \mathcal{X} = \mathcal{K} \otimes \mathcal{Y} = \mathcal{N}$. A unitary transformation $U_\Phi \in \mathbf{U}(\mathcal{N})$ is a *unitary realization* of Φ if $\mathrm{tr}_\mathcal{Y} U_\Phi (\rho \otimes |0_\mathcal{X}\rangle\langle 0_\mathcal{X}|) U_\Phi^\dagger = \Phi(\rho)$ for any $\rho \in \mathbf{D}(\mathcal{H})$.

Quantum circuits. It is assumed that any quantum circuit Q in this paper is unitary and is composed of gates in some reasonable, universal, finite set of unitary quantum gates. For convenience, we may identify a circuit Q with the unitary operator it induces. Since non-unitary and unitary quantum circuits are equivalent in computational power [1], it is sufficient to treat only unitary quantum circuits, which justifies the above assumption. For avoiding unnecessary complication, however, the descriptions of procedures often include non-unitary operations in the subsequent sections. Even in such cases, it is always possible to construct unitary quantum circuits that essentially achieve the same procedures described. When proving statements concerning quantum perfect zero-knowledge proofs or proofs having perfect completeness, we assume that the Hadamard transformation and any classical reversible transformations are exactly implementable in our gate set. This condition may not hold with an arbitrary universal gate set, but is satisfied by most of the standard gate sets including the Shor basis [23], and thus, the author believes that it is not restrictive. These subtle issues regarding choices of the universal gate set is discussed in the full version of this paper [18]. It is stressed, however, that all of our statements not concerning quantum perfect zero-knowledge proofs nor proofs having perfect completeness do hold with an arbitrary choice of the universal gate set (the completeness and soundness conditions may become worse by negligible amounts in some of the claims, which does not matter for the final main statements).

A quantum circuit Q is q_{in}-*in* q_{out}-*out* if it exactly implements a unitary realization U_Φ of some q_{in}-in q_{out}-out admissible transformation Φ. For convenience, we may identify a circuit Q with Φ in such a case. As a special case of this, a quantum circuit Q is a *generating circuit* of a quantum state ρ of q

qubits if it exactly implements a unitary realization of a zero-in q-out admissible transformation that always outputs ρ. A family $\{Q_x\}$ of quantum circuits is *polynomial-time uniformly generated* if there exists a deterministic procedure that, on every input x, outputs a description of Q_x and runs in time polynomial in $|x|$. It is assumed that the number of gates in any circuit is not more than the length of the description of that circuit, which assures that Q_x has size polynomial in $|x|$. An ensemble $\{\rho_x\}$ of quantum states is *polynomial-time preparable* if there exists a polynomial-time uniformly generated family $\{Q_x\}$ of quantum circuits such that each Q_x is a generating circuit of ρ_x. In what follows, we may use the notation $\{\rho(x)\}$ instead of $\{\rho_x\}$ for ensembles of quantum states simply for descriptional convenience.

Quantum computational indistinguishability. We use the notions of quantum computational indistinguishability introduced by Watrous [27]: polynomially quantum indistinguishable ensembles of quantum states and polynomially quantum indistinguishable ensembles of admissible transformations.

Definition 9. *Let $S \subseteq \{0,1\}^*$ be an infinite set and let $m \in$ poly. For each $x \in S$, let ρ_x and σ_x be mixed states of $m(|x|)$ qubits. The ensembles $\{\rho_x : x \in S\}$ and $\{\sigma_x : x \in S\}$ are polynomially quantum indistinguishable if it holds for all but finitely many $x \in S$ that, for every choice of $k, p, s \in$ poly, an ensemble $\{\xi_x : x \in S\}$ where ξ_x is a mixed state of $k(|x|)$ qubits, and an $(m(|x|) + k(|x|))$-in one-out quantum circuit Q of size at most $s(|x|)$,*

$$|\langle 1|Q(\rho_x \otimes \xi_x)|1\rangle - \langle 1|Q(\sigma_x \otimes \xi_x)|1\rangle| < \frac{1}{p(|x|)}.$$

Definition 10. *Let $S \subseteq \{0,1\}^*$ be an infinite set and let $l, m \in$ poly. For each $x \in S$, let Φ_x and Ψ_x be $l(|x|)$-in $m(|x|)$-out admissible transformations. The ensembles $\{\Phi_x : x \in S\}$ and $\{\Psi_x : x \in S\}$ are polynomially quantum indistinguishable if it holds for all but finitely many $x \in S$ that, for every choice of $k, p, s \in$ poly, an ensemble $\{\xi_x : x \in S\}$ where ξ_x is a mixed state of $l(|x|) + k(|x|)$ qubits, and an $(m(|x|) + k(|x|))$-in one-out quantum circuit Q of size at most $s(|x|)$,*

$$|\langle 1|Q((\Phi_x \otimes I_{2^{k(|x|)}})(\xi_x))|1\rangle - \langle 1|Q((\Psi_x \otimes I_{2^{k(|x|)}})(\xi_x))|1\rangle| < \frac{1}{p(|x|)}.$$

In what follows, we will often use the term "computationally indistinguishable" instead of "polynomially quantum indistinguishable" for simplicity. Also, we will often informally say that mixed states ρ_x and σ_x or admissible transformations Φ_x and Ψ_x are computationally indistinguishable when $x \in S$ to mean that the ensembles $\{\rho_x : x \in S\}$ and $\{\sigma_x : x \in S\}$ or $\{\Phi_x : x \in S\}$ and $\{\Psi_x : x \in S\}$ are polynomially quantum indistinguishable.

Quantum zero-knowledge proofs. For readability, in what follows, the arguments x and $|x|$ are often dropped in various functions. It is assumed that operators acting on subsystems of a given system are extended to the entire system by tensoring with the identity, as it will be clear from the context upon what part of a system a given operator acts. Although all the statements in this paper can be proved only in terms of languages without using promise problems [5], in what follows we define models and prove statements in terms of promise problems, for generality and for the compatibility with some other studies on quantum zero-knowledge proofs [25,17,27,3]. This paper follows a manner in Ref. [25] when defining various honest-verifier quantum zero-knowledge proofs, and that in Ref. [27] when defining various general quantum zero-knowledge proofs.

We start with formally defining quantum verifiers and quantum provers. An m-*message* quantum verifier V is a mapping of the form $V\colon \{0,1\}^* \to \{0,1\}^*$. For every input $x \in \{0,1\}^*$, the string $V(x)$ is interpreted as a $\lceil (m(|x|) + 1)/2 \rceil$-tuple $(V(x)_1, \ldots, V(x)_{\lceil (m(|x|)+1)/2 \rceil})$, with each $V(x)_j$ a description of a polynomial-size quantum circuit acting over the qubits in the verifier's private space and message qubits. A quantum verifier V is *uniform* if the corresponding mapping V is polynomial-time computable, and is *non-uniform* if no restrictions are placed on the complexity of the mapping V (but each circuit $V(x)_j$ must have size polynomial in $|x|$). Similarly, an m-*message* quantum prover P is a mapping of the form $P\colon \{0,1\}^* \to \{0,1\}^*$. For every input $x \in \{0,1\}^*$, the string $P(x)$ is interpreted as a $\lceil m(|x|)/2 \rceil$-tuple $(P(x)_1, \ldots, P(x)_{\lceil m(|x|)/2 \rceil})$, with each $P(x)_j$ a description of a quantum circuit acting over the qubits in the prover's private space and message qubits. No restrictions are placed on the complexity of the mapping P, and each $P(x)_j$ can be an arbitrary unitary transformation.

First we define the notions of various *honest-verifier* quantum zero-knowledge proofs. Given a quantum verifier V and a quantum prover P, let $\mathrm{view}_{V,P}(x,j)$ be the quantum state that V possesses immediately after the jth transformation of P during an execution of the protocol between V and P. Now we define the classes $\mathrm{HVQPZK}(m,c,s)$, $\mathrm{HVQSZK}(m,c,s)$, and $\mathrm{HVQZK}(m,c,s)$ of problems having m-message honest-verifier quantum perfect, statistical, and computational zero-knowledge proof systems, respectively, with completeness at least c and soundness at most s.

Definition 11. *Given functions $m \in \mathrm{poly}$ and $c, s\colon \mathbb{Z}^+ \to [0,1]$, a problem $A = \{A_{\mathrm{yes}}, A_{\mathrm{no}}\}$ is in $\mathrm{HVQPZK}(m,c,s)$ ($\mathrm{HVQSZK}(m,c,s)$) [$\mathrm{HVQZK}(m,c,s)$] iff there exist an m-message uniform honest quantum verifier V and an m-message honest quantum prover P such that*

(Completeness and Soundness) *(V, P) forms an m-message quantum interactive proof system with completeness at least c and soundness at most s,*

(Honest-Verifier Zero-Knowledge) *there exists a polynomial-time preparable ensemble $\{S_V(x,j)\}$ of quantum states such that $S_V(x,j) = \mathrm{view}_{V,P}(x,j)$ for every $x \in A_{\mathrm{yes}}$ and $j \in T$ ($\|S_V(x,j) - \mathrm{view}_{V,P}(x,j)\|_{\mathrm{tr}}$ is negligible with respect to $|x|$ for all but finitely many $(x,j) \in A_{\mathrm{yes}} \times T$) [the ensembles $\{S_V(x,j)\colon (x,j) \in A_{\mathrm{yes}} \times T\}$ and $\{\mathrm{view}_{V,P}(x,j)\colon (x,j) \in A_{\mathrm{yes}} \times T\}$ are polynomially quantum indistinguishable], where $T = \{1, \ldots, \lceil \frac{m(|x|)}{2} \rceil\}$.*

Using these, we define the classes HVQPZK, HVQSZK, and HVQZK of problems having honest-verifier quantum perfect, statistical, and computational zero-knowledge proof systems, respectively.

Definition 12. *A problem* $A = \{A_{\text{yes}}, A_{\text{no}}\}$ *is in* HVQPZK *(*HVQSZK*)* *[*HVQZK*] if there exists a function* $m \in$ poly *such that* A *is in* HVQPZK $\left(m, \frac{2}{3}, \frac{1}{3}\right)$ *(*HVQSZK $\left(m, \frac{2}{3}, \frac{1}{3}\right)$*)* *[*HVQZK $\left(m, \frac{2}{3}, \frac{1}{3}\right)$*].*

Note that it is easy to see that we can amplify the success probability of honest-verifier quantum perfect/statistical/computational zero-knowledge proof systems by sequential repetition, which justifies Definition 12.

Next we define the notions of various quantum zero-knowledge proofs. Let V be an arbitrary non-uniform quantum verifier. Suppose that V possesses some auxiliary quantum state in $\mathbf{D}(\mathcal{A})$ at the beginning and possesses some quantum state in $\mathbf{D}(\mathcal{Z})$ after having received the last message from the prover, for some Hilbert spaces $\mathcal{A}$ and $\mathcal{Z}$. For such V, for any quantum prover P, and for every $x \in \{0,1\}^*$, let $\langle V, P \rangle(x)$ denote the admissible transformation in $\mathbf{T}(\mathcal{A}, \mathcal{Z})$ induced by the interaction between V and P on input x. Note that the last transformation of V is *not* considered as a part of the interaction, since we want to focus on the state V would possess immediately after having received the last message from P. We call this $\langle V, P \rangle(x)$ the *induced admissible transformation* from V, P, and x. We define the classes $\mathrm{QPZK}(m, c, s)$, $\mathrm{QSZK}(m, c, s)$, and $\mathrm{QZK}(m, c, s)$ of problems having m-message quantum perfect, statistical, and computational zero-knowledge proof systems, respectively, with completeness at least c and soundness at most s, as follows.

Definition 13. *Given functions* $m \in$ poly *and* $c, s \colon \mathbb{Z}^+ \to [0, 1]$, *a problem* $A = \{A_{\text{yes}}, A_{\text{no}}\}$ *is in* $\mathrm{QPZK}(m, c, s)$ *(*$\mathrm{QSZK}(m, c, s)$*)* *[*$\mathrm{QZK}(m, c, s)$*] iff there exist an* m-*message uniform honest quantum verifier* V *and an* m-*message honest quantum prover* P *such that*

(Completeness and Soundness) (V, P) *forms an* m-*message quantum interactive proof system with completeness at least* c *and soundness at most* s,

(Zero-Knowledge) *there exists a polynomial-time uniformly generated family* $\{Q_{x,y}\}$ *of quantum circuits such that, for any* m-*message non-uniform quantum verifier* V', *the circuit* $Q_{x,V'(x)}$ *exactly implements an admissible transformation* $S_{V'}(x)$ *such that* $S_{V'}(x) = \langle V', P \rangle(x)$ *for every* $x \in A_{\text{yes}}$ *(*$\| S_{V'}(x) - \langle V', P \rangle(x) \|_\diamond$ *is negligible with respect to* $|x|$ *for all but finitely many* $x \in A_{\text{yes}}$*)* *[the ensembles* $\{S_{V'}(x) \colon x \in A_{\text{yes}}\}$ *and* $\{\langle V', P \rangle(x) \colon x \in A_{\text{yes}}\}$ *are polynomially quantum indistinguishable]*, *where* $\langle V', P \rangle(x)$ *is the induced admissible transformation from* V', P, *and* x.

Using these, we define the classes QPZK, QSZK, and QZK of problems having quantum perfect, statistical, and computational zero-knowledge proof systems, respectively.

Definition 14. *A problem* $A = \{A_{\text{yes}}, A_{\text{no}}\}$ *is in* QPZK *(*QSZK*)* *[*QZK*] if there exists a function* $m \in$ poly *such that* A *is in* QPZK $\left(m, \frac{2}{3}, \frac{1}{3}\right)$ *(*QSZK $\left(m, \frac{2}{3}, \frac{1}{3}\right)$*)* *[*QZK $\left(m, \frac{2}{3}, \frac{1}{3}\right)$*].*

Again note that it is not hard to see that we can amplify the success probability of quantum perfect/statistical/computational zero-knowledge proof systems by sequential repetition, which justifies Definition 14.

In the classical case, the most common definition of perfect zero-knowledge proofs would be the one that allows the simulator to output "FAIL" with small probability, say, with probability at most $\frac{1}{2}$ [7,22]. Adopting this convention leads to alternative definitions of honest-verifier and general quantum perfect zero-knowledge proof systems. At a glance, the two types of definitions seem likely to form different complexity classes of quantum perfect zero-knowledge proofs. Fortunately, it is proved in Section 6 that the two types of definitions result in the same complexity class of quantum perfect zero-knowledge proofs. Such equivalence is not known in the classical case.

3   Computational Zero-Knowledge Case

We start with showing that any *honest-verifier* quantum computational zero-knowledge proof system with two-sided bounded error can be transformed into one with perfect completeness (if the completeness error in the original proof system is negligible, which may be assumed without loss of generality since the success probability can be amplified by sequential repetition). This can be basically proved by using a method for usual quantum interactive proofs due to Kitaev and Watrous (Theorem 2 of Ref. [16]), but now it is necessary for the honest-verifier zero-knowledge property to carefully and explicitly construct a protocol for the honest prover. The proof is found in the full version of this paper [18].

Lemma 15. *Let $m \in$ poly, let $\varepsilon \colon \mathbb{Z}^+ \to [0,1]$ be any negligible function such that there exists a polynomial-time uniformly generated family $\{Q_x\}$ of quantum circuits such that Q_{1^n} exactly performs the unitary transformation $U_{\varepsilon(n)} = \begin{pmatrix} \sqrt{\varepsilon(n)} & \sqrt{1-\varepsilon(n)} \\ \sqrt{1-\varepsilon(n)} & -\sqrt{\varepsilon(n)} \end{pmatrix}$, and let $\delta \colon \mathbb{Z}^+ \to [0,1]$ be any function that satisfies $\delta > \varepsilon$. Then, $\mathrm{HVQZK}(m, 1-\varepsilon, 1-\delta) \subseteq \mathrm{HVQZK}(m+2, 1, 1-(\delta-\varepsilon)^2)$.*

Next we show that any honest-verifier quantum computational zero-knowledge proof system that involves polynomially many messages can be parallelized to one that involves only three messages. This can be achieved again by applying a method in usual quantum interactive proofs due to Kitaev and Watrous (Theorem 4 of Ref. [16]). The main idea in their parallelization protocol is that the verifier receives each snapshot state of the underlying proof system as the first message, and then checks if the following three properties are satisfied: (i) the first snapshot state is a legal state in the underlying proof system after the first message, (ii) the last snapshot state can make the original verifier accept, and (iii) any two consecutive snapshot states are indeed transformable with each other by one round of communication. The verifier first checks if the conditions (i) and (ii) really hold for the received snapshot states. He then randomly chooses a consecutive pair of the snapshot states and challenges the prover to show the transformability from one to the other. It is straightforward to show that their construction preserves the honest-verifier zero-knowledge property.

Lemma 16. *Let* $m \in \mathrm{poly}$ *and let* $\delta \colon \mathbb{Z}^+ \to [0,1]$ *be any function. Then,* $\mathrm{HVQZK}(m, 1, 1 - \delta) \subseteq \mathrm{HVQZK}\left(3, 1, 1 - \frac{\delta^2}{4m^2}\right).$

Finally we show that any three-message honest-verifier quantum computational zero-knowledge proof system can be transformed into a three-message public-coin one in which the message from the verifier consists of only one classical bit. Marriott and Watrous (Theorem 5.4 of Ref. [19]) showed such a transformation in the case of usual quantum interactive proofs. In their construction, the verifier first receives a state that is supposed to be the reduced state in the verifier's private space after the second message in the original proof system, and then challenges the prover to recover either the state the original verifier would have after the first message or that after the third message, depending on the outcome of the public coin-flip. It is easy to show that their construction preserves the honest-verifier zero-knowledge property.

Lemma 17. *Let* $c, s \colon \mathbb{Z}^+ \to [0,1]$ *be any functions that satisfy* $c^2 > s$. *Then, any problem in* $\mathrm{HVQZK}(3, c, s)$ *has a three-message public-coin honest-verifier quantum computational zero-knowledge proof system with completeness at least* $\frac{1+c}{2}$ *and soundness at most* $\frac{1+\sqrt{s}}{2}$ *in which the message from the verifier consists of only one classical bit.*

Now we can use the quantum rewinding technique due to Watrous [27] to show that any three-message public-coin honest-verifier quantum computational zero-knowledge proof system in which the message from the verifier consists of only one classical bit is computational zero-knowledge even against any dishonest non-uniform quantum verifier.

Lemma 18. *Any three-message public-coin honest-verifier quantum computational zero-knowledge proof system such that the message from the verifier consists of only one classical bit is computational zero-knowledge against any non-uniform quantum verifier.*

Proof. Let $A = \{A_{\mathrm{yes}}, A_{\mathrm{no}}\}$ be a problem having a three-message public-coin honest-verifier quantum computational zero-knowledge proof system such that the message from the verifier consists of only one classical bit. Let V and P be the corresponding honest quantum verifier and honest quantum prover, respectively. Let M and N be the quantum registers sent to V at the first message and at the third message, respectively, and let R and S be the single-qubit registers that are used to store the classical information representing the outcome b of a public coin flipped by V, where R is inside the private space of V and S is sent to P. Let S_V be the simulator for V such that, if x is in A_{yes}, the states $S_V(x, 1)$ and $\mathrm{view}_{V,P}(x, 1)$ consisting of qubits in M are computationally indistinguishable and the states $S_V(x, 2)$ and $\mathrm{view}_{V,P}(x, 2)$ consisting of qubits in (M, N, R) are also computationally indistinguishable.

Simulator for General Verifier W

1. Store the auxiliary quantum state ρ in the quantum register X. Prepare the quantum registers S, W, M, N, R, and A, and further prepare a single-qubit quantum register F. Initialize all the qubits in F, S, W, M, N, R, and A to state $|0\rangle$.
2. Apply the generating circuit Q of the quantum state $S_V(x, 2)$ to the qubits in (M, N, R, A).
3. Apply W_1 to the qubits in (S, W, X, M), where W_1 is the first transformation of the simulated verifier W.
4. Compute the exclusive-or of the contents of R and S and write the result in F.
5. Measure the qubit in F in the $\{|0\rangle, |1\rangle\}$ basis. If this results in $|0\rangle$, output the qubits in (W, X, M, N, R), otherwise apply $W_1^\dagger$ to the qubits in (S, W, X, M) and then apply $Q^\dagger$ to the qubits in (M, N, R, A).
6. Apply the phase-flip if all the qubits in F, S, W, M, N, R, and A are in state $|0\rangle$, apply Q to the qubits in (M, N, R, A), and apply W_1 to the qubits in (S, W, X, M). Output the qubits in (W, X, M, N, R).

Fig. 1. Simulator for a general verifier W

Consider a generating circuit Q of the quantum state $S_V(x, 2)$. Without loss of generality, it is assumed that Q acts over the qubits in (M, N, R, A), where A is the quantum register consisting of q_A qubits for some $q_A \in$ poly. For any non-uniform quantum verifier W and any auxiliary quantum state ρ for W stored in the quantum register X inside the private space of W, we construct an efficiently implementable admissible mapping Φ that corresponds to a simulator T_W for W. Without loss of generality it is assumed that the message from W consists of a single *classical* bit, since the honest prover can easily enforce this constraint by measuring the message from the verifier before responding to it. Let W be the quantum register consisting of all the qubits in the private space of W except for those in X and M after the second message having been sent. We consider the procedure described in Fig. 1, which is the implementation of Φ.

Suppose that the input x is in A_{yes}. We shall show that (i) the gap between $\frac{1}{2}$ and the probability of obtaining $|0\rangle$ as the measurement result in Step 5 must be negligible regardless of the auxiliary quantum state ρ, and (ii) the output state in Step 5 in the construction conditioned on the measurement result being $|0\rangle$ must be computationally indistinguishable from the state W would possess after the third message. With these two properties, the quantum rewinding technique due to Watrous [27] works well, by using the amplification lemma for the case with negligible perturbations, which is also due to Watrous [27]. This ensures the computational zero-knowledge property against W.

For the generating circuit Q' of the quantum state $\text{view}_{V,P}(x, 2)$ (here no restrictions are placed on the size of Q'), consider the "ideal" construction of the simulator such that Q' is applied instead of Q in Step 2 of the "real" simulator construction.

We first show the property (i).

Since the state $\text{view}_{V,P}(x,2)$ can be written of the form $\text{view}_{V,P}(x,2) = \frac{1}{2}(\sigma_0 \otimes |0\rangle\langle 0| + \sigma_1 \otimes |1\rangle\langle 1|)$ for some quantum states σ_0 and σ_1 in (M, N), the probability of obtaining $|0\rangle$ as the measurement result in Step 5 in the "ideal" construction is exactly equal to $\frac{1}{2}$ regardless of the auxiliary quantum state ρ, because $\text{tr}_{\mathcal{N}}\sigma_0 = \text{tr}_{\mathcal{N}}\sigma_1$ necessarily holds in this case, where $\mathcal{N}$ is the Hilbert space corresponding to N.

Now, from the honest-verifier computational zero-knowledge property, the states $S_V(x,2)$ and $\text{view}_{V,P}(x,2)$ in $(\mathsf{M}, \mathsf{N}, \mathsf{R})$ are computationally indistinguishable. Since the circuit implementing W_1 is of size polynomial with respect to $|x|$, it follows that the gap between $\frac{1}{2}$ and the probability of obtaining $|0\rangle$ as the measurement result in Step 5 in the "real" construction must be negligible regardless of the auxiliary quantum state ρ, which proves the property (i).

Now we show the property (ii).

Let $\xi_i = \Pi_i W_1 (|0_{\mathcal{S} \otimes \mathcal{W}}\rangle\langle 0_{\mathcal{S} \otimes \mathcal{W}}| \otimes \rho \otimes \sigma_i \otimes |i\rangle\langle i|) W_1^\dagger \Pi_i$ be an unnormalized state in $(\mathsf{S}, \mathsf{W}, \mathsf{X}, \mathsf{M}, \mathsf{N}, \mathsf{R})$ for each $i \in \{0,1\}$, where $\Pi_i = |i\rangle\langle i|$ is the projection operator over the qubit in S, and $\mathcal{S}$ and $\mathcal{W}$ are the Hilbert spaces corresponding to S and W, respectively. Then, in the "ideal" construction, conditioned on the measurement result being $|0\rangle$ in Step 5, the output is the state $\text{tr}_{\mathcal{S}}(\xi_0 + \xi_1)$.

Noticing that $\text{tr}_{\mathcal{S}}\frac{\xi_i}{\text{tr}\xi_i}$ is exactly the state the verifier W would possess after the third message when the second message from W is i and that the probability of the second message from W being i is exactly equal to $\text{tr}\xi_i$ for each $i \in \{0,1\}$, $\text{tr}_{\mathcal{S}}(\xi_0 + \xi_1) = \text{tr}\xi_0 \cdot \text{tr}_{\mathcal{S}}\frac{\xi_0}{\text{tr}\xi_0} + \text{tr}\xi_1 \cdot \text{tr}_{\mathcal{S}}\frac{\xi_1}{\text{tr}\xi_1}$ is exactly the state W would possess after the third message.

Towards a contradiction, suppose that the output state in Step 5 in the "real" construction conditioned on the measurement result being $|0\rangle$ is computationally distinguishable from $\text{tr}_{\mathcal{S}}(\xi_0 + \xi_1)$. Let D be the corresponding distinguisher that uses the auxiliary quantum state ρ'. We construct a distinguisher D' for $S_V(x,2)$ and $\text{view}_{V,P}(x,2)$ from D.

On input quantum state η that is either $S_V(x,2)$ or $\text{view}_{V,P}(x,2)$, D' uses the auxiliary quantum state $\rho \otimes \rho'$, where ρ is the auxiliary quantum state the verifier W would use. D' prepares the quantum registers $\mathsf{S}, \mathsf{W}, \mathsf{M}, \mathsf{N}, \mathsf{R}$ and another quantum register Y. D' stores ρ in the register X, η in the register $(\mathsf{M}, \mathsf{N}, \mathsf{R})$, and ρ' in Y. All the qubits in S and W are initialized to state $|0\rangle$. Now D' applies W_1 to the qubits in $(\mathsf{S}, \mathsf{W}, \mathsf{X}, \mathsf{M})$, and then applies D to the qubits in $(\mathsf{W}, \mathsf{X}, \mathsf{M}, \mathsf{N}, \mathsf{R}, \mathsf{Y})$.

It is obvious from this construction that D' with the auxiliary quantum state $\rho \otimes \rho'$ forms a distinguisher for $S_V(x,2)$ and $\text{view}_{V,P}(x,2)$ if D with the auxiliary quantum state ρ' forms a distinguisher for the output state in Step 5 in the "real" simulator construction conditioned on the measurement result being $|0\rangle$ and the state $\text{tr}_{\mathcal{S}}(\xi_0 + \xi_1)$. This contradicts the computational indistinguishability between $S_V(x,2)$ and $\text{view}_{V,P}(x,2)$, and thus the property (ii) follows. $\square$

Now we are ready to show Theorem 1 that states HVQZK = QZK.

Proof (of Theorem 1). It is trivial that HVQZK $\supseteq$ QZK, and we show that HVQZK $\subseteq$ QZK. From Lemma 15, we can start with an m-message honest-verifier quantum computational zero-knowledge proof system of perfect completeness with soundness at most $1 - \delta$ for some $m \in$ poly and δ such that $1 - \delta$ is polynomially bounded away from one. Now from Lemmas 16 and 17 together with Lemma 18, we have that $\mathrm{HVQZK}(m, 1, 1 - \delta) \subseteq \mathrm{HVQZK}(3, 1, 1 - \delta') \subseteq \mathrm{QZK}\left(3, 1, \frac{1+\sqrt{1-\delta'}}{2}\right)$, where $\delta' = \frac{\delta^2}{4m^2}$. Finally, the sequential repetition establishes $\mathrm{HVQZK} \subseteq \mathrm{QZK}$.   □

This simultaneously shows Theorem 2, the equivalence of public-coin and general quantum computational zero-knowledge proofs, and Theorem 3, the equivalence of quantum computational zero-knowledge proofs of perfect completeness and general ones.

To show Theorem 4, we need another two properties. First, it is trivial that parallel repetition of honest-verifier quantum zero-knowledge proofs preserves the honest-verifier zero-knowledge property. Together with the perfect parallel repetition theorem for three-message quantum interactive proofs due to Kitaev and Watrous (Theorem 6 of Ref. [16]), this implies the following.

Lemma 19. *Let $c, s \colon \mathbb{Z}^+ \to [0, 1]$ be any functions such that $c > s$. Then, for any $k \in$ poly, $\mathrm{HVQZK}(3, c, s) \subseteq \mathrm{HVQZK}(3, c^k, s^k)$.*

Second, it is easy to extend Lemma 18 to the following more general statement.

Lemma 20. *Any three-message public-coin honest-verifier quantum computational zero-knowledge proof system such that the message from the verifier consists of $O(\log n)$ bits for every input of length n is computational zero-knowledge against any non-uniform quantum verifier.*

Now Theorem 4 can be proved as follows.

Proof (of Theorem 4). For any $p \in$ poly, take $q \in$ poly such that $2^{\frac{q}{2}} \geq \log p + 2$. Then, from Lemmas 15, 16, and 19, we have that $\mathrm{HVQZK} \subseteq \mathrm{HVQZK}(3, 1, 2^{-q})$. With Lemma 17, this further implies that any problem in HVQZK has a three-message public-coin honest-verifier quantum computational zero-knowledge proof system of perfect completeness with soundness at most $\frac{1}{2} + 2^{-\frac{q}{2}-1}$ in which the message from the verifier consists of only one classical bit. For every input of length n, we run this proof system $\lceil \log p(n) \rceil + 2$ times in parallel. From Lemma 19, this results in a three-message public-coin honest-verifier quantum computational zero-knowledge proof system of perfect completeness with soundness at most $\frac{1}{4p(n)}\left(1 + 2^{-\frac{q(n)}{2}}\right)^{\lceil \log p(n) \rceil + 2} \leq \frac{1}{p(n)}$ in which the message from the verifier consists of $\lceil \log p(n) \rceil + 2$ bits. Now Lemma 20 ensures that this proof system is computational zero-knowledge even against any dishonest quantum verifier.   □

4   Statistical Zero-Knowledge Case

All the properties shown for the computational zero-knowledge case also hold for the statistical zero-knowledge case. The proofs are essentially same as in the computational zero-knowledge case. This proves Theorems 5 and 6, and also gives alternative proofs of the facts that HVQSZK = QSZK and that public-coin quantum statistical zero-knowledge equals general quantum statistical zero-knowledge, which were first shown by Watrous [27] using his previous results [25].

5 Perfect Zero-Knowledge Case

Now we move to the perfect zero-knowledge case. Although our approach for the computational and statistical zero-knowledge cases basically works well even for the perfect zero-knowledge case, some of our transformations do not preserve the *perfect* zero-knowledge property. In particular, our method of making proof systems perfectly complete no longer works well for quantum perfect zero-knowledge case, and we need to use a slightly modified parallelization method.

As mentioned in Section 3, the verifier in the Kitaev-Watrous parallelization protocol checks if the last snapshot state can make the original verifier accept *before* proceeding to the test for consecutivity. The problem arises here, in the check for the last snapshot state, when parallelizing an honest-verifier quantum perfect zero-knowledge proof system with *imperfect* completeness. Because of imperfect completeness, the verifier's check can fail even if the honest prover prepares every snapshot state honestly, which means that the verifier's check causes a small perturbation to the snapshot states. Now we have difficulty in *perfectly* simulating the behavior of the honest prover with respect to these perturbed states, which spoils the perfect zero-knowledge property.

To avoid this difficulty, we modify the parallelization protocol as follows. Our basic idea is to postpone the verifier's check for the last snapshot state until after the third message. At the final verification of the verifier, with equal probability he either carries out the postponed check for the last snapshot state or just carries out the original final verification procedure. Now the honest-verifier perfect zero-knowledge property becomes straightforward, since there is no perturbation to all the snapshot states until after the last transformation of the verifier. The completeness accepting probability cannot be worse than that in the original protocol. However, the soundness condition now becomes a bit harder to prove, because we can no longer assume that the last snapshot state prepared by a dishonest prover makes the original verifier accept, when analyzing the probability to pass the transformability test for two consecutive snapshot states. Nevertheless, we can show that our modified parallelization protocol above indeed works well, and we have the following lemma. The proof is found in the full version of this paper [18].

Lemma 21. *Let $m \in$ poly be such that $m \geq 4$ and let $\varepsilon, \delta \colon \mathbb{Z}^{+} \to [0, 1]$ be any functions such that $\varepsilon < \frac{\delta^2}{16(m+1)^2}$. Then,*
$$\mathrm{HVQPZK}(m, 1 - \varepsilon, 1 - \delta) \subseteq \mathrm{HVQPZK}\big(3, 1 - \tfrac{\varepsilon}{2}, 1 - \tfrac{\delta^2}{32(m+1)^2}\big).$$

For Lemmas 17 and 18, exactly the same constructions can be used to show their perfect zero-knowledge versions. Putting things together, we have Theorem 7 that states HVQPZK = QPZK, and Theorem 8, the equivalence of public-coin and general quantum perfect zero-knowledge proofs.

6 Equivalence of Two Definitions of Quantum Perfect Zero-Knowledge

In the classical case, the most common definition of perfect zero-knowledge proofs would be the one that allows the simulator to output "FAIL" with small probability [7,22]. Adopting this convention leads to the following alternative definitions of honest-verifier and general quantum perfect zero-knowledge proof systems.

Definition 22. *Given functions $m \in$ poly and $c, s \colon \mathbb{Z}^+ \to [0,1]$, a problem $A = \{A_{\mathrm{yes}}, A_{\mathrm{no}}\}$ is in $\mathrm{HVQPZK}'(m, c, s)$ iff there exist an m-message uniform honest quantum verifier V and an m-message honest quantum prover P such that*

(Completeness and Soundness) *(V, P) forms an m-message quantum interactive proof system with completeness at least c and soundness at most s,*
(Honest-Verifier Perfect Zero-Knowledge) *there exists a polynomial-time preparable ensemble $\{S_V(x, j)\}$ of quantum states such that $S_V(x, j) = p_{x,j}|0\rangle\langle 0| \otimes |0_{\mathcal{H}_j}\rangle\langle 0_{\mathcal{H}_j}| + (1 - p_{x,j})|1\rangle\langle 1| \otimes \mathrm{view}_{V,P}(x, j)$ for some $0 \leq p_{x,j} \leq \frac{1}{2}$, for every $x \in A_{\mathrm{yes}}$ and for each $1 \leq j \leq \lceil \frac{m(|x|)}{2} \rceil$, where $\mathcal{H}_j$ is the Hilbert space such that $\mathrm{view}_{V,P}(x, j)$ is in $\mathbf{D}(\mathcal{H}_j)$.*

Definition 23. *Given functions $m \in$ poly and $c, s \colon \mathbb{Z}^+ \to [0,1]$, a problem $A = \{A_{\mathrm{yes}}, A_{\mathrm{no}}\}$ is in $\mathrm{QPZK}'(m, c, s)$ iff there exist an m-message uniform honest quantum verifier V and an m-message honest quantum prover P such that*

(Completeness and Soundness) *(V, P) forms an m-message quantum interactive proof system with completeness at least c and soundness at most s,*
(Perfect Zero-Knowledge) *there exists a polynomial-time uniformly generated family $\{Q_{x,y}\}$ of quantum circuits such that, for any m-message non-uniform quantum verifier V', the circuit $Q_{x,V'(x)}$ exactly implements an admissible transformation $S_{V'}(x)$ such that, for every $x \in A_{\mathrm{yes}}$, $S_{V'}(x) = p_x(\Phi_0 \otimes \Psi_{\mathrm{fail}}) + (1 - p_x)(\Phi_1 \otimes \langle V', P\rangle(x))$ for some $0 \leq p_x \leq \frac{1}{2}$, where $\langle V', P\rangle(x) \in \mathbf{T}(\mathcal{A}, \mathcal{Z})$ is the induced admissible transformation from V', P, and x for some Hilbert spaces $\mathcal{A}$ and $\mathcal{Z}$, $\Psi_{\mathrm{fail}} \in \mathbf{T}(\mathcal{A}, \mathcal{Z})$ is the admissible transformation that always outputs $|0_{\mathcal{Z}}\rangle\langle 0_{\mathcal{Z}}|$, and Φ_b is the admissible transformation that takes nothing as input and outputs $|b\rangle\langle b|$, for each $b \in \{0, 1\}$.*

In Definitions 22 and 23, the first qubit of the output of the simulator indicates whether or not the simulation succeeds — $|0\rangle\langle 0|$ is interpreted as failure and $|1\rangle\langle 1|$ as success.

Definition 24. *A problem $A = \{A_{\mathrm{yes}}, A_{\mathrm{no}}\}$ is in HVQPZK' (QPZK') if there exists a function $m \in$ poly such that A is in $\mathrm{HVQPZK}'\left(m, \frac{2}{3}, \frac{1}{3}\right)$ ($\mathrm{QPZK}'\left(m, \frac{2}{3}, \frac{1}{3}\right)$).*

It is not obvious at a glance that HVQPZK = HVQPZK' and QPZK = QPZK', i.e., that the definitions of honest-verifier and general quantum perfect zero-knowledge proof systems using Definitions 11 and 13 are equivalent to those using Definitions 22 and 23. Fortunately, with Theorem 7, we can show that HVQPZK = HVQPZK' and QPZK = QPZK'. It is stressed that such equivalence is not known in the classical case.

Theorem 25. HVQPZK = HVQPZK' *and* QPZK = QPZK'.

Note that QPZK $\subseteq$ QPZK' $\subseteq$ HVQPZK' is obvious. From Theorem 7, we have HVQPZK = QPZK. Therefore, to show Theorem 25, it is sufficient to show that HVQPZK' $\subseteq$ HVQPZK. Now, the idea is to modify the protocol of the honest prover for the HVQPZK' system so that the honest prover "adjusts" his behavior to that of the simulator, i.e., he privately runs the simulator and intentionally fails to return the correct response whenever the simulator fails. The detailed proof is found in the full version of this paper [18].

Acknowledgement. The author is grateful to John Watrous for his helpful comments on the choice of the universal gate set. The author would also like to thank anonymous referees for their helpful suggestions for improving this paper. This work is supported by the Strategic Information and Communications R&D Promotion Programme No. 031303020 of the Ministry of Internal Affairs and Communications of Japan.

References

1. Aharonov, D., Kitaev, A.Yu., Nisan, N.: Quantum circuits with mixed states. In: Proceedings of the Thirtieth Annual ACM Symposium on Theory of Computing, pp. 20–30 (1998)
2. Aiello, W., Håstad, J.: Statistical zero-knowledge languages can be recognized in two rounds. Journal of Computer and System Sciences 42(3), 327–345 (1991)
3. Ben-Aroya, A., Ta-Shma, A.: Quantum expanders and the quantum entropy difference problem. arXiv.org e-Print archive, quant-ph/0702129 (2007)
4. Damgård, I., Fehr, S., Salvail, L.: Zero-knowledge proofs and string commitments withstanding quantum attacks. In: Franklin, M. (ed.) CRYPTO 2004. LNCS, vol. 3152, pp. 254–272. Springer, Heidelberg (2004)
5. Even, S., Selman, A.L., Yacobi, Y.: The complexity of promise problems with applications to public-key cryptography. Information and Control 61(2), 159–173 (1984)
6. Fortnow, L.J.: The complexity of perfect zero-knowledge. In: Micali, S. (ed.) Randomness and Computation. Advances in Computing Research, vol. 5, pp. 327–343. JAI Press (1989)
7. Goldreich, O.: Foundations of Cryptography – Volume 1 Basic Tools. Cambridge University Press, Cambridge (2001)
8. Goldreich, O.: Zero-knowledge twenty years after its invention. Electronic Colloquium on Computational Complexity, Report No. 63 (2002)
9. Goldreich, O., Goldwasser, S.: On the limits of nonapproximability of lattice problems. Journal of Computer and System Sciences 60(3), 540–563 (2000)

10. Goldreich, O., Micali, S., Wigderson, A.: Proofs that yield nothing but their validity or all languages in NP have zero-knowledge proof systems. Journal of the ACM 38(3), 691–729 (1991)
11. Goldreich, O., Sahai, A., Vadhan, S.P.: Honest-verifier statistical zero-knowledge equals general statistical zero-knowledge. In: Proceedings of the Thirtieth Annual ACM Symposium on Theory of Computing, pp. 399–408 (1998)
12. Goldreich, O., Vadhan, S.P.: Comparing entropies in statistical zero knowledge with applications to the structure of SZK. In: Fourteenth Annual IEEE Conference on Computational Complexity, pp. 54–73 (1999)
13. Goldwasser, S., Micali, S., Rackoff, C.W.: The knowledge complexity of interactive proof systems. SIAM Journal on Computing 18(1), 186–208 (1989)
14. van de Graaf, J.: Towards a formal definition of security for quantum protocols. PhD thesis, Département d'Informatique et de Recherche Opérationnelle, Université de Montréal (December 1997)
15. Kitaev, A.Yu., Shen, A.H., Vyalyi, M.N.: Classical and Quantum Computation. Graduate Studies in Mathematics, vol. 47. American Mathematical Society (2002)
16. Kitaev, A.Yu., Watrous, J.H.: Parallelization, amplification, and exponential time simulation of quantum interactive proof systems. In: Proceedings of the Thirty-Second Annual ACM Symposium on Theory of Computing, pp. 608–617 (2000)
17. Kobayashi, H.: Non-interactive quantum perfect and statistical zero-knowledge. In: Ibaraki, T., Katoh, N., Ono, H. (eds.) ISAAC 2003. LNCS, vol. 2906, pp. 178–188. Springer, Heidelberg (2003)
18. Kobayashi, H.: General properties of quantum zero-knowledge proofs. arXiv.org e-Print archive, arXiv:0705.1129 [quant-ph] (2007)
19. Marriott, C., Watrous, J.H.: Quantum Arthur-Merlin games. Computational Complexity 14(2), 122–152 (2005)
20. Nielsen, M.A., Chuang, I.L.: Quantum Computation and Quantum Information. Cambridge University Press, Cambridge (2000)
21. Okamoto, T.: On relationships between statistical zero-knowledge proofs. Journal of Computer and System Sciences 60(1), 47–108 (2000)
22. Sahai, A., Vadhan, S.P.: A complete problem for statistical zero knowledge. Journal of the ACM 50(2), 196–249 (2003)
23. Shor, P.W.: Fault-tolerant quantum computation. In: 37th Annual Symposium on Foundations of Computer Science, pp. 56–65 (1996)
24. Vadhan, S.P.: An unconditional study of computational zero-knowledge. SIAM Journal on Computing 36(4), 1160–1214 (2006)
25. Watrous, J.H.: Limits on the power of quantum statistical zero-knowledge. In: 43rd Annual Symposium on Foundations of Computer Science, pp. 459–468 (2002)
26. Watrous, J.H.: PSPACE has constant-round quantum interactive proof systems. Theoretical Computer Science 292(3), 575–588 (2003)
27. Watrous, J.H.: Zero-knowledge against quantum attacks. In: Proceedings of the 38th Annual ACM Symposium on Theory of Computing, pp. 296–305 (2006)

The Layered Games Framework
for Specifications and Analysis of Security
Protocols

Amir Herzberg and Igal Yoffe

Computer Science Department, Bar Ilan University,
Ramat Gan, 52900, Israel
{herzbea,ioffei}@cs.biu.ac.il

Abstract. The layered games framework provides a solid foundation to
the accepted methodology of building complex distributed systems, as a
'stack' of independently-developed protocols. Each protocol in the stack,
realizes a corresponding 'layer' model, over the 'lower layer'. We define
layers, protocols and related concepts. We then prove the *fundamental
lemma of layering*. The lemma shows that given a stack of protocols
$\{\pi_i\}_{i=1}^u$, s.t. for every $i \in \{1, \ldots u\}$, protocol π_i realizes layer L_i over
layer L_{i-1}, then the entire stack can be composed to a single protocol
$\pi_{u||\ldots||1}$, which realizes layer L_u over layer L_0.

The fundamental lemma of layering allows precise specification, design
and analysis of each layer independently, and combining the results to
ensure properties of the complete system. This is especially useful when
considering (computationally-bounded) adversarial environments, as for
security and cryptographic protocols.

Our specifications are based on *games*, following many works in applied cryptography. This differs from existing frameworks allowing compositions of cryptographic protocols, which are based on *simulatability*
of ideal functionality.

1   Introduction

The design and analysis of complex distributed systems, such as the Internet and
applications using it, is an important and challenging goal. Such systems are designed in modular fashion, typically by decomposing the system into multiple
layers (or modules-). Some of the well known layered network architectures include the 'OSI 7-layers reference model' and the 'IETF 5-layers reference model'
(also referred to as the Internet or TCP/IP model); see e.g. [30]. The present
work is part of an effort, described in [25], to extend such layered networking
architectures, to support secure e-commerce applications. Figure 1 shows the five
IETF layers, together with two optional security sub-layers, and the four secure
e-commerce layers of [25].

Layered (or modular) architectures allow to specify, design, analyze, implement and test protocols for each layer, independently of protocols for other layers.
This is based on the paradigm of *lower layers abstraction*: when discussing and

R. Canetti (Ed.): TCC 2008, LNCS 4948, pp. 125–141, 2008.
© International Association for Cryptologic Research 2008

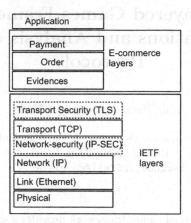

Fig. 1. IETF and e-commerce layers; (optional) security sub-layers marked with dotted contour

analyzing a protocol π_i for layer i, running in multiple nodes, we abstract the satisfactory behaviors of the lower layers by a single abstract *layer model* L_{i-1}, and the satisfactory behaviors of layer i into abstract layer model L_i. Protocol π_i *realizes layer model* L_i *over layer model* L_{i-1}, if the behavior of (multiple instances of) π_i running over layer model L_{i-1}, satisfies layer model L_i (except with negligible probability). We write this as: $\mathsf{L}_i \vdash \left[\mathsf{L}_{i-1}^{\pi_i} \right]$.

A pair of protocols π_i and π_{i-1}, of layers $i, i+1$, can be composed into a single protocol, which we denote as $\pi_{i||i-1}$. Our main result is the *fundamental lemma of layering*, showing that by composing protocols of multiple layers, we can implement a high-layer model directly over a low-layer model. Given layer models $\{\mathsf{L}_i\}_{i=0}^{l}$, and protocols $\pi_1, \ldots, \pi_l$, where $\mathsf{L}_i \vdash \left[\mathsf{L}_{i-1}^{\pi_i} \right]$ for $i = 1, \ldots, l$, their layered composition $\pi_{1||\ldots||l}$ implements L_l over L_0, i.e. $\mathsf{L}_l \vdash \left[\mathsf{L}_0^{\pi_{1||\ldots||l}} \right]$. This provides firm foundations to the security of modular and layered architectures, as in Figure 1.

For example, in [27] we define the *delivery evidences layer* model L_{DE}, and the lower *communication layer* model L_{Comm}; and we show a protocol π_{DE} s.t. $\mathsf{L}_{DE} \vdash \left[\mathsf{L}_{Comm}^{\pi_{DE}} \right]$. Similarly, in [26] we define the *orders layer* model L_{Orders}, and show protocol π_{Order} s.t. $\mathsf{L}_{Orders} \vdash \left[\mathsf{L}_{DE}^{\pi_{Order}} \right]$. Using the fundamental lemma of layering, the composite protocol $\pi_{DE||O}$ realizes the orders layer directly over the communication layer, i.e. $\mathsf{L}_{Orders} \vdash \left[\mathsf{L}_{Comm}^{\pi_{DE||O}} \right]$. This is illustrated in Figure 2, where we outline the games each of the protocols (π_{DE}, π_{Order} and their composition

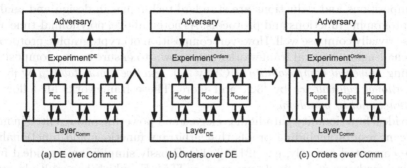

Fig. 2. Layering of realizations of the Order and Delivery Evidences (DE) layers

$\pi_{DE\|O}$, the two lower layers (Comm and DE), the two experiments protocols (DE and Orders), and the adversary protocol.

The layered games framework provides solid foundations to the accepted methodology, of using layered architectures (also called reference models), to specify, design, analyze, implement and test each layer independently. In spite of the extensive use of layered architectures, such foundations did exist prior to this work. For example, the IP (Internet Protocol) layer is essentially only required to provide a vaguely-described 'best effort' service. Existing proposals and standard of specifications of layers are only stated informally, often by partial-specification for the *operation* of the protocols, rather than to the *service* the higher layer can rely on. Composition of protocols is also used without formal definition or proof.

A possible explanation for the fact that layering was not yet based on formal foundations, in spite of its wide use, is the fact that similar compositions work as expected for many models, often trivially. For example, the composition of two polynomial time algorithms is trivially also a polynomial time algorithm. However, as [2] argue, composition properties require proof, and may not hold for all (natural) models. For example, the composition of two polynomial time interactive Turing machines (ITM), or of an (infinite) state machine with polynomial-time transition function, may not be polynomial-time, in the natural setting where the outputs of each machine is considered part of the inputs of the other. Indeed, in developing the layered games framework, we found that some definitional choices could have subtle but critical impact on composability. Details within.

Precise specifications of models for network layers can be hard to write and analyze, since they depend on many implementation and environment aspects. However, such rigorous specifications, and analysis, are critical, at least for security and cryptographic protocols, which must resist adversarial attacks. The layered games framework allows meaningful models, and analysis of implementations (protocols), using standard reduction techniques and composition of protocols (layers).

Compositions and reductions are standard techniques in design and analysis of cryptographic functions and protocols. As noted above, polynomial-time algorithms trivially compose well. However, composition of cryptographic protocols is more challenging. Several frameworks were shown to ensure secure composition, including *universal composability* (UC) by [14], *reactive simulatability* by [5, 34], *observational equivalence* by [32], and more. These frameworks all follow the *ideal functionality paradigm*.

The ideal functionality paradigm is elegant and powerful, and resulted in many significant results, including proofs that arbitrary functions and functionalities can be computed securely, e.g. [21, 12, 14]. Grossly simplifying, an 'ideal functionality' for layer i is a single program or ITM F_i, which has multiple copies of the interfaces to layer $i + 1$. Protocol π_i is considered secure, if executions of multiple copies of it over F_{i-1}, are *indistinguishable* from executions of F_i.

However, it may not always be feasible to define an ideal functionality capturing the possible behaviors of a realistic network layer. In fact, even defining the behaviors of each layer is challenging; transforming this into a program, would be impractical or impossible, and may result in over-specification. Note that over-specification of layers (or protocols) is usually considered harmful by practitioners, see e.g. [9].

This inability to use ideal functionalities as specifications for networking and e-commerce layer models, is our motivation in developing the layered games framework. The layered games framework allows protocol compositions with realistic specifications for network and e-commerce layer models, and with emphasis on simplicity and usability, even at some reduction in scope and generality.

As the name implies, the layered games framework is based on the *game playing paradigm*, instead of following the ideal functionality paradigm. The game playing paradigm is central to the theory of cryptography, see e.g. [21, 20]. Game playing supports strong analytical tools, e.g. [8], and may facilitate the use of (semi) automated proof-checking tools, see e.g. [24].

In the game-playing paradigm, one specifies an interactive game between a component and an adversary, where security is defined by the probability of the adversary winning in the game. With information-theoretic games the adversarial entity is allowed unbounded computational resources, while *concrete* and *probabilistic polynomial time* games assume certain limitations on adversarial resources, e.g. available time. Game-based specifications are widely used, and available for many cryptographic primitives such as digital signature and encryption schemes, pseudo-random functions, and much more, e.g., [22, 23, 20].

Some primitives have secure implementations for game-based specifications, where the corresponding ideal functionalities are not realizable, see [17, 11, 13]. This provides another motivation for investigating compositions of protocols satisfying game-playing specifications. However, our focus is different: allowing realistic models for network layers, without trying to define them as 'ideal functionality'.

Further related works. Our execution model is closely related to the execution models of I/O Automata of [33], especially the Probabilistic I/O Automata model of Canetti et al. [15], and to the Reactive Simulatability framework [5, 6, 35]. In an especially related work, Backes et al. [4] define a relaxed notion of *conditional reactive simulatability*, where simulation is required only if the environment fulfills some constraints; however, there are significant differences between the works, most notably their constraints are on the environment and not on the lower layers.

The layered games framework follows the *computational* approach to cryptography, which treats protocols and cryptographic schemes as programs/machines, operating on arbitrary stings (bits). This is in contrast to the *symbolic* approach, where cryptographic operations are seen as functions on a space of symbolic (formal) expressions, and security properties are stated as symbolic expressions; see [18, 10]. Several works investigate compositions of cryptographic protocols with the symbolic approach, e.g. Datta et al. [16] and Backes at al. [3]. We believe that it may be possible and beneficial, to extend the layered games framework to support symbolic/formal analysis, possibly building on recent results on the relationships between the two approaches, such as [1]. This may facilitate the use of verification tools; notice also that we use state machines as the basic computational model, which can also be helpful in applying verification tools.

Organization. In Section 2 we define protocols, configurations (of protocols), and executions (of configurations). In Section 3 we define layer games, models and realizations. In Section 4 we present and prove the fundamental lemma of layering. We conclude and discuss future work in Section 5.

For space limitations, the proof and detailed examples of applications of the framework are deferred to the full version of this paper [28]; see also [27, 26].

2   Protocols, Configurations and Executions

2.1   Protocols

Our basic element of computation is a *protocol*. We use protocols to model all the entities compromising the systems we investigate, including even adversarial entities ('the adversary'). Protocols are state machines[1] that accept input on one of few *input interfaces*, and produce output on one or more *output interfaces*. The transition function δ maps the input (interface and value), current state and random bits, to a new state and to outputs on the different output interfaces. We use $\perp$ to denote a special value which is not a binary string ($\perp \notin \{0,1\}^*$); a protocol outputs $\perp$ on some output interface to signal 'no output'.

---

[1] We use state machines, rather than e.g. ITM as in Universal Composability [14], since we found it simpler, and easier to ensure that an execution involving multiple protocols, some of which are adversarial, will have well-defined scheduling and distribution of events. Also, in many cases protocols may be represented by *finite* state machines, which may have advantages including possible use of automated verification tools.

The transition function δ can depend on two additional inputs: random bits and a security parameter. The random bits may be ignored to define deterministic protocols, including analysis of protocols using pseudo-random bits. The (unary) security parameter, allows to define computational properties of the protocol and of specifications, such as security against computationally-bounded adversary. Specifically, we use the security parameter to define a *polynomial* protocol.

Definition 1 (Protocol). *A protocol π is a tuple $\langle S, I_{IN}, I_{OUT}, \delta \rangle$ where:*

1. *S is a set of states, where $\bot \in S$ is the initial state,*
2. *I_{IN} is a set of input interface identifiers,*
3. *I_{OUT} is a set of output interface identifiers,*
4. *$\delta : IN \to OUT$ is a transition function, with:*
   - *Domain $IN = 1^* \times S \times I_{IN} \times \{0,1\}^* \times \{0,1\}^*$ (security parameter, current state, input interface, input value, random bits).*
   - *Range $OUT = S \times \prod_{i \in I_{OUT}} (\{0,1\}^* \cup \{\bot\})$. The outputs consist of a new state, denoted $\delta.S \in S$, and output values $\delta.ov[\iota] \in \{0,1\}^* \cup \{\bot\}$ for each interface $\iota \in I_{OUT}$.*

The protocol is polynomial *if δ is polynomial-time computable, and if the length of the outputs is the same as the length of the inputs[2], plus a polynomial in the security parameter, i.e. $\exists c \in \mathbb{N}$ s.t. $\forall (1^k, s, \iota_i, x, r) \in IN, \iota_o \in I_{OUT}$: $|\delta.ov[\iota_o](k, s, \iota_i, x, r)| \le |x| + |k|^c$.*

Notations

Π, Π_{poly}: Let Π denote the set of all protocols, and Π_{poly} denote the set of polynomial protocols.

Dot notation: the range of δ is a set of pairs $(s, ov[\iota])$, where $s \in S$ is the new state and $ov[\iota] \in \{0,1\}^* \cup \{\bot\}$ is the output on each output interface $\iota \in I_{OUT}$. To refer directly to the state or the outputs, we use dot notation as in $\delta.s(\cdot)$ and $\delta.ov[\iota](\cdot)$ respectively. We similarly use dot notation in other places, i.e. $\alpha.\beta$ refers to element β of a record or tuple α.

We can connect protocols, via their interfaces, in different *configurations*, as we define next. We can also connect from an output interface of a protocol, to an input interface of the same protocol; this makes it trivial to compose several protocols into a single protocol, which is useful (see Section 4). Note that if we

---

[2] This restriction of the output length to be the same as input length, plus some 'overhead' which depends only on the security parameter, is a simple method to prevent exponential blow-up in input and output lengths, as outputs of one protocol become inputs to another protocol during execution. This restriction is reasonable in practice, and sufficient for our needs; for example, it allows a protocol to 'duplicate' input from one interface, to multiple output interfaces, but maintains a polynomial bound on the length of the inputs and outputs on each interface during the execution. More elaborate ways to to prevent exponential blow-up were presented by [31] describing a general model for systems which satisfy certain acyclic conditions, [14] and [29] for UC, and [6] for reactive simulatability.

compose several polynomial protocols in this manner, then the resulting protocol is also polynomial.

2.2   Configuration

We study interactions of multiple protocols, connected via their interfaces; we call the set of interconnected protocols a *configuration*. Configuration are a *directed graph*, whose nodes P are identifiers for protocols, and whose edges are defined by mappings $p' = \mathsf{nP}(p, \iota)$ (for 'next protocol') and $\iota' = \mathsf{nI}(p, \iota)$ (for 'next interface'), mapping *output interface* $\iota \in \mathsf{oI}(p)$ of node p, to *input interface* $\iota' \in \mathsf{iI}(p')$ of node p'. Identification of the input and output interfaces, corresponds to the awareness of the network-layer, e.g. of router or firewall, to the identification of the network interface card on which a packet was received. For example, Figure 2, shows three (homomorphic) configurations. The definition follows.

Definition 2 (Configuration). *A configuration is a tuple* $C = \langle \mathsf{P}, \mathsf{iI}, \mathsf{oI}, \mathsf{nP}, \mathsf{nI} \rangle$, *where:*

P *is a set of protocol instance identifiers,*
iI, oI *map identifiers in* P *to input and output interfaces, respectively,*
nP *maps from instance identifier* $p \in \mathsf{P}$ *and an output interface* $\iota \in \mathsf{oI}(p)$, *to*
   $p' = \mathsf{nP}(p, \iota)$, *where either* $p' = \bot$ *or* $p' \in \mathsf{P}$ *(another instance),*
nI *maps from instance identifier* $p \in \mathsf{P}$ *and an output interface* $\iota \in \mathsf{oI}(p)$, *to*
   input interface ι', *where if* $\mathsf{nP}(p, \iota) \in \mathsf{P}$ *then* $\iota' \in \mathsf{iI}(\mathsf{nP}(p, \iota))$,

Above, we defined configurations without any 'size' parameter, as required e.g. to analyze protocols and distributed algorithms designed for networks with a variable number of parties (and where complexities may depend on the number of parties). This is for simplicity and to avoid clutter; the extensions to (uniform or non-uniform) 'configuration families' seem quite obvious. Notice that for many applications, e.g. in [27, 26], it may be sufficient to consider a small fixed set of parties.

   Still, configurations as defined above, are quite general. In particular, we intentionally avoided assuming any specific communication or synchronization mechanisms. This allows use of the framework in diverse scenarios, e.g. with or without assumptions on synchronization, communication and failures.

2.3   Executions

An *execution* is a sequence of events, each event corresponding to one transition of a protocol π running in one node $p \in \mathsf{P}$ inside a configuration $C = \langle \mathsf{P}, \mathsf{iI}, \mathsf{oI}, \mathsf{nP}, \mathsf{nI} \rangle$; to define the execution, we use a mapping $\pi = \Gamma(p)$ from the protocol identifiers P to the protocols realizing each node.

   An important design goal, is that the set of executions of a given configuration C, with a specific mapping to protocols Γ, would be a well-defined random variable. This makes it easier to use an execution as a 'subroutine', to facilitate reduction-based reasoning and proofs. To further simplify such reductions, we require that executions be a *deterministic* function of explicit random-tape inputs. Specifically, the ith event in the execution, denoted ξ_i, is defined by the (deterministic) transition function of the protocol $\Gamma(p_i)$ invoked at this event (where p_i is the identifier of that node). We allow the protocol to make random choices, but only using uniformly-selected random bits $R_i \in_R \{0,1\}^*$, provided as input to the transition function. Let $\mathbb{R} = \{R_i \equiv \{0,1\}^*\}_{i=1,2,\dots}$ be the sequence whose elements are the sets of all binary strings $\{0,1\}^*$; each execution is a deterministic function of the specific sequence $R \in \mathbb{R}$ used in that execution (i.e. $R = \{R_i\}_{i=1,2,\dots}$ s.t. $(\forall i)R_i = \{0,1\}^*$).

Each protocol instance has its own state, and in each round may decide to invoke interfaces of multiple other protocol instances; see for example the configurations in Figure 2. Therefore, some scheduling mechanism for events is required. To ensure well-defined executions, without any non-deterministic choice (except for the explicit use of the random input strings $R \in \mathbb{R}$), we use a deterministic *schedule* $\mathcal{S}$ (cf. [15]).

A schedule $\mathcal{S}$ of configuration $C = \langle \text{P, il, ol, nP, nl} \rangle$, is a sequence of pairs $\mathcal{S} = \{\langle p_i, \iota_i \rangle\}_{i \in \mathbb{N}}$ where $p_i \in \text{P}$. We (later) require protocols to perform correctly for *any* schedule, therefore, the schedule can be considered as adversarial (and not even limited by computational assumptions). On the other hand, the schedule, is defined in advance and cannot depend on the execution (or on the random bits $R \in \mathbb{R}$); in a sense, we separated the adversarial mechanisms into a non-adaptive, computationally-unlimited element (the schedule), and an adaptive, usually computationally-limited element (modeled as a protocol, or multiple protocols, in the configuration, and aware of only inputs on its interfaces). A schedule could, of course, prevent events from happening; to prevent this from being a trivial method to cause executions where the adversary wins, our definitions of games (later) consider the adversary as winning only if some event happens, rather than by the absence of some event.

A similar issue, where we tried to avoid non-determinism, involves how we handle multiple pending inputs, submitted on the same input interface. Our definition delivers inputs on an interface, in the order in which they were submitted. We do this by keeping a *FIFO queue* $Q[p, \iota]$, for protocol instance p and input interface ι, with regular semantics for the *enqueue, dequeue,* and *is_non_empty* operations. Other choices may be possible.

Definition 3 (Execution). *Let* $C = \langle \text{P, il, ol, nP, nl} \rangle$ *be a configuration. Let* $\mathcal{S} = \{\langle p_i \in \text{P}, \iota_i \in \text{il}(p_i) \rangle\}_{i \in \mathbb{N}}$ *be a schedule of* C. *Let* $\Gamma : \text{P} \to \Pi$ *be a mapping of the protocol identifiers* P *to specific protocols.*

The execution $X_k(C, \Gamma, \mathcal{S}; R)$ *of security parameter* $k \in 1^*$, *configuration* C, *protocol mapping* Γ, *schedule* $\mathcal{S}$ *and sequence (of random bits)* $R = \{R_i\} \in \mathbb{R}$, *is the sequence of execution events* $\{\xi_i\} = \{\langle p_i \in \text{P}, \iota_i \in \text{il}(p_i), iv_i, ov_i[\cdot] \rangle\}$ *resulting from the following process:*

FOR ALL $p \in \mathsf{P}$: $s[p] := \bot$;
$Q[p_1, \iota_1].\text{ENQUEUE}(0)$; $X := \{\}$

FOR $i := 1$ TO ∞ DO:

IF $(p_i \in \mathsf{P}, \iota_i \in I_{IN}(p_i)$ AND $Q[p_i, \iota_i].\text{IS\_NON\_EMPTY}())$ THEN:
1. $iv_i := Q[p_i, \iota_i].\text{DEQUEUE}()$;
2. $\langle S, I_{IN}, I_{OUT}, \delta \rangle := \Gamma(p_i)$.
3. $\langle s[p_i], ov_i[\iota \in I_{OUT}] \rangle := \delta(k, s[p_i], \iota_i, iv_i; R_i)$;
4. $\forall \iota \in I_{OUT}$: IF $ov_i[\iota] \neq \bot$
    THEN: $Q[\mathsf{nP}(p_i, \iota), \mathsf{nl}(p_i, \iota].\text{ENQUEUE}(ov_i[\iota])$;

Let $X_k(C, \Gamma, \mathcal{S})$ be the random variable $X_k(C, \Gamma, \mathcal{S}; R)$ for $R \in_R \mathbb{R}$.

If all protocols in the range of Γ are *polynomial*, we say that Γ is *polynomial*. If Γ is polynomial, then $X_k(C, \Gamma, \mathcal{S})[l]$ is sampleable in time polynomial in k and l, where $X_k(C, \Gamma, \mathcal{S})[l]$ denotes the l first events of $X_k(C, \Gamma, \mathcal{S})$. This allows a polynomial protocol to run polynomial number of steps of an execution containing polynomial protocols, as part of its computational process (e.g. for reduction proofs). We restate this observation in the following proposition.

Proposition 1 (Executions of polynomial protocols are efficiently sampleable). *Let $C = \langle \mathsf{P}, \mathsf{il}, \mathsf{ol}, \mathsf{nP}, \mathsf{nl} \rangle$ be a configuration and $\Gamma : \mathsf{P} \to \Pi_{\mathsf{poly}}$ be a mapping of the protocol identifiers P to specific polynomial protocols. Then $X_k(C, \Gamma, \mathcal{S})[l]$ is sampleable in probabilistic polynomial time (as a function of k and l).*

3 Layer Games, Models and Realizations

From this section, our discussion is focused, for simplicity, on *layered architectures*, as in Figure 1. We believe that it is not too difficult to generalize our concepts and results, but that this will cause (mostly technical) complexities, that may make the resulting definitions less easy to understand and use.

The basic idea of layered architectures, is *abstraction*. Namely, the designer of protocol π_i for layer i, is oblivious to details of lower layers, and only cares about the *layer model* of layer $i - 1$, denoted L_{i-1}. The layer model L_{i-1} defines all possible behaviors observable to layer i, resulting from the operation of layer $i - 1$ protocols and of all lower layers. The goal of the designer of protocol π_i, for layer i, is to ensure that when instances of π_i operate over any instantiation of Γ_{i-1} of layer model L_{i-1}, the resulting operation satisfies layer model L_i.

In the first subsection below, we give a game-based definition of a *layer model*, with conditions on the outcomes of the game, defining when a protocol Γ_L is considered to satisfy layer model L; we denote this by $\mathsf{L} \models \Gamma_\mathsf{L}$. In the second subsection, we define the *realization* relation, denoted $\mathsf{L}_U \vdash \begin{bmatrix} \pi_U \\ \mathsf{L}_L \end{bmatrix}$, indicating that protocol π_U, when running over lower layer L_L, realizes layer model L_U.

3.1   Layer Models

We define the layer model L, by a simple zero-sum (win-lose) game between an *adversary protocol*, with identifier A, and a *layer protocol*, with identifier I_L. These protocols interact only via a third protocol, the *experiment protocol*, with identifier Exp, as shown in Figure 3. The experiment protocol defines the 'rules of the game', and in particular the outcome, which Exp produces on a designated output interface outcome. Specifically, in every execution, Exp outputs a value on outcome (at most) once, and this value is a single bit: 1 if the adversary wins (protocol failed the game), and 0 if the adversary losses (protocol passed the game). The game includes an *expected winning rate* $\alpha \in [0,1]$ (typically $\alpha = 0$ or $\alpha = \frac{1}{2}$), defining the expected (or permitted) probability that the adversary will win, i.e. eventually have 1 on outcome.

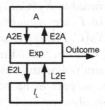

Fig. 3. Layer Model Configuration. If for every Γ_A holds Pr(outcome = 1) $\le \alpha +$ +negl(k), then the layer protocol Γ_L satisfies L = (Γ_{Exp}, α), or: L $\models \Gamma_L$.

We later implement layer i over layer $i - 1$, by multiple instances of protocol π_i, one in each processor in the network. For simplicity, we assume a constant number of instances n; it seems straightforward to extend the results to allow n to be a parameter. It is convenient to define a separate input and output interfaces between the experiment and each instance. Namely, for $j \in \{1, \ldots, n\}$, the configuration includes interface E2L$_j$ from Exp to I_L, and interface L2E$_j$ from I_L to Exp. Finally, we use a single interface E2A from Exp to A, and a single interface A2E from A to Exp. This completes the definition of the *layer modeling game configuration* C_{LM} (for some constant number n of instances).

For $\phi \in \{Exp, A\}$, let $\Gamma(\phi) = \Gamma_\phi$ be the protocol instantiating node ϕ; similarly, let $\Gamma(I_L) = \Gamma_L$ be a protocol realizing I_L. Given schedule S, let $Exp_{\Gamma_A, \Gamma_L, S}^{\Gamma_{Exp}}(k, l; R)$ denote the output of outcome after l events in the execution $X_k(C_{LM}, \Gamma, S; R)$, for $R \in \mathbb{R}$, or $\perp$ if there was no such output.

Definition 4 (Layer model). *A (polynomial) layer model is a pair* L = (Γ_{Exp}, α), *where* Γ_{Exp} *is a (polynomial) protocol and* $\alpha \in [0, 1]$. *We say that protocol* $\Gamma_L \in \Pi_{poly}$ *computationally satisfies layer model* L, *and write* L $\models_{poly} \Gamma_L$, *if for every* $\Gamma_A \in \Pi_{poly}$, *schedule* S, *polynomial* l *and large enough* k, *holds:*

$$\Pr_{R \in \mathbb{R}} \left(Exp_{\Gamma_A, \Gamma_L, S}^{\Gamma_{Exp}}(k, l(k); R) = 1 \right) \le \alpha + negl(k)$$

where **negl** *is some negligible function (asymptotically smaller than any strictly positive polynomial), and* $\mathsf{Exp}_{\Gamma_A,\Gamma_L,\mathcal{S}}^{\Gamma_{\mathsf{Exp}}}(k,l;R)$ *is defined as above.*

Protocol Γ_L statistically satisfies L, *if the above holds when protocols are not required to be polynomial, and perfectly satisfies* L *if this holds even when we remove the* **negl**(k) *term. These notions are denoted* $L \models_{\mathsf{stat}} \Gamma_L$ *and* $L \models_{\mathsf{perf}} \Gamma_L$, *respectively.*

We observe the trivial relation among the three notions of satisfaction.

Proposition 2. *For any layer model* L *and any protocol* Γ_L *holds:*

$$L \models_{\mathsf{perf}} \Gamma_L \Rightarrow L \models_{\mathsf{stat}} \Gamma_L \Rightarrow L \models_{\mathsf{poly}} \Gamma_L$$

Notation: we may write $L \models \Gamma_L$, when it is obvious that we refer to $\models_{\mathsf{poly}}$.

3.2 Layer Realization Indistinguishability Game

We now define and investigate another game, which we call *indistinguishable layer realization games*, which is similar to indistinguishability games used in many cryptographic definitions, e.g. pseudo-random functions [19], and especially to the 'left-or-right indistinguishability' (LOR) of [7]. Layer realization games are convenient for the common layered and modular ('top-down') design methodologies. As in previous sections, we had to tradeoff generality for simplicity and ease-of-use.

The configuration of layer realization indistinguishability games is illustrated in Figure 4. Like in layer model games, the configuration contains nodes A, Exp and I_L, where A and I_L are connected only via Exp. There are $n + 1$ additional nodes, where n is the (constant) number of instances: n *realization nodes* (instances) $\{R_j\}_{j=1,\dots,n}$, and one *lower layer node* I_{LL}.

As in the layer model games, without loss of generality, we use a single input and output interface from the experiment (or 'higher layer') to each instance in I_L, and therefore we will have the interfaces E2L$_j$, L2E$_j$, E2A and A2E as before. The configuration also includes interfaces E2R$_j$, R2E$_j$, R2L$_j$ and L2R$_j$, connecting between Exp and R, and between R and I_{LL}. This completes the definition of the *layer realization configuration* C_{LR} (for a fixed number n of instances).

All the realization nodes are instantiated by (mapped to) the same protocol π, which is tested for realization of layer L over lower layer LL. Namely, $(\forall j \in \{1, \dots, n\})\Gamma(R_j) = \pi$, where Γ is the mapping we will use in the execution of the game (with n instances).

In layer realization indistinguishability games, we use a specific experiment protocol $\mathsf{Exp}^{\mathsf{IND}}$, which we define below, i.e. $\Gamma(\mathsf{Exp}) = \mathsf{Exp}^{\mathsf{IND}}$. Here are some basic details about $\mathsf{Exp}^{\mathsf{IND}}$. Upon initialization, $\mathsf{Exp}^{\mathsf{IND}}$ flips a fair coin $b \in_R \{L, R\}$, where L stands for either Layer or Left, and R stands for either Realization or Right. The game ends when $\mathsf{Exp}^{\mathsf{IND}}$ receives a guess b' of either L or R from the adversary A, which arrives on a dedicated Guess input interface. Upon receiving the guess b', $\mathsf{Exp}^{\mathsf{IND}}$ outputs on its outcome output interface 1 if $b = b'$, and 0 otherwise.

Given adversary protocol $\Gamma(\mathsf{A}) = \Gamma_\mathsf{A}$, protocols for the two layers $\Gamma(I_\mathsf{L}) = \Gamma_\mathsf{L}$, $\Gamma(I_\mathsf{LL}) = \Gamma_\mathsf{LL}$, sequence of random bit sequences $R \in \mathbb{R}$ and schedule $\mathcal{S}$, let $\mathsf{Exp}^{\mathsf{IND}}{}_{\Gamma_\mathsf{A},\Gamma_\mathsf{L},\Gamma_\mathsf{LL},\pi,\mathcal{S}}(k,l;R)$ denote the output of outcome after l events in the execution $X_k(C_{LR}, \Gamma, \mathcal{S}; R)$, or $\bot$ if there was no such output.

Definition 5 (Layer realization). *Let* L, LL *be two polynomial layer models. Protocol* π *computationally realizes* layer model L *over layer model* LL, *which we denote by* $\mathsf{L} \vdash_{\mathsf{poly}} \left[\begin{smallmatrix} \pi \\ \mathsf{LL} \end{smallmatrix} \right]$, *if for every polynomial algorithm* Γ_LL *s.t.* $\mathsf{LL} \models \Gamma_\mathsf{LL}$, *there exists a polynomial algorithm* Γ_L *s.t.* $\mathsf{L} \models \Gamma_\mathsf{L}$, *s.t. every polynomial algorithm* Γ_A *and for every schedule* $\mathcal{S}$ *and every polynomial* l, *for sufficiently large* k *holds*

$$\Pr_{R \in \mathbb{R}} \left(\mathsf{Exp}^{\mathsf{IND}}{}_{\Gamma_\mathsf{A},\Gamma_\mathsf{L},\Gamma_\mathsf{LL},\pi,\mathcal{S}}(k, l(k); R) = 1 \right) \leq \frac{1}{2} + \mathsf{negl}(k)$$

Protocol π *statistically realizes* layer model L *over layer model* LL, *which we denote by* $\mathsf{L} \vdash_{\mathsf{stat}} \left[\begin{smallmatrix} \pi \\ \mathsf{LL} \end{smallmatrix} \right]$, *if the above holds when protocols are not required to be polynomial, and* perfectly realizes L *over* LL, *which we denote by* $\mathsf{L} \vdash_{\mathsf{perf}} \left[\begin{smallmatrix} \pi \\ \mathsf{LL} \end{smallmatrix} \right]$ *if this holds even when we remove the* $\mathsf{negl}(k)$ *term.*

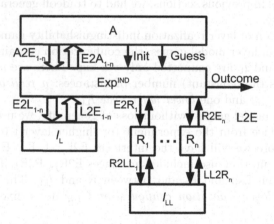

Fig. 4. The Layer Realization Indistinguishability game. Protocol π realizes layer L over layer LL, if for every adversary Γ_A and every lower-layer protocol Γ_LL, there is some protocol Γ_L satisfying layer model L, s.t. the adversary cannot distinguish between Γ_L and between the composition of n instances of π over Γ_LL.

In summary, protocol π *realizes* layer model Lover layer model LL, if for every adversary protocol Γ_A and every lower-layer protocol Γ_LL, there is some protocol Γ_L satisfying layer L, s.t. the Γ_A cannot distinguish between interacting with Γ_L and interacting with π operating over Γ_LL, where Γ_A interacts only via $\mathsf{Exp}^{\mathsf{IND}}$. Intuitively, $\left[\begin{smallmatrix} \pi \\ \Gamma_\mathsf{LL} \end{smallmatrix} \right]$ is a good implementation of L, if the adversary A cannot distinguish between it and between some protocol Γ_L which satisfies L, when interacting via $\mathsf{Exp}^{\mathsf{IND}}$, better than the trivial winning rate of $\frac{1}{2}$. To complete the description, we now present the *indistinguishability experiment* $\mathsf{Exp}^{\mathsf{IND}}$.

Definition 6 (Layer realization indistinguishability experiment). *Let* $\mathsf{Exp}^{\mathsf{IND}} = \langle S, I_{IN}, I_{OUT}, \delta \rangle$ *be the following protocol:*

$S = \{\bot, \mathsf{testing}, \mathsf{done}\}$
$I_{IN} = \{\mathsf{Init}, \mathsf{Guess}\} \cup \{\mathsf{A2E}_j\}_{j=1,\ldots,n} \cup \{\mathsf{L2E}_j\}_{j=1,\ldots,n} \cup \{\mathsf{R2E}_j\}_{j=1,\ldots,n}$
$I_{OUT} = \{\mathsf{outcome}\} \cup \{\mathsf{E2A}_j\}_{j=1,\ldots,n} \cup \{\mathsf{E2L}_j\}_{j=1,\ldots,n} \cup \{\mathsf{E2R}_j\}_{j=1,\ldots,n}$
δ:

　　1. In initialization state $\bot$, *upon any input, select randomly* $b \in_R \{\mathsf{L}, \mathsf{R}\}$, *and move to* testing *state.*

　　2. In testing *state, pass all input events on interface* $\mathsf{A2E}_i$, *for* $i \in \{1, \ldots, n\}$, *to corresponding output event on output interface* $\mathsf{E2L}_i$ *(if* $b = \mathsf{L}$*) or* $\mathsf{E2R}_i$ *(if* $b = \mathsf{R}$*), and all input events on interfaces* $\mathsf{L2E}_i$ *(if* $b = \mathsf{L}$*) or* $\mathsf{R2E}_i$ *(if* $b = \mathsf{R}$*), to corresponding output events on interface* $\mathsf{E2A}_i$.

　　3. When, in testing *state, the guess input interface* Guess *is invoked with input (guess)* $b' \in \{\mathsf{L}, \mathsf{R}\}$, *output on* outcome *the value 1 if* $b = b'$, *and 0 otherwise* $(b \neq b')$. *Move to the* done *state (and ignores all further inputs).*

4　The Fundamental Lemma of Layering

We now show the fundamental lemma of layering, allowing compositions of protocols of multiple layers. This provides firm foundations to the accepted methodology of designing, implementing, analyzing and testing of each layer independently, yet relying on their composition to ensure expected properties.

　　We first need to define layering of *protocols*. We actually consider two different variants of protocol layering:

- Layering of two realization protocols π_L, π_{LL}. As discussed, we assumed (for simplicity) that there are n instantiations of the realization protocol of each layer; each of these has two input interfaces and two output interfaces, one for the higher layer and one for the lower layer. We define $\pi_{LL||L} = \begin{bmatrix} \pi_L \\ \pi_{LL} \end{bmatrix}$ in the obvious way.

- Layering of the n instances of the realization protocol π_L, on top of a protocol realizing the lower-layer model Γ_{LL}. We define $\Gamma_{LL||L} = \begin{bmatrix} \pi_L \\ \Gamma_{LL} \end{bmatrix}$ in the obvious way.

Note our convention of using π_x for protocols instantiating realizations (of n instances), and Λ_x for instantiations of a (lower) layer model. Also, note that if π_L and π_{LL} (or Γ_{LL}) are polynomial, then $\Gamma_{LL||L}$ is also polynomial.

　　We first present the *'composition preserves satisfaction' lemma*, which justifies considering abstraction of all lower layers, into a single 'virtual protocol'. For both this and the fundamental lemma of layering (below), we present only the computational version (the statistical and perfect versions are similar).

Lemma 1 (Composition preserves satisfaction). *Let* L, LL *be two polynomial layer models, and* $\pi_\mathsf{L}, \Gamma_\mathsf{LL}$ *be polynomial protocols, such that* π_L *computationally realizes* L *over* LL*, namely* $\mathsf{L} \vdash_\mathsf{poly} \left[\begin{smallmatrix}\pi_\mathsf{L}\\\mathsf{LL}\end{smallmatrix}\right]$*, and and* Γ_LL *computationally satisfies* LL*, namely* $\mathsf{LL} \models_\mathsf{poly} \Gamma_\mathsf{LL}$*. Then the composite protocol* $\Gamma_{\mathsf{LL}||\mathsf{L}}$ *satisfies* L*, namely* $\mathsf{L} \models_\mathsf{poly} \Gamma_{\mathsf{LL}||\mathsf{L}}$*. Or, as a formula:*

$$\left(\mathsf{L} \vdash_\mathsf{poly} \left[\begin{smallmatrix}\pi_\mathsf{L}\\\mathsf{LL}\end{smallmatrix}\right]\right) \bigwedge (\mathsf{LL} \models_\mathsf{poly} \Gamma_\mathsf{LL}) \Rightarrow (\mathsf{L} \models_\mathsf{poly} \Gamma_{\mathsf{LL}||\mathsf{L}})$$

The *composite realization* lemma shows that we can prove realization of each layer separately, and the composition of the realizations will be a realization of the highest layer over the lowest layer. We state the lemma for only three layers - generalization for an arbitrary stack is immediate.

Lemma 2 (The Fundamental Lemma of Layering). *Let* $\mathsf{L}_3, \mathsf{L}_2, \mathsf{L}_1$ *be three polynomial layer models, and* π_2, π_3 *be polynomial protocols, such that* π_3 *computationally realizes* L_3 *over* L_2*, and* π_2 *computationally realizes* L_2 *over* L_1*. Then* $\pi_{2||3} = \left[\begin{smallmatrix}\pi_3\\\pi_2\end{smallmatrix}\right]$ *computationally realizes* L_3 *over* L_1*.*

Furthermore, let Γ_{L_1} *be a polynomial protocol that computationally satisfies* L_1*, namely* $\mathsf{L}_1 \models_\mathsf{poly} \Gamma_{\mathsf{L}_1}$*. Then* $\Gamma_{1||2||3} = \left[\begin{smallmatrix}\pi_{2||3}\\\Gamma_1\end{smallmatrix}\right]$ *satisfies* L_3*, i.e.* $\mathsf{L}_3 \models_\mathsf{poly} \Gamma_{1||2||3}$*.*

5   Conclusions and Research Directions

In this work, we try to lay solid, rigorous foundations, to the important methodology of layered decomposition of distributed systems and network protocols, particularly concerning security in adversarial settings. The framework is built on previous works on modeling and analysis of (secure) distributed systems, as described in the introduction, but it is clearly a very ambitious goal, possibly overambitious, and certainly beyond the reach of a single publication. There are many directions that require further research. Here are some:

- The best way to test and improve such a framework, is simply by using it to analyze different problems and protocols; there are many interesting and important problems, that can benefit from such analysis. As one important example, consider the *secure channel layer* problem. Many protocols and applications assume they operate over 'secure, reliable connections'. In practice, this is often done using the standard layers in Figure 1, in one of two methods. In the first method, we use TLS (for security) over TCP (for reliability) over the 'best effort' service of IP. In the second method, we use TCP (for reliability) over IP-Sec (for security), again over 'best effort' (IP). It would be interesting to define a 'secure, reliable connection' layer, and to analyze these two methods with respect to it.
- There are many desirable extensions to the framework, including: support for corruptions of nodes, including adaptive and/or mobile corruptions (proactive security and forward security); adaptive control of the number of nodes; support for side channels such as timing and power.

- In this work, we focused on layered configurations. These are sufficient for many scenarios. However, there are other scenarios. It would be interesting to identify important non-layered scenarios, and find appropriate games, specifications and composition properties, which will support them, possibly as generalizations of our definitions and results.
- It would be interested to explore the relationships between the layered games framework, and other formal frameworks for study of distributed algorithms and protocols (see introdcution).
- The framework is based on the computational approach to security, where attackers can compute arbitrary functions on information available to it (e.g. ciphertext). Many results and tools are based on symbolic analysis, see introduction (and [18, 10, 1]). It can be very useful to find how to apply such techniques and tools, within the framework.

Acknowledgments

We would like to thank Yehuda Lindell, Ran Canetti, Dominique Unruh, Alejandro Hevia, Mark Manulis and Dennis Hofheinz for interesting discussions and helpful comments.

References

[1] Abadi, Rogaway: Reconciling two views of cryptography (the computational soundness of formal encryption). JCRYPTOL: Journal of Cryptology 15 (2002)
[2] Abadi, M., Lamport, L.: Composing specifications. ACM Trans. Program. Lang. Syst. 15(1), 73–132 (1993)
[3] Backes, Datta, Derek, Mitchell, Turuani: Compositional analysis of contract-signing protocols. TCS: Theoretical Computer Science 367 (2006)
[4] Backes, M., Dürmuth, M., Hofheinz, D., Küsters, R.: Conditional Reactive Simulatability. In: Gollmann, D., Meier, J., Sabelfeld, A. (eds.) ESORICS 2006. LNCS, vol. 4189, pp. 424–443. Springer, Heidelberg (2006)
[5] Backes, M., Pfitzmann, B., Waidner, M.: A General Composition Theorem for Secure Reactive Systems. In: Naor, M. (ed.) TCC 2004. LNCS, vol. 2951, pp. 336–354. Springer, Heidelberg (2004)
[6] Backes, M., Pfitzmann, B., Waidner, M.: Secure Asynchronous Reactive Systems. Cryptology ePrint Archive, Report, 2004/082 (2004), http://eprint.iacr.org/
[7] Bellare, M., Desai, A., Jokipii, E., Rogaway, P.: A concrete security treatment of symmetric encryption. In: Proceedings of the 38th Annual Symposium on Foundations of Computer Science (FOCS 1997), October 20–22, IEEE Computer Society Press, Los Alamitos (1997)
[8] Bellare, M., Rogaway, P.: The security of triple encryption and a framework for code-based game-playing proofs. In: Vaudenay, S. (ed.) EUROCRYPT 2006. LNCS, vol. 4004, pp. 3–540. Springer, Heidelberg (2006), http://dx.doi.org/10.1007/11761679_25
[9] Bradner, S.: Key words for use in RFCs to Indicate Requirement Levels. RFC (Best Current Practice) (March 1997), http://www.ietf.org/rfc/rfc2119.txt

[10] Burrows, Abadi, Needham: A logic of authentication. ACMTCS: ACM Transactions on Computer Systems 8 (1990)

[11] Canetti, Kushilevitz, Lindell: On the limitations of universally composable two-party computation without set-up assumptions. In: JCRYPTOL: Journal of Cryptology, 19th edn. (2006)

[12] Canetti, R.: Security and Composition of Multiparty Cryptographic Protocols. Journal of Cryptology 13(1), 143–202 (2000)

[13] Canetti, R., Fischlin, M.: Universally Composable Commitments. In: Kilian, J. (ed.) CRYPTO 2001. LNCS, vol. 2139, pp. 19–40. Springer, Heidelberg (2001)

[14] Canetti, R.: Universally Composable Security: A New Paradigm for Cryptographic Protocols. In: IEEE Symposium on Foundations of Computer Science, pp. 136–145 (2001) updated version: Cryptology ePrint Archive, Report 2000/067

[15] Canetti, R., Cheung, L., Kaynar, D.K., Liskov, M., Lynch, N.A., Pereira, O., Segala, R.: Time-bounded task-PIOAs: A framework for analyzing security protocols. In: Dolev, S. (ed.) DISC 2006. LNCS, vol. 4167, pp. 3–540. Springer, Heidelberg (2006), http://dx.doi.org/10.1007/11864219_17

[16] Datta, A., Derek, A., Mitchell, J.C., Pavlovic, D.: A derivation system and compositional logic for security protocols. J. Comput. Secur. 13(3), 423–482 (2005)

[17] Datta, A., Derek, A., Mitchell, J.C., Ramanathan, A., Scedrov, A.: Games and the impossibility of realizable ideal functionality. In: Halevi, S., Rabin, T. (eds.) TCC 2006. LNCS, vol. 3876, pp. 360–379. Springer, Heidelberg (2006)

[18] Dolev, D., Yao, A.: On the security of public key protocols. IEEE Transactions on Information Theory 29(2), 198–208 (1983)

[19] Goldreich, Goldwasser, Micali: How to construct random functions. JACM: Journal of the ACM 33 (1986)

[20] Goldreich, O.: Foundations of Cryptography. Basic Applications, vol. 2. Cambridge University Press, New York (2004)

[21] Goldreich, O., Micali, S., Wigderson, A.: How to play any mental game or A completeness theorem for protocols with honest majority. In: STOC, pp. 218–229. ACM, New York (1987)

[22] Goldwasser, S., Micali, S.: Probabilistic encryption & how to play mental poker keeping secret all partial information. In: STOC 1982: Proceedings of the fourteenth annual ACM symposium on Theory of computing, pp. 365–377. ACM Press, New York, USA (1982)

[23] Goldwasser, S., Micali, S., Yao, A.: Strong signature schemes. In: STOC 1983: Proceedings of the fifteenth annual ACM symposium on Theory of computing, pp. 431–439. ACM Press, New York, USA (1983)

[24] Halevi, S.: A plausible approach to computer-aided cryptographic proofs. Report, 2005/181, Cryptology ePrint Archive (June 2005), http://eprint.iacr.org/2005/181.pdf

[25] Herzberg, A., Yoffe, I.: Layered Architecture for Secure E-Commerce Applications. In: SECRYPT 2006 - International Conference on Security and Cryptography, pp. 118–125. INSTICC Press (2006)

[26] Herzberg, A., Yoffe, I.: On Secure Orders in the Presence of Faults. In: De Prisco, R., Yung, M. (eds.) SCN 2006. LNCS, vol. 4116, pp. 126–140. Springer, Heidelberg (2006) New version: Foundations of Secure E-Commerce: The Order Layer, in Cryptology ePrint Archive, Report 2006/352.

[27] Herzberg, A., Yoffe, I.: The delivery and evidences layer. Cryptology ePrint Archive, Report 2007/139 (2007), http://eprint.iacr.org/

[28] Herzberg, A., Yoffe, I.: Layered specifications, design and analysis of security protocols. Cryptology ePrint Archive, Report 2006/398 (2006)

[29] Hofheinz, D., Müller-Quade, J., Unruh, D.: Polynomial Runtime in Simulatability Definitions. In: CSFW 2005: Proceedings of the 18th IEEE Computer Security Foundations Workshop (CSFW 2005), Washington, DC, USA, pp. 156–169. IEEE Computer Society, Los Alamitos (2005)

[30] Kurose, J.F., Ross, K.W.: Computer networking: a top-down approach featuring the Internet. Addison-Wesley, Reading (2003)

[31] Küsters, R.: Simulation-Based Security with Inexhaustible Interactive Turing Machines. In: CSFW 2006: Proceedings of the 19th IEEE Workshop on Computer Security Foundations, Washington, DC, USA, pp. 309–320. IEEE Computer Society Press, Los Alamitos (2006)

[32] Lincoln, P., Mitchell, J., Mitchell, M., Scedrov, A.: A probabilistic poly-time framework for protocol analysis. In: CCS 1998: Proceedings of the 5th ACM conference on Computer and communications security, pp. 112–121. ACM Press, New York (1998)

[33] Lynch, N.A., Tuttle, M.R.: Hierarchical correctness proofs for distributed algorithms. In: PODC 1987: Proceedings of the sixth annual ACM Symposium on Principles of distributed computing, pp. 137–151. ACM Press, New York (1987)

[34] Pfitzmann, B., Waidner, M.: Composition and integrity preservation of secure reactive systems. In: CCS 2000: Proceedings of the 7th ACM conference on Computer and communications security, pp. 245–254. ACM Press, New York (2000)

[35] Pfitzmann, B., Waidner, M.: A Model for Asynchronous Reactive Systems and its Application to Secure Message Transmission. In: SP 2001: Proceedings of the 2001 IEEE Symposium on Security and Privacy, Washington, DC, USA, pp. 184–200. IEEE Computer Society Press, Los Alamitos (2001)

Universally Composable Multi-party Computation with an Unreliable Common Reference String*

Vipul Goyal[1],** and Jonathan Katz[2],***

[1] Department of Computer Science, UCLA
vipul@cs.ucla.edu
[2] Department of Computer Science, University of Maryland
jkatz@cs.umd.edu

Abstract. Universally composable (UC) multi-party computation has been studied in two settings. When a majority of parties are honest, UC multi-party computation is possible without any assumptions. Without a majority of honest parties, UC multi-party computation is impossible in the plain model, but feasibility results have been obtained in various augmented models. The most popular such model posits a *common reference string* (CRS) available to parties executing the protocol.

In either of the above settings, some *assumption* regarding the protocol execution is made: i.e., that many parties are honest in the first case, or that a legitimately-chosen string is available in the second. If this assumption is incorrect then all security is lost.

A natural question is whether it is possible to design protocols secure if *either one* of these assumptions holds, i.e., a protocol which is secure if *either* at most s players are dishonest *or* if up to $t > s$ players are dishonest but the CRS is chosen in the prescribed manner. We show that such protocols exist if and only if $s + t < n$.

1 Introduction

Protocols proven to satisfy the definition of *universal composability* [5] offer strong and desirable security guarantees. Informally speaking, such protocols remain secure even when executed concurrently with arbitrary other protocols running in some larger network, and can be used as sub-routines of larger protocols in a modular fashion.

Universally composable (UC) multi-party computation of arbitrary functionalities has been investigated in two settings. When a majority of the parties running a protocol are assumed to be honest, UC computation of arbitrary functionalities is possible without any cryptographic assumptions. (This is claimed

---

* This work was done in part while the authors were visiting IPAM.
** Research supported in part by NSF ITR and Cybertrust programs (including grants #0430254, #0627781, #0456717, and #0205594).
*** Research supported in part by the U.S. Army Research Laboratory, NSF CAREER award #0447075, and US-Israel Binational Science Foundation grant #2004240.

R. Canetti (Ed.): TCC 2008, LNCS 4948, pp. 142–154, 2008.
© International Association for Cryptologic Research 2008

in [5], building on [3,17].) This result holds in the so-called "plain model" which assumes only pairwise private and authenticated channels between each pair of parties. (A broadcast channel or a PKI are not needed [10], since fairness and output delivery are not guaranteed in the UC framework.)

In contrast, when the honest players *cannot* be assumed to be in the majority, it is known that UC computation of general functions is not possible in the plain model regardless of any cryptographic assumptions made. Canetti and Fischlin [7] showed the impossibility of two-party protocols for commitment and zero knowledge, and Canetti, Kushilevitz, and Lindell [8] ruled out UC two-party computation of a wide class of functionalities.

To circumvent these far-reaching impossibility results, researchers have investigated various *augmented* models in which UC computation without honest majority might be realizable [5,7,9,1,12,6,15]. The most widely-used of these augmented models is the one originally suggested by Canetti and Fischlin [7], in which a *common reference string* (CRS) is assumed to be available to all parties running a given execution of a protocol. (The use of a common reference string in cryptographic protocols has a long history that can be traced back to [4].) Canetti and Fischlin show that UC commitments and zero knowledge are possible in the two-party setting when a CRS is available, and later work of Canetti et al. [9] shows that (under suitable cryptographic assumptions) a CRS suffices for UC multi-party computation of arbitrary functionalities.

In summary, there are two types of what we might term "assumptions about the world" under which UC multi-party computation is known to be possible:

- When a strict minority of players are dishonest.
- When an arbitrary number of players may be dishonest, but a trusted CRS (or some other setup assumption) is available.

Our contribution. Known protocols designed under one of the assumptions listed above are *completely insecure* in case the assumption turns out to be false. For example, the BGW protocol [3] — which is secure when a majority of the parties are honest — is completely insecure in case half or more of the parties are dishonest. Similarly, the CLOS protocol [9] — which is secure for an arbitrary number of corrupted parties when a trusted CRS is available — is completely insecure in the presence of even a *single* corrupted party if the protocol is run using a CRS σ that is taken from the wrong distribution or, even worse, adversarially generated. Given this state of affairs, a natural question is whether it is possible to design a *single* protocol Π that uses a common reference string σ and simultaneously guarantees the following:

- *Regardless of how σ is generated* (and, in particular, even if σ is generated adversarially), Π is secure as long as at most s parties are corrupted.
- If σ is generated "honestly" (i.e., by a trusted third party according to the specification), then Π is secure as long as at most t parties are corrupted.

In this case, we will call the protocol Π an "(s, t)-secure protocol". It follows from [7,8] that (s, t)-security for general functionalities is only potentially achievable if

$s < n/2$, where n is the total number of parties running the protocol. A priori, we might hope to achieve the "best possible" result that $(\lfloor(n-1)/2\rfloor, n-1)$-secure protocols exist for arbitrary functionalities.

Here, we show tight positive and negative answers to the above question. First, we show that for any $s + t < n$ (and $s < n/2$) there exists an (s,t)-secure protocol realizing any functionality. We complement this by showing that this is, unfortunately, the best possible: if $s + t = n$ then there is a large class of functionalities (inherited, in some sense, from [8]) for which no (s,t)-secure protocol exists. We prove security under adaptive corruptions for our positive result, while our negative result holds even for the case of non-adaptive corruptions.

For n odd, the extremes of our positive result (i.e., $s = t = \lfloor(n-1)/2\rfloor$, or $s = 0$, $t = n-1$) correspond to, respectively, a protocol secure for honest majority (but relying on cryptographic assumptions) or one secure against an arbitrary number of malicious parties but requiring a CRS. (For n even we obtain a protocol that tolerates $s = \lfloor(n-1)/2\rfloor$ corruptions regardless of how the CRS is constructed, and $t = s + 1$ corruptions if the CRS is honestly-generated.) Our results also exhibit new protocols in between these extremes. Choice of which protocol to use reflects a tradeoff between the level of confidence in the CRS and the number of corruptions that can be tolerated: e.g., choosing $s = 0$ represents full confidence in the CRS, while setting $s = t = \lfloor(n-1)/2\rfloor$ means that there is effectively no confidence in the CRS at all.

Related work. Another suggestion for circumventing the impossibility results of [7,8] has been to use a definition of security where the ideal-model simulator is allowed to run in *super-polynomial* time [16,2]. This relaxation is sufficient to bypass the known impossibility results and leads to constructions of protocols for any functionality without setup assumptions. While these constructions seem to supply adequate security for certain applications, they require stronger (sub-exponential time) complexity assumptions and can be problematic when used as sub-routines within larger protocols.

Some other recent work has also considered the construction of protocols having "two tiers" of security. Barak, Canetti, Nielsen, and Pass [1] show a protocol relying on a key-registration authority: if the key-registration authority acts honestly the protocol is universally composable, while if this assumption is violated the protocol still remains secure in the stand-alone sense. Ishai et al. [13] and Katz [14], in the stand-alone setting, studied the question of whether there exist protocols that are "fully-secure" (i.e., guaranteeing privacy, correctness, and fairness) in the presence of a dishonest minority, yet still "secure-with-abort" otherwise. While the motivation in all these cases is similar, the problems are different and, in particular, a solution to our problem does not follow from (or rely on) any of these prior results.

Groth and Ostrovsky [11] recently introduced the *multi-CRS model* for universally composable multi-party computation. In this model, roughly speaking, the parties have access to a set of k common reference strings, some k' of which are "good" (i.e., guaranteed to have been chosen honestly). The remaining $k - k'$ strings are "bad", and can be chosen in an arbitrary manner. (Of course, it is not known which strings are "good" and which are "bad".) Groth and Ostrovsky explore conditions on k, k' under which UC multi-party computation is still possible. Although in both their case and our own the question boils down to what security guarantees can be achieved in the presence of a "bad" CRS, our end results are very different. In the work of Groth and Ostrovsky the number of corruptions to be tolerated is fixed and there are assumed to be some minimal number k' of "good" strings among the k available ones. In our work, in contrast, it is possible that *no* "good" CRS is available at all; even in this case, though, we would still like to ensure security against some (necessarily) smaller set of corrupted parties. On the other hand, we do rely on the Groth-Ostrovsky result as a building block for our positive result.

2 Preliminaries

2.1 Review of the UC Framework

We give a brief overview of the UC framework, referring the reader to [5] for further details. The UC framework allows for defining the security properties of cryptographic tasks so that security is maintained under general composition with an unbounded number of instances of arbitrary protocols running concurrently. In the UC framework, the security requirements of a given task are captured by specifying an ideal functionality run by a "trusted party" that obtains the inputs of the participants and provides them with the desired outputs. Informally, then, a protocol securely carries out a given task if running the protocol in the presence of a real-world adversary amounts to "emulating" the desired ideal functionality.

The notion of emulation in the UC framework is considerably stronger than that considered in previous models. As usual, the real-world model includes the parties running the protocol and an adversary $\mathcal{A}$ who controls their communication and potentially corrupts parties, while the ideal-world includes a simulator $\mathcal{S}$ who interacts with an ideal functionality $\mathcal{F}$ and dummy players who simply send input to/receive output from $\mathcal{F}$. In the UC framework, there is also an additional entity called the *environment* $\mathcal{Z}$. This environment generates the inputs to all parties, observes all their outputs, and interacts with the adversary in an arbitrary way throughout the computation. A protocol Π is said to *securely realize* an ideal functionality $\mathcal{F}$ if for any real-world adversary $\mathcal{A}$ that interacts with $\mathcal{Z}$ and real players running Π, there exists an ideal-world simulator $\mathcal{S}$ that interacts with $\mathcal{Z}$, the ideal functionality $\mathcal{F}$, and the "dummy" players communicating with $\mathcal{F}$, such that *no* poly-time environment $\mathcal{Z}$ can distinguish whether it is interacting with $\mathcal{A}$ (in the real world) or $\mathcal{S}$ (in the ideal world). $\mathcal{Z}$ thus serves as an "interactive distinguisher" between a real-world execution of the protocol Π and an ideal execution of functionality $\mathcal{F}$. A key point is that $\mathcal{Z}$ cannot be re-wound by $\mathcal{S}$; in other words, $\mathcal{S}$ must provide a so-called "straight-line" simulation.

The following *universal composition theorem* is proven in [5]. Consider a protocol Π that operates in the $\mathcal{F}$-hybrid model, where parties can communicate as usual and in addition have ideal access to an unbounded number of copies of the functionality $\mathcal{F}$. Let ρ be a protocol that securely realizes $\mathcal{F}$ as sketched above, and let Π^ρ be identical to Π with the exception that the interaction with *each copy* of $\mathcal{F}$ is replaced with an interaction with a *separate instance* of ρ. Then Π and Π^ρ have essentially the same input/output behavior. In particular, if Π securely realizes some functionality $\mathcal{G}$ in the $\mathcal{F}$-hybrid model then Π^ρ securely realizes $\mathcal{G}$ in the standard model (i.e., without access to any functionality).

2.2   Definitions Specific to Our Setting

We would like to model a single protocol Π that uses a CRS σ, where σ either comes from a trusted functionality $\mathcal{F}_{CRS}$ (defined as in [7] and all subsequent work on UC computation in the CRS model) or is chosen in an arbitrary manner by the environment $\mathcal{Z}$. A technical detail is that parties running Π can trivially "tell" where σ comes from depending on which incoming communication tape σ is written on (since an ideal functionality would write inputs to a different tape than $\mathcal{Z}$ would). Because this does not correspond to what we are attempting to model in the real world, we need to effectively "rule out" protocols that utilize this additional knowledge. The simplest way to do this is to define a "malicious CRS" functionality $\mathcal{F}_{mCRS}$ that we now informally describe. Functionality $\mathcal{F}_{mCRS}$ takes input σ from the adversary $\mathcal{A}$ and then, when activated by any party P_i, sends σ to that party. The overall effect of this is that $\mathcal{A}$ (and hence $\mathcal{Z}$) can set the CRS to any value of its choice; however, it is forced to provide the *same* value to all parties running protocol Π. When the parties interact with $\mathcal{F}_{CRS}$, this (intuitively) means that the CRS is "good"; when they interact with $\mathcal{F}_{mCRS}$ the CRS is "bad". We refer to this setting, where parties interact with either $\mathcal{F}_{CRS}$ or $\mathcal{F}_{mCRS}$ but do not know which, as the *mixed CRS model*. We can now define an (s,t)-secure protocol.

Definition 1. *We say a protocol Π (s,t)-securely realizes a functionality $\mathcal{F}$ in the mixed CRS model if*

(a) *Π securely realizes $\mathcal{F}$ in the $\mathcal{F}_{mCRS}$-hybrid model when at most s parties are corrupted.*

(b) *Π securely realizes $\mathcal{F}$ in the $\mathcal{F}_{CRS}$-hybrid model when at most t parties are corrupted.*

We stress that Π itself does not "know" in which of the two hybrid models it is being run. $\mathcal{S}$, however, may have this information hard-wired in. More concretely: although Π is a fixed protocol, two different ideal-world adversaries $\mathcal{S}$, $\mathcal{S}'$ may be used in proving each part of the definition above.

3   Positive Result for $s + t < n$

We begin by showing our positive result: if $s+t < n$ and $s < n/2$ (where n is the total number of parties running the protocol), then essentially any functionality

$\mathcal{F}$ can be (s, t)-securely realized in the mixed CRS model. This is subject to two minor technical conditions [9] we discuss briefly now.

Non-trivial protocols. The ideal process does not require the ideal-process adversary to deliver the messages that are sent between the ideal functionality and the parties. A corollary of the above fact is that a protocol that "hangs" (i.e., never sends any messages and never generates output) securely realizes any ideal functionality. However, such a protocol is uninteresting. Following [9], we therefore let a non-trivial protocol be one for which all parties generate output if the real-life adversary delivers all messages and all parties are honest.

Well-formed functionalities. A well-formed functionality is oblivious of the corruptions of parties, runs in polynomial time, and reveals the internal randomness used by the functionality to the ideal-process adversary in case all parties are corrupted [9]. This class contains all functionalities we can hope to securely realize from a non-trivial protocol in the presence of adaptive corruptions, as discussed in [9].

We can now formally state the result of this section:

Theorem 1 *Fix s, t, n with $s + t < n$ and $s < n/2$. Assume that enhanced trapdoor permutations, augmented non-committing encryption schemes, and dense cryptosystems exist. Then for every well-formed n-party functionality $\mathcal{F}$, there exists a non-trivial protocol Π which (s, t)-securely realizes $\mathcal{F}$ against adaptive adversaries in the mixed CRS model.*

The cryptographic assumptions of the theorem are inherited directly from [9], and we refer the reader there for formal definitions of each of these. Weaker assumptions suffice to achieve security against static corruptions; see [9].

To prove the above theorem, we rely on the results of Groth and Ostrovsky regarding the multi-CRS model [11]. Informally, they show the following result: Assume parties $P_1, \ldots, P_n$ having access to $k \geq 1$ strings $\sigma_1, \ldots, \sigma_k$. As long as $k' > k/2$ of these strings are honestly generated according to some specified distribution $\mathcal{D}$ (and assuming the same cryptographic assumptions of the theorem stated above), then for every well-formed functionality $\mathcal{F}$ there exists a non-trivial protocol Π securely realizing $\mathcal{F}$. We stress that the remaining $k - k'$ strings can be generated arbitrarily (i.e., adversarially), even possibly depending on the k' honestly-generated strings.

Building on the above result, we now describe our construction. We assume there are n parties $P_1, \ldots, P_n$ who wish to run a protocol to realize a (well-formed) functionality $\mathcal{F}$. Construct a protocol Π as follows:

1. All parties begin with the same string σ^* provided as input. (Recall the parties do not know whether this is a "good" CRS or a "bad" CRS.) $P_1, \ldots, P_n$ first "amplify" the given string σ^* to m CRSs $\sigma_1^*, \ldots, \sigma_m^*$, where m is a parameter which is defined later on. The requirements here are simply that if σ^* is "good", then each of $\sigma_1^*, \ldots, \sigma_m^*$ should be "good" also. (If σ^* is "bad" then we impose no requirements on $\sigma_1^*, \ldots, \sigma_m^*$.)

The above can be accomplished by using the CLOS protocol [9] as follows. Define an ideal functionality $\mathcal{F}_{m\_new\_CRS}$ which generates m new CRSs from the appropriate distribution $\mathcal{D}$ (where $\mathcal{D}$ refers to the the distribution used in the Groth-Ostrovsky result mentioned above) and outputs these to all parties. We use the CLOS protocol to realize the functionality $\mathcal{F}_{m\_new\_CRS}$. When running the CLOS protocol, use the given string σ^* as the CRS.

Note that when σ^* was produced by $\mathcal{F}_{CRS}$, security of the CLOS protocol guarantees that the m resulting CRSs are all chosen appropriately. On the other hand, there are *no* guarantees in case σ^* was produced by $\mathcal{F}_{mCRS}$, but recall that we do not require anything in that case anyway.

2. Following the above, each party P_i chooses a string σ_i according to distribution $\mathcal{D}$ (where, again, $\mathcal{D}$ is the distribution used in the Groth-Ostrovsky result mentioned above), and broadcasts σ_i to all other parties.[1]

3. Each party receives $\sigma_1, \ldots, \sigma_n$, and sets $\sigma^*_{m+i} = \sigma_i$ for $i = 1$ to n.

4. All parties now have $n + m$ strings $\sigma^*_1, \ldots, \sigma^*_{n+m}$. These strings are used to run the Groth-Ostrovsky protocol for $\mathcal{F}$.

We claim that for any s, t satisfying the conditions of Theorem 1, it is possible to set m so as to obtain a protocol Π that (s, t)-securely realizes $\mathcal{F}$. The conditions we need to satisfy are as follows:

– When Π is run in the F_{CRS}-hybrid model, σ^* is a "good" CRS and so the strings $\sigma^*_1, \ldots, \sigma^*_m$ are also "good". The $n - t$ honest parties contribute another $n - t$ "good" strings in step 2, above, for a total of $m + n - t$ "good" strings in the set of strings $\sigma^*_1, \ldots, \sigma^*_{n+m}$. At most t of the strings in this set (namely, those contributed by the t malicious parties) can be "bad". For the Groth-Ostrovsky result to apply, we need $m + n - t > t$ or

$$m > 2t - n. \tag{1}$$

– When Π is run in the F_{mCRS}-hybrid model, σ^* is adversarially-chosen and so we must assume that the strings $\sigma^*_1, \ldots, \sigma^*_m$ are also "bad". In step 2, the malicious parties contribute another s "bad" strings (for a total of $m + s$ "bad" strings), while the $n - s$ honest parties contribute $n - s$ "good" strings. For the Groth-Ostrovsky result to apply, we now need $n - s > m + s$ or

$$m < n - 2s. \tag{2}$$

Since m, t, n are all integers, Equations (1) and (2) imply

$$2t - n \leq n - 2s - 2$$

or $s + t \leq n - 1$. When this condition holds, the equations can be simultaneously satisfied by setting $m = n - 2s - 1$, which gives a positive solution if $s < n/2$.

The security of the above construction follows from the security of the Groth-Ostrovsky protocol [11] (the details are omitted).

---

[1] The "broadcast" used here is the UC broadcast protocol from [10] (which achieves a weaker definition than "standard" broadcast, but suffices for constructing protocols in the UC framework).

4 Impossibility Result for $s + t \geq n$

In this section, we state and prove our main impossibility result which shows that the results of the previous section are tight.

Theorem 2 *Let n, t, s be such that $s + t \geq n$. Then there exists a well-formed deterministic functionality for which no non-trivial n-party protocol exists that (s, t)-securely realizes $\mathcal{F}$ in the mixed CRS model.*

We in fact show that the above theorem holds for a large class of functionalities. That is, there exists a large class of functionalities for which no such non-trivial protocol exists.

The proof of Theorem 2 relies on ideas from the impossibility result of Canetti, Kushilevitz, and Lindell [8] that applies to 2-party protocols in the plain model. Since ours is inherently a multi-party scenario, our proof proceeds in two stages. In the first stage of our proof, we transform any n-party protocol Π that securely computes a function f in the mixed CRS model, into a two-party protocol Σ in the mixed CRS model that computes a related function g (derived from f). Protocol Σ guarantees security in the $\mathcal{F}_{CRS}$-hybrid model when either party is corrupted, and security in the $\mathcal{F}_{mCRS}$-hybrid model when the *second* party is corrupted. In the second stage of our proof, we show that one of the parties running Σ can run a successful *split simulator strategy* [8] against the other. As in [8], the existence of a split simulator strategy means that the class of functionalities that can be securely realized by the two-party protocol Σ is severely restricted. This also restricts the class of functionalities f which can be realized using the original n-party protocol.

We now give the details. Let $x \| y$ denote the concatenation of x and y. We first define the t-*division* of a function f.

Definition 2. *Let $f = (f_1, \ldots, f_n)$ be a function taking n inputs $x_1, \ldots, x_n$ and returning n (possibly different) outputs. Define the two-input/two-output function $g = (g_1, g_2)$, the t-division of f via:*

$$g_1 \left(\overbrace{(x_1 \| \cdots \| x_t)}^{I_1}, \overbrace{(x_{t+1} \| \cdots \| x_n)}^{I_2} \right) = f_1(x_1, \ldots, x_n) \| \cdots \| f_t(x_1, \ldots, x_n)$$

$$g_2 \left((x_1 \| \cdots \| x_t), (x_{t+1} \| \cdots \| x_n) \right) = f_{t+1}(x_1, \ldots, x_n) \| \cdots \| f_n(x_1, \ldots, x_n).$$

Lemma 1. *Let n, t, s be such that $s + t = n$ and $s < n/2$. Say Π is an (s, t)-secure protocol by which parties $P_1, \ldots, P_n$ holding inputs $x_1, \ldots, x_n$ can evaluate a function $f(x_1, \ldots, x_n)$. Then there exists a two-party protocol Σ by which parties p_1, p_2 holding inputs $I_1 = x_1 \| \ldots \| x_t$ and $I_2 = x_{t+1} \| \ldots \| x_n$ can evaluate the t-division function $g(I_1, I_2)$. Furthermore, Σ is secure when either parties is corrupted in the $\mathcal{F}_{CRS}$-hybrid model, and secure against a dishonest p_2 in the $\mathcal{F}_{mCRS}$-hybrid model.*

Proof. We construct the protocol Σ using the protocol Π. The basic idea is as follows. The parties p_1 and p_2 break their input I_1, I_2 into several parts and start emulating n parties running the protocol Π to compute f on those inputs. Some of these parties in Π are controlled and emulated by p_1 and others by p_2. Finally when Π finishes, p_1 and p_2 get several outputs f_i meant for parties controlled by them. Using these outputs, p_1 and p_2 then individually reconstruct their final output g_1 and g_2. More details follow.

The parties p_1, p_2 hold inputs $I_1 = x_1 \| \ldots \| x_t$ and $I_2 = x_{t+1} \| \ldots \| x_n$ and wish to compute the function g. Party p_1 internally starts emulating parties $P_1, \ldots, P_t$ on inputs $x_1, \ldots, x_t$, respectively, to compute the function f. Similarly, p_2 starts emulating parties $P_{t+1}, \ldots, P_n$ on inputs $x_{t+1}, \ldots, x_n$. Whenever Π requires party P_i to send a message M to party P_j, this is handled in the natural way: If $i, j \leq t$ (resp., $i, j > t$), then p_1 (resp., p_2) internally delivers M from P_i to P_j. If $i \leq t$ and $j > t$, then p_1 sends the message (i, j, M) to p_2 who then internally delivers M to P_j as if it were received from P_i. The case $i > t$ and $j \leq t$ is handled similarly. After Π finishes, $P_1, \ldots, P_t$ halt outputting $f_1, \ldots, f_t$ and hence p_1 obtains $g_1 = f_1 \| \ldots \| f_t$. Similarly, p_2 obtains $g_2 = f_{t+1} \| \ldots \| f_n$.

As for the security claims regarding Σ, recall that Π is t-secure in the $\mathcal{F}_{CRS}$-hybrid model. This means that Π securely computes f in the presence of any coalition of up to t corrupted parties. This in particular means that Π remains secure if all of $P_1, \ldots, P_t$ are corrupted. Thus, Σ remains secure against a dishonest p_1 (who controls $P_1, \ldots, P_t$) in the $\mathcal{F}_{CRS}$-hybrid model. Also since $s \leq t$ (because $s < n/2$), protocol Π is secure even if $P_{t+1}, \ldots, P_n$ are corrupted and hence Σ is secure against a dishonest p_2 in the $\mathcal{F}_{CRS}$-hybrid model. Furthermore, Π is s-secure in the $\mathcal{F}_{mCRS}$-hybrid model. This means that Π remains secure even if $P_{t+1}, \ldots, P_n$ are corrupted. Hence Σ is secure against a dishonest p_2 (but not necessarily against a dishonest p_1) in the $\mathcal{F}_{mCRS}$-hybrid model.

We now show that a malicious p_2 can run a successful *split simulator strategy* [8] against an honest p_1 in protocol Σ when run in the $\mathcal{F}_{mCRS}$-hybrid model. This shows that even if p_1 remains honest, there is a large class of functionalities that cannot be securely realized by Σ.[2] Using the previous lemma, this in turn shows the existence of a class of functionalities which cannot be (s, t)-securely realized by Π (when $t + s \geq n$).

Showing the existence of a successful split simulator strategy for p_2 amounts to reproving the main technical lemma of [8] in our setting. We start by recalling a few definitions and notations from [9,8]. Part of our proof is taken almost verbatim from [8].

Notation. Let $g : D_1 \times D_2 \to \{0,1\}^* \times \{0,1\}^*$ be a deterministic, polynomial-time computable function, where $D_1, D_2 \subseteq \{0,1\}^*$ are arbitrary (possibly infinite) domains of inputs. Function g is denoted by $g = (g_1, g_2)$ where g_1 and g_2 denote the outputs of p_1 and p_2, respectively. The following definition corresponds to [8, Def. 3.1].

---

[2] In [8], it was shown that *either* party p_1 or p_2 could run a split simulator strategy against the other. In our case, we only show that p_2 can do so against p_1. Hence, the class of functionalities which we prove are impossible to realize is smaller than that in [8].

Definition 3. *Let Σ be a protocol securely computing g. Let $D_\kappa \subseteq D_2$ be a polynomial-size subset of inputs (i.e., $|D_\kappa| = \text{poly}(\kappa)$, where κ is a security parameter). Then a corrupted party p_2 is said to run a* split *adversarial strategy if it consists of machines p_2^a and p_2^b such that:*

1. *On input $(1^\kappa, D_\kappa, I_2)$, with $I_2 \in D_\kappa$, party p_2 internally gives machine p_2^b the input $(1^\kappa, D_\kappa, I_2)$.*
2. *An execution between (an honest) p_1 running Σ and $p_2 = (p_2^a, p_2^b)$ works as follows:*
   (a) *p_2^a interacts with p_1 according to some specified strategy.*
   (b) *At some stage of the execution p_2^a hands p_2^b a value I_1'.*
   (c) *When p_2^b receives I_1' from p_2^a, it computes $J_1' = g_1(I_1', I_2')$ for some $I_2' \in D_\kappa$ of its choice.*
   (d) *p_2^b hands p_2^a the value J_1', and p_2^a continues interacting with p_1.*

We define a successful strategy as in [8, Def. 3.2].

Definition 4. *Let Σ, g, κ be as in Definition 3. Let $\mathcal{Z}$ be an environment who hands input I_1 to p_1 and a pair (D_κ, I_2) to p_2 where $D_\kappa \subseteq D_2$, $|D_\kappa| = \text{poly}(\kappa)$, and I_2 is chosen uniformly in D_κ. Then a split adversarial strategy for p_2 is said to be* successful *if for every $\mathcal{Z}$ as above and every input z to $\mathcal{Z}$, the following conditions hold in a real execution of p_2 with $\mathcal{Z}$ and honest p_1:*

1. *The value I_1' output by p_2^a in step 2b of Definition 3 is such that for every $I_2 \in D_\kappa$, it holds that $g_2(I_1', I_2) = g_2(I_1, I_2)$.*
2. *The honest party p_1 outputs $g_1(I_1, I_2')$, where I_2' is the value chosen by p_2^b in step 2c of Definition 3.*

We now prove a lemma akin to [8, Lem. 3.3].

Lemma 2. *Let Σ be a non-trivial, two-party protocol computing g, which is secure in the $\mathcal{F}_{CRS}$-hybrid model when either party is corrupted, and secure in the $\mathcal{F}_{mCRS}$-hybrid model when p_2 is corrupted. Then there exists a machine p_2^a such that for every machine p_2^b of the form described in Definition 3, the split adversarial strategy $p_2 = (p_2^a, p_2^b)$ is successful in the $\mathcal{F}_{mCRS}$-hybrid model, except with negligible probability.*

Proof. The proof in our setting is very similar to the proof of the main technical lemma in [8]. Here we only sketch a proof, highlighting the main differences. We refer the reader to [8] for complete details.

In the proof of [8], they first consider the real-world execution where party p_1 is controlled by the environment $\mathcal{Z}$ through a dummy adversary $\mathcal{A}_D$ who simply forwards messages received from the environment to party p_2 and vice versa. Parties p_1 and p_2 have inputs I_1 and I_2, respectively, and execute Σ; we assume that Σ securely computes g. Thus, there exists a simulator $\mathcal{S}$ that interacts with the ideal process and such that $\mathcal{Z}$ cannot distinguish an execution of a real-world process from an execution of the ideal process. Notice that in the ideal world, $\mathcal{S}$ must send an input I_1' to the ideal functionality computing g, and receives an output J_1' from this functionality such that I_1' and J_1' are functionally equivalent to I_1 and $g_1(I_1, I_2')$ respectively. (Here, I_2' is chosen by p_2.) This implies that if $\mathcal{Z}$ simply runs the code of an honest p_1, the ideal-world simulator $\mathcal{S}$ is able to *extract* the inputs of the honest player p_1 and also force its output to be J_1'.

In our setting, in the $\mathcal{F}_{\text{CRS}}$-hybrid model (i.e., if the string σ is an honestly-generated CRS), protocol Σ is secure regardless of which party is corrupted. This means that there exists a simulator $\mathcal{S}$ who generates a CRS σ and is then able to extract the input of the honest player p_1.

Now consider the case of the $\mathcal{F}_{mCRS}$-hybrid model, i.e., when Σ is run with an adversarially-generated string σ. In this case, a malicious p_2 can just run $\mathcal{S}$ to generate a CRS and interact with p_1. At a high level, the machine p_2^a just consists of running $\mathcal{S}$ with the honest p_1. Machine p_2^a forwards every message that it receives from p_1 to $\mathcal{S}$ as if it came from $\mathcal{Z}$. Similarly, every message that $\mathcal{S}$ sends to $\mathcal{Z}$ is forwarded by p_2^a to p_1 in the real execution. When $\mathcal{S}$ outputs a value I_1' that it intends to send to the ideal functionality computing g, then p_2^a gives this value to p_2^b. Later, when p_2^b gives a value J_1' to p_2^a, then p_2^a gives it to $\mathcal{S}$ as if it came from the ideal functionality computing g. Hence, a malicious p_2 is able to use the simulator $\mathcal{S}$ to do whatever the simulator $\mathcal{S}$ was doing in the $\mathcal{F}_{\text{CRS}}$-hybrid model. This in particular means that p_2 is able to extract the input of the honest p_1 and run a *successful* split simulator strategy. This completes our proof sketch.

Completing the proof of Theorem 2. As shown by [8], the existence of a successful split simulator strategy for p_2 against an honest p_1 rules out the realization of several interesting well-formed functionalities. This, in turn, rules out several n-input functionalities f whose secure computation implies secure computation of g by Lemma 1. We give a concrete example in what follows.

We consider *single-input functions which are not efficiently invertible* [8]. The definition of an efficiently-invertible function is given as in [8]:

Definition 5. *A polynomial-time function $g : D \to \{0,1\}^*$ is efficiently invertible if there exists a* PPT *machine M such that for every distribution $\hat{D} = \{\hat{D}_\kappa\}$ over D that is sampleable by a non-uniform,* PPT *Turing machine, the following is negligible:*

$$\Pr_{x \leftarrow \hat{D}_\kappa} \left[M(1^\kappa, g(x)) \notin g^{-1}(g(x)) \right].$$

Let t, s, n be such that $t + s = n$ and $s < n/2$. We consider the following functionality $\mathcal{F}$: Let parties $P_1, \ldots, P_t$ hold inputs $x_1, \ldots, x_t$, while $P_{t+1}, \ldots, P_n$ have no inputs. The output of $P_1, \ldots, P_t$ is $\perp$ while the output of $P_{t+1}, \ldots, P_n$ is $f(x_1 \| \cdots \| x_t)$ for an function f which is *not* efficiently invertible.

If there exists an n-party protocol Π that (s, t)-securely realizes $\mathcal{F}$, then there exists a 2-party protocol Σ computing the function $g(I_1, \perp) = (\perp, f(I_1))$, which is secure against corruption of either party in the $\mathcal{F}_{\text{CRS}}$-hybrid model and secure against corruption of the second party in the $\mathcal{F}_{mCRS}$-hybrid model. Lemma 2, however, implies that p_2 can run a successful split simulator strategy and extract an input I_1' such that $g(I_1, \perp) = g(I_1', \perp)$, or equivalently $f(I_1) = f(I_1')$. Since all the information computable by p_2 during an execution of Σ should follow from its output $f(I_1)$ alone, it follows that I_1' is computable given $f(I_1)$. This contradicts the assumption that f is not efficiently invertible.

Hence, we conclude that there does not exist such a protocol Π to evaluate the functionality $\mathcal{F}$. This impossibility result can be extended to include a large class of functionalities as in [8].

References

1. Barak, B., Canetti, R., Nielsen, J.B., Pass, R.: Universally composable protocols with relaxed set-up assumptions. In: 45th Annual Symposium on Foundations of Computer Science (FOCS), pp. 186–195. IEEE, Los Alamitos (2004)
2. Barak, B., Sahai, A.: How to play almost any mental game over the net — concurrent composition using super-polynomial simulation. In: 46th Annual Symposium on Foundations of Computer Science (FOCS), IEEE, Los Alamitos (2005)
3. Ben-Or, M., Goldwasser, S., Wigderson, A.: Completeness theorems for non-cryptographic fault-tolerant distributed computation. In: 20th Annual ACM Symposium on Theory of Computing (STOC), pp. 1–10. ACM, New York (1988)
4. Blum, M., Feldman, P., Micali, S.: Non-interactive zero-knowledge and its applications. In: 20th Annual ACM Symposium on Theory of Computing (STOC), pp. 32–42. ACM, New York (1988)
5. Canetti, R.: Universally composable security: A new paradigm for cryptographic protocols. In: 42nd Annual Symposium on Foundations of Computer Science (FOCS), pp. 136–147. IEEE, Los Alamitos (2001) Preliminary full version available as Cryptology ePrint Archive Report 2000/067
6. Canetti, R., Dodis, Y., Pass, R., Walfish, S.: Universally composable security with global setup. In: Vadhan, S.P. (ed.) TCC 2007. LNCS, vol. 4392, pp. 61–85. Springer, Heidelberg (2007)
7. Canetti, R., Fischlin, M.: Universally composable commitments. In: Kilian, J. (ed.) CRYPTO 2001. LNCS, vol. 2139, pp. 19–40. Springer, Heidelberg (2001)
8. Canetti, R., Kushilevitz, E., Lindell, Y.: On the limitations of universally composable two-party computation without set-up assumptions. J. Cryptology 19(2), 135–167 (2006)
9. Canetti, R., Lindell, Y., Ostrovsky, R., Sahai, A.: Universally composable two-party and multi-party secure computation. In: 34th Annual ACM Symposium on Theory of Computing (STOC), pp. 494–503 (2002)
10. Goldwasser, S., Lindell, Y.: Secure multi-party computation without agreement. J. Cryptology 18(3), 247–287 (2005)
11. Groth, J., Ostrovsky, R.: Cryptography in the multi-string model. In: Menezes, A. (ed.) CRYPTO 2007. LNCS, vol. 4622, pp. 323–341. Springer, Heidelberg (2007)
12. Hofheinz, D., Müller-Quade, J., Unruh, D.: Universally composable zero-knowledge arguments and commitments from signature cards. In: Proc. 5th Central European Conference on Cryptology (2005)
13. Ishai, Y., Kushilevitz, E., Lindell, Y., Petrank, E.: On combining privacy with guaranteed output delivery in secure multiparty computation. In: Dwork, C. (ed.) CRYPTO 2006. LNCS, vol. 4117, pp. 483–500. Springer, Heidelberg (2006)

14. Katz, J.: On achieving the best of both worlds in secure multiparty computation. In: 39th Annual ACM Symposium on Theory of Computing (STOC), pp. 11–20. ACM, New York (2007)
15. Katz, J.: Universally composable multi-party computation using tamper-proof hardware. In: Naor, M. (ed.) EUROCRYPT 2007. LNCS, vol. 4515, pp. 115–128. Springer, Heidelberg (2007)
16. Prabhakaran, M., Sahai, A.: New notions of security: Achieving universal composability without trusted setup. In: 36th Annual ACM Symposium on Theory of Computing (STOC), pp. 242–251 (2004)
17. Rabin, T., Ben-Or, M.: Verifiable secret sharing and multi-party protocols with honest majority. In: 21st Annual ACM Symposium on Theory of Computing (STOC), pp. 73–85. ACM, New York (1989)

Efficient Protocols for Set Intersection and Pattern Matching with Security Against Malicious and Covert Adversaries*

Carmit Hazay and Yehuda Lindell

Department of Computer Science
Bar-Ilan University, Israel
{harelc,lindell}@cs.biu.ac.il

Abstract. In this paper we construct efficient secure protocols for *set intersection* and *pattern matching*. Our protocols for securely computing the set intersection functionality are based on secure pseudorandom function evaluations, in contrast to previous protocols that used secure polynomial evaluation. In addition to the above, we also use secure pseudorandom function evaluation in order to achieve secure pattern matching. In this case, we utilize specific properties of the Naor-Reingold pseudorandom function in order to achieve high efficiency.

Our results are presented in two adversary models. Our protocol for secure pattern matching and one of our protocols for set intersection achieve security against *malicious adversaries* under a relaxed definition where one corruption case is simulatable and for the other only privacy (formalized through indistinguishability) is guaranteed. We also present a protocol for set intersection that is fully simulatable in the model of covert adversaries. Loosely speaking, this means that a malicious adversary can cheat, but will then be caught with good probability.

1 Introduction

In the setting of secure two-party computation, two parties wish to jointly compute some function of their private inputs while preserving a number of security properties. In particular, the parties wish to ensure that nothing is revealed beyond the output (privacy), that the output is computed according to the specified function (correctness) and more. The standard definition today (cf. [5] following [13,4,17]) formalizes security by comparing a real protocol execution to an "ideal execution" where an incorruptible trusted party helps the parties compute the function. Specifically, in the ideal world the parties just send their inputs (over perfectly secure communication lines) to the trusted party, who computes the function honestly and sends the output to the parties. A real protocol (in which parties interact arbitrarily) is said to be secure if any adversarial attack on a real protocol can essentially be carried out also in the ideal world (of course,

---

* This research was supported by an Eshkol scholarship and Infrastructures grant from the Israel Ministry of Science and Technology.

R. Canetti (Ed.): TCC 2008, LNCS 4948, pp. 155–175, 2008.
© International Association for Cryptologic Research 2008

in the ideal world the adversary can do almost nothing and this guarantees that the same is true also in the real world). This definition of security is often called *simulation-based* because security is demonstrated by showing that a real protocol execution can be "simulated" in the ideal world.

This setting has been widely studied, and it has been shown that any efficient two-party functionality can be securely computed [24,12,11]. These feasibility results demonstrate the wide applicability of secure computation, in principle. However, they fall short of what is needed in implementations because they are far from efficient enough to be used in practice (with a few exceptions). This is not surprising because the results are general and do not utilize any special properties of the specific problem being solved. The focus of this paper is the development of efficient protocols for specific problems of interest.

Relaxed notions of security. Recently, the field of data mining has shown great interest in secure computation, for the purpose of "privacy-preserving data mining". However, most of the protocols that have been constructed with this aim in mind are only secure in the presence of *semi-honest adversaries* who follow the protocol specification (but may try to examine the messages they receive to learn more than they should). Unfortunately, in many cases, this level of security is not sufficient. Rather, adversarial parties are willing to behave maliciously – meaning that they may divert arbitrarily from the protocol specification – in their aim to cheat. It seems that it is hard to obtain highly efficient protocols that are secure in the presence of malicious adversaries (under the standard simulation-based definitions), and two decades after the foundational feasibility results of [12] we only know of very few non-trivial secure computation problems that can be solved with high efficiency in this model. In this paper, we consider two different relaxations in order to achieve higher efficiency:

- *One-sided simulatability:* According to this notion of security, full simulation is provided for one of the corruption cases, while only privacy (via computational indistinguishability) is guaranteed for the other corruption case. This notion of security is useful when considering functionalities for which only one party receives output. In this case, privacy is guaranteed when the party not receiving output is corrupted (and this is formalized by saying that the party cannot distinguish between different inputs used by the other party), whereas full simulation via the ideal/real paradigm is guaranteed when the party receiving output is corrupted. This notion of security has been considered in the past; see [19,8] for example.
- *Security in the presence of covert adversaries:* This notion of security provides the following guarantee. A malicious adversary may be able to cheat (e.g., learn the other party's private input). However, if it follows such a strategy, it is guaranteed to be caught with probability at least ϵ, where ϵ is called the "deterrence factor" (in this paper, we use $\epsilon = 1/2$). This definition is formalized within the ideal/real simulation paradigm and so has all the advantages offered by it. This definition was recently introduced in [2].

We stress that both notions are relaxations and are not necessarily sufficient for all applications. For example, security in the presence of covert adversaries would not suffice when the computation relates to highly sensitive data or when there are no repercussions to a party being caught cheating. Likewise, the guarantee of privacy alone (as in one-sided simulatability for one of the corruption cases) is sometimes not sufficient. For example, the properties of independence of inputs and correctness are not achieved, and they are sometimes needed. Nevertheless, in many cases, such relaxations are acceptable. Furthermore, using these relaxations, we are able to construct protocols that are much more efficient than anything known that achieves full security in the presence of malicious adversaries (where security is formalized via the ideal/real simulation paradigm).

Secure set intersection. The bulk of this paper is focused on solving the set intersection problem. In this problem, two parties with private sets wish to learn the intersection of their sets and nothing more. There are many cases where such a computation is useful. For example, two health insurance companies may wish to ensure that no one has taken out the same insurance with both of them (if this is forbidden), or the government may wish to ensure that no one receiving social welfare is currently employed and paying income tax. By running secure protocols for these tasks, sensitive information about law-abiding citizens is not unnecessarily compromised.

We present two protocols for this task. The first achieves security in the presence of malicious adversaries with *one-sided simulatability* while the second is secure in the presence of *covert adversaries*. Both protocols take a novel approach. Specifically, instead of using protocols for secure polynomial evaluation [18], our protocols are based on running secure subprotocols for pseudorandom function evaluation. In addition, we use only standard assumptions (e.g., the decisional Diffie-Hellman assumption) and do not resort to random oracles.

In order to get a feel of how our protocol works we sketch the general idea underlying it. The parties run many executions of a protocol for securely computing a pseudorandom function, where one party inputs the key to the pseudorandom function and the other inputs the elements of its set. Denoting the pseudorandom function by F, the input of party P_1 by X and the input of party P_2 by Y, we have that at the end of this stage party P_2 holds the set $\{F_k(y)\}_{y \in Y}$ while P_1 has learned nothing. Then, P_1 just needs to locally compute the set $\{F_k(x)\}_{x \in X}$ and send it to P_2. By comparing which elements appear in both sets, P_2 can learn the intersection (but nothing more). This is a completely different approach to that taken until now that has defined polynomials based on the sets and used secure polynomial evaluations to learn the intersection. We stress that the "polynomial approach" has only been used successfully to achieve security in the presence of semi-honest adversaries [14,9], or together with random oracles when malicious adversaries are considered [9]. (We exclude the use of techniques that use general zero-knowledge proofs because these are not efficient.)

Secure pattern matching. We present an efficient secure protocol for solving the basic problem of *pattern matching* [3,15]. In this problem, one party holds a text T and the other a pattern p. The aim is for the party holding the pattern to learn all the locations of the pattern in the text (and there may be many) while the other learns nothing about the pattern. As with our protocols for secure set intersection, the use of secure pseudorandom function evaluation lies at the heart of our solution. However, here we also utilize specific properties of the Naor-Reingold pseudorandom function [20], enabling us to obtain a simple protocol that is significantly more efficient than that obtained by running known general protocols. Our protocol is secure in the presence of *malicious adversaries with one-sided simulatability*, and is the first to address this specific problem.

Related work. The problem of secure set intersection was studied in [9] who presented protocols for both the semi-honest and malicious cases. However, their protocol for the case of malicious adversaries assumes a random oracle. This problem was also studied in [14] whose main focus was the semi-honest model; their protocols for the malicious case use multiple zero-knowledge proofs for proving correct behavior and as such are not very efficient. As we have mentioned, both of the above works use oblivious polynomial evaluation as the basic building block in their solutions.

2   Definitions and Tools

2.1   Definitions

We denote the security parameter by n and computational indistinguishability of ensembles X and Y by $X \stackrel{c}{\equiv} Y$; see [11] for formal definitions. We adopt the convention whereby a machine is said to run in polynomial-time if its number of steps is polynomial in its *security parameter* alone. We use the shorthand PPT to denote probabilistic polynomial-time. Two basic building blocks that we utilize in our constructions are ensembles of pseudorandom functions, denoted by F_{PRF}, and ensembles of pseudorandom permutations, denoted by F_{PRP}, as defined in [10]. We also denote the ensemble of truly random functions by H_{Func} and the ensemble of truly random permutations by H_{Perm}.

One sided simulation for two-party protocols. Two of our protocols achieve a level of security that we call one-sided simulation. In these protocols, P_2 receives output while P_1 should learn nothing. In one-sided simulation, *full simulation* is possible when P_2 is corrupted. However, when P_1 is corrupted we only guarantee *privacy*, meaning that it learns nothing whatsoever about P_2's input (this is straightforward to formalize because P_1 receives no output). This is a relaxed level of security and does not achieve everything we want; for example, independence of inputs and correctness are not guaranteed. Nevertheless, for this level of security we are able to construct highly efficient protocols that are secure in the presence of malicious adversaries. The formal definition appears in the full version; we present it very briefly here. Let $\mathrm{REAL}_{\pi,\mathcal{A}(z),i}(x,y,n)$ denote the output of the honest party and the adversary $\mathcal{A}$ (controlling party P_i) after a real execution of protocol π, where P_1 has input x, P_2 has input y, $\mathcal{A}$ has auxiliary input z, and the security parameter is n. Let $\mathrm{IDEAL}_{f,\mathcal{S}(z),i}(x,y,n)$ be the analogous distribution in an ideal execution with a trusted party who computes f for the parties. Finally, let $\mathrm{VIEW}^{\mathcal{A}}_{\pi,\mathcal{A}(z),i}(x,y,n)$ denote the view of the adversary after a real execution of π as above. Then, we have the following definition:

Definition 1. *Let f be a two-party functionality where only P_2 receives output. We say that a protocol π securely computes f with one-sided simulation if the following holds:*

1. *For every non-uniform* PPT *adversary $\mathcal{A}$ in the real model, there exists a non-uniform* PPT *adversary $\mathcal{S}$ for the ideal model, such that for every $x, y, z \in \{0,1\}^*$*

$$\left\{ \text{REAL}_{\pi,\mathcal{A}(z),2}(x,y,n) \right\}_{n \in N} \stackrel{c}{\equiv} \left\{ \text{IDEAL}_{f,\mathcal{S}(z),2}(x,y,n) \right\}_{n \in N}$$

2. *For every non-uniform* PPT *adversary $\mathcal{A}$, all pairs of inputs $y, y' \in \{0,1\}^*$ with $|y| = |y'|$, and all inputs $x, z \in \{0,1\}^*$,*

$$\left\{ \text{VIEW}^{\mathcal{A}}_{\pi,\mathcal{A}(z),1}(x,y,n) \right\}_{n \in N} \stackrel{c}{\equiv} \left\{ \text{VIEW}^{\mathcal{A}}_{\pi,\mathcal{A}(z),1}(x,y',n) \right\}_{n \in N}$$

Security in the presence of covert adversaries. In this setting, the adversary may deviate from the protocol specification in an attempt to cheat, and as such is malicious. However, if it follows a strategy which enables it to achieve something that is not possible in the ideal model (like learning the honest party's input), then its cheating is guaranteed to be detected by the honest party with probability at least ϵ, where ϵ is a deterrent parameter. This definition is formalized in three ways in [2]; we consider their strongest definition here. In this definition, the ideal model is modified so that the adversary may send a special cheat message to the trusted party. In such a case, the trusted party tosses coins so that with probability ϵ the adversary is caught and a message corrupted is sent to the honest party (indicating that the other party attempted to cheat). However, with probability $1 - \epsilon$, the ideal-model adversary is allowed to cheat and so the trusted party sends it the honest party's full input and also allows it to set the output of the honest party. We refer the reader to [2] and the full version of this paper for further details. The output distribution of an execution of this modified ideal model for a given ϵ and parameters as above is denoted $\text{IDEALSC}^{\epsilon}_{f,\mathcal{S}(z),i}(x,y,n)$. We have the following:

Definition 2. *Let f, π and ϵ be as above. Protocol π is said to securely compute f in the presence of covert adversaries with ϵ-deterrent if for every non-uniform* PPT *adversary $\mathcal{A}$ for the real model, there exists a non-uniform probabilistic polynomial-time adversary $\mathcal{S}$ for the ideal model such that for every $i \in \{1,2\}$, every $x, y \in \{0,1\}^*$ with $|x| = |y|$, and every auxiliary input $z \in \{0,1\}^*$:*

$$\left\{ \text{IDEALSC}^{\epsilon}_{f,\mathcal{S}(z),i}(x,y,n) \right\}_{n \in \mathbb{N}} \stackrel{c}{\equiv} \left\{ \text{REAL}_{\pi,\mathcal{A}(z),i}(x,y,n) \right\}_{n \in \mathbb{N}}$$

The two notions of security. We remark that one-sided simulatability and security in the presence of covert adversaries are incomparable notions. On the one hand, the guarantees provided by security under one-sided simulation cannot be breached, even by a malicious adversary. This is not the case for security in the presence of covert adversaries where it is possible for a malicious adversary to successfully cheat. On the other hand, the formalization of security for covert adversaries is such that any deviation from what can be achieved in the ideal model is considered cheating (and so will result in the adversary being caught with probability ϵ). This is not the case for one-sided simulatability where one of the parties can make its input depend on the other, or cause the result to not be correctly computed, without ever being caught.

2.2   Tools

In this section, we describe the basic tools used in our constructions. Full descriptions and proofs are provided in the full version of this paper.

Oblivious transfer. We use oblivious transfer in order to achieve secure pseudorandom function evaluation (see below), which in turn is used for our set intersection protocols. For our protocols that achieve one-sided simulatability, we need an oblivious transfer protocol that achieves one-sided simulatability. Such a protocol can be constructed using homomorphic encryption, based on the protocol of [1]. The protocol needs some modifications in order to obtain simulatability in the case that the receiver is corrupted. We can instantiate our protocol with either the El-Gamal [6] or Paillier [21] homomorphic encryptions schemes. However, our instantiation using El-Gamal is considerably more efficient; see the full version. We remark that our protocols actually need to run multiple oblivious transfers in parallel. For the sake of this, we define the *multi-oblivious transfer functionality* with m executions, denoted $\mathcal{F}_{\mathrm{OT}}^m$ as follows:

$$((x_1^0, x_1^1), \ldots, (x_m^0, x_m^1), (\sigma_1, \ldots, \sigma_m)) \to (\lambda, (x_1^{\sigma_1}, \ldots, x_m^{\sigma_m}))$$

Our protocol for computing this functionality works by running the basic protocol in parallel, using the same homomorphic encryption key in each execution. This yields higher efficiency and the number of asymmetric operations per transfer is essentially two. We denote a protocol that securely realizes $\mathcal{F}_{\mathrm{OT}}^m$ with one-sided simulation by π_{OT}^m.

Our protocol that achieves security for covert adversaries needs an oblivious transfer protocol that is secure for covert adversaries. Such a protocol was presented in [2] and essentially requires 4 exponentiations only per execution.

Oblivious pseudorandom function evaluation. Let $(I_{\mathrm{PRF}}, F_{\mathrm{PRF}})$ be an ensemble of pseudorandom functions, where I_{PRF} is a probabilistic polynomial-time algorithm that generates keys (or more exactly, that samples a function from the ensemble). The task of oblivious pseudorandom function evaluation with F_{PRF} is that of securely computing the functionality $\mathcal{F}_{\mathrm{PRF}}$ defined by

$$(k, x) \mapsto (\lambda, F_{\mathrm{PRF}}(k, x)) \tag{1}$$

where $k \leftarrow I_{\mathrm{PRF}}(1^n)$ and $x \in \{0, 1\}^n$.[1] We will use the Naor-Reingold [20] pseudorandom function ensemble F_{PRF} (with some minor modifications). For every n,

---

[1] If k is not a "valid" key in the range of $I_{\mathrm{PRF}}(1^n)$, then we allow the function to take any arbitrary value. This simplifies our presentation.

the function's key is the tuple $k = (p, q, g^{a_0}, a_1, \ldots, a_n)$, where p is a prime, q is an n-bit prime divisor of $p - 1$, $g \in Z_p^*$ is of order q, and $a_0, a_1, \ldots, a_n \in_R Z_q^*$. (This is slightly different from the description in [20] but makes no difference to the pseudorandomness of the ensemble.) The function itself is defined by

$$F_{\mathrm{PRF}}(k, x) = g^{a_0 \cdot \prod_{i=1}^n a_i^{x_i}} \bmod p$$

We remark that this function is not pseudorandom in the classic sense of it being indistinguishable from a random function whose range is composed of all strings of a given length. Rather, it is indistinguishable from a random function whose range is the group generated by g as defined above. This suffices for our purposes. A protocol for oblivious pseudorandom function evaluation of this function was presented in [8] and involves the parties running an oblivious transfer execution for every bit of the input x. In the full version we prove that the protocol of [8] preserves the security level of the oblivious transfer used (whether it be full security, one-sided simulatability, or security in the presence of covert adversaries). Using the oblivious transfer of [2] we therefore have that for $x \in \{0, 1\}^\ell$, the cost of securely computing $\mathcal{F}_{\mathrm{PRF}}$ in the presence of covert adversaries is essentially 4ℓ exponentiations. We remark that by using a multi-oblivious transfer protocol, we can run many executions of π_{PRF} simultaneously. This is of great importance for efficiency.

3   Secure Set-Intersection

In this section we present our main result. We show how to securely compute the two-party set-intersection functionality $\mathcal{F}_\cap$, where each party enters a *set* of values from some predetermined domain. If the input sets are legal, i.e. they are made up of distinct values, then the functionality sends the intersection of these inputs to P_2 and nothing to P_1. Otherwise P_2 is given $\perp$. Let X and Y denote the respective input sets of P_1 and P_2, and let the domain of elements be $\{0, 1\}^{p(n)}$ for some known polynomial $p(n)$. We assume that $p(n) = \omega(\log n)$; this is needed for proving security and can always be achieved by padding the elements if necessary. Functionality $\mathcal{F}_\cap$ is defined by:

$$(X, Y) \mapsto \begin{cases} (\lambda, X \cap Y), & \text{if } X, Y \subseteq \{0, 1\}^{p(n)} \text{ and are legal sets} \\ (\lambda, \perp), & \text{otherwise} \end{cases}$$

We present two protocols in this section: the first achieves one-sided simulatability in the presence of malicious adversaries, and the second achieves security in the presence of covert adversaries with deterrent $\epsilon = 1/2$.

3.1   Secure Set Intersection with One-Sided Simulatability

The basic idea behind this protocol was described in the introduction. We therefore proceed directly to the protocol, which uses a subprotocol π_{PRF} that securely computes $\mathcal{F}_{\mathrm{PRF}}$ with one-sided simulatability (functionality $\mathcal{F}_{\mathrm{PRF}}$ was defined in Eq. (1) above).

Protocol π_{INT}

- **Inputs:** The input of P_1 is X where $X \subseteq \{0,1\}^{p(n)}$ contains m_1 items, and the input of P_2 is Y where $Y \subseteq \{0,1\}^{p(n)}$ contains m_2 items.
- **Auxiliary inputs:** Both parties have the security parameter 1^n and the polynomial p bounding the lengths of all elements in X and Y. In addition, P_1 is given m_2 (the size of Y) and P_2 is given m_1 (the size of X).
- **The protocol:**
    1. Party P_1 chooses a key $k \leftarrow I_{\mathrm{PRF}}(1^{p(n)})$ for the pseudorandom function. Then, the parties run m_2 parallel executions of π_{PRF}. P_1 enters the key k chosen above in all of the executions, whereas P_2 enters a different value $y \in Y$ in each execution. The output of P_2 from these executions is the set $U = \{(F_{\mathrm{PRF}}(k,y))\}_{y \in Y}$.
    2. P_1 sends P_2 the set $V = \{F_{\mathrm{PRF}}(k,x)\}_{x \in X}$ in a randomly permuted order, where k is the same key P_1 used in Protocol π_{PRF} in the previous step.
    3. P_2 outputs all y's for which $F_{\mathrm{PRF}}(k,y) \in V$. I.e., for every y let f_y be the output of P_2 from π_{PRF} when it used input y. Then, P_2 outputs the set $\{y \mid f_y \in V\}$.

Theorem 3. *Assume that π_{PRF} securely computes $\mathcal{F}_{\mathrm{PRF}}$ with one-sided simulation. Then π_{INT} securely computes $\mathcal{F}_{\cap}$ with one-sided simulation.*

Proof Sketch: In the case that P_1 is corrupted we need only show that P_1 learns nothing about P_2's inputs. This follows from the fact that the only messages that P_1 receives are in the executions of π_{PRF} which also reveals nothing about P_2's input to P_1. The formal proof of this follows from a standard hybrid argument.

We now proceed to the case that P_2 is corrupted; here we must present a simulator but can also rely on the fact that the π_{PRF} subprotocol is simulatable. Thus, we can analyze the security of π_{INT} in a hybrid model where a trusted party computes $\mathcal{F}_{\mathrm{PRF}}$ for the parties. In this model, P_1 and P_2 just send their inputs to π_{PRF} to the trusted party. Thus, the simulator $\mathcal{S}$ for $\mathcal{A}$ who controls P_2 receives $\mathcal{A}$'s inputs $y_1, \ldots, y_{m_2}$ to the pseudorandom function evaluations. $\mathcal{S}$ chooses a unique random value z_i for each distinct y_i and hands it to $\mathcal{A}$ as its output in the ith evaluation. $\mathcal{S}$ then sends $y_1, \ldots, y_{m_2}$ to the trusted party computing $\mathcal{F}_{\cap}$ and receives back a subset of the values (this is the output $X \cap Y$); let t be the number of values in the subset. $\mathcal{S}$ completes $X \cap Y$ with a set of $m_1 - t$ random values of length $p(n)$ each, computes the set V from this set as an honest P_1 would and hands it to $\mathcal{A}$.[2] Finally, $\mathcal{S}$ outputs whatever $\mathcal{A}$ outputs. The proof is completed by proving that the ability to distinguish the simulation from a real execution can be converted into the ability to distinguish the pseudorandom function from random. ∎

Efficiency. Note first that since π_{PRF} can be run in parallel and has only a constant number of rounds, protocol π_{INT} also has only a constant number of rounds. Next, the number of exponentiations is $O(m_2 \cdot p(n) + m_1)$. This is due to the fact that each local computation of the Naor-Reingold pseudorandom function can be carried out with just one modular exponentiation and n modular multiplications

---

[2] Since $p(n)$ is superlogarithmic, the probability that any of the random values sent by $\mathcal{S}$ are in P_1's input set is negligible.

(which are equivalent to another exponentiation). Thus, computing the set V requires $O(m_1)$ exponentiations. In addition, for inputs of length $p(n)$, Protocol π_{PRF} consists of running $p(n)$ oblivious transfers (each requiring $O(1)$ exponentiations). Thus m_2 such executions require $O(m_2 \cdot p(n))$ exponentiations. We remark that since $p(n)$ is the size of the input elements it is typically quite small (e.g., the size of an SSN). If this is not the case, then the input can be hashed to a fixed size using a collision-resistant hash function. Thus, $m_2 \cdot p(n) + m_1$ will typically be much smaller than $m_1 \cdot m_2$. (Recall that we do need to assume that $p(n)$ is large enough so that a randomly chosen string does not intersect with any of the sets except with very small probability. However, this can still be quite small.)

We remark that our protocol is much more efficient than that of [14] (although they achieve full simulatability). This is due to the fact that in their protocol every party P_i is required to execute $O(m_1 \cdot m_2)$ zero-knowledge proofs of knowledge, and a similar number of asymmetric computations. (Many of these proofs can be made efficient but not all. In particular, their protocol is only secure as long as the players prove that they do not send the all-zero polynomial. However, no efficient protocols for proving this are known.)

3.2   Secure Set Intersection in the Presence of Covert Adversaries

In this section we present a protocol for securely computing set-intersection in the presence of covert adversaries. Our protocol is based on the high-level idea demonstrated in protocol π_{INT} (achieving one-sided simulation for malicious adversaries). In order to motivate this protocol, we explain why π_{INT} cannot be simulated in the case that P_1 is corrupted. The problem arises from the fact that P_1 may use different keys in the different evaluations of π_{PRF} and in the computation of V. In such a case, the simulator cannot construct a set of values X that corresponds with P_1's behavior. Another problem that arises is that if P_1 can choose the key k by itself, then it can make it so that for some distinct values y and y' it holds that $F_{\mathrm{PRF}}(k, y) = F_{\mathrm{PRF}}(k, y')$. This enables P_2 to effectively make its set X larger, affecting the size of the intersection. Needless to say, this strategy cannot be carried out in the ideal model. Thus, the main objective of the additional steps in our protocol below is to ensure that P_1 uses the same *randomly chosen* k in *all* of the π_{PRF} evaluations as well as in the construction V. This is achieved in the following ways. First, the parties run two series of executions of the π_{PRF} protocol where in one execution real values are used and in the other dummy values are used. Party P_2 then checks that P_1 used the same key in all of dummy executions. This check is carried out by having P_1 and P_2 generate the randomness that P_1 should use in these subprotocols by coin tossing (where P_1 receives coins and P_2 receives a commitment to those coins). Then, P_1 simply reveals the coins used in the dummy series and P_2 can fully verify its behavior. Second, P_1 and P_2 first apply a pseudorandom permutation to their inputs and then a pseudorandom function. Then, P_1 sends two sets V_0 and V_1, and opens one of them to P_2 in order to prove that it was constructed by applying the pseudorandom function with the *same key* as used in

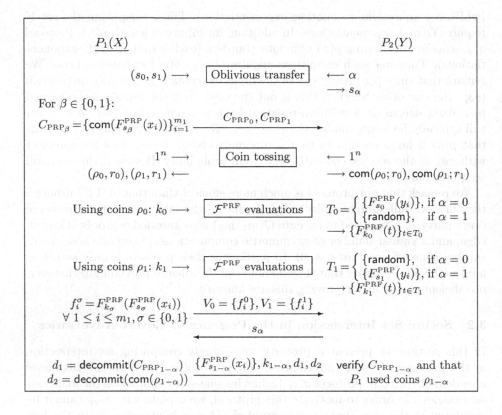

Fig. 1. A high-level diagram of our protocol

the dummy evaluations. The reason that the pseudorandom permutation is first applied is to hide P_1's values from P_2 when one of the sets V_0, V_1 is "opened". The difficulty in implementing this idea is to devise a way that P_2 can compute the intersection and check all of the above, without revealing more about P_1's input than allowed. Technically, this is achieved by having V_0 equal the set of values $F_{\mathrm{PRF}}(k_0, F_{\mathrm{PRP}}(s_0, x))$ and having V_1 equal the values $F_{\mathrm{PRF}}(k_1, F_{\mathrm{PRP}}(s_1, x))$. Then, P_2 learns either (k_0, s_1) or (k_1, s_0). In this way, it cannot derive any information from the sets (it only knows one of the keys). However, it is enough to check P_1's behavior. A high-level overview of the protocol appears in Figure 1 and the full description (starting with the tools that we use) follows below.

Tools: Our protocol uses the following primitives and subprotocols:

– A pseudorandom permutation with sampling algorithm I_{PRP}. We denote a sampled key by s and the computation of the permutation with key s and input x by $F_{\mathrm{PRP}}(s, x)$.
– A pseudorandom function with sampling algorithm I_{PRF}. We denote a sampled key by k and the computation of the permutation with key k and input x by $F_{\mathrm{PRF}}(k, x)$.

- A perfectly-binding commitment scheme com; we denote by $\text{com}(x; r)$ the commitment to a string x using random coins r.
- An oblivious transfer protocol that is secure in the presence of covert adversaries with deterrent $\epsilon = 1/2$ and can be run in parallel. An efficient protocol that achieves this was presented in [2]. We denote this protocol by π_{OT}.
- An efficient coin-tossing protocol that is secure in the presence of covert adversaries with deterrent $\epsilon = 1/2$. Such a protocol can be constructed by using the protocol of [16], with commitments based on El-Gamal encryption [6] (this enables highly efficient zero-knowledge proofs; see the full version). The exact functionality we need is not plain coin-tossing but rather $(1^n, 1^n) \mapsto ((\rho, r), \text{com}(\rho; r))$ where $\rho \in_R \{0,1\}^n$ and r is random and of sufficient length for committing to ρ. We denote this protocol by π_{CT}.
- A protocol π_{PRF} for computing $\mathcal{F}_{\text{PRF}}$ as defined in Eq. (1), that is secure in the presence of covert adversaries with $\epsilon = 1/2$; see Section 2.2.

We are now ready to present our protocol.

Protocol $\pi_\cap$

- **Inputs:** The input of P_1 is X where $X \subseteq \{0,1\}^{p(n)}$ contains m_1 items, and the input of P_2 is Y where $Y \subseteq \{0,1\}^{p(n)}$ contains m_2 items.
- **Auxiliary inputs:** Both parties have the security parameter 1^n and the polynomial p bounding the lengths of all elements in X and Y. In addition, P_1 is given m_2 (the size of Y) and P_2 is given m_1 (the size of X).
- **The protocol:**
  1. *Oblivious transfer (secure in the presence of covert adversaries):*
     (a) Party P_1 chooses a pair of keys $s_0, s_1 \leftarrow \mathcal{I}_{\text{PRP}}(1^{p(n)})$ for a PRP.
     (b) Party P_2 chooses a random bit $\alpha \in_R \{0,1\}$.
     (c) P_1 and P_2 execute the oblivious transfer protocol π_{OT}. P_1 inputs the keys s_0 and s_1 and plays the sender, and P_2 inputs α and plays the receiver. If one of the parties receives corrupt_i or abort_i as output, it outputs it and halts. Otherwise P_2 receives s_α.
  2. P_1 computes $C_{\text{PRP}_0} = \{\text{com}(F_{\text{PRP}}(s_0, x))\}_{x \in X}$, $C_{\text{PRP}_1} = \{\text{com}(F_{\text{PRP}}(s_1, x))\}_{x \in X}$ and sends C_{PRP_0} and C_{PRP_1} to P_2.
  3. The parties run the coin-tossing protocol π_{CT} computing $(1^{q(n)}, 1^{q(n)}) \to ((\rho, r), \text{com}(\rho; r))$ twice, where $q(n)$ is the number of random bits needed to both choose a key $k \leftarrow \mathcal{I}_{\text{PRF}}(1^{p(n)})$ and run m_2 executions of the PRF protocol (see below). Party P_1 receives for output (ρ_0, r_0) and (ρ_1, r_1), and P_2 receives $c_{\rho_0} = \text{com}(\rho_0; r_0)$ and $c_{\rho_1} = \text{com}(\rho_1; r_1)$, where ρ_0, ρ_1 are each of length $q(n)$.
  4. *Run oblivious PRF evaluations:*
     (a) The parties run m_2 executions of the oblivious PRF evaluation protocol π_{PRF} *in parallel*, in which P_1 inputs the same randomly chosen key $k_0 \leftarrow \mathcal{I}_{\text{PRF}}(1^{p(n)})$ in each execution, and P_2 enters the elements of the set
     $$T_0 = \{F_{\text{PRP}}(s_0, y)\}_{y \in Y} \text{ (if } \alpha = 0\text{), and } m_2 \text{ random values of size } p(n) \text{ (if } \alpha = 1\text{). Let } U_0 \text{ be the set of outputs received by } P_2 \text{ in these executions. The randomness used by } P_1 \text{ in all of the executions (and for choosing the key } k_0\text{) is the string } \rho_0 \text{ from the coin-tossing above.}$$

    (b) The parties run another m_2 executions of π_{PRF} *in parallel,* in which P_1 inputs the same randomly chosen key $k_1 \leftarrow \mathcal{I}_{\mathrm{PRF}}(1^{p(n)})$ each time, and P_2 enters m_2 random values of size $p(n)$ (if $\alpha = 0$), and the elements of the set $T_1 = \{F_{\mathrm{PRP}}(s_1, y)\}_{y \in Y}$ (if $\alpha = 1$). Let U_1 be the set of outputs received by P_2 in these executions. The randomness used by P_1 in all of the executions (and for choosing the key k_1) is the string ρ_1 from the coin-tossing above.

5. P_1 computes and sends P_2 the sets of values $V_0 = \{F_{\mathrm{PRF}}(k_0, F_{\mathrm{PRP}}(s_0, x))\}_{x \in X}$ and $V_1 = \{F_{\mathrm{PRF}}(k_1, F_{\mathrm{PRP}}(s_1, x))\}_{x \in X}$, in randomly permuted order.

6. *Run checks:*

    (a) If either $|V_0|$ or $|V_1|$ are smaller than m_1 or not distinct, P_2 outputs corrupted$_1$, otherwise it sends P_1 the key s_α.

    (b) If P_2 sends s such that $s \notin \{s_0, s_1\}$, then P_1 halts. Otherwise, P_1 sets α such that $s = s_\alpha$. Then, P_1 sends P_2 the decommitments for all values in the set $C_{\mathrm{PRP}_{1-\alpha}}$, and the decommitment of $c_{\rho_{1-\alpha}}$.

    (c) Let $W_{1-\alpha}$ denote the opening of $C_{\mathrm{PRP}_{1-\alpha}}$ and $\rho_{1-\alpha}$ the opening of $c_{\rho_{1-\alpha}}$. First, P_2 checks that the responses of P_1 to its messages in the m_2 executions of the PRF evaluations in which it input random strings are exactly the responses of an honest P_1 using random coins $\rho_{1-\alpha}$ to generate $k_{1-\alpha}$ and run the subprotocols. Furthermore, P_2 checks that $V_{1-\alpha} = \{F_{\mathrm{PRF}}(k_{1-\alpha}, w)\}_{w \in W_{1-\alpha}}$ using $k_{1-\alpha}$ as above. In case the above does not hold, P_2 outputs corrupted$_1$. Otherwise, let f_y be the output received by P_2 from the PRF evaluation in which it input $F_{\mathrm{PRP}}(s_\alpha, y)$. Party P_2 outputs the set $\{y \mid f_y \in V_\alpha\}$.

We now prove the security of the protocol:

Theorem 4. *Assume that $\pi_{\mathrm{OT}}, \pi_{\mathrm{CT}}, \pi_{\mathrm{PRF}}$ are secure in the presence of covert adversaries with deterrent $\epsilon = \frac{1}{2}$, and assume that* com *is a perfectly-binding commitment scheme and that F_{PRF} and F_{PRP} are pseudorandom function and permutation families, respectively. Then Protocol $\pi_\cap$ securely computes the set-intersection functionality $\mathcal{F}_\cap$ in the presence of covert adversaries with $\epsilon = \frac{1}{2}$.*

Proof: We will separately consider the case that P_1 is corrupted and the case that P_2 is corrupted. The case where both parties are honest is straightforward and therefore omitted. We present the proof in a hybrid model in which a trusted party is used to compute the oblivious transfer and coin-tossing computations. We denote these functionalities by $\mathcal{F}_{\mathrm{OT}}$ and $\mathcal{F}_{\mathrm{CT}}$. (Unfortunately, we cannot do the same for π_{PRF} because P_1 needs to use the coins ρ_0, ρ_1 in the protocol.)

Party P_1 is corrupted. Let $\mathcal{A}$ be an adversary controlling the party P_1; we construct a simulator $\mathcal{S}$ as follows:

1. $\mathcal{S}$ receives X and z, and invokes $\mathcal{A}$ on this input.
2. $\mathcal{S}$ plays the trusted party for the oblivious transfer execution with $\mathcal{A}$ as the sender, and receives the input that $\mathcal{A}$ sends to the trusted party:

(a) If this input is abort_1 or $\mathsf{corrupted}_1$, then $\mathcal{S}$ sends abort_1 or $\mathsf{corrupted}_1$ (respectively) to the trusted party computing $\mathcal{F}_\cap$, simulates P_2 aborting and halts (outputting whatever $\mathcal{A}$ outputs).

(b) If the input is cheat_1, then $\mathcal{S}$ sends cheat_1 to the trusted party. If it receives back $\mathsf{corrupted}_1$, then it hands $\mathcal{A}$ the message $\mathsf{corrupted}_1$ as if it received it from the trusted party, simulates P_2 aborting and halts (outputting whatever $\mathcal{A}$ outputs). If it receives back $\mathsf{undetected}$ (and the input set Y of the honest P_2) then $\mathcal{S}$ proceeds as follows. First, it hands $\mathcal{A}$ the message $\mathsf{undetected}$ together with a random α that $\mathcal{A}$ expects to receive (as P_2's input to π_{OT}). Next, it uses the input Y of P_2 that it obtained in order to perfectly emulate P_2 in the rest of the execution. That is, it runs P_2's honest strategy with input Y while interacting with $\mathcal{A}$ playing P_1 for the rest of the execution. Let Z be the output for P_2 that it receives. $\mathcal{S}$ sends Z to the trusted party (for P_2's output) and outputs whatever $\mathcal{A}$ outputs. The simulation ends here in this case.

(c) If the input is a pair of keys s_0, s_1, $\mathcal{S}$ proceeds with the simulation below.[3]

3. $\mathcal{S}$ receives from $\mathcal{A}$ two sets of commitments C_{PRP_0} and C_{PRP_1}.

4. $\mathcal{S}$ receives from $\mathcal{A}$ its input for $\mathcal{F}_{\mathrm{CT}}$. In case it equals abort_1, $\mathsf{corrupted}_1$, or cheat_1, then $\mathcal{S}$ behaves exactly as above in the OT execution. Otherwise $\mathcal{S}$ chooses random (ρ_0, r_0) and (ρ_1, r_1) of the appropriate length and hands them to $\mathcal{A}$.

5. $\mathcal{S}$ runs the simulator $\mathcal{S}_{\mathrm{PRF}}$ guaranteed to exist for the protocol π_{PRF} (by the assumption that it is secure) on the residual $\mathcal{A}$ at this point (i.e., $\mathcal{S}$ defines an adversary $\mathcal{A}'$ that is just $\mathcal{A}$ with the messages sent until now hardwired into it). If $\mathcal{S}_{\mathrm{PRF}}$ wishes to send abort_1, $\mathsf{corrupted}_1$ or cheat_1 in any of the executions, then $\mathcal{S}$ acts exactly as above. Otherwise, $\mathcal{S}$ proceeds. Let t be the transcript of messages sent by $\mathcal{A}$ in the simulated view of π_{PRF} as generated by $\mathcal{S}_{\mathrm{PRF}}$ (we define the residual $\mathcal{A}$ so that it outputs this transcript and so this is also what is output by $\mathcal{S}_{\mathrm{PRF}}$).

6. $\mathcal{S}$ receives from $\mathcal{A}$ two sets of computed values V_0 and V_1. If they are not of size m_1 or not distinct, $\mathcal{S}$ sends $\mathsf{corrupted}_1$ to the trusted party, simulates P_2 aborting and halts (outputting whatever $\mathcal{A}$ outputs).

7. Otherwise, $\mathcal{S}$ hands $\mathcal{A}$ the key s_0 and receives back $\mathcal{A}$'s decommitments of C_{PRP_1} and c_{ρ_1}. $\mathcal{S}$ then rewinds $\mathcal{A}$, hands it s_1 and receives back its decommitments of C_{PRP_0} and c_{ρ_0}. Simulator $\mathcal{S}$ runs the same checks as an honest P_2 would run (it uses the transcript t to check that $\mathcal{A}$ acted honestly using the randomness ρ_0, ρ_1). We have two cases:

(a) *Case 1 – all of the checks carried by $\mathcal{S}$ in both rewindings pass:* Let k_0 and k_1 denote the keys that an honest P_1 would have used in the PRF evaluations when its coins are ρ_0 and ρ_1, respectively (where ρ_b is value committed to in c_{ρ_b}). Then, $\mathcal{S}$ chooses a random bit $\alpha \in_R \{0,1\}$ and sends the trusted party the set $\{F^{-1}_{\mathrm{PRP}}(s_\alpha, w)\}_{w \in W_\alpha}$.

(b) *Case 2 – there exists a bit $\alpha \in \{0,1\}$ so that the checks when $\mathcal{S}$ sent $s_{1-\alpha}$ failed:* Simulator $\mathcal{S}$ sends cheat_1 to the trusted party. If it receives

---

[3] We assume a mapping from *any* string to a valid key for the pseudorandom permutation.

back corrupted$_1$ then it rewinds $\mathcal{A}$ and sends it $s_{1-\alpha}$ again. If it receives back undetected then it rewinds $\mathcal{A}$ and sends it s_α. Then, it runs the last step of the protocol exactly as P_2 would, using P_2's real input. $\mathcal{S}$ then sends the trusted party whatever P_2 would output in the ideal model.

8. $\mathcal{S}$ outputs whatever $\mathcal{A}$ outputs and halts.

Let $\epsilon = \frac{1}{2}$. We prove that for every $X \subseteq \{0,1\}^{p(n)}$ of size m_1 and $Y \subseteq \{0,1\}^{p(n)}$ of size m_2, and every $z \in \{0,1\}^*$

$$\left\{\text{IDEALSC}^\epsilon_{\mathcal{F}_\cap, \mathcal{S}(z), 1}(X, Y, n)\right\}_{n \in N} \stackrel{c}{\equiv} \left\{\text{HYBRID}^{\text{OT,CT}}_{\pi_\cap, \mathcal{A}(z), 1}(X, Y, n)\right\}_{n \in N}$$

Recall that in the above $\{\mathcal{F}_{\text{OT}}, \mathcal{F}_{\text{CT}}\}$-hybrid model, the view of P_1 includes its output from $\mathcal{F}_{\text{CT}}$, the messages sent during the π_{PRF} executions, and the value s_α that P_2 sends after receiving V_0 and V_1. Thus the only difference between the hybrid and ideal executions is within the π_{PRF} executions. This is due to the fact that $\mathcal{S}$ invokes $\mathcal{S}_{\text{PRF}}$ whereas in a hybrid execution a real π_{PRF} execution is run between P_1 and P_2. Clearly, the views of $\mathcal{A}$ in these executions are computationally indistinguishable. The more interesting challenge is thus to prove that the joint output distributions of P_2 and these views are computationally indistinguishable.

We consider three different cases. In the first case $\mathcal{A}$'s input to $\mathcal{F}_{\text{OT}}$ or $\mathcal{F}_{\text{CT}}$ is either corrupted$_1$, abort$_1$ or cheat$_1$. Let bad$_1$ denote this event. In this case, the execution is either aborted (with P_2 receiving abort$_1$ or corrupted$_1$) or $\mathcal{S}$ receives the honest P_2's full input with which to perfectly complete the simulation. Thus,

$$\{\text{IDEALSC}^\epsilon_{\mathcal{F}_\cap, \mathcal{S}(z), 1}(X, Y, n) \mid \text{bad}_1\} \equiv \{\text{HYBRID}^{\text{OT,CT}}_{\pi, \mathcal{A}(z), 1}(X, Y, n) \mid \text{bad}_1\}$$

In the second case, $\mathcal{A}$ provides valid inputs for $\mathcal{F}_{\text{OT}}$ and $\mathcal{F}_{\text{CT}}$, yet there exists an $\alpha \in \{0,1\}$ value for which $\mathcal{A}$ does not provide a valid response in Step 6 of the protocol; denote this event by bad$_2$. Now, if P_2 sent α to $\mathcal{F}_{\text{OT}}$ then $\mathcal{A}$ cannot deviate from the protocol within the π_{PRF} executions on $T_{1-\alpha}$ without *definitely* getting caught by P_2 (and the simulator). Thus, in both the hybrid and ideal executions, P_2 outputs corrupted$_1$ with the same probability. Furthermore, when it does not output corrupted$_1$, simulator $\mathcal{S}$ concludes the simulation with P_2's real input (note that although these inputs are already used earlier in π_{PRF}, since $\mathcal{S}$ knows the values k_0, k_1 it can conclude the simulation even when receiving P_2's inputs later). Thus, the only difference is that in the real protocol, the π_{PRF} executions are run with P_2's inputs whereas in the simulation $\mathcal{S}_{\text{PRF}}$ is used. By the security of $\mathcal{S}_{\text{PRF}}$ we have:

$$\{\text{IDEALSC}^\epsilon_{\mathcal{F}_\cap, \mathcal{S}(z), 1}(X, Y, n) \mid \text{bad}_2\} \stackrel{c}{\equiv} \{\text{HYBRID}^{\text{OT,CT}}_{\pi, \mathcal{A}(z), 1}(X, Y, n) \mid \text{bad}_2\}$$

The last case we need to consider is when neither bad$_1$ nor bad$_2$ occur; denote this event by $\neg$bad. Let k_0 and k_1 be the keys that $\mathcal{A}$ used in all of the π_{PRF} executions, and let s_0 and s_1 be the values that $\mathcal{A}$ input to the oblivious transfer. Then we have the following claim:

Claim 5. *Let* $X_\alpha = \{F_{\mathrm{PRP}}^{-1}(s_\alpha, w)\}_{w \in W_\alpha}$ *and consider the event* $\neg$bad *where neither* bad_1 *nor* bad_2 *occur. Then, for every* $\alpha \in \{0,1\}$ *and set* $Y \subseteq \{0,1\}^{p(n)}$, *it holds that* $z \in X_\alpha \cap Y$ *if and only if* $F_{\mathrm{PRF}}(k_\alpha, F_{\mathrm{PRP}}(s_\alpha, z)) \in V_\alpha \cap U_\alpha$, *except with negligible probability.*

Proof Sketch: If $z \in X_\alpha \cap Y$, then $F_{\mathrm{PRF}}(k_\alpha, F_{\mathrm{PRP}}(s_\alpha, z)) \in V_\alpha \cap U_\alpha$ because $\mathcal{A}$ uses the same key k_α for the PRF evaluation that defines U_α and for computing V_α. If this were not the case, then $\mathcal{A}$ would be caught cheating with probability at least $1/2$ (whereas here we are dealing with the case that $\mathcal{A}$ provides answers that never result in it being caught cheating).

As for the other direction, assume that $F_{\mathrm{PRF}}(k_\alpha, F_{\mathrm{PRP}}(s_\alpha, z)) \in V_\alpha \cap U_\alpha$. Then a problem can arise if there exist $y \in Y$ and $x \in X$ such that $x \neq y$ and yet $F_{\mathrm{PRF}}(k_\alpha, F_{\mathrm{PRP}}(s_\alpha, x)) = F_{\mathrm{PRF}}(k_\alpha, F_{\mathrm{PRP}}(s_\alpha, y))$. If $\mathcal{A}$ could choose X after k_α is known, then it could indeed cause such an event to happen. However, notice that $\mathcal{A}$ is committed to its inputs (in C_{PRP_0} and C_{PRP_1}) before k_α is chosen in the coin tossing. Thus, the probability that such a "collision" occurs, where the probability is taken over the choice of k_α and the sets X and Y are already fixed, is negligible (or else F_{PRF} can be distinguished from random). $\blacksquare$

This implies that the output received by P_2 in the hybrid and ideal executions is the same (except with negligible probability). Combining this with the fact that the view of $\mathcal{A}$ is clearly indistinguishable in both executions, we have:

$$\{\mathrm{IDEALSC}_{\mathcal{F}_\cap, \mathcal{S}(z),1}^\epsilon(X,Y,n) \mid \neg\mathsf{bad}\} \stackrel{c}{\equiv} \{\mathrm{HYBRID}_{\pi,\mathcal{A}(z),1}^{\mathrm{OT},\mathrm{CT}}(X,Y,n) \mid \neg\mathsf{bad}\}$$

Combining the above three cases, and noting that the events bad_1 and bad_2 happen with probability that is negligibly close in the hybrid and ideal executions, we have that the output distributions are computationally indistinguishable, as required.

Party P_2 is corrupted. Let $\mathcal{A}$ be an adversary controlling party P_2. We construct a simulator $\mathcal{S}$ as follows:

1. $\mathcal{S}$ receives Y and z, and invokes $\mathcal{A}$ on this input.
2. $\mathcal{S}$ plays the trusted party for the oblivious transfer execution with $\mathcal{A}$ as the receiver. $\mathcal{S}$ receives the input that $\mathcal{A}$ sends to the trusted party. If this input is abort_2, $\mathsf{corrupted}_2$ or cheat_2, then $\mathcal{S}$ works in an analogous way as when this occurs in the simulation when P_1 is corrupted.
   If the input equals a bit α, then $\mathcal{S}$ samples a key $s_\alpha \leftarrow \mathcal{I}_{\mathrm{PRP}}(1^{p(n)})$ as the honest P_1 does, and hands it to $\mathcal{A}$ emulating $\mathcal{F}_{\mathrm{OT}}$'s answer. $\mathcal{S}$ samples a second key $s_{1-\alpha} \leftarrow \mathcal{I}_{\mathrm{PRP}}(1^{p(n)})$ as above, and keeps it for later.
3. $\mathcal{S}$ sends $\mathcal{A}$ two sets of m_2 commitments C_{PRP_0} and C_{PRP_1} to distinct random values of length $p(n)$.
4. $\mathcal{S}$ receives from $\mathcal{A}$ its input for $\mathcal{F}_{\mathrm{CT}}$. In case it equals abort_2, $\mathsf{corrupted}_2$, or cheat_2, then $\mathcal{S}$ behaves exactly as above in the OT execution. Otherwise $\mathcal{S}$ chooses random (ρ_0, r_0) and (ρ_1, r_1) of the appropriate length and hands $c_{\rho_0} = \mathrm{com}(\rho_0; r_0)$ and $c_{\rho_1} = \mathrm{com}(\rho_1; r_1)$ to $\mathcal{A}$.

5. S simulates the PRF evaluations as follows. If $\alpha = 0$ (where α is $\mathcal{A}$'s input to the oblivious transfer), then S runs the simulator S_{PRF} on the residual $\mathcal{A}$ for the first m_2 executions, and follows the honest P_1's instructions using random coins ρ_1 for the second m_2 executions (where the "first" and "second" set is as in the order described in the protocol). In contrast, if $\alpha = 1$, then S follows the honest P_1's instructions using random coins ρ_0 for the first m_2 executions and runs the simulator S_{PRF} on the residual $\mathcal{A}$ for the second m_2 executions.

In the m_2 executions simulated by S_{PRF}, simulator S receives the input that S_{PRF} wishes to send to the trusted party as its input in the PRF executions:

(a) If any of these inputs is abort_2, $\mathsf{corrupted}_2$, or cheat_2, then S behaves exactly as above in the OT execution.

(b) Else, let T' denote the set of m_2 elements (with length bounded by $p(n)$) that S_{PRF} wishes to send as $\mathcal{A}$'s inputs to π_{PRF}. Then S hands S_{PRF} the set $\{F_{\mathrm{PRF}}(k_\alpha, t)\}_{t \in T'}$ as its output from the trusted party, where $k_\alpha \leftarrow I_{\mathrm{PRF}}(1^{p(n)})$ is a randomly generated key. In addition, S defines the set $Y' = \{F_{\mathrm{PRP}}^{-1}(s_\alpha, t)\}_{t \in T'}$. (If Y' is not exactly of size m_2, then S adds $m_2 - |Y'|$ random elements of size $p(n)$; recall that $p(n) = \omega(\log n)$ and so random values are in the intersection with only negligible probability.)

6. S sends the trusted party computing $\mathcal{F}_\cap$ the set Y' that it recorded and receives back for output the set Z (note $Z = X \cap Y'$). Then it chooses $m_2 - |Z|$ distinct random elements and adds them to Z. Finally, S computes and sends $\mathcal{A}$ the sets $V_\alpha = \{F_{\mathrm{PRF}}(k_\alpha, F_{\mathrm{PRP}}(s_\alpha, z))\}_{z \in Z}$ and $V_{1-\alpha} = \{F_{\mathrm{PRF}}(k_{1-\alpha}, w)\}_{\mathsf{com}(w) \in C_{\mathrm{PRP}_{1-\alpha}}}$. We remark that the elements of V_α are randomly permuted before being sent.

7. S receives from $\mathcal{A}$ the value s_α and responds with the decommitments of $C_{\mathrm{PRP}_{1-\alpha}}$ and the decommitment of $c_{\rho_{1-\alpha}}$. If $\mathcal{A}$ did not send s_α, then S halts.

8. S outputs whatever $\mathcal{A}$ outputs.

Let $\epsilon = \frac{1}{2}$. We prove that for every $X \subseteq \{0,1\}^{p(n)}$ of size m_1 and $Y \subseteq \{0,1\}^{p(n)}$ of size m_2, and every $z \in \{0,1\}^*$

$$\left\{ \mathrm{IDEALSC}_{\mathcal{F}_\cap, S(z), 2}^{\epsilon}(X, Y, n) \right\}_{n \in N} \stackrel{c}{\equiv} \left\{ \mathrm{HYBRID}_{\pi_\cap, \mathcal{A}(z), 2}^{\mathrm{OT,CT}}(X, Y, n) \right\}_{n \in N}$$

Note first that the simulation differs from a real execution with respect to how the sets C_{PRF_α} and V_α are generated, and with respect to the decommitments of $C_{\mathrm{PRF}_{1-\alpha}}$ (recall that in the real execution P_1 uses its input X for these computations whereas the simulator does not know X). Nevertheless, the views cannot be distinguished due to the hiding property of F_{PRF}, F_{PRP} and com. As in the previous analysis, we begin with the case where $\mathcal{A}$ sends abort_2, cheat_2 or $\mathsf{corrupted}_2$ to $\mathcal{F}_{\mathrm{OT}}$ or $\mathcal{F}_{\mathrm{CT}}$. Due to the similarity to the case were P_1 is corrupted we omit the details here. Let bad denote the event where $\mathcal{A}$ sends abort_2, $\mathsf{corrupted}_2$ or cheat_2. Then relying on the above discussion it holds that,

$$\{\mathrm{IDEALSC}_{\mathcal{F}_\cap, S(z), 2}^{\epsilon}(X, Y, n) \mid \mathsf{bad}\} \equiv \{\mathrm{HYBRID}_{\pi, \mathcal{A}(z), 2}^{\mathrm{OT,CT}}(X, Y, n) \mid \mathsf{bad}\}$$

Next we analyze the security of P_1 in case $\mathcal{A}$ provides valid inputs to $\mathcal{F}_{\mathrm{OT}}$ and $\mathcal{F}_{\mathrm{CT}}$, and prove through the following series of games that the output distributions are computationally indistinguishable. For lack of space in this abstract, we only sketch this part of the proof.

Game H_1: In the first game the simulator has access to an oracle $\mathcal{O}_{F_{\mathrm{PRP}}}$ for computing F_{PRP} such that instead of computing F_{PRP} using $s_{1-\alpha}$, it queries the oracle. Clearly the output distribution of the current and original simulation is identical.

Game H_2: In this game we replace $\mathcal{O}_{F_{\mathrm{PRP}}}$ with an oracle $\mathcal{O}_{H_{\mathrm{Perm}}}$ computing a truly random permutation while the rest of the execution is as above. Indistinguishability holds using a standard reduction.

Game H_3: The next game is identical to the previous one except that the simulator knows the real input X of P_1 but uses it only for the computation of $V_{1-\alpha}$ and $C_{\mathrm{PRP}_{1-\alpha}}$. Since the oracle is a truly random permutation, the distribution here is identical (note that X is a set and thus all items are distinct).

Game H_4: In this game the simulator is given an oracle $\mathcal{O}_{F_{\mathrm{PRF}}}$ for computing F_{PRF} (with a random key) which it uses instead of computing F_{PRF} using k_α. The only difference is that in H_3, the coins used to generate k_α are committed to in c_{ρ_α} whereas in H_4 the oracle uses a random key that is independent of those coins. The fact that these games are indistinguishable therefore follows from the hiding property of the commitment scheme. Note that the executions using $k_{1-\alpha}$ remain the same.

Game H_5: Next we replace $\mathcal{O}_{F_{\mathrm{PRF}}}$ with a truly random function $\mathcal{O}_{H_{\mathrm{Func}}}$; indistinguishability here follows from the pseudorandomness of F_{PRF}.

Game H_6: In this game we let the simulator query its PRF oracle on the real input set X of P_1. That is, the simulator uses X for the entire computation as the real party P_1. Now, since $\mathcal{O}_{H_{\mathrm{Func}}}$ is a truly random function, we have the same output distribution in both games.

Game H_7: Here we modify $\mathcal{O}_{H_{\mathrm{Perm}}}$ back into $\mathcal{O}_{F_{\mathrm{PRP}}}$. This replacement affects the PRP computation for the $(1 - \alpha)$th set of PRP evaluations.

Game H_8: In this game we modify $\mathcal{O}_{H_{\mathrm{Func}}}$ back into $\mathcal{O}_{F_{\mathrm{PRF}}}$.

Game H_{10}: Finally, we let the simulator conduct the PRF and PRP computations by itself. This does not affect the outputs of these functions, but as above a reduction to the hiding property of the commitment c_{ρ_α} is needed because now the coins used to generate the key k_α are committed to in c_{ρ_α}. In addition we let $\mathcal{S}$ compute C_{PRP_α} as the honest P_1 would. Since $\mathcal{S}$ is not required to decommit these commitments, it is again easy to reduce indistinguishability here to the hiding property of these commitments.

We therefore conclude that H_{10} is computationally indistinguishable from the (original) ideal simulation by $\mathcal{S}$. However, H_{10} is identical to the real execution in the hybrid model, completing the proof. ∎

Efficiency. We analyze the complexity of protocol $\pi_\cap$. We first count the number of asymmetric operations; in particular, modular exponentiations. Note that each invocation of π_{PRF} with inputs of length $p(n)$ requires $4p(n)+1$ exponentiations, because every invocation of the covert oblivious transfer requires at most 4 such computations, and π_{PRF} runs an oblivious transfer for every bit of P_2's input (one additional exponentiation is used for obtaining the final result). Given that there are $2m_2$ executions of π_{PRF}, we have that the number of exponentiations is approximately $8m_2 \cdot (p(n)+1) + m_1$. As we have already mentioned, $p(n)$ is expected to be quite small in most cases (and a collision-resistant hash function can be used when not). We note that our protocol is completely modular meaning that any protocol π_{PRF} for any pseudorandom function F_{PRF} can be used. Thus, the development of a more efficient protocol π_{PRF} will automatically result in our protocol also being more efficient. In terms of round efficiency, $\pi_{\cap}$ has a constant number of rounds due to the round efficiency of π_{OT} in the covert model, and the fact that all these executions are run in parallel.

4   Secure Pattern Matching

The basic problem of *pattern matching* is the following one: given a text T of length N and a pattern p of length m, find all the locations in the text where pattern p appears in the text. Stated differently, for every $i = 1, \ldots, N - m + 1$, let T_i be the substring of length m that begins at the ith position in T. Then, the basic problem of pattern matching is to return the set $\{i \mid T_i = p\}$. This problem has been intensively studied and can be solved optimally in time that is linear in size of the text [3,15].

In this section, we address the question of how to securely compute the above basic pattern matching functionality. The functionality, denoted $\mathcal{F}_{\mathrm{PM}}$, is defined by

$$((T, m), p) \mapsto \begin{cases} (\lambda, \{i \mid T_i = p\}) & \text{if } |p| \le m \\ (\lambda, \{i \mid T_i = p_1 \ldots p_m\}) & \text{otherwise} \end{cases}$$

where T_i is defined as above, T and p are binary strings and p_i is the ith bit in p. Note that P_1 who holds the text learns nothing about the pattern held by P_2, and the only thing that P_2 learns about the text held by P_1 is the locations where its pattern appears.

Although similar questions have been considered in the past (e.g., keyword search [8]), to the best of our knowledge, this is the first work considering the basic problem of pattern matching as described above. The main difference between keyword search and the problem that we consider here is that in keyword search, each keyword is assumed to appear only once. However, here the text is viewed as a stream and a pattern can appear multiple times. Furthermore, the strings $T_i, T_{i+1}, \ldots$ are dependent on each other (adjacent T_i's only differ in their first and last characters). Thus, it is not possible to apply a pseudorandom function to each T_i and use a protocol to securely compute $\mathcal{F}_{\mathrm{PRP}}$ on p as in the case of keyword search. Thus it seems that finding a secure solution for this problem is harder.

We present a protocol for securely computing $\mathcal{F}_{\mathrm{PM}}$ in the presence of *malicious adversaries* with *one-sided simulatability*. The basic idea behind our protocol is for P_1 and P_2 to run a *single* execution of π_{PRF} for securely computing a pseudo-random function with one-sided simulatability; let $f = F_{\mathrm{PRF}}(k,p)$ be the output received by P_2. Then, P_1 locally computes the pseudorandom function on T_i for every i and sends the results $\{F_{\mathrm{PRF}}(k,T_i)\}$ to P_2. Party P_2 can then find all the matches by just seeing where f appears in the series sent by P_1. Unfortunately, within itself, this is insufficient because P_2 can then detect repetitions within T. That is, if $T_i = T_j$ then P_2 will learn this because this implies that $F_{\mathrm{PRF}}(k,T_i) = F_{\mathrm{PRF}}(k,T_j)$. However, if $T_i \neq p$, this should not be revealed. We therefore include the index i of the subtext T_i in the computation and have P_1 send the values $F_{\mathrm{PRF}}(k,T_i\|\langle i\rangle)$ where $\langle i\rangle$ denotes the binary representation of i. This in turns generates another problem because now it is not possible for P_2 to see where p appears given only $F_{\mathrm{PRF}}(k,p)$; this is solved by having P_2 obtain $F_{\mathrm{PRF}}(k,p\|\langle i\rangle)$ for every i. Although this means that P_2 obtains n different outputs of F_{PRF} (because there are n different indices i), we utilize specific properties of the Naor-Reingold pseudorandom function, and the protocol π_{PRF} for computing it, in order to have P_2 obtain all of these values while running only a *single* execution of π_{PRF}. Due to lack of space, we defer the description of how this is achieved to the full version.

Protocol π_{PM}

- **Inputs:** The input of P_1 is a binary string T of size N, and the input of P_2 is a binary pattern p of size m.
- **Auxiliary Inputs:** the security parameter 1^n, and the input sizes N and m.
- **The protocol:**
  1. Party P_1 chooses a key for computing the Naor-Reingold function on inputs of length $m+\log N$; denote the key $k = (p,q,g^{a_0},a_1,\ldots,a_{m+\log N})$.
  2. The parties execute a modified version of π_{PRF} for computing the Naor-Reingold function, where P_1 enters the key k and P_2 enters its pattern p of length m. The modification is such that P_2's output is the *set* $\{f_i = F_{\mathrm{PRF}}(k,p\|\langle i\rangle)\}_{i=1}^{N-m+1}$, rather than just a single value.
  3. For every i, let $t_i = F_{\mathrm{PRF}}(k,T_i\|\langle i\rangle)$. Then, P_1 sends P_2 the set $\{(i,t_i)\}_{i=1}^{N-m+1}$.
  4. P_2 outputs the set of indices $\{i\}$ for which $f_i = t_i$.

Theorem 6. *Let F_{PRF} denote the Naor-Reingold function and assume that it is pseudorandom. Furthermore, assume that protocol π_{PRF} securely computes the functionality $(k,p) \mapsto (\lambda, \{F_{\mathrm{PRF}}(k,p\|\langle i\rangle)\}_{i=1}^{N-m+1})$ in the presence of malicious adversaries with one-sided simulatability. Then protocol π_{PM} securely computes $\mathcal{F}_{\mathrm{PM}}$ in the presences of malicious adversaries with one-sided simulatability.*

Proof Sketch: For the case that P_1 is corrupted, we need to show that P_1 learns nothing about P_2's input. This follows immediately from the fact that π_{PRF} is secure with one-sided simulatability and P_1 receives no other messages. For the case that party P_2 is corrupted we need to present a simulator. Very briefly, the simulator $\mathcal{S}$ works by obtaining the pattern p that $\mathcal{A}$ inputs to π_{PRF}

and generating values t_i that are completely random when $p \neq T_i$ and that equal f_i when $p = T_i$ ($\mathcal{S}$ knows when $p \neq T_i$ and when $p = T_i$ because this is given by the output received from the trusted party). The security is thus reduced to the pseudorandomness of the Naor-Reingold function. ∎

Efficiency. π_{PM} has a constant number of rounds, and each parties carries out approximately $2N$ exponentiations where N is the length of the text.

References

1. Aiello, W., Ishai, Y., Reingold, O.: Priced Oblivious Transfer: How to Sell Digital Goods. In: Pfitzmann, B. (ed.) EUROCRYPT 2001. LNCS, vol. 2045, pp. 110–135. Springer, Heidelberg (2001)
2. Aumann, Y., Lindell, Y.: Security Against Covert Adversaries: Efficient Protocols for Realistic Adversaries. In: Vadhan, S.P. (ed.) TCC 2007. LNCS, vol. 4392, pp. 137–156. Springer, Heidelberg (2007)
3. Boyer, R.S., Moore, J.S.: A Fast String Searching Algorithm. Comm ACM 20, 762–772 (1977)
4. Beaver, D.: Foundations of Secure Interactive Computing. In: Feigenbaum, J. (ed.) CRYPTO 1991. LNCS, vol. 576, pp. 377–391. Springer, Heidelberg (1992)
5. Canetti, R.: Security and Composition of Multiparty Cryptographic Protocols. Journal of Cryptology 13(1), 143–202 (2000)
6. El-Gamal, T.: A Public-Key Cryptosystem and a Signature Scheme Based on Discrete Logarithms. In: Blakely, G.R., Chaum, D. (eds.) CRYPTO 1984. LNCS, vol. 196, pp. 10–18. Springer, Heidelberg (1985)
7. Feige, U., Shamir, A.: Zero Knowledge Proofs of Knowledge in Two Rounds. In: Brassard, G. (ed.) CRYPTO 1989. LNCS, vol. 435, pp. 526–544. Springer, Heidelberg (1990)
8. Freedman, M.J., Ishai, Y., Pinkas, B., Reingold, O.: Keyword Search and Oblivious Pseudorandom Functions. In: Kilian, J. (ed.) TCC 2005. LNCS, vol. 3378, pp. 303–324. Springer, Heidelberg (2005)
9. Freedman, M.J., Nissim, K., Pinkas, B.: Efficient Private Matching and Set Intersection. In: Cachin, C., Camenisch, J.L. (eds.) EUROCRYPT 2004. LNCS, vol. 3027, pp. 1–19. Springer, Heidelberg (2004)
10. Goldreich, O.: Foundations of Cryptography. Basic tools, vol. 1. Cambridge University Press, Cambridge (2001)
11. Goldreich, O.: Foundations of Cryptography. Basic Applications, vol. 2. Cambridge University Press, Cambridge (2004)
12. Goldreich, O., Micali, S., Wigderson, A.: How to Play any Mental Game – A Completeness Theorem for Protocols with Honest Majority. In: 19th STOC, pp. 218–229 (1987)
13. Goldwasser, S., Levin, L.: Fair Computation of General Functions in Presence of Immoral Majority. In: Menezes, A., Vanstone, S.A. (eds.) CRYPTO 1990. LNCS, vol. 537, pp. 77–93. Springer, Heidelberg (1991)
14. Kissner, L., Song, D.X.: Privacy-Preserving Set Operations. In: Shoup, V. (ed.) CRYPTO 2005. LNCS, vol. 3621, pp. 241–257. Springer, Heidelberg (2005)
15. Knuth, D.E., Morris, J.H., Pratt, V.R.: Fast Pattern Matching in Strings. SIAM Journal on Computing 6, 323–350 (1977)

16. Lindell, Y.: Parallel Coin-Tossing and Constant-Round Secure Two-Party Computation. Journal of Cryptology 16(3), 143–184 (2003)
17. Micali, S., Rogaway, P.: Secure Computation. In: Feigenbaum, J. (ed.) CRYPTO 1991. LNCS, vol. 576, pp. 392–404. Springer, Heidelberg (1992)
18. Naor, M., Pinkas, B.: Oblivious Transfer and Polynomial Evaluation. In: 31st STOC, pp. 245–254 (1999)
19. Naor, M., Pinkas, B.: Efficient Oblivious Transfer Protocols. In: 12th SODA, pp. 448–457 (2001)
20. Naor, M., Reingold, O.: Number-Theoretic Constructions of Efficient Pseudo-Random Functions. In: 38th FOCS, pp. 231–262 (1997)
21. Paillier, P.: Public-key Cryptosystems Based on Composite Degree Residuosity Classes. In: Stern, J. (ed.) EUROCRYPT 1999. LNCS, vol. 1592, pp. 223–238. Springer, Heidelberg (1999)
22. Pedersen, T.P.: Non-Interactive and Information-Theoretical Secure Verifiable Secret Sharing. In: Feigenbaum, J. (ed.) CRYPTO 1991. LNCS, vol. 576, pp. 129–140. Springer, Heidelberg (1992)
23. Rabin, M.: How to Exchange Secrets by Oblivious Transfer. Tech. Memo TR-81, Aiken Computation Laboratory, Harvard U. (1981)
24. Yao, A.: How to Generate and Exchange Secrets. In: 27th FOCS, pp. 162–167 (1986)

Fast Private Norm Estimation and Heavy Hitters

Joe Kilian[1,*], André Madeira[2], Martin J. Strauss[3,**], and Xuan Zheng[4,***]

[1] Department of Computer Science, Rutgers University, Piscataway, NJ 08854 USA
jkilian@cs.rutgers.edu
[2] Department of Computer Science, Rutgers University, Piscataway, NJ 08854 USA
amadeira@cs.rutgers.edu
[3] Departments of Math and EECS, University of Michigan, Ann Arbor,
MI 48109 USA
martinjs@umich.edu
[4] Department of EECS, University of Michigan, Ann Arbor, MI 48109 USA
xuanzh@eecs.umich.edu

Abstract. We consider the problems of computing the Euclidean norm of the difference of two vectors and, as an application, computing the large components (Heavy Hitters) in the difference. We provide protocols that are approximate but private in the semi-honest model and efficient in terms of time and communication in the vector length N. We provide the following, which can serve as building blocks to other protocols:

- *Euclidean norm problem:* we give a protocol with quasi-linear local computation and polylogarithmic communication in N leaking only the true value of the norm. For processing massive datasets, the intended application, where N is typically huge, our improvement over a recent result with quadratic runtime is significant.
- *Heavy Hitters problem:* suppose, for a prescribed B, we want the B largest components in the difference vector. We give a protocol with quasi-linear local computation and polylogarithmic communication leaking only the set of true B largest components and the Euclidean norm of the difference vector. We justify the leakage as (1) desirable, since it gives a measure of goodness of approximation; or (2) inevitable, since we show that there are contexts where linear communication is required for approximating the Heavy Hitters.

1 Introduction

Secure Multiparty Computation (SMC) has been studied for decades since [6,22]. Any protocol for computing a function can be converted, gate-by-gate, to a *private* protocol, in which no party learns anything from the protocol messages

* Supported in part by NSF grant CCF-0728937
** Supported in part by NSF grant DMS-0354600.
*** Supported in part by NSF grant DMS-0354600.

R. Canetti (Ed.): TCC 2008, LNCS 4948, pp. 176–193, 2008.
© International Association for Cryptologic Research 2008

other than what can be inferred from the function's input/output relation. The computational overhead is at most polynomial in the size of the inputs.

In recent years, however, input sizes in many problems have grown to the point where "polynomial computational overhead" is too coarse a measure; both computation and communication should be minimized. For example, absent privacy concerns, applications may require that a protocol use at most polylogarithmic communication—this occurs in processing distributed internet traffic at line speeds or in performing data mining algorithms in very large datasets. General-purpose SMC may blow up communication exponentially, so additional techniques are needed. In one theoretical approach, individual protocols are designed for functions of interest such as database lookup [9,18,7] and building decision trees [19]. Another important approach [21], converts any protocol into a private one with little communication blowup, but imposes a computational blowup that may be exponential in the communication complexity.

The approach we follow, which was introduced in [11], is to substitute an approximate function for the desired exact function. Many functions of interest have good approximations that can be computed efficiently both in terms of computation and communication. A caveat is that the traditional definition of privacy is no longer appropriate. Instead, a protocol π computing an approximation g to a function f is a private approximation protocol [11] for f if

- π is a private protocol for g in the traditional sense that the messages of π leak nothing beyond what is implied by inputs and g; and,
- the output g leaks nothing beyond what is implied by the inputs and f.

Several examples were given in [11]. Another important example was given in [15], where the authors provided an estimate $\|a - b\|_\sim$ (for integer-valued vectors a and b held by Alice and Bob respectively) as the first non-trivial example of polylogarithmic communication and polynomial computation. We will analyze and make use of this protocol for our results.

1.1 Our Results

Consider the general problem where Alice and Bob hold vectors, a and b, of dimension N, and they want an efficient summary for the vector sum $c = a + b$. We analyze two problems in this setting.

First, we consider the Euclidean norm estimation problem, in which we wish to output a tight approximation to $\|c\|_2 = \left(\sum_i c_i^2\right)^{1/2}$, the l_2 norm of vector c. The problem is a well-known building block for other protocols in the non-private setting, since it is used to estimate the skew of the data. A private protocol approximating $\|c\|_2$ using polylogarithmic communication in the vector length first appeared in [15]. Our results are based on their protocol; we strengthen its privacy guarantees and decrease its local computational costs. Specifically, we obtain a $O(N \log N)$ local computational cost versus the implied $\Omega(N^2)$, while keeping equivalent communication and round complexity costs.

Second, we consider the Euclidean approximate Heavy Hitters problem, in which there is a parameter, B, and the players ideally want c_{opt}, the B largest terms in c; i.e., the B biggest values together with the corresponding indices.

Unfortunately, finding c_{opt} exactly requires linear communication. Instead, the players use polylogarithmic communication (and polynomial work and $O(1)$ rounds) to output a vector $\tilde{c}$ with $\|\tilde{c} - c\|_2 \le (1 + \epsilon)\|c_{opt} - c\|_2$. In our protocol, the players learn nothing more than what can be deduced from c_{opt} and $\|c\|_2$. (We discuss below the significance of leaking $\|c\|_2$.) We can immediately use this result as a black box for *taxicab* approximate heavy hitters, *i.e.*, finding $\tilde{c}$ with $\|\tilde{c} - c\|_1 \le (1 + \epsilon)\|c_{opt} - c\|_1$, leaking c_{opt} and $\|c\|_2$. We omit the development of this extension in the interest of space.

In the basic result, we give an at-most-B-term representation that is nearly as good (in the Euclidean sense) as the best B-term representation and leaks no more than the best B-term representation and *the exact Euclidean norm*. Although leaking the Euclidean norm represents a weaker result than not leaking it, actually (i) leaking $\|c\|_2$ is necessary and (ii) computing or approximating $\|c\|_2$ is desirable in some circumstances. First, we sketch a straightforward lower bound showing that, for some (reasonable) values of parameters $M, N, \dots$, computing $\tilde{c}$ leaking only c_{opt} requires $\Omega(N)$ communication. In fact, for some (artificial) classes of inputs, $\Omega(N)$ communication is needed unless $\|c\|_2$ itself is not only potentially leaked, but also actually computed exactly. On the other hand, one can regard the Euclidean norm as semantically interesting, so that we can regard the top B terms *together with the Euclidean norm* as a compound, extended summary. In particular, since $\tilde{c}$ is computed, leaking $\|c\|_2$ is equivalent to leaking $\|c\|_2^2 - \|\tilde{c}\|_2^2 = \|\tilde{c} - c\|_2^2$, *i.e.*, the error in our representation, which is a useful and common desired result. Our protocol indeed can be modified to output an approximation $\|\tilde{c} - c\|_\sim$ with $\|\tilde{c} - c\|_2 \le \|\tilde{c} - c\|_\sim \le (1 + \epsilon)\|\tilde{c} - c\|_2$, so we can regard the protocol as solving two cascaded approximation problems: find a near-best representation $\tilde{c}$, then find an approximation $\|\tilde{c} - c\|_\sim$ to $\|\tilde{c} - c\|_2$. It is natural to expect a protocol for $\tilde{c}$ to leak c_{opt} and a protocol for $\|\tilde{c} - c\|_\sim$ to leak $\|\tilde{c} - c\|_2$; while lower bounds prevent that, we can compute $\tilde{c}$ and $\|\tilde{c} - c\|_\sim$ *simultaneously* and guarantee that, *overall*, we leak only c_{opt} and $\|\tilde{c} - c\|_2$.

1.2 Related Work

Other works in private communication-efficient protocols include the Private Information Retrieval problem [9,18,7], building decision trees [19], the set intersection and matching problem [12], and computing the kth-ranked element [2]. The breakthrough work of [21] gives a general technique for converting any protocol into a private protocol with little communication overhead. However, this comes at the expense of local computational costs, which may increase exponentially. Thus, other general or application-specific techniques are needed.

The seminal work of [11] introduced the notion of private approximations and gave several protocols. Some negative results followed in [14] for approximations to NP-hard functions and more on NP-hard search problems appears in [5]. Recently, [15] gave a private approximation to the Euclidean norm that is central to our paper. Statistical work such as [8] also addresses approximate summaries over large databases, but differs from our work in many parameters, such as the number of players and the allowable communication.

Several papers address the Heavy Hitters problem, in a variety of contexts. Many of the needed ideas can be seen in [16] as well as in [3,4,10,13]. However, none are directly suitable when privacy is a concern.

Road Map

This paper is organized as follows. In Section 2, we present some necessary definitions used throughout the paper. We review private approximations in Section 3. In Section 4 we present our results for the private Euclidean norm estimation. Finally, in Section 5, we present our private approximate Euclidean Heavy Hitters protocol and some suitable lower bounds that motivate our results.

2   Preliminaries

Fix parameters N, M, B, k, and a distortion parameter ϵ. In this paper, we consider only two players, Alice and Bob, holding input vectors a and b respectively, each of dimension N, and taking integer values in the range $[-M, +M]$. Let k be a security and failure probability parameter and $neg(k, N)$ be an arbitrary negligible function of k and N, i.e. a function that shrinks faster than any inverse polynomial in k and N. We guarantee summaries whose error is at most the factor $(1 + \epsilon)$ times the error of the best possible summary; and we will be interested in protocols that use communication $\text{poly}(B, \log(N), k, \log(M), 1/\epsilon)$, local computation $\text{poly}(B, N, k, \log(M), 1/\epsilon)$, and $O(1)$ of rounds.

The Euclidean norm of a vector c is $\|c\|_2 = \left(\sum_i c_i^2 \right)^{1/2}$. For the Heavy Hitters protocol, we are interested in summaries of size B for the combined vector $c = a + b$. For example, we are interested ideally in the largest B terms of c. A vector c is written $c = (c_0, c_1, c_2, \ldots, c_{N-1}) = \sum_j c_j \delta_j$, where j is an *index*, c_j is a *value*, δ_j is the vector that is 1 at index j and 0 elsewhere, and $c_j \delta_j$, which can be implemented compactly and equivalently written as the pair (j, c_j), is a *term*, in which c_j is the *coefficient*. We compare terms by the *magnitudes* of their coefficients, breaking ties by the indices. That is, we will say that $(j, c_j) < (k, c_k)$ if $|c_j| < |c_k|$ or both $|c_j| = |c_k|$ and $j < k$. Thus all terms are strictly comparable. A heavy hitter summary is an expression of the form $\sum_{i \in \Lambda} \eta_i \delta_i$. If $|\Lambda|$ must be at most B, then the best heavy hitter summary c_{opt} for a vector c occurs where $\{(i, \eta_i) : i \in \Lambda\}$ consists of the B largest terms.

2.1   Approximate Data Summaries

A function g is said to be an $\langle \epsilon, \delta \rangle$-approximation of f if, for all inputs x, $\Pr[(1 - \epsilon)f(x) \leq g(x) \leq (1 + \epsilon)f(x)] \geq 1 - \delta$ holds for an approximation error $\epsilon \in (0, 1)$ and confidence parameter $\delta \in (0, 1)$. The probabilistic guarantee is over the randomness of g.

In the exact Heavy Hitters problem, we are given parameters B and N and the goal is to find the B largest terms in a vector of dimension N. In the approximate Heavy Hitters problem, however, we want a summary $\tilde{c} = \sum_{i \in \Lambda} \eta_i \delta_i$ such that $\|\tilde{c} - c\| \leq (1 + \epsilon)\|c_{\text{opt}} - c\|$, where the norms are all Euclidean norms.

In order to describe previous relevant algorithms, we first need some definitions. Fix a vector $c = (c_0, c_1, c_2, \ldots, c_{N-1}) = \sum_{0 \le i < N} c_i \delta_i$, whose terms are $t_0 = (0, c_0), t_1 = (1, c_1), \ldots, t_{N-1} = (N-1, c_{N-1})$. Suppose the sequence $i'_0, i'_1, \ldots$ is a decreasing rearrangement of c, i.e., $t_{i'_0} > t_{i'_1} > \cdots > t_{i'_{N-1}}$.

Definition 1 (Significant index). *Let $I \subseteq [0, N)$ be a set of indices. Then i is a (I, θ)-significant index for c if and only if $c_i^2 \ge \theta \sum_{j \in I} |c_j|^2$.*

That is, an index is significant if the corresponding value is large compared with all the other values. In some of the algorithms below, we will find the largest term (if it is sufficiently large), subtract it off, then recurse on the residual signal. This motivates the following definitions.

Definition 2 (Qualified index set). *Fix parameters ℓ and θ. The set $Q = \{i'_0, i'_1, \ldots, i'_{m-1}\}$ is a (ℓ, θ)-qualified index set for c if and only if (a) $m \le \ell$; (b) $\forall j \in [0, m-1]$, i'_j is a $(\{i'_j, i'_{j+1}, \ldots, i'_{N-1}\}, \theta)$-significant index; and (c) i'_m is NOT a $(\{i'_m, i'_{m+1}, \ldots, i'_{N-1}\}, \theta)$-significant index.*

That is, a qualified index set consists of the largest possible dimension m for a prefix of $i'_0, i'_1, \ldots, i'_{m-1}$ such that, for each $j < m$, we have $c_{i'_j}^2 \ge \theta(c_{i'_j}^2 + c_{i'_{j+1}}^2 + c_{i'_{j+2}}^2 + \cdots + c_{i'_{N-1}}^2)$. In particular, if the terms happen to be in decreasing order to begin with, i.e., if $|c_0| > |c_1| > \cdots$, then a qualified index set is $\{0, 1, 2, \ldots, m-1\}$ for the largest m such that, for each $j < m$, we have $c_j^2 \ge \theta(c_j^2 + c_{j+1}^2 + c_{j+2}^2 + \cdots c_{N-1}^2)$. Note that for each ℓ, θ, and vector c, there is only one (ℓ, θ)-qualified index set for c. We use $Q_{c,\ell,\theta}$ to denote it and sometimes write $Q_{\ell,\theta}$ when c is understood. The following Proposition is then straightforward.

Proposition 3. *For any $\theta_1 < \theta_2$, Q_{ℓ,θ_2} set is a subset of Q_{ℓ,θ_1}.*

Proposition 4. *Fix parameters N, M, B, k, ϵ and the vector c as above. If $\tilde{c} = \sum_{i \in Q_{c,B,\frac{\epsilon}{B(1+\epsilon)}}} c_i \delta_i$, then $\|\tilde{c} - c\|_2^2 \le (1+\epsilon)\|c_{\mathrm{opt}} - c\|_2^2$.*

Proof. Assume without loss of generality that $|c_0| > |c_1| > \cdots$ and let $q = |Q_{c,B,\frac{\epsilon}{B(1+\epsilon)}}|$. If $q = B$, then $\tilde{c} = c_{\mathrm{opt}}$ and we are done. Otherwise we have $\|\tilde{c} - c\|_2^2 = \sum_{q \le i < B} |c_i|^2 + \|c_{\mathrm{opt}} - c\|_2^2 \le B|c_q|^2 + \|c_{\mathrm{opt}} - c\|_2^2 \le \frac{\epsilon}{1+\epsilon}\|\tilde{c} - c\|_2^2 + \|c_{\mathrm{opt}} - c\|_2^2$, whence $(1 - \epsilon/(1+\epsilon))\|\tilde{c} - c\|_2^2 \le \|c_{\mathrm{opt}} - c\|_2^2$. The result follows.   □

The algorithms below work from a linear *sketch* of a vector.

Definition 5 (Sketch of a vector). *Given a vector c, a linear sketch of c is Rc, where R is a random matrix, called the measurement matrix, generated from a prescribed distribution*

In our case, as is typical, the matrix R is a pseudorandom matrix that can be generated from a short pseudorandom seed. We use sketching for the NORM_ESTIMATION protocol (Protocol 1 in Section 4), in which the generator needs to be secure against small space, and a different measurement matrix in the non-private Euclidean Heavy Hitters protocol, where, *e.g.*, pairwise independence suffices for the pseudorandom number generator.

An algorithm in connection with the approximate Euclidean Heavy Hitters problem satisfying the following is known [13]:

Theorem 6. *Fix* N, M, B, k, ϵ *as above, and* $\theta \geq$ poly$(\log(N), \log(M),$ $B, k, 1/\epsilon)^{-1}$. *There is a distribution on sketch matrices* R *and a corresponding algorithm that, from* R *and sketch* Rc *of a vector* c, *outputs a superset of* $Q_{c,B,\theta}$, *in time* poly$(\log(N), \log(M), B, k, 1/\epsilon)$.

In particular, the number or rows in R and the size of the output is bounded by the expression poly$(\log(N), \log(M), B, k, 1/\epsilon)$ in accordance with the time bound on the algorithm. The algorithm admits efficient Secure Function Evaluation protocols, and can be modified to run privately in poly$(\log(N), \log(M), B, k, 1/\epsilon)$ time. Note that the algorithm returns a superset of $Q_{c,B,\theta}$ but that even $Q_{c,B,\theta}$ itself suffices for a good approximation.

2.2   Private Two-Player Protocol

SMC allows two or more parties to evaluate a previously-agreed-upon function of their inputs, while hiding their inputs from each other. Here, we assume that all parties are computationally bounded and semi-honest, meaning they follow the protocol but may keep message histories in an attempt to learn more than is prescribed. The adversary is thus passive and can't modify the behavior of corrupted parties. In [21], the authors have shown how to transform a semi-honest protocol into a protocol secure in the malicious model, where parties deviate from the protocol arbitrarily using a different input or outputing the wrong answer, or even exiting from the protocol prematurely. Therefore, we assume parties are semi-honest for the remainder of the paper.

Formally, a two-party computation is specified by a (possibly randomized) mapping g from a pair of inputs $(a, b) \in \{0,1\}^* \times \{0,1\}^*$ to a pair of outputs $(c, d) \in \{0,1\}^* \times \{0,1\}^*$. Let $\pi = (\pi_A, \pi_B)$ be a two-party protocol computing g. Consider the probability space induced by the execution of π on input $\mathbf{x} = (a, b)$ (induced by the independent choices of random inputs r_A, r_B). Let view$_A^\pi(\mathbf{x})$ (resp., view$_B^\pi(\mathbf{x})$) denote the entire view of Alice (resp., Bob) in this execution, including her input, random input, and all messages she has received. Let output$_A^\pi(\mathbf{x})$ (resp., output$_B^\pi(\mathbf{x})$) denote Alice's (resp., Bob's) output. Note that the above four random variables are defined over the same probability space. Two distributions (or ensembles) $\mathcal{D}_1$ and $\mathcal{D}_2$ are said to be *computationally indistinguishable* with security parameter k, $\mathcal{D}_1 \stackrel{c}{\equiv} \mathcal{D}_2$, if, for any $X_1 \sim \mathcal{D}_1$ and $X_2 \sim \mathcal{D}_2$ and, for any family of polynomial-size circuits $\{C_k\}$, we have $|\Pr(C_k(X_1) = 1) - \Pr(C_k(X_2) = 1)| \leq \text{neg}(k)$.

Definition 7 (Private two-party protocol). *Let* X *be the set of all valid inputs* $\mathbf{x} = (a, b)$. *A protocol* π *is a* private protocol *computing* g *if the following properties hold:*

Correctness. *The joint outputs are distributed according to $g(a,b)$. Formally,*

$$\{(\text{output}_A^\pi(\mathbf{x}), \text{output}_B^\pi(\mathbf{x}))\}_{\mathbf{x}\in X} \equiv \{(g_A(\mathbf{x}), g_B(\mathbf{x}))\}_{\mathbf{x}\in X},$$

where $(g_A(\mathbf{x}), g_B(\mathbf{x}))$ is the joint distribution of the outputs of $g(\mathbf{x})$.

Privacy. *There exist probabilistic polynomial-time algorithms $\mathcal{S}_A, \mathcal{S}_B$, also known as* simulators, *such that:*

$$\{(\mathcal{S}_A(a, g_A(\mathbf{x})), g_B(\mathbf{x}))\}_{\mathbf{x}=(a,b)\in X} \stackrel{c}{\equiv} \{(\text{view}_A^\pi(\mathbf{x}), \text{output}_B^\pi(\mathbf{x}))\}_{\mathbf{x}\in X}$$

$$\{(g_A(\mathbf{x}), \mathcal{S}_B(b, g_B(\mathbf{x})))\}_{\mathbf{x}=(a,b)\in X} \stackrel{c}{\equiv} \{(\text{output}_A^\pi(\mathbf{x}), \text{view}_B^\pi(\mathbf{x}))\}_{\mathbf{x}\in X}$$

Yao, in its seminal work [22], provided a general technique:

Proposition 8 (General-Purpose SMC [22]). *Two parties holding inputs x and y can privately compute any circuit C with communication and computation $O(k(|C| + |x| + |y|))$, where k is a security parameter, in $O(1)$ rounds.*

We also require the following notion of evaluating a circuit with ROM securely. In this context, the ith party has a table $R_i \in (\{0,1\}^r)^s$, a function of its inputs. Then, the circuit has *lookup gates*, which on inputs (i, j) returns $R_i[j]$.

Proposition 9 (Secure Circuit with ROM [21]). *If C is a circuit with ROM, then it can be securely evaluated with $O(k|C|T(r,s))$ communication in $O(1)$ rounds, where $T(r,s)$ is the communication of 1-out-of-s Oblivious Transfer (OT) protocol on words of size r.*

We will need the following standard definitions for our results in Section 5.

Definition 10 (Additive Secret Sharing). *An intermediate value x of a joint computation is said to be* secret shared *between Alice and Bob if Alice holds r and Bob holds $x - r$, modulo some large prime, where r is a random number independent of all inputs and outputs.*

Definition 11 (Private Sample Sum). *At the start, Alice holds a vector a of dimension N and Bob holds a vector b. Alice and Bob also hold a secret sharing of an index i. At the end, Alice and Bob hold a secret sharing of $a_i + b_i$.*

That is, neither the index i nor the value $a_i + b_i$ becomes known to the parties. Efficient protocols for this problem can be found (or can be constructed immediately from related results) in [21,11], under various assumptions about the existence of Private Information Retrieval, such as in [7].

Proposition 12. *There is a protocol* PRIVATE-SAMPLE-SUM *for the Private Sample Sum problem that requires* poly(N, k) *computation,* poly$(\log(N), k)$ *communication, and $O(1)$ rounds.*

3   Private Approximations

In this Section, we review the notion of private approximations introduced in [11].

Definition 13 (Private Approximation Protocol (*strict* sense) [11]). *A two-party private approximation protocol for a deterministic, common-output function g on inputs a and b is* strict *if it computes an approximation $\widetilde{g}$ to g such that: (a) $\widetilde{g}$ is a good approximation to g (in the appropriate sense); (b) π is a private protocol for $\widetilde{g}$ in the traditional sense (Definition 7); and (c) (Functional Privacy) there exists a probabilistic polynomial-time (PPT) simulator S s.t.:*

$$\{\mathcal{S}(g(\mathbf{x}))\}_{\mathbf{x}=(a,b)\in X} \stackrel{c}{\equiv} \widetilde{g}(\mathbf{x}).$$

In the case the output to both parties is a deterministic function, a (weakly) equivalent definition is as follows, known as the "liberal" definition in [11]:

Definition 14 (Private Approximation Protocol (*liberal* sense) [11]). *A two-party private approximation protocol for a deterministic, common-output function g on inputs a and b is* liberal *if it computes an approximation $\widehat{g}$ to g such that: (a) $\widehat{g}$ is a good approximation to g (in the appropriate sense); (b) π is a private protocol for $\widehat{g}$, with correctness as in Definition 7 and privacy as in existing PPT simulators $\mathcal{S}_A$ and $\mathcal{S}_B$ such that:*

$$\{\mathcal{S}_A(a, g(\mathbf{x}))\}_{\mathbf{x}=(a,b)\in X} \stackrel{c}{\equiv} \{\text{view}_A^\pi(\mathbf{x})\}_{\mathbf{x}\in X}$$

$$\{\mathcal{S}_B(b, g(\mathbf{x}))\}_{\mathbf{x}=(a,b)\in X} \stackrel{c}{\equiv} \{\text{view}_B^\pi(\mathbf{x})\}_{\mathbf{x}\in X};$$

and (c) (Functional Privacy) there exists a PPT simulator S such that:

$$\{\mathcal{S}(g(\mathbf{x}))\}_{\mathbf{x}=(a,b)\in X} \stackrel{c}{\equiv} \widehat{g}(\mathbf{x}).$$

We elaborate on a general technique, originally sketched in [11], to construct a private protocol in the *strict* sense *given* a private protocol in the *liberal* sense. We also show the intuitive fact that the converse always holds.

Proposition 15 (equivalency between *liberal* and *strict* definitions). *Any private approximation protocol in the* liberal *sense requiring only polylogarithmic communication complexity can be transformed into a private approximation protocol in the* strict *sense with the same asymptotic communication complexity, local computational costs, and rounds. The converse holds true as well.*

Proof. Let $\widehat{g}$ and $\widetilde{g}$ be $\langle\epsilon, \delta\rangle$-approximations of g; and $\widehat{\pi}$ and $\widetilde{\pi}$ be private protocols computing $\widehat{g}$ and $\widetilde{g}$ in the *liberal* and *strict* sense respectively. Now, suppose there are simulators in the *strict* sense. Then, putting $\widehat{g} = \widetilde{g}$, a simulator for the *liberal* definition can be constructed by simulating $\widehat{g}(a, b) = \widetilde{g}(a, b)$ from $g(a, b)$ using the hypothesized simulator for functional privacy, then simulating Alice's (or Bob's) view from $\widehat{g}(a, b)$ and a (or b) using the hypothesized *strict* simulator.

In the other direction, suppose there is a simulator in the *liberal* definition. Let τ be a transcript of Alice's view except for input a. Define $\widetilde{g} = \widehat{g}.\tau$ to be $\widehat{g}$ with τ encoded into its low-order bits. We assume that this kind of encoding into approximations can be accomplished without significantly affecting the goodness of approximation; in fact, we will assume that the value represented does not change at all, even if the "approximate" value is zero—that is, τ is auxiliary data rather than an actual part of the value of $\widetilde{g}$. Furthermore, since τ is polyloga-rithmically bounded in the input size, the communication overhead of $\widetilde{g}$ over $\widehat{g}$ is at most the size of τ, since a protocol for $\widehat{g}$ also serves as a protocol for $\widetilde{g}$. It is trivial to simulate the protocol messages given a and $\widetilde{g}$. Use the hypothesized simulator in the *liberal* definition to show functional privacy of $\widetilde{g}$.   □

In Section 4, we apply the technique above of encoding the transcript into the low-order bits to the NORM_ESTIMATION protocol from [15], originally presented in the *liberal* definition, to achieve a more secure version abiding by the *strict* definition. Furthermore, our Heavy Hitters result in Section 5 is formally proven in the *strict* sense using the same idea.

4   Private Euclidean Norm Estimation

We consider the setting in which Alice and Bob hold integer-valued vectors a and b respectively, each of dimension N. In [15], the authors provided a protocol for privately approximating the Euclidean norm of the vector difference $\|c\| = \|a - b\|$ as well as the similar vector sum. Before we present our enhancements, it is instructive to review the inner workings of their protocol and its guarantees, given in Protocol 1 and Proposition 16 respectively.

NORM_ESTIMATION

Inputs: N-dimensional vectors a and b with integer values in the range $[-M, M]$.
Output: An $\langle \epsilon, \delta \rangle$-approximation of $\|c\|^2$, where $c = a - b$.

1. Alice and Bob exchange a seed of a pseudorandom generator G and generate a pseudorandom orthonormal matrix A.
2. Set $T = T_{\max} = NM^2$
3. Repeat ({Assertion: $\|c\|^2 \leq T$})
   (a) $\forall j \in [l]$, a secure circuit with ROM (with lookup tables on Aa and Ab) independently generates random coordinates i_j, computes $(Ax)_{i_j}^2$, and independently generates z_j from a Bernoulli($N(Ax)_{i_j}^2/(TB)$) distribution.
   (b) $T = T/2$
4. Until $\sum_i z_i \geq l/(4B)$ or $T < 1$
5. Output $E = (2TB)/l \cdot \sum_i z_i$ as an estimate of $\|c\|^2$.

Protocol 1. Private approximation protocol of the square l_2 difference [15]

Proposition 16. *(Private l_2 approximation [15]) Suppose Alice and Bob have integer-valued vectors a and b in $[-M, M]^N$ and let $c = a - b$. Fix distortion ϵ and security parameter k. There is a protocol* NORM_ESTIMATION *that computes an approximation $\|c\|_\sim$ to the Euclidean norm of the vector difference, $\|c\|_2$, such that it (a) outputs $\frac{1}{1+\epsilon}\|c\|_2 \leq \|c\|_\sim \leq \|c\|_2$; (b) requires* poly$(k \log(M)N/\epsilon)$ *local computation,* poly$(k \log(M) \log(N)/\epsilon)$ *communication, and $O(1)$ rounds; and (c) is a private approximation protocol for $\|c\|_2$ in the liberal sense.*

  Furthermore, the protocol's only access to a and b is through the matrix-vector products Aa and Ab, where A is a pseudorandom matrix known to both players.

The access is possible through evaluating a circuit with ROM securely; i.e. a circuit with lookup gates on inputs Aa and Ab (see Proposition 9).

Observe that although the *communication* complexity of this protocol is low, the *computational* complexity of their protocol is quadratic in the vector dimension N. The protocol multiplies the matrix A, which has $\Theta(N^2)$ degrees of freedom by the input vectors a and b, thus requiring $\Omega(N^2)$ computations. Before we present our enhancements to Protocol 1, we first sketch the intuition behind its construction, correctness and privacy guarantees.

In [20], the authors have shown that picking a random $N \times N$ orthonormal matrix A from a distribution defined by the Haar measure ensures that each component of Ax, for any vector x, is tightly concentrated around its root mean square, $\|x\|/\sqrt{N}$. Formally, there exists a $c > 0$ such that

$$\Pr\left[|(Ax)_i| \geq t\|x\|/\sqrt{N}\right] \leq e^{-ct^2} \tag{1}$$

holds for any $i = 1, \ldots, N$, any $t > 1$ and any $x \in \mathbb{R}^N$. This transformation ensures that the "mass" of vector x is uniformly spread among the N coordinates while preserving the vector norm, i.e. $\|x\| = \|Ax\|$. Protocol 1 uses this fact and constructs A using pseudorandom generators instead, guaranteeing nonetheless that $(1 - 2^{-\Theta(k)})\|x\|^2 \leq \|Ax\|^2 \leq \|x\|^2$ holds except with $\operatorname{neg}(k, N)$ probability. Note that with each component tightly concentrated around the root mean square, one can construct an unbiased sample estimator which is an $\langle \epsilon, \delta \rangle$-approximation by straightforward application of Chernoff bounds. However, to achieve privacy, the protocol must sample the coordinates $(Ax)_i$ *obliviously* as to prevent either party from learning the sampled values (it does so by using a secure circuit with ROM; see Section 2.2). Furthermore, the protocol ensures that the final estimate E depends only on $\|x\|$ by using Bernoulli trials to squash the higher moments of E, thus preventing non-simulatable information from leaking. In particular, this also achieves *Functional Privacy* as needed in Definition 14. For its correctness argument, the protocol guarantees that each z_j has enough information to approximate $l_2(x)$ tightly by scaling the Bernoulli trials by a loop variable T and exiting the loop when the sum of the trials is large enough for tight estimation. We refer the reader to [15] for complete analysis of Protocol 1.

4.1   Faster Approximation

As argued in the last Section, the computation bottleneck of Protocol 1 is the multiplication of the pseudorandom matrix A by the input vectors a and b. Computing Aa (and Ab) requires $\Omega(N^2)$ due to the $\Theta(N^2)$ degrees of freedom of matrix A. We recall that this multiplication step is crucial for both the *correctness* and *privacy* guarantees. The matrix transformation ensures that the "mass" of the vector is uniformly spread among all coordinates while preserving the norm. Such process allows a circuit to sample logarithmic many coordinates for a tight estimation of the same norm. Preserving the norm ensures that the Bernouilli trials can be simulated for the privacy proof.

We perform a similar but faster matrix transformation on Alice and Bob's input vectors. The transformation also spreads the "mass" of the vector uniformly and preserves the vector norms as required by the correctness and privacy arguments of Protocol 1. However, our matrix multiplication takes only $O(N \log N)$ time as opposed to $\Omega(N^2)$.

Our approach, based on the technique of Ailon and Chazelle [1], is to randomly choose from a "sufficiently random" family of easily computable orthonormal transformations, as follows. Given a vector $x = (x_1, x_2, \ldots, x_N)$, we flip the sign of each x_i independently with probability $1/2$ and then apply a Hadamard transform to it, yielding a new vector x^*. Thus we choose uniformly from a family of 2^N linear transformations, each corresponding to a choice of which variables to sign-flip. Since the Hadamard transform is orthonormal and can be computed in $O(N \log N)$ time, it follows that each transformation in our family is orthonormal and computable in $O(N \log N)$ time, as flipping the sign of a variable is an orthonormal transformation with trivial computational overhead.

Next, we observe that each x_j^*, viewed in isolation, is a random linear combination of signed and unsigned x_i's, scaled by $1/\sqrt{N}$. We prove that each x_i^* is not larger than the root mean square of x, or $\|x\|/\sqrt{N}$, with high probability. Thus, we achieve a similar bound for each coordinate $(Ax)_i$ as in equation Eq. (1), which suffices for the *correctness* and *privacy* proofs of the original protocol.

The following lemma summarizes the above discussion and claims.

Lemma 1. *Let x and x' be vectors of dimension N, with each x_i' being the result of flipping the signal of the corresponding x_i with probability $1/2$. Then, for any $\lambda > 0$, applying a Hadamard transform to vector x', yielding $x^* = \frac{1}{\sqrt{N}} H_N x'$, where H_N is the $N \times N$ Hadamard matrix, we have that*

$$\Pr\left[|x_i^*| \geq \lambda \frac{|x|}{\sqrt{N}}\right] \leq 2e^{-\lambda^2/2}. \tag{2}$$

Proof. We analyze the case for a particular x_j^*, for $j \in [1, N]$. Let $Z_1, \ldots, Z_N$ be independent variables such that $Z_i = (\zeta_i x_i)/\sqrt{N}$, where $\zeta_i \in_R \{+1, -1\}$. Here, $\in_R$ denotes drawing each ζ_i independently and uniformly at random. Note that $E[Z_i] = 0$. Now, let $S = \sum_i^N Z_i$. We then define a martingale sequence $X_0, X_1, \ldots, X_N$ by setting $X_0 = E[S]$ and, for $i \in [1, N]$, $X_i = E[S|Z_1, \ldots, Z_i]$. We now apply Azuma's inequality as follows. Recall that for a martingale sequence $X_0, X_1, \ldots, X_N$ s.t. $|X_k - X_{k-1}| \leq c_k$, $\Pr[|X_t - X_0| \geq \lambda] \leq 2 \exp\left(-\frac{\lambda^2}{2\sum_{k=1}^t c_k^2}\right)$ for any $t \geq 0$ and any $\lambda > 0$. For our martingale difference sequence let $c_k = |X_k - X_{k-1}|$ and thus we get

$$\Pr\left[|X_N - X_0| \geq \lambda \frac{|x|}{\sqrt{N}}\right] \leq 2 \exp\left(-\frac{\lambda^2}{2} \frac{|x|^2}{N} \frac{1}{\sum_{k=1}^N c_k^2}\right). \tag{3}$$

Thus, to prove Eq. (2) it suffices to show that $\sum_{k=1}^N c_k^2 \leq |x|^2/N$. Note that $X_k - X_{k-1} = Z_k = (\zeta_k x_k)/\sqrt{N}$, and thus $\sum_{k=1}^N c_k^2 = \sum_{k=1}^N (X_k - X_{k-1})^2 = \sum_{k=1}^N \left(\frac{\zeta_k x_k}{\sqrt{N}}\right)^2 = \frac{|x|^2}{N}$. Therefore, applying it to Eq. (3) guarantees Eq. (2). $\square$

4.2   More Secure Approximation

In [15], the authors have shown that Protocol 1 is secure in the *liberal* sense. They provided the NORM_ESTIMATION simulator that guarantees both functional privacy and private computation of protocol $\widehat{\pi}$ (Protocol 1) computing an approximation $\widehat{g}$ of $g = \|x\|^2$. Their simulator receives the *exact* output $\|x\|^2$ for generating the protocol transcripts. To be secure in the *strict* sense, besides showing functional privacy, one must provide a simulator that is able to produce computationally indistingishable views from Alice's and Bob's *without access* to the exact output, but *only* to the approximation output $\widehat{g}$ (see Definition 13). The original NORM_ESTIMATION simulator from [15] is shown next.

NORM_ESTIMATION simulator

Input: $\|x\|^2$
Output: a computationally indistinguishable distribution from Protocol 1

1. Generate a random seed of G
2. Set $T = T_{\max} = nM^2$
3. Repeat:
   (a) $\forall j \in [l]$, independently generate z_j from a Bernoulli($\|x\|^2/(TB)$) distribution
   (b) $T = T/2$
4. Until $\sum_i z_i \geq l/(4B)$ or $T < 1$
5. Output $E = (2TB)/l \cdot \sum_i z_i$

Simulator 1. The NORM_ESTIMATION simulator from [15]

Simulator 1 above guarantees that the probabilities of the Bernoulli trials from the real and simulated views differ only by neg(k, N). Thus, given access to the exact and approximate outputs, $g(x)$ and $\widehat{g}(x)$ respectively, all messages exchanged —the seed, the oblivious transfer (OT) invocations by the secure circuit, and the output—are simulatable. Specifically, the final value of T is also simulatable. Note that simulating the final value of T is crucial to the privacy argument. If the number of invocations made by the secure circuit differ between the real and simulated views, the distribution on the resulting transcripts will no longer be indistinguishable. Furthermore, observe that the *exact* output $g(x)$ is necessary for simulating such number of steps since its magnitude dictates the loop exit condition. Clearly, using $\widehat{g}(x) = (1 \pm \epsilon)\|x\|^2$ to replace $g(x) = \|x\|^2$ in the Bernoulli trials would make the probabilities differ by a factor in the order of $O(\epsilon)$, a non-negligible factor in our security setting; i.e. we expect $O(2^{-\Theta(k)})$.

Nonetheless, we show how can we transform this *liberal* protocol into a *strict* one, by using the general technique outlined in Section 3. We define a new approximation function $\widetilde{g}$ based on $\widehat{g}$. Let $\widehat{\tau}$ denote the transcript of protocol $\widehat{\pi}$ and let $\widetilde{g} = \widehat{g}.\widehat{\tau}$, meaning that the output of the new approximation function is the output of the original approximation function $\widehat{g}$ concatenated to the entire transcript of protocol $\widehat{\pi}$ (one can view $\widehat{\tau}$ as encoded into the low-order bits of the approximation—in which case we assume the goodness of approximation is not substantially changed—or, alternatively, as auxiliary data and not part of the output itself). The transcript $\hat{\tau}$ in this case is just a concatenation of the seed used for the pseudorandom generator, all OT invocations by the secure circuit, and the final approximate output. Thus, the communication costs at most doubled; and thus still remain asymptotically $\mathrm{poly}(k \log(M) \log(N)/\epsilon)$.

Let the new protocol $\tilde{\pi}$ computing $\tilde{g}$ be identical to $\hat{\pi}$ with the additional output of $\hat{\tau}$ along with $\hat{g}$. It remains to show that $\tilde{\pi}$ can be privately computed. We thus create Simulator 2, which clearly generates indistinguishable views for Alice and Bob, since all messages exchanged in protocol $\tilde{\pi}$ are simulated properly: the random seed messages, a matching number of OT calls as well as the final output. Finally, it is clear that $\tilde{g}$ is *functionally private* to g since one can use the NORM_ESTIMATION simulator to output $\hat{\tau}$ along with $\hat{g}(x)$ *given only* $g(x) = \|x\|$.

$\tilde{\pi}$ simulator

Input: $\tilde{g}(x) = \hat{g}(x).\hat{\tau}$
Output: a computationally indistinguishable transcript from $\tilde{\pi}(x)$

1. Extract $\hat{g}(x)$ and $\hat{\tau}$ from the input $\tilde{g}(x) = \hat{g}(x).\hat{\tau}$
2. Extract the random seed for the pseudorandom generator from $\hat{\tau}$ and send it to the other party.
3. Simulate the OT calls from Step 3 in Protocol 1 by playing back the messages exchanged in $\hat{\tau}$.
4. Output $\tilde{g}(x)$

Simulator 2. Simulator for $\tilde{\pi}$

5   Private Euclidean Heavy Hitters

Consider the same input setting from the previous Section. Here, both parties want to learn a representation $\tilde{c} = \sum_{t \in T_{out}} t$ such that $\|c - \tilde{c}\|_2^2 \leq (1 + \epsilon)\|c - c_{opt}\|_2^2$ and such that at most c_{opt} and $\|c\|_2$ is revealed. Unless otherwise stated, we consider the private Euclidean Heavy Hitters problem as simply the private Heavy Hitters problem. A protocol is given in Figure 2.

5.1   Analysis

First, to gain intuition, we consider some easy special cases of the protocol's operation. For our analysis, assume that the terms in c are already positive and in decreasing order, $c_0 > c_1 > \cdots > c_{N-1} > 0$. We will be able to find the coefficient value of any desired term, so we focus on the set of indices. Let $I_{opt} = \{0, 1, 2, \ldots, B - 1\}$ denote the set of indices for the optimal B terms. The set I of indices is defined in Figure 2. Thus $Q_{c,B,\theta} \subseteq Q_{c,B,\frac{\theta}{1+\epsilon}} \subseteq I_{opt}$ and $Q_{c,B,\frac{\theta}{1+\epsilon}} \subseteq I$.

The ideal output is I_{opt}, though any superset of $Q_{c,B,\theta}$ suffices to get an approximation with error at most $(1 + \epsilon)$ times optimal. This includes the set

$I \supseteq Q_{c,B,\theta}$ that the non-private algorithm has recovered. The set I_B of the largest B terms indexed by I contains $Q_{c,B,\theta}$, so I_B is a set of at most B terms with error at most $(1+\epsilon)$ times optimal. If $|Q_{c,B,\theta}| = B$, then $I_B = Q_{c,B,\theta} = I_{\text{opt}}$, and I_B is a private and correct output.

PRIVATE_HEAVY_HITTERS

- *Known parameters*: N, M, B, ϵ, k, which determine $\theta = \frac{\epsilon}{B(1+\epsilon)}$ and B'.
- *Inputs*: N-dimensional vectors a and b with integer values in the range $[-M, M]$.
- *Output*: With probability at least $1 - 2^{-k}$, a set T_{out} of at most B terms, such that $\left\| c - \sum_{t \in T_{\text{out}}} t \right\|_2^2 \leq (1+\epsilon) \left\| c - \sum_{t \in T_{\text{opt}}} t \right\|_2^2$.

---

1. Exchange pseudorandom seeds (in the clear). Generate measurement matrices R_1 and R_2. Alice locally constructs sketches $R_1 a$ and $R_2 a = (R_2^0 a, R_2^1 a, \ldots R_2^{B-1} a)$, where the matrix R_1 is used for a non-private Euclidean Heavy Hitters and the matrix $R_2 = (R_2^0, R_2^1, \ldots, R_2^{B-1})$ is used for B independent repetitions of NORM_ESTIMATION. Bob similarly constructs $R_1 b$ and $R_2 b$.

2. Using general-purpose SMC, do
   - Use an existing (non-private) Euclidean Heavy Hitters protocol to get, from $R_1 a$ and $R_1 b$, a secret-sharing of a superset I of $Q_{c,B,\frac{\theta}{1+\epsilon}}$, in which I has exactly $B' \leq \text{poly}(\log(N), \log(M), B, k, 1/\epsilon)$ indices. (Pad, if necessary.)

3. Use PRIVATE-SAMPLE-SUM to compute, from I, a, and b, secret-shared values for each index in I. Let T denote the corresponding set of secret-shared terms. (Both the index and value of each term in T is secret shared.) Enumerate I as $I = \{i_0, i_1, \ldots\}$ with $t_{i_0} > t_{i_1} > \cdots$.

4. Using SMC, do
   - for $j = 0$ to $B - 1$
   (a) From $R_2^j, R_2^j a, R_2^j b, t_0, t_1, \ldots, t_{i_{j-1}}$, sketch $r_j = c - (t_{i_0} + t_{i_1} + \cdots + t_{i_{j-1}})$ as $R_2^j r_j = (R_2^j a + R_2^j b - R_2^j (t_{i_0} + t_{i_1} + \cdots + t_{i_{j-1}}))$.
   (b) use NORM_ESTIMATION to estimate $\|r_j\|_2^2$ as $\|r_j\|_\sim^2$, satisfying $\frac{1}{1+\epsilon}\|r_j\|_2^2 \leq \|r_j\|_\sim^2 \leq \|r_j\|_2^2$.
   (c) If $|c_{i_j}|^2 < \theta \|r_j\|_\sim^2$, break (out of for-loop)
   (d) Output t_j

5. Encode the pseudorandom seeds for R_1 and R_2 into the low-order bits of the output or (as we assume here) provide R_1 and R_2 as auxiliary output.

Protocol 2. Protocol for the Euclidean Heavy Hitters problem

The difficulty arises when $|Q_{c,B,\theta}| < B$, in which case some of I_B may be arbitrary and should not be allowed to leak. So the algorithm needs to find a private subset I_{out} with $Q_{c,B,\theta} \subseteq I_{\text{out}} \subseteq I_B$. The challenge is subtle. Let s denote $|Q_{c,B,\theta}|$. If the algorithm knew s, the algorithm could easily output $Q_{c,B,\theta}$, which is the indices of the top s terms, a correct and private output. Unfortunately, determining $Q_{c,B,\theta}$ or $s = |Q_{c,B,\theta}|$ requires $\Omega(N)$ communication (see Section 5.2), so we cannot hope to find $Q_{c,B,\theta}$ exactly. Non-private norm estimation can be used to find a subset I_{out} with $Q_{c,B,\theta} \subseteq I_{\text{out}} \subseteq Q_{c,B,\frac{\theta}{1+\epsilon}} \subseteq I_{\text{opt}}$, which is correct, but not quite private. Given $|I_{out}|$, the contents of $I_{out} \subseteq I_{opt}$ are indeed private, but the *size* of I_{out} is, generally, non-private. Fortunately, if we use a *private* protocol for norm estimation, $|I_{out}|$ remains private. We now proceed to a formal analysis.

Theorem 17. *Protocol* PRIVATE_HEAVY_HITTERS *requires* $\mathrm{poly}(N, \log(M), B, k, 1/\epsilon)$ *local computation,* $\mathrm{poly}(\log(N), \log(M), B, k, 1/\epsilon)$ *communication, and* $O(1)$ *rounds.*

Proof. By existing work, all costs of Steps 1 to 3 are as claimed. Now consider Step 4. Observe that the function being computed there has inputs and outputs of size bounded by $\mathrm{poly}(\log(N), \log(M), B, k, 1/\epsilon)$ and takes time polynomial in the size of its inputs. In particular, the instances of NORM_ESTIMATION do *not* start from scratch with respect to a or b; rather, they pick up from the precomputed short sketches $R_2 a$ and $R_2 b$. It follows that this function can be wrapped with SMC, preserving the computation and communication up to polynomial blowup in the size of the input and keeping the round complexity to $O(1)$. □

We now turn to correctness and privacy. Let I_{out} denote the set of indices corresponding to the set T_{out} of output terms.

Theorem 18. *Protocol* PRIVATE_HEAVY_HITTERS *is correct.*

Proof. The correctness of Steps 2 and 3 follows from previous work. In Step 4, we first show that $Q_{B, \frac{\epsilon}{B(1+\epsilon)}} \subseteq I_{out}$. We assume that $\frac{1}{1+\epsilon}\|r_j\|_2^2 \leq \|r_j\|_2^2 \lesssim \|r_j\|_2^2$ always holds; by Proposition 16, this happens with high probability. Thus, if $|c_{i_j}|^2 \geq \frac{\epsilon}{B(1+\epsilon)}\|r_j\|_2^2$, then $|c_{i_j}|^2 \geq \frac{\epsilon}{B(1+\epsilon)}\|r_j\|_2^2 \geq \frac{\epsilon}{B(1+\epsilon)}\|r_i\|_2^2$. By construction, $Q_{B, \frac{\epsilon}{B(1+\epsilon)}} \subseteq I$. A straightforward induction shows that, if $j \in Q_{B, \frac{\epsilon}{B(1+\epsilon)}}$, then iteration j outputs t_{i_j} and the previous iterations output exactly the set of the j larger terms in I. By Proposition 4, since I_{out} is a superset of $Q_{B, \frac{\epsilon}{B(1+\epsilon)}}$, if $\tilde{c} = \sum_{j \in I_{out}} c_{i_j} \delta_{i_j}$, then $\|\tilde{c} - c\|_2^2 \leq (1+\epsilon)\|c_{opt} - c\|_2^2$, as desired. □

Before giving the complete privacy argument, we give a lemma, similar to the above. Suppose a set P of indices is a subset of another set Q of indices. We will say that P is a *prefix* of Q if $i \in P, t_j > t_i$, and $j \in Q$ imply $j \in P$.

Lemma 2. *Output set* I_{out} *is a prefix of* $Q_{B, \frac{\epsilon}{B(1+\epsilon)^2}}$ *except with probability* 2^{-k}.

Proof. Note that $Q_{B, \frac{\epsilon}{B(1+\epsilon)^2}}$ is a subset of I and $Q_{B, \frac{\epsilon}{B(1+\epsilon)^2}}$ is a prefix of the universe, so $Q_{B, \frac{\epsilon}{B(1+\epsilon)^2}}$ is a prefix of I. The set I_{out} is also a prefix of I. Thus, of the sets I_{out} and $Q_{B, \frac{\epsilon}{B(1+\epsilon)^2}}$, one is a prefix of the other (or they are equal).

So suppose, toward a contradiction, that $Q_{B, \frac{\epsilon}{B(1+\epsilon)^2}}$ is a proper prefix of I_{out}. Let $q = \left|Q_{B, \frac{\epsilon}{B(1+\epsilon)^2}}\right|$, so q is the least number such that i_q is *not* in $Q_{B, \frac{\epsilon}{B(1+\epsilon)^2}}$. If the protocol halts before considering q, then $I_{out} \subseteq Q_{B, \frac{\epsilon}{B(1+\epsilon)^2}}$, a contradiction. So we may assume that $q < B$ (so the for-loop doesn't terminate). Then, by definition of $Q_{B, \frac{\epsilon}{B(1+\epsilon)^2}}$, we have $|c_{i_q}|^2 < \frac{\epsilon}{B(1+\epsilon)^2}\sum_{j \geq q}|c_{i_j}|^2$. It follows that

$|c_{i_q}|^2 < \frac{\epsilon}{B(1+\epsilon)^2} \sum_{i \geq q} |c_i|^2 = \frac{\epsilon}{B(1+\epsilon)^2} \|r_q\|_2^2 \leq \frac{\epsilon}{B(1+\epsilon)} \|r_q\|_\sim^2$. Thus the protocol halts without outputting t_q, after outputting exactly $Q_{B,\frac{\epsilon}{B(1+\epsilon)^2}}$. $\qquad\square$

Finally, Theorem 19 ensures privacy and Theorem 20 summarizes our results.

Theorem 19. *Protocol* PRIVATE_HEAVY_HITTERS *leaks only* $\|c\|_2^2$ *and* c_{opt}.

Proof. With the random inputs R_1 and R_2 encoded into the output, it is straightforward to show that Protocol PRIVATE_HEAVY_HITTERS is a private protocol in the traditional sense that the protocol messages leak no more than the inputs and outputs. This is done by composing simulators for PRIVATE-SAMPLE-SUM and SMC. It remains only to show only that we can simulate the joint distribution on $(\widetilde{c}, R_1, R_2)$ given as simulator-input c_{opt} and $\|c\|$. We will show that R_1 is indistinguishable from independent of the joint distribution of $(\widetilde{c}, R_2)$, which we will simulate directly.

First, we show that R_1 is independent. Except with probability $2^{-\Omega(k)}$, the intermediate set I is a superset of $Q_{B,\frac{\epsilon}{B(1+\epsilon)^2}}$ and the norm estimation is correct. In that case, the protocol outputs a prefix of $Q_{B,\frac{\epsilon}{B(1+\epsilon)^2}}$ and we get identical output if I is replaced by $Q_{B,\frac{\epsilon}{B(1+\epsilon)^2}}$. Also, $Q_{B,\frac{\epsilon}{B(1+\epsilon)^2}}$ can be constructed from c_{opt} and $\|c\|_2$. Since the protocol proceeds without further reference to R_1, we have shown that the pair $(\widetilde{c}, R_2)$ is indistinguishable from being independent of R_1. It remains only to simulate $(\widetilde{c}, R_2)$.

Note that the output $\widetilde{c}$ does depend non-negligibly on R_2. If $|c_{i_j}|^2$ is very close to $\theta \|r_j\|_2^2$, then the test $|c_{i_j}|^2 < \theta \|r_j\|_\sim^2$ in the protocol may succeed with probability non-negligibly far from 0 and from 1, depending on R_2, since the distortion guarantee on $\|r_j\|_\sim^2$ is only the factor $(1 \pm \epsilon)$.

The simulator is as follows. Assume that the terms in c_{opt} are $t_0, t_1, \ldots, t_{B-1}$ with decreasing order, $t_0 > t_1 > \cdots > t_{B-1}$. For each $j \leq B$, compute $E_j = \|c - (t_0 + t_1 + \cdots + t_{j-1})\|_2^2 = \|c\|_2^2 - \|t_0 + t_1 + \cdots + t_{j-1}\|_2^2$ and then run the NORM_ESTIMATION simulator on input E_j and ϵ to get a sample from the joint distribution $(\widetilde{E}_j, \overline{R}_2)$, where $\widetilde{E}_j$ is a good estimate to E_j. Our simulator then outputs t_{i_j} if $|c_{i_j}|^2 \geq \frac{\epsilon}{B(1+\epsilon)} \widetilde{E}_j$, and halts, otherwise, following the final for-loop of the protocol. Call the output of the simulator $\widetilde{s} = \sum_j t_{i_j} \delta_{i_j}$.

Again using the fact that a prefix of $Q_{B,\frac{\epsilon}{B(1+\epsilon)^2}}$ is output, if $j \in Q_{B,\frac{\epsilon}{B(1+\epsilon)^2}}$, then $i_j = j$; *i.e.*, the jth largest output term is the jth largest overall, so that, if j is output, $E_j = \|r_j\|_2^2$. Thus $(\widetilde{E}_j, \overline{R}_2)$ is distributed indistinguishably from $(\|r_j\|_\sim^2, R_2)$. The protocol finishes deterministically using I and $\|r_j\|_\sim^2$ and the simulator finishes deterministically using $Q_{B,\frac{\epsilon}{B(1+\epsilon)^2}}$ and $\widetilde{E}_j$, but, since the protocol output is identical if I is replaced by $Q_{B,\frac{\epsilon}{B(1+\epsilon)^2}}$, the distributions on output $(\widetilde{c}, R_2)$ of the protocol and $(\widetilde{s}, \overline{R}_2)$ of the simulator are indistinguishable. $\qquad\square$

Theorem 20. *Suppose Alice and Bob hold integer-valued vectors* a *and* b *in* $[-M, M]^N$, *respectively. Let* B, k *and* ϵ *be user-defined parameters. Let* $c = a + b$. *Let* T_{opt} *be the set of the largest* B *terms in* c. *There is a protocol, taking* a, b, B k *and* ϵ *as input, that computes a representation* $\widetilde{c}$ *of at most* B

terms such that it: (a) outputs $\widetilde{c}$ with $\|\widetilde{c} - c\|_2 \leq (1 + \epsilon)\|c_{\text{opt}} - c\|_2$; (b) uses poly$(N, \log(M), B, k, 1/\epsilon)$ time, poly$(\log(N), \log(M), B, k, 1/\epsilon)$ communication, and $O(1)$ rounds; and (c) succeeds with probability $1 - 2^{-k}$ and leaks only c_{opt} and $\|c\|_2$ on security parameter k.

Corollary 21. *With the same hyptotheses and resource bounds, there is a protocol that computes $\widetilde{c}$ and an approximation $\|\widetilde{c} - c\|_\sim$ to $\|\widetilde{c} - c\|_2$ such that $\frac{1}{1+\epsilon}\|\widetilde{c} - c\|_2 \leq \|\widetilde{c} - c\|_\sim \leq \|\widetilde{c} - c\|_2$ and the protocol leaks only c_{opt} and $\|\widetilde{c} - c\|_2$.*

Proof. Run the main protocol and output also $\|\widetilde{c} - c\|_\sim$, computed in the course of the main protocol. Note that $\|\widetilde{c} - c\|_2^2 = \|c\|_2^2 - \|\widetilde{c}\|_2^2$ and both $\|c\|_2$ and $\widetilde{c}$ are available to the main simulator (as input and output, resp.), so we can modify the main simulator to compute $\|\widetilde{c} - c\|_2^2$ as well.   □

5.2   Lower Bounds

In this Section, we state some lower bounds for problems related to our main problem in this Section, such as computing an approximation to c_{opt} without leaking $\|c\|_2$. The results are straightforward, but we include Theorem 22 to motivate the approximation and Theorem 23 to motivate leakage of the Euclidean norm in protocols we present. The proofs, based on the set disjointness problem, will appear in the journal version of this article.

Theorem 22. *There is an infinite family of settings of parameters M, N, B, k such that any protocol that computes the Euclidean norm exactly on the sum c of individually-held inputs a and b, uses communication $\Omega(N)$. Similarly, any protocol that computes the exact Heavy Hitters or computes the qualified set $Q_{c,1,1}$ exactly uses communication $\Omega(N)$.*

Theorem 23. *There is an infinite family of settings of parameters M, N, B, k, ϵ such that any protocol that solves the Euclidean Heavy Hitters problem on the sum c of individually-held inputs a and b, leaking only c_{opt}, uses communication $\Omega(N)$. Furthermore, for an infinite class of inputs in which $\|c\|_2$ is not constant, any such protocol either computes $\|c\|_2$ or uses communication $\Omega(N)$.*

References

1. Ailon, N., Chazelle, B.: Approximate nearest neighbors and the fast Johnson-Lindenstrauss transform. In: Proc. 38th Annual ACM STOC, pp. 557–563 (2006)
2. Aggarwal, G., Mishra, N., Pinkas, B.: Secure computation of the k th-ranked element. In: Cachin, C., Camenisch, J.L. (eds.) EUROCRYPT 2004. LNCS, vol. 3027, pp. 40–55. Springer, Heidelberg (2004)
3. Alon, N., Gibbons, P.B., Matias, Y., Szegedy, M.: Tracking join and self-join sizes in limited storage. J. Comput. Syst. Sci. 64(3), 719–747 (2002)
4. Alon, N., Matias, Y., Szegedy, M.: The space complexity of approximating the frequency moments. J. Comput. Syst. Sci. 58(1), 137–147 (1999)

5. Beimel, A., Carmi, P., Nissim, K., Weinreb, E.: Private approximation of search problems. In: Proc. 38th Annual ACM STOC, pp. 119–128 (2006)
6. Ben-Or, M., Goldwasser, S., Wigderson, A.: Completeness theorems for non-cryptographic fault-tolerant distributed computation. In: Proc. 20th Annual ACM STOC, pp. 1–10. ACM Press, New York (1988)
7. Cachin, C., Micali, S., Stadler, M.: Computationally private information retrieval with polylogarithmic communication. In: Stern, J. (ed.) EUROCRYPT 1999. LNCS, vol. 1592, pp. 404–414. Springer, Heidelberg (1999)
8. Chawla, S., Dwork, C., McSherry, F., Smith, A., Wee, H.: Toward privacy in public databases. In: Kilian, J. (ed.) TCC 2005. LNCS, vol. 3378, pp. 363–385. Springer, Heidelberg (2005)
9. Chor, B., Goldreich, O., Kushilevitz, E., Sudan, M.: Private information retrieval. Journal of the ACM 45, 965–981 (1998)
10. Cormode, G., Muthukrishnan, S.: What's hot and what's not: Tracking most frequent items dynamically. In: Proc. ACM PODS, pp. 296–306 (2003)
11. Feigenbaum, J., Ishai, Y., Malkin, T., Nissim, K., Strauss, M., Wright, R.N.: Secure multiparty computation of approximations. Transactions on Algorithms (2006). An extended abstract appeared in ICALP 2001
12. Freedman, M., Nissim, K., Pinkas, B.: Efficient private matching and set intersection. In: Cachin, C., Camenisch, J.L. (eds.) EUROCRYPT 2004. LNCS, vol. 3027, pp. 1–19. Springer, Heidelberg (2004)
13. Gilbert, A., Guha, S., Indyk, P., Kotidis, Y., Muthukrishnan, S., Strauss, M.: Fast, small-space algorithms for approximate histogram maintenance. In: Proc. 34th Annual ACM STOC, pp. 389–398 (2002)
14. Halevi, S., Kushilevitz, E., Krauthgamer, R., Nissim, K.: Private approximations of NP-hard functions. In: Proc. 33th Annual ACM STOC, pp. 550–559 (2001)
15. Indyk, P., Woodruff, D.P.: Polylogarithmic private approximations and efficient matching. In: Proc. Third Theory of Cryptography Conference, pp. 245–264 (2006)
16. Kushilevitz, E., Mansour, Y.: Learning decision trees using the fourier sprectrum. In: Proc. 23th Annual ACM STOC, pp. 455–464 (1991)
17. Kushilevitz, E., Nisan, N.: Communication complexity. Cambridge University Press, Cambridge (1997)
18. Kushilevitz, E., Ostrovsky, R.: Replication is NOT needed: SINGLE database, computationally-private information retrieval. In: Proc. 38th IEEE FOCS, pp. 364–373 (1997)
19. Lindell, Y., Pinkas, B.: Privacy preserving data mining. J. Cryptology 15(3), 177–206 (2002)
20. Milman, V.D., Schechtman, G.: Asymptotic Theory of Finite Dimensional Normed Spaces. Lecture Notes in Mathematics, vol. 1200 (1986)
21. Naor, M., Nissim, K.: Communication preserving protocols for secure function evaluation. In: Proc. 33th Annual ACM STOC, pp. 590–599 (2001)
22. Yao, A.: Protocols for secure computation. In: Proc. 23rd IEEE FOCS, pp. 160–164 (1982)

Matroids Can Be Far from Ideal Secret Sharing

Amos Beimel[1,*], Noam Livne[2,**], and Carles Padró[3,***]

[1] Dept. of Computer Science
Ben-Gurion University
Beer-Sheva, Israel
beimel@cs.bgu.ac.il
[2] Weizmann Institute of Science
Rehovot, Israel
noam.livne@weizmann.ac.il
[3] Dept. of Applied Mathematics 4
Universitat Politècnica de Catalunya
Barcelona, Spain
cpadro@ma4.upc.edu

Abstract. In a secret-sharing scheme, a secret value is distributed among a set of parties by giving each party a share. The requirement is that only predefined subsets of parties can recover the secret from their shares. The family of the predefined authorized subsets is called the access structure. An access structure is ideal if there exists a secret-sharing scheme realizing it in which the shares have optimal length, that is, in which the shares are taken from the same domain as the secrets. Brickell and Davenport (J. of Cryptology, 1991) proved that ideal access structures are induced by matroids. Subsequently, ideal access structures and access structures induced by matroids have received a lot of attention. Seymour (J. of Combinatorial Theory, 1992) gave the first example of an access structure induced by a matroid, namely the Vamos matroid, that is non-ideal. Beimel and Livne (TCC 2006) presented the first non-trivial lower bounds on the size of the domain of the shares for secret-sharing schemes realizing an access structure induced by the Vamos matroid.

In this work, we substantially improve those bounds by proving that the size of the domain of the shares in every secret-sharing scheme for those access structures is at least $k^{1.1}$, where k is the size of the domain of the secrets (compared to $k + \Omega(\sqrt{k})$ in previous works). Our bounds are obtained by using non-Shannon inequalities for the entropy function. The importance of our results are: (1) we present the first proof that there exists an access structure induced by a matroid which is not nearly ideal, and (2) we present the first proof that there is an access structure whose information rate is strictly between 2/3 and 1. In addition, we present a better lower bound that applies only to *linear* secret-sharing schemes realizing the access structures induced by the Vamos matroid.

* Partially supported by the Frankel Center for Computer Science at the Ben-Gurion University.
** Partially supported by the Israel Science Foundation (grant No. 460/05).
*** Partially supported by the Spanish Ministry of Education and Science under project TSI2006-02731.

R. Canetti (Ed.): TCC 2008, LNCS 4948, pp. 194–212, 2008.
© International Association for Cryptologic Research 2008

1   Introduction

1.1   Ideal Secret-Sharing Schemes and Matroids

Secret-sharing schemes, which were introduced by Shamir [31] and Blakley [5] nearly 30 years ago, are nowadays used in many cryptographic protocols. In these schemes there is a finite set of parties, and a collection $\mathcal{A}$ of subsets of the parties (called the access structure). A secret-sharing scheme for $\mathcal{A}$ is a method by which a dealer distributes shares of a secret value to the parties such that (1) any subset in $\mathcal{A}$ can reconstruct the secret from its shares, and (2) any subset not in $\mathcal{A}$ cannot reveal any partial information about the secret in the information-theoretic sense. Clearly, the access structure $\mathcal{A}$ must be monotone, that is, all supersets of a set in $\mathcal{A}$ are also in $\mathcal{A}$.

Ito, Saito, and Nishizeki [18] proved that there exists a secret-sharing scheme for every monotone access structure. Their proof is constructive, but the obtained schemes are very inefficient: the ratio between the length in bits of the shares and that of the secret is exponential in the number of parties. Nevertheless, some access structures admit secret-sharing schemes with much shorter shares. A secret-sharing scheme is called *ideal* if the shares of every participant are taken from the same domain as the secret. As proved in [20], this is the optimal size for the domain of the shares. The access structures which can be realized by ideal secret-sharing schemes are called *ideal access structures*.

The exact characterization of ideal access structures is a longstanding open problem, which has interesting connections to combinatorics and information theory. The most important result towards giving such characterization is by Brickell and Davenport [8], who proved that every ideal access structure is induced by a matroid, providing a necessary condition for an access structure to be ideal. A sufficient condition is obtained as a consequence of the linear construction of ideal secret-sharing schemes due to Brickell [7]. Namely, an access structure is ideal if it is induced by a matroid that is representable over some finite field. However, there is a gap between the necessary condition and the sufficient condition. Seymour [30] proved that the access structures induced by the Vamos matroid are not ideal. Other examples of non-ideal access structures induced by matroids have been presented by Matúš [26]. Hence, the necessary condition above is not sufficient. Moreover, Simonis and Ashikmin [33] constructed ideal secret-sharing schemes for the access structures induced by the non-Pappus matroid, which is not representable over any field. This means that the sufficient condition is not necessary. Therefore, the study of the access structures that are induced by matroids is useful in the search of new results about the characterization of ideal access structures.

Another motivation in studying access structures induced by matroids arises from the separation result of Martí-Farré and Padró [24]. Namely, by using an old result by Seymour [29], they generalized the result by Brickell and Davenport [8], proving that in every secret-sharing scheme whose access structure is *not* induced by a matroid there is at least one participant whose domain of shares has size at least $k^{1.5}$, where k is the size of the domain of secrets. In other words, by proving that an access structure is not induced by a matroid, we prove a lower bound of $k^{1.5}$ for the size of the shares' domain. Therefore, the access structures that are not induced by matroids are clearly far from being ideal.

We rephrase the above result using the notion of information rate of [9]. The *information rate* of a secret-sharing scheme is $\log k / \log s$, where k is the size of the domain of the secrets and s is the maximum size of the domains of shares. That is, the information rate is the relation between the length in bits of the secret and the maximum length of the shares. Ideal secret-sharing schemes are those having information rate equal to 1. The information rate of an access structure $\mathcal{A}$ is the supremum of the information rates of all secret-sharing schemes realizing the access structure with a finite domain of shares. Stating the aforementioned result in the new notation, if $\mathcal{A}$ is not induced by a matroid, the information rate of every secret-sharing scheme for $\mathcal{A}$ is at most $2/3$, hence the information rate of $\mathcal{A}$ is at most $2/3$. This is not the case for the non-ideal access structures induced by matroids, which can be very close to ideal. An access structure $\mathcal{A}$ is *nearly ideal* if its information rate is 1. A non-ideal but nearly-ideal access structure is presented in [22,27].

At this point, two natural open questions arise. First, which matroids induce ideal access structures? And second, what can be said about the optimal size of the shares' domain for access structures induced by matroids?

Even though several interesting results have been given in [33,26,27], the first question is far from being solved. Since an ideal secret-sharing scheme can be seen as a representation of the corresponding matroid, this question can be thought of as a representability problem. Very little is known about the second question. For instance, the only known non-trivial lower bound on the optimal size of the shares' domain for access structures induced by matroids has been presented by Beimel and Livne [2]. Specifically, for an access structure induced by the Vamos matroid, they prove a lower bound of $k + \Omega(\sqrt{k})$, where k is the size of the domain of the secrets.

The best constructions of secret-sharing realizing access structures induced by matroids are the constructions for general access structures, e.g., in [4,32,7,19]; in these constructions most access structures induced by matroids require shares of exponential length. However, prior to this work, even the following question was open.

Question 1. *Does there exist a matroid such that its induced access structures are not nearly ideal?*

Observe that the lower bound given in [2] for an access structure induced by the Vamos matroid does not imply that it is not nearly ideal. For comparison, for general access structures the best known lower bound is given by Csirmaz [13] who proves that for every n there is an access structure $\mathcal{A}_n$ with n participants such that for every secret-sharing scheme realizing $\mathcal{A}_n$ there is at least one participant whose share has length at least $(n/\log n)\log k$.

Moreover, the following open problem, which was posed by Martí-Farré and Padró [23], was unsolved.

Question 2. *Does there exist an access structure whose optimal share size is $\Theta(k^\alpha)$ for some constant $1 < \alpha < 3/2$?*

That is, Martí-Farré and Padró ask if there is an access structure whose information rate is strictly between $2/3$ and 1. As a consequence of the result of [24], if such an access structure exists, it must be induced by a matroid.

1.2   Our Results

In this paper we answer the above two questions about access structures induced by matroids. Specifically, we prove new lower bounds on the size of the domains of shares in secret-sharing schemes for the access structures induced by the Vamos matroid, substantially improving the bound given in [2]. The Vamos matroid induces two non-isomorphic access structures. We prove for them lower bounds on the size of the domains of shares of, respectively, $k^{10/9}$ and $k^{11/10}$, where k is the size of the domain of the secrets (compared to $k + \Omega(\sqrt{k})$ in [2]).

Therefore, we present here the first examples of access structures induced by matroids that are not nearly ideal, resolving Question 1. Moreover, we solve Question 2 in the affirmative: As a consequence of our lower bound and the upper bound of $k^{4/3}$ that was proved in [25], the access structures induced by the Vamos matroid are the required examples.

The interest of our result is increased by the use of the so called non-Shannon inequalities in our proof. By using the basic properties of the entropy function, namely, the so-called Shannon inequalities, Csirmaz [13] proved the best known lower bounds for secret-sharing schemes mentioned above. On the negative side, Csirmaz proved that using only Shannon inequalities one cannot improve his lower bounds by a factor larger than $\log n$. More relevant to this work, several bounds on the joint entropy of the shares of subsets of parties for access structures induced by matroids were proved in [2] using Shannon inequalities (see Theorem 14 and Theorem 15 in Section 2 below). However, these bounds are only on the joint entropy of the shares and the authors of [2] could not use them to prove lower bounds for access structures induced by matroids. This is not a coincidence as in [24] it is proved that it is not possible to obtain bounds for access structures induced by matroids by using only this technique (since the rank function of the matroid satisfies the Shannon inequalities).

Nevertheless, there exist several inequalities for the entropies of a set of random variables that cannot be deduced from the Shannon inequalities. These are the so-called non-Shannon inequalities. The first examples of such inequalities were given by Zhang and Yeung [36], and other examples have been found subsequently [15]. In this paper, we combine the entropy inequalities of [2] and the non-Shannon inequality of Zhang and Yeung [36] to obtain a simple and elegant proof of our result. The inequality of [36] was previously used related to the Vamos matroid in [16] for proving lower bounds for network coding and in [27] for proving that this matroid is not asymptotically entropic (the latter result gives an alternative proof that the access structures induced by the Vamos matroid are not ideal). We believe that non-Shannon inequalities will be used for proving new lower bounds for secret-sharing schemes, possibly improving the best known lower bound given by Csirmaz [13].

In addition, by applying a similar technique to the Ingleton's inequality [17,28], which applies only to linear random variables, we obtain a lower bound of $k^{5/4}$ for the size of the shares' domains for *linear* secret-sharing schemes whose access structures are induced by the Vamos matroid.

2   Preliminaries

In this section we define secret-sharing schemes, review some background on matroids, and discuss the connection between secret-sharing schemes and matroids. The definition of secret-sharing presented in this paper uses the entropy function; in the appendix we review the relevant definitions from information theory.

2.1   Secret Sharing

Definition 1 (Access Structure). *Let P be a finite set of parties. A collection $\mathcal{A} \subseteq 2^P$ is monotone if $B \in \mathcal{A}$ and $B \subseteq C$ imply that $C \in \mathcal{A}$. An access structure is a monotone collection $\mathcal{A} \subseteq 2^P$ of non-empty subsets of P. Sets in $\mathcal{A}$ are called* authorized, *and sets not in $\mathcal{A}$ are called* unauthorized.

Definition 2 (Distribution Scheme). *Let $P = \{p_1, \ldots, p_n\}$ be a set of parties, and $p_0 \notin P$ be a special party called the dealer. An n-party distribution scheme $\Sigma = \langle \Pi, \mu \rangle$ with domain of secrets K is a pair where μ is a probability distribution on some finite set R (the set of random strings) and Π is a mapping from $K \times R$ to a set of n-tuples $K_1 \times K_2 \times \ldots \times K_n$, where K_i is called the* share-domain *of p_i. A dealer distributes a secret $s \in K$ according to Σ by first sampling a string $r \in R$ according to μ, computing a vector of shares $\Pi(s, r) = (s_1, \ldots, s_n)$, and then privately communicating each share s_i to the party p_i.*

We next give a definition of secret-sharing scheme using the entropy function. This definition is the same as that of [20,10] and is equivalent to the definition of [11,1,3]. Before stating the definition, we present some notations. Let $\mathcal{A}$ be an access structure on the set of parties P. We defined a distribution scheme Σ as a probabilistic mapping that given a secret s generates a vector of shares. It will be convenient to view the secret as the share of the dealer, and for every $T \subseteq P \cup \{p_0\}$ to consider the vector of shares of T. Any probability distribution on the domain of secrets, together with the distribution scheme Σ, induces, for any $T \subseteq P \cup \{p_0\}$, a probability distribution on the vector of shares of the parties in T. We denote the random variable taking values according to this probability distribution on the vector of shares of T by S_T, and by S the random variable denoting the secret (i.e., $S = S_{\{p_0\}}$). Note that for disjoint subsets T_1, T_2, the random variable denoting the vector of shares of $T_1 \cup T_2$ can be written either as $S_{T_1 \cup T_2}$ or as $S_{T_1} S_{T_2}$. For a singleton $\{b\}$, we will write S_b instead of $S_{\{b\}}$.

Definition 3 (Secret-Sharing Scheme). *We say that a distribution scheme is a secret-sharing scheme realizing an access structure $\mathcal{A}$ with respect to a given probability distribution on the secrets, denoted by a random variable S, if the following conditions hold.*

CORRECTNESS. *For every authorized set $T \in \mathcal{A}$, the shares of the parties in T determine the secret, that is,*

$$H(S|S_T) = 0. \tag{1}$$

PRIVACY. *For every unauthorized set $T \notin \mathcal{A}$, the shares of the parties in T do not disclose any information on the secret, that is,*

$$H(S|S_T) = H(S). \tag{2}$$

Remark 4. Although the above definition considers a specific distribution on the secrets, Blundo et al. [6] proved that its correctness and privacy are actually independent of this distribution: If a scheme realizes an access structure with respect to one distribution on the secrets, then it realizes the access structure with respect to any distribution with the same support.

Karnin et al. [20] have showed that the size of the domain of shares of each non-redundant party (that is, a party that appears in at least one minimal authorized set) is at least the size of the domain of secrets. This motivates the definition of ideal secret sharing.

Definition 5 (Ideal Secret-Sharing Scheme and Ideal Access Structure). *A secret-sharing scheme with domain of secrets K is ideal if the domain of shares of each party is K. An access structure $\mathcal{A}$ is ideal if there exists an ideal secret-sharing scheme realizing it over some finite domain of secrets.*

2.2   Matroids

A matroid is an axiomatic abstraction of linear independence. There are several equivalent axiomatic systems to describe matroids: by independent sets, by bases, by the rank function, or, as done here, by circuits. For more background on matroid theory the reader is referred to [35,28].

Definition 6 (Matroid). *A matroid $\mathcal{M} = \langle V, \mathcal{C} \rangle$ is a finite set V and a collection $\mathcal{C}$ of subsets of V that satisfy the following three axioms:*

(C0) $\emptyset \notin \mathcal{C}$.
(C1) *If $X \neq Y$ and $X, Y \in \mathcal{C}$, then $X \nsubseteq Y$.*
(C2) *If C_1, C_2 are distinct members of $\mathcal{C}$ and $x \in C_1 \cap C_2$, then there exists $C_3 \in \mathcal{C}$ such that $C_3 \subseteq (C_1 \cup C_2) \setminus \{x\}$.*

The elements of V are called points, *or simply* elements, *and the subsets in $\mathcal{C}$ are called* circuits.

For example, let $G = (V, E)$ be an undirected simple graph and $\mathcal{C}$ be the collection of simple cycles in G. Then, $(E, \mathcal{C})$ is a matroid.

Definition 7 (Rank, Independent and Dependent Sets). *A subset of V is dependent in a matroid $\mathcal{M}$ if it contains a circuit. If a subset is not dependent, it is* independent. *The* rank *of a subset $T \subseteq V$, denoted $\text{rank}(T)$, is the size of the largest independent subset of T.*

Definition 8 (Connected Matroid). *A matroid is* connected *if for every pair of distinct elements x and y there is a circuit containing x and y.*

2.3   Matroids and Secret Sharing

In this section we describe the results relating ideal secret-sharing schemes and matroids. We first define access structures induced by matroids.

Definition 9. *Let $\mathcal{M} = \langle V, \mathcal{C} \rangle$ be a connected matroid and $p_0 \in V$. The* induced access structure *of $\mathcal{M}$ with respect to p_0 is the access structure $\mathcal{A}$ on $P = V \setminus \{p_0\}$ defined by*

$$\mathcal{A} \stackrel{def}{=} \{T :\ \text{there exists } C_0 \in \mathcal{C} \text{ such that } p_0 \in C_0 \text{ and } C_0 \setminus \{p_0\} \subseteq T\}.$$

That is, a set T is a minimal authorized set of $\mathcal{A}$ if by adding p_0 to it, it becomes a circuit of $\mathcal{M}$. We think of p_0 as the dealer. We say that an access structure is induced by $\mathcal{M}$, *if it is obtained by setting some arbitrary element of $\mathcal{M}$ as the dealer. In this case, we say that $\mathcal{M}$ is the* appropriate matroid *of $\mathcal{A}$, and that $\mathcal{A}$ is induced by $\mathcal{M}$ with respect to p_0.*

Remark 10. The term *the appropriate matroid* is justified, as if some access structure is induced by a matroid, this matroid is unique.

The following fundamental result, proved by Brickell and Davenport [8], gives a necessary condition for an access structure to have an ideal secret-sharing scheme.

Theorem 11 ([8]). *If an access structure is ideal, then it has an appropriate matroid.*

The following result of [21] shows a connection between the rank function of the appropriate matroid and the joint entropy of the collections of shares.

Lemma 12 ([21]). *Assume that the access structure $\mathcal{A} \subseteq 2^P$ is ideal, and let $\langle P \cup \{p_0\}, \mathcal{C} \rangle$ be its appropriate matroid where $p_0 \notin P$. Let Σ be an ideal secret-sharing scheme realizing $\mathcal{A}$ where S is the random variable denoting the secret. Then $H(S_T) = \text{rank}(T) \cdot H(S)$ for any $T \subseteq P \cup \{p_0\}$, where $\text{rank}(T)$ is the rank of T in the matroid.*

Example 13. Consider the threshold access structure $\mathcal{A}_t$, which consists of all subsets of participants of size at least t, and Shamir's scheme [31] which is an ideal secret-sharing scheme realizing it. In this scheme, to share a secret s, the dealer randomly chooses a random polynomial $p(x)$ of degree $t-1$ such that $p(0) = s$, and the the the share of the ith participant is $p(i)$. The appropriate matroid of $\mathcal{A}_t$ is the uniform matroid with $n+1$ points, whose circuits are the sets of size $t+1$ and $\text{rank}(T) = \min\{|T|, t\}$. Since every t points determine a unique polynomial of degree $t-1$, in Shamir's scheme $H(S_T) = \min\{|T|, t\} H(S)$, as implied by Lemma 12.

We next quote results from [2] proving lower and upper bounds on the size of shares' domains of subsets of parties in matroid-induced access structures. These results generalize the results of [21] on ideal secret-sharing schemes to non-ideal secret-sharing schemes for matroid-induced access structures.

Theorem 14 ([2]). *Let $\mathcal{M} = \langle V, \mathcal{C} \rangle$ be a connected matroid where $|V| = n+1$, and $p_0 \in V$. Furthermore, let $\mathcal{A}$ be the induced access structure of $\mathcal{M}$ with respect to p_0, and let Σ be any secret-sharing scheme realizing $\mathcal{A}$. For every $T \subseteq V$,*

$$H(S_T) \geq \text{rank}(T) \cdot H(S).$$

Theorem 15 ([2]). *Let $\mathcal{M} = \langle V, \mathcal{C} \rangle$ be a connected matroid where $|V| = n+1$, $p_0 \in V$ and let $\mathcal{A}$ be the induced access structure of $\mathcal{M}$ with respect to p_0. Furthermore, let Σ be any secret-sharing scheme realizing $\mathcal{A}$, and let $\lambda \geq 0$ be such that $H(S_v) \leq (1+\lambda)H(S)$ for every $v \in V \setminus \{p_0\}$. Then, for every $T \subseteq V$*

$$H(S_T) \leq \text{rank}(T)(1 + \lambda)H(S) + (|T| - \text{rank}(T))\lambda n H(S). \tag{3}$$

2.4   The Vamos Matroid

In this paper we prove lower bounds on the size of shares in secret-sharing schemes realizing the access structures induced by the Vamos matroid. The Vamos matroid [34] is the smallest known matroid that is non-representable over any field, and is also non-algebraic (for more details on these notions see [35,28]; we will not need these notions in this paper).

Definition 16 (The Vamos Matroid). *The Vamos matroid $\mathcal{V}$ is defined on the set $V = \{v_1, v_2, \ldots, v_8\}$. Its independent sets are all the sets of cardinality ≤ 4 except for five: $\{v_1, v_2, v_3, v_4\}$, $\{v_1, v_2, v_5, v_6\}$, $\{v_3, v_4, v_5, v_6\}$, $\{v_3, v_4, v_7, v_8\}$, and $\{v_5, v_6, v_7, v_8\}$.*

Note that these 5 sets are all the unions of two pairs from $\{v_1, v_2\}$, $\{v_3, v_4\}$, $\{v_5, v_6\}$, and $\{v_7, v_8\}$, excluding $\{v_1, v_2, v_7, v_8\}$. The five sets listed in Definition 16 are circuits in $\mathcal{V}$ while the set $\{v_1, v_2, v_7, v_8\}$ is independent; these facts will be used later.

There are two non-isomorphic access structures induced by the Vamos matroid. First, the access structures obtained by setting v_1, v_2, v_7, or v_8 as the dealer are isomorphic. The other access structure is obtained by setting v_3, v_4, v_5, or v_6 as the dealer.

Definition 17 (The Access Structures $\mathcal{V}_6$ and $\mathcal{V}_8$). *The access structure $\mathcal{V}_8$ is the access structure induced by the Vamos matroid with respect to v_8. That is, in this access structure the parties are $\{v_1, \ldots, v_7\}$ and a set of parties is a minimal authorized set if this set together with v_8 is a circuit in $\mathcal{V}$. The access structure $\mathcal{V}_6$ is the access structure induced by the Vamos matroid with respect to v_6. That is, in this access structure the parties are $\{v_1, \ldots, v_5, v_7, v_8\}$ and a set of parties is a minimal authorized set if this set together with v_6 is a circuit in $\mathcal{V}$.*

Example 18. We next give examples of authorized and non-authorized sets in $\mathcal{V}_6$.

1. The set $\{v_5, v_7, v_8\}$ is authorized, since $\{v_5, v_6, v_7, v_8\}$ is a circuit.
2. The circuit $\{v_1, v_2, v_3, v_4\}$ is unauthorized, since the set $\{v_1, v_2, v_3, v_4, v_6\}$ does not contain a circuit that contains v_6. To check this, we first note that this 5-set itself cannot be a circuit, since it contains the circuit $\{v_1, v_2, v_3, v_4\}$. Second, the only circuit it contains is $\{v_1, v_2, v_3, v_4\}$, which does not contain v_6.
3. The set $\{v_1, v_2, v_7, v_8\}$ is a minimal authorized set, since $\{v_1, v_2, v_6, v_7, v_8\}$ is a circuit (as it is dependent, and no circuit of size 4 is contained in it).

3   Lower Bounds for the Vamos Access Structure

In this section we prove our main result, stating that the access structures induced by the Vamos matroid cannot be close to ideal. That is, their information rate is bounded away from 1.

We will use a non-Shannon information inequality proved by Zhang and Yeung [36]. This inequality was used related to the Vamos matroid in [16] for proving lower bounds for network coding and in [27] for proving that a function is not asymptotically entropic.

Theorem 19 ([36, Theorem 3]). *For every four discrete random variables $A, B, C,$ and D the following inequality holds:*

$$3[H(CD) + H(BD) + H(BC)] + H(AC) + H(AB)$$
$$\geq H(D) + 2[H(C) + H(B)] + H(AD) + 4H(BCD) + H(ABC). \quad (4)$$

Seymour [30] proved that $\mathcal{V}_6$ and $\mathcal{V}_8$ are not ideal. Inequality (4) was used in [27] to give an alternative proof of this fact. We next present the proof of [27]. Assume there is an ideal secret-sharing scheme realizing the Vamos access structure $\mathcal{V}_6$. Define the following random variables

$$A \stackrel{\text{def}}{=} S_{\{v_1, v_2\}},$$
$$B \stackrel{\text{def}}{=} S_{\{v_3, v_4\}},$$
$$C \stackrel{\text{def}}{=} S_{\{v_5, v_6\}},$$
$$D \stackrel{\text{def}}{=} S_{\{v_7, v_8\}}. \quad (5)$$

By Lemma 12 $H(S_T) = \text{rank}(T)H(S)$ for every set $T \subseteq \{v_1, \ldots, v_8\}$. Since all sets of size 2 are independent in the Vamos matroid, $H(A) = H(B) = H(C) = H(D) = 2H(S)$. Furthermore, by the definition of the circuits of size 4 in the Vamos matroid $H(AB) = H(AC) = H(BC) = H(BD) = H(CD) = 3H(S)$ while $H(AD) = 4H(S)$. Finally, $H(BCD) = H(ABC) = 4H(S)$. Under the above definition of $A, B, C,$ and D we notice that the l.h.s. of (4) is $33H(S)$ while the r.h.s. of (4) is $34H(S)$, a contradiction. Note that this proof strongly exploits the fact that the random variable AD, which corresponds to the shares of the independent set $\{v_1, v_2, v_7, v_8\}$, appears in the r.h.s. of (4), while the random variables appearing in the l.h.s. of (4) correspond to the shares of circuits in the matroid.

Applying Theorem 14 and Theorem 15, we can generalize the above proof and prove that V_6 cannot be close to ideal. That is, we can prove that in every secret-sharing scheme realizing V_6, the size of the entropy of the share of at least one party is at least $(1 + 1/110)H(S)$. Using direct arguments, we prove that the size of the entropy of the share of at least one party is at least $(1 + 1/9)H(S)$. Before we formally state our result, we prove two lemmas. First, to aid us in proving the better lower bound, we rearrange Inequality (4):

Lemma 20. *For every four discrete random variables A, B, C, and D the following inequality holds:*

$$3H(C|D) + 2H(C|B) + H(B|C) + H(A|C)$$
$$\geq H(A|D) + 3H(C|BD) + H(BC|D) + H(C|AB). \quad (6)$$

Proof. The claim is proved by a simple manipulation of (4). By (28), $3H(BCD) = 3H(C|BD) + 3H(BD)$ and $H(ABC) = H(C|AB) + H(AB)$. Substituting these expressions in (4) and rearranging the terms, we get

$$3H(CD) + 3H(BC) + H(AC)$$
$$\geq H(D) + 2[H(C) + H(B)]$$
$$+ H(AD) + 3H(C|BD) + H(BCD) + H(C|AB). \quad (7)$$

By (28), $2H(BC) = 2H(B) + 2H(C|B)$, $H(BC) = H(C) + H(B|C)$, and $H(AC) = H(C) + H(A|C)$. Substituting these expressions in (7) and rearranging the terms, we get

$$3H(CD) + 2H(C|B) + H(B|C) + H(A|C)$$
$$\geq H(D) + H(AD) + 3H(C|BD) + H(BCD) + H(C|AB). \quad (8)$$

By (28), $3H(CD) = 3H(D) + 3H(C|D)$, $H(AD) = H(D) + H(A|D)$, and $H(BCD) = H(D) + H(BC|D)$. Substituting these expressions in (8) and rearranging the terms, we get (6). $\quad \square$

To prove our lower bounds, we need the following simple lemma whose proof can be found in [2]. For completeness we present its proof here. Informally, this lemma states that if a set T is unauthorized and $T \cup \{b\}$ is authorized for some participant b, then guessing b's share given the shares of T is at least as hard as guessing the secret. Otherwise, the unauthorized set T can guess the share of b, and via the share compute the secret. Since, by the privacy requirement, the unauthorized set T cannot have any information on the secret, the entropy of the share must be at least $H(S)$.

Lemma 21. *Let $T \subseteq V \setminus \{p_0\}$ and $b \notin T$ such that $T \cup \{b\} \in \mathcal{A}$ and $T \notin \mathcal{A}$. Then, $H(S_b|S_T) \geq H(S)$.*

Proof. By applying (33) twice,

$$H(S, S_b|S_T) = H(S_b|S_T) + H(S|S_b, S_T) = H(S|S_T) + H(S_b|S, S_T).$$

The proof is straightforward from the second equality by taking into account that $H(S|S_T) = H(S)$, $H(S|S_b, S_T) = 0$, and that the conditional entropy function is nonnegative.   □

3.1   Proving the Lower Bound for $\mathcal{V}_6$

We next state and prove our main result.

Theorem 22. *In any secret-sharing scheme realizing $\mathcal{V}_6$ with respect to a distribution on the secrets denoted by a random variable S, the entropy of the shares of at least one party is at least $(1 + 1/9)H(S)$.*

Proof. We fix any scheme realizing $\mathcal{V}_6$ and define λ as

$$\lambda \stackrel{\text{def}}{=} \frac{\max_{1 \leq i \leq 8}(H(S_{v_i}))}{H(S)} - 1.$$

In particular, for $1 \leq i \leq 8$:

$$H(S_{v_i}) \leq (1 + \lambda)H(S). \tag{9}$$

Recall that $H(S_{v_6}) = H(S)$ as v_6 is the dealer. We use the same random variables A, B, C, and D as defined in (5). We will show that Lemma 20 implies that $\lambda \geq 1/9$.

We start with giving upper-bounds on the terms on the left hand side of (6). Recall that v_6 is the dealer, $C = S_{\{v_5, v_6\}}$, and $D = S_{\{v_7, v_8\}}$. Thus, since $\{v_5, v_7, v_8\}$ is authorized,

$$\begin{aligned} H(C|D) &= H(S_{v_5}|S_{v_7}, S_{v_8}) + H(S_{v_6}|S_{v_5}, S_{v_7}, S_{v_8}) \quad \text{(from (33))} \\ &\leq H(S_{v_5}) \leq (1 + \lambda)H(S). \end{aligned} \tag{10}$$

Similarly,

$$H(C|B) \leq (1 + \lambda)H(S). \tag{11}$$

Next, recall that $B = S_{\{v_3, v_4\}}$. By applying (29) and (33),

$$H(B|C) = H(S_{v_4}|C) + H(S_{v_3}|S_{v_4}, S_{v_5}, S_{v_6})$$
$$\leq H(S_{v_4}) + H(S_{v_3}, S_{v_6}|S_{v_4}, S_{v_5}) - H(S_{v_6}|S_{v_4}, S_{v_5})$$
$$= H(S_{v_4}) + H(S_{v_3}|S_{v_4}, S_{v_5}) + H(S_{v_6}|S_{v_3}, S_{v_4}, S_{v_5}) - H(S_{v_6}|S_{v_4}, S_{v_5}).$$

Therefore, since $\{v_3, v_4, v_5\}$ is authorized and $\{v_4, v_5\}$ is unauthorized,

$$H(B|C) \leq H(S_{v_4}) + H(S_{v_3}) - H(S) \leq (1 + 2\lambda)H(S).$$

Similarly,

$$H(A|C) \leq (1 + 2\lambda)H(S). \tag{12}$$

So, the l.h.s. of (6) is at most $(7 + 9\lambda)H(S)$.

We continue by giving lower-bounds on the terms in the right hand side of (6). First, by using (32) and (33),

$$H(A|D) = H(S_{v_1}|D) + H(S_{v_2}|D, S_{v_1})$$
$$\geq H(S_{v_1}|D, S_{v_2}) + H(S_{v_2}|D, S_{v_1})$$
$$\geq 2H(S), \tag{13}$$

where the last inequality is obtained from Lemma 21 as $\{v_1, v_2, v_7, v_8\}$ is a minimal authorized set. Second, from (33) and (2) as BD is unauthorized

$$H(C|BD) \geq H(S_{v_6}|BD) \geq H(S). \tag{14}$$

Third, by (33), (32), and Lemma 21,

$$H(BC|D) = H(B|D) + H(C|BD)$$
$$\geq H(S_{v_3}|D) + H(S)$$
$$\geq H(S_{v_3}|D, S_{v_1}) + H(S).$$

From Lemma 21 and the fact that $\{v_1, v_3, v_7, v_8\}$ is a minimal authorized set,

$$H(BC|D) \geq 2H(S).$$

Fourth, from (33) and (2) as AB is unauthorized,

$$H(C|AB) \geq H(S_{v_6}|AB) \geq H(S). \tag{15}$$

So, the r.h.s. of (6) is at least $8H(S)$.

To conclude, we have proved that the l.h.s. of (6) is at most $(7+9\lambda)H(S)$ and the r.h.s. of (6) is at least $8H(S)$. As the l.h.s. of (6) should be at least the r.h.s. of (6), we deduce that $(7 + 9\lambda)H(S) \geq 8H(S)$, which implies that $\lambda \geq 1/9$. $\square$

By Remark 4, we can assume without loss of generality that the distribution on the secrets is uniform, that is, if the domain of secrets is K, then $H(S) = \log|K|$. Furthermore, by (27), if the domain of shares of v_i is K_i, then $H(S_{v_i}) \leq \log|K_i|$. Thus, we can reformulate Theorem 22 as follows.

Corollary 23. *In any secret-sharing scheme realizing $\mathcal{V}_6$ with respect to a distribution on the secrets with support K, the size of the domain of shares of at least one party is at least $|K|^{1+1/9}$.*

3.2 Proving the Lower Bound for $\mathcal{V}_8$

In a similar manner to the proof of the lower bound for $\mathcal{V}_6$, we prove a slightly weaker lower-bound for $\mathcal{V}_8$. As before, we begin by rearranging Inequality (4). The next lemma is proved similarly to Lemma 20.

Lemma 24. *For every four discrete random variables A, B, C, and D the following inequality holds:*

$$3H(D|C) + 2H(D|B) + H(BD) + H(B|A)$$
$$\geq H(D) + H(D|A) + H(B|C) + 4H(D|BC) + H(B|AC). \qquad (16)$$

Theorem 25. *In any secret-sharing scheme realizing $\mathcal{V}_8$ with respect to a distribution on the secrets denoted by a random variable S, the entropy of the shares of at least one party is at least $(1 + 1/10)H(S)$.*

Proof. We fix any scheme realizing $\mathcal{V}_8$ and we define λ as in the proof of Theorem 22. Then $H(S_{v_i}) \leq (1 + \lambda)H(S)$ for every $i = 1, \ldots, 8$. Recall that $H(S_{v_8}) = H(S)$ as v_8 is the dealer. We use the same random variables A, B, C, and D as defined in (5). In a similar way as in Theorem 22, we find bounds on the terms of (16) to obtain a bound on λ.

Claim. $H(B|A) \leq (1 + 3\lambda)H(S)$.

To prove this claim, we first observe that

$$\begin{aligned} H(B|A) &= H(S_{v_3}, S_{v_4}|S_{v_1}, S_{v_2}) \\ &\leq H(S_{v_3}) + H(S_{v_4}|S_{v_1}, S_{v_2}, S_{v_3}) \\ &\leq (1 + \lambda)H(S) + H(S_{v_4}|S_{v_1}, S_{v_2}, S_{v_3}). \end{aligned} \qquad (17)$$

We now bound $H(S_{v_4}|S_{\{v_1, v_2, v_3\}})$. By applying (33) twice,

$$\begin{aligned} H(S_{v_4}, S_{v_5}|S_{\{v_1, v_2, v_3\}}) &= H(S_{v_4}|S_{\{v_1, v_2, v_3\}}, S_{v_5}) + H(S_{v_5}|S_{\{v_1, v_2, v_3\}}) \\ &= H(S_{v_5}|S_{\{v_1, v_2, v_3\}}, S_{v_4}) + H(S_{v_4}|S_{\{v_1, v_2, v_3\}}). \end{aligned} \qquad (18)$$

Thus, by (18)

$$\begin{aligned} H(S_{v_4}|S_{\{v_1, v_2, v_3\}}) &= H(S_{v_4}|S_{\{v_1, v_2, v_3, v_5\}}) + H(S_{v_5}|S_{\{v_1, v_2, v_3\}}) \\ &\quad - H(S_{v_5}|S_{\{v_1, v_2, v_3, v_4\}}). \end{aligned} \qquad (19)$$

We next bound each of the elements of the above sum, and get the desired result. First,

$$H(S_{v_5}|S_{\{v_1, v_2, v_3, v_4\}}) \leq H(S_{v_5}) \leq (1 + \lambda)H(S).$$

Second, from Lemma 21 we have

$$H(S_{v_5}|S_{\{v_1, v_2, v_3, v_4\}}) \geq H(S).$$

Next observe that $\{v_1, v_2, v_3, v_5\}$ is authorized in $\mathcal{V}_8$, and hence $H(S_{v_8} | S_{\{v_1,v_2,v_3,v_5\}}) = 0$, thus,

$$
\begin{aligned}
H(S_{v_4} | S_{\{v_1,v_2,v_3,v_5\}}) &= H(S_{\{v_1,v_2,v_3,v_5\}}, S_{v_4}) - H(S_{\{v_1,v_2,v_3,v_5\}}) \\
&= H(S_{\{v_1,v_2,v_3,v_4,v_5\}}) \\
&\quad - [H(S_{v_8} | S_{\{v_1,v_2,v_3,v_5\}}) + H(S_{\{v_1,v_2,v_3,v_5\}})] \\
&= H(S_{\{v_1,v_2,v_3,v_4,v_5\}}) - H(S_{\{v_1,v_2,v_3,v_5,v_8\}}) \\
&\leq H(S_{\{v_1,v_2,v_3,v_4,v_5,v_8\}}) - H(S_{\{v_1,v_2,v_3,v_5,v_8\}}) \\
&= H(S_{v_4} | S_{\{v_1,v_2,v_3,v_5,v_8\}}) \\
&\leq H(S_{v_4} | S_{\{v_1,v_2,v_5,v_8\}}) \\
&= H(S_{v_4} S_{v_8} | S_{\{v_1,v_2,v_5\}}) - H(S_{v_8} | S_{\{v_1,v_2,v_5\}}) \\
&= [H(S_{v_4} | S_{\{v_1,v_2,v_5\}}) + H(S_{v_8} | S_{\{v_1,v_2,v_4,v_5\}})] \\
&\quad - H(S_{v_8} | S_{\{v_1,v_2,v_5\}}) \\
&\leq H(S_{v_4}) + 0 - H(S) \\
&\leq \lambda H(S).
\end{aligned}
\tag{20}
$$

In the last steps we used that $\{v_1, v_2, v_4, v_5\}$ is a minimal authorized subset. Now, by summing up the bounds,

$$
H(S_{v_4} | S_{\{v_1,v_2,v_3\}}) \leq \lambda H(S) + (1 + \lambda) H(S) - H(S) = 2\lambda H(S). \tag{21}
$$

Thus, by (17) and (21), $H(B|A) \leq (1 + 3\lambda) H(S)$, which concludes the proof of our claim.

Since $\{v_5, v_6, v_7\}$ is an authorized set,

$$
\begin{aligned}
H(D) - H(D|C) &= (H(S_{v_7}) + H(S_{v_8} | S_{v_7})) \\
&\quad - (H(S_{v_7} | S_{\{v_5,v_6\}}) + H(S_{v_8} | S_{\{v_5,v_6,v_7\}})) \\
&= H(S_{v_7}) + H(S) - H(S_{v_7} | S_{\{v_5,v_6\}}) - 0 \\
&\geq H(S).
\end{aligned}
\tag{22}
$$

Thus, by (16) and (22),

$$
\begin{aligned}
&2H(D|C) + 2H(D|B) + H(BD) + H(B|A) \\
&\geq H(D|A) + H(B|C) + 4H(D|BC) + H(B|AC) + H(S).
\end{aligned}
\tag{23}
$$

We next give upper bounds for the terms in the l.h.s. of (23). We proved before that $H(B|A) \leq (1 + 3\lambda) H(S)$. For the rest of the terms in the l.h.s. we use straightforward bounds. First,

$$
H(D|C) = H(S_{v_7} S_{v_8} | C) \leq H(S_{v_8} | S_{v_7} C) + H(S_{v_7}) \leq (1 + \lambda) H(S)
$$

because $\{v_5, v_6, v_7\}$ is authorized, and similarly $H(D|B) \leq (1 + \lambda) H(S)$. Second, $H(BD) = H(S_{\{v_3,v_4,v_7,v_8\}}) = H(S_{v_8} | S_{\{v_3,v_4,v_7\}}) + H(S_{\{v_3,v_4,v_7\}}) \leq 3(1 + \lambda) H(S)$, since $\{v_3, v_4, v_7\}$ is authorized. Thus, the l.h.s. of (23) is less than $(8 + 10\lambda) H(S)$.

We continue by giving lower bounds for the terms in the r.h.s. of (23). First, by Lemma 21,

$$H(D|A) = H(S_{v_8}|S_{\{v_1,v_2,v_7\}}) + H(S_{v_7}|S_{\{v_1,v_2\}}) \geq 2H(S),$$

since $\{v_1, v_2, v_7\}$ is unauthorized and $\{v_1, v_2, v_5, v_7\}$ is authorized. Second,

$$H(B|C) \geq H(B|AC) \geq H(S) \qquad (24)$$

since $\{v_1, v_2, v_5, v_6\}$ is unauthorized and $\{v_1, v_2, v_3, v_4, v_5, v_6\}$ is authorized. Next, $H(D|BC) \geq H(S)$ since the set $\{v_3, v_4, v_5, v_6\}$ is unauthorized, while $\{v_7, v_8\}$ contains the dealer v_8. Finally, $H(B|AC) \geq H(S)$ by (24). Thus, we conclude that the r.h.s. of (23) is at least $9H(S)$.

Finally, the bounds we obtained for both sides of Inequality (23) imply that $\lambda \geq 1/10$. $\qquad \Box$

Corollary 26. *In any secret-sharing scheme realizing $\mathcal{V}_8$ with respect to a distribution on the secret with support K, the size of the domain of shares of at least one party is at least $|K|^{1+1/10}$.*

3.3   Lower Bounds for Linear Secret-Sharing Schemes

In the following, we present a lower bound for the size of the shares' domain that applies only to *linear* secret-sharing schemes with access structure $\mathcal{V}_6$ or $\mathcal{V}_8$. Nearly all known secret-sharing schemes are linear. A secret-sharing scheme is linear if the distribution scheme is such that the domain of secrets K, the domain of random strings R, and the domains of shares of the i-th party K_i, for every i, are vector spaces over some finite field, Π is a linear mapping, and the distribution on random strings μ is uniform. This bound is obtained in a very similar way as the previous ones by using an inequality due to Ingleton [17], which applies only to linear random variables, that is, random variables defined by linear mappings.

Theorem 27 ([17,28]). *For every four* linear *discrete random variables A, B, C, and D the following inequality holds:*

$$H(CD) + H(BD) + H(BC) + H(AC) + H(AB)$$
$$\geq H(C) + H(B) + H(AD) + H(BCD) + H(ABC). \qquad (25)$$

The proof of the next lemma is very similar to the one of Lemma 20.

Lemma 28. *For every four linear discrete random variables A, B, C, and D the following inequality holds:*

$$H(C|D) + H(C|B) + H(A|C)$$
$$\geq H(A|D) + H(C|BD) + H(C|AB). \qquad (26)$$

The following result is proved in a similar way to the proof of Theorem 22.

Theorem 29. *In any* linear *secret-sharing scheme realizing* $\mathcal{V}_6$ *with respect to a distribution on the secrets denoted by a random variable S, the entropy of the shares of at least one party is at least $(1 + 1/4)H(S)$.*

Proof. We fix any *linear* scheme realizing $\mathcal{V}_6$ and define

$$\lambda \stackrel{\text{def}}{=} \max_{1 \le i \le 8}(H(S_{v_i}))/H(S) - 1.$$

We use the same random variables A, B, C, and D as defined in (5). Note that all bounds proved in Section 3.1 apply, in particular, to linear secret-sharing realizing $\mathcal{V}_6$. Thus, by (10), (11), and (12), the l.h.s. of (26) is at most $(3 + 4\lambda)H(S)$. By (13), (14), and (15), the r.h.s. of (26) is at least $4H(S)$. This implies that $(3 + 4\lambda)H(S) \ge 4H(S)$, which implies that $\lambda \ge 1/4$.

Corollary 30. *In any* linear *secret-sharing scheme realizing* $\mathcal{V}_6$ *with respect to a distribution on the secrets with support K, the size of the domain of shares of at least one party is at least $|K|^{1+1/4}$.*

Finally, the same bound applies to the linear secret-sharing schemes with access structure $\mathcal{V}_8$ by duality. The *dual* of an access structure $\mathcal{A}$ is the access structure

$$\mathcal{A}^* \stackrel{\text{def}}{=} \{T \subseteq P : P \setminus T \notin \mathcal{A}\}.$$

It is well known that, for every linear secret-sharing scheme Σ with access structure $\mathcal{A}$, there exists a linear secret sharing scheme Σ^* for $\mathcal{A}^*$ such that the domain of the shares of every participant is the same for Σ and for Σ^* (see [14], for instance). Therefore, since $\mathcal{V}_8^*$ is isomorphic to $\mathcal{V}_6$, the bounds in Theorem 29 and Corollary 30 apply also to the access structure $\mathcal{V}_8$.

References

1. Beimel, A., Chor, B.: Universally ideal secret sharing schemes. IEEE Trans. on Information Theory 40(3), 786–794 (1994)
2. Beimel, A., Livne, N.: On matroids and non-ideal secret sharing. In: Halevi, S., Rabin, T. (eds.) TCC 2006. LNCS, vol. 3876, pp. 482–501. Springer, Heidelberg (2006)
3. Bellare, M., Rogaway, P.: Robust computational secret sharing and a unified account of classical secret-sharing goals. In: Proc. of the 14th conference on Computer and communications security, pp. 172–184 (2007)
4. Benaloh, J., Leichter, J.: Generalized secret sharing and monotone functions. In: Goldwasser, S. (ed.) CRYPTO 1988. LNCS, vol. 403, pp. 27–35. Springer, Heidelberg (1990)
5. Blakley, G.R.: Safeguarding cryptographic keys. In: Proc. of the 1979 AFIPS National Computer Conference, pp. 313–317 (1979)
6. Blundo, C., De Santis, A., Vaccaro, U.: On secret sharing schemes. Inform. Process. Lett. 65(1), 25–32 (1998)

7. Brickell, E.F.: Some ideal secret sharing schemes. Journal of Combin. Math. and Combin. Comput. 6, 105–113 (1989)
8. Brickell, E.F., Davenport, D.M.: On the classification of ideal secret sharing schemes. J. of Cryptology 4(73), 123–134 (1991)
9. Brickell, E.F., Stinson, D.R.: Some improved bounds on the information rate of perfect secret sharing schemes. J. of Cryptology 5(3), 153–166 (1992)
10. Capocelli, R.M., De Santis, A., Gargano, L., Vaccaro, U.: On the size of shares for secret sharing schemes. J. of Cryptology 6(3), 157–168 (1993)
11. Chor, B., Kushilevitz, E.: Secret sharing over infinite domains. J. of Cryptology 6(2), 87–96 (1993)
12. Cover, T.M., Thomas, J.A.: Elements of Information Theory. John Wiley & Sons, Chichester (1991)
13. Csirmaz, L.: The size of a share must be large. J. of Cryptology 10(4), 223–231 (1997)
14. van Dijk, M., Jackson, W.A., Martin, K.M.: A note on duality in linear secret sharing schemes. Bull. of the Institute of Combinatorics and its Applications 19, 98–101 (1997)
15. Dougherty, R., Freiling, C., Zeger, K.: Six new non-Shannon information inequalities. In: IEEE International Symposium on Information Theory (ISIT), pp. 233–236 (2006)
16. Dougherty, R., Freiling, C., Zeger, K.: Networks, matroids, and non-Shannon information inequalities. IEEE Trans. on Information Theory 53(6), 1949–1969 (2007)
17. Ingleton, A.W.: Conditions for representability and transversability of matroids. In: Proc. Fr. Br. Conf 1970, pp. 62–67. Springer, Heidelberg (1971)
18. Ito, M., Saito, A., Nishizeki, T.: Secret sharing schemes realizing general access structure. In: Proc. of the IEEE Global Telecommunication Conf., Globecom 1987, pp. 99–102 (1987)
19. Karchmer, M., Wigderson, A.: On span programs. In: Proc. of the 8th IEEE Structure in Complexity Theory, pp. 102–111 (1993)
20. Karnin, E.D., Greene, J.W., Hellman, M.E.: On secret sharing systems. IEEE Trans. on Information Theory 29(1), 35–41 (1983)
21. Kurosawa, K., Okada, K., Sakano, K., Ogata, W., Tsujii, S.: Nonperfect secret sharing schemes and matroids. In: Helleseth, T. (ed.) EUROCRYPT 1993. LNCS, vol. 765, pp. 126–141. Springer, Heidelberg (1994)
22. Livne, N.: On matroids and non-ideal secret sharing. Master's thesis, Ben-Gurion University, Beer-Sheva (2005)
23. Martí-Farré, J., Padró, C.: Secret sharing schemes with three or four minimal qualified subsets. Designs, Codes and Cryptography 34(1), 17–34 (2005)
24. Martí-Farré, J., Padró, C.: On secret sharing schemes, matroids and polymatroids. In: Vadhan, S.P. (ed.) TCC 2007. LNCS, vol. 4392, pp. 253–272. Springer, Heidelberg (2007)
25. Martí-Farré, J., Padró, C.: On secret sharing schemes, matroids and polymatroids. Journal version of [24]. Technical Report 2006/077, Cryptology ePrint Archive (2006), http://eprint.iacr.org/
26. Matúš, F.: Matroid representations by partitions. Discrete Mathematics 203, 169–194 (1999)
27. Matúš, F.: Two constructions on limits of entropy functions. IEEE Trans. on Information Theory 53(1), 320–330 (2007)

28. Oxley, J.G.: Matroid Theory. Oxford University Press, Oxford (1992)
29. Seymour, P.D.: A forbidden minor characterization of matroid ports. Quart. J. Math. Oxford Ser. 27, 407–413 (1976)
30. Seymour, P.D.: On secret-sharing matroids. J. of Combinatorial Theory, Series B 56, 69–73 (1992)
31. Shamir, A.: How to share a secret. Communications of the ACM 22, 612–613 (1979)
32. Simmons, G.J., Jackson, W., Martin, K.M.: The geometry of shared secret schemes. Bulletin of the ICA 1, 71–88 (1991)
33. Simonis, J., Ashikhmin, A.: Almost affine codes. Designs, Codes and Cryptography 14(2), 179–197 (1998)
34. Vamos, P.: On the representation of independence structures (1968)(unpublished manuscript)
35. Welsh, D.J.A.: Matroid Theory. Academic press, London (1976)
36. Zhang, Z., Yeung, R.W.: On characterization of entropy function via information inequalities. IEEE Trans. on Information Theory 44(4), 1440–1452 (1998)

A   Basic Definitions from Information Theory

In this appendix, we review the basic concepts of information theory used in this paper. For a complete treatment of this subject see, e.g., [12]. All the logarithms here are of base 2.

Given a finite random variable X, we define the *entropy* of X, denoted $H(X)$, as

$$H(X) \stackrel{\text{def}}{=} - \sum_{x, \Pr[X=x]>0} \Pr[X = x] \log \Pr[X = x].$$

It can be proved that

$$0 \leq H(X) \leq \log |\operatorname{supp}(X)|, \tag{27}$$

where $|\operatorname{supp}(X)|$ is the size of the support of X (the number of values with probability greater than zero). The upper bound is obtained if and only if the distribution of X is uniform.

Given two finite random variables X and Y (possibly dependent), we define the *conditioned entropy of X given Y* as

$$H(X|Y) \stackrel{\text{def}}{=} H(XY) - H(Y). \tag{28}$$

For convenience, when dealing with the entropy function, XY will denote $X \cup Y$. From the definition of the conditional entropy, the following properties can be proved:

$$0 \leq H(X|Y) \leq H(X), \tag{29}$$

$$H(Y) \leq H(XY), \tag{30}$$

and

$$H(XY) \leq H(X) + H(Y). \tag{31}$$

Given three finite random variable X, Y and Z (possibly dependent), the following properties hold:

$$H(X|Y) \geq H(X|YZ),$$ (32)

$$H(XY|Z) = H(X|YZ) + H(Y|Z) \geq H(Y|Z),$$ (33)

and

$$H(XY|Z) \leq H(X|Z) + H(Y|Z).$$ (34)

Perfectly-Secure MPC
with Linear Communication Complexity*

Zuzana Beerliová-Trubíniová and Martin Hirt

ETH Zurich, Department of Computer Science, CH-8092 Zurich
{bzuzana,hirt}@inf.ethz.ch

Abstract. Secure multi-party computation (MPC) allows a set of n players to securely compute an agreed function, even when up to t players are under the control of an adversary. Known *perfectly secure* MPC protocols require communication of at least $\Omega(n^3)$ field elements per multiplication, whereas cryptographic or unconditional security is possible with communication linear in the number of players. We present a perfectly secure MPC protocol communicating $\mathcal{O}(n)$ field elements per multiplication. Our protocol provides perfect security against an active, adaptive adversary corrupting $t < n/3$ players, which is optimal. Thus our protocol improves the security of the most efficient information-theoretically secure protocol at no extra costs, respectively improves the efficiency of perfectly secure MPC protocols by a factor of $\Omega(n^2)$. To achieve this, we introduce a novel technique – constructing detectable protocols with the help of so-called hyper-invertible matrices, which we believe to be of independent interest. Hyper-invertible matrices allow (among other things) to perform efficient correctness checks of many instances in parallel, which was until now possible only if error-probability was allowed.

Keywords: Multi-party computation, efficiency, perfect security, hyper-invertible matrix.

1 Introduction

1.1 Secure Multi-party Computation

Secure multi-party computation (MPC) enables a set of n players to securely evaluate an agreed function even when t of the players are corrupted by a central adversary. A *passive adversary* can read the internal state of the corrupted players, trying to obtain some information he is not entitled to. An *active adversary* can additionally make the corrupted players deviate from the protocol, trying to falsify the outcome of the computation. In this work, we consider active adversaries.

The MPC problem dates back to Yao [Yao82]. The first generic solutions presented in [GMW87, CDvdG87, GHY87] (based on cryptographic intractability assumptions) and later [BGW88, CCD88, RB89, Bea91b] (with information-theoretic security) are rather inefficient and thus of theoretical interest mainly.

---

* This work was partially supported by the Zurich Information Security Center. It represents the views of the authors.

R. Canetti (Ed.): TCC 2008, LNCS 4948, pp. 213–230, 2008.
© International Association for Cryptologic Research 2008

1.2   Efficiency of MPC Protocols

In the recent years lots of research concentrated on designing protocols with lower communication complexity. In this paper we concentrate on bit-complexity, measured in bits sent by honest players. The following table gives an overview on the currently most efficient MPC protocols (in the respective security model), where κ denotes the bit-length of a field element (resp. the security parameter).

Thresh.	Security	Bits/Mult.	Reference
$t < n/3$	perfect	$\mathcal{O}(n^3\kappa)$	[HMP00]
$t < n/2$	unconditional	$\mathcal{O}(n^2\kappa)$	[BH06]
$t < n/2$	cryptographic	$\mathcal{O}(n\kappa)$	[HN06]
$t < n/3$	unconditional	$\mathcal{O}(n\kappa)$	[DN07]

All above protocols use "player elimination" (or its generalization "dispute control") – a technique that enables converting non-robust (but detectable) protocols into robust protocols, essentially without any efficiency loss. Furthermore, all but the perfectly secure protocol use circuit randomization [Bea91a], which reduces the multiplication of two shared values to two reconstructions, given a precomputed sharing of a random multiplication triple (a, b, c) with $c = ab$. Such triples can be non-robustly generated and checked in advance – making use of parallelization. Checking the correctness of many instances in parallel can be done very efficiently when negligible error-probability is allowed, however until now no perfectly secure efficient parallel correctness-checks are known.

1.3   Contributions

In this paper, we present a novel technique which, at the same time, allows to perfectly and very efficiently verify a bunch of sharings and (if the check says that they are correct) to extract a set of (new) correct random sharings given that a sub-set of the original sharings is random.

More precisely, given n supposedly random sharings, up to t of them distributed by corrupted players (and thus possibly of a wrong degree, non-random, etc), we can check whether they are all correct and if so (locally) compute $n - 2t$ correct and uniform random sharings. The check is (despite of being perfectly secure) highly efficient; it only requires the reconstruction of $2t$ sharings, each towards a single player.

In other words, we can non-robustly but detectably generate $\Omega(n)$ uniform random sharings, unknown to the adversary, with perfect security and communicating $\mathcal{O}(n^2)$ field elements. By now, similarly efficient protocols to generate random sharings are known only with probabilistic checks, which provides a lower level of security and is less elegant.

The novel technique is based on so-called *hyper-invertible matrices*, i.e., matrices whose every square sub-matrix is invertible. Applying n sharings to such a matrix results in n sharings with the property that (i) if *any* (up to t) of the inputs sharings are broken, then this can be seen in *every* subset of t output sharings, and (ii) if *any* $n - t$ input sharings are uniform random, then *every* subset of size $n - t$ of output sharings is uniform random.

Using hyper-invertible matrices and some techniques from [Bea91a, HMP00, DN07], we construct a perfectly secure multi-party protocol with optimal resilience and linear communication complexity. This can be seen as an efficiency improvement (the most efficient known MPC protocol with perfect security communicates $\mathcal{O}(n^3)$ field elements per multiplication [HMP00]), or alternatively as a security improvement (the most secure known MPC protocol with linear communication provides error probability [DN07]). In either case, we consider the new protocol to be more elegant, as it employs neither two-dimensional sharings (like all previous perfectly-secure MPC protocols) nor probabilistic checks (like all previous MPC protocols with linear communication complexity).

2 Preliminaries

2.1 Model

We consider a set $\mathcal{U}$ of users, who can give input and receive output, and a set $\mathcal{P}$ of n players, $\mathcal{P} = \{P_1, \ldots, P_n\}$, who perform the computation. The players and users are connected by a complete network of secure (private and authentic) synchronous channels.

The function to be computed is specified as an arithmetic circuit over a finite field $\mathcal{F}$ (with $|\mathcal{F}| > 2n$), with input, addition, multiplication, random, and output gates. We denote the number of gates of each type by c_I, c_A, c_M, c_R, and c_O, respectively.

The faultiness of players or users is modeled in terms of a central adversary corrupting players and users. The adversary can corrupt up to t players for any fixed t with $t < n/3$ and any number of users, and make them deviate from the protocol in any desired manner. The adversary is computationally unbounded, active, adaptive, and rushing. The security of our protocols is perfect, i.e., information-theoretic without any error probability.

To every player $P_i \in \mathcal{P}$ a unique, non-zero element $\alpha_i \in \mathcal{F} \setminus \{0\}$ is assigned.

For the ease of presentation, we always assume that the messages sent through the channels are from the right domain — if a player receives a message which is not in the right domain (e.g., no message at all), he replaces it with an arbitrary message from the specified domain.

2.2 Byzantine Agreement

In our multi-party protocol we use Byzantine agreement in both its shapes, broadcast and consensus. Broadcast allows a sender to distribute a value x, such that all players receive the same value x' (even if the sender is faulty), and $x = x'$ if the sender is honest. Consensus allows the players, each holding an input x_i, to reach agreement on a value x', where $x = x'$ if every honest players holds $x_i = x$. For $t < n/3$, both broadcast and consensus can be simulated with perfect security by a sub-protocol communicating $\mathcal{O}(n^2)$ bits [BGP92, CW92]. We denote the communication complexity needed for agreeing on a k bit message as $\mathcal{BA}(k) = n^2 k$.

2.3   Player-Elimination Framework

Player Elimination [HMP00] is a general technique, used for constructing efficient MPC protocols. It allows to transform (typically very efficient) non-robust protocols into robust protocols at essentially no additional costs.

The basic idea is to divide the computation into segments and repeat the non-robust evaluation of each segment until it succeeds, whereby limiting the total number of times the adversary can cause a segment to fail. Each evaluation of a segment proceeds in three steps: (1.) detectable computation (2.) fault detection and (3.) fault localization.

Definition 1. *A detectable protocol is a passively secure protocol that can (in the presence of an active adversary) produce incorrect output, however this will be detected by at least one honest player. We say that after detecting a fault the player gets* unhappy *(sets his happy-bit to unhappy).*

In the detectable computation, the actual non-robust (but detectable) protocol is invoked to compute the segment. In the fault detection the players agree on whether or not there are some unhappy players. If all players are happy the computation of the segment was successful, the players keep the output and proceed to the next segment. Otherwise the segment failed, the output is discarded and a pair of players $E = \{P_i, P_j\}$ containing at least one corrupted player is localized in the fault localization, eliminated from the actual player set and the segment is repeated with the new player set.[1] We denote the original player set as $\mathcal{P}$ (containing n players, up to t of them faulty), and the actual (reduced) player set as $\mathcal{P}'$ (containing n' players, up to t' of them faulty).

By selecting the size of a segment such that there are t segments, the overall costs of the resulting robust protocol are at most twice the costs of the non-robust protocol (plus the overhead costs for the fault detection and the player elimination).

Special care needs to be taken such that the computation after a (sequence of) player elimination is "compatible" with the outputs of previous segments. We ensure this compatibility be fixing the degree of all sharings to t, independent of the actual threshold t'. Note that a sharing (among $\mathcal{P}'$) of degree t can be reconstructed as long as $t + 2t' < n'$, what is clearly satisfied when $t < n/3$.

Technically, a player-elimination protocol proceeds as follows:

---

[1] Note that we eliminate *players* and not *users*. If a party playing the role of a player as well as the role of a user is eliminated from the player set, it still keeps its user role – can give input and receive output.

Protocol with Player-Elimination

Let $\mathcal{P}' \leftarrow \mathcal{P}$, $n' \leftarrow n$, $t' \leftarrow t$. Divide computation into t segments of similar size, and do the following for each segment:

0. Every $P_i \in \mathcal{P}'$ sets his happy-bit to happy (i.e., P_i did not observe a fault).

1. DETECTABLE COMPUTATION: Compute the actual segment in detectable manner, such that (i) if all players in $\mathcal{P}'$ follow their protocol, then the computation succeeds and all players remain happy, and (ii) if the output is incorrect, then at least one honest player in $\mathcal{P}'$ detects so and gets unhappy.

2. FAULT DETECTION: Reach agreement on whether or not all players in $\mathcal{P}'$ are happy (involves Byzantine Agreement). If all players are happy, proceed with the next segment. If at least one player is unhappy, proceed with the following fault-localization procedure.

3. FAULT LOCALIZATION: Find $E \subseteq \mathcal{P}'$ with $|E| = 2$, containing at least one corrupted player.

4. PLAYER ELIMINATION: Set $\mathcal{P}' \leftarrow \mathcal{P}' \setminus E$, $n' \leftarrow n' - 2$, $t' \leftarrow t' - 1$, and repeat the segment.

2.4 Circuit Randomization

Circuit randomization [Bea91a] allows to compute a sharing $[z]$ of the product z of two factors x and y, shared as $[x]$ and $[y]$, at the costs of two public reconstructions, when a pre-shared random triple $([a], [b], [c])$ with $c = ab$ is available. This technique allows to first prepare c_M shared multiplication triples $([a], [b], [c])$, and then to evaluate a circuit with c_M multiplication by a sequence of public reconstructions.

The trick of circuit randomization is that $z = xy$ can be expressed as $z = ((x-a)+a)((y-b)+b)$, hence $z = de+db+ae+c$, where (a, b, c) is a multiplication triple and $d = x - a$ and $e = y - b$. For a random multiplication triple, d and e are random values independent of x and y, hence a sharing $[z]$ can be linearly computed as $[z] = [de] + d[b] + e[a] + [c]$, by reconstructing $[d] = [x] - [a]$ and $[e] = [y] - [b]$.

3 Hyper-invertible Matrices

3.1 Definition

A hyper-invertible matrix is a matrix of which every (non-trivial) square submatrix is invertible.

Definition 2. *An r-by-c matrix M is* hyper-invertible *if for any index sets $R \subseteq \{1, \ldots, r\}$ and $C \subseteq \{1, \ldots, c\}$ with $|R| = |C| > 0$, the matrix M_R^C is invertible, where M_R denotes the matrix consisting of the rows $i \in R$ of M, M^C denotes the matrix consisting of the columns $j \in C$ of M, and $M_R^C = \left(M_R\right)^C$.*

3.2   Construction

We present a construction of a hyper-invertible n-by-n matrix M over a finite field $\mathcal{F}$ with $|\mathcal{F}| \geq 2n$. A hyper-invertible r-by-c matrix can be extracted as a sub-matrix of such a matrix with $n = \max(r, c)$.

Construction 1. Let $\alpha_1, \ldots, \alpha_n, \beta_1, \ldots, \beta_n$ denote fixed distinct elements in $\mathcal{F}$, and consider the function $f : \mathcal{F}^n \to \mathcal{F}^n$, mapping $(x_1, \ldots, x_n)$ to $(y_1, \ldots, y_n)$ such that the points $(\beta_1, y_1), \ldots, (\beta_n, y_n)$ lie on the polynomial $g(\cdot)$ of degree $n-1$ defined by the points $(\alpha_1, x_1), \ldots, (\alpha_n, x_n)$. Due to the linearity of Lagrange interpolation, f is linear and can be expressed as a matrix $M = \{\lambda_{i,j}\}_{i=1,\ldots,n}^{j=1,\ldots n}$, where $\lambda_{i,j} = \prod_{\substack{k=1 \\ k \neq j}}^{n} \frac{\beta_i - \alpha_k}{\alpha_j - \alpha_k}$.

Lemma 1. *Construction 1 yields a hyper-invertible n-by-n matrix M.*

Proof. We have to show that for any index sets $R, C \subseteq \{1, \ldots, n\}$ with $|R| = |C| > 0$, M_R^C is invertible. As $|R| = |C|$, it is sufficient to show that the mapping defined by M_R^C is surjective, i.e., for every $\vec{y}_R$ there exists an $\vec{x}_C$ such that $\vec{y}_R = M_R^C \vec{x}_C$. Equivalently, we show that for every $\vec{y}_R$ there exists an $\vec{x}$ such that $\vec{y}_R = M_R \vec{x}$ and $\vec{x}_{\overline{C}} = \vec{0}$, where $\overline{C} = \{1, \ldots, n\} \setminus C$. Remember that M is defined such that the points $(\alpha_1, x_1), \ldots, (\alpha_n, x_n), (\beta_1, y_1), \ldots, (\beta_n, y_n)$ lie on a polynomial $g(\cdot)$ of degree $n - 1$. Given the n points $\{(\alpha_j, 0)\}_{j \notin C}$ and $\{(\beta_i, y_i)\}_{i \in R}$, the polynomial $g(\cdot)$ can be determined by Lagrange interpolation, and $\vec{x}_C$ can be computed linearly from $\vec{y}_R$. Hence, M_R^C is invertible.   □

3.3   Properties

The mappings defined by hyper-invertible matrices have a very nice symmetry property: Any subset of n input/output values can be expressed as a linear function of the remaining n input/output values:

Lemma 2. *Let M be a hyper-invertible n-by-n matrix and $(y_1, \ldots, y_n) = M(x_1, \ldots, x_n)$. Then for any index sets $A, B \subseteq \{1, \ldots, n\}$ with $|A| + |B| = n$, there exists an invertible linear function $f : \mathcal{F}^n \to \mathcal{F}^n$, mapping the values $\{x_i\}_{i \in A}, \{y_i\}_{i \in B}$ onto the values $\{x_i\}_{i \notin A}, \{y_i\}_{i \notin B}$.*

Proof. We have $\vec{y} = M\vec{x}$ and $\vec{y}_B = M_B \vec{x} = M_B^A \vec{x}_A + M_B^{\overline{A}} \vec{x}_{\overline{A}}$. Due to hyper-invertibility, $M_B^{\overline{A}}$ is invertible, and $\vec{x}_{\overline{A}} = \left(M_B^{\overline{A}}\right)^{-1} (\vec{y}_B - M_B^A \vec{x}_A)$. $\vec{y}_{\overline{B}}$ can be computed similarly.   □

4   Protocol Overview

The new MPC protocol proceeds in two phases: the preparation phase and the computation phase.

In the preparation phase, degree-t sharings of random (a, b, c)-triples are generated (in parallel), one for every multiplication gate. Furthermore, for every random gate as well as for every input gate, a t-sharing of a random r is generated. For the sake of simplicity, we generate $c_M + c_R + c_I$ random triples, where for random and input gates, only the first component is used. The preparation phase makes use of the player-elimination technique.

In the computation phase, the actual circuit is computed. Input gates are evaluated with help of a pre-shared random value r. Due to the linearity of the used secret-sharing, the linear gates can be computed locally – without communication. Random gates are evaluated simply by picking an unused pre-shared random value r. Multiplication gates are evaluated with help of one prepared (a, b, c)-triple, using Beaver's circuit randomization technique [Bea91a]. Output gates involve a (robust) secret reconstruction.

5 Secret Sharing

5.1 Definitions and Notation

As secret-sharing scheme, we use the standard Shamir sharing scheme [Sha79].

Definition 3. *We say that a value s is (correctly) d-shared (among the players in $\mathcal{P}'$) if every honest player $P_i \in \mathcal{P}'$ is holding a share s_i of s, such that there exists a degree-d polynomial $p(\cdot)$ with $p(0) = s$ and $p(\alpha_i) = s_i$ for every $P_i \in \mathcal{P}'$.[2] The vector $(s_1, \ldots, s_{n'})$ of shares is called a d-sharing of s, and is denoted by $[s]_d$. A (possibly incomplete) set of shares is called d-consistent if these shares lie on a degree d polynomial.*

Most of the sharings used in our protocol are t-sharings – denoted as $[\cdot]_t$. In the preparation phase we also temporarily use t'- and $2t'$-sharings (denoted by $[\cdot]_{t'}$ and $[\cdot]_{2t'}$, respectively).

By saying that the players in $\mathcal{P}'$ compute (locally) $([y^{(1)}]_{d'}, \ldots, [y^{(m')}]_{d'}) = f([x^{(1)}]_d, \ldots, [x^{(m)}]_d)$ (for any function $f : \mathcal{F}^m \to \mathcal{F}^{m'}$) we mean that every player P_i applies this function to his shares, i.e. computes $(y_i^{(1)}, \ldots, y_i^{(m')}) = f(x_i^{(1)}, \ldots, x_i^{(m)})$. Note that by applying any linear function to correct d-sharings we get a correct d-sharing of the output. However, by multiplying two correct d-sharings we get a correct $2d$-sharing of the product, i.e. $[a]_d[b]_d = [ab]_{2d}$.

5.2 The Share Protocol

The following (trivial) Share protocol allows an honest dealer P_D to correctly d-share a secret s among the players in $\mathcal{P}'$, while communicating $\mathcal{O}(n\kappa)$ bits. We stress that this protocol does not ensure that the resulting sharing is consistent; a corrupted dealer might distribute totally inconsistent shares. The consistency of sharings must be verified separately.

Protocol Share($P_D \in (\mathcal{P} \cup \mathcal{U}), s, d$)
1. P_D chooses a random degree-d polynomial $p(\cdot)$ with $s = p(0)$ and sends $s_i = p(\alpha_i)$ to every $P_i \in \mathcal{P}'$.

---

[2] Where α_i denotes the unique fixed value assigned to P_i.

5.3   The Reconstruct Protocols

We use two reconstruction protocols: one for private and one for public reconstruction. Both can be either robust or only detectable – depending on the degree of the sharings to be reconstructed.

In the private reconstruction protocol the players simply send their shares to the receiver P_R (a player or a user) who interpolates the secret (if possible).

Protocol ReconsPriv($P_R \in (\mathcal{P} \cup \mathcal{U}), d, [s]_d$)
1. Every player $P_i \in \mathcal{P}'$ sends his share s_i of s to P_R.
2. If there exists a degree-d polynomial $p(\cdot)$ such that at least $d + t' + 1$ of the received shares lie on it, then P_R computes the secret $s = p(0)$. Otherwise P_R gets unhappy.

Lemma 3. *For $d < n' - 2t'$, the protocol* ReconsPriv *robustly reconstructs $[s]_d$ towards P_R. For $d < n' - t'$,* ReconsPriv *detectably reconstructs $[s]_d$ towards P_R (i.e., P_R either outputs s or gets unhappy, where the latter only happens when some players are faulty).* ReconsPriv *communicates $\mathcal{O}(n\kappa)$ bits.*

The public reconstruction protocol ReconsPubl takes $T = n' - 2t' = n - 2t = \Omega(n)$ correct d-sharings $[s_1]_d, \ldots, [s_T]_d$ and publicly (to all players in $\mathcal{P}'$) outputs the (correct) values $s_1, \ldots, s_T$ or fails (with at least one honest player being unhappy). In ReconsPubl we use the idea of [DN07]: first the T sharings $[s_1]_d, \ldots, [s_T]_d$ are expanded (using a linear error-correcting code) to n' sharings $[u_1]_d, \ldots, [u_{n'}]_d,$[3] each of which is reconstructed towards *one* player in $\mathcal{P}'$ (using ReconsPriv). Then every $P_i \in \mathcal{P}'$ sends his reconstructed value u_i to every other player in $\mathcal{P}'$, who tries to decode (with error correction) the received code word $(u_1, \ldots, u_{n'})$ to $s_1, \ldots, s_T$. ReconsPubl communicates $\mathcal{O}(n^2\kappa)$ bits to reconstruct $T = \Omega(n)$ sharings.

Protocol ReconsPubl($d, [s_1]_d, \ldots, [s_T]_d$)
1. For every $j = 1, \ldots, n'$ the players in $\mathcal{P}'$ (locally) compute $[u_j]_d$ as:
$$[u_j]_d = [s_1]_d + [s_2]_d\beta_j + [s_3]_d\beta_j^2 + \ldots + [s_T]_d\beta_j^{T-1}$$
2. For every $P_i \in \mathcal{P}'$, ReconsPriv is invoked to reconstruct $[u_i]_d$ towards P_i.
3. Every $P_i \in \mathcal{P}'$ sends u_i (or $\perp$ if unhappy) to every $P_j \in \mathcal{P}'$.
4. $\forall P_i \in \mathcal{P}'$: If P_i received at least $T + t'$ ($T - 1$)-consistent values (in the previous step), he computes $s_1, \ldots, s_T$ from any T of them. Otherwise he gets unhappy.

Lemma 4. *For $d < n' - 2t'$, the protocol* ReconsPubl *robustly reconstructs $[s_1]_d, \ldots, [s_T]_d$ towards all players in $\mathcal{P}'$. For $d < n' - t'$,* ReconsPubl *detectably reconstructs $[s_1]_d, \ldots, [s_T]_d$ towards all players in $\mathcal{P}'$ (i.e., every $P_i \in \mathcal{P}'$ either outputs $s_1, \ldots, s_T$ or gets unhappy, where the latter only happens when some players are faulty).* ReconsPubl *communicates $\mathcal{O}(n^2\kappa)$ bits.*

---

[3] For this we interpret $s_1, \ldots, s_T$ as coefficients of a degree $T - 1$ polynomial and $u_1, \ldots, u_{n'}$ as evaluations of this polynomial at n' fixed positions $\beta_1, \ldots, \beta_{n'}$.

6 Preparation Phase

6.1 Overview

The goal of the preparation phase is to generate correct t-sharings of $c_M + c_R + c_I$ secret random triples (a_k, b_k, c_k), such that $c_k = a_k b_k$ for $k = 1, \ldots, c_M + c_R + c_I$. We stress that all resulting sharings must be t-sharings (rather than t'-sharings) among the player set $\mathcal{P}'$.[4]

The preparation phase uses player elimination, i.e. the generation of the triples is divided into t segments of length $\ell = \lceil \frac{c_M + c_R + c_I}{t} \rceil$. In every segment the non-robust protocol GenerateTriples is invoked, which either generates correct triples, or fails with at least one honest player being unhappy.

The generation of the triples follows the approach of [DN07]: First, the players generate random a and b values, both simultaneously shared with degree t (for outputting) and degree t' (for multiplication). Additionally, the players generate random value r, simultaneously shared with degree t and degree $2t'$. Then, they locally compute the $2t'$-sharing $[ab]_{2t'}$ (by every player multiplying his respective shares), publicly reconstruct the difference $[ab]_{2t'} - [r]_{2t'}$ and add it (locally) to $[r]_t$, resulting in $[ab]_t$. Finally, the players output the triple $([a]_t, [b]_t, [ab]_t)$.

Definition 4. *A value x is (d, d')-shared among the players $\mathcal{P}'$, denoted as $[x]_{d,d'}$, if x is both d-shared and d'-shared. We denote such a sharing as a* double-sharing, *and the pair of shares held by each player as his* double-share.

We (trivially) observe that the sum of correct $(d; d')$-sharings is a correct (d, d')-sharing of the sum.

6.2 Generating Random Double-Sharings

The following non-robust protocol DoubleShareRandom(d, d') either generates T independent secret random values $r_1, \ldots, r_T$, each independently (d, d')-shared among $\mathcal{P}'$, or fails with at least one honest player being unhappy.

The generation of the random double-sharings employs hyper-invertible matrices: First, every player $P_i \in \mathcal{P}'$ selects and double-shares a random value s_i. Then, the players compute double-sharings of the values r_i, defined as $(r_1, \ldots, r_{n'}) = M(s_1, \ldots, s_{n'})$, where M is a hyper-invertible n'-by-n' matrix. $2t'$ of the resulting double-sharings are reconstructed, each towards a different player, who verify the correctness of the double-sharings (and gets unhappy in case of a fault). The remaining $n' - 2t' = T$ double-sharings are outputted. This procedure guarantees that if all honest players are happy, then at least n' double-sharings are correct (the $n' - t'$ double-sharings inputted by honest players, as well as the t' double-sharings verified by honest players), and due to the hyper-invertibility of M, all $2n'$ double-sharings must be correct (the remaining double-sharings can be computed linearly from the good double-sharings).

---

[4] Remember that as $t \le n' - 2t'$ (according to Lemma 3 and 4), such sharings can be robustly reconstructed (regardless of the actual player set $\mathcal{P}'$).

Furthermore, the outputted double-sharings are random and unknown to the adversary, as there is a bijective mapping from any T double-sharings inputted by honest players to the outputted double-sharings.

Protocol DoubleShareRandom(d, d')

1. SECRET SHARE: Every $P_i \in \mathcal{P}'$ chooses a random s_i and acts (twice in parallel) as a dealer in Share to distribute the shares among the players in $\mathcal{P}'$, resulting in $[s_i]_{d,d'}$.

2. APPLY M: The players in $\mathcal{P}'$ (locally) compute $([r_1]_{d,d'}, \ldots, [r_{n'}]_{d,d'}) = M([s_1]_{d,d'}, \ldots, [s_{n'}]_{d,d'})$. In order to do so, every P_i computes his double-share of each r_j as linear combination of his double-shares of the s_k-values.

3. CHECK: For $i = T+1, \ldots, n'$, every $P_j \in \mathcal{P}'$ sends his double-share of $[s_i]_{d,d'}$ to P_i, who checks that *all* n' double-shares define a correct double-sharing of some value s_i. More precisely, P_i checks that all d-shares indeed lie on a polynomial $g(\cdot)$ of degree d, and that all d'-shares indeed lie on a polynomial $g'(\cdot)$ of degree d', and that $g(0) = g'(0)$. If any of the checks fails, P_i gets unhappy.

4. OUTPUT: The remaining T double-sharings $[r_1]_{d,d'}, \ldots, [r_T]_{d,d'}$ are outputted.

Lemma 5. *If* DoubleShareRandom(d, d') *succeeds (i.e., all honest players are happy), it outputs* $T = n' - 2t'$ *correct and random* (d, d')-*sharings (among* $\mathcal{P}'$), *unknown to the adversary.* DoubleShareRandom *communicates* $\mathcal{O}(n^2\kappa)$ *bits to generate* $\Omega(n)$ *double-sharings.*

Proof. CORRECTNESS: Assume that all honest players remain happy during the protocol. Then for all honest P_i with $i \in \{T+1, \ldots, n'\}$, the sharing of r_i checked by P_i in Step 3 is a correct (d, d')-sharing. As $T = n' - 2t'$, there are at least t' correct sharings of the values r_k. Furthermore, every sharing of an s_i distributed by an honest P_i in Step 1 is a correct (d, d')-sharing. Thus there are at least $n' - t'$ correct sharings of the values s_k. Given these (at least) n' correct (d, d')-sharings, the sharings of *all* other values s_k and r_k can be computed linearly. As a linear combination of a correct (d, d')-sharing is again a (d, d')-sharing, it follows that all values $s_1, \ldots, s_{n'}, r_1, \ldots, r_{n'}$ are correctly (d, d')-shared.

PRIVACY: The adversary knows (at most) t' of the input sharings s_k (those provided by corrupted players), and t' of the output sharings r_k (with $k > T$, those reconstructed towards corrupted players). When fixing these $2t'$ sharings, then there exists a bijective mapping between any other (honest) T input sharings and the first T output sharings (Lemma 2), hence the sharings $[r_1]_{d,d'}, \ldots, [r_T]_{d,d'}$ are uniformly at random, unknown to the adversary.

COMMUNICATION: The stated communication can easily be verified by inspecting the protocol.   □

6.3   Generating Random Triples

Now we present the non-robust protocol GenerateTriples that either generates $T = n' - 2t'$ correctly t-shared (a, b, c)-triples, or fails (with at least one honest player being unhappy). The idea of the protocol GenerateTriples is the following: First DoubleShareRandom is invoked 3 times to generated the random double-sharings $[a_1]_{t,t'}, \ldots, [a_T]_{t,t'}$, $[b_1]_{t,t'}, \ldots, [b_T]_{t,t'}$, and $[r_1]_{t,2t'}, \ldots, [r_T]_{t,2t'}$, respectively. Then for every pair a_k, b_k, a t-sharing of the product $c_k = a_k b_k$ is computed by reducing the locally computed $2t'$-sharing $[c_k]_{2t'} = [a_k]_{t'}[b_k]_{t'}$ to a t-sharing $[c_k]_t$ using the t-sharing $[r_k]_t$ and the $2t'$-sharing $[r_k]_{2t'}$ of the random value r_k.

Protocol GenerateTriples

1. GENERATE DOUBLE-SHARINGS: Invoke DoubleShareRandom three times in parallel to generate the double-sharings $[a_1]_{t,t'}, \ldots, [a_T]_{t,t'}$, $[b_1]_{t,t'}, \ldots, [b_T]_{t,t'}$, and $[r_1]_{t,2t'}, \ldots, [r_T]_{t,2t'}$.

2. MULTIPLY:

   2.1 For $k = 1, \ldots, T$, the players in $\mathcal{P}'$ compute (locally) the $2t'$-sharing $[c_k]_{2t'}$ of $c_k = a_k b_k$ as $[c_k]_{2t'} = [a_k]_{t'}[b_k]_{t'}$ (by every player computing the product of his shares).

   2.2 For $k = 1, \ldots, T$, the players in $\mathcal{P}'$ compute (locally) a $2t'$-sharing of the difference $[d_k]_{2t'} = [c_k]_{2t'} - [r_k]_{2t'}$

   2.3 Invoke ReconsPubl ($\mathcal{R} = \mathcal{P}', d = 2t', [d_1]_{2t'}, \ldots, [d_T]_{2t'}$) to reconstruct $d_1, \ldots, d_T$ towards every player in $\mathcal{P}'$.

   2.4 For $k = 1, \ldots, T$, the players in $\mathcal{P}'$ compute (locally) the t-sharing $[c_k]_t = [r_k]_t + [d_k]_0$, where $[d_k]_0$ denotes the constant sharing $[d_k]_0 = (d_k, \ldots, d_k)$.

3. OUTPUT: The t-shared triples $([a_1]_t, [b_1]_t, [c_1]_t), \ldots, ([a_T]_t, [b_T]_t, [c_T]_t)$ are outputted.

Lemma 6. *If GenerateTriples succeeds (i.e., all honest players are happy), it outputs independent random t-sharings of $T = \Omega(n)$ random triples $(a_1, b_1, c_1), \ldots, (a_T, b_T, c_T)$ with a_k, b_k independent uniform random values and $c_k = a_k b_k$ for $k = 1, \ldots, T$. GenerateTriples communicates $\mathcal{O}(n^2 \kappa)$ bits.*

Proof. The security of GenerateTriples follows directly from the security of DoubleShareRandom.   □

6.4   Preparation Phase — Main Protocol

The following protocol PreparationPhase divides the generation of the $c_M + c_R + c_I$ triples into t segments of length $\ell = \lceil \frac{c_M + c_R + c_I}{t} \rceil$. In each segment the triples are generated invoking the non-robust protocol GenerateTriples (as often as necessary), then the players reach agreement on whether or not all players are happy. If yes, they proceed to the next segment. Otherwise, a pair of players is identified in FaultLocalization, excluded from the actual player set $\mathcal{P}'$ and the segment is repeated (with the new $\mathcal{P}'$ and all players setting their happy-bit to happy).

Protocol PreparationPhase

For each segment $k = 1, \ldots, t$ do:

0. Every $P_i \in \mathcal{P}'$ sets his happy-bit to happy.

1. TRIPLE GENERATION: Invoke GenerateTriples $\lceil \frac{\ell}{T} \rceil$ times in parallel.

2. FAULT DETECTION: Reach agreement whether or not at least one player is unhappy:

    2.1 Every $P_i \in \mathcal{P}'$ sends his happy-bit to every $P_j \in \mathcal{P}'$, who gets unhappy if at least one P_i claims to be unhappy.

    2.2 The players in $\mathcal{P}'$ run a consensus protocol on their respective happy-bits. If the consensus outputs "happy", then the generated triples are outputted and the segment is finished. Otherwise, the following Fault-Localization step is executed.

3. FAULT LOCALIZATION: Localize $E \subseteq \mathcal{P}'$ with $|E| = 2$ and at least one player in E being corrupted:

    3.0 Denote the player $P_r \in \mathcal{P}'$ with the smallest index r as the referee.[5]

    3.1 Every $P_i \in \mathcal{P}'$ sends everything he received and all random values he chose during the computation of the actual segment (including fault detection) to P_r.

    3.2 Given the values received in Step 3.1, P_r can reproduce every message that should have been sent (by applying the respective protocol instructions of the sender), and compare it with the value that the recipient claims to have received. Then P_r broadcasts (l, i, j, x, x'), where l is the index of a message where P_i should have sent x to P_j, but P_j claims to have received $x' \neq x$.

    3.3 The accused players broadcast whether they agree with P_r. If P_i disagrees, set $E = \{P_r, P_i\}$, if P_j disagrees, set $E = \{P_r, P_j\}$, otherwise set $E = \{P_i, P_j\}$.

4. PLAYER ELIMINATION: Set $\mathcal{P}' \leftarrow \mathcal{P}' \setminus E$, $n' \leftarrow n' - 2$, $t' \leftarrow t' - 1$, and repeat the segment.

Lemma 7. *The protocol* PreparationPhase *generates independent random t-sharings of $c_M + c_R + c_I$ secret triples (a_k, b_k, c_k) with a_k, b_k independent uniform random values and $c_k = a_k b_k$ for $k = 1, \ldots, c_M + c_R + c_I$.* PreparationPhase *communicates $\mathcal{O}\big((c_M + c_R + c_I)n\kappa + n^2\kappa + t\,\mathcal{BA}(\kappa)\big)$ bits, which amounts to $\mathcal{O}\big((c_M + c_R + c_I)n\kappa + n^3\kappa\big)$ bits overall.*

7   Computation Phase

In the computation phase, the circuit is robustly evaluated, whereby all intermediate values are t-shared among the players in $\mathcal{P}'$.

Input gates are realized by reconstructing a pre-shared random value r towards the input-providing user, who then broadcasts the difference of this r and his input.

---

[5] The communication can be balanced by selecting a player who has not yet been referee in a previous segment.

Due to the linearity of the secret-sharing scheme, linear gates can be computed locally simply by applying the linear function to the shares, i.e. for any linear function $f(\cdot,\cdot)$, a sharing $[c] = [f(a,b)]$ is computed by letting every player P_i compute $c_i = f(a_i, b_i)$.

With every random gate, one random sharing $[r]$ (from the preparation phase) is associated and $[r]_t$ is directly used as outcome of the random gate.

With every multiplication gate, one $([a],[b],[c])$-triple (from the preparation phase) is associated, which is used to compute a sharing of the product at the cost of two public reconstruction. For the sake of efficiency, we evaluate $T/2$ multiplication gates at once (such that we can publicly reconstruct T sharings at once). This of course requires that these multiplication gates do not depend on each other, i.e., that they all have the same multiplicative depth in the circuit.[6]

Output gates involve a (robust) secret reconstruction.

Protocol ComputationPhase

Evaluate the gates of the circuit as follows:

- INPUT GATE (USER U INPUTS s):
  1. Reconstruct the associated sharing $[r]_t$ towards U with ReconsPriv$(U, t, [r])$. This is robust because $t < n' - 2t'$.
  2. User U computes and broadcasts the difference $d = s - r$.
  3. Every $P_i \in \mathcal{P}'$ computes his share s_i of s locally as $s_i = d + r_i$.

- ADDITION/LINEAR GATE: Every $P_i \in \mathcal{P}'$ applies the linear function on his respective shares.

- RANDOM GATE: Pick the sharing $[r]_t$ associated with the gate.

- MULTIPLICATION GATE: Up to $\lfloor T/2 \rfloor$ (where $T = n - 2t$) multiplication gates are processed simultaneously. Denote the factor sharings as $([x_1],[y_1]),\ldots,([x_{T/2}],[y_{T/2}])$, and the associated triples as $([a_1],[b_1],[c_1]),\ldots,([a_{T/2}],[b_{T/2}],[c_{T/2}])$. The products $[z_1],\ldots,[z_{T/2}]$ are computed as follows:
  1. For $k = 1,\ldots,T/2$, the players compute $[d_k] = [x_k] - [a_k]$ and $[e_k] = [y_k] - [b_k]$.
  2. Invoke ReconsPubl to publicly reconstruct the T t-sharings $(d_1, e_1),\ldots,(d_{T/2}, e_{T/2})$. Note that this is robust, as $t < n' - 2t'$.
  3. For $k = 1,\ldots,T/2$, the players compute the product sharings $[z_k]_t = [de]_0 + d[b]_t + e[a]_t + [c]_t$, where $[de]_0$ denotes the (implicitly defined) 0-sharing of de.

- OUTPUT GATE (OUTPUT $[s]$ TO USER U): Invoke ReconsPriv$(U, t, [s]_t)$.

Lemma 8. *The protocol* ComputationPhase *perfectly securely evaluates a circuit with c_I input, c_R random, c_M multiplication, and c_O output gates, given $c_I + c_R + c_M$ pre-shared random multiplication triples, with communicating*

---

[6] The multiplicative depth of a gate is the maximum number of multiplication gates on any path from input/random gates to this gate.

$\mathcal{O}\big((c_I n + c_M n + c_O n + D_M n^2)\kappa + c_I\,\mathcal{BA}(\kappa)\big)$ bits, where D_M denotes the multiplicative depth of the circuit.

Theorem 1. *The MPC protocol consisting of* PreparationPhase *and* ComputationPhase *evaluates a circuit with c_I input, c_R random, c_M multiplication, and c_O output gates, with communicating $\mathcal{O}\big((c_I n + c_R n + c_M n + c_O n + D_M n^2)\kappa + (c_I + n)\,\mathcal{BA}(\kappa)\big)$ bits, which amounts to $\mathcal{O}\big((c_I n^2 + c_R n + c_M n + c_O n + D_M n^2)\kappa + n^3 \kappa\big)$ bits, where D_M denotes the multiplicative depth of the circuit. The protocol is perfectly secure against an active adversary corrupting $t < n/3$ players.*

The communication complexity for giving input can be improved from $\mathcal{O}(n^2 \kappa)$ per input to $\mathcal{O}(n\kappa)$. Details can be found in Appendix A.

Theorem 2. *The MPC protocol given in Appendix A evaluates a circuit with c_I input, c_R random, c_M multiplication, and c_O output gates, with communicating $\mathcal{O}\big((c_I n + c_R n + c_M n + c_O n + D_M n^2)\kappa + n\,\mathcal{BA}(\kappa)\big)$ bits, which amounts to $\mathcal{O}\big((c_I n + c_R n + c_M n + c_O n + D_M n^2)\kappa + n^3 \kappa\big)$ bits, where D_M denotes the multiplicative depth of the circuit. The protocol is perfectly secure against an active adversary corrupting $t < /n/3$ players.*

8   Conclusions

We have presented a perfectly secure multi-party computation protocol with optimal security $(t < n/3)$, which communicates only $\mathcal{O}(n)$ field elements per multiplication.

Compared with the previously most efficient perfectly-secure MPC protocol [HMP00], this is a speedup of $\theta(n^2)$ with the same level of security.

Compared with the previously "most secure" MPC protocol with linear communication complexity [DN07], this improves the security from unconditional to perfect, and at the same time slightly improves the communication overhead (from $\mathcal{O}(n^4 \kappa)$ in [DN07] to $\mathcal{O}(n^3 \kappa)$ here).

This speed-up was possible due to a new technique, so-called hyper-invertible matrices. Such matrices allow to detectably generate $\Omega(n)$ random sharings at costs $\mathcal{O}(n^2)$, with perfect security (i.e., without any probabilistic checks as used in all previous highly-efficient MPC protocols). We believe that this approach is much more natural than the previous approach with probabilistic checks (for example, [DN07] needs to work in an extension field to keep the error-probability small).

References

[Bea91a]   Beaver, D.: Efficient multiparty protocols using circuit randomization. In: Feigenbaum, J. (ed.) CRYPTO 1991. LNCS, vol. 576, pp. 420–432. Springer, Heidelberg (1992)

[Bea91b]   Beaver, D.: Secure multiparty protocols and zero-knowledge proof systems tolerating a faulty minority. Journal of Cryptology, 75–122 (1991)

[BGP92]   Berman, P., Garay, J.A., Perry, K.J.: Bit optimal distributed consensus.
             In: Computer Science Research, Preliminary version has appeared in
             Proc. 21st STOC, pp. 313–322 (1992)

[BGW88]   Ben-Or, M., Goldwasser, S., Wigderson, A.: Completeness theorems for
             non-cryptographic fault-tolerant distributed computation. In: Proc. 20th
             ACM Symposium on the Theory of Computing (STOC), pp. 1–10 (1988)

[BH06]   Beerliova-Trubiniova, Z., Hirt, M.: Efficient multi-party computation
             with dispute control. In: Halevi, S., Rabin, T. (eds.) TCC 2006. LNCS,
             vol. 3876, pp. 305–328. Springer, Heidelberg (2006)

[CCD88]   Chaum, D., Crépeau, C., Damgård, I.: Multiparty unconditionally secure
             protocols (extended abstract). In: Proc. 20th ACM Symposium on the
             Theory of Computing (STOC), pp. 11–19 (1988)

[CDvdG87]   Chaum, D., Damgård, I., van de Graaf, J.: Multiparty computations
             ensuring privacy of each party's input and correctness of the result.
             In: Pomerance, C. (ed.) CRYPTO 1987. LNCS, vol. 293, pp. 87–119.
             Springer, Heidelberg (1988)

[CW92]   Coan, B.A., Welch, J.L.: Modular construction of a Byzantine agree-
             ment protocol with optimal message bit complexity. Information and
             Computation 97(1), 61–85 (1992); Preliminary version has appeared in
             Proc. 8th PODC (1989)

[DN07]   Damgård, I., Nielsen, J.B.: Robust multiparty computation with linear
             communication complexity. In: Menezes, A. (ed.) CRYPTO 2007. LNCS,
             vol. 4622, Springer, Heidelberg (2007)

[GHY87]   Galil, Z., Haber, S., Yung, M.: Cryptographic computation: Secure fault-
             tolerant protocols and the public-key model. In: Pomerance, C. (ed.)
             CRYPTO 1987. LNCS, vol. 293, pp. 135–155. Springer, Heidelberg
             (1988)

[GMW87]   Goldreich, O., Micali, S., Wigderson, A.: How to play any mental
             game — a completeness theorem for protocols with honest majority.
             In: Proc. 19th ACM Symposium on the Theory of Computing (STOC),
             pp. 218–229 (1987)

[HMP00]   Hirt, M., Maurer, U., Przydatek, B.: Efficient secure multi-party compu-
             tation. In: Okamoto, T. (ed.) ASIACRYPT 2000. LNCS, vol. 1976, pp.
             143–161. Springer, Heidelberg (2000)

[HN06]   Hirt, M., Nielsen, J.B.: Robust multiparty computation with linear com-
             munication complexity. In: Dwork, C. (ed.) CRYPTO 2006. LNCS,
             vol. 4117, pp. 463–482. Springer, Heidelberg (2006)

[RB89]   Rabin, T., Ben-Or, M.: Verifiable secret sharing and multiparty protocols
             with honest majority. In: Proc. 21st ACM Symposium on the Theory of
             Computing (STOC), pp. 73–85 (1989)

[Sha79]   Shamir, A.: How to share a secret. Communications of the ACM 22,
             612–613 (1979)

[Yao82]   Yao, A.C.: Protocols for secure computations. In: Proc. 23rd IEEE Sym-
             posium on the Foundations of Computer Science (FOCS), pp. 160–164.
             IEEE, Los Alamitos (1982)

Appendix

A   Totally Linear Protocol

To construct a totally linear MPC protocol we propose a more efficient input protocol. For the sake of simpler presentation we assume that all inputs are given at the beginning of the computation stage.

We first present the input protocol LinearInput that allows a set of dealers $D \subset \mathcal{P} \cup \mathcal{U}$ each having T inputs to (robustly) share these inputs among the players in $\mathcal{P}'$ (using pre-computed t-sharings of random values). If there is a user with more than T inputs, he plays a role of more dealers.

Protocol LinearInput (every $D_k \in D$ having inputs $s_k^{(1)}, \ldots, s_k^{(T)}$ with associated random
t-**sharings $[r_k^{(1)}]_t, \ldots, [r_k^{(T)}]_t$)**
1. RECONSTRUCT: For every $D_k \in D$ and every $l = 1, \ldots, T$ invoke ReconsPriv($D_k, [r_k^{(l)}]_t$) to reconstruct the secret random value $r_k^{(l)}$ towards D_k.
2. COMPUTE DIFFERENCE: Every $D_k \in D$ computes for every $l = 1, \ldots, T$ the difference $d_k^{(l)} = s_k^{(l)} - r_k^{(l)}$.
3. BROADCAST: Invoke Broadcast to let every dealer $D_k \in D$ broadcast (towards the players in $\mathcal{P}'$) the T computed differences $d_k^{(1)}, \ldots, d_k^{(T)}$.
4. COMPUTE LOCALLY AND OUTPUT: For every $D_k \in D$ and every $l = 1, \ldots, T$ the players in $\mathcal{P}'$ (locally) compute the sharing of the input $s_k^{(l)}$ as $[s_k^{(l)}]_t = [d_k^{(l)}]_0 + [r_k^{(l)}]_t$.

The robust protocol Broadcast is constructed in three steps.

We first present a non-robust broadcast protocol for $\mathcal{P}'$ PE − Broadcast.

Note, that broadcasting a value can be interpreted as sharing this value with degree zero, thus checking whether every player distributed his value consistently is the same as checking the correctness of sharings with degree zero, which we can easily do applying HIM.

Protocol PE − Broadcast(every $P_i \in \mathcal{P}'$ has input x_i)
1. DISTRIBUTE VALUES: Every P_i shares his input with Share $(P_i, x_i, d = 0)$, i.e. sends x_i to every $P_j \in \mathcal{P}'$. Resulting in n' (supposed) 0-sharings

$$[x_1]_0, \ldots, [x_{n'}]_0$$

2. APPLY HIM M: The players in $\mathcal{P}'$ compute locally the 0-sharings $[\hat{x}_1]_0, \ldots, [\hat{x}_{n'}]_0$ as

$$([\hat{x}_1]_0, \ldots, [\hat{x}_{n'}]_0) = M([x_1]_0, \ldots, [x_{n'}]_0)$$

3. CHECK: Every $P_i \in \mathcal{P}'$ checks the correctness of $[\hat{x}_i]_0$. For this every $P_j \in \mathcal{P}'$ sends his share of $\hat{x}_i$ to P_i. If the values received by P_i are not 0-consistent (equal), P_i gets unhappy.

4. OUTPUT: Every $P_j \in \mathcal{P}'$ outputs the values received in Step 1.)

Now we construct a robust broadcast protocol for $\mathcal{P}'$ BroadcastFor$\mathcal{P}'$ using PE − Broadcast, player elimination and segmentation. BroadcastFor$\mathcal{P}'$ allows the players in $\mathcal{P}'$, each holding ℓ values $x_i^{(1)}, \ldots, x_i^{(\ell)}$ to broadcast this values among the players in $\mathcal{P}'$.

Protocol BroadcastFor$\mathcal{P}'$
For each segment $k = 1, \ldots, t$ (of length $\ell' = \lceil \frac{\ell}{t} \rceil$) do:
0. Every $P_i \in \mathcal{P}'$ sets his happy-bit to happy.
1. PE-BROADCAST: Invoke PE − Broadcast $\ell' = \lceil \frac{\ell}{t} \rceil$ times in parallel, i.e. for $l = 1, \ldots, \ell'$ invoke PE − Broadcast to let every $P_i \in \mathcal{P}'$ broadcast his input $x_i = x_i^{(l+(k-1)\ell')}$.
2. FAULT DETECTION: Reach agreement whether or not at least one player is unhappy:
   2.1 Every $P_i \in \mathcal{P}'$ sends his happy-bit to every $P_j \in \mathcal{P}'$, who gets unhappy if at least one P_i claims to be unhappy.
   2.2 The players in $\mathcal{P}'$ run a consensus protocol on their respective happy-bits. If the consensus outputs "happy", then the generated triples are outputted and the segment is finished. Otherwise, the following Fault-Localization step is executed.
3. FAULT LOCALIZATION: Localize $E \subsetneq \mathcal{P}'$ with $|E| - 2$ and at least one player in E being corrupted:
   3.0 Denote the player $P_r \in \mathcal{P}'$ with the smallest index r as the referee.[7]
   3.1 Every $P_i \in \mathcal{P}'$ sends everything he received and all random values he chose during the computation of the actual segment (including fault detection) to P_r.
   3.2 Given the values received in Step 3.1, P_r can reproduce every message that should have been sent (by applying the respective protocol instructions of the sender), and compare it with the value that the recipient claims to have received. Then P_r broadcasts (l, i, j, x, x'), where l is the index of a message where P_i should have sent x to P_j, but P_j claims to have received $x' \neq x$.
   3.3 The accused players broadcast whether they agree with P_r. If P_i disagrees, set $E = \{P_r, P_i\}$, if P_j disagrees, set $E = \{P_r, P_j\}$, otherwise set $E = \{P_i, P_j\}$.
4. PLAYER ELIMINATION: Set $\mathcal{P}' \leftarrow \mathcal{P}' \setminus E$, $n' \leftarrow n' - 2$, $t' \leftarrow t' - 1$, and repeat the segment.

Finally we present the protocol Broadcast that enables a set of dealers D (players or users), each holding T values to robustly broadcast this values, among the players in $\mathcal{P}'$.

---

[7] The communication can be balanced by selecting a player who has not yet been referee in a previous segment.

The idea of the protocol is to let every dealer expand his T values to n' values (using an error-correcting code tolerating t' errors) and to send each of these values to one player in $\mathcal{P}'$. Then the players in $\mathcal{P}'$ invoke BroadcastForP' to broadcast the received values and final (locally) compute the original values from the broadcasted values using error-correction.

Protocol Broadcast(every dealer D_k holding $a_k^{(0)}, \ldots, a_k^{(T-1)}$)

1. EXPAND AND DISTRIBUTE: For every dealer D_k denote the polynomial defined by the values $a_k^{(0)}, \ldots, a_k^{(T-1)}$ as $p_k(x)$, i.e.

$$p_k(x) = a_k^{(0)} + a_k^{(1)} x + \ldots + a_k^{(T-1)} x^{T-1}$$

. The dealer D_k computes for every player $P_i \in \mathcal{P}'$ the point $p_k(\alpha_i)$ and sends it to P_i.

2. BROADCAST: The players in $\mathcal{P}'$ invoke BroadcastForP' with P_i having input $p_1(\alpha_i), \ldots, p_{|D|}(\alpha_i)$.

3. COMPUTE AND OUTPUT: For every dealer D_k every $P_i \in \mathcal{P}'$ locally computes the values $a_k^{(0)}, \ldots, a_k^{(T-1)}$ from the broadcasted values $p_k(\alpha_1), \ldots, p_k(\alpha_{n'})$ (using error-correction).

MPC vs. SFE: Perfect Security in a Unified Corruption Model*

Zuzana Beerliová-Trubíniová, Matthias Fitzi, Martin Hirt, Ueli Maurer, and Vassilis Zikas

Department of Computer Science, ETH Zurich
{bzuzana,fitzi,hirt,maurer,vzikas}@inf.ethz.ch

Abstract. Secure function evaluation (SFE) allows a set of players to compute an arbitrary agreed function of their private inputs, even if an adversary may corrupt some of the players. Secure multi-party computation (MPC) is a generalization allowing to perform an arbitrary on-going (also called reactive or stateful) computation during which players can receive outputs and provide new inputs at intermediate stages.

At Crypto 2006, Ishai *et al.* considered mixed threshold adversaries that either passively corrupt some fixed number of players, or, alternatively, actively corrupt some (smaller) fixed number of players, and showed that for certain thresholds, cryptographic SFE is possible, whereas cryptographic MPC is not.

However, this separation does not occur when one considers *perfect* security. Actually, past work suggests that no such separation exists, as all known general protocols for perfectly secure SFE can also be used for MPC. Also, such a separation does not show up with *general adversaries*, characterized by a collection of corruptible subsets of the players, when considering passive and active corruption.

In this paper, we study the most general corruption model where the adversary is characterized by a collection of adversary classes, each specifying the subset of players that can be actively, passively, or fail-corrupted, respectively, and show that in this model, perfectly secure MPC separates from perfectly secure SFE. Furthermore, we derive the exact conditions on the adversary structure for the existence of perfectly secure SFE resp. MPC, and provide efficient protocols for both cases.

1 Introduction

1.1 Secure Function Evaluation and Secure Multi-party Computation

Secure function evaluation (SFE) allows a set $\mathcal{P} = \{p_1, \ldots, p_n\}$ of n players to compute an arbitrary agreed function f of their inputs $x_1, \ldots, x_n$ in a secure way. Security means that dishonest players can neither falsify the output of the computation, nor obtain information about the honest players' inputs (except what they can derive from their own inputs and outputs). (Reactive) secure multi-party computation (MPC) is a generalization of SFE. Here, the function to be computed is reactive, meaning that players

---

* This research was partially supported by the Swiss National Science Foundation (SNF), project no. 200020-113700/1 and by the Zurich Information Security Center (ZISC).

© International Association for Cryptologic Research 2008

can give inputs and get outputs several times during the course of the computation, and every output can depend on all inputs given so far.

A bit more formally, SFE and MPC can be best described by considering a hypothetical trusted party which performs the specified task on behalf of the players. In SFE, the trusted party is non-reactive: it takes inputs from the players, evaluates the function, and announces the outputs (and disappears). In MPC, the trusted party is reactive: it continuously interacts with the players, taking inputs and sending outputs. It maintains an internal state which is updated with every input, and every output is computed based on this state. The goal of SFE and MPC is to *simulate* this trusted party among the set $\mathcal{P}$ of players. The potential dishonesty of players is modeled by a central adversary corrupting players, where players can be actively corrupted (the adversary takes full control over them), passively corrupted (the adversary can read their internal state), or fail-corrupted (the adversary can make them crash at any suitable time). A crashed player stops sending any messages, but the adversary cannot read the internal state of the player (unless he is actively or passively corrupted at the same time).

Typical examples of SFE include e-voting, i.e., the computation of the sum of the players' secret votes, or the double-agent problem, i.e., the identification of identical entries in several confidential databases. An example of MPC is the simulation of a fair stock market, where inputs (e.g. new trading orders) are given and outputs (e.g. current stock prices) are provided while the computation proceeds.

SFE (and MPC) was introduced by Yao [Yao82]. The first general solutions were given by Goldreich, Micali, and Wigderson [GMW87]; these protocols are secure under some intractability assumptions. Later solutions [BGW88, CCD88, RB89, Bea91b] provide information-theoretic security.

1.2 Summary of Known Results

In the seminal papers solving the general SFE and MPC problems, the adversary is specified by a single corruption type (active or passive) and a threshold t on the tolerated number of corrupted players. Goldreich, Micali, and Wigderson [GMW87] proved that, based on cryptographic intractability assumptions, general secure MPC is possible if and only if $t < n/2$ players are actively corrupted, or, alternatively, if and only if $t < n$ players are passively corrupted. In the information-theoretic model, Ben-Or, Goldwasser, and Wigderson [BGW88] and independently Chaum, Crépeau, and Damgård [CCD88] proved that unconditional security is possible if and only if $t < n/3$ for active corruption, and for passive corruption if and only if $t < n/2$.

These results were unified and extended by fail-corruption in [FHM98] by proving that perfectly secure MPC is achievable if and only if $3t_a + 2t_p + t_f < n$, where t_a, t_p, and t_f denote the upper bounds on the number of actively, passively, and fail-corrupted players, respectively.

Another line of generalization is concerned with so-called general adversaries: Here, the adversary is not characterized by a threshold, but rather by an enumeration of the possible subsets of players that the adversary can corrupt.[1] In [HM97] (see also [HM00])

---

[1] This allows to model non-symmetric settings where not every player's potential dishonesty is modeled in exactly the same way. Some coalitions of colluding players might be more likely than others, and some players might have a higher level of dishonesty than others.

it was proved that perfect security is possible if and only if no two corruptible subsets cover the full players set (passive adversary), respectively no three corruptible subsets cover the full player set (active adversary). These results naturally generalize the threshold results of $2t < n$, respectively $3t < n$. These results were unified to a mixed general adversary in [FHM99], where the adversary is characterized by an enumeration of classes, each class consisting of an actively corruptible subset of players and of a passively corruptible subset of players. Fail-corruption was not considered. The bounds on the existence of perfectly secure MPC are a natural combination of the bounds in the threshold model.

A similar development of generalizations (from threshold to general adversaries) can be observed in the area of Byzantine agreement protocols [LSP82, DS82, LF82, MP91, GP92, FM98, AFM99].

Recently, Ishai et al. [IKLP06] considered a mixed model in which the adversary can either corrupt t_a players actively, or, alternatively, t_p players passively (in contrast to previous work [FHM98], where the adversary could corrupt t_a players actively, and, simultaneously, t_p players passively). They showed that for $t_p < n$ and $t_a < n/2$ cryptographically secure SFE is possible, whereas, for $t_p = n - 1$ and $t_a \geq 1$, cryptographically secure (reactive) MPC is not possible.

1.3 Contributions of This Paper

The original motivation for this paper was to determine the exact conditions for SFE and MPC in the natural and most general adversary model where all corruption types can occur. We characterize the adversary's corruption capability by an *adversary structure* $\mathcal{Z} = \{(A_1, E_1, F_1), \ldots, (A_m, E_m, F_m)\}$, where $A_k, E_k, F_k \subseteq \mathcal{P}$ and $A_k \subseteq E_k$ and $A_k \subseteq F_k$. The adversary can (secretly) choose an arbitrary *adversary class* $Z_k = (A_k, E_k, F_k) \in \mathcal{Z}$ and actively corrupt the players in A_k, passively corrupt the players in E_k, and fail-corrupt the players in F_k. In the technical sections of this paper, we present and prove exact conditions on the adversary structure to allow perfectly secure MPC and perfectly secure SFE. This unifies all previously considered models, where either not all three types of corruption were considered, or where the corruption capability was specified in terms of thresholds.

Interestingly, the conditions for SFE and MPC are different. This is surprising since all known results on perfectly secure protocols suggest no such separation. In fact, a first separating example was observed by Almann [Alt99]. In particular, when considering active, passive, and fail-corruption (but only *threshold* type), then no such separation has been observed [FHM98]. When considering general adversaries (with active and passive corruption, but *without fail-corruption*), no separation can be observed neither [FHM99]. However, in the combination of both these models, the separation shows up. This indicates that the most general adversary model considered here is both natural and appropriate, since all restricted models hide the fact that SFE and MPC separate.

We describe a simple example of an adversary structure which separates, i.e., for which SFE with perfect security is possible but MPC is not. Let $\mathcal{P} = \{p_1, p_2, p_3, p_4\}$ and $\mathcal{Z} = \{Z_1, Z_2, Z_3\}$, where $Z_1 = (\emptyset, \{p_1\}, \emptyset)$, $Z_2 = (\{p_2\}, \{p_2\}, \{p_2, p_4\})$, and

$Z_3 = (\{p_3\}, \{p_3\}, \{p_3, p_4\})$. In other words, the adversary can either corrupt p_1 passively, or corrupt p_2 actively and fail-corrupt p_4, or corrupt p_3 actively and fail-corrupt p_4.[2]

A protocol for SFE works as follows: First use p_4 as trusted party with the constraint that p_4 sends the output of the function first to p_1 and then to p_2 and p_3. If p_4 crashes, then restart the protocol using p_1 as trusted party (the crashing of p_4 guarantees that the adversary did not choose $Z_1 \in \mathcal{Z}$ and hence that p_1 is uncorrupted). If p_1 has received the output from p_4 before p_4 crashed, then he forwards it to p_2 and p_3, otherwise he evaluates the function on the inputs received by p_2 and p_3 and sends them the output. The security of this protocol is trivial to verify. The impossibility of MPC for this example follows from the observation that if some intermediate value v — part of the state of an MPC protocol — is not known to p_1, then there is no protocol that always reveals it to him. Indeed, if in such a protocol the adversary crashes p_4 and forces p_2 or p_3 to send random messages whenever he is instructed to send something (she can do so by choosing Z_2 or Z_3), then with non-zero probability, p_1 will not be able to decide whether p_2 or p_3 is misbehaving and will accept a value different than v, contradicting perfect security.

2   The Model

We consider the standard secure-channels model introduced in [BGW88, CCD88]: The players $p_1, \ldots, p_n$ are connected by a complete network of bilateral synchronous secure channels. The computation is described as an arithmetic circuit over some finite field $\mathbb{F}$, consisting of addition (or linear) gates and multiplication gates.

The security of our protocols is information-theoretic without error probability, which is called *perfect* security and is the strongest possible security notion. A protocol is defined to be secure if it realizes a trusted functionality (computing the function f), where the term "realize" is defined via the simulation paradigm [Can00, MR91, Bea91a, DM00, PW01] which, in a nutshell, guarantees that whatever the adversary can achieve in the real world where the protocol is executed, she could also achieve in the ideal setting with the trusted functionality.[3] This security notion implies in particular that the adversary cannot obtain any information about the players' inputs beyond what is implied by the outputs (secrecy), and that she cannot influence the outputs other than by choosing the inputs of the corrupted players (correctness).

The adversary's corruption capability is characterized by an adversary structure $\mathcal{Z} = \{(A_1, E_1, F_1), \ldots, (A_m, E_m, F_m)\}$ (for some m). The adversary chooses a triple in $\mathcal{Z}$ non-adaptively,[4] i.e., before the beginning of the protocol; this triple is denoted as

---

[2] Additionally, $Z_4 = (\{p_4\}, \{p_4\}, \{p_4\})$ could be tolerated, but this would unnecessarily complicate the example.

[3] While our protocols can be proven secure in any of these simulation-based frameworks, with perfect indistinguishability of the real and the ideal world, we will in this paper not give full-fledged simulation-based security proofs; this is consistent with the previous literature on secure SFE and MPC.

[4] In contrast, an *adaptive* adversary can corrupt more and more players during the protocol execution, subject only to the constraint that the corrupted sets are within one of the triples in $\mathcal{Z}$. We do not consider the adaptive setting in this paper, but our results could be generalized to it.

$Z^* = (A^*, E^*, F^*)$ and is called the *actual adversary class* or simply the actual adversary. The players in A^*, E^*, and F^* are actively, passively, and fail-corrupted, respectively. Note that Z^* is not known to the honest players and appears only in the security analysis. A protocol is called $\mathcal{Z}$-*secure* if it is secure against an adversary with corruption power characterized by $\mathcal{Z}$.

For notational simplicity we assume that $A \subseteq E$ and $A \subseteq F$ for any $(A, E, F) \in \mathcal{Z}$ (anyway, an actively corrupted player can behave as being passively or fail-corrupted). Furthermore, as most constructions only need to consider the maximal classes of a structure, we define the maximal structure $\overline{\mathcal{Z}} = \{(A, E, F) \in \mathcal{Z} : \nexists (A', E', F') \in \mathcal{Z}$ with $(A, E, F) \neq (A', E', F')$ and $A \subseteq A', E \subseteq E', F \subseteq F'\}$.

To simplify the description, we adopt the following convention: Whenever a player does not receive a message (when expecting one), or receives a message outside of the expected range, then the special symbol $\perp \notin \mathbb{F}$ is taken for this message. Note that after a player has crashed, he only sends $\perp$. If a player has followed the protocol instructions correctly up to a certain point, he is called *correct* at that point, independently of whether he is actually corrupted. A player who has deviated from the protocol (e.g., has crashed or has sent inconsistent messages) is called *incorrect*.

3   Tools (Sub-protocols)

In this section we present some protocols that are used as building blocks in the main sections. Several of these protocols are non-robust, i.e., they might abort when faults occur. In case of abortion, all (correct) players agree on a non-empty set $B \subseteq \mathcal{P}$ of incorrect players; we say then that *the protocol aborts with B*.

3.1   Broadcast and Consensus

A *broadcast protocol* allows a sender p with input value v to distribute v among a set $\mathcal{P}$ of players, where it is guaranteed that all correct players in $\mathcal{P}$ output the same value v' (consistency), and that $v' = v$ when the sender is correct during the execution of the protocol (correctness). Similarly, a *consensus protocol* allows a set $\mathcal{P}$ of players, each holding an input value v_i, to reach agreement, such that every correct player in $\mathcal{P}$ outputs the same value v' (consistency), and that $v' = v$ if all (correct) players hold as input v (correctness).

In [AFM99] a tight condition for the existence of perfectly-secure broadcast and consensus is given for the model with active and fail-corruption. Those protocols assume pairwise authenticated (but not necessarily private) channels, hence they remain secure even when the adversary is allowed to passively corrupt any number of players. Therefore these conditions immediately translate to our model:

Lemma 1. *In the secure channels model, perfectly $\mathcal{Z}$-secure broadcast and consensus among a set $\mathcal{P}$ of players is possible if and only if $C_{BC}(\mathcal{P}, \mathcal{Z})$ holds, where*

$$C_{BC}(\mathcal{P}, \mathcal{Z}) \iff \begin{cases} \forall (A_1, E_1, F_1), (A_2, E_2, F_2), (A_3, E_3, F_3) \in \mathcal{Z} : \\ A_1 \cup A_2 \cup A_3 \cup (F_1 \cap F_2 \cap F_3) \neq \mathcal{P}. \end{cases}$$

We denote the broadcast and the consensus protocol of [AFM99] by Broadcast and Consensus, respectively.

3.2 Crash Detection

We present a protocol which allows the players in $\mathcal{P}$ to commonly detect whether a specific player $p \in \mathcal{P}$ is alive or has crashed. Such a decision cannot be sharp, as an actively corrupted player can always behave as having crashed, i.e., not send any messages during the execution of the protocol. However, we require that correct players are always identified as "alive", and crashed players are always identified as "crashed".

Protocol CDP($\mathcal{P}, \mathcal{Z}, p$)

1. p sends a 1-bit to every $p_j \in \mathcal{P}$.
2. Every $p_j \in \mathcal{P}$ sets $b_j := 1$ if he received a 1-bit, and $b_j := 0$ otherwise.
3. The players in $\mathcal{P}$ invoke Consensus on inputs $b_1, \ldots, b_n$.
4. Every $p_j \in \mathcal{P}$ outputs "alive" when the output of the consensus protocol is 1, and "crashed" otherwise.

Lemma 2. *If $C_{\mathrm{BC}}(\mathcal{P}, \mathcal{Z})$ holds, then the protocol* CDP($\mathcal{P}, \mathcal{Z}, p$) *has the following properties: Consistency: The (correct) players agree on the output. Correctness: If p is correct until the end of* CDP, *then every (correct) player outputs "alive" and if p has crashed before the invocation of* CDP, *then every (correct) player outputs "crashed".*[5]

Proof. Correctness: When p is correct, then every (correct) $p_j \in \mathcal{P}$ sets $b_j := 1$ and, by definition of consensus, all correct players decide on 1 and output "alive". When p has crashed before CDP is invoked, then every correct $p_j \in \mathcal{P}$ sets $b_j := 0$, and hence all correct players output "crashed". Consistency: As the output is decided by using consensus, the output of all correct players is identical.   □

3.3 Strong Broadcast

Intuitively, a fail-corrupted player never sends a "wrong" message; in the worst case, he sends no message at all. This intuition does not apply to broadcast (according to the standard definition): When the sender of a broadcast protocol crashes, only consistency of the output is guaranteed. But the output value can be arbitrary.[6]

We lift the intuition that fail-corrupted players never send "wrong" messages to broadcast by introducing the notion of *strong broadcast*: A protocol with sender p, holding input v, achieves strong broadcast when it achieves broadcast and additionally ensures that the output is in $\{v, \perp\}$ when the sender is not actively corrupted. We show how to construct a protocol for p to strongly broadcast v, given a protocol for broadcast (e.g., Broadcast) and CDP.

---

[5] Note that in any case the adversary learns the output of CDP.

[6] In [AFM99], the output of broadcast can even be chosen by the adversary, when the sender crashes.

Protocol StrongBroadcast($\mathcal{P}, \mathcal{Z}, p, v$)
1. Invoke Broadcast to have p broadcast his input v. For each $p_j \in \mathcal{P}$, let v_j denote p_j's output in Broadcast.
2. Invoke CDP to detect whether p is alive or has crashed.
3. Every $p_j \in \mathcal{P}$ outputs v_j when p is alive, and $\perp$ when p has crashed.

Lemma 3. *If $C_{BC}(\mathcal{P}, \mathcal{Z})$ holds, then the protocol StrongBroadcast($\mathcal{P}, \mathcal{Z}, p, v$) has the following properties: Consistency: All (correct) players output the same value v'. Correctness: If the sender p is correct, then $v' = v$; if p crashed before the invocation of the protocol, then $v' = \perp$; if p crashes during the protocol, then $v' \in \{v, \perp\}$.*

Proof. Consistency follows immediately from the consistency property of Broadcast and the consistency property of CDP. For correctness we consider 3 cases: (a) If the sender p is correct through the whole protocol, then the consistency property of Broadcast implies that for all correct p_j's, $v_j = v$ and the correctness property of CDP implies that all correct players will output "alive" in CDP, hence they will all output v in StrongBroadcast. (b) If p has already crashed *before* the invocation of StrongBroadcast, then this is detected in Step 2 (by CDP) and the protocol outputs $\perp$. (c) If p crashes during the protocol but is correct up to that point, then either this is detected in Step 2 and the protocol outputs $\perp$, or p is still alive at the beginning of Step 2 and has correctly broadcast his input v. Since, when p is not actively-corrupted one of the above 3 cases must hold, the output of StrongBroadcast for such a p is always in $\{v, \perp\}$. $\qquad\square$

3.4 Secret Sharing

A secret-sharing scheme allows a player (called the dealer) to distribute a secret, in such a way that only qualified sets of players can reconstruct it. As secret-sharing scheme, we employ a sum sharing (i.e., the secret is split into summands that add up to the secret), folded with a replication sharing (i.e., every summand is given to a subset of the players): Such a sharing is characterized by a *sharing specification* $\mathcal{S}$, which is a vector of subsets of the player set $\mathcal{P}$. A value s is *shared* with respect to a sharing specification $\mathcal{S} = (S_1, \ldots, S_m)$, when there exist summands $s_1, \ldots, s_m$ with $s = \sum s_k$, and s_k is given to every $p_i \in S_k$. For a player $p_i \in \mathcal{P}$, we consider the vector $(s_{i_1}, \ldots, s_{i_\ell})$ of summands held by p_i to be p_i's *share* of s, denoted as $\langle s \rangle_i$. The vector of all shares, denoted as $\langle s \rangle = (\langle s \rangle_1, \langle s \rangle_2, \ldots, \langle s \rangle_n)$, is a *sharing* of s. We say that $\langle s \rangle$ is a (consistent) sharing of s according to $(\mathcal{P}, \mathcal{S})$, if for each $S_k \in \mathcal{S}$ all (correct) players in S_k have the same view on s_k and $s = \sum_{k=1}^{m} s_k$.

For an adversary structure $\mathcal{Z}$, we say that a sharing specification $\mathcal{S}$ is $\mathcal{Z}$-*private* if for any sharing $\langle s \rangle$ according to $\mathcal{S}$ and for any adversary in $\mathcal{Z}$, there exists a summand s_k which this adversary does not know. Formally, $\mathcal{S}$ is $\mathcal{Z}$-private if $\forall (A, E, F) \in \mathcal{Z} \ \exists S \in \mathcal{S} : \ S \cap E = \emptyset$. For an adversary structure $\mathcal{Z}$ with maximal classes $\overline{\mathcal{Z}} = \{(\cdot, E_1, \cdot), \ldots, (\cdot, E_m, \cdot)\}$, we denote the natural $\mathcal{Z}$-private sharing specification by $S_{\mathcal{Z}} = (\mathcal{P} \backslash E_1, \ldots, \mathcal{P} \backslash E_m)$.

In the following, we describe the protocol Share which allows a dealer p to share a value s among the players in $\mathcal{P}$ according to a sharing specification $\mathcal{S}$. The protocol is a modification to tolerate fail-corruption of the sharing protocol from [Mau02]. It may abort when p is incorrect.

Protocol Share($\mathcal{P}, \mathcal{Z}, \mathcal{S}, p, s$)

1. Dealer p chooses the summands $s_2, \ldots, s_{|\mathcal{S}|}$ randomly and sets $s_1 := s - \sum_{k=2}^{|\mathcal{S}|} s_k$.

2. Execute the following steps for $k = 1, \ldots, |\mathcal{S}|$:

   (a) p sends s_k to every $p_i \in S_k$, who denotes the received value as $s_k^{(i)}$ ($\perp$ when no value is received).

   (b) Every $p_i \in S_k$ sends $s_k^{(i)}$ to every $p_j \in S_k$, who denotes the received value as $s_k^{(i,j)}$.

   (c) For each $p_j \in S_k$ StrongBroadcast is invoked to have p_j broadcast a complaint bit $b_{k,j}$, where $b_{k,j} = 1$ when $s_k^{(j)} = \perp$ or $s_k^{(i,j)} \notin \{s_k^{(j)}, \perp\}$ for some i, and $b_{k,j} = 0$ otherwise.

   (d) If a complaint was reported (i.e., $b_{k,j} = 1$ for some j), then StrongBroadcast is invoked to have p broadcast s_k, and every $p_j \in S_k$ sets $s_k^{(j)}$ to the broadcasted value.

3. If p broadcasts $\perp$ in Step 2d, then Share aborts with $B = \{p\}$.

Lemma 4. *If $C_{BC}(\mathcal{P}, \mathcal{Z})$ holds and $\mathcal{S}$ is a $\mathcal{Z}$-private sharing specification, then the protocol Share $(\mathcal{P}, \mathcal{Z}, \mathcal{S}, p, s)$ has the following properties. Correctness: Share either outputs a consistent sharing of some s', where $s' = s$ unless the dealer is actively corrupted, or it aborts with $B = \{p\}$; it does not abort if p is correct. Secrecy: No information on s leaks to the adversary.*

Proof. Correctness: The consistency of the sharing is guaranteed because correct players either hold the same value for a common summand, or they complain and get a consistent value for the summand by strong broadcast. Because all sent and broadcasted summands are s_k such that $s = \sum s_k$ it is clear that the shared value is s when the dealer is correct. Lastly, the protocol only aborts when the dealer is incorrect in an invocation of StrongBroadcast. Secrecy: Because $\mathcal{S}$ is $\mathcal{Z}$-private we know that the summands of corrupted players do not reveal information on s. On the other hand, the dealer only broadcasts summands for which a complaint is broadcast, i.e., two players (claim to) have different values for that summand. This only happens when the dealer or one of the disputing players is actively corrupted, or when the dealer has crashed. In the first case, the adversary is entitled to know the summand, and in the second case, the summand will not be broadcasted (the dealer has crashed). □

Reconstructing a shared value towards a player is straight-forward: All players send the summands they know (i.e., their share) to the output player, who tries to find the correct value for each summand and computes the secret as the sum of the summands. However, finding the correct value of a summand is not always possible when corrupted players send wrong values or no value to the output player. So we need an extra condition on the adversary structure to ensure that the output player can always decide on the value of every summand. We can slightly relax this condition when a sharing is reconstructed publicly (rather than towards a dedicated output player): In this case, the players can decide, depending on the published values, whether a summand is uniquely defined or not, and if not, agree on a set $B \subseteq \mathcal{P}$ of incorrect players.

In the sequel, we present the protocols Announce and Reconstruct to announce a summand, respectively reconstruct a sharing, towards a dedicated player, and the protocols PublicAnnounce and PublicReconstruct to announce a summand, respectively to reconstruct a sharing, towards all players. The latter protocols are non-robust; they might abort with a non-empty set $B \subseteq \mathcal{P}$ of incorrect players. The abortion of the protocol PublicAnnounce will allow to derive information on the actual adversary class, which will be helpful in the output protocol of SFE.

Protocol Announce($\mathcal{P}, \mathcal{Z}, S_k, s_k, p$)

1. Every $p_i \in S_k$ sends s_k to p, who denotes the received value as $s_k^{(i)}$ ($\perp$ when no value is received).
2. Let $V \subseteq \mathbb{F}$ denote the set of values v that are "explainable" with some adversary in $\mathcal{Z}$, i.e., for which there is an adversary class $(A, E, F) \in \mathcal{Z}$, such that $\{p_i \in S_k : s_k^{(i)} = \perp\} \subseteq F$ and $\{p_i \in S_k : s_k^{(i)} \notin \{v, \perp\}\} \subseteq A$.
3. p sets s_k to be the smallest element in V.

Lemma 5. *If* $\forall (A_1, E_1, F_1), (A_2, E_2, F_2) \in \mathcal{Z}: S_k \not\subseteq A_1 \cup A_2 \cup (F_1 \cap F_2)$, *then the protocol* Announce *robustly announces* s_k *to* p.

Proof. We have to prove that (i) the set V contains the correct summand s_k and (ii) the set V contains no other values. (i) Observe that the summands $s_k^{(i)}$ received by p satisfy that $\{p_i \in S_k : s_k^{(i)} = \perp\} \subseteq F^\star$ and $\{p_i \in S_k : s_k^{(i)} \notin \{s_k, \perp\}\} \subseteq A^\star$, where $(A^\star, E^\star, F^\star)$ denotes the actual adversary class. As $(A^\star, E^\star, F^\star) \in \mathcal{Z}$, it follows that $s_k \in V$. (ii) Consider any value $v \in V$. There exists an adversary class $(A, E, F) \in \mathcal{Z}$ such that $\{p_i \in S_k : s_k^{(i)} = \perp\} \subseteq F$ and $\{p_i \in S_k : s_k^{(i)} \notin \{v, \perp\}\} \subseteq A$. By assumption we know that $S_k \not\subseteq A \cup A^\star \cup (F \cap F^\star)$, hence there exists a player $p_i \in S_k$ with $s_k^{(i)} \neq \perp$, $p_i \notin A$ and $p_i \notin A^\star$. This implies that $v = s_k^{(i)} = s_k$. $\qquad\square$

Protocol Reconstruct($\mathcal{P}, \mathcal{Z}, \mathcal{S}, \langle s \rangle, p$)

1. For every $S_k \in \mathcal{S}$, Announce is invoked to have the correct summand s_k announced towards p.
2. p computes $s := \sum_{k=1}^{|\mathcal{S}|} s_k$ and outputs s.

Lemma 6. *If* $\forall k = 1, \ldots, |\mathcal{S}|$, $\forall (A_1, E_1, F_1), (A_2, E_2, F_2) \in \mathcal{Z}: S_k \not\subseteq A_1 \cup A_2 \cup (F_1 \cap F_2)$, *then the protocol* Reconstruct *robustly reconstructs* s *towards* p.

The proof follows immediately from Lemma 5.

Protocol PublicAnnounce($\mathcal{P}, \mathcal{Z}, S_k, s_k$)

1. Every $p_i \in S_k$ publishes his value for s_k (denoted as $s_k^{(i)}$) using StrongBroadcast.
2. Every $p_j \in \mathcal{P}$: determine the set $V \subseteq \mathbb{F}$ of values that are "explainable" with some adversary in $\mathcal{Z}$ (see protocol Announce).
3. Every $p_j \in \mathcal{P}$: output $s_k \in V$ if $|V| = 1$, otherwise abort with $B = \{p_i \subseteq S_k : s_k^{(i)} = \perp\}$.

Lemma 7. *If $C_{\mathrm{BC}}(\mathcal{P}, \mathcal{Z})$ holds and $\forall (A_1, \cdot, \cdot), (A_2, \cdot, \cdot) \in \mathcal{Z}: S_k \not\subseteq A_1 \cup A_2$, then the protocol PublicAnnounce either publicly announces s_k, or aborts with a non-empty set $B \subseteq \mathcal{P}$ of incorrect players. When it aborts, then there exists an adversary class $(A, E, F) \in \mathcal{Z}$ such that $S_k \subseteq A^\star \cup A \cup (F^\star \cap F)$.*

Proof. As V contains at least the correct summand s_k (see proof of Lemma 5), it is clear that PublicAnnounce either outputs s_k or aborts. It remains to be shown that when it aborts with B, then $|B| > 0$ and there exists an adversary class $(A, E, F) \in \mathcal{Z}$ such that $S_k \subseteq A^\star \cup A \cup (F^\star \cap F)$. Note that $s_k \in V$, hence PublicAnnounce aborts only when there exists a value $v \neq s_k$ with $v \in V$. This implies that there is an adversary class $(A, E, F) \in \mathcal{Z}$ with $\{p_i \in S_k : s_k^{(i)} = \perp\} \subseteq F$ and $\{p_i \in S_k : s_k^{(i)} \notin \{v, \perp\}\} \subseteq A$. Because $v \neq s_k$, we need $\{p_i \in S_k : s_k^{(i)} \neq \perp\} \subseteq A \cup A^\star$, which implies that $S_k \subseteq A^\star \cup A \cup (F^\star \cap F)$. Furthermore, B must be non-empty, because otherwise $S_k \subseteq (A^\star \cup A)$ would hold, contradicting the assumption in the lemma. $\qquad\square$

Protocol PublicReconstruct($\mathcal{P}, \mathcal{Z}, \mathcal{S}, \langle s \rangle$)

1. For every $S_k \in \mathcal{S}$, PublicAnnounce is invoked to have the correct summand s_k announced. If an invocation of PublicAnnounce aborts with B, then also PublicReconstruct aborts with B.
2. Every $p_j \in \mathcal{P}$ computes $s := \sum_{k=1}^{|\mathcal{S}|} s_k$ and outputs s.

Lemma 8. *If $C_{\mathrm{BC}}(\mathcal{P}, \mathcal{Z})$ holds and $\forall k = 1, \ldots, |\mathcal{S}|$, $\forall (A_1, \cdot, \cdot), (A_2, \cdot, \cdot) \in \mathcal{Z}: S_k \not\subseteq A_1 \cup A_2$, then the protocol PublicReconstruct either publicly reconstructs s, or aborts with a non-empty set B of incorrect players.*

The proof follows immediately from Lemma 7.

3.5 Multiplication

We present a protocol for securely computing a sharing of the product of two shared values. The protocol is a variation of the multiplication protocol of [Mau02], capturing fail-corruptions. The multiplication protocol may abort when faults occur, with outputting a set $B \subseteq \mathcal{P}$ of incorrect players.

The idea of the protocol is the following: As s and t are shared according to $\mathcal{S}$, we can use the summands $s_1, \ldots, s_{|\mathcal{S}|}$ and $t_1, \ldots, t_{|\mathcal{S}|}$ to compute the product st as $st := \sum_{k,\ell=1}^{|\mathcal{S}|} s_k t_\ell$. To do so, each term $x_{k,\ell} = s_k t_\ell$ of this sum is shared by every player knowing both s_k and t_ℓ. Then the players perform consistency checks on the shared summands and compute the sum of the shared terms $x_{k,\ell}$, which results in a sharing of st.

Protocol Mult$(\mathcal{P}, \mathcal{Z}, \mathcal{S}, \langle s \rangle, \langle t \rangle)$

1. For every $(S_k, S_\ell) \in \mathcal{S} \times \mathcal{S}$, the following steps are executed:
   (a) Every $p_i \in (S_k \cap S_\ell)$ computes the products $x_{k,\ell} := s_k t_\ell$ and invokes
   Share$(\mathcal{P}, \mathcal{Z}, \mathcal{S}, p_i, x_{k,\ell})$; denote the resulting sharing as $\langle x_{k,\ell}^{(i)} \rangle$.
   (b) Let p_i denote the player with the smallest index in $(S_k \cap S_\ell)$. For every
   $p_j \in (S_k \cap S_\ell)$, the difference $\langle x_{k,\ell}^{(j)} \rangle - \langle x_{k,\ell}^{(i)} \rangle$ is computed and, by invoking
   PublicReconstruct, reconstructed.
   (c) If all differences are 0, then the sharing $\langle x_{k,\ell}^{(i)} \rangle$ of p_i is adopted as sharing
   of $x_{k,\ell}$, i.e., $\langle x_{k,\ell} \rangle := \langle x_{k,\ell}^{(i)} \rangle$. Otherwise (i.e., some difference is non-zero),
   PublicAnnounce is invoked to have both s_k and t_ℓ announced, and a default
   sharing $\langle x_{k,\ell} \rangle$ of $x_{k,\ell} = s_k t_\ell$ is created (e.g., the first summand is set to $x_{k,\ell}$
   and the other summands are set to 0).
2. Each player in $\mathcal{P}$ (locally) computes his share of the product st as the sum of his
   shares of all terms $x_{k,\ell}$.
3. If any of the invoked sub-protocols aborts with B, then also Mult aborts with B.

Lemma 9. *Assume that $\mathcal{S}$ is a $\mathcal{Z}$-private sharing specification, $\langle s \rangle$ and $\langle t \rangle$ are consistent sharings according to $\mathcal{S}$, $C_{\mathrm{BC}}(\mathcal{P}, \mathcal{Z})$ holds, and the following two conditions hold: $\forall S_k, S_\ell \in \mathcal{S}, \forall (A, \cdot, \cdot) \in \mathcal{Z} : S_k \cap S_\ell \not\subseteq A$ and $\forall S_k \in \mathcal{S}, \forall (A_1, \cdot, \cdot), (A_2, \cdot, \cdot) \in \mathcal{Z} : S_k \not\subseteq A_1 \cup A_2$. Then the protocol Mult$(\mathcal{P}, \mathcal{Z}, \mathcal{S}, \langle s \rangle, \langle t \rangle)$ has the following properties. Correctness: It either outputs a sharing of st according to $(\mathcal{P}, \mathcal{S})$ or it aborts with a non-empty set $B \subseteq \mathcal{P}$ of incorrect players. Secrecy: No information on the inputs (i.e., on $\langle s \rangle$ and $\langle t \rangle$) leaks to the adversary.*

Proof. Correctness: The conditions in the lemma are sufficient for all the invoked sub-protocols (Share, PublicReconstruct, PublicAnnounce). The condition $\forall S_k, S_\ell \in \mathcal{S}$, $\forall (A, \cdot, \cdot) \in \mathcal{Z} : S_k \cap S_\ell \not\subseteq A$ ensures that every $x_{k,\ell}$ is known to at least one player p_i who is not actively corrupted; hence if no invocation of Share aborts and all differences are zero, then the shared values are correct. Privacy: Due to the security of Share, the invocations of Share do not leak information to the adversary. Furthermore, PublicAnnounce is only invoked on summands s_k, t_ℓ when two players in $S_k \cap S_\ell$ contradict each other; at least one of these players is actively corrupted, hence the adversary already knows s_k, t_ℓ before PublicAnnounce is invoked. □

3.6 Resharing

In the context of MPC, we will need to reshare shared values according to a different sharing specification. The key idea is to have every summand s_k in the original sharing being reshared according to the new sharing specification, and then distributively add the sharings of the summand, resulting in a new sharing of the original value.

In the following we describe the protocol Reshare$(\mathcal{P}, \mathcal{Z}, \mathcal{S}, \mathcal{S}', \langle s \rangle)$.

Lemma 10. *Assume that $\mathcal{S}'$ is a $\mathcal{Z}$-private sharing specification, $\langle s \rangle$ is a consistent sharing according to $\mathcal{S}$, $C_{\mathrm{BC}}(\mathcal{P}, \mathcal{Z})$ holds, and for all $(A_1, \cdot, \cdot), (A_2, \cdot, \cdot) \in \mathcal{Z}$ the following two conditions hold: $\forall S_k \in \mathcal{S} : S_k \not\subseteq A_1 \cup A_2$ and $\forall S'_k \in \mathcal{S}' : S'_k \not\subseteq A_1 \cup A_2$.*

Protocol Reshare($\mathcal{P}, \mathcal{Z}, \mathcal{S}, \mathcal{S}', \langle s \rangle$)

1. For every $S_k \in \mathcal{S}$, the following steps are executed:
   (a) Every $p_i \in S_k$ invokes Share($\mathcal{P}, \mathcal{Z}, \mathcal{S}', p_i, s_k$); denote the resulting sharing as $\langle s_k^{(i)} \rangle$.
   (b) Let p_i denote the player with the smallest index in S_k. For every $p_j \in S_k$, the difference $\langle s_k^{(j)} \rangle - \langle s_k^{(i)} \rangle$ is computed and, by invoking PublicReconstruct, reconstructed.
   (c) If all differences are 0, then the sharing $\langle s_k^{(i)} \rangle$ of p_i is adopted as sharing of s_k, i.e., $\langle s_k \rangle := \langle s_k^{(i)} \rangle$. Otherwise (i.e., some difference is non-zero), PublicAnnounce is invoked to have s_k announced, and a default sharing $\langle s_k \rangle$ of s_k according to $\mathcal{S}'$ is created.
2. Every $p_i \in \mathcal{P}$ (locally) computes the sum of his shares of all summands s_k.
3. If any of the invoked sub-protocols aborts with B, then also Reshare aborts with B.

Then the protocol Reshare($\mathcal{P}, \mathcal{Z}, \mathcal{S}, \mathcal{S}', \langle s \rangle$) has the following properties. Correctness: It either outputs a sharing of s according to $(\mathcal{P}, \mathcal{S}')$ or it aborts with a non-empty set $B \subseteq \mathcal{P}$ of incorrect players. Secrecy: No information on the inputs (i.e., on $\langle s \rangle$) leaks to the adversary.

Proof. Correctness: The conditions in the lemma are sufficient for all the invoked sub-protocols (Share, PublicReconstruct, PublicAnnounce). The condition $\forall S_k \in \mathcal{S}$, $\forall (A_1, \cdot, \cdot), (A_2, \cdot, \cdot) \in \mathcal{Z} : S_k \not\subseteq A_1 \cup A_2$ implies that $\forall S_k \in \mathcal{S}, \forall (A, \cdot, \cdot) \in \mathcal{Z} : S_k \not\subseteq A$, which ensures that every s_k is known to at least one player p_i who is not actively corrupted; hence if no invocation of Share aborts and all differences are zero, then the shared values are correct. Privacy: Due to the security of Share, the invocations of Share do not leak information to the adversary. Furthermore, PublicAnnounce is only invoked on the summand s_k when two players in S_k contradict each other; at least one of these players is actively corrupted, hence the adversary already knows s_k before PublicAnnounce is invoked.   □

4   (Reactive) Multi-party Computation

In this section we prove the necessary and sufficient condition on the adversary structure $\mathcal{Z}$ for the existence of perfectly $\mathcal{Z}$-secure multi-party computation protocols. The sufficiency of the condition is proved by constructing an MPC protocol.

Theorem 1. *A set $\mathcal{P}$ of players can perfectly $\mathcal{Z}$-securely compute any (reactive) computation when $C_{\mathrm{MULT}}(\mathcal{P}, \mathcal{Z})$ and $C_{\mathrm{REC}}(\mathcal{P}, \mathcal{Z})$ hold, where*

$$C_{\mathrm{MULT}}(\mathcal{P}, \mathcal{Z}) \iff \begin{cases} \forall (A_1, E_1, F_1), (A_2, E_2, F_2), (A_3, E_3, F_3) \in \mathcal{Z} : \\ \quad E_1 \cup E_2 \cup A_3 \cup (F_1 \cap F_2 \cap F_3) \neq \mathcal{P} \end{cases}$$

$$C_{\mathrm{REC}}(\mathcal{P}, \mathcal{Z}) \iff \begin{cases} \forall (A_1, E_1, F_1), (A_2, E_2, F_2), (A_3, E_3, F_3) \in \mathcal{Z} : \\ \quad E_1 \cup A_2 \cup A_3 \cup (F_2 \cap F_3) \neq \mathcal{P} \end{cases}$$

The condition C_{MULT} is needed for (non-robust) multiplication. The condition C_{REC} is needed for robust reconstruction.

4.1 The MPC Protocol

The circuit C to be computed consists of input, addition, multiplication, and output gates.[7] The reactiveness of the computation is modeled by assigning to each gate a point in time when it should be evaluated.

The circuit is evaluated in a gate-by-gate fashion, where for input, multiplication, and output gates, the corresponding sub-protocol Share, Mult, and Reconstruct, respectively, is invoked. Due to the linearity of the sharing, addition (or linear) gates can be evaluated locally by the players.

The non-robustness of the used sub-protocols is addressed differently depending on the type of the gate: When in an input gate the input player does not share his input, the players just pick a default sharing of some pre-agreed default value. The reconstruction protocol of the output gate is robust under the necessary condition for MPC. The multiplication of shared values can abort (with a set $B \subseteq P$ of incorrect players). If this happens, the multiplication is retried in a smaller setting, namely with the player set $P' = P \setminus B$ and the adversary structure Z' which contains only those adversary classes which are compatible with the fact that the players in B are incorrect. More precisely, first both factors are re-shared to the new setting with P' and Z', then the multiplication sub-protocol is invoked within this setting, and upon success, the resulting sharing of the product is re-shared to the original setting with P and Z. This process is repeated until the multiplication succeeds, and with each repetition, the active player set P' becomes smaller.

For the sake of clarity, we introduce two operators on adversary structures: For a set $B \subseteq P$, we denote by $Z|^{B \subseteq F}$ the sub-structure of Z that contains only adversaries who can fail-corrupt all the players in B, i.e., $Z|^{B \subseteq F} = \{(A, E, F) \in Z : B \subseteq F\}$. Furthermore, for a set $P' \subseteq P$, we denote by $Z|_{P'}$ the adversary structure with all classes in Z restricted to the player set P', i.e., $Z|_{P'} = \{(A \cap P', E \cap P', F \cap P') : (A, E, F) \in Z\}$. As syntactic sugar, we write $Z|^{B \subseteq F}_{P'}$ for $(Z|^{B \subseteq F})|_{P'}$.

It immediately follows from the above definitions that when the players in B have been detected to be incorrect, then the actual adversary $Z^\star$ is in $Z|^{B \subseteq F}$. Furthermore, we exclude the players in B from the multiplication protocol, and the new setting is $P' = P \setminus B$ and $Z' = Z|^{B \subseteq F}_{P \setminus B}$. One can easily verify that the conditions C_{BC}, C_{MULT}, and C_{REC} hold in $(P \setminus B, Z|^{B \subseteq F}_{P \setminus B})$ when they hold in (P, Z), for an arbitrary $B \subseteq P$. This results in the MPC protocol described below.

Lemma 11. *The protocol* MPC *is perfectly Z-secure if $C_{\text{MULT}}(P, Z)$ and $C_{\text{REC}}(P, Z)$ hold.*

Proof (sketch). One can easily verify that the conditions in the lemma imply all conditions required in the sub-protocols, hence the security of the MPC protocol follows from the security of the sub-protocols.   □

---

[7] This does not exclude probabilistic circuits, as a random gate can be simulated by having each player input a random value and take the sum of those values as the output.

Protocol MPC($\mathcal{P}, \mathcal{Z}, C$)

1. Initialize the set of players detected as incorrect to $\mathcal{P}_\perp := \emptyset$. Set the default sharing specification $\mathcal{S} := \mathcal{S}_{\mathcal{Z}}$.
2. For every gate to be evaluated, do the following:
   - *Input gate for p:* Invoke Share to have p share his input according to $(\mathcal{P}, \mathcal{S})$. If Share aborts, then a default sharing of some pre-agreed default value is taken.
   - *Addition gate:* Every $p_i \in \mathcal{P}$ locally computes the sum of his respective shares.
   - *Multiplication gate:* Denote the sharings of the factors as $\langle s \rangle$ and $\langle t \rangle$, respectively, and denote the set of active players as $\mathcal{P}' = \mathcal{P} \setminus \mathcal{P}_\perp$, the adversary structure compatible with $\mathcal{P}_\perp$ being incorrect as $\mathcal{Z}' = \mathcal{Z}|_{\mathcal{P} \setminus \mathcal{P}_\perp}^{\mathcal{P}_\perp \subseteq F}$, and the corresponding ($\mathcal{Z}'$-private) sharing specification as $\mathcal{S}' = \mathcal{S}_{\mathcal{Z}'}$. Invoke Reshare($\mathcal{P}', \mathcal{Z}', \mathcal{S}, \mathcal{S}', \langle s \rangle$) and Reshare($\mathcal{P}', \mathcal{Z}', \mathcal{S}, \mathcal{S}', \langle t \rangle$) to obtain the sharings $\langle s \rangle'$ and $\langle t \rangle'$ for $(\mathcal{P}', \mathcal{S}')$, respectively. Invoke Mult($\mathcal{P}', \mathcal{Z}', \langle s \rangle', \langle t \rangle'$) to obtain a sharing $\langle st \rangle'$ of the product, according to $(\mathcal{P}', \mathcal{S}')$. Invoke Reshare($\mathcal{P}', \mathcal{Z}', \mathcal{S}', \mathcal{S}, \langle st \rangle'$) to reshare this product according to $(\mathcal{P}, \mathcal{S})$.[a] If any of the sub-protocols aborts with set B then set $P_\perp := P_\perp \cup B$ and repeat the gate.
   - *Output gate for p:* Invoke Reconstruct to have the output reconstructed towards p.

---

[a] Reshare outputs a sharing according to $(\mathcal{P}', \mathcal{S})$, which is trivially also a sharing according to $(\mathcal{P}, \mathcal{S})$ since all players in $\mathcal{P} \setminus \mathcal{P}'$ are incorrect.

4.2 Impossibility of MPC

In this section we prove that perfectly secure (reactive) MPC is not possible for some circuits when $C_{\text{MULT}}(\mathcal{P}, \mathcal{Z})$ or $C_{\text{REC}}(\mathcal{P}, \mathcal{Z})$ is violated. We first prove that when $C_{\text{MULT}}(\mathcal{P}, \mathcal{Z})$ is violated, then even non-reactive computations cannot be securely evaluated (Lemma 12). Secondly, we prove that when $C_{\text{REC}}(\mathcal{P}, \mathcal{Z})$ is violated, then the players in $\mathcal{P}$ cannot hold a secret joint state, which excludes the evaluation of (non-trivial) reactive circuit (Lemma 13).

Lemma 12. *If $C_{\text{MULT}}(\mathcal{P}, \mathcal{Z})$ is violated, then there exist (even non-reactive) circuits which cannot be evaluated perfectly $\mathcal{Z}$-securely.*

Proof. Consider $\mathcal{P}$ and $\mathcal{Z}$ with $C_{\text{MULT}}(\mathcal{P}, \mathcal{Z})$ violated, and assume, to arrive at a contradiction, that for every circuit C a perfectly $\mathcal{Z}$-secure protocol exists. There exist $(A_1, E_1, F_1), (A_2, E_2, F_2), (A_3, E_3, F_3) \in \mathcal{Z}$ with $E_1 \cup E_2 \cup A_3 \cup (F_1 \cap F_2 \cap F_3) = \mathcal{P}$. Let $F = F_1 \cap F_2 \cap F_3$, $\mathcal{P}' = \mathcal{P} \setminus F$, and for $i = 1, 2, 3$, let $A_i' = A_i \setminus F$ and $E_i' = E_i \setminus F$. The alleged protocol must also be perfectly secure for the player set $\mathcal{P}'$ and the adversary structure (with only active and passive corruption) $\mathcal{Z}' = \{(A_1', E_1'), (A_2', E_2'), (A_3', E_3')\}$, because one particular strategy of the adversary is to fail-corrupt the players in F and make them crash at the very beginning of the protocol. However, for $(\mathcal{P}', \mathcal{Z}')$ perfectly secure (non-reactive) MPC protocols do not exist for all circuits, as proven in [FHM99, Thm. 1]. $\square$

Lemma 13. *If $C_{\text{REC}}(\mathcal{P}, \mathcal{Z})$ is violated, then the players cannot hold a secret joint state with perfect security.*

Proof. Consider $\mathcal{P}$ and $\mathcal{Z}$ with $C_{\text{REC}}(\mathcal{P}, \mathcal{Z})$ violated, hence there exist (A_1, E_1, F_1), (A_2, E_2, F_2), $(A_3, E_3, F_3) \in \mathcal{Z}$ with $E_1 \cup A_2 \cup A_3 \cup (F_2 \cap F_3) = \mathcal{P}$. Without loss of generality assume that $E_1 = \{p_1\}$, $A_2 = \{p_2\}$, $A_3 = \{p_3\}$, and $F_2 = F_3 = \{p_4\}$. We denote the view of p_i as v_i. To arrive at a contradiction, assume that these views define a secret joint state v. Privacy requires that v_1 does not determine v, hence there exists a different state $v' \neq v$ which could be represented by the views (v_1, v_2', v_3', v_4'). Now consider the following two cases: (i) The secret state is v, and the adversary corrupts (A_2, E_2, F_2) and makes p_4 crash and p_2 take a random view, which (with perhaps negligible probability) could be v_2'. (ii) The secret state is v', and the adversary corrupts (A_3, E_3, F_3) and makes p_4 crash and p_3 take a random view, which (with perhaps negligible probability) could be v_3. In both cases, the views of the players are $(v_1, v_2', v_3, \perp)$, but the joint state is once v and once $v' \neq v$, contradicting perfect security. $\square$

5 Secure Function Evaluation

In this section we prove the sufficient and necessary condition on the adversary structure $\mathcal{Z}$ for the existence of perfectly $\mathcal{Z}$-secure function evaluation protocols. The sufficiency of the condition is proved by constructing an SFE protocol. Note that the condition for SFE is weaker than the condition for MPC.

Theorem 2. *A set $\mathcal{P}$ of players can perfectly $\mathcal{Z}$-securely compute any function if and only if $C_{\text{MULT}}(\mathcal{P}, \mathcal{Z})$ and $C_{\text{NREC}}(\mathcal{P}, \mathcal{Z})$ hold, where*

$$C_{\text{MULT}}(\mathcal{P}, \mathcal{Z}) \Longleftrightarrow \begin{cases} \forall (A_1, E_1, F_1), (A_2, E_2, F_2), (A_3, E_3, F_3) \in \mathcal{Z} : \\ \quad E_1 \cup E_2 \cup A_3 \cup (F_1 \cap F_2 \cap F_3) \neq \mathcal{P} \end{cases}$$

$$C_{\text{NREC}}(\mathcal{P}, \mathcal{Z}) \Longleftrightarrow \begin{cases} \exists \text{ an ordering } ((A_1, E_1, F_1), \ldots, (A_m, E_m, F_m)) \text{ of } \overline{\mathcal{Z}} \text{ s.t.}[8] \\ \forall i, j, k \in \{1, \ldots, m\}, i \leq k : E_k \cup A_i \cup A_j \cup (F_i \cap F_j) \neq \mathcal{P} \end{cases}$$

The condition C_{MULT} is needed for (non-robust) multiplication. The condition C_{NREC} is needed for non-robust reconstruction. Essentially, the latter condition allows for a reconstruction protocol in which the actual adversary gets information on the output only once it cannot disturb the protocol anymore.

5.1 The SFE Protocol

Our SFE protocol follows the standard approach of SFE protocols, namely to first secret-share all inputs, then to evaluate the circuit gate by gate, and then to reconstruct the output. However, the protocol employs sharings which are not robustly reconstructible. This means that the adversary can break down the computation in such a

---

[8] Remember that $\overline{\mathcal{Z}}$ denotes the maximum classes in $\mathcal{Z}$. One can verify that such an ordering exists for $\overline{\mathcal{Z}}$ exactly if it exists for $\mathcal{Z}$.

way that all sharings are lost. As the circuit is non-reactive, we can handle such an abortion by repeating the whole protocol, including the input stage. The correct players will give the same inputs in every iteration, but the adversary might give different inputs. However, in a failed iteration, the adversary does not get any information about any secrets (more precisely, the adversary could perfectly simulate all messages received within a failed iteration already beforehand), so the inputs chosen by the adversary in the successful iteration are independent of the other players' inputs.

Termination is guaranteed by the fact that whenever an iteration aborts, then a non-empty set $B \subseteq \mathcal{P}$ of incorrect players is identified, and the next iteration will proceed without these players. Hence the number of iterations is bounded by n.

The delicate task is the output protocol. For simplicity, we describe the protocol only for a single public output s; however, it naturally extends to a vector s of several public outputs, which then can be extended to capture private outputs with standard techniques (the output player inputs a one-time pad used for perfectly blinding the private element of the output vector).

The intuition of the output protocol is as follows: First observe that in our sharing, the privacy against each adversary is protected by a particular summand. More precisely, for every adversary class $(A_k, E_k, F_k) \in \mathcal{Z}$ there exists a summand s_k which is given only to the players in $S_k \in \mathcal{S}$ with $S_k \cap E_k = \emptyset$ (we even have $S_k = \mathcal{P} \setminus E_k$). As long as this summand is not published, an adversary of class (A_k, E_k, F_k) does not obtain information about the output (from the point of view of the adversary, s_k is a perfect blinding of the output, and all other summands s_i are either known to the adversary or are distributed uniformly). Second, observe that whenever the publishing of some summand s_k fails (i.e. the protocol PublicAnnounce aborts), then a set $B \subseteq \mathcal{P}$ of incorrect players is identified. The information that the players in B are incorrect leaks information about the actual adversary $(A^\star, E^\star, F^\star)$, namely that $B \subseteq F^\star$. The key idea of the output protocol is to publish the summands in such an order that whenever PublicAnnounce aborts with B, then the information that the players in B are incorrect excludes the possibility that the actual adversary is from a class whose summand has already being published. In other words: If an adversary of class (A_i, E_i, F_i) could potentially abort the announcing of the summand s_k associated with the adversary class (A_k, E_k, F_k), then the summand s_k must be announced strictly before the summand s_i is announced.

Let $((A_1, E_1, F_1), \ldots, (A_m, E_m, F_m))$ denote an ordering of the maximum structure $\overline{\mathcal{Z}}$ satisfying

$$\forall 1 \leq i, j, k \leq m, i \leq k : E_k \cup A_i \cup A_j \cup (F_i \cap F_j) \neq \mathcal{P},$$

and let S denote the induced sharing specification $\mathcal{S} = (S_1, \ldots, S_m)$ with $S_k = \mathcal{P} \backslash E_k$. Then the following protocol perfectly $\mathcal{Z}$-securely publicly reconstructs a sharing $\langle s \rangle$ according to $\mathcal{S}$, or aborts with a non-empty set $B \subseteq \mathcal{P}$ of incorrect players. Privacy of the protocol is guaranteed under the assumption that those summands of $\langle s \rangle$ that are unknown to the adversary are uniformly distributed. This is the case for all sharings in our protocols.

Protocol OutputGeneration($\mathcal{P}, \mathcal{Z}, \mathcal{S} = (S_1, \ldots, S_m), \langle s \rangle$)

1. For $k = 1, \ldots, m$, the following steps are executed *sequentially*:
   (a) PublicAnnounce($\mathcal{P}, \mathcal{Z}, S_k, s_k$) is invoked to have the correct summand s_k published.
   (b) If PublicAnnounce aborts with B, then OutputGeneration *immediately* aborts with B.
2. Every $p_j \in \mathcal{P}$ (locally) computes $s := \sum_{k=1}^m s_k$ and outputs s.

Lemma 14. *Assume that $\mathcal{S}$ is a $\mathcal{Z}$-private sharing specification constructed as explained, $C_{BC}(\mathcal{P}, \mathcal{Z})$ holds, the condition $\forall S_k \in \mathcal{S}, (A_1, \cdot, \cdot), (A_2, \cdot, \cdot) \in \mathcal{Z} : S_k \not\subseteq A_1 \cup A_2$ holds, and $\langle s \rangle$ is a consistent sharing according to $\mathcal{S}$ with the property that those summands that are unknown to the adversary are randomly chosen. Then the protocol OutputGeneration either publicly reconstructs s, or it aborts with a non-empty set $B \subseteq \mathcal{P}$ of incorrect players. If OutputGeneration aborts, then the protocol does not leak any information on s to the actual adversary.*

Proof. First observe that the pre-conditions of PublicAnnounce are satisfied. Second, observe that by construction of $\mathcal{S}$, we have $\forall i, j, k \in \{1, \ldots, m\}, i \leq k : (\mathcal{P} \backslash S_k) \cup A_i \cup A_j \cup (F_i \cap F_j) \neq \mathcal{P}$. Now assume that the invocation of PublicAnnounce($\mathcal{P}, \mathcal{Z}, S_k, s_k$) aborts with $B \subseteq \mathcal{P}$. It follows from Lemma 7 that the actual adversary $(A^\star, E^\star, F^\star)$ satisfies the property that there exists $(A_j, E_j, F_j) \in \mathcal{Z}$ such that $S_k \subseteq A^\star \cup A_j \cup (F^\star \cap F_j)$. By the construction of $\mathcal{S}$, no adversary class $(A_i, E_i, F_i) \in \mathcal{Z}$ with $i \leq k$ satisfies this condition, hence the summand associated with the actual adversary has not yet been announced. □

In the following we describe the SFE protocol.

Protocol SFE($\mathcal{P}, \mathcal{Z}, \mathcal{C}$)

0. Let $\mathcal{S} = (\mathcal{P} \backslash E_1, \ldots, \mathcal{P} \backslash E_m)$ for the assumed ordering $((A_1, E_1, F_1), \ldots, (A_m, E_m, F_m))$ of $\mathcal{Z}$.
1. *Input stage:* For every input gate in $\mathcal{C}$, Share is invoked to have the input player p_i share his input x_i according to $\mathcal{S}$.[a]
2. *Computation stage:* The gates in $\mathcal{C}$ are evaluated as follows:
   – *Addition gate:* Every $p_i \in \mathcal{P}$ locally computes the sum of his respective shares.
   – *Multiplication gate:* Invoke Mult to compute a sharing of the product according to $\mathcal{S}$.
3. *Output stage:* Invoke OutputGeneration($\mathcal{P}, \mathcal{Z}, \mathcal{S}, \langle s \rangle$) for the sharing $\langle s \rangle$ of the public output.
4. If any of the subprotocols aborts with B, then set $\mathcal{P} := \mathcal{P} \backslash B$, and set $\mathcal{Z}$ to the adversary structure which is compatible with B being incorrect, i.e., $\mathcal{Z} := \mathcal{Z}|_{\mathcal{P}'}^{B \subseteq F}$, and go to Step 1.

---

[a] If in a later iteration a player $p_i \notin \mathcal{P}$ should give input, then the players in $\mathcal{P}$ pick the default sharing of a default value.

Lemma 15. *The above SFE protocol is perfectly $\mathcal{Z}$-secure if $C_{\mathrm{MULT}}(\mathcal{P}, \mathcal{Z})$ and $C_{\mathrm{NREC}}(\mathcal{P}, \mathcal{Z})$ hold.*

Proof (sketch). One can easily verify that the conditions in the lemma imply all conditions required in the sub-protocols, hence the security of the SFE protocol follows from the security of the sub-protocols.

Special care needs to be taken for the fact that the adversary can abort the protocol and provoke repetitions. Termination of this process is obvious, as in every repetition the player set shrinks. Also correctness is straight-forward. Privacy is argued as follows: The adversary can perfectly simulate his view in every iteration which aborts (even without knowing the public output), hence his capability to abort an iteration does not give him any additional power. □

5.2 Impossibility of SFE

In this section we prove that perfectly $\mathcal{Z}$-secure SFE is not possible for some circuits when $C_{\mathrm{MULT}}(\mathcal{P}, \mathcal{Z})$ or $C_{\mathrm{NREC}}(\mathcal{P}, \mathcal{Z})$ is violated. The necessity of $C_{\mathrm{MULT}}(\mathcal{P}, \mathcal{Z})$ follows immediately from Lemma 12. It remains to show that $C_{\mathrm{NREC}}(\mathcal{P}, \mathcal{Z})$ is necessary:

Lemma 16. *If $C_{\mathrm{NREC}}(\mathcal{P}, \mathcal{Z})$ is violated, then there exist functions which cannot be evaluated perfectly $\mathcal{Z}$-securely.*

Proof. Consider $\mathcal{P}$ and $\mathcal{Z}$ with $C_{\mathrm{NREC}}(\mathcal{P}, \mathcal{Z})$ violated, i.e., for every ordering $((A_1, E_1, F_1), \ldots, (A_m, E_m, F_m))$ of $\overline{\mathcal{Z}}$ there exists $i, j, k \in \{1, \ldots, m\}$ such that $i \leq k$ and $E_k \cup A_i \cup A_j \cup (F_i \cap F_j) = \mathcal{P}$. Consider the identity function, where every player $p_i \in \mathcal{P}$ inputs some value x_i, and the public output is the vector $(x_1, \ldots, x_n)$. To arrive at a contradiction, assume that there exists a perfectly $\mathcal{Z}$-secure SFE protocol for this function. This protocol implicitly defines for every set $L \subseteq \mathcal{P}$ the protocol round in which the players in L obtain full joint information about the output. We denote the index of this round as $\phi(L)$, i.e., the joint view of the players in L in round $\phi(L)$ gives full information on $(x_1, \ldots, x_n)$, but their joint view in round $\phi(L) - 1$ does not give full information. The function ϕ implies an ordering $((A_1, E_1, F_1), \ldots, (A_m, E_m, F_m))$ on the adversary classes in $\overline{\mathcal{Z}}$ such that for every $1 \leq i \leq k \leq m : \phi(E_i) \leq \phi(E_k)$. Denote by i, j, k those indices that satisfy $i \leq k$ and $E_k \cup A_i \cup A_j \cup (F_i \cap F_j) = \mathcal{P}$ (which are assumed to exist for contradiction). The adversary corrupts (A_i, E_i, F_i) and behaves as follows: Up to round $\phi(E_i) - 1$, the adversary lets the corrupted players behave correctly. In round $\phi(E_i)$, the adversary crashes the players in $F_i \cap F_j$, and has the players in $A_i \setminus (F_i \cap F_j)$ send random values (also in all subsequent rounds). Still, the adversary obtains full information on the output in round $\phi(E_i)$ (she knows all correct messages that were sent, respectively should have been sent to the players in E_i). However, the players in E_k do not have full information *before* round $\phi(E_k) \geq \phi(E_i)$. Hence these players cannot with certainty distinguish the current situation from the situation when the output vector would be different, the players in class (A_j, E_j, F_j) would be corrupted, those in $F_j \cap F_i$ would be crashed, and those in $A_j \setminus (F_j \cap F_i)$ would send random messages. Hence the adversary has obtained full information about the output vector, but some uncorrupted players do not, contradicting perfect security. □

6   Conclusions and Open Problems

We have considered an adversary whose corruption capability is described by a collection $\mathcal{Z}$ of adversary classes (A, E, F), where the adversary may actively corrupt the players in A, passively corrupt the players in E, and fail-corrupt the players in F. This model unifies all corruption models considered in the literature. Indeed, all these models are special cases of our model, in the sense that they consider either not all corruption types, or only threshold corruption.

For this general adversary model, we have derived exact conditions for the existence of perfectly secure multi-party computation (MPC) and secure function evaluation (SFE). It turned out that the condition for SFE is strictly weaker than the condition for MPC. In fact, there are simple adversary structures for which perfectly secure SFE is possible, but perfectly secure MPC and verifiable secret sharing are not possible. This separation does not show up in the restricted models considered so far. The following theorem states this separation. It follows immediately from the separating example in the introduction with $\mathcal{P} = \{p_1, p_2, p_3, p_4\}$ and $\overline{\mathcal{Z}} = \{(\emptyset, \{p_1\}, \emptyset), (\{p_2\}, \{p_2\}, \{p_2, p_4\}),$ $(\{p_3\}, \{p_3\}, \{p_3, p_4\})\}$.

Theorem 3. *Perfectly secure MPC and SFE separate, i.e., there exist $\mathcal{P}$ and $\mathcal{Z}$ such that perfectly $\mathcal{Z}$-secure SFE among the players in $\mathcal{P}$ is possible, whereas perfectly $\mathcal{Z}$-secure MPC is not.*

This paper considers only protocols with perfect security, and does not handle the cases of unconditional (i.e., information theoretic with error probability) or cryptographic security. In particular, the proofs of Lemmata 13 and 16 exploit the fact that not even small error probability is allowed. Hence, the proof of separation does not carry over to unconditional or cryptographic security. Moreover, the exact bounds for $\mathcal{Z}$-secure MPC and SFE in these models are not known yet.

Acknowledgments

We would like to thank Bernd Altmann for interesting discussions. In particular, a first separating example was observed by Altmann [Alt99].

References

[Alt99]   Altmann, B.: Constructions for efficient multi-party protocols secure against general adversaries. Diploma Thesis, ETH Zurich (1999)

[AFM99]   Altmann, B., Fitzi, M., Maurer, U.: Byzantine agreement secure against general adversaries in the dual failure model. In: Jayanti, P. (ed.) DISC 1999. LNCS, vol. 1693, pp. 123–137. Springer, Heidelberg (1999)

[Bea91a]   Beaver, D.: Foundations of secure interactive computing. In: Feigenbaum, J. (ed.) CRYPTO 1991. LNCS, vol. 576, pp. 377–391. Springer, Heidelberg (1992)

[Bea91b]   Beaver, D.: Secure multiparty protocols and zero-knowledge proof systems tolerating a faulty minority. Journal of Cryptology 4(2), 370–381 (1991)

[BGW88]   Ben-Or, M., Goldwasser, S., Wigderson, A.: Completeness theorems for non-cryptographic fault-tolerant distributed computation. In: STOC 1988, pp. 1–10 (1988)

[Can00]   Canetti, R.: Security and composition of multiparty cryptographic protocols. Journal of Cryptology 13(1), 143–202 (2000)

[CCD88]   Chaum, D., Crépeau, C., Damgård, I.: Multiparty unconditionally secure protocols (extended abstract). In: STOC 1988, pp. 11–19 (1988)

[DM00]   Dodis, Y., Micali, S.: Parallel reducibility for information-theoretically secure computation. In: Bellare, M. (ed.) CRYPTO 2000. LNCS, vol. 1880, pp. 74–92. Springer, Heidelberg (2000)

[DS82]   Dolev, D., Strong, H.R.: Polynomial algorithms for multiple processor agreement. In: STOC 1982, pp. 401–407 (1982)

[FHM98]   Fitzi, M., Hirt, M., Maurer, U.: Trading correctness for privacy in unconditional multi-party computation. In: Krawczyk, H. (ed.) CRYPTO 1998. LNCS, vol. 1462, pp. 121–136. Springer, Heidelberg (1998)

[FHM99]   Fitzi, M., Hirt, M., Maurer, U.: General adversaries in unconditional multi-party computation. In: Lam, K.-Y., Okamoto, E., Xing, C. (eds.) ASIACRYPT 1999. LNCS, vol. 1716, pp. 232–246. Springer, Heidelberg (1999)

[FM98]   Fitzi, M., Maurer, U.: Efficient Byzantine agreement secure against general adversaries. In: Kutten, S. (ed.) DISC 1998. LNCS, vol. 1499, pp. 134–148. Springer, Heidelberg (1998)

[GMW87]   Goldreich, O., Micali, S., Wigderson, A.: How to play any mental game — a completeness theorem for protocols with honest majority. In: STOC 1987, pp. 218–229 (1987)

[GP92]   Garay, J.A., Perry, K.J.: A continuum of failure models for distributed computing. In: Segall, A., Zaks, S. (eds.) WDAG 1992. LNCS, vol. 647, pp. 153–165. Springer, Heidelberg (1992)

[HM97]   Hirt, M., Maurer, U.: Complete characterization of adversaries tolerable in secure multi-party computation. In: PODC 1997, pp. 25–34 (1997); Full version appeared in Journal of Cryptology,13(1), 31–60 (2000) [HM00]

[HM00]   Hirt, M., Maurer, U.: Player simulation and general adversary structures in perfect multiparty computation. Journal of Cryptology 13(1), 31–60 (2000)

[IKLP06]   Ishai, Y., Kushilevitz, E., Lindell, Y., Petrank, E.: On combining privacy with guaranteed output delivery in secure multiparty computation. In: Dwork, C. (ed.) CRYPTO 2006. LNCS, vol. 4117, pp. 483–500. Springer, Heidelberg (2006)

[LF82]   Lamport, L., Fischer, M.J.: Byzantine generals and transaction commit protocols. Technical Report Opus 62, SRI International (Menlo Park, CA), TR (1982)

[LSP82]   Lamport, L., Shostak, R., Pease, M.: The Byzantine generals problem. ACM Transactions on Programming Languages and Systems 4(3), 382–401 (1982)

[Mau02]   Maurer, U.: Secure multi-party computation made simple. In: Cimato, S., Galdi, C., Persiano, G. (eds.) SCN 2002. LNCS, vol. 2576, pp. 14–28. Springer, Heidelberg (2003); Full version appeared in Discrete Applied Mathematics, 154(2), 370–381 (2006)

[MP91]   Meyer, F.J., Pradhan, D.K.: Consensus with dual failure modes. IEEE Transactions on Parallel and Distributed Systems 2(2), 214–222 (1991)

[MR91]   Micali, S., Rogaway, P.: Secure computation. In: Feigenbaum, J. (ed.) CRYPTO 1991. LNCS, vol. 576, pp. 392–404. Springer, Heidelberg (1992)

[PW01]   Pfitzmann, B., Waidner, M.: A model for asynchronous reactive systems and its application to secure message transmission. In: IEEE Symposium on Security and Privacy, pp. 184–200 (2001)

[RB89]   Rabin, T., Ben-Or, M.: Verifiable secret sharing and multiparty protocols with honest majority. In: STOC 1989, pp. 73–85 (1989)

[Yao82]   Yao, A.C.: Protocols for secure computations. In: FOCS 1982, pp. 160–164 (1982)

Bridging Game Theory and Cryptography: Recent Results and Future Directions

Jonathan Katz*

Department of Computer Science
University of Maryland
jkatz@cs.umd.edu

Abstract. Motivated by the desire to develop more realistic models of, and protocols for, interactions between mutually distrusting parties, there has recently been significant interest in combining the approaches and techniques of game theory with those of cryptographic protocol design. Broadly speaking, two directions are currently being pursued:

Applying cryptography to game theory: Certain game-theoretic equilibria are achievable if a trusted *mediator* is available. The question here is: *to what extent can this mediator be replaced by a distributed cryptographic protocol run by the parties themselves?*

Applying game-theory to cryptography: Traditional cryptographic models assume some honest parties who faithfully follow the protocol, and some arbitrarily malicious players against whom the honest players must be protected. Game-theoretic models propose instead that all players are simply *self-interested* (i.e., rational), and the question then is: *how can we model and design meaningful protocols for such a setting?*

In addition to surveying known results in each of the above areas, I suggest some new definitions along with avenues for future research.

1 Introduction

The fields of *game theory* and *cryptographic protocol design* are both concerned with the study of "interactions" among mutually distrusting parties. These two subjects have, historically, developed almost entirely independently within different research communities and, indeed, they tend to have a very different flavor. Recently, however, motivated by the desire to develop more realistic models of (and protocols for) such interactions, there has been significant interest in combining the techniques and approaches of both fields.

Current research at the intersection of game theory and cryptography can be classified into two broad categories: applying cryptographic protocols to game-theoretic problems, and applying game-theoretic models and definitions to the general area of cryptographic protocol design. In a bit more detail:

---

* Research supported in part by the U.S. Army Research Laboratory, NSF CAREER award #0447075, and US-Israel Binational Science Foundation grant #2004240.

R. Canetti (Ed.): TCC 2008, LNCS 4948, pp. 251–272, 2008.
© International Association for Cryptologic Research 2008

- Certain game-theoretic equilibria are possible if parties rely on the existence of an external trusted party called a *mediator*. (All the relevant definitions are given in Section 2.) This naturally motivates a cryptographer[1] to ask: *can the trusted mediator be replaced by a protocol that is run by the parties themselves?* Research aimed at understanding the conditions under which the answer is positive, and developing appropriate protocols in such cases, is described in Section 3.
- Traditionally, cryptographic protocols are designed under the assumption that some parties are *honest* and faithfully follow the protocol, while some parties are *malicious* and behave in an arbitrary fashion. The game-theoretic perspective, however, is that all parties are simply *rational* and behave in their own best interests. This viewpoint is incomparable to the cryptographic one: although no one can be trusted to follow the protocol (unless it is in their own best interests), the protocol need not prevent "irrational" behavior. The general question here is: *what models and protocols are appropriate for this setting?* This work is discussed in Section 4.

This paper surveys recent work in both the directions listed above, with a cryptographic audience in mind. This survey focuses more on the problems being addressed than on the solutions that have been proposed, and will thus emphasize definitions rather than concrete results. I also propose new definitional approaches to some of the problems under discussion, and have made a particular effort to highlight promising directions for future research.

Dodis and Rabin have recently written an excellent survey [16] that covers very similar ground as the present work. The present survey is perhaps a bit more technical, and somewhat more opinionated. Surveys more tangentially related to the topics considered here include those by Linial [33] and Halpern [25].

It is fascinating to observe that the recent growth of interest in blending game theory and cryptography has paralleled a surge of attention focused on game theory by computer scientists in general, most notably (for the purposes of this work) in the fields of computational complexity (see, e.g., [38, Chap. 2]), networking and distributed algorithms (see, e.g., [38, Chap. 14]), network security (see, e.g., [10] and [38, Chaps. 23, 27]), information security economics [38, Chap. 25], and more. These are all well beyond the scope of the present work.

Note: Due to space limitations, this survey has been shortened somewhat. A full version will be posted and maintained at http://eprint.iacr.org. Comments and corrections are very much appreciated.

2   A Crash Course in Game Theory

This section reviews some central game-theoretic concepts. I have tried to simplify things when, in my view, nothing of essence is lost (vis-a-vis the results presented here). For extensive further details, the reader is referred to [39,19].

---

[1] Although, interestingly, the question was first asked in the economics community.

We begin by introducing the notion of *normal form games*. A n-player game $\Gamma = (\{A_i\}_{i=1}^n, \{u_i\}_{i=1}^n)$, presented in normal form, is determined by specifying, for each player P_i, a set of possible *actions* A_i and a *utility function* $u_i \colon A_1 \times \cdots \times A_n \mapsto \mathbb{R}$. Letting $A \overset{\text{def}}{=} A_1 \times \cdots \times A_n$, we refer to a tuple of actions $\boldsymbol{a} = (a_1, \ldots, a_n) \in A$ as an *outcome*. The utility function u_i of party P_i expresses this player's preferences over outcomes: P_i prefers outcome $\boldsymbol{a}$ to outcome $\boldsymbol{a}'$ iff $u_i(\boldsymbol{a}) > u_i(\boldsymbol{a}')$. (We also say that P_i *weakly* prefers $\boldsymbol{a}$ to $\boldsymbol{a}'$ if $u_i(\boldsymbol{a}) \geq u_i(\boldsymbol{a}')$.) We assume that the $\{A_i\}, \{u_i\}$ are common knowledge among the players, although the assumption of known utilities seems rather strong and it is preferable to avoid it (or assume only limited knowledge).

The game is played by having each party P_i select an action $a_i \in A_i$, and then having all parties play their actions *simultaneously*. The "payoff" to P_i is given by $u_i(a_1, \ldots, a_n)$ and, as noted above, P_i is trying to maximize this value.

Two-player games (for reasonably sized A_1, A_2) can be represented conveniently in matrix form by labeling the rows (resp., columns) of the matrix with the actions in A_1 (resp., A_2). The entry in the cell at row $a_1 \in A_1$ and column $a_2 \in A_2$ contains a tuple (u_1, u_2) indicating the payoffs to P_1 and P_2, respectively, given the outcome $\boldsymbol{a} = (a_1, a_2)$. For example, the following represents a game where $A_1 = \{C, D\}$, $A_2 = \{C', D'\}$, and, e.g., $u_1(C, D') = 1$ and $u_2(C, D') = 3$:

Table 1. A two-player game

	C'	D'
C	$(2, 2)$	$(1, 3)$
D	$(3, 1)$	$(0, 0)$

Types, and games of incomplete information. The above definition corresponds to so-called games of *perfect* (or *complete*) information. One can also consider extensions that model different features of "real-world" interactions, such as inputs provided to the parties at the beginning of the game whose values affect players' utilities. (In the game theory literature these inputs are said to determine the *type* of each party.) We now provide a simplified definition incorporating this situation; see [39,19] for the general case.

Let $\Gamma = (\{A_i\}, \{u_i\})$ be as above, where the $\{u_i\}$ are now functions from $(\{0, 1\}^*)^n \times A$ to the reals. Let $\mathcal{D}$ be a distribution over vectors $(t_1, \ldots, t_n)$, where each t_i is a binary string. A game is now played as follows: first, $(t_1, \ldots, t_n)$ is sampled according to $\mathcal{D}$, and P_i is given t_i. Next, each player P_i plays an action $a_i \in A_i$ as before; once again, these are all assumed to played simultaneously. Then, each player P_i receives payoff $u_i(t_1, \ldots, t_n, a_1, \ldots, a_n)$.

2.1 Nash Equilibria

If parties play a game (of perfect information), what can we expect to happen? Say P_1 knows the actions $a_2, \ldots, a_n$ that the other parties are going to take. Then it will choose the action $a_1 \in A_1$ that maximizes $u_1(a_1, \ldots, a_n)$; we call this a_1 a *best response* of P_1 to the actions of the other players. (A best response need not be unique.) Given this action chosen by the first player, P_2 will then choose a best response $a_2' \in A_2$, and so on. We see that a tuple $\boldsymbol{a} = (a_1, \ldots, a_n)$ is "self-enforcing" only if each a_i represents P_i's best response to $\boldsymbol{a}_{-i} \stackrel{\text{def}}{=} (a_1, \ldots, a_{i-1}, a_{i+1}, \ldots, a_n)$. A tuple with this property is called a *Nash equilibrium*, and this serves as the starting point for all further analysis of the game. Formally, if we let $(a_i', \boldsymbol{a}_i)$ denote $(a_1, \ldots, a_{i-1}, a_i', a_{i+1}, \ldots, a_n)$, we have:

Definition 1. *Let* $\Gamma = (\{A_i\}_{i=1}^n, \{u_i\}_{i=1}^n)$ *be a game presented in normal form, and let* $A = A_1 \times \cdots \times A_n$. *A tuple* $\boldsymbol{a} = (a_1, \ldots, a_n) \in A$ *is a* (pure-strategy) Nash equilibrium *if for all* i *and any* $a_i' \in A_i$ *it holds that* $u_i(a_i', \boldsymbol{a}_{-i}) \leq u_i(\boldsymbol{a})$.

Another way of expressing this is to say that $a_i \in A_i$ *weakly dominates* $a_i' \in A_i$ *relative to* $\boldsymbol{a}_{-i}$ if $u_i(a_i, \boldsymbol{a}_{-i}) \geq u_i(a_i', \boldsymbol{a}_{-i})$. Then $\boldsymbol{a}$ is a Nash equilibrium if, for all i, the action a_i weakly dominates all actions in A_i relative to $\boldsymbol{a}_{-i}$.

In the example of Table 1, (C, D') is a pure-strategy Nash equilibrium: given that P_1 plays C, the second player prefers to play D'; given that P_2 plays D', the first player prefers to play C. A second Nash equilibrium is given by (D, C').

In the above definition of a pure-strategy Nash equilibrium, the "strategy" of P_i was to deterministically play a_i (hence the name *pure* strategy). If we limit players to such strategies, a Nash equilibrium may not exist in a given game. To remedy this, we allow players to follow *randomized* strategies as well. Specifically, if σ_i is a probability distribution over A_i then we also let σ_i represent the strategy in which P_i samples $a_i \in A_i$ according to σ_i and then plays this action. (We recover deterministic strategies by letting σ_i assign probability 1 to some action.) Given a strategy vector $\boldsymbol{\sigma} = (\sigma_1, \ldots, \sigma_n)$, we overload notation and let $u_i(\boldsymbol{\sigma})$ denote the *expected* utility of P_i given that all parties play according to $\boldsymbol{\sigma}$. (We remark that although this is the standard way to assign utilities to distributions over outcomes, doing so makes the generally unrealistic assumption that players are *risk neutral* in that they care only about their expected utility.) The strategy σ_i is a *best response* to $\boldsymbol{\sigma}_{-i}$ if it maximizes $u_i(\sigma_i, \boldsymbol{\sigma}_{-i})$. Then:

Definition 2. *Let* $\Gamma = (\{A_i\}_{i=1}^n, \{u_i\}_{i=1}^n)$ *be as above, and let* σ_i *be a distribution over* A_i. *Then* $\boldsymbol{\sigma} = (\sigma_1, \ldots, \sigma_n)$ *is a* (mixed-strategy) Nash equilibrium *if for all* i *and any distribution* σ_i' *over* A_i *it holds that* $u_i(\sigma_i', \boldsymbol{\sigma}_{-i}) \leq u_i(\boldsymbol{\sigma})$.

One can verify that in the two-party game of Table 1, the strategy vector in which P_1 plays C with probability $1/2$, and in which P_2 plays C' with probability $1/2$ is a (mixed-strategy) Nash equilibrium.

The celebrated theorem of Nash [37] is that any game of perfect information where the $\{A_i\}$ are finite has a (mixed-strategy) Nash equilibrium. The finiteness assumption is necessary, as there are examples of two-player games with countably-infinite action sets where no mixed-strategy Nash equilibrium exists.

Nash equilibria for games of incomplete information can be defined in the natural way based on the above. Here the strategy of player P_i corresponds to a *function* mapping its received input t_i to an action $a_i \in A_i$; pure strategies correspond to deterministic functions. Note that here we must take into account parties' *expected* utilities even when considering pure-strategy Nash equilibria, since the utility of P_i may depend on the types of the other players, and these are unknown at the time P_i chooses its action.

2.2 Other Equilibrium Concepts

Nash equilibria are considered by many to be *the* fundamental equilibrium notion for games. Nevertheless, it is of interest to explore various refinements and strengthenings of this concept.

Dominated strategies and iterated deletion. Given a game $\Gamma = (\{A_i\}, \{u_i\})$, we say that action $a_i \in A_i$ is *strictly dominated* with respect to A_{-i} if there exists a randomized strategy $\sigma_i \in \Delta(A_i)$ such that $u_i(\sigma_i, \boldsymbol{a}_{-i}) > u_i(a_i, \boldsymbol{a}_{-i})$ for all $\boldsymbol{a}_{-i} \in A_{-i}$ (where $A_{-i} \stackrel{\text{def}}{=} \times_{j \neq i} A_j$). I.e., a_i is strictly dominated if P_i can always improve its situation by *not* playing a_i. An action $a_i \in A_i$ is *weakly dominated* with respect to A_{-i} if there exists a randomized strategy $\sigma_i \in \Delta(A_i)$ such that (1) $u_i(\sigma_i, \boldsymbol{a}_{-i}) \geq u_i(a_i, \boldsymbol{a}_{-i})$ for all $\boldsymbol{a}_{-i} \in A_{-i}$, and (2) there exists $\boldsymbol{a}_{-i} \in A_{-i}$ such that $u_i(\sigma_i, \boldsymbol{a}_{-i}) > u_i(a_i, \boldsymbol{a}_{-i})$. I.e., P_i can never improve its situation by playing a_i, and can sometimes improve its situation by not playing a_i.

It seems that a rational player will never choose a strictly dominated action. In fact, it is not hard to show that in any Nash equilibrium, no player assigns positive probability to any strictly dominated action. Arguably, a rational player should also never choose a weakly dominated action (although the argument in this case is less clear). If we accept this assumption, then a Nash equilibrium in which some party plays a weakly dominated action with positive probability is not expected to occur in practice. For example, consider the following game:

	C'	D'
C	$(10, 10)$	$(1, 1)$
D	$(10, 0)$	$(2, 2)$

(C, C') is a Nash equilibrium. However, action C of player P_1 is weakly dominated by action D. Thus, we may expect that P_1 plays D — but this forces us to the Nash equilibrium (D, D'). Note that both players now end up doing worse! Intuitively, both players prefer the Nash equilibrium (C, C'), but this is not "stable" in a sense we will define below.

Say we are given a game Γ^0, and we have eliminated the weakly dominated actions of each player from consideration. This leaves us with "effective" action sets $\{A_i^1\}$ for each player. We may now iterate the process, and remove any actions that are weakly dominated in the "reduced game" $\Gamma^1 = (\{A_i^1\}, \{u_i\})$, etc. This leads to the following definition.

Definition 3. *Given $\Gamma = (\{A_i\}, \{u_i\})$ and $\hat{A} \subseteq A$, let $\mathsf{DOM}_i(\hat{A})$ denote the set of strategies in $\hat{A}_i$ that are weakly dominated with respect to $\hat{A}_{-i}$. For $k \geq 1$, set $A_i^k \stackrel{\text{def}}{=} A_i^{k-1} \setminus \mathsf{DOM}_i(A^{k-1})$. Set $A_i^\infty \stackrel{\text{def}}{=} \cap_k A_i^k$. A Nash equilibrium $\boldsymbol{\sigma}$ of Γ survives iterated deletion of weakly dominated strategies if $\sigma_i \in \Delta(A_i^\infty)$ for all i.*

Stability with respect to trembles. Another means to distinguish among a set of Nash equilibria is to ask how stable each such equilibrium is to "mistakes" (or *trembles*) of the other players. Such mistakes might correspond to a real mistake on the part of some player (e.g., a player chooses an irrational strategy by accident), some "out-of-band" event (e..g, a network failure), or the fact that a player's utility is slightly different than originally thought.

To define stability with respect to trembles, we must first define a metric d on the strategy space $\Delta(A)$ of the players. Assuming A is finite, a natural candidate is statistical difference and we assume this in the definition that follows. Various notions of stability with respect to trembles have been considered in the game theory literature, although some of them seem problematic in a cryptographic setting. The following seems best for our context:

Definition 4. *Let $\Gamma = (\{A_i\}, \{u_i\})$, and let σ be a Nash equilibrium in Γ. Then σ is* stable with respect to trembles *if there exists an $\epsilon > 0$ such that for all i and every $\sigma'_{-i} \in \Delta(A_{-i})$ with $d(\sigma_{-i}, \sigma'_{-i}) < \epsilon$, the strategy σ_i is a best response to σ'_{-i}. I.e., for every $\sigma'_i \in \Delta(A_i)$ it holds that $u_i(\sigma'_i, \sigma'_{-i}) \leq u_i(\sigma_i, \sigma'_{-i})$.*

That is, even if P_i believes there is some small probability that the other players will make a mistake (and not play according to σ_{-i}), it is still in P_i's best interests to play according to σ_i.

As an example, consider the following two-player game:

	A'	B'	C'
A	$(10, 2)$	$(1, 0)$	$(0, 1)$
B	$(10, 0)$	$(0, 0)$	$(100, 100)$

(A, A') is a Nash equilibrium, but it is not stable with respect to trembles: if P_1 believes that P_2 might play C' with any positive probability ϵ (but still plays B' with probability 0), then P_1 will prefer to play B rather than A. On the other hand, (C, C') is a Nash equilibrium that *is* stable with respect to trembles: for small enough $\epsilon > 0$, even if P_1 believes that P_2 might play something other than C' with probability ϵ, it is still in P_1's best interest to play C.

I am not aware of any results stating conditions under which stable equilibria are guaranteed to exist.

Coalitions. Thus far, we have only been considering single-player deviations, i.e., whether it is in any single player's best interests to deviate from some prescribed strategy. Cryptographers generally prefer to think in terms of coalitions of players acting together. In general, a Nash equilibrium provides no "protection" against such coalitions.

What does it mean for a coalition C to prefer one outcome to another? There are at least four natural possibilities:

- C prefers σ to σ' only if *every* player in C weakly prefers σ to σ', and some player in C strictly prefers σ to σ'.
- C prefers σ to σ' only if the sum of the utilities of the parties in C improves; i.e., if $\sum_{i \in C} u_i(\sigma) > \sum_{i \in C} u_i(\sigma')$. (Note that for this to make sense, we must

assume that the utility functions of all players in C are measured in the same units.) This definition can be viewed as capturing the ability of players in C to make "side payments" to each other before or after the game.

- C prefers σ to σ' if *any* player in C prefers σ to σ', i.e., if $u_i(\sigma) > u_i(\sigma')$ for some $i \in C$. The definition makes sense if we think of one adversarial party corrupting other parties and taking complete control over their actions. Note also that preference of σ to σ' with respect to this definition implies preference with respect to the previous two definitions.

- Another possibility is to simply assume utility functions u_C for each possible coalition C, and then define preference in the obvious way. This is the most general approach (it subsumes the previous three), but requires additional assumptions about players' utilities.

We adopt the third definition here.

Given a set $C = \{i_1, \ldots, i_t\} \subset [n]$ and a vector $\sigma = (\sigma_1, \ldots, \sigma_n)$, we let $A_C \overset{\text{def}}{=} \times_{i \in C} A_i$, $\sigma_C \overset{\text{def}}{=} (\sigma_{i_1}, \ldots, \sigma_{i_t})$, and $\sigma_{-C} \overset{\text{def}}{=} \sigma_{[n] \setminus C}$. Then:

Definition 5. *Let* $\Gamma = (\{A_i\}_{i=1}^n, \{u_i\}_{i=1}^n)$. *Then for* $1 \leq t < n$ *the strategy vector* $\sigma = (\sigma_1, \ldots, \sigma_n)$ *is a* t-resilient equilibrium *if for all* $C \subset [n]$ *with* $|C| \leq t$, *all* $i \in C$, *and any* $\sigma_C' \in \Delta(A_C)$, *it holds that* $u_i(\sigma_C', \sigma_{-C}) \leq u_i(\sigma)$.

That is, for every coalition C of size at most t, no member of the coalition improves its situation no matter how the members of C coordinate their actions.

Observe that a 1-resilient equilibrium is a Nash equilibrium. Extending other equilibrium concepts to the case of coalitions seems not to have been explored significantly.

Mixed models. It is standard in game theory to assume that all players are rational. Recent work [1,35,2] has explored models where most parties are rational, but some players are *malicious* and behave arbitrarily. Treating players as malicious can be viewed (to some extent) as treating their utilities as completely unknown. It is also possible to assume that some players honestly follow the proscribed protocol — perhaps out of altruism or laziness — rather than seeking to improve their utility (although it should be in these players' interests to run the protocol altogether). These are interesting directions that are not discussed any further in this survey.

2.3   Correlated Equilibria

Correlated equilibria [3] offer another solution concept with some advantages relative to Nash equilibria. In some games, there may exist a correlated equilibrium that, for every party P_i, gives a better payoff to P_i than any Nash equilibrium (see [36] for an example). More generally, correlated equilibria have payoffs outside the convex hull of all Nash equilibria, and therefore give more options to the players. Finally, correlated equilibria of any game can be computed in polynomial time, something not believed to be the case for Nash equilibria.

Given a game $\Gamma = (\{A_i\}, \{u_i\})$, we define a *mediated* version of Γ which relies on a trusted, external party M called the *mediator*. The game is now played in two stages: first, the mediator chooses a vector of actions $\boldsymbol{a} \in A$ according to some known distribution $\mathcal{M}$, and then hands the *recommendation* a_i to player P_i. The players then play Γ as before by choosing any action in their respective action sets. Players are "supposed" to follow the recommendation of the mediator, and a *correlated equilibrium* is one in which it is in each player's best interests to do so. To formally define this notion, let $u_i(a'_i, \boldsymbol{a}_{-i} \mid a_i)$ denote the expected utility of P_i, given that it plays action a'_i after having received recommendation a_i and all other parties play their recommended actions $\boldsymbol{a}_{-i}$. (The expectation here is over $\boldsymbol{a}$ sampled according to $\mathcal{M}$.)

Definition 6. *Let* $\Gamma = (\{A_i\}, \{u_i\})$. *A distribution* $\mathcal{M} \in \Delta(A)$ *is a* correlated equilibrium *if for all* $\boldsymbol{a} = (a_1, \ldots, a_n)$ *in the support of* $\mathcal{M}$, *all* i, *and all* $a'_i \in A_i$, *it holds that* $u_i(a'_i, \boldsymbol{a}_{-i} \mid a_i) \leq u_i(\boldsymbol{a} \mid a_i)$.

Any Nash equilibrium is a correlated equilibrium, but Nash equilibria correspond to the special case where $\mathcal{M}$ is a *product distribution* over the A_i.

Let $u_i(\mathcal{M})$ denote the expected utility of P_i when all parties follow their actions as recommended by $\mathcal{M}$. A definition equivalent to the previous one, but better suited for extensions to coalitions as well as the computational setting, is:

Definition 7. *Let* $\Gamma = (\{A_i\}, \{u_i\})$. *A distribution* $\mathcal{M} \in \Delta(A)$ *is a* correlated equilibrium *if for all* i *and any* $f_i : A_i \to A_i$ *it holds that*

$$u_i(f_i(a_i), \boldsymbol{a}_{-i}) \leq u_i(\mathcal{M}),$$

where $\boldsymbol{a}$ *is sampled according to* $\mathcal{M}$.

As an example of a game with a correlated equilibrium that is not a Nash equilibrium, consider the two-party game of Table 1 and the distribution that assigns probability $1/3$ to each of (C, D'), (D, C'), and (C, C'). One can check that neither party has any incentive to deviate from their recommended action, and each player has expected utility 2 (an improvement on the mixed-strategy Nash equilibrium described in Section 2.1).

In games of incomplete information (as we have defined them in Section 2), a mediated game is played as follows: first, a vector $(t_1, \ldots, t_n)$ is sampled according to a distribution $\mathcal{D}$, and t_i is given to P_i. Then, each party P_i sends some t'_i to the mediator. Based on the vector $\boldsymbol{t}' = (t'_1, \ldots, t'_n)$ received, the mediator samples a vector $\boldsymbol{a} \in A$ according to a distribution $\mathcal{M}(\boldsymbol{t}')$, and recommends action a_i to player P_i. The parties then play as before, choosing whether or not to follow the mediator's recommendation. Correlated equilibria in this situation are defined as the natural extension of the above, through we stress that a player's strategy now determines *both* what value t'_i it sends to the mediator (as a function of the received input t_i) as well as what action it plays in the game. A correlated equilibrium is said to be *truthful* if it is in each party's best interest to send $t'_i = t_i$ to the mediator. The *revelation principle* characterizes when truthful correlated equilibria exist.

Correlated equilibria in the presence of coalitions. The basic approach used to handle coalitions in Definition 5 can be extended in the natural way to the case of correlated equilibria. Two variants of the definition are obtained, however, depending on the details of how the mediated game is played. If *ex ante* collusion is allowed, the parties in a coalition C may coordinate their strategies in advance, but are assumed unable to communicate after the mediator provides them with their recommended actions. If *ex post* collusion is allowed, the parties in C can communicate even after receiving their recommendations from the mediator.

Definition 8. *Let* $\Gamma = (\{A_i\}, \{u_i\})$ *be an* n-party game, and let $1 \leq k < n$. A *distribution* $\mathcal{M} \in \Delta(A)$ *is an* ex ante t-resilient correlated equilibrium *if for all* $C \subset [n]$ *with* $|C| \leq k$, *any functions* $\{f_i : A_i \to A_i\}_{i \in C}$, *and all* $i \in C$ *it holds that* $u_i(\{f_i(a_i)\}_{i \in C}, a_{-C}) \leq u_i(\mathcal{M})$, *where* a *is sampled according to* $\mathcal{M}$.

$\mathcal{M}$ *is an* ex post t-resilient correlated equilibrium *if for all* C *as above, any function* $f_C : A_C \to A_C$, *and any* $i \in C$ *it holds that* $u_i(f_C(a_C), a_{-C}) \leq u_i(\mathcal{M})$, *where* a *is sampled according to* $\mathcal{M}$.

2.4 Extensive Form Games

Extensive form games remove the assumption that players act simultaneously. Such games are best described as occurring in a sequence of rounds, where in any given round the game might specify that all parties play simultaneously (as in a normal form game) or that some subset of designated parties plays. Play of the game thus defines a *history* of the actions taken by the players thus far, and a player P_i's strategy σ_i now specifies, for each round in which it is P_i's turn to move, a (randomized) function mapping possible histories to actions. Players' utilities are now functions of terminal histories (i.e., histories that occur at the end of the game), rather than functions of the strategy vector of the players. We rely on the above intuitive description rather than present a formal definition.

We provide a simple example of an extensive form game, which also demonstrates how introducing alternation can affect the outcome of a game. Consider a seller P_1 and a buyer P_2, where P_1 can either sell high (H) or low (L), and P_2 can choose either to buy (B) or not (N). Payoffs are given by the matrix on the left, but we will assume that the seller announces its action first. This gives an extensive form game in which the buyer can follow any of four (pure) strategies; we let XY denote the strategy where P_2 chooses X if the seller chooses H, and P_2 chooses Y if the seller chooses L. This extensive form game is represented in normal form in the matrix on the right.

Table 2. An extensive form game presented in normal form

	B	N		BB	BN	NB	NN
H	$(10,1)$	$(0,0)$	H	$(10,1)$	$(10,1)$	$(0,0)$	$(0,0)$
L	$(5,6)$	$(0,0)$	L	$(5,6)$	$(0,0)$	$(5,6)$	$(0,0)$

Looking at the game on the right, we see that (L, NB) is a Nash equilibrium. The strategy being followed by P_2 is to refuse to buy if the buyer charges the higher price, and if the seller knows that P_2 will follow this strategy then it is in the seller's best interest to charge a low price. In contrast, the game on the left (where parties move simultaneously) has the unique Nash equilibrium (H, B).

Something odd about the Nash equilibrium (L, NB) of the extensive form game is that P_2 is, in essence, threatening to play *irrationally* if P_1 plays H (since, conditioned on P_1 playing H, the buyer is better off playing B than N). Another way to say this is that P_2 plays rationally given any realizable history (where history h is *realizable with respect to σ* if this history occurs with positive probability when all parties play according to σ), but P_2 threatens to play irrationally at some non-realizable history. In the following section, we will discuss a refinement of Nash equilibria that eliminates such "empty threat" strategies.

2.5   Equilibrium Concepts in Extensive Form Games

Any game in extensive form can be viewed as a normal form game by letting the set of allowable actions correspond to the players' strategies. Thus, all the equilibrium concepts we have discussed previously can be applied to extensive form games as well. However, it is often more natural to view certain games in extensive form, and thinking of games in this way motivates new equilibrium concepts. In particular, a question that arises with regard to extensive form games is whether we need to "pay attention" to players' strategies at non-realizable histories. In some cases paying attention to such strategies makes intuitive sense, while in other cases the situation is less clear.

Subgame perfect equilibria. As noticed in the previous section, certain strategy vectors may be Nash equilibria but contain "empty threats" by one or more of the players. Subgame perfect Nash equilibria eliminate this possibility. To define this concept, we introduce (informally) the notion of the *reduced game Γ^h* of an extensive form game Γ. Basically, Γ^h corresponds to Γ where some initial history h is fixed; we may view Γ^h as the continuation of Γ conditioned on the fact that history h has been observed thus far. A strategy σ_i in Γ naturally induces a strategy σ_i^h in Γ^h by setting $\sigma_i^h(h') = \sigma_i(h\|h')$.

Definition 9. *Let Γ be an extensive form game, and let σ be a Nash equilibrium in Γ. Then σ is subgame perfect if for all possible histories h of Γ, the strategy vector σ^h is a Nash equilibrium of the reduced game Γ^h.*

Recall that a history h is realizable (with respect to σ) if it occurs with positive probability when all parties follow σ. If the definition above only quantified over realizable histories, then every Nash equilibrium would satisfy the definition.

In the game of Table 2 the Nash equilibrium (L, NB) is not subgame perfect because, conditioned on the (non-realizable) history in which P_1 plays H, player P_2 prefers to play B instead of N. Equilibrium (H, BB) is subgame perfect.

Stability with respect to trembles. There are two possible ways to extend the definition of stability with respect to trembles to extensive form games, depending on whether or not subgame perfection is also required. The following definition does *not* take subgame perfection into account. Say two strategies σ_i, σ_i' of P_i *yield equivalent play with respect to σ* if for every history h realizable with respect to σ it holds that $\sigma_i(h) = \sigma_i'(h)$. (This just means that, assuming all other parties play σ_{-i}, play proceeds identically whether P_i plays σ_i or σ_i'.)

Definition 10. *Let Γ be an extensive form game, and let σ be a Nash equilibrium in Γ. Then σ is stable with respect to trembles (for realizable histories) if there exists an $\epsilon > 0$ such that for all i and every σ_{-i}' with $d(\sigma_{-i}, \sigma_{-i}') < \epsilon$ there exists a σ_i' that is a best response to σ_{-i}' and such that σ_i and σ_i' yield equivalent play with respect to σ.*

2.6   Cryptographic Considerations

In a cryptographic setting, it is natural to modify the way games are treated and the way various equilibrium notions are defined. We give an example of how this might be done for the specific case of parties running a protocol in the standard cryptographic sense, though it can be easily extended for more general scenarios (for examples, parties running a protocol and then taking some action as in Section 3, or parties who receive some initial input as in Section 4).

As usual in the cryptographic setting, we introduce a security parameter k provided to all parties at the beginning of the game. The action of a player P_i now corresponds to running an interactive Turing machine (ITM) M_i. This ITM M_i takes as input some current state and incoming messages from the other parties, and outputs the next message of player P_i along with updated state. The message m_i is then sent to all other parties (we are assuming here that communication is over a broadcast channel). We require M_i to run in probabilistic polynomial-time, which we take to mean that the next message function is computed in time polynomial in k. This definition allows M_i to run for an unbounded number of rounds and, if desired, we can additionally require that the expected number of rounds for which M_i runs is also polynomial.

Utility functions take the security parameter k as input, and are functions mapping transcripts of a protocol execution to the reals that can be computed in time polynomial in k. We stress that, as in extensive form games, utilities depend only on the "observable outcome" of the game play.

For the purposes of this section, we define a *computational game Γ* to be one in which the actions of each player correspond to the set of probabilistic polynomial-time ITMs, and where the utilities of each player are polynomial-time computable. We remark that we no longer need to consider mixed strategies, since a mixed strategy that can be implemented in polynomial time corresponds to a pure strategy (since pure strategies correspond to randomized ITMs).

An important difference between the cryptographic setting and the setting we have considered until now is that *now parties are assumed to be indifferent to negligible changes in their utilities.* For example:

Definition 11. *Let* $\Gamma = (\{A_i\}, \{u_i\})$ *be a computational game. A strategy vector* $M = (M_1, \ldots, M_n)$ *is a* computational Nash equilibrium *if for all* i *and any probabilistic polynomial-time ITM* M'_i *there is a negligible function* ϵ *such that*

$$u_i(k, M'_i, M_{-i}) - u_i(k, M) \leq \epsilon(k).$$

I am unaware of any result characterizing conditions under which Nash equilibria exist in computational games.

Subgame perfection and related notions. Execution of a protocol can naturally be regarded as an extensive form game. Extending equilibrium notions for extensive form games to the computational setting is, however, less obvious. For example, a first approach to extending the notion of subgame perfection to the computational setting would be to say that the strategy vector σ of the game Γ is subgame perfect if for all possible histories h of Γ, the strategy vector σ^h is a computational Nash equilibrium of the reduced game Γ^h. However, this ignores the probability with which history h is reached! On the other hand, it is unclear how to assign a probability to a non-realizable history. We are not aware of any definition of computational subgame perfection that deals with these issues.

A recent definition suggested by Kol and Naor [29] explicitly rejects the idea of "weighting" the utility of strategies according to the probability with which a given history is reached. Instead, informally, they require that conditioned on reaching *any* history that occurs with positive probability, players' strategies should remain in equilibrium. In their definition of a computational game, they allow players to use ITMs which run in time polynomial in $k + r$, where r is the number of rounds that have been played thus far. (Thus, the next-message function in their case may be viewed as a function from the entire history/transcript thus far to a next message, rather than from some internal state and a set of incoming messages to a next message, as defined above.) For lack of any better name, we refer to ITMs of this sort as running in *liberal* polynomial time and refer to the notion of t-resilient* equilibria for strategy vectors that remain in equilibrium even with respect to this stronger class of machines. Finally, we let $u_i(\cdot \mid h)$ denote the expected utility of P_i conditioned on history h. We now give the definition of Kol and Naor:[2]

Definition 12. *Let* Γ *be a computational game. A strategy vector* $M = (M_1, \ldots, M_n)$ *is* computationally t-immune[3] *if for every history h realizable with respect to* M *and every* i, *the strategy vector* M^h *is a t-resilient* Nash equilibrium in the reduced game* Γ^h. *I.e., for every* $C \subset [n]$ *with* $|C| \leq t$ *and every liberal polynomial-time ITM* M'_C *there is a negligible function* ϵ *such that*

$$u_i(k, M'_C, M_{-C} \mid h) - u_i(k, M \mid h) \leq \epsilon(k).$$

---

[2] One change we introduce is to condition on observable histories rather than on players' random coins (which may be private).

[3] Note that immunity refers to an entirely different concept in [1].

2.7   Critiques of Game Theory

Without going into much detail here, I will simply say that it is not at all clear whether game theory provides the "best" way of modeling interactions, both in general as well as specifically in a cryptographic setting. (All of the critiques I mention here are well-known, and not in any way novel.) For starters, it is unclear the extent to which the behavior of most people can be modeled as rational. (*Social economists* study exactly this issue.) Even if we are willing to believe that people act rationally, it is not always clear when a protocol designer can assume any knowledge of their utilities.

Irrespective of the above, many of the solution concepts are unsatisfying. The notion of a Nash equilibrium is perhaps the most intuitively appealing one, but in cases where multiple Nash equilibria exist it is unclear which one the parties will settle on or even if they can agree to settle on one at all. Other notions have been introduced in an effort to distinguish among various Nash equilibria, but it seems that for every such notion there exists a game in which applying the notion goes against one's intuition. (See, e.g., [19, pp. 462–463] for an example in the context of iterated deletion of weakly dominated strategies where it is to one party's advantage to publicly burn their money.)

3   Implementing Mediators Using Cryptography

As we have seen in Section 2.3, if parties are willing to assume the existence of a trusted mediator then they can potentially achieve certain equilibria that may be "preferable" to any of the available Nash equilibria. If a trusted mediator is *not* available, the question becomes: *to what extent can the parties themselves run a protocol in place of the mediator?*

This question was first explored in the economics community [14,6,18,9,42,43,4] (see [2] for a summary of these results), where researchers suggested "cheap talk" protocols by which parties could communicate amongst themselves to implement a correlated equilibrium. (As the terminology suggests, communication among the players is "cheap" in the sense that it costs nothing; it is also "worth nothing" in the sense that players are not "bound" to any statements they make; e.g., there is no legal recourse if someone lies). In the cryptography community, the question was first addressed by Dodis, Halevi, and Rabin [15].

3.1   Defining the Problem

Let us begin by defining the basic problem. (Other variants and extensions will be explored below.) We are given some n-party game $\Gamma = (\{A_i\}, \{u_i\})$ in normal form, along with a correlated equilibrium $\mathcal{M}$. We then define the extensive form game Γ_{CT} in which all parties hold a common security parameter k and first communicate in a "cheap talk" phase. The parties then play Γ, making their moves simultaneously (as always). Following the game-theoretic convention, all parties must play some action in Γ. (I.e., we do not allow player P_i to "abort" in

Γ unless this is an action in A_i.) On the other hand, following the cryptographic convention we *do* allow players to abort (and refuse to send any more messages) during the cheap talk phase.

We make no assumptions regarding the exact communication model during the cheap talk phase. For now, however, we assume that colluding parties can communicate "out of band" throughout the entire game. (This assumption is removed in Section 3.4.) Thus, for now we focus on *ex post* correlated equilibria which are resilient to coalitions even when such communication is allowed.

A player's strategy in Γ_{CT} determines both the protocol it runs in the cheap talk phase as well as the action it plays in Γ. We may now define the basic goal:

Definition 13. *Let Γ be a game, and let $\mathcal{M}$ be an ex post t-resilient correlated equilibrium in Γ. Let Γ_{CT} be the cheap talk extension of Γ, and let σ be an efficient strategy vector in Γ_{CT}. Then σ is a t-resilient implementation of $\mathcal{M}$ if (1) σ is a t-resilient computational equilibrium in Γ_{CT}, and (2) for all i, it holds that $u_i(k, \sigma) = u_i(k, \mathcal{M})$.*

One might strengthen the definition to require that the *distribution* of payoffs in Γ_{CT} (both for each party as well as when considering joint distributions among multiple parties) is close to the distribution of payoffs in the original mediated game. A stronger requirement of a different flavor is given by Lepinski et al. [30], who require (informally) that any vector of expected payoffs achievable by $\mathcal{C}$ in Γ_{CT} (i.e., even ones that are sub-optimal for $\mathcal{C}$) can also be achieved by $\mathcal{C}$ in the original mediated game. We do not impose such requirements here.

3.2 A Simple Observation

It is instructive to begin with a relatively simple observation: if t, n, and the communication model are such that *completely fair* secure multi-party computation [20, Def. 7.5.4] is possible, then *any* correlated equilibrium $\mathcal{M}$ of any game Γ has a t-resilient implementation: During the cheap talk phase the parties run a completely fair protocol Π computing $a \leftarrow \mathcal{M}$, where P_i receives a_i as output. Following the cheap talk phase, each party plays the action it received as output in Π. It is not hard to see that the strategy vector thus specified (i.e., "run Π and then play the result") is a t-resilient (computational) equilibrium with expected payoffs identical to those in the original mediated game.

Applying the above observation to the standard communication model, we see that if parties are connected by pairwise point-to-point channels then a t-resilient implementation of any correlated equilibrium exists when $t < n/3$. If a broadcast channel or a PKI is additionally assumed, then t-resilient implementations exist whenever $t < n/2$. The above all follow from standard results in secure multi-party computation [11,8,40,7]. Lepinski et al. [30] show how to achieve completely fair secure computation for any $t < n$ — and hence show t-resilient implementations of any correlated equilibrium for $t < n$ — in a non-standard communication model where "secure envelopes" are assumed. (Completely fair secure multi-party computation using point-to-point channels and broadcast is, in general, impossible for $t \geq n/2$ [13].) Assuming secure envelopes may be reasonable in some settings, but such envelopes seem impossible to realize (without assuming trusted parties) in a distributed setting such as the Internet.

3.3 Implementing Mediators without Completely Fair MPC

The natural next question is: when can a correlated equilibrium be implemented even though completely fair secure computation is ruled out? The initial result in this direction is due to Dodis, Halevi, and Rabin [15], who examine the case $t = 1, n = 2$. Before explaining their solution, we first introduce some terminology. Let Γ be a game in normal form. Then the *minimax profile* against player P_i is an action $\boldsymbol{a}_{-i} \in A_{-i}$ (or, more generally, in the product distribution $\times_{j \neq i} \Delta(A_i)$) minimizing $\max_{a_i \in A_i} \{u_i(a_i, \boldsymbol{a}_{-i})\}$. In other words, a minimax profile $\boldsymbol{a}_{-i}$ "punishes" P_i by giving P_i its lowest possible utility, assuming P_i plays a best response to the strategy of the other parties.

The basic idea of Dodis, Halevi, and Rabin is as follows: Let $\mathcal{M}$ be a correlated equilibrium in some two-party game Γ. In Γ_{CT}, the two parties run a protocol Π for computing $(a_1, a_2) \leftarrow \mathcal{M}$, where party P_i receives a_i as output. This protocol Π is "secure-with-abort" (cf. [20, Def. 7.2.6]), which informally means that privacy and correctness hold but fairness does not; in particular, we assume it is possible for P_1 to receive its output even though P_2 does not. After running Π, each party plays the action it received as output in Π; if P_2 does not receive output from Π then it plays the minimax profile against P_1.

It is not hard to see that this is a 1-resilient implementation of $\mathcal{M}$. First, it is immediate that if both parties play the indicated strategy, then the payoffs of both parties in Γ_{CT} are exactly the payoffs they would receive by playing $\mathcal{M}$ in Γ. Let us now argue that this is a computational Nash equilibrium in Γ_{CT}. We first observe that P_2 has no incentive to deviate: no matter how P_2 plays when running Π, party P_1 receives correctly-distributed output and plays according to the correlated equilibrium. Given this, P_2's best action to play in Γ is given by its own output from Π. We remark that here we are relying on the assumption that P_2 can only run *polynomial-time* strategies, and that P_2 is indifferent to negligible differences in its expected utility, exactly as we have defined things in Definition 11.

As for P_1, the only way it can (effectively) deviate during the cheap talk phase is by running Π until it receives its own output a_1 and then possibly aborting the protocol so that P_2 does not receive any output. We claim that it is never to P_1's advantage to abort. (Note that the analysis in [15] seems to assume that P_1 either *never* aborts or *always* aborts, but of course P_1 can determine whether to abort based on its output.) If P_1 allows P_2 to receive its output, this induces some mixed strategy σ_2 that will be played by P_2. (I.e., σ_2 represents the marginal distribution on P_2's recommended action according to $\mathcal{M}$, conditioned on the fact that P_1's recommended action is a_1.) Since $\mathcal{M}$ is a correlated equilibrium, a_1 is a best response to σ_2. If P_1 aborts, then P_2 will play a minimax profile σ_2' against P_1. By definition of a minimax profile, P_1's best response to σ_2' cannot give P_1 better utility than its best response to σ_2.

We conclude that P_1 always does worse by aborting the execution of Π. Given that both parties receive output in Π, it is obviously to P_1's advantage to play its recommended action. This completes the proof.

Extensions. The above ideas do not extend easily to a more general setting. For example, consider the case $t = 1, n = 3$ (with point-to-point communication). If these parties run a protocol Π in which a single deviating party can abort the computation *without being identified*, then the remaining parties do not know which player to "punish". In fact, essentially this situation is inherent in general [2, Theorem 4]. On the other hand, specific correlated equilibria may be implementable using the general approach discussed below.

Next look at the case $t = 2, n < 5$. Observe that even if one party, say P_1, is identified as cheating, the naive approach of having the remaining parties play a minimax profile against P_1 may not work. For one thing, although such a profile might result in a worse payoff for P_1, it may actually lead to a better payoff for a second player, say P_2, colluding with P_1. (And recall that 2-resilience only holds if deviations help *no one* in the coalition.) Moreover, if players play a minimax profile against P_1, it may be possible for P_2 (who, recall, is colluding with P_1) to deviate from the minimax profile and thus benefit P_1.

We are thus motivated to define a stronger notion of "punishment". Following [1, Def. 5], though differing in some respects, we define:

Definition 14. *Let Γ be an n-party game with correlated equilibrium $\mathcal{M}$. A strategy vector $\boldsymbol{\sigma}$ is a t-punishment strategy with respect to $\mathcal{M}$ if for all $\mathcal{C} \subset [n]$ with $|\mathcal{C}| \leq t$, all $\boldsymbol{\sigma}'_C$, and all $i \in \mathcal{C}$ it holds that $u_i(\boldsymbol{\sigma}'_C, \boldsymbol{\sigma}_{-C}) \leq u_i(\mathcal{M})$.*

That is, any coalition would be better off following the recommendations of $\mathcal{M}$ rather than playing against $\boldsymbol{\sigma}_{-C}$.

If a t-punishment strategy is available for a given correlated equilibrium $\mathcal{M}$, then this gives hope that a variant of the Dodis-Halevi-Rabin approach can be used to give a t-resilient implementation of $\mathcal{M}$. See [1,2] for work along these lines. Also relevant is the work of [31,27,28], discussed in more detail in the following section. As we have mentioned earlier, a partial converse [2] of the positive result just mentioned shows that, in general, if a t-punishment strategy is *not* available for a given correlated equilibrium $\mathcal{M}$, then this equilibrium cannot be implemented. Further work is need to better characterize the exact conditions under which a given correlated equilibrium can or cannot be implemented.

3.4   Implementing *ex ante* Equilibria (and More)

This section provides a brief discussion of work aimed at a slightly different aspect of the problem. Assume now that colluding parties *cannot* communicate "out of band" once Γ_{CT} begins; i.e., during the cheap talk phase of Γ_{CT} all communication is done over a public channel, and after the cheap talk phase — when it is time for the parties to play Γ — there is no inter-party communication at all. (Colluding parties can try to communicate over the broadcast channel, but if they are obvious about it then this will be detected by the other parties and punished.) It is then meaningful to ask whether it is possible to implement an *ex ante* correlated equilibrium of Γ in the cheap talk extension Γ_{CT}.

This problem is not immediately solved even if completely fair secure computation is possible. The problem is that *covert channels* may exist in the protocol itself. If such covert communication is possible, an ex ante correlated equilibrium may no longer remain an equilibrium. Informally, say a protocol is *collusion free* if covert communication is impossible. (We remark, however, that it seems sufficient here to prevent covert communication only *after* the parties have learned their output, since communication between the colluding parties before they learn their recommended actions will not affect an ex ante equilibrium.) Lepinski et al. [31] show how to construct a collusion-free protocol assuming the existence of "secure envelopes"; their work is further developed in [27,28]. Some impossibility results for collusion-free protocols are shown in [31], though it is not clear what are the implications of these results for the specific problem of implementing ex ante correlated equilibria.

Collusion freeness may also be interesting in other contexts; see [31,27,28] for further discussion. Recent work [27,28] has looked at stronger notions of collusion freeness, with the aim of achieving game-theoretic guarantees such as *strategic equivalence* between a mediated game and the cheap talk implementation of it. In that work, it is assumed that parties cannot communicate "out of band" *even before the protocol begins*; furthermore, a protocol should not only prevent covert communication between parties but should also prevent parties from agreeing on a common bit. We do not give further discussion here.

3.5 Future Directions

The immediate open question is to further characterize when a given *ex post* correlated equilibrium of a game is implementable (in, say, the standard communication model, either with or without broadcast). One direction to explore is when using a *partially fair* protocol [34,17,13,12,21] might suffice. Also, recent results [22] show that complete fairness for $t \geq n/2$ is achievable *for certain functions* in the standard communication model, thus giving hope that for certain restricted classes of correlated equilibria a cheap talk implementation might be possible even when general fair computation is not. Yet another direction is to explore other communication models, e.g., when a simultaneous broadcast channel is available. Or, taking a cue from the work on collusion-free protocols, we may ask what can be achieved under the assumption that colluding parties cannot communicate once the protocol begins. (Cleve's impossibility proof [13] fails in both the aforementioned settings.) These questions are interesting both in the current context as well as in a purely cryptographic sense.

In another direction, we can strengthen Definition 13 to require cheap talk protocols to satisfy stronger game-theoretic notions such as subgame perfection.

(See also the following section.) The Dodis-Halevi-Rabin approach, in particular, will usually not yield a subgame perfect equilibrium.

4   Rational Multi-party Computation

We now briefly discuss the second research direction mentioned in the Introduction. Here, there is no underlying game; rather, the protocol itself *is* the game, in the sense that parties' utilities are now functions of the inputs and outputs of the parties running the protocol. The difference between this setting and the standard setting of secure computation is that, in contrast to the standard setting where some parties are assumed to follow the protocol and other may behave arbitrarily, in the current setting we only guarantee that all players are *rational*. (Thus, the models are incomparable.) The questions here are: *how can we construct "meaningful" protocols in this setting?* and (more tantalizingly) *does this setting enable us to circumvent impossibility results that hold under the standard definition of secure computation?*

Let us jump right in with a "straw man" definition that, as far as I know, is new. Assume a set of parties $P_1, \ldots, P_n$ where party P_i begins holding input x_i. We assume the vector of inputs $\boldsymbol{x} = (x_1, \ldots, x_n)$ is chosen according to some known distribution $\mathcal{D}$. The parties want to compute a possibly probabilistic function f, where $f(\boldsymbol{x})$ outputs a vector $\boldsymbol{y} = (y_1, \ldots, y_n)$ and P_i receives y_i. The parties run some protocol $\boldsymbol{\Pi} = (\Pi_1, \ldots, \Pi_n)$, and we assume this protocol is *correct* in the sense that it yields the correct output if run honestly. (However, we do not assume the parties use their given inputs; see below.) The utility function of P_i is now a polynomial-time function of its view during the execution of $\boldsymbol{\Pi}$, the initial inputs $\boldsymbol{x}$, and the outputs $\boldsymbol{y}_{-i}$ of all other parties. (Note that inputs may be viewed as *types* in the sense defined in Section 2.) For treating coalitions, it seems best to define, for each possible coalition $\mathcal{C}$, a utility function $u_\mathcal{C}$ that is a function of the coalition's view, the inputs $\boldsymbol{x}$, and the outputs $\boldsymbol{y}_{-\mathcal{C}}$ of the other parties. We let Γ_{real} denote the real-world game thus defined.

In an ideal world computation of f (see [20]), a party P_i receiving input x_i can replace its input with some other value $x_i' = \delta_i(x_i)$; we allow δ_i to be probabilistic and allow $x_i' = \perp$, which is treated as an abort. After parties hand their inputs to the ideal functionality, the functionality computes $\boldsymbol{y} = f(\boldsymbol{x}')$ and gives y_i to P_i. Each party then outputs an arbitrary (polynomial-time) function $\pi_i(\cdot)$ of its view; this is left implicit in what follows, and we thus let δ_i stand for the entire strategy of P_i in the ideal world game Γ_f. The utility functions u_i are as above, except that these are now applied to the output of P_i, the inputs $\boldsymbol{x}$, and the outputs $\boldsymbol{y}_{-i}$ of the other parties (and analogously for coalitions).

Shoham and Tennenholtz [41] define the class of *NCC functions* for which, roughly speaking, setting δ_i to the identity function is a Nash equilibrium for all $\mathcal{D}$. Focusing on NCC functions appears to be a mistake that unnecessarily limits the class of functions under study.

Let $\Pi_i \circ \delta_i$ denote the real-world strategy where P_i changes its input x_i to $x_i' = \delta_i(x_i)$, and then runs Π_i using input x_i'. Then:

Definition 15. *Let $\delta = (\delta_1, \ldots, \delta_n)$ be a Nash equilibrium of Γ_f with respect to utilities $\{u_i\}$ and input distribution $\mathcal{D}$. Then Π is a* Nash protocol *for f (with respect to δ, $\{u_i\}$, and $\mathcal{D}$) if (1) $\Pi \circ \delta = (\Pi_1 \circ \delta_1, \ldots, \Pi_n \circ \delta_n)$ is a computational Nash equilibrium in Γ_{real}, and (2) for all i, it holds that $u_i(k, \Pi \circ \delta) = u_i(k, \delta)$.*

A definition of t-resilience may be derived from the above. Note that privacy, etc. are *not* explicitly required; it is our belief that questions of rationality should be separated from questions of security against malicious behavior.

An easy observation is that any protocol for completely fair secure computation tolerating t malicious parties is a t-resilient protocol for any δ, $\{u_i\}$, and $\mathcal{D}$. We also conjecture that if a protocol Π is resilient for *all* δ, $\{u_i\}$, and $\mathcal{D}$, then it is completely fair. Thus, things only become interesting if (1) we are in a setting where completely fair secure computation is impossible; and/or (2) we look at equilibrium concepts *stronger* than a Nash equilibrium. We briefly discuss these issues now. More extensive discussion will appear in the full version of this paper.

Constructing Nash protocols without completely fair MPC. This relates to the question, raised earlier, as to when relying on rationality of the parties might enable *circumvention* of impossibility results. As one example, depending on the utilities assumed it is possible to achieve complete fairness (which, note, is attained in the ideal model used in Definition 15) even in the presence of coalitions consisting of half or more of the parties [1,35,24,29]. Similarly, it is possible to implement Byzantine agreement over point-to-point channels even in the presence of coalitions controlling 1/3 or more of the parties [23].

Rational secret sharing and stronger notions of equilibrium. Halpern and Teague [26] were the first to suggest that Nash protocols do not suffice but, instead, stronger notions are needed. As a motivating example [26], consider t-out-of-n secret sharing (here, $t < n$) under the assumption that each party (1) prefers to learn the secret above all else; and (2) otherwise, prefers other parties not learn the secret. Consider the naive protocol in which each party simply broadcasts their share. (We assume authenticated shares, so each party can choose either to broadcast the correct value or nothing.) This is clearly a Nash protocol, since no matter what any particular party does at least t parties broadcast their share and everyone reconstructs the secret. Nevertheless, it appears that each P_i would prefer *not* to broadcast: if at least t other parties broadcast, then everyone (including P_i) gets the secret as before; however, if fewer than t parties broadcast then only P_i recovers the secret. That is, following the protocol is weakly dominated by *not* following the protocol, and we might expect that no one follows the protocol. (and hence the protocol is not very useful).

To address this, Halpern and Teague suggest to look for Nash protocols where players' strategies survive iterated deletion of weakly dominated strategies. Such protocols were constructed in [26,1,35,24].

Kol and Naor [29] argue that the requirement of surviving iterated deletion does not suffice to rule out protocols that are, intuitively, irrational. The notion is also difficult to work with and does not seem to capture intuition very well;

moreover, it leads to other undesirable consequences such as the fact that, if we do not assume simultaneous channels (and thus allow rushing), then protocols in which two parties are supposed to speak in the same round are inherently problematic. (Since each party will simply wait for the other to go first.) Kol and Naor thus suggest another notion that we have given as Definition 12. Their definition rules out protocols that, intuitively, seem rational to follow.

We suggest to explore using the notion of resistance to trembles. (cf. Definition 10). This requirement rules out the naive protocol mentioned above as well as the counterexample of Kol-Naor; on the other hand, the protocols of [26,1,35,24] appear to satisfy it.

The work of [27,28] offers other definitions of rational MPC.

4.1 Future Directions

The community has not yet settled on a definition for rational MPC, and finding the "right" definition seems important for further progress in this area. Looking at constructions, we note that almost all positive results for rational MPC thus far assume the utility functions inherited from [26] (an exception is [23]); a natural step is to characterize when rational MPC is possible for other classes of utilities. One can also look for closer connections between the questions considered in Sections 3 and 4.

More broadly, one might explore applications of the ideas described here to scenarios that are more complicated than function evaluation; trust inference in distributed systems serves as one compelling example. Another direction is to realize that secure computation does not happen in a vacuum, but instead may occur within an existing legal framework; given this, game theory might be profitably applied to analyze protocols satisfying the definitions of [5,32].

Acknowledgments. I am very grateful to Ran Canetti and the TCC '08 program committee for inviting me to write this survey, and to Ran for helpful discussions and suggestions.

References

1. Abraham, I., Dolev, D., Gonen, R., Halpern, J.: Distributed computing meets game theory: Robust mechanisms for rational secret sharing and multiparty computation. In: Proc. 25th ACM Symposium on Principles of Distributed Computing (PODC), pp. 53–62. ACM Press, New York (2006)
2. Abraham, I., Dolev, D., Halpern, J.: Lower bounds on implementing robust and resilient mediators. In: 5th Theory of Cryptography Conference (TCC) (2008), http://arxiv.org/abs/0704.3646
3. Aumann, R.: Subjectivity and correlation in randomized strategies. J. Mathematical Economics 1, 67–96 (1974)
4. Aumann, R., Hart, S.: Long cheap talk. Econometrica 71(6), 1619–1660 (2003)
5. Aumann, Y., Lindell, Y.: Security against covert adversaries: Efficient protocols for realistic adversaries. In: Vadhan, S.P. (ed.) TCC 2007. LNCS, vol. 4392, Springer, Heidelberg (2007)

6. Barany, I.: Fair distribution protocols or how the players replace fortune. Mathematics of Operations Research 17(2), 327–340 (1992)
7. Beaver, D.: Multiparty protocols tolerating half faulty processors. In: Brassard, G. (ed.) CRYPTO 1989. LNCS, vol. 435, pp. 560–572. Springer, Heidelberg (1990)
8. Ben-Or, M., Goldwasser, S., Wigderson, A.: Completeness theorems for non-cryptographic fault-tolerant distributed computation. In: 20th ACM Symposium on Theory of Computing (STOC), pp. 1–10 (1988)
9. Ben-Porath, E.: Cheap talk in games with incomplete information. J. Economic Theory 108(1), 45–71 (2003)
10. Buttyán, L., Hubaux, J.-P.: Security and Cooperation in Wireless Networks. Cambridge University Press, Cambridge (2007)
11. Chaum, D., Crépeau, C., Damgård, I.: Multiparty unconditionally secure protocols. In: 20th ACM Symposium on Theory of Computing, pp. 11–19 (1988)
12. Cleve, R.: Controlled gradual disclosure schemes for random bits and their applications. In: Brassard, G. (ed.) CRYPTO 1989. LNCS, vol. 435, Springer, Heidelberg (1990)
13. Cleve, R.: Limits on the security of coin flips when half the processors are faulty. In: 18th ACM Symposium on Theory of Computing (STOC), pp. 364–369 (1986)
14. Crawford, V., Sobel, J.: Strategic information transmission. Econometrica 50(6), 1431–1451 (1982)
15. Dodis, Y., Halevi, S., Rabin, T.: A cryptographic solution to a game theoretic problem. In: Dwork, C. (ed.) CRYPTO 2006. LNCS, vol. 4117, Springer, Heidelberg (2006)
16. Dodis, Y., Rabin, T.: Cryptography and game theory. In: Nisan, et al. [38]
17. Even, S., Goldreich, O., Lempel, A.: A randomized protocol for signing contracts. Comm. ACM 28(6), 637–647 (1985)
18. Forges, F.: Universal mechanisms. Econometrica 58(6), 1341–1364 (1990)
19. Fudenberg, D., Tirole, J.: Game Theory. MIT Press, Cambridge (1991)
20. Goldreich, O.: Foundations of Cryptography, Basic Applications, vol. 2. Cambridge University Press, Cambridge (2004)
21. Goldwasser, S., Levin, L.: Fair computation of general functions in presence of immoral majority. In: Menezes, A., Vanstone, S.A. (eds.) CRYPTO 1990. LNCS, vol. 537, pp. 77–93. Springer, Heidelberg (1991)
22. Gordon, S.D., Hazay, C., Katz, J., Lindell, Y.: Complete fairness in secure two-party computation. Manuscript (2007)
23. Gordon, S.D., Katz, J.: Byzantine agreement with a rational adversary. In: Rump session presentation, Crypto (2006)
24. Gordon, S.D., Katz, J.: Rational secret sharing, revisited. In: Security and Cryptography for Networks (SCN), pp. 229–241 (2006)
25. Halpern, J.: Computer science and game theory: A brief survey. In: The New Palgrave Dictionary of Economics, 2nd edn. (2008), (Anticipated publication date: 2008) http://www.cs.cornell.edu/home/halpern/
26. Halpern, J., Teague, V.: Rational secret sharing and multiparty computation. In: 36th Annual ACM Symp. on Theory of Computing (STOC), pp. 623–632 (2004)
27. Izmalkov, S., Lepinski, M., Micali, S.: Rational secure computation and ideal mechanism design. In: IEEE Symposium on Foundations of Computer Science (FOCS), See also technical report MIT-CSAIL-TR-2007-040 (2005), http://hdl.handle.net/1721.1/38208
28. Izmalkov, S., Lepinski, M., Micali, S.: Verifiably secure devices. In: 5th Theory of Cryptography Conference (TCC) (2008)

29. Kol, G., Naor, M.: Cryptography and game theory: Designing protocols for exchanging information. In: 5th Theory of Cryptography Conference (TCC) (2008)
30. Lepinski, M., Micali, S., Peikert, C., Shelat, A.: Completely fair SFE and coalition-safe cheap talk. In: 23rd Annual ACM Symposium on Principles of Distributed Computing (PODC), pp. 1–10 (2004)
31. Lepinski, M., Micali, S., Shelat, A.: Collusion-free protocols. In: ACM Symposium on Theory of Computing (STOC) (2005)
32. Lindell, A.Y.: Legally-enforceable fairness in secure multiparty computation. In: CT-RSA (to appear 2008)
33. Linial, N.: Game-theoretic aspects of computer science. In: Aumann, R., Hart, S. (eds.) Handbook of Game Theory with Economic Applications, vol. 2, pp. 1340–1395. North-Holland, Amsterdam (1994)
34. Luby, M., Micali, S., Rackoff, C.: How to simultaneously exchange a secret bit by flipping a symmetrically-biased coin. In: 24th Annual IEEE Symposium on Foundations of Computer Science (FOCS), pp. 11–21 (1983)
35. Lysyanskaya, A., Triandopoulos, N.: Rationality and adversarial behavior in multiparty computation. In: Dwork, C. (ed.) CRYPTO 2006. LNCS, vol. 4117, Springer, Heidelberg (2006)
36. Moulin, H., Vial, J.-P.: Strategically zero sum games. Intl. J. Game Theory 7(3/4), 201–221 (1978)
37. Nash, J.: Non-cooperative games. Annals of Mathematics 54(2), 286–295 (1951)
38. Nisan, N., Roughgarden, T., Tardos, É., Vazirani, V.: Algorithmic Game Theory. Cambridge University Press, Cambridge (2007)
39. Osborne, M., Rubinstein, A.: A Course in Game Theory. MIT Press, Cambridge (2004)
40. Rabin, T., Ben-Or, M.: Verifiable secret sharing and multiparty protocols with honest majority. In: 21st ACM Symp. on Theory of Computing, pp. 73–85 (1989)
41. Shoham, Y., Tennenholtz, M.: Non-cooperative computation: Boolean functions with correctness and exclusivity. Theoretical Comp. Sci. 343(1–2), 97–113 (2005)
42. Urbano, A., Villa, J.: Computational complexity and communication: Coordination in two-player games. Econometrica 70(5), 1893–1927 (2002)
43. Urbano, A., Villa, J.: Computationally restricted unmediated talk under incomplete information. Economic Theory 23(2), 283–320 (2004)

Verifiably Secure Devices

Sergei Izmalkov[1], Matt Lepinski[2], and Silvio Micali[3]

[1] MIT Department of Economics
izmalkov@mit.edu
[2] BBN Technologies
mlepinski@bbn.com
[3] MIT CSAIL
silvio@csail.mit.edu

Abstract. We put forward the notion of a *verifiably secure device*, in essence a stronger notion of secure computation, and achieve it in the ballot-box model. Verifiably secure devices

1. Provide a perfect solution to the problem of *achieving* correlated equilibrium, an important and extensively investigated problem at the intersection of game theory, cryptography and efficient algorithms; and
2. Enable the secure evaluation of multiple interdependent functions.

1 Introduction

FROM GMW TO ILM1 SECURITY. As put forward by Goldreich, Micali and Wigderson [11] (improving on two-party results of Yao [17]), secure computation consists of capturing **crucial aspects** of an abstract computation aided by a trusted party, by means of a concrete **implementation** that does not trust anyone. However, what is deemed crucial to capture and what constitutes an implementation have been changing over time. In order to achieve fundamental desiderata in a game theoretic setting, where incentives are added to the mixture and players are assumed to behave rationally, in [13] we put forward a stronger notion of secure computation, and achieved it in the *ballot-box model*. In essence, this is a physical model using ballots and a ballot randomizer, that is, the same "hardware" utilized from time immemorial for running a lottery and tallying secret votes. We refer to our 2005 notion as *ILM1 security*. Our main reason for introducing ILM1 security was implementing normal-form mechanisms in the stand-alone setting.

ILM2 SECURITY. In this paper, we put forward a yet stronger notion of secure computation, herein referred to as *ILM2 security*, and achieve it in a variant of the ballot-box model. ILM2 security enables us to attain even more sophisticated applications. In particular, it enables us to (1) perfectly achieve *correlated equilibrium*, a crucial desideratum at the intersection of cryptography, game theory, and efficient algorithms; and (2) securely implement interdependent functions and mechanisms.

R. Canetti (Ed.): TCC 2008, LNCS 4948, pp. 273–301, 2008.
© International Association for Cryptologic Research 2008

SETTING UP A COMPARISON. In an fairer world, we could have come up with our present security notion in 2005. But the world is not fair, and we could not even conceive ILM2's security requirements back then. To ease their comparison, we sketch both approaches in the next two subsections, initially focussing on the secure evaluation of a single, probabilistic, finite function f from n inputs to $n + 1$ outputs. Without loss of generality, $f : (\{0,1\}^a)^n \to (\{0,1\}^b)^{n+1}$.

We break each approach into the following components: (1) *The Ideal Evaluation*, describing the "target computation", that is, an evaluation of f with the help of a trusted party; (2) *The Concrete Model*, highlighting the "mechanics" of an evaluation of f without a trusted party; and (3) *The Security Notion*, describing the extent to which the concrete model captures the ideal evaluation.

Traditionally, in summarizing secure computation, the second and third components are merged. In the original GMW definition, there was certainly no compelling need to treat the concrete model as a "variable." Subsequently, their concrete model (i.e., a communication protocol among all players) persisted in all variants of secure computation, thus possibly generating a sense of "inevitability." We instead highlight the concrete model as an independent component because one of our contributions is indeed a change of scenery of this aspect of secure computation.

1.1   The ILM1 Approach

THE IDEAL EVALUATION. In the ILM1 approach, an ideal evaluation of f proceeds in three stages.

1. In the input stage, each player i either (1.1) publicly aborts, in which case the entire evaluation ends, or (1.2) privately and independently chooses an a-bit string x_i and privately sends it to T.
2. In the computation stage, T publicizes a random string σ, and then privately evaluates f on all x_i's so as to obtain the b-bit values $y, y_1, \ldots, y_n$.
3. In the output stage, T publicizes y and privately hands back y_i to each player i.

(Note that the players do not communicate in an ILM1 ideal evaluation. By contrast, in the GMW approach, the ideal evaluation of f cannot be precisely matched by a GMW-secure protocol unless it offers the players the option of communicating to each other prior to aborting or sending T their inputs.)

THE CONCRETE MODEL. In the ILM1 approach, the concrete model for mimicking an ideal evaluation of f continues to be that of the original GMW approach: namely, a multi-round communication protocol P_f executed by all players, where each player secretly holds and updates a share of the global state of the evaluation. The only difference is in the communication model. That is, rather than relying on broadcasting and/or private channels (as traditional GMW-secure protocols do), ILM1 protocols rely on ballots and a ballot randomizer.

THE SECURITY NOTION. At the highest level, for P_f to be an ILM1-secure protocol for a function f, there must be an "output-preserving" bijection between the players' non-aborting strategies in P_f and the players' non-aborting strategies in an ideal evaluation of f.[1]

Furthermore, to match the privacy of f's ideal evaluation, it is required that

1. During the input-commitment stage the only information collectively available to any set of the players about the inputs of the other players consists of a fixed string —which is no information at all, and strictly less than observing a random string;[2]

2. During the computation stage the only information collectively available to any set of the players about the inputs of the other players consists of a common random string; and

3. During the output stage, the only information collectively available to any set of the players about the inputs of the other players consists of the desired outcome —in a pre-specified, deterministic encoding.

(Note that our use of the term "collectively available information" in relation to a set of the players does not refer to information known to at least a member of the set. Rather, it refers to the information that one would obtain were he able to join together the information available to each member of the set. Barring the existence of external and undetected means of communication, an individual member of the set can only get such "collective" information by querying the other members after the protocol P_f terminates. As deducible by the above sketchy description, an ILM1-secure protocol does not enable any inter-player communication, as demanded in the ideal evaluation.)

Main Properties of ILM1 Security

- *Perfect implementation of Normal-Form Mechanisms in the Stand-Alone Setting.* Our original reason for introducing the ILM1 notion of security was to be able to perfectly implement normal form mechanisms in the stand-alone setting, something that was not implied by GMW-secure computation. We refer the reader to [13] for the definition of such a perfect implementation (and actually to MIT-CSAIL-TR-2007-040 for a more precise explanation). Here we are happy to quickly recall what a normal-form mechanism and the stand-alone setting are.

---

[1] By contrast, a player has "much more to do" in a GMW-secure protocol P_f than in an ideal evaluation of f. In the latter setting, in fact, there are exactly $2^a + 1$ "strategies" for a player i, one for each possible i-input to f plus aborting. Accordinly, if f operates on 10-bit inputs, the total number of strategies is roughly one thousand. By contrast, letting for concreteness P_f be the original protocol of [11], player i not only has to decide between 2^a inputs or aborting, but can also decide which encryptions to broadcast, which query bits to use in a zero-knowledge proof and so on. Thus, while each player has roughly 1000 strategies in an ideal evaluation of h, he may easily have more than 2^{1000} strategies in P_f. Such richness of strategies severely limits the relevance of GMW-secure protocols to game-theoretic applications.

[2] Quite differently, in a GMW-secure protocol for f, the players —by broadcasting and/or privately exchanging all kinds of strings— can make available plenty of additional information.

Very informally, a (finite) normal-form mechanism is a function f : $(\{0,1\}^a)^n \to (\{0,1\}^a)^{n+1}$. The mechanism play coincides with an ideal evaluation of f as in the ILM1 approach. However, such play is analyzed as a game in a context specifying players' preferences and thus the players get different utilities (think of "dollars prizes") for different outputs of f.

Very informally too, by a normal-form mechanism in the stand-alone setting we mean that "nothing else happens after a single play of the mechanism."

- *Fairness and Perfect Security without Honest Majority.* Informally fairness means that either all (honest) players receive their right outputs, or nobody does. It is a notable property of ILM1 security that it simultaneously guarantees fairness and perfect information-theoretic security without relying on the majority of the players to be honest. Here by a "honest" player we mean one sticking to his protocol instructions no matter what.

  (GMW-secure protocols —relying on broadcasting and/or private channels— do not guarantee fairness unless the majority of the players are honest. Indeed, Cleve [7] shows that not even the the probabilistic function that, on no input, outputs a random bit can be fairly and efficiently computed when half of the players can abort the protocol —let alone maliciously deviate from their instructions. Remarkably, in 2004, Lepinski, Micali, Peikert and Shelat [15] put forward a protocol guaranteeing fairness without any honest majority in a mixed model involving broadcasting and regular ballots.[3] The security of their protocol, however, was only computational.)

- *Perfect Security and Universal Composibility Without Honest Majority.* ILM1-secure protocols satisfy "composibility" as defined in 2000 by Dodis and Micali [9]. It is by now recognized that their notion actually coincides, in the perfect information-theoretic setting, with universal composibility as defined in 2001 by Canetti [6]. Indeed, Halevi and Rabin show that *perfect* simulatability via *straight-line* simulators (as demanded in [9]) implies universal composibility.

  The universal composibility of ILM1-secure protocols is remarkable because it is achieved together with perfect information-theoretic security and without relying on any honest majority, something that was not known to be possible in the past.

1.2 The ILM2 Approach

In the ILM2 approach, an ideal evaluation of f continues to proceed in three stages. There are two possibilities for the first stage: one with aborts and one without. We refer to the first one as *weak* ideal evaluation, and the second one as *strong* ideal evaluation. With start by presenting the latter, simpler and more provocative notion.

---

[3] I.e., their ballots needed not to be identical, nor was a ballot randomizer needed for their construction.

STRONG IDEAL EVALUATION

1. In the input stage, each player i —independently from all others— secretly chooses an input x_i and gives it to T in an envelope,

2. In the computation stage, T privately opens all received envelopes and then privately computes $y, y_1, \ldots, y_n = f(x_1, \ldots, x_n)$.

3. In the output stage, T publicizes y and publicly hands back to each player i an envelope containing y_i.

THE CONCRETE MODEL. In the ILM2 approach, the concrete model continues to rely on the same hardware of the ILM1 approach (i.e., identical ballots and a ballot randomizer), but no longer coincides with a communication protocol among the players. Instead, the encoding of the global state is now fully contained in a sequence of envelopes publicly manipulated by a *verifiable* entity T'. Let us explain.

In the ILM2 ideal evaluation of a function f, T is *trusted*. Indeed, he could change the outputs and/or reveal undue information about the players' inputs (e.g., via the envelopes he hands back to the players or after the evaluation) without any fear of being "caught." By contrast, to securely evaluate f, T' is called to perform a *unique sequence of ballot operations* such that the players *can verify that the right operations have been performed*. For instance, if privacy did not matter, upon receiving the envelopes containing the players' inputs, T' could be required to open all of them. In which case it would be trivial to check whether he has done what was required of him. (To be sure, such a verifiable T' would still be trusted not to —say— publicly open half of the envelopes and destroy the other half. But such trust is much milder: because any deviation from opening all envelopes would become of public record, T' can be kept accountable.)

Because the actions required from T' are uniquely determined and verifiable, human or not, we think of T' as a *verifiable device*.

THE SECURITY NOTION. The ILM2 security notion is the most stringent we can (currently) conceive. Before sketching it, it should be realized that, in whatever model used (encrypt-and-broadcast, private-channel, ballot-box, etc.) each "action" essentially has a public and a private informational component.[4] For instance, in the ballot-box model, the action of "publicly opening envelope j" generates only a public component: namely, "(PUBLICLYOPEN,j,c)" where c is the now exposed content of (former) envelope j. As for another example, "party i privately opens envelope j" generates a public component "(PRIVATELYOPEN, i,j)" together with the private component c for party i, where c is the content of the (former) envelope j. The correctness requirements of an ILM2 concrete evaluation of f are not surprising (and are formally presented later on). The privacy requirements are instead surprising. Namely, in a verified computation,

---

[4] In the encrypt-and-broadcast model, when player i sends a message m to player j encrypted with j's key, the public component is "ciphertext C from i to j" while the private component to j is "m".

1. In a correct execution of the verifiable device, the public history generated (i.e., the concatenation of all public informational components) is a *fixed* string R (depending only on f) and *vice versa* whenever the public history of an execution of the verifiable device is R, then f has been evaluated correctly and privately on the actual inputs, no matter what they may be;
2. The private history of each player i consists only of his own input x_i; and
3. The private history of the verifiable device consists only of a random string.

Remarks

- *Perfect Security and Universal Composibility without Honest Majority.* As for the ILM1 case, this is due to the fact that ILM2 security satisfies the Dodis-Micali conditions.
- *Perfect Determinism, and Perfect Independence.* ILM2 security is the only notion of secure computation that is totally deterministic to the players. They do not observe any common randomness and do not even generate any local randomness. The situation is not too different for the verifiable device. Namely, he does not generate any local randomness, but individually observes a random string ρ. Such string however cannot be communicated by the device to the players by means of the operations available to him (which are verifiable anyway). And should the device reveal ρ to some players afterwards, it would be "too late." During the executions the players have been *absolutely isolated from one another*.
- *Hidden Aborts.* Let us now explain in what sense it is meaningful to consider ideal executions in which players cannot abort. In a typical secure protocol, it makes no sense to worry about player i learning j's input by pointing a gun at j's head. The analysis of any protocol should be relative only to the actions available within the protocol's model. Nonetheless, aborting is quite possible within the confines of a protocol's actions. For instance, a player who is required to broadcast the decryption of a given cipher-text might be able to cause an abort by broadcasting a different string. That is, one ability of aborting arises when the set of available actions is richer than that "handleable" by the protocol. This source of aborts, however, may not be always present, and is not be present in the ILM2 case. Nonetheless, there is one more source of aborts: namely, taking "no action." That is, aborts may also occur in models for which "doing nothing" is distinguishable from "the prescribed actions". In the broadcast model, if a player is envisaged to broadcast a bit, broadcasting nothing is quite a different thing. Doing nothing is also distinguishable from all possible actions in the version of the ballot-box model envisaged in ILM1-secure protocols, and easily enables a player to halt the joint computation. Indeed, in the ILM1 approach, the global state of the concrete computation is shared in n pieces, each known to a different player, and each necessary for the progress of the computation. Thus, if a player suicides carrying his own piece to the other world, the computation cannot be continued in any meaningful way.

  By contrast, in an ILM2-secure protocol, after the verifiable device starts taking public actions on the envelopes encoding the players' inputs, all envelopes are in the hands of the device, and the global state of the computation is contained fully in those envelopes. Thus, from that point on, in a properly verified execution the computation goes on (correctly and privately) independently of what the players may or may not do. Device abort is not an issue either. Because our devices are called to do a single verifiable action at every point of their executions, only device no-action is meaningful to analyze, and it is a notable property of our construction that, if needed, the verifiable device can be substituted at any moment with another verifiable device without any loss.

In an ILM2-secure protocol, therefore, the only stage in which the issue of abort might possibly arise is the input stage; because there the players' participation is crucial to ensure that envelopes properly encoding their inputs are handed to the device. There, however, we have engineered players' input commitment so as to "hide aborts." Conceptually, a player contributes an input bit to the evaluation of f as follows. First, he publicly receives from the device two envelopes, both publicly generated with different prescribed contents. Then the player is asked to secretly permute them: leaving them in the same order corresponds to secretly inputting 0, flipping their order corresponds to secretly inputting 1. Formally speaking, therefore, aborting is indistinguishable from secretly inputting 0.

- *Public Aborts.* Practically speaking, however, enforcing "abort indistinguishability" requires building and analyzing some ad hoc simple gadget with an associated protocol. (If you fail in designing them, just ask us!) Such a gadget, of course, would be a physical assumption not only *additional* to ballots and ballot-boxes, but also *quite new,* while ballots and ballot-randomizers have been around for time immemorial and are thus easy to accept as "physical axioms."[5] Altogether, most readers would prefer to stick with the more intuitive operation of "player i secretly permutes two envelopes, or publicly aborts." In this case, we can only achieve the following type of ideal evaluation.

WEAK IDEAL EVALUATION

1′. In the input stage, each player i —independently from all others— either secretly chooses an input x_i and gives it to T in an envelope, or publicly gives T the input $x_i = 0^a$ —i.e., the concatenation of a 0s.

2. In the computation stage, T privately opens all received envelopes and then privately computes $y, y_1, \ldots, y_n = f(x_1, \ldots, x_n)$.

3. In the output stage, T publicizes y and publicly hands back to each player i an envelope containing y_i.

Definitionally, it is clear that that the strong and weak versions of ILM2 security are both stronger than ILM1 security. Let us now present two specific concrete

---

[5] Put it this way: if you think that it is not possible to randomize identical ballots, you should also explain (1) why people have been shuffling decks of cards for ever and for real money; *and* (2) why electoral precincts do not simplify voting procedures by always adopting roll-call voting.

settings requiring ILM2 security. The first one actually involves the implementation of function with no inputs. Therefore it is does not matter whether the players can or cannot abort! Thus, weak or strong, ILM2 security wins anyway.

1.3   Perfect Achievement of Correlated Equilibrium

Consider the following trivial probabilistic function f that, on no inputs, outputs a pair of strings:

$f() = (C, G)$, (C, G), or (D, G) —each with probability $1/3$.

Obviously, using proper binary encodings, an ILM1-secure protocol P_f for f exists. Ideally, such a P_f should successfully replace an ideal evaluation for f in any setting. However, *this is not true for the following setting.*

	E	F	G	H
A	100, 0	$-\infty, -\infty$	$-\infty, -\infty$	$-\infty, -\infty$
B	$-\infty, -\infty$	0, 100	$-\infty, -\infty$	$-\infty, -\infty$
C	$-\infty, -\infty$	$-\infty, -\infty$	4, 4	1, 5
D	$-\infty, -\infty$	$-\infty, -\infty$	5, 1	0, 0

Let Row and Column be the two players of the normal-form game $\mathbb{G}$ described by the above pay-off matrix. Accordingly, Row and Column are confined to two separate rooms: the first facing a keyboard with 4 buttons A,B,C and D; and the second with a keyboard whose buttons are E,F,G, and H. Without any external help, only the following (reasonable) Nash equilibria exist in $\mathbb{G}$: (A,E), which is very good for Row, (B,F), which is very good for Column, (C, H), (D, G), and the mixed equilibrium $(\frac{1}{2}C + \frac{1}{2}D, \frac{1}{2}G + \frac{1}{2}H)$, which yields payoff 2.5 to each player. Mankind's interest, however, is that the outcome of $\mathbb{G}$ is either (C, G), (C, H), or (D, G), each with probability $\frac{1}{3}$. Accordingly, an angel in all his splendor descends from the sky, privately evaluates the above function $f()$ to obtain a pair (X, Y), tells Row and Column that he has done so, and then provides Row with an envelope containing X, and Column with an envelope containing Y. Technically, this angelic intervention puts Row and Column in a *correlated equilibrium* $\mathbb{A}$, for "angelic". (Correlated equilibria were proposed by Aumann [1].) In essence, each player is better off playing the recommendation received from the angel if he believes the other will do so. It is evident that the expected utility of each player in $\mathbb{A}$ is $10/3$. Clearly, however, Row and Column would prefer to play a different correlated equilibrium: namely $\mathbb{E}$, the correlated equilibrium corresponding to selecting (A,E) or (B,F), each with probability $1/2$. Unfortunately, they cannot reach such equilibrium without external help. Nor can they use the X and Y they respectively received by the angel (which is external help indeed!) in order to "translate" their angelic recommendations into their preferable ones. The point is that if X=C, then Row has no idea whether Y=G or Y=H. Each is equally probable to him. (Symmetrically, if Y=G, then Column has no idea whether X=C or X=D, both are equally probable to him.) If X=D, then Row knows that Y=G, but he also know that Column has no idea whether X=C or X=D. Accordingly, the expected payoff provided to the players by any "translation strategy" is $-\infty$. This must be very frustrating, since (1) Row and Column receive an expected utility of 50 each in $\mathbb{E}$, and (2) Row and Column only need a *single* random bit to coordinate their actions to achieve $\mathbb{E}$!

Consider now replacing the angel for f with the ILM1-secure protocol P_f. (This is indeed possible because ILM1 security does not require any honest majority.) Then, after correctly and privately computing their P_f-recommendations, Row and Column can simply ignore them and use instead the first common random bit generated by P_f to coordinate and play equilibrium $\mathbb{E}$ instead. (Indeed, at some point of any execution of P_f, 5 envelopes previously randomized by the ballot-box are publicly opened, thus publicly revealing a random permutation of 5 elements, which can be trivially translated into a single random bit.)

How is this possible in view of the claim that ILM1 successfully evaluates any finite function in the stand-alone setting? The answer is that *the above setting is not stand alone*. Indeed, Row and Column are not just asked to play P_f and get immediately rewarded. They are asked to play first P_f and *then* $\mathbb{G}$, and they ultimately get $\mathbb{G}$'s rewards. Thus, even a minimal deviation from the stand-alone model, together with the minimal presence of common random bit, sets apart what can happen with the help of an angel and what can happen with traditional secure computation. In other words, so far in secure computation we have been conditioned to think (and the last author agrees to take some blame) that "randomness and efficient computation are for free." Unfortunately, this is not true in game-theoretic settings. In general,

> Common randomness *as a side product of secure computation is akin to* pollution *as a side product of energy transformation.*

A Perfect Solution to a Beautiful Problem. As introduced by Aumann [1], a correlated equilibrium for a normal-form game G is a probability distribution f over the actions available to the players such that if —somehow!— a profile of "recommended actions" $(a_1, \ldots, a_n)$ is chosen according to f, and each player i learns a_i without gaining any other information about a_{-i}, then no single player can deviate from his recommendation in a play of G and improve his expected utility. Given the description of any equilibrium E, the obvious problem consists of finding a concrete way that precisely simulates a trusted party correctly sampling and privately handing out to the players E's recommendations.

Despite much work [2,5,10,8,16,15,12], all prior solutions to this problem were imperfect. Specific deficiencies included limited rationality, strategic-option alteration, limitations on the numbers of players, infeasible computation and resources, and imposing a pre-specified beliefs to yet-to-be-determined players.

By contrast ILM2 security provides a perfect solution to the achieving correlated equilibrium. (Details will be given in the final paper.) Note that the main concerns here is orthogonal to composibility, that does not care about preserving strategic opportunities. Indeed, generating a common random bit is OK vis à vis composibility, but alters the achievement of correlated equilibrium.

1.4   Interdependent Secure Computations and Mechanisms

More generally, we now consider evaluating multiple *interdependent* functions, each not only receiving fresh inputs from the players and producing public and private outputs, but also receiving an additional secret input consisting of state information from the evaluation of another function, and passing some state information of its own evaluation as an additional secret input to another function. For simplicity sake, below we formalize just the case of a pre-defined linear sequence of such functions. (In particular, we could handle trusted parties can operate more than once, each time passing their state to different trusted parties, operate simultaneously, recursively, et cetera. Of course, the more complex the setting, the richer are the strategic opportunities of the players. We are not interested in analyzing them, but rather to match them exactly, whatever they may be —without envisaging concurrently executing extraneous secure protocols.)

(WEAK) IDEAL EVALUATION OF INTERDEPENDENT FUNCTIONS. Let F be a sequence of functions, $F = f_1, \ldots, f_k$, where $f_i : (\{0,1\}^a)^{n+1} \to (\{0,1\}^a)^{n+2}$. Letting s_0 and s_{k+1} be empty strings, an ideal evaluation of F proceeds in k phases, with the help of k separate trusted parties: $T_1, \ldots, T_k$. The jth phase consists of 3 stages.

1. In the input stage, T_j secretly receives state information s_{j-1} and publicly receives the identities of all aborting players. Further, each non-aborting player i independently and secretly chooses an input x_i^j and gives it to T in an envelope, or publicly aborts. Each aborting player i publicly gives T_j the input $x_i^j = 0^a$ —i.e., the concatenation of a 0s.
2. In the computation stage, T_j privately opens all received envelopes and then privately computes
$$s_j, y^j, y_1^j, \ldots, y_n^j = f_j(s_{j-1}, x_1^j, \ldots, x_n^j).$$
3. In the output stage, T_j publicizes y^j, privately hands to each player i an envelope containing y_i^j, and privately hands s_j to T_{j+1}.

Note that this weak evaluation can be changed in several ways if desired or necessary to model the situation at hand. For instance, rather than forcing aborting players to always contribute the input 0^a in the future, one may envisage a player aborting in the input stage of phase j as contributing the input 0^a in just that phase, but give him an opportunity to fully participate in future phases. This may be a natural choice, but of course it enlarges the players' signalling abilities. As for another possible change, one may demand that no envelope containing the private output of an aborting player be ever given to him. A third alternative may consist of banning the aborting players from future participation, thus changing the functions of future phases so as to have fewer inputs (although this technically is already beyond the simpler setting envisaged above). And so on. It all depends on the setting we wish to model.

The "strong" version of the above definition can be obtained by removing the possibility of aborting altogether.

To go from sequences of functions to the analysis of sequences of normal-form mechanisms one needs only to specify the players' preferences over the outcomes.

The Inachievability of Secure Interdependency with Prior Notions.
ILM1 security does not suffice for meaningfully implementing a sequence of interdependent functions/mechanisms. (GMW security would be even less meaningful, particularly if relying on private channels.) The main problem is as follows.
Assume that player i aborts in the jth phase. In the ideal setting, i necessarily
aborts in the input stage. Accordingly, T_j uses 0^a as i's input for the function
f_j, completes the computation of f_j, returns all public and private outputs to
the right players, and finally provides T_{j+1} with the correct state information
s_j. This is the case because in the ideal setting T_j has already received from
T_{j-1} the state information s_{j-1}. By contrast, when i aborts the execution of an
ILM1-secure protocol P_{f_j} for f_j, he makes his own share of the global computation disappear with him, causing the entire sequence of evaluations to grind
to a halt. (In fact, any way of continuing would proceed with an incorrect state:
the other players do have their own shares of the current global state, but their
shares alone are consistent with any feasible global state at that point.) If the
whole evaluation were just a mental game, endowing a player with the ability of
halting the entire sequence of future evaluations by his aborting in specific phase
might not matter. But causing the entire sequence of evaluations to abort may be
disastrous in other settings, where "real incentives" are associated to the whole
enterprize. For instance, assume that the government of a major country is pri-
vatizing its national resources (it has happened before!) by means of a complex
sequence of inter-dependent normal-form mechanisms, so as to achieve complex
social objectives. Gas has been allocated first, oil second, and so on, with the
players paying real money for these resources (and possibly selling off assets they
previously owned in order to raise the necessary cash). And then, suddenly, one
of players commits suicide. What should the government do? Sending every one
home, as if nothing ever happened, and demanding that the allocated resources
be returned is not an option: who is going to return the assets some players
had to sell (possibly in different countries) in order to win some of the present
resources? Nor is it an option to recognize all allocations already made and stop
the future ones. In fact, due to the interdependency of the mechanisms in the
sequence, a player may have chosen to acquire a given resource in one of the early
phases (by strategically choosing his secret inputs to the first functions in the
sequence) only in order to improve his chance to win resources more attractive to
him in the future phases. Nor is it an option to allocate the remaining resources
by means of a different sequence of mechanisms. The players' past strategies
depended on that very evaluation to continue with the right state.[6]

---

[6] Notice that although exogenous incentives, such as fines, may discourage abortions,
they are incapable of perfectly solving the problem. On one hand, fining players will
not resurrect the right computation state. On the other, finding the right fine to
impose is not an easy theoretical problem. Assume that a player i aborts because,
based on his received private information, he realizes that the rest of the mechanisms
—if continued— would cause him immense financial harm. Then, to induce him not
to do so, a mechanism designer must impose a fine greater than i's expected loss.
But, in general, the designer may be unaware of the players' preferences.

In essence, not every thing in life is a mental game without incentives, and to properly deal with these incentives one needs to preserve the originally envisaged strategic opportunities. It should be clear that weak ILM2 security is instead capable of matching the above ideal scenario. Indeed, in ILM2-secure computation, the global state continues to remain available to each verifiable device D_j (corresponding to trusted party T_j) whether the players abort or not. Moreover, the players do not have the ability to signal in any additional way from one mechanism to the next. Not even random strings will enable to alter the strategic opportunities available to the players in a weak-ILM2-secure protocol.

Finally, note that one may also consider *strong* ideal evaluations of interdependent functions, and that strong-ILM2-secure protocols will be able to match these more stringent requirements.

2 Notation

Basics. We denote by $\mathbb{R}^+$ the set of non-negative reals; by Σ the alphabet consisting of English letters, arabic numerals, and punctuation marks; by Σ^* the set of all finite strings over Σ; by $\bot$ a symbol not in Σ; by SYM_k the group of permutations of k elements; by $x := y$ the operation that assigns value y to variable x; by $\emptyset$ the empty set, and by ϕ the empty string/sequence/vector.

If x is a sequence, by either x^i or x_i we denote x's ith element,[7] and by $\{x\}$ the set $\{z : x^i = z$ for some $i\}$. If x is a sequence of k integers, and m is an integer, by $x + m$ we denote the sequence $x^1 + m, \ldots, x^k + m$. If x and y are sequences, respectively of length j and k, by $x \circ y$ we denote their concatenation (i.e., the sequence of $j + k$ elements whose ith element is x^i if $i \leq j$, and y^{i-j} otherwise). If x and y are strings (i.e., sequences with elements in Σ), we denote their concatenation by xy.

Players and profiles. We always denote by N the (finite) set of players, and by n its cardinality. If i is a player, $-i$ denotes the set of the other $n - 1$ players, that is, $-i = N \setminus \{i\}$. Similarly, if $C \subset N$, then $-C$ denotes $N \setminus C$. A profile is a vector indexed by N. If x is a profile, then, for all $i \in N$ and $C \subset N$, x_i is i's component of x and x_C is the sub-profile of x indexed by C; thus: $x = (x_i, x_{-i}) = (x_C, x_{-C})$.

Probability distributions. All distributions considered in this paper are over finite sets. If $X : S \to \mathbb{R}^+$ is a distribution over a set S, we denote its support by $[X]$, that is, $[X] = \{s \in S : X(s) > 0\}$. We denote by $rand(S)$ the uniform distribution over S.

If A is a probabilistic algorithm, the distribution over A's outputs on input x is denoted by $A(x)$. A probabilistic function $f : X \to Y$ is *finite* if X and Y are both finite sets and, for every $x \in X$ and $y \in Y$, the probability that $f(x) = y$ has a finite binary representation.

---

[7] For any given sequence, we shall solely use superscripts, or solely subscripts, to denote all of its elements.

3   The Ballot-Box Model

The ballot-box model ultimately is an abstract model of communication, but possesses a quite natural physical interpretation. The physical setting is that of a group of players, 1 through n, and a distinguished "player" 0, the device, seated around a table together and acting on a set of *ballots* with the help of a randomizing device, the *ballot-box*. Within this physical setting, one has considerable latitude in choosing reasonable actions. Indeed, in this paper, we envisage more actions than in [13].

3.1   Intuition

BALLOTS. There are two kinds of ballots: *envelopes* and *super-envelopes*. Externally, all ballots of the same kind are identical, but super-envelopes are slightly larger than envelopes. An envelope may contain a symbol from a finite alphabet, and a super-envelope may contain a sequence of envelopes. (Our constructions actually needs only envelopes containing an integer between 1 and 5, and super-envelopes capable of containing at most 5 envelopes.) An envelope perfectly hides and guarantees the integrity of the symbol it contains until it is opened. A super-envelope tightly packs the envelopes it contains, and thus keeps them in the same order in which they were inserted. Initially, all ballots are empty and in sufficient supply.

BALLOT-BOX ACTIONS. There are 10 classes of ballot-box actions. Each action in the first 7 classes is referred to as a *public action*, because it is performed in plain view, so that all players know exactly which action has been performed, and its consequences are the same no matter who performs it. These 7 classes are: (1) publicly write a symbol on a piece of paper and seal it into a new, empty envelope; (2) publicly open an envelope to reveal its content to all players; (3) publicly seal a sequence of envelopes into a new super-envelope; (4) publicly open a super-envelope to expose its inner envelopes; (5) publicly reorder a sequence of envelopes; (6) publicly destroy a ballot; and (7) do nothing. The last three classes just simplify the description of our construction.

An action in the eighth class is referred to as an *action of Nature*. Such an action consists of "ballot boxing" a publicly chosen sequence of ballots, that is, reordering the chosen ballots according to a permutation randomly chosen by —and solely known to— Nature.

Each action of 9th and 10th classes is referred to as a *private action*, because some details about either its inputs or outputs are known solely to the player (or device) performing it. These two classes are: (9) privately open, read the content, and reseal an envelope; and (10) secretly reorder a sequence of envelopes. We imagine that the players observe what ballots these actions are performed upon, but the actions themselves are performed outside of public view. For instance, to perform an action of class 10, a player can shuffle the envelopes behind his back or within a box, so that only he knows in what order the envelopes are returned on the table.

PUBLIC INFORMATION. Conceptually, the players observe which actions have been performed on which ballots. Formally, (1) we associate to each ballot a unique identifier, a positive integer that is common information to all players (these identifiers correspond to the order in which the ballots are placed on the table for the first time or returned to the table —e.g., after being ballot-boxed); and (2) we have each action generate, when executed, a public string of the form "A, i, j, k, l, ..."; where A is a string identifying the action, i is the number corresponding to the player performing the action, and $j, k, l, ...$ are the identifiers of the ballots involved. If the action is public, for convenience, the identity of the player performing it is not recorded, since the effect of the action is the same no matter by whom the action is performed. The *public history* is the concatenation of the public strings generated by all actions executed thus far. Similarly, the *private history* of each player is the concatenation of the private strings generated by the private actions performed by, respectively, the player. The private string is the content of the opened envelope for a "private read" action and the actual permutation for a "secret permute" action.

3.2 Formalization

An *envelope* is a triple $(j, c, 0)$, where j is a positive integer, and c a symbol of Σ. A *super-envelope* is a triple (j, c, L), where both j and L are positive integers, and $c \in \Sigma^L$. A *ballot* is either an envelope or a super-envelope. If (j, c, L) is a ballot, we refer to j as its *identifier*, to c as its *content*, and to L as its *level*. (As we shall see, L represents the number of inner envelopes contained in a ballot.)

A set of ballots B is *well-defined* if distinct ballots have distinct identifiers. If B is a well-defined set of ballots, then I_B denotes the set of identifiers of B's ballots. For $j \in I_B$, B_j (or the expression *ballot j*) denotes the unique ballot of B whose identifier is j. For $J \subset I_B$, B_J denotes the set of ballots of B whose identifiers belong to J. To emphasize that ballot j actually is an envelope (super-envelope) we may use the expression *envelope j* (*super-envelope j*).

Relative to a well-defined set of ballots B: if j is an envelope in B, then $cont_B(j)$ denotes the content of j; if $x = j^1, \ldots, j^k$ is a sequence of envelope identifiers in I_B, then $cont_B(x)$ denotes the concatenation of the contents of these envelopes, that is, the string $cont_B(j^1) \cdots cont_B(j^k)$.

A *global memory* for a set of players N consists of a triple (B, R, H), where

- B is a well defined set of ballots;
- R is a sequence of strings in Σ^*, $R = R^1, R^2, \ldots$; and
- H a tuple of sequences of strings in Σ^*, $H = H_0, H_1, \ldots, H_n$.

We refer to B as the *ballot set;* to R as the *public history;* to each element of R as a *record;* to H as the *private history;* to H_0 as the *private history of the device;* and to each H_i as the *private history of player i*. The *empty global memory* is the global memory for which the ballot set, the public history, and all the private histories are all empty. We denote the set of all possible global memories by GM.

Ballot-box actions are functions from GM to GM. The subset of ballot-box actions available at a given global memory gm is denoted by $\mathcal{A}_{gm}$. The actions in $\mathcal{A}_{gm}$ are described below, grouped in 10 classes. For each $a \in \mathcal{A}_{gm}$ we provide a formal identifier; an informal reference (to facilitate the high-level description of our constructions); and a functional specification. If $gm = (B, R, H)$, we actually specify $a(gm)$ as a program acting on variables B, R, and H. For convenience, we include in R the auxiliary variable ub, the *identifier upper-bound*: a value equal to 0 for an empty global memory, and always greater than or equal to any identifier in I_B.

1. $(\textsc{NewEn}, c)$ —where $c \in \Sigma$.
   "Make a new envelope with public content c."
   $\mathsf{ub} := \mathsf{ub} + 1$; $B := B \cup \{(\mathsf{ub}, c, 0)\}$; and $R := R \circ (\textsc{NewEn}, c, \mathsf{ub})$.

2. $(\textsc{OpenEn}, j)$ —where j is an envelope identifier in I_B.
   "Publicly open envelope j to reveal content $cont_B(j)$."
   $B := B \setminus \{B_j\}$ and $R := R \circ (\textsc{OpenEn}, j, cont_B(j), \mathsf{ub})$.

3. $(\textsc{NewSup}, j_1, \ldots, j_L)$ —where $L \le 5$, and $j_1, \ldots, j_L \in I_B$ are distinct envelope identifiers.
   "Make a new super-envelope containing the envelopes $j_1, \ldots, j_L$."
   $\mathsf{ub} := \mathsf{ub} + 1$; $B := B \cup \{(\mathsf{ub}, (cont_B(j_1)), \ldots, (cont_B(j_L)), L)\}$;
   $B := B \setminus \{B_{j_1}, \ldots, B_{j_L}\}$; and $R := R \circ (\textsc{NewSup}, j_1, \ldots, j_L, \mathsf{ub})$.

4. $(\textsc{OpenSup}, j)$ —where $j \in I_B$ is the identifier of a level-L super-envelope.[8]
   "Open super-envelope j."
   letting $cont_B(j) = (c_1, \ldots, c_L)$, $B := B \cup \{(\mathsf{ub}+1, c_1, 0), \ldots, (\mathsf{ub}+L, c_L, 0)\}$;
   $B := B \setminus \{B_j\}$; $\mathsf{ub} := \mathsf{ub} + L$; and $R := R \circ (\textsc{OpenSup}, j, \mathsf{ub})$.

5. $(\textsc{PublicPermute}, j_1, \ldots, j_k, p)$ —where $k \le 5$, $j_1, \ldots j_k \in I_B$ are distinct identifiers of ballots of the same level L, and $p \in \mathrm{SYM}_k$.
   "Publicly permute $j_1, \ldots, j_k$ according to p."
   $B := B \cup \{(\mathsf{ub}+1, cont_B(j_{p(1)}), L), \ldots, (\mathsf{ub}+k, cont_B(j_{p(K)}), L)\}$; $\mathsf{ub} := \mathsf{ub}+k$; $B := B \setminus \{B_{j_1}, \ldots, B_{j_k}\}$; and $R := R \circ (\textsc{PublicPermute}, j_1, \ldots, j_k, p, \mathsf{ub})$.

6. $(\textsc{Destroy}, j)$ —where j is a ballot identifier in I_B.
   "Destroy ballot j"
   $B := B \setminus \{B_j\}$ and $R := R \circ (\textsc{Destroy}, j, \mathsf{ub})$.

7. $(\textsc{DoNothing})$.
   "Do nothing"
   $B := B$ and $R := R \circ (\textsc{DoNothing}, \mathsf{ub})$.

8. $(\textsc{BallotBox}, j_1, \ldots, j_k)$ — where $k \le 5$ and $j_1, \ldots j_k \in I_B$ are distinct identifiers of ballots of the same level L.
   "Ballotbox $j_1, \ldots, j_k$"
   $p \leftarrow rand(\mathrm{SYM}_k)$; $B := B \cup \{(\mathsf{ub}+p(1), cont_B(j_1), L), \ldots, (\mathsf{ub}+p(k), cont_B(j_k), L)\}$; $B := B \setminus \{B_{j_1}, \ldots, B_{j_k}\}$; $\mathsf{ub} := \mathsf{ub}+k$; and $R := R \circ (\textsc{BallotBox}, j_1, \ldots, j_k, \mathsf{ub})$.

9. $(\textsc{PrivRead}, i, j)$ —where $i \in [0, n]$ and j is an envelope identifier in I_B.
   "i privately reads and reseals envelope j."
   $R := R \circ (\textsc{PrivRead}, i, j, \mathsf{ub})$ and $H_i := H_i \circ cont_B(j)$.

---

[8] All the ballot-box actions involving multiple super-envelopes require as inputs and produce as outputs the ballots of the same level (see below). Thus, the level of any ballot can be deduced from the public history.

10. (SECRETPERMUTE, $i, j_1, \ldots, j_k, p$)   —where $i \in [0, n]$, $k \leq 5$, $p \in \mathrm{SYM}_k$, and $j_1, \ldots j_k \in I_B$ are distinct identifiers of ballots with the same level L. "i secretly permutes $j_1, \ldots, j_k$ (according to p)."
$B := B \cup \{(\mathrm{ub} + 1, cont_B(j_{p(1)}), L), \ldots, (\mathrm{ub} + k, cont_B(j_{p(K)}), L)\}$; $B := B \backslash \{B_{j_1}, \ldots, B_{j_k}\}$; $\mathrm{ub} := \mathrm{ub} + k$; $R := R \circ (\mathrm{SECRETPERMUTE}, i, j_1 \ldots, j_k, \mathrm{ub})$, and $H_i := H_i \circ p$.

REMARKS

- All ballot-box actions are deterministic functions, except for the actions of Nature.
- The variable ub never decreases and coincides with the maximum of all identifiers "ever in existence." Notice that we never re-use the identifier of a ballot that has left, temporarily or for ever, the table. This ensures that different ballots get different identifiers.
- Even though we could define the operations NEWSUP, PUBLICPERMUTE, BALLOTBOX, and SECRETPERMUTE to handle an arbitrary number of ballots, it is a strength of our construction that we never need to operate on more than 5 ballots at a time. We thus find it convenient to define such bounded operations to highlight the practical implementability of our construction.

Definition 1. *A global memory gm is* feasible *if there exists a sequence of global memories $gm^0, gm^1, \ldots, gm^k$, such that gm^0 is the empty global memory; $gm^k = gm$; and, for all $i \in [1, k]$, $gm^i = a^i(gm^{i-1})$ for some $a^i \in \mathcal{A}_{gm^{i-1}}$.*
If (B, R, H) is a feasible memory, we refer to R as a feasible public history.

REMARK. If $gm = (B, R, H)$ is feasible, then $\mathcal{A}_{gm}$ is easily computable from R alone (and so is ub). Indeed, what ballots are in play, which ballots are envelopes and which are super-envelopes, *et cetera*, are all deducible from R. Therefore, different feasible global memories that have the same public history also have the same set of available actions. This motivates the following definition.

Definition 2. *If R is a feasible public history, by $\mathcal{A}_R$ we denote the set of available actions for any feasible global memory with public history R.*

4   The Notion of a (Not Necessarily Verifiable) Device

Definition 3. *Let $\mathcal{D}$ be a sequence of K functions. We say that $\mathcal{D}$ is a ballot-box device (of length K) if, for all $k \in [1, K]$, public histories R and private histories H_0, $\mathcal{D}^k(R, H_0)$ specifies a single action. If a private action is specified, then it has $i = 0$.*
An execution of $\mathcal{D}$ on an initial feasible global memory (B^0, R^0, H^0) is a sequence of global memories $(B^0, R^0, H^0), \ldots, (B^K, R^K, H^K)$ such that $(B^k, R^k, H^k) = a^k(B^{k-1}, R^{k-1}, H^{k-1})$ for all $k \in [1, K]$, where $a^k = \mathcal{D}^k(R^{k-1}, H_0^{k-1})$.
If e is an execution of $\mathcal{D}$, by $B^k(e)$, $R^k(e)$, and $H^k(e)$ we denote, respectively, the ballot set, the public history, and the private history of e at round k.

By $R_{\mathcal{D}}^k(e)$ and $H_{0,\mathcal{D}}^k(e)$ we denote, respectively, the last k records of $R^k(e)$ and $H_0^k(e) \setminus H_0^0$ (i.e., "the records appended to R^0 and H_0^0 by executing $\mathcal{D}$").

The executions of $\mathcal{D}$ on initial memory gm^0 constitute a distribution,[9] which we denote by $EX_{\mathcal{D}}(gm^0)$.

REMARKS

- Note that if $\mathcal{D} = \mathcal{D}^1, \ldots, \mathcal{D}^K$ and $\mathcal{T} = \mathcal{T}^1, \ldots, \mathcal{T}^L$ are ballot-box devices, then their concatenation, that is, $\mathcal{D}^1, \ldots, \mathcal{D}^K, \mathcal{T}^1, \ldots, \mathcal{T}^L$ is a ballot-box device too.

5   The Notion of a Verifiably Secure Computer

Definition 4. *An address is a finite sequence x of distinct positive integers. An address vector x is a vector of mutually disjoint addresses, that is, $\{x_i\} \cap \{x_j\} = \phi$ whenever $i \neq j$. The identifier set of an address vector $x = (x_1, \ldots, x_k)$ is denoted by I_x and defined to be the set $\bigcup_{i=1}^{k} \{x_i\}$. If B is a set of ballots, then we define $cont_B(x)$ to be the vector $(cont_B(x_1), \ldots, cont_B(x_k))$. If i is a positive integer, then $x + i$ is the address vector whose jth component is $x_j + i$ (i.e., each element of sequence x_j is increased by i).*

As usual, an address profile is an address vector indexed by the set of players.

A computer $\mathcal{D}$ for a function f is a special ballot-box device. Executed on an initial global memory in which specific envelopes (the "input envelopes") contain an input x for f, $\mathcal{D}$ replaces such envelopes with new ones (the "output envelopes") that will contain the corresponding output $f(x)$. Of course, no property is required from $\mathcal{D}$ if the initial memory is not of the proper form.

Definition 5. *Let $f : X^a \to Y^b$ be a finite function, where $X, Y \subset \Sigma^*$; and let $x = x_1, \ldots, x_a$ be an address vector. We say that a feasible global memory $gm = (B, R, H)$ is proper for f and x if $I_x \subset I_B$ and $cont_B(x) \in X^a$.*

With modularity in mind, we actually envision that an execution of a computer $\mathcal{D}$ may be preceded and/or followed by the execution of other computers. We thus insist that $\mathcal{D}$ does not "touch" any ballots of the initial memory besides its input envelopes. This way, partial results already computed, if any, will remain intact.

Definition 6. *Let $f : X^a \to Y^b$ be a finite function, where $X, Y \subset \Sigma^*$; let x and y be two address vectors. We say that a ballot-box device $\mathcal{D}$ is a verifiably secure computer for f, with input address vector x and output address vector y, if there exist a constant sequence U and a straight-line no-input simulator SIM such that, for any execution e of $\mathcal{D}$ on an initial memory $gm^0 = (B^0, R^0, H^0)$, proper*

---

[9] Indeed, although each function $\mathcal{D}^k$ is deterministic, $\mathcal{D}^k(R)$ may return an action of nature.

for f and x and with identifier upper-bound ub_0, the following three properties hold:

    1. Correctness: $cont_{B^K(e)}(y + \mathsf{ub}) = f(cont_{B^0}(x))$.

    2. Privacy: $R_{\mathcal{D}}^K(e) = U$ *and* $H_{0,\mathcal{D}}^K(e) = SIM()$.

    3. Clean Operation: $B^K(e) = B_{\{y+\mathsf{ub}\}} \cup B^0 \setminus B_{\{x\}}$.

We refer to SIM as $\mathcal{D}$'s simulator; to $B_{\{x\}}$ as the input *envelopes; and to $B_{\{y+\mathsf{ub}\}}$ as the* output *envelopes. For short, when no confusion may arise, we refer to $\mathcal{D}$ as a* computer.

REMARKS

- *Correctness.* Semantically, Correctness states that the output envelopes will contain f evaluated on the contents of the input envelopes. Syntactically, Correctness implies that each integer of each address $y_j + \mathsf{ub}$ is the identifier of an envelope in $B^K(e)$.
- *Privacy.* By running a computer $\mathcal{D}$ for f, the only *additional* information about f's inputs or outputs gained by the players consists of $R_{\mathcal{D}}^K$, the portion of the public history generated by $\mathcal{D}$'s execution. Privacy guarantees that this additional information is constant, thus the players neither learn anything about each other inputs or outputs nor receive any residual information. At the same time, in any execution the internal information of the device is the random string that can be generated with the same odds by a straight-line no-input simulator. Thus, the device also does not learn anything about the players' inputs or outputs.
- *Clean Operation.* Clean Operation guarantees that $\mathcal{D}$
    1. Never touches an initial ballot that is not an input envelope (in fact, if a ballot is acted upon, then it is either removed from the ballot set, or receives a new identifier), and
    2. Eventually replaces all input envelopes with the output envelopes (i.e., other ballots generated by $\mathcal{D}$ are temporary, and will not exist in the final ballot set).
- *Simplicity.* Note that, since the public history generated by computer $\mathcal{D}$ is fixed, $\mathcal{D}$'s functions do not depend on public history R. Also, as we shall see, the private actions of the devices we construct depend only on at most 5 last records of H_0. Thus, one can interpret $\mathcal{D}$ as a simple automaton, that keeps in its internal memory last 5 private records, and reads the fixed string U record-by-record to find actions it has to perform.
- *Straight-line Simulators.* Our simulators are straight-line in the strictest possible sense. In essence, SIM is run independently many times, and each time outputs a random permutation in SYM_5 and its inverse. (The simulator is in fact called only after the device "privately opens a sequence of 5 envelopes" whose content is guaranteed —by construction— to be a random permutation of the integers 1 through 5.)

6   Three Elementary Ballot-Box Computers

In this section we first provide verifiably secure computers for three elementary functions. (These computers will later on be used as building blocks for constructing computers for arbitrary finite functions.) Our three elementary functions are:

1. *Permutation Inverse,* mapping a permutation $p \in \text{SYM}_5$ to p^{-1}.
2. *Permutation Product,* mapping a pair of permutations $(p, q) \in (\text{SYM}_5)^2$ to pq —i.e., the permutation of SYM_5 so defined: $pq(i) = p(q(i))$.
3. *Permutation Clone,* mapping a permutation $p \in \text{SYM}_5$ to the pair of permutations (p, p).

ENCODINGS. Note that the notion of a verifiably secure computer $\mathcal{D}$ for f applies to functions f from strings to strings. (Indeed, f's inputs and outputs must be represented as the concatenation of, respectively, the symbols contained in $\mathcal{D}$'s input and output envelopes.) Thus we need to encode the inputs and outputs of Permutation Inverse, Product and Clone as strings of symbols. This is naturally done as follows.

Definition 7. *We identify a permutation s in SYM_5 with the 5-long string $s_1 s_2 s_3 s_4 s_5$, such that $s_j = s(j)$. Relative to a well-defined set of ballots B, we say that a sequence σ of 5 envelope identifiers is an envelope encoding of a permutation if $cont_B(\sigma) \in SYM_5$.*

If σ is an envelope encoding of a permutation in SYM_5, we refer to this permutation by $\hat{\sigma}$. We consistently use lower-case Greek letters to denote envelope encodings.

DEVICE CONVENTIONS. To simplify our description of a device $\mathcal{D}$ we adopt the following conventions.

- Rather than describing $\mathcal{D}$ as a sequence of K functions that, on input a public history R and a private history H_0, output a ballot-box action feasible for any global memory with public history R, we present $\mathcal{D}$ as a list of K actions $a^1, \ldots, a^K$ (to be performed no matter what the public history may be). Should any such a^k be infeasible for a particular global memory, we interpret it as the "do nothing" action, which is always feasible.
- We describe each action a^k via its informal reference (as per Definition 3.2), using an explicit and convenient reference to the identifiers it generates. For instance, when we say "Make a new envelope x with public content c", we mean (1) "Make a new envelope with public content c" and (2) "refer to the identifier of the newly created envelope as x" —rather than $ub + 1$.
- We (often) collapse the actions of several rounds into a single *conceptual round,* providing convenient names for the ballot identifiers generated in the process. For instance, if p is a permutation in SYM_5, the conceptual round "Create an envelope encoding σ of p" stands for the following 5 actions:

Make a new envelope σ_1 with public content p_1.
Make a new envelope σ_2 with public content p_2.
Make a new envelope σ_3 with public content p_3.
Make a new envelope σ_4 with public content p_4.
Make a new envelope σ_5 with public content p_5.

6.1   A Verifiably Secure Computer for Permutation Inverse

Device $\mathcal{INV}_\sigma$

Input address: σ —an envelope encoding of a permutation in SYM$_5$.

(1) Create an envelope encoding α of the identity permutation $\mathcal{I} = 12345$.

(2) For $\ell = 1$ to 5: make a new super-envelope A_ℓ containing the pair of envelopes $(\sigma_\ell, \alpha_\ell)$.

(3) Ballotbox $A_1, \ldots, A_5$ to obtain $A'_1, \ldots, A'_5$.

(4) For $\ell = 1$ to 5: open super-envelope A'_ℓ to expose envelope pair (ν_ℓ, μ_ℓ).

(5) For $\ell = 1$ to 5: privately read and reseal ν_ℓ, and denote its content by $\hat{\nu}_\ell$. Set $\hat{\nu} = \hat{\nu}_1 \circ \cdots \circ \hat{\nu}_5$.

(6) For $\ell = 1$ to 5: make a new super-envelope B_ℓ containing the pair of envelopes (ν_ℓ, μ_ℓ).

(7) Secretly permute $B_1, \ldots, B_5$ according to $\hat{\nu}^{-1}$ to obtain $B'_1, \ldots, B'_5$.

(8) For $\ell = 1$ to 5: open super-envelope B'_ℓ to expose envelope pair (β_ℓ, ρ_ℓ). Set $\rho = \rho_1, \ldots, \rho_5$.

(9) For $\ell = 1$ to 5: open envelope β'_ℓ and denote its content by $\hat{\beta}_\ell$.

Output address: $37, 39, 41, 43, 45$.

Lemma 1. *For any 5-long address σ, $\mathcal{INV}_\sigma$ is a verifiably secure computer for permutation inverse, with input address σ and output address $37, 39, 41, 43, 45$.*

Proof. As per Definition 6, let us establish Correctness, Privacy and Clean Operation for $\mathcal{INV}_\sigma$. Consider an execution of $\mathcal{INV}_\sigma$ on any initial memory gm^0 proper for permutation inverse and σ, and let $\mathbf{ub}^0$ be the identifier upper-bound of gm^0.

CORRECTNESS. Step 1 generates 5 new identifiers (increasing $\mathbf{ub}^0$ by 5). Step 2 binds together, in the same super-envelope A_ℓ, the ℓth envelope of σ and α. It generates 5 new identifiers, and all of its actions are feasible since $\sigma \in I_B$. Step 3 applies the same, random and secret, permutation to both $\hat{\sigma}$ and $\hat{\alpha}$, generating 5 new identifiers. Letting x be this secret permutation, Step 4 "puts on the table" the envelope encodings $\nu = \nu_1, \ldots, \nu_5$ and $\mu = \mu_1, \ldots, \mu_5$, where $\hat{\nu} = x\hat{\sigma}$ and $\hat{\mu} = x\mathcal{I} = x$, and generates 10 new identifiers. At the end of Step 4, both $\hat{\nu}$ and $\hat{\mu}$ are totally secret. In Step 5, however, the device learns $\hat{\nu}$ and reseals envelope encoding ν. Step 6 puts ν and μ back into super-envelopes $B_1, \ldots, B_5$, generating 5 new identifiers. In Step 7, the device secretly applies permutation $\hat{\nu}^{-1}$ to both $\hat{\nu}$ and $\hat{\mu}$, generating 5 new identifiers. The action of

Step 7 is feasible because $\hat{\sigma} \in \mathrm{SYM}_5$, thus $\hat{\nu} \in \mathrm{SYM}_5$. Step 8 "puts on the table" the envelope encodings β and ρ, where $\hat{\beta} = \hat{\nu}^{-1}\hat{\nu} = Id$ and $\hat{\rho} = \hat{\nu}^{-1}x$, and generates 10 new identifiers. Step 9 reveals contents of β, which are $\hat{\beta} = 12345$. Thus, $\rho = \mathrm{ub}^0 + 37, \mathrm{ub}^0 + 39 \ldots, \mathrm{ub}^0 + 45$; and $\hat{\rho} = \hat{\sigma}^{-1}x^{-1}x = \hat{\sigma}^{-1}$ as desired.

PRIVACY. It is clear that the public history generated by $\mathcal{D}$ is a fixed constant. And the very fact that the contents of β revealed in Step 9 are 1, 2, $\ldots$, 5 in the fixed order serves as a proof that the device had used the correct permutation to invert the contents of $\hat{\nu}$ and $\hat{\mu}$. Constructing the required simulator is also trivial, as the contents of $H_{0,D}$ are a random permutation $\hat{\nu}$ and its inverse. Thus, SIM consists of: (1) generating a random permutation $r = r_1 \ldots r_5 \in \mathrm{SYM}_5$; (2) for $\ell = 1$ to 5: writing a string r_ℓ; and (3) writing a string r^{-1}.

CLEAN OPERATION. Trivially follows by construction.

6.2 A Verifiably Secure Computer for Permutation Product

Device $\mathcal{MULT}_{\sigma,\tau}$

Input addresses: σ and τ —each an envelope encoding of a permutation in SYM_5.

(1) Execute computer $\mathcal{INV}_\sigma$ to obtain the envelope encoding α.

(2) For $\ell = 1$ to 5: make a new super-envelope A_ℓ containing the pair of envelopes (α_ℓ, τ_ℓ).

(3) Ballotbox $A_1, \ldots, A_5$ to obtain $A'_1, \ldots, A'_5$.

(4) For $\ell = 1$ to 5: open super-envelope A'_ℓ to expose envelope pair (ν_ℓ, μ_ℓ).

(5) For $\ell = 1$ to 5: privately read and reseal ν_ℓ, and denote its content by $\hat{\nu}_\ell$. Set $\hat{\nu} = \hat{\nu}_1 \circ \cdots \circ \hat{\nu}_5$.

(6) For $\ell = 1$ to 5: make a new super-envelope B_ℓ containing the pair of envelopes (ν_ℓ, μ_ℓ).

(7) Secretly permute $B_1, \ldots, B_5$ according to $\hat{\nu}^{-1}$ to obtain $B'_1, \ldots, B'_5$.

(8) For $\ell = 1$ to 5: open super-envelope B'_ℓ to expose envelope pair (β_ℓ, ρ_ℓ). Set $\rho = \rho_1, \ldots, \rho_5$.

(9) For $\ell = 1$ to 5: open envelope β'_ℓ and denote its content by $\hat{\beta}_\ell$.

Output address: $77, 79, 81, 83, 85$.

Lemma 2. *For any two, disjoint, 5-long addresses σ and τ, $\mathcal{MULT}_{\sigma,\tau}$ is a verifiably secure computer for permutation product, with input addresses σ and τ and output address $77, 79, 81, 83, 85$.*

Proof. To establish Correctness, note that envelopes α generated in Step 1 contain $\hat{\alpha} = \hat{\sigma}^{-1}$; contents of ν and μ in Step 4 are $\hat{\nu} = x\hat{\sigma}^{-1}$ and $\hat{\mu} = x\hat{\tau}$ for a random $x \in \mathrm{SYM}_5$; and contents of β and ρ in Step 8 are $\hat{\nu}^{-1}\hat{\nu} = \mathcal{I}$ and $\hat{\rho} = (x\hat{\sigma}^{-1})^{-1}x\hat{\tau} = \hat{\sigma}\hat{\tau}$. Privacy and Clean Operation trivially follow. By construction, the public history generated by $\mathcal{MULT}_{\sigma,\tau}$ is fixed, and the SIM has to generate a random permutation and its inverse twice (for $\mathcal{INV}$ in Step 1 and for Steps 5 and 7.)

6.3   A Verifiably Secure Computer for Permutation Clone

Device $CLONE_\sigma$

Input address: σ —an envelope encoding of a permutation in SYM_5.

(1) Execute computer INV_σ to obtain the envelope encoding α.

(2) Create two envelope encodings, β and γ, of the identity permutation I.

(3) For $\ell = 1$ to 5: make a new super-envelope A_ℓ containing the triple of envelopes $(\alpha_\ell, \beta_\ell, \gamma_\ell)$.

(4) Ballotbox $A_1, \ldots, A_5$ to obtain $A'_1, \ldots, A'_5$.

(5) For $\ell = 1$ to 5: open super-envelope A'_ℓ to expose envelope triple $(\nu_\ell, \mu_\ell, \eta_\ell)$.

(6) For $\ell = 1$ to 5: privately read and reseal ν_ℓ, and denote its content by $\hat{\nu}_\ell$. Set $\hat{\nu} = \hat{\nu}_1 \circ \cdots \circ \hat{\nu}_5$.

(7) For $\ell = 1$ to 5: make a new super-envelope B_ℓ containing the pair of envelopes $(\nu_\ell, \mu_\ell, \eta_\ell)$.

(8) Secretly permute $B_1, \ldots, B_5$ according to $\hat{\nu}^{-1}$ to obtain $B'_1, \ldots, B'_5$.

(9) For $\ell = 1$ to 5: open super-envelope B'_ℓ to expose envelope triple $(\delta_\ell, \psi_\ell, \rho_\ell)$.

(10) For $\ell = 1$ to 5: open envelope δ_ℓ.[10]

Output addresses: $92, 95, 98, 101, 104$ and $93, 96, 99, 102, 105$.

Lemma 3. *For any 5-long address σ, $CLONE_\sigma$ is a verifiably secure computer for permutation clone, with input address σ and output addresses $92, 95, 98, 101, 104$ and $93, 96, 99, 102, 105$.*

7   General Verifiably Secure Computers

Recall that any finite function $f : \{0,1\}^a \to \{0,1\}^b$ can be easily (and quite efficiently) represented as a combinatorial circuit, and thus as a fixed sequence of the following basic functions:

- $COIN$, the probabilistic function that, on no input, returns a random bit;
- $DUPLICATE$, the function that, on input a bit b, returns a pair of bits (b, b);
- AND, the function that, on input a pair of bits (b_1, b_2), returns 1 if and only if $b_1 = b_2 = 1$; and
- NOT, the function that, on input a bit b, returns the bit $1 - b$.

We thus intend to prove that each of these basic functions has a ballot-box computer, and then obtain a ballot-box computer for any desired $f : \{0,1\}^a \to \{0,1\}^b$ by utilizing these 4 basic computers. To this end, we must first decide how to encode binary strings and binary functions.

---

[10] Note that Steps 2–10, in essence, correspond to a protocol for permutation inverse that on input α produces two identical envelope encodings, each encoding $\hat{\alpha}^{-1}$.

Definition 8. *We define the SYM_5 encoding of a k-bit binary string $x = b_1, \ldots,$ b_k, denoted by $\bar{x}$, to be $\bar{b}_1 \cdots \bar{b}_k$, where*

$$\bar{0} = 12345; \quad \bar{1} = 12453.$$

The SYM_5 encoding is immediately extended to binary functions as follows: $\bar{f}(\bar{x}) = \overline{f(x)}$ for all x.

One of our basic computer has already been constructed: namely,

Lemma 4. *For any envelope encoding σ, $\mathcal{CLONE}$ is a ballot-box computer for DUPLICATE with input address σ.*

Proof. Because $\mathcal{CLONE}$ duplicates any permutation in SYM_5, in particular it duplicates 12345 and 12453.

We thus proceed to build the other 3 basic computers

7.1 A Verifiably Secure Computer for $\overline{COIN}$

Device $\mathcal{COIN}$

(1) Create an envelope encoding α of $\mathcal{I}$ and an envelope encoding β of a.

(2) Make new super-envelopes A and B containing envelopes $\alpha_1, \ldots, \alpha_5$ and $\beta_1, \ldots, \beta_5$, respectively.

(3) Ballotbox A and B to obtain super-envelopes C and D.

(4) Open C to expose an envelope encoding γ. Destroy D.

Output address: $15, 16, 17, 18, 19$.

Lemma 5. *$\mathcal{COIN}$ is a verifiably secure computer for $\overline{COIN}$, with no input address and output address $15, \ldots, 19$.*

Proof. The only non-trivial part to prove Correctness is to demonstrate that contents of γ are random and belong to $\{\mathcal{I}, a\}$. Indeed, at the end of Step 2, A contains a sequence of 5 envelopes encoding $\mathcal{I}$, and B contains a sequence of 5 envelopes encoding permutation a. At the end of Step 3, the contents of C are either those of A or of B with equal probabilities. Thus, at the end of Step 4, the content of address γ is random and is either $\mathcal{I}$ or a.

Clean Operation is trivial, and Privacy straightforwardly follows by noting that the public history is fixed and there is no private history of the device generated by $\mathcal{COIN}$.

7.2 Verifiably Secure Computers for $\overline{NOT}$ and $\overline{AND}$

In proving the existence of ballot-box computers for $\overline{NOT}$ and $\overline{AND}$ we rely on the result of [3] that the Boolean functions NOT and AND can be realized as sequences of group operations in SYM_5.[11] Here is our rendition of it.

Let $\mathcal{I} = 12345, a = 12453, b = 25341, c_1 = 34125, c_2 = 12354,$ and $c_3 = 42153$; and let $\tilde{x}, x'$ and x^* be the operators defined to act on a permutation $x \in \text{SYM}_5$ as follows:

$$\tilde{x} = c_1^{-1}xc_1, \quad x' = c_2^{-1}xc_2, \quad \text{and} \quad x^* = c_3^{-1}xc_3.$$

Then, recalling that $\bar{0} = \mathcal{I}$ and $\bar{1} = a$, the following lemma can be verified by direct inspection.

Barrington's Lemma. *If $x_1 = \overline{b_1}$ and $x_2 = \overline{b_2}$, where b_1 and b_2 are bits, then*

$$\overline{\neg b_1} = (x_1 a^{-1})^*, \quad \text{and} \quad \overline{b_1 \wedge b_2} = (x_1 \tilde{x}_2 x_1^{-1} \tilde{x}_2^{-1})'.$$

Lemma 6. *There exist ballot-box computers $\mathcal{NOT}$ and $\mathcal{AND}$ for, respectively, $\overline{NOT}$ and $\overline{AND}$.*

Proof. The lemma follows by combining Barrington's lemma and our Lemmas 1,2 and 3. That is, each of $\mathcal{NOT}$ and $\mathcal{AND}$ is obtained in four steps. First, by expanding the operators of the formulas of Lemma 6 so as to show all relevant constants a, c_1, c_2 and c_3. Second, by generating envelope encodings for each occurrence of each constant. Third, in the case of $\mathcal{AND}$, by using our elementary computer $\mathcal{CLONE}$ so as to duplicate x_1 and x_2. Forth, by replacing each occurrence of permutation inverse and permutation product in the formulas of Lemma 6 with, respectively, our elementary computers $\mathcal{INV}$ and $\mathcal{MULT}$. Accordingly, the simulators for $\mathcal{NOT}$ and $\mathcal{AND}$ can be obtained by running the simulators of their individual elementary computers in the proper order.

7.3 Verifiably Secure Computers for Arbitrary Finite Functions

Theorem 1. *Every finite function has a verifiably secure computer.*

Proof. Let $f : \{0,1\}^a \to \{0,1\}^b$ be a finite function. Then (by properly ordering the "gates" of a combinatorial circuit for f) there exists a fixed sequence $C_f = F_1, F_2, \ldots, F_K$ such that:

---

[11] Note that neither $DUPLICATE$ nor $COIN$ can be realized in SYM_5, and thus one cannot compute arbitrary functions in SYM_5. Indeed, the result of [3] was solely concerned with implementing a restricted class of finite functions called NC^1. At high level, we bypass this limitation by (1) representing permutations in SYM_5 as sequences of 5 envelopes and (2) using these *physical* representations and our ballot-box operations for implementing $DUPLICATE$ and $COIN$ (in addition to NOT and AND). That is, rather than viewing a permutation in SYM_5 as a single, 5-symbol string, we view it a sequence of 5 distinct symbols, and put each one of them into its own envelope, which can then be manipulated separately by our ballot-box operations. Such "segregation" of permutations of SYM_5 into separate envelopes is crucial to our ability of performing general computation, and in a private way too.

- Each F_i is either $COIN$, $DUPLICATE$, NOT or AND;
- Each input bit of F_i is either one of the original input bits or one of the output bits of F_j for $j < i$; and
- For each a-bit input x, the b-bit output $f(x)$ can be computed by evaluating (in order) all functions F_i on their proper inputs, and then concatenating (in order) all their output bits not used as inputs by some F_j. (Such bits are guaranteed to be exactly b in total.)

Define now $\mathcal{D}_i$ as follows:

$\mathcal{D}_i = \mathcal{COIN}$ if $F_i = COIN$;
$\mathcal{D}_i = \mathcal{CLONE}$ if $F_i = DUPLICATE$;
$\mathcal{D}_i = \mathcal{NOT}$ if $F_i = NOT$;
$\mathcal{D}_i = \mathcal{AND}$ if $F_i = AND$.

Let $\mathcal{D}$ be the concatenation of $\mathcal{D}_1, \ldots, \mathcal{D}_K$, with appropriately chosen input addresses, so that the lth input address of $\mathcal{D}_i$ matches the mth output address of $\mathcal{D}_j$ whenever the lth input bit of F_i is the mth output bit of F_j. Then, $\mathcal{D}$ is a ballot-box computer for f. In fact, $\mathcal{D}$'s correctness follows from the correctness of each computer $\mathcal{D}_i$. Furthermore, $\mathcal{D}$'s privacy follows from the fact that each $\mathcal{D}_i$ has a simulator SIM_i, and thus a simulator SIM for $\mathcal{D}$ can be obtained by executing (in order) $SIM_1, SIM_2, \ldots, SIM_K$. Finally, the clean operation of $\mathcal{D}$ follows from the clean operation of each $\mathcal{D}_i$.

REMARKS

- Note that our ballot-box computer $\mathcal{D}_f$ is "as efficient as" f itself. Indeed, the description of $\mathcal{D}_f$ is linear in the description of C_f. This is so because, letting $C_f = F_1, F_2, \cdots$, $\mathcal{D}_f$ replaces each F_i with a ballot-box computer for $\overline{F_i}$ that has constant number of actions and generates a constant number of identifiers. Moreover, assuming each function F_i is executable in constant time (since it operates on at most two bits) and assuming that each ballot-box action is executable in constant time (since it operates on at most 5 ballots), the time needed to run $\mathcal{D}_f$ is also linear in the time needed to run C_f.
- If $\mathcal{D}$ is executed on an initial global memory whose ballot set coincides with just the input envelopes, then the ballot set of $\mathcal{D}$'s final global memory of coincides with just the output envelopes.

8   The Input Stage (and the Output Stage)

Having described the concrete computation stage of ILM2 security, we just need to describe its input and output stages. The output stage is trivial implemented. In essence, the device publicly opens the sequence of envelopes containing the public output y, and hands over to each player i the sequence of envelopes containing y_i. For the latter, one must formally enrich the ballot-box model with the basic public action "Hand over envelope j to player i." We omit to do so formally in this extended abstract.[12]

In this section we thus limit ourselves to defining and implementing the input stage.

8.1   Verifiably Secure Input Committers

Intuitively, a verifiably secure input committer $\mathcal{IC}$ is a protocol that enables each player i to give to the verifiably secure device $\mathcal{D}$ a separate sequence of envelopes S_i whose contents encode i's chosen input m_i so that: (1) in any execution of $\mathcal{IC}$ the contents of the envelopes in S_i properly encode m_i; and (2) only player i knows m_i, while the other players and the device learn only a fixed public history.

Since in the input stage the players must interact with the device, we start with formalizing the notion of a protocol (restricted to our needs) and then the notion of a committer. Afterwards, we proceed to construct a committer.

Definition 9. *A (tight) protocol $\mathcal{P}$ is a triple (K, PS, AF), where*

- *K, the* length *of the protocol, is a positive integer;*
- *PS, the* player sequence, *is a sequence of K integers each from 0 to n; and*
- *AF, the* action function, *is a mapping from $K \times \mathcal{R}$ —where $\mathcal{R}$ is the set of all feasible public histories— to sets of ballot-box actions such that, for all $k \in [1, K]$ and $R \in \mathcal{R}$, $AF(k, R)$ specifies for player $i = PS^k$ either a single action or a pair of actions as follows:*

  * *a single action a^k if $i = 0$;*
  * *a pair of SECRETPERMUTE actions $\{a_0^k, a_1^k\}$, where a_0^k and a_1^k permute the same ballots $j_0, j_1 \in I_B$ with, respectively, permutations 12 and 21.[13]*

If $a = a_b^k$, we refer to b as a's *hidden bit* and refer to a as the *0-action* (of the pair) if $b = 0$, and as the *1-action* otherwise.

For a given $\mathcal{P}$, let ℓ_i be the number of times player i is asked to act by $\mathcal{P}$, that is $\ell_i = \#\{k \in [1, K] : PS^k = i\}$. Let $z_{\mathcal{P}}$ be a profile of strings, such that z_i is an ℓ_i-bit string.

---

[12] If we want to capture additional security desiderata, such as deniability, then we should ensure that the players privately read and destroy the envelopes they finally receive. To this effect, it may be advantageous to enrich the ballot-box model with the operation "flash the content to envelope j to just player i." (In essence this is the operation corresponding to raising a card facing down just in the direction of player i.) The device will then destroy the envelope flushed to i. Again, the destroy action is not essential, and can be simulated by a verifiable device by means of a fixed sequence of the other ballot-box actions. Additional desiderata can also demand the "simultaneous execution" of some of the actions.

[13] Note that the set of the current ballot identifiers I_B can be fully obtained from R. For more general protocols, the actions of player i are allowed to depend on H_i too.

An *execution of* $\mathcal{P}$ with *associated profile* $z_{\mathcal{P}}$ on an initial feasible global memory (B^0, R^0, H^0) is a sequence of global memories $(B^0, R^0, H^0), \ldots, (B^K, R^K, H^K)$ such that $(B^k, R^k, H^k) = a^k(B^{k-1}, R^{k-1}, H^{k-1})$ for all $k \in [1, K]$, where:

- $a^k = AF(k, PS, gm^{k-1})$, if $PS^k = 0$; and
- $a^k = a_b^k \in AF(k, PS, gm^{k-1})$, where b is the mth bit of z_i, where $m = \#\{j \in [1, k] : PS^k = i\}$, if $PS^k = i$.

Since the execution of $\mathcal{P}$ in all respects is similar to the execution of a device, we will retain all the notation defined for executions of devices.

Definition 10. *Let* $\mathcal{IC}$ *be a ballot-box protocol,* $(\bar{0}, \bar{1})$ *a bit-by-bit encoding, and* x *an address profile,* $x = x_1, \ldots, x_n$. *We say that* $\mathcal{IC}$ *is a L-bit* verifiably secure input committer *for* $(\bar{0}, \bar{1})$ *with output address profile* x *if there exists a unique sequence* U *such that, for every execution* e *of* $\mathcal{IC}$ *with associated profile* z, *whose initial global memory is empty, the following three properties hold:*

1. Correctness: $\forall i \in [1, n], \ H_i^K(e) = z_i \in \{0, 1\}^L$ *and* $cont_{B^K(e)}(x_i) = \overline{z_i}$.

2. Privacy: $R^K(e) = U$ *and* $H_0^K(e) = \emptyset$.

3. Clean Termination: $I_{B^K(e)} = I_x$.

REMARKS

- *Correctness.* Correctness requires that, once $\mathcal{IC}$ is executed, the private history string of each player is of length L, and that the ballots corresponding to the output address profile $x_1, \ldots, x_n$ contain the bit-by-bit encoding of the players' intended inputs. This requirement has both syntactic and semantic implications. Syntactically, Correctness implies that each x_i is of length $5L$ and each element of x_i is the identifier of an envelope in $B^K(e)$. Further, it requires that the number of times each player acts in $\mathcal{IC}$ is equal to $5L$ and that the length of his private history is also $5L$. Semantically, Correctness implies that the envelopes of address x_i encode a message z_i freely chosen by player i alone. (I.e., the other players have no control over the value of z_i.) This is so because the envelopes in x_i are guaranteed to contain the bit-by-bit encoding of the final private record of player i. Recall that i's private history, H_i, grows, by a bit at a time, in only one case: when i himself secretly chooses one of two complementary actions (in our physical interpretation, when i temporarily holds two envelopes behind his back, and then returns them in the order he chooses). This is an "atomic" (in the sense of "indivisible") choice of i, and therefore the other players have no control over it.
- *Privacy.* Privacy requires that, at each round k of $\mathcal{IC}$, the public information available to the acting player always consists of the fixed sequence U^{k-1}, no matter what are the intended inputs. Thus, while Correctness implies that any control over the choice of a player's message solely rests with that player, Privacy implies that all messages are independently chosen, as demanded in the ideal normal-form mechanism, and totally secret at the end of the committer's execution.

Privacy thus rules out any possibility of signaling among players during the execution of a committer.

Note that the information available to a player i consists of both his private history H_i and the public history R. In principle, therefore, the privacy condition should guarantee that no information about other players' strategies is deducible from H_i and R *jointly*. As argued above, however, H_i^K depends on i's strategy alone. Thus, formulating Privacy in terms of the public history alone is sufficient.

- *Clean Termination.* Clean termination ensures that only the envelopes containing the desired encoding of the players' private messages remain on the table.

8.2   Our Verifiably Secure Committer

Protocol $Commit_L$

For player $i = 1$ to n DO: For $t = 1$ to L and bit b_{it} Do:

(1) Create an envelope encoding $\alpha^{(i,t)}$ of permutation $\mathcal{I} = 12345$.

(2) Create an envelope encoding $\beta^{(i,t)}$ of permutation $a = 12453$.

(3) Make a new super-envelope $A^{(i,t)}$ containing envelopes $\alpha_1^{(i,t)}, \ldots, \alpha_5^{(i,t)}$.

(4) Make a new super envelope $B^{(i,t)}$ containing envelopes $\beta_1^{(i,t)}, \ldots, \beta_5^{(i,t)}$.

(5) Player i secretly permutes $A^{(i,t)}$ and $B^{(i,t)}$ according to 12, if $b_{it} = 0$, and to 21, otherwise, to obtain the super-envelopes $C^{(i,t)}$ and $D^{(i,t)}$.

(6) Open $C^{(i,t)}$ to expose envelopes $\gamma_1^{(i,t)}, \ldots, \gamma_5^{(i,t)}$. Set $\gamma^{(i,t)} = \gamma_1^{(i,t)}, \ldots, \gamma_5^{(i,t)}$.

(7) Destroy $D^{(i,t)}$.

Output Addresses: For each $i \in N$, the sequence $x_i = \gamma^{(i,1)}, \ldots, \gamma^{(i,L)}$.

Lemma 7. *Protocol $Commit_L$ is an L-bit verifiably secure committer for the SYM_5 encoding.*

PROOF. At the end of Step 3, $A^{(i,t)}$ contains a sequence of 5 envelopes encoding the identity permutation $\mathcal{I}$, and, at the end of Step 4, $B^{(i,t)}$ contains a sequence of 5 envelopes encoding permutation a. Thus, recalling that in the SYM_5 encoding $\mathcal{I} = \bar{0}$ and $a = \bar{1}$, at the end of Step 5, $C^{(i,t)}$ contains an envelope encoding of $\bar{b}$ if player i "chooses the bit b". Thus, at the end of Step 6, the content of address x_i is z_i, where H_i is a binary string of length L, as demanded by Definition 10. All other properties are trivially established.

References

1. Aumann, R.: Subjectivity and correlation in randomized strategies. J. Math. Econ. 1, 67–96 (1974)
2. Bárány, I.: Fair distribution protocols or how the players replace fortune. Mathematics of Operation Research 17, 327–341 (1992)

3. Barrington, D.: Bounded-width polynomial-size branching programs recognize exactly those languages in NC1. In: Proceedings of STOC (1986)
4. Ben-Or, M., Goldwasser, S., Wigderson, A.: Completeness Theorems for Non-Cryptographic Fault-Tolerant Distributed Computation. In: Proceedings of STOC (1988)
5. Ben-Porath, E.: Correlation without mediation: Expanding the set of equilibria outcomes by "cheap" pre-play procedures. Journal of Economic Theory 80, 108–122 (1998)
6. Canetti, R.: Universally composable security. A new paradaigm for cryptographic protocols. In: Proceedings of FOCS (2001)
7. Cleve, R.: Limits on the Security of Coin Flips When Half the Processors are Faulty. In: Proceedings of STOC (1986)
8. Dodis, Y., Halevi, S., Rabin, T.: A Cryptographic Solution to a Game Theoretic Problem. In: Bellare, M. (ed.) CRYPTO 2000. LNCS, vol. 1880, Springer, Heidelberg (2000)
9. Dodis, Y., Micali, S.: Parallel Reducibility for Information-Theoretically Secure Computation. In: Bellare, M. (ed.) CRYPTO 2000. LNCS, vol. 1880, Springer, Heidelberg (2000)
10. Gerardi, D.: Unmediated communication in games with complete and incomplete information. Journal of Economic Theory 114, 104–131 (2004)
11. Goldreich, O., Micali, S., Wigderson, A.: How to play any mental game. In: Proceedings of STOC (1987)
12. Halpern, J., Teague, V.: Rational secret sharing and multiparty computation. In: Proceedings of STOC (2004)
13. Izmalkov, S., Lepinski, M., Micali, S.: Rational secure function evaluation and ideal mechanism design. In: Proceedings of FOCS (2005)
14. Kushilevitz, E., Lindell, Y., Rabin, T.: Information-Theoretically Secure Protocols and Security under Composition. In: Proceedings of STOC (2006)
15. Lepinksi, M., Micali, S., Peikert, C., Shelat, A.: Completely fair SFE and coalition-safe cheap talk. In: Proceedings of PODC (2004)
16. Urbano, A., Vila, J.E.: Computational complexity and communication: Coordination in two-player games. Econometrica 70(5), 1893–1927 (2002)
17. Yao, A.: A proof of Yao's protocol for secure two-party computation (2004) (Never published. The result is presented in Lindell and Pinkas), http://www.cs.biu.ac.il/ lindell/abstracts/yao_abs.html

Lower Bounds on Implementing Robust and Resilient Mediators

Ittai Abraham[1], Danny Dolev[2,*], and Joseph Y. Halpern[3,**]

[1] Hebrew University
ittaia@cs.huji.ac.il
[2] Hebrew University
dolev@cs.huji.ac.il
[3] Cornell University
halpern@cs.cornell.edu

Abstract. We provide new and tight lower bounds on the ability of players to implement equilibria using cheap talk, that is, just allowing communication among the players. One of our main results is that, in general, it is impossible to implement three-player Nash equilibria in a bounded number of rounds. We also give the first rigorous connection between Byzantine agreement lower bounds and lower bounds on implementation. To this end we consider a number of variants of Byzantine agreement and introduce reduction arguments. We also give lower bounds on the running time of two player implementations. All our results extended to lower bounds on (k, t)-*robust* equilibria, a solution concept that tolerates deviations by coalitions of size up to k and deviations by up to t players with unknown utilities (who may be malicious).

1 Introduction

The question of whether a problem in a multiagent system that can be solved with a trusted mediator can be solved by just the agents in the system, without the mediator, has attracted a great deal of attention in both computer science (particularly in the cryptography community) and game theory. In cryptography, the focus on the problem has been on *secure multiparty computation*. Here it is assumed that each agent i has some private information x_i. Fix functions $f_1, \ldots, f_n$. The goal is have agent i learn $f_i(x_1, \ldots, x_n)$ without learning anything about x_j for $j \neq i$ beyond what is revealed by the value of $f_i(x_1, \ldots, x_n)$. With a trusted mediator, this is trivial: each agent i just gives the mediator its private value x_i; the mediator then sends each agent i the value $f_i(x_1, \ldots, x_n)$. Work on

* Part of the work was done while the author visited Cornell university. The work was funded in part by ISF, ISOC, NSF, CCR, and AFOSR.

** Supported in part by NSF under grants CCR-0208535, ITR-0325453, and IIS-0534064, by ONR under grant N00014-01-10-511, by the DoD Multidisciplinary University Research Initiative (MURI) program administered by the ONR under grants N00014-01-1-0795 and N00014-04-1-0725, and by AFOSR under grant FA9550-05-1-0055.

© International Association for Cryptologic Research 2008

multiparty computation (see [18] for a survey) provides conditions under which this can be done. In game theory, the focus has been on whether an equilibrium in a game with a mediator can be implemented using what is called *cheap talk*— that is, just by players communicating among themselves (see [28] for a survey).

There is a great deal of overlap between the problems studied in computer science and game theory. But there are some significant differences. Perhaps the most significant difference is that, in the computer science literature, the interest has been in doing multiparty computation in the presence of possibly malicious adversaries, who do everything they can to subvert the computation. On the other hand, in the game theory literature, the assumption is that players have preference and seek to maximize their utility; thus, they will subvert the computation iff it is in their best interests to do so. Following [1], we consider here both rational adversaries, who try to maximize their utility, and possibly malicious adversaries (who can also be considered rational adversaries whose utilities we do not understand).

1.1 Our Results

In this paper we provide new and optimal lower bounds on the ability to implement mediators with cheap talk. Recall that a *Nash equilibrium* σ is a tuple of strategies such that given that all other players play their corresponding part of σ then the best response is also to play σ. Given a Nash equilibrium σ we say that a strategy profile ρ is a k-*punishment strategy for* σ if, when all but k players play their component of ρ, then no matter what the remaining k players do, their payoff is strictly less than what it is with σ. We now describe some highlights of our results in the two simplest settings: (1) where rational players cannot form coalitions and there are no malicious players (this gives us the solution concept of Nash equilibrium) and (2) where there is at most one malicious player. We describe our results in a more general setting in Section 1.2.

No bounded implementations: In [1] it was shown that any Nash equilibrium with a mediator for three-player games with a 1-punishment strategy can be implemented using cheap talk. The expected running time of the implementation is constant. It is natural to ask if implementations with a bounded number of rounds exist for all three-player games. Theorem 2 shows this is not the case, implementations must have infinite executions and cannot be bounded for all three-player games. This lower bound highlights the importance of using randomization. An earlier attempt to provide a three-player cheap talk implementation [8] uses a bounded implementation, and hence cannot work in general. The key insight of the lower bound is that when the implementation is bounded, then at some point the punishment strategy must become ineffective. The details turn out to be quite subtle. The only other lower bound that we are aware of that has the same flavor is the celebrated FLP result [15] for reaching agreement in asynchronous systems, which also shows that no bounded implementation exists. However, we use quite different proof techniques than FLP.

Byzantine Agreement and Game Theory: We give the first rigorous connection between Byzantine agreement lower bounds and lower bounds on implementation. To get the lower bounds, we need to consider a number of variants of Byzantine agreement, some novel. The novel variants require new impossibility results. We have four results of this flavor:

1. Barany [6] gives an example to show that, in general, to implement an equilibrium with a mediator in a three-player game, it is necessary to have a 1-punishment strategy. Using the power of randomized Byzantine agreement lower bounds we strengthen his result and show in Theorem 4 that we cannot even get an ϵ-implementation in this setting.
2. Using the techniques of [7] or [17], it is easy to show that any four-player game Nash equilibrium with a mediator can be implemented using cheap talk even if no 1-punishment strategy exists. Moreover, these implementations are *universal*; they do not depend on the players' utilities. In Theorem 3 we prove that universal implementations do not exist in general for three-player games. Our proof uses a nontrivial reduction to the weak Byzantine agreement (WBA) problem [24]. To obtain our lower bound, we need to prove a new impossibility result for WBA, namely, that no protocol with a finite expected running time can solve WBA.
3. In [1] we show that for six-player games with a 2-punishment strategy, any Nash equilibrium can be implemented even in the presence of at most one malicious player. In Theorem 5 we show that for five players even ϵ-implementation is impossible. The proof uses a variant of Byzantine agreement; this is related to the problem of *broadcast with extended consistency* introduced by Fitzi et al. [16]. Our reduction maps the rational player to a Byzantine process that is afraid of being detected and the malicious player to a standard Byzantine process.
4. In Theorem 8, we show that for four-player games with at most one malicious player, to implement the mediator, we must have a PKI setup in place, even if the players are all computationally bounded and even if we are willing to settle for ϵ-implementations. Our lower bound is based on a reduction to a novel relaxation of the Byzantine agreement problem.

Bounds on running time: We provide bounds on the number of rounds needed to implement two-player games. In Theorem 9(a) we prove that the expected running time of any implementation of a two-player mediator equilibrium must depend on the utilities of the game, even if there is a 1-punishment strategy. This is in contrast to the three-player case, where the expected running time is constant. In Theorem 9(b) we prove that the expected running time of any ϵ-implementation of a two-player mediator equilibrium for which there is no 1-punishment strategy must depend on ϵ. Both results are obtained using a new two-player variant of the secret-sharing game. The only result that we are aware of that has a similar spirit is that of Boneh and Naor [9], where it is shown that two-party protocols with "bounded unfairness" of ϵ must have running time that depends on the value of ϵ. The implementations given by Urbano and Vila [31,32] in the two-player case are independent of the utilities; the above results show that their implementation cannot be correct in general.

1.2 Our Results for Implementing Robust and Resistant Mediators

In [1] (ADGH from now on), we argued that it is important to consider deviations by both rational players, who have preferences and try to maximize them, and players that can be viewed as malicious, although it is perhaps better to think of them as rational players whose utilities are not known by the other players or mechanism designer. We considered equilibria that are (k, t)-*robust*; roughly speaking, this means that the equilibrium tolerates deviations by up to k rational players, whose utilities are presumed known, and up to t players with unknown utilities (i.e., possibly malicious players). We showed how (k, t)-robust equilibria with mediators could be implemented using cheap talk, by first showing that, under appropriate assumptions, we could implement secret sharing in a (k, t)-robust way using cheap talk. These assumptions involve standard considerations in the game theory and distributed systems literature, specifically, (a) the relationship between k, t and n, the total number of players in the system; (b) whether players know the exact utilities of other players; (c) whether there are broadcast channels or just point-to-point channels; (d) whether cryptography is available; and (e) whether the game has a $(k + t)$-*punishment strategy*; that is, a strategy that, if used by all but at most $k + t$ players, guarantees that every player gets a worse outcome than they do with the equilibrium strategy. Here we provide a complete picture of when implementation is possible, providing lower bounds that match the known upper bounds (or improvements of them that we have obtained). The following is a high-level picture of the results. (The results discussed in Section 1.1 are special cases of the results stated below. Note that all the upper bounds mentioned here are either in ADGH, slight improvements of results in ADGH, or are known in the literature; see Section 3 for the details. The new results claimed in the current submission are the matching lower bounds.)

- If $n > 3k + 3t$, then mediators can be implemented using cheap talk; no punishment strategy is required, no knowledge of other agents' utilities is required, and the cheap-talk strategy has bounded running time that does not depend on the utilities (Theorem 1(a) in Section 3).
- If $n \leq 3k + 3t$, then we cannot, in general, implement a mediator using cheap talk without knowledge of other agents' utilities (Theorem 3). Moreover, even if other agents' utilities are known, we cannot, in general, implement a mediator without having a punishment strategy (Theorem 4) nor with bounded running time (Theorem 2).
- If $n > 2k + 3t$, then mediators can be implemented using cheap talk if there is a punishment strategy (and utilities are known) in finite expected running time that does not depend on the utilities (Theorem 1(b) in Section 3).
- If $n \leq 2k + 3t$, then we cannot, in general, ϵ-implement a mediator using cheap talk, even if there is a punishment strategy and utilities are known (Theorem 5).

- If $n > 2k + 2t$ and we can simulate broadcast then, for all ϵ, we can ϵ-implement a mediator using cheap talk, with bounded expected running time that does not depend on the utilities in the game or on ϵ (Theorem 1(c) in Section 3). (Intuitively, an ϵ-implementation is an implementation where a player can gain at most ϵ by deviating.)
- If $n \leq 2k+2t$, we cannot, in general, ϵ-implement a mediator using cheap talk even if we have broadcast channels (Theorem 7). Moreover, even if we assume cryptography and broadcast channels, we cannot, in general, ϵ-implement a mediator using cheap talk with expected running time that does not depend on ϵ (Theorem 9(b)); even if there is a punishment strategy, then we still cannot, in general, ϵ-implement a mediator using cheap talk with expected running time independent of the utilities in the game (Theorem 9(a)).
- If $n > k + 3t$ then, assuming cryptography, we can ϵ-implement a mediator using cheap talk; moreover, if there is a punishment strategy, the expected running time does not depend on ϵ (Theorem 1(e) in Section 3).
- If $n \leq k + 3t$, then even assuming cryptography, we cannot, in general, ϵ-implement a mediator using cheap talk (Theorem 8).
- If $n > k + t$, then assuming cryptography and that a PKI (Public Key Infrastructure) is in place, [1] we can ϵ-implement a mediator (Theorem 1(d) in Section 3); moreover, if there is a punishment strategy, the expected running time does not depend on ϵ (Theorem 1(e) in Section 3).

The lower bounds are existential results; they show that if certain conditions do not hold, then there exists an equilibrium that can be implemented by a mediator that cannot be implemented using cheap talk. There are other games where these conditions do not hold but we can nevertheless implement a mediator.

1.3   Related Work

There has been a great deal of work on implementing mediators, both in computer science and game theory. The results above generalize a number of results that appear in the literature. We briefly discuss the most relevant work on implementing mediators here. Other work related to this paper is discussed where it is relevant.

In game theory, the study of implementing mediators using cheap talk goes back to Crawford and Sobel [11]. Barany [6] shows that if $n \geq 4$, $k = 1$, and $t = 0$ (i.e., the setting for Nash equilibrium), a mediator can be implemented in a game where players do not have private information. Forges [17] provides what she calls a *universal mechanism* for implementing mediators; essentially, when combining her results with those of Barany, we get the special case of Theorem 1(a) where $k = 1$ and $t = 0$. Ben-Porath [8] considers implementing a mediator with cheap talk in the case that $k = 1$ if $n \geq 3$ and there is a 1-punishment strategy. He seems to have been the first to consider punishment strategies (although his notion is different from ours: he requires that there be an equilibrium that

---

[1] We can replace the assumption of a PKI here and elsewhere by the assumption that there is a trusted preprocessing phase where players may broadcast.

is dominated by the equilibrium that we are trying to implement). Heller [22] extends Ben-Porath's result to allow arbitrary k. Theorem 1(b) generalizes Ben-Porath and Heller's results. Although Theorem 1(b) shows that the statement of Ben-Porath's result is correct, Ben-Porath's implementation takes a bounded number of rounds; Theorem 2 shows it cannot be correct. [2] Heller proves a matching lower bound; Theorem 5 generalizes Heller's lower bound to the case that $t > 0$. (This turns out to require a much more complicated game than that considered by Heller.) Urbano and Vila [31,32] use cryptography to deal with the case that $n = 2$ and $k = 1$; [3] Theorem 1(e)) generalizes their result to arbitrary k and t. However, just as with Ben-Porath, Urbano and Vila's implementation takes a bounded number of rounds; As we said in Section 1.1, Theorem 9(a) shows that it cannot be correct.

In the cryptography community, results on implementing mediators go back to 1982 (although this terminology was not used), in the context of *(secure) multiparty computation*. Since there are no utilities in this problem, the focus has been on essentially what we call here *t-immunity*: no group of t players can prevent the remaining players from learning the function value, nor can they learn the other players' private values. Results of Yao [33] can be viewed as showing that if $n = 2$ and appropriate computational hardness assumptions are made, then, for all ϵ, we can obtain 1-immunity with probability greater than $1 - \epsilon$ if appropriate computational hardness assumptions hold. Goldreich, Micali, and Wigderson [19] extend Yao's result to the case that $t > 0$ and $n > t$. Ben-Or, Goldwasser, and Wigderson [7] and Chaum, Crépeau, and Damgard [10] show that, without computational hardness assumptions, we can get t-immunity if $n > 3t$; moreover, the protocol of Ben-Or, Goldwasser, and Wigderson does not need an ϵ "error" term. Although they did not consider utilities, their protocol actually gives a (k,t)-robust implementation of a mediator using cheap talk if $n > 3k + 3t$; that is, they essentially prove Theorem 1(a). (Thus, although these results predate those of Barany and Forges, they are actually stronger.) Rabin and Ben-Or [29] provide a t-immune implementation of a mediator with "error" ϵ if broadcast can be simulated. Again, when we add utilities, their protocol actually gives an $\epsilon-(k,t)$-robust implementation. Thus, they essentially prove Theorem 1(c). Dodis, Halevi, and Rabin [12] seem to have been the first to apply cryptographic techniques to game-theoretic solution concepts; they consider the case that $n = 2$ and $k = 1$ and there is no private information (in which case the equilibrium in the mediator game is a *correlated equilibrium* [5]); their result is essentially that of Urbano and Vila [32] (although their protocol does not suffer form the problems of that of Urbano and Vila).

Halpern and Teague [21] were perhaps the first to consider the general problem of multiparty computation with rational players. In this setting, they essentially prove Theorem 1(d) for the case that $t = 0$ and $n \geq 3$. However, their focus is on

---

[2] Although Heller's implementation does not take a bounded number of rounds, it suffers from problems similar to those of Ben-Porath.

[3] However, they make somewhat vague and nonstandard assumptions about the cryptographic tools they use.

the solution concept of *iterated deletion*. They show that there is no Nash equilibrium for rational multiparty computation with rational agents that survives iterated deletion and give a protocol with finite expected running time that does survive iterated deletion. If $n \leq 3(k + t)$, it follows easily from Theorem 2: that there is no multiparty computation protocol that is a Nash equilibrium, we do not have to require that the protocol survive iterated deletion to get the result if $n \leq 3(k + t)$. Various generalizations of the Halpern and Teague results have been proved. We have already mentioned the work of ADGH. Lysanskaya and Triandopoulos [27] independently proved the special case of Theorem 1(c) where $k = 1$ and $t + 1 < n/2$ (they also consider survival of iterated deletion); Gordon and Katz [20] independently proved a special case of Theorem 1(d) where $k = 1$, $t = 0$, and $n \geq 2$.

In this paper we are interested in implementing equilibrium by using standard communication channels. An alternate option is to consider the possibility of simulating equilibrium by using much stronger primitives. Izmalkov, Micali, and Lepinski [23] show that, if there is a punishment strategy and we have available strong primitives that they call *envelopes* and *ballot boxes*, we can implement arbitrary mediators perfectly (without an ϵ error) in the case that $k = 1$, in the sense that every equilibrium of the game with the mediator corresponds to an equilibrium of the cheap-talk game, and vice versa. In [26,25], these primitives are also used to obtain implementation that is perfectly collusion proof in the model where, in the game with the mediator, coalitions cannot communicate. (By way of contrast, we allow coalitions to communicate.) Unfortunately, envelopes and ballot boxes cannot be implemented under standard computational and systems assumptions [25].

The rest of this paper is organized as follows. In Section 2, we review the relevant definitions. In Section 3, we briefly discuss the upper bounds, and compare them to the results of ADGH. In Section 4, we prove the lower bounds.

2   Definitions

We give a brief description of the definitions needed for our results here. More detailed definitions and further discussion can be found in [3].

We are interested in implementing mediators. Formally, this means we need to consider three games: an *underlying game* Γ, an extension Γ_d of Γ with a mediator, and a cheap-talk extension Γ_{CT} of Γ. Our underlying games are *(normal-form) Bayesian games*. These are games of incomplete information, where players make only one move, and these moves are made simultaneously. The "incomplete information" is captured by assuming that nature makes the first move and chooses for each player i a *type* in some set T_i, according to some distribution that is commonly known. Formally, a Bayesian game Γ is defined by a tuple (N, T, A, u, μ), where N is the set of players, $T = \times_{i \in N} T_i$ is the set of possible types, μ is the distribution on types, $A = \times_{i \in N} A_i$ is the set of action profiles, and $u_i : T \times A$ is the utility of player i as a function of the types prescribed by nature and the actions taken by all players.

Given an underlying Bayesian game Γ as above, a game Γ_d with a mediator d that extends Γ is, informally, a game where players can communicate with the mediator and then perform an action from Γ. The utility of player i in Γ_d depends just on its type and the actions performed by all the players. Although we think of a *cheap-talk* game as a game where players can communicate with each other (using point-to-point communication and possibly broadcast), formally, it is a game with a special kind of mediator that basically forwards all the messages it receives to their intended recipients. We assume that mediators and players are just interacting Turing machines with access to an unbiased coin (which thus allows them to choose uniformly at random from a finite set of any size). Γ_{CT} denotes the cheap-talk extension of Γ.

When considering a deviation by a coalition K, one may want to allow the players in K to communicate with each other. If Γ' is an extension of an underlying game Γ (including Γ itself) and $K \subseteq N$, let $\Gamma' + CT(K)$ be the extension of Γ where the mediator provides private cheap-talk channels for the players in K in addition to whatever communication there is in Γ'. Note that $\Gamma_{CT} + CT(K)$ is just Γ_{CT}; players in K can already talk to each other in Γ_{CT}.

A *strategy* for player i in a Bayesian game Γ is a function from i's type to an action in A_i; in a game with a mediator, a strategy is a function from i's type and message history to an action. We allow behavior strategies (i.e., randomized strategies); such a strategy gets an extra argument, which is a sequence of coin flips (intuitively, what a player does can depend on its type, the messages it has sent and received if we are considering games with mediators, and the outcome of some coin flips). We use lower-case Greek letters such as σ, τ, and ρ to denote a strategy profile; σ_i denotes the strategy of player i in strategy profile σ; if $K \subseteq N$, then σ_K denotes the strategies of the players in K and σ_{-K} denotes the strategies of the players not in K. Given a strategy profile σ a player $i \in N$ and a type $t_i \in T_i$ let $u_i(t_i, \sigma)$ be the expected utility of player i given that his type is t_i and each player $j \in N$ is playing the strategy σ_j. Note that a strategy profile— whether it is in the underlying game, or in a game with a mediator extending the underlying game (including a cheap-talk game)—induces a mapping from type profiles to distributions over action profiles. If Γ_1 and Γ_2 are extension of some underlying game Γ, then strategy σ_1 in Γ_1 *implements* a strategy σ_2 in Γ_2 if both σ and σ' induce the same function from types to distributions over actions. Note that although our informal discussion in the introduction talked about *implementing mediators*, the formal definitions (and our theorems) talk about implementing strategies. Our upper bounds show that, under appropriate assumptions, for *every* (k, t)-robust equilibrium σ in a game Γ_1 with a mediator, there exists an equilibrium σ' in the cheap-talk game Γ_2 corresponding to Γ_1 that implements σ; the lower bounds in this paper show that, if these conditions are not met, there exists a game with a mediator and an equilibrium in that game that cannot be implemented in the cheap-talk game. Since our definition of games with a mediator also allow arbitrary communication among the agents, it can also be shown that every equilibrium in a cheap-talk game can be implemented in the mediator game: the players simply ignore the mediator and communicate with each other.

The utility function in the games we consider is defined on type and action profiles. Note that we use the same utility function both for an underlying game Γ and all extensions of it. As usual, we want to talk about the expected utility of a strategy profile, or of a strategy profile conditional on a type profile. We abuse notation and continue to use u_i for this, writing for example, $u_i(t_K, \sigma)$ to denote the expected utility to player i if the strategy profile σ is used, conditional on the players in K having the types t_K. Since the strategy σ here can come from the underlying game or some extension of it, the function u_i is rather badly overloaded. We sometimes include the relevant game as an argument to u_i to emphasize which game the strategy profile σ is taken from, writing, for example, $u_i(t_K, \Gamma', \sigma)$.

We now define the main solution concept used in this paper: (k, t)-robust equilibrium. The k indicates the size of coalition we are willing to tolerate, and the t indicates the number of players with unknown utilities. These t players are analogues of faulty players or adversaries in the distributed computing literature, but we can think of them as being perfectly rational. Since we do not know what actions these t players will perform, nor do we know their identities, we are interested in strategies for which the payoffs of the remaining players are immune to what the t players do.

Definition 1. *A strategy profile σ in a game Γ is t-immune if, for all $T \subseteq N$ with $|T| \leq t$, all strategy profiles τ, all $i \notin T$, and all types $t_i \in T_i$ that occur with positive probability, we have $u_i(t_i, \Gamma + CT(T), \sigma_{-T}, \tau_T) \geq u_i(t_i, \Gamma, \sigma)$.*

Intuitively, σ is t-immune if there is nothing that players in a set T of size at most t can do to give the remaining players a worse payoff, even if the players in T can communicate.

Our notion of (k, t)-robustness requires both t-immunity and the fact that, no matter what t players do, no subset of size at most k can all do better by deviating, even with the help of the t players, and even if all $k + t$ players share their type information.

Definition 2. *Given $\epsilon \geq 0$, σ is an ϵ-(k, t)-robust equilibrium in game Γ if σ is t-immune and, for all $K, T \subseteq N$ such that $|K| \leq k$, $|T| \leq t$, and $K \cap T = \emptyset$, and all types $t_{K \cup T} \in T_{K \cup T}$ that occur with positive probability, it is not the case that there exists a strategy profile τ such that*

$$u_i(t_{K \cup T}, \Gamma + CT(K \cup T), \tau_{K \cup T}, \sigma_{-(K \cup T)}) > u_i(t_i, \Gamma + CT(T), \tau_T, \sigma_{-T}) + \epsilon$$

for all $i \in K$. A (k, t)-robust equilibrium is just a 0-(k, t)-robust equilibrium.

Note that a $(1, 0)$-robust equilibrium is just a Nash equilibrium, and an ϵ-$(1, 0)$-robust equilibrium is what has been called an ϵ-Nash equilibrium in the literature. The notion $(k, 0)$-robust equilibrium is essentially Aumann's [4] notion of resilience to coalitions, except that we allow communication by coalition members (see [3] for a discussion of the need for such communication). Heller [22] used essentially this notion. The notion $(0, t)$-robustness is somewhat in the spirit of Eliaz's [13] notion of t fault-tolerant implementation. Both our notion of $(0, t)$-robustness and Eliaz's notion of t-fault tolerance require that what the players not in T do is a best response to whatever the players in T do (given that all the players not in T follow the recommended strategy); however, Eliaz does not require an analogue of t-immunity.

In [1] we considered a stronger version of robust equilibrium. Roughly speaking, in this stronger version, we require that, if a coalition deviates, only one coalition member need be better off, rather than all coalition members. In [3] we formally define this stronger notion and discuss its motivation. We note that all our lower and upper bounds works for both notions; we focus on Definition 1 here because it is more standard in the game theory literature. (Other notions of equilibrium have been considered in the literature; see the appendix for discussion.)

In this paper, we are interested in the question of when a (k, t)-robust equilibrium σ in a game Γ_d with a mediator extending an underlying game Γ can be implemented by an ϵ-(k, t)-robust equilibrium σ' in the cheap-talk extension Γ_{CT} of Γ. If this is the case, we say that σ' is an ϵ-(k, t)-robust implementation of σ. (We sometimes say that (Γ_{CT}, σ') is an ϵ-(k, t)-robust implementation of (Γ_d, σ) if we wish to emphasize the games.)

3   The Possibility Results

Definition 3. *If Γ_d is an extension of an underlying game Γ with a mediator d, a strategy profile ρ in Γ is a k-punishment strategy with respect to a strategy profile σ in Γ_d if for all subsets $K \subseteq N$ with $|K| \leq k$, all strategies ϕ in $\Gamma + CT(K)$, all types $t_K \in T_K$, and all players $i \in K$:*

$$u_i(t_K, \Gamma_d, \sigma) > u_i(t_K, \Gamma + CT(K), \phi_K, \rho_{-K}).$$

If the inequality holds with $\geq$ replacing $>$, ρ is a weak k-punishment strategy with respect to σ.

Intuitively, ρ is k-punishment strategy with respect to σ if, for any coalition K of at most k players, even if the players in K share their type information, as long as all players not in K use the punishment strategy in the underlying game, there is nothing that the players in K can do in the underlying game that will give them a better expected payoff than playing σ in Γ_d.

The notion of utility variant is used to make precise that certain results do not depend on knowing the players' utilities (see [3] for details).

Theorem 1. *Suppose that Γ is Bayesian game with n players and utilities u, d is a mediator that can be described by a circuit of depth c, and σ is a (k, t)-robust equilibrium of a game Γ_d with a mediator d.*

(a) If $3(k + t) < n$, then there exists a strategy σ_{CT} in $\Gamma_{CT}(u)$ such that for all utility variants $\Gamma(u')$, if σ is a (k, t)-robust equilibrium of $\Gamma_d(u')$, then $(\Gamma_{CT}(u'), \sigma_{CT})$ implements $(\Gamma_d(u'), \sigma)$. The running time of σ_{CT} is $O(c)$.

(b) If $2k + 3t < n$ and there exists a $(k + t)$-punishment strategy with respect to σ, then there exists a strategy σ_{CT} in Γ_{CT} such that σ_{CT} implements σ. The expected running time of σ_{CT} is $O(c)$.

(c) If $2(k + t) < n$ and broadcast channels can be simulated, then, for all $\epsilon > 0$, there exists a strategy $\sigma_{\mathrm{CT}}^{\epsilon}$ in Γ_{CT} such that $\sigma_{\mathrm{CT}}^{\epsilon}$ ϵ-implements σ. The running time of $\sigma_{\mathrm{CT}}^{\epsilon}$ is $O(c)$.

(d) If $k + t < n$ then, assuming cryptography and that a PKI is in place, there exists a strategy $\sigma_{\mathrm{CT}}^{\epsilon}$ in Γ_{CT} such that $\sigma_{\mathrm{CT}}^{\epsilon}$ ϵ-implements σ. The expected running time of $\sigma_{\mathrm{CT}}^{\epsilon}$ is $O(c) \cdot f(u) \cdot O(1/\epsilon)$ where $f(u)$ is a function of the utilities.

(e) If $k + 3t < n$ or if $k + t < n$ and a trusted PKI is in place, and there exists a $(k + t)$-punishment strategy with respect to σ, then, assuming cryptography, there exists a strategy $\sigma_{\mathrm{CT}}^{\epsilon}$ in Γ_{CT} such that $\sigma_{\mathrm{CT}}^{\epsilon}$ ϵ-implementers σ. The expected running time of $\sigma_{\mathrm{CT}}^{\epsilon}$ is $O(c) \cdot f(u)$ where $f(u)$ is a function of the utilities but is independent of ϵ.

We briefly comment on the differences between Theorem 1 and the corresponding Theorem 4 of ADGH. In ADGH, we were interested in finding strategies that were not only (k, t)-robust, but also survived iterated deletion of weakly dominated strategies. For part (a), in ADGH, a behavioral strategy was used that had no upper bound on running time. This was done in order to obtain a strategy that survived iterated deletion. However, it is observed in ADGH that, without this concern, a strategy with a known upper bound can be used. As we observed in the introduction, part (a), as stated, actually follows from [7]. Part (b) here is the same as in ADGH. In part (c), we assume here the ability to simulate broadcast; ADGH assumes cryptography. As we have observed, in the presence of cryptography, we can simulate broadcast, so the assumption here is weaker. In any case, as observed in the introduction, part (c) follows from known results [29]. Parts (d) and (e) are new, and will be proved in [2]. The proof uses ideas from [19] on multiparty computation. For part (d), where there is no punishment strategy, ideas from [14] on getting ϵ-fair protocols are also required. Our proof of part (e) shows that if $n > k + 3t$, then we can essentially set up a PKI on the fly. These results strengthen Theorem 4(d) in ADGH, where punishment was required and n was required to be greater than $k + 2t$.

4   The Impossibility Results

No Bounded Implementations

We prove that it is impossible to get an implementation with bounded running time in general if $2k + 3t < n \leq 3k + 3t$. This is true even if there is a punishment strategy. This result is optimal. If $3k + 3t < n$, then there does exist a bounded implementation; if $2k + 3t < n \leq 3k + 3t$ there exists an unbounded implementation that has constant *expected* running time.

Theorem 2. *If $2k + 3t < n \le 3k + 3t$, there is a game Γ and a strong (k,t)-robust equilibrium σ of a game Γ_d with a mediator d that extends Γ such that there exists a $(k+t)$-punishment strategy with respect to σ for which there do not exist a natural number c and a strategy σ_{CT} in the cheap talk game extending Γ such that the running time of σ_{CT} on the equilibrium path is at most c and σ_{CT} is a (k,t)-robust implementation of σ.*

Proof. We first assume that $n = 3$, $k = 1$, and $t = 0$. We consider a family of 3-player games $\Gamma_3^{n,k+t}$, where $2k + 3t < n \le 3k + 3t$, defined as follows. Partition $\{1, \ldots, n\}$ into three sets B_1, B_2, and B_3, such that B_1 consists of the first $\lfloor n/3 \rfloor$ elements in $\{1, \ldots, n\}$, B_3 consists of the last $\lceil n/3 \rceil$ elements, and B_2 consists of the remaining elements.

Let p be a prime such that $p > n$. Nature chooses a polynomial f of degree $k + t$ over the p-element field $GF(p)$ uniformly at random. For $i \in \{1,2,3\}$, player i's type consists of the set of pairs $\{(h, f(h)) \mid h \in B_i\}$. Each player wants to learn $f(0)$ (the secret), but would prefer that other players do not learn the secret. Formally, each player must play either 0 or 1. The utilities are defined as follows:

- if all players output $f(0)$ then all players get 1;
- if player i does not output $f(0)$ then he gets -3;
- otherwise players i gets 2.

Consider the mediator game where each player is supposed to tell the mediator his type. The mediator records all the pairs (h, v_h) it receives. If at least $n-t$ pairs are received and there exists a unique degree $k + t$ polynomial that agrees with at least $n - t$ of the pairs then the mediator interpolates this unique polynomial f' and sends $f'(0)$ to each player; otherwise, the mediator sends 0 to each player.

Let σ_i be the strategy where player i truthfully tells the mediator his type and follows the mediator's recommendation. It is easy to see that σ is a $(1,0)$-robust equilibrium (i.e., a Nash equilibrium). If a player i deviates by misrepresenting or not telling the mediator up to t of his shares, then everyone still learns; if the player misrepresents or does not tell the mediator about more of his shares, then the mediator sends the default value 0. In this case i is worse off. For if 0 is indeed the secret, which it is with probability $1/2$, i gets 1 if he plays 0, and -3 if he plays 1. On the other hand, if 1 is the secret, then i gets 2 if he plays 1 and -3 otherwise. Thus, no matter what i does, his expected utility is at most $-1/2$. This argument also shows that if ρ_i is the strategy where i decides 0 no matter what, then ρ is a 1-punishment strategy with respect to σ.

Suppose, by way of contradiction, that there is a cheap-talk strategy σ' in the game Γ_{CT} that implements σ such that any execution of σ' takes at most c rounds. We say that a player i *learns the secret by round b* of σ' if, for all executions (i.e., plays) r and r' of σ' such that i has the same type and the same message history up to round b, the secret is the same in r and r'. Since we have assumed that all plays of σ' terminate in at most c rounds, it must be the case that all players learn the secret by round c of σ'. For if not, there are two executions r and r' of σ' that i cannot distinguish by round c, where the secret is different in r and r'. Since i must play the same move in r and r', in one case he is not playing the secret, contradicting the assumption that σ' implements σ. Thus, there must exist a round $b \leq c$ such that all three players learn the secret at round b of σ' and, with nonzero probability, some player, which we can assume without loss of generality is player 1, does not learn the secret at round $b - 1$ of σ'. This means that there exists a type t_1 and message history h_1 for player 1 of length $b - 1$ that occurs with positive probability when player 1 has type t_1 such that, after $b-1$ rounds, if player 1 has type t_1 and history h_1, player 1 considers it possible that the secret could be either 0 or 1. Thus, there must exist type profiles t and t' that correspond to polynomials f and f' such that $t_1 = t_1'$, $f(0) \neq f'(0)$ and, with positive probability, player 1 can have history h_1 with both t and t', given that all three players play σ'.

Let h_2 be a history for player 2 of length $b - 1$ compatible with t and h_1 (i.e., when the players play σ', with positive probability, player 1 has h_1, player 2 has h_2, and the true type profile is t); similarly, let h_3 be a history of length $b-1$ for player 3 compatible with t' and h_1. Note that player i's action according to σ_i is completely determined by his type, his message history, and the outcome of his coin tosses. Let $\sigma_2'[t_2, h_2]$ be the strategy for player 2 according to which player 2 uses σ_2' for the first $b - 1$ rounds, and then from round b on, player 2 does what it would have done according to σ_2' if its type had been t_2 and its message history for the first $b-1$ rounds had been h_2 (that is, player 2 modifies his actual message history by replacing the prefix of length $b-1$ by h_2, and leaving the rest of the message history unchanged). We can similarly define $\sigma_3'[t_3', h_3]$. Consider the strategy profile $(\sigma_1', \sigma_2'[t_2, h_2], \sigma_3'[t_3', h_3])$. Since $\sigma_i'[t_i, h_i]$ is identical to σ_i' for the first $b - 1$ steps, for $i = 2, 3$, there is a positive probability that player 1 will have history h_1 and type t_1 when this strategy profile is played. It should be clear that, conditional on this happening, the probability that player 1 plays 0 or 1 is independent of the actual types and histories of players 2 and 3. This is because players 2 and 3's messages from time b depend only on i's messages, and not on their actual type and history. Thus, for at least one of 0 and 1, it must be the case that the probability that player 1 plays this value is strictly less than 1. Suppose without loss of generality that the probability of playing $f(0)$ is less than 1.

We now claim that $\sigma_3'[t_3', h_3]$ is a profitable deviation for player 3. Notice that player 3 receives the same messages for the first b rounds of σ' and $(\sigma_1', \sigma_2', \sigma_3'[t_3', h_3])$. Thus, player 3 correctly plays the secret no matter what the type profile is, and gets payoff of at least 1. Moreover, if the type profile is t, then, by construction, with positive probability, after $b-1$ steps, player 1's history will be h_1 and player 2's history will be h_2. In this case, σ_2' is identical to $\sigma_2'[t_2, h_2]$, so the play will be identical to $(\sigma_1', \sigma_2'[t_2, h_2], \sigma_3'[t_3', h_3])$. Thus, with positive probability, player 1 will not output $f(0)$, and player 3 will get payoff 2. This means player 3's expected utility is greater than 1.

For the general case, suppose that $2k+3t < n \leq 3k+3t$. Consider the n-player game $\Gamma^{n,k,t}$, defined as follows. Partition the players into three groups, B_0, B_1, and B_2, as above. As in the 3-player game, nature chooses a polynomial f of degree $k + t$ over the field $GF(p)$ with a prime $p > n$ uniformly at random, but now player i's type is just the pair $(i, f(i))$. Again, the players want to learn $f(0)$, but would prefer that other players do not learn the secret, and must output a value in F. The payoffs are similar in spirit to the 3-player game:

- if at least $n - t$ players output $f(0)$ then all players that output $f(0)$ get 1;
- if player i does not output $f(0)$ then he gets -3;
- otherwise player i gets 2.

The mediator's strategy is essentially identical to that in the 3-player game (even though now it is getting one pair (h, v_h) from each player rather than a set of such pairs from a single player). Similarly, each player i's strategy in $\Gamma_d^{n,k,t}$, which we denote σ_i^n, is essentially identical to the strategy in the 3-player game with the mediator. Again, if ρ_i^n is the strategy in the n-player game where i plays 0 no matter what his type, then it is easy to check that ρ^n is a $(k+t)$-punishment strategy with respect to σ^n.

Now suppose, by way of contradiction, that there exists a strategy σ' in the cheap-talk extension $\Gamma_{CT}^{n,k,t}$ of $\Gamma^{n,k,t}$ that is a (k, t)-robust implementation of σ^n such that all executions of σ' take at most c rounds. We show in [3] that we can use σ' to get a $(1, 0)$-robust implementation in the 3-player mediator game $\Gamma_{3,d}^{n,k+t}$, contradicting the argument above.   □

Byzantine Agreement and Game Theory

In [1] it is shown that if $n > 3k + 3t$, we can implement a mediator in a way that does not depend on utilities and does not need a punishment strategy. Using novel connections to randomized Byzantine agreement lower bounds, we show that neither of these properties hold in general if $n \leq 3k + 3t$.

We start by showing that we cannot handle all utilities variants if $n \leq 3k+3t$. Our proof exposes a new connection between utility variants and the problem of *Weak Byzantine Agreement* [24]. Lamport [24] showed that there is no deterministic protocol with bounded running time for *weak Byzantine agreement* if $t \geq n/3$. We prove a stronger lower bound for any randomized protocol that only assumes that the running time has finite expectation.

Proposition 1. *If* $\max\{2, k + t\} < n \leq 3k + 3t$, *all* 2^n *input values are equally likely, and* P *is a (possibly randomized) protocol with finite expected running time (that is, for all protocols* P'' *and sets* $|T| \leq k + t$, *the expected running time of processes* P_{N-T} *given* (P_{N-T}, P_T'') *is finite), then there exists a protocol* P' *and a set* T *of players with* $|T| \leq k + t$ *such that an execution of* (P_{N-T}, P_T') *is unsuccessful for the weak Byzantine agreement problem with nonzero probability.*

The idea of our impossibility result is to construct a game that captures weak Byzantine agreement. The challenge in the proof is that, while in the Byzantine agreement problem, nature chooses which processes are faulty, in the game, the players decide whether or not to behave in a faulty way. Thus, we must set up the incentives so that players gain by choosing to be faulty iff Byzantine agreement cannot be attained, while ensuring that a (k, t)-robust cheap-talk implementation of the mediator's strategy in the game will solve Byzantine agreement.

Theorem 3. *If $2k + 2t < n \leq 3k + 3t$, there is a game $\Gamma(u)$ and a strong (k, t)-robust equilibrium σ of a game Γ_d with a mediator d that extends Γ such that there exists a $(k + t)$-punishment strategy with respect to σ and there does not exist a strategy σ_{CT} such that for all utility variants $\Gamma(u')$ of $\Gamma(u)$, if σ is a (k, t)-robust equilibrium of $\Gamma_d(u')$, then $(\Gamma_{\mathrm{CT}}(u'), \sigma_{\mathrm{CT}})$ is a (k, t)-robust implementation of $(\Gamma_d(u'), \sigma)$.*

Theorem 3 shows that we cannot, in general, get a *uniform* implementation if $n \leq 3k + 3t$. As shown in Theorem 1(b)–(e), we can implement mediators if $n \leq 3k + 3t$ by taking advantage of knowing the players' utilities.

We next prove that if $2k + 3t < n \leq 3k + 3t$, although mediators can be implemented, they cannot be implemented without a punishment strategy. In fact we prove that they cannot even be ϵ–implemented without a punishment strategy. Barany [6] proves a weaker version of a special case of this result, where $n = 3$, $k = 1$, and $t = 0$. It is not clear how to extend Barany's argument to the general case, or to ϵ–implementation. We use the power of randomized Byzantine agreement lower bounds for this result.

Theorem 4. *If $2k + 2t < n \leq 3k + 3t$, then there exists a game Γ, an $\epsilon > 0$, and a strong (k, t)-robust equilibrium σ of a game Γ_d with a mediator d that extends Γ, for which there does not exist a strategy σ_{CT} in the CT game that extends Γ such that σ_{CT} is an ϵ–(k, t)-robust implementation of σ.*

We now show that the assumption that $n > 2k + 3t$ in Theorem 1 is necessary. More precisely, we show that if $n \leq 2k + 3t$, then there is a game with a mediator that has a (k, t)-robust equilibrium that does not have a (k, t)-robust implementation in a cheap-talk game. We actually prove a stronger result: we show that there cannot even be an ϵ–(k, t)-robust implementation, for sufficiently small ϵ.

Theorem 5. *If $k + 2t < n \leq 2k + 3t$, there exists a game Γ, a strong (k, t)-robust equilibrium σ of a game Γ_d with a mediator d that extends Γ, a $(k+t)$-punishment strategy with respect to σ, and an $\epsilon > 0$, such that there does not exist a strategy σ_{CT} in the CT extension of Γ such that σ_{CT} is an ϵ–(k, t)-robust implementation of σ.*

The proof of Theorem 5 splits into two cases: (1) $2k + 2t < n \leq 2k + 3t$ and $t \geq 1$ and (2) $k + 2t < n \leq 2k + 2t$. For the first case, we use a reduction to a generalization of the Byzantine agreement problem called the (k, t)-Detect/Agree *problem.* This problem is closely related to the problem of *broadcast with extended consistency* introduced by Fitzi et al. [16].

Theorem 6. *If $2k + 2t < n \leq 2k + 3t$ and $t \geq 1$, there exists a game Γ, an $\epsilon > 0$, a strong (k, t)-robust equilibrium σ of a game Γ_d with a mediator d that extends Γ, and a $(k + t)$-punishment strategy with respect to σ, such that there does not exist a strategy σ_{CT} in the CT extension of Γ which is an ϵ–(k, t)-robust implementation of σ.*

We then consider the second case of Theorem 5, where $k + 2t < n \leq 2k + 2t$. Since we do not assume players know when other players have decided in the underlying game, our proof is a strengthening of the lower bounds of [30,22].

Theorem 7. *If $k + 2t < n \leq 2k + 2t$, there exist a game Γ, an $\epsilon > 0$, a mediator game Γ_d extending Γ, a strong (k,t)-robust equilibrium σ of Γ_d, and a $(k+t)$-punishment strategy ρ with respect to σ, such that there is no strategy σ_{CT} that is an ϵ-(k,t)-robust implementation of σ in the cheap-talk extension of Γ, even with broadcast channels.*

Our last lower bound using Byzantine agreement impossibility results gives a lower bound that matches the upper bound of Theorem 1(e) for the case that $n > k + 3t$. We show that a PKI cannot be set up on the fly if $n \leq k + 3t$. Our proof is based on a reduction to a lower bound for the *(k,t)-partial broadcast problem*, a novel variant of Byzantine agreement that can be viewed as capturing minimal conditions that still allow us to prove strong randomized lower bounds.

Theorem 8. *If $\max(2, k + t) < n \leq k + 3t$, then there is a game Γ, a strong (k,t)-robust equilibrium σ of a game Γ_d with a mediator d that extends Γ for which there does not exist a strategy σ_{CT} in the CT game that extends Γ such that σ_{CT} is an ϵ-(k,t)-robust implementation of σ even if players are computationally bounded and we assume cryptography.*

Tight Bounds on Running Time

We now turn our attention to running times. We provide tight bounds on the number of rounds needed to ϵ-implement equilibrium when $k + t < n \leq 2(k+t)$. When $2(k + t) < n$ then the expected running time is independent of the game utilities and independent of ϵ. We show that for $k + t < n \leq 2(k + t)$ this is not the case. The expected running time must depend on the utilities, and if punishment does not exist then the running time must also depend on ϵ.

Theorem 9. *If $k + t < n \leq 2(k + t)$ and $k \geq 1$, then there exists a game Γ, a mediator game Γ_d that extends Γ, a strategy σ in Γ_d, and a strategy ρ in Γ such that*

(a) for all ϵ and b, there exists a utility function $u^{b,\epsilon}$ such that σ is a (k,t)-robust equilibrium in $\Gamma_d(u^{b,\epsilon})$ for all b and ϵ, ρ is a (k,t)-punishment strategy with respect to σ in $\Gamma(u^{b,\epsilon})$ if $n > k + 2t$, and there does not exist an ϵ-(k,t)-robust implementation of σ that runs in expected time b in the cheap-talk extension $\Gamma_{\mathrm{CT}}(u^{b,\epsilon})$ of $\Gamma(u^{b,\epsilon})$;

(b) there exists a utility function u such that σ is a (k,t)-robust equilibrium in $\Gamma_d(u)$ and, for all b, there exists ϵ such that there does not exist an ϵ-(k,t)-robust implementation of σ^i that runs in expected time b in the cheap-talk extension $\Gamma_{\mathrm{CT}}(u)$ of $\Gamma(u)$.

This is true even if players are computationally bounded, we assume cryptography and there are broadcast channels.

Note that, in part (b), it is not assumed that there is a (k, t)-punishment strategy with respect to σ in $\Gamma(u)$. With a punishment strategy, for a fixed family of utility functions, we can implement an ϵ-(k, t)-robust strategy in the mediator game using cheap talk with running time that is independent of ϵ; with no punishment strategy, the running time depends on ϵ in general.

References

1. Abraham, I., Dolev, D., Gonen, R., Halpern, J.Y.: Distributed computing meets game theory: Robust mechanisms for rational secret sharing and multiparty computation. In: Proc. 25th ACM Symp. Principles of Distributed Computing, pp. 53–62 (2006)
2. Abraham, I., Dolev, D., Gonen, R., Halpern, J.Y.: Distributed computing meets game theory: Robust mechanisms for rational secret sharing and multiparty computation (unpublished manuscript, 2007)
3. Abraham, I., Dolev, D., Halpern, J.Y.: Lower bounds on implementing robust and resilient mediators. arXiv:0704.3646v2
4. Aumann, R.J.: Acceptable points in general cooperative n-person games. Contributions to the Theory of Games, Annals of Mathematical Studies IV, 287–324 (1959)
5. Aumann, R.J.: Correlated equilibrium as an expression of Bayesian rationality. Econometrica 55, 1–18 (1987)
6. Barany, I.: Fair distribution protocols or how the players replace fortune. Mathematics of Operations Research 17, 327–340 (1992)
7. Ben-Or, M., Goldwasser, S., Wigderson, A.: Completeness theorems for noncryptographic fault-tolerant distributed computation. In: Proc. 20th ACM Symp. Theory of Computing, pp. 1–10 (1988)
8. Ben-Porath, E.: Cheap talk in games with incomplete information. J. Economic Theory 108(1), 45–71 (2003)
9. Boneh, D., Naor, M.: Timed commitments. In: Bellare, M. (ed.) CRYPTO 2000. LNCS, vol. 1880, pp. 236–254. Springer, Heidelberg (2000)
10. Chaum, D., Crépeau, C., Damgard, I.: Multiparty unconditionally secure protocols. In: Proc. 20th ACM Symp. Theory of Computing, pp. 11–19 (1988)
11. Crawford, V.P., Sobel, J.: Strategic information transmission. Econometrica 50(6), 1431–1451 (1982)
12. Dodis, Y., Halevi, S., Rabin, T.: A cryptographic solution to a game theoretic problem. In: Bellare, M. (ed.) CRYPTO 2000. LNCS, vol. 1880, pp. 112–130. Springer, Heidelberg (2000)
13. Eliaz, K.: Fault-tolerant implementation. Review of Economic Studies 69(3), 589–610 (2002)
14. Even, S., Goldreich, O., Lempel, A.: A randomized protocol for signing contracts. Commun. ACM 28(6), 637–647 (1985)
15. Fischer, M.J., Lynch, N.A., Paterson, M.S.: Impossibility of distributed consensus with one faulty processor. Journal of the ACM 32(2), 374–382 (1985)
16. Fitzi, M., Hirt, M., Holenstein, T., Wullschleger, J.: Two-threshold broadcast and detectable multi-party computation. In: Biham, E. (ed.) EUROCRYPT 2003. LNCS, vol. 2656, pp. 51–67. Springer, Heidelberg (2003)
17. Forges, F.: Universal mechanisms. Econometrica 58(6), 1341–1364 (1990)

18. Goldreich, O.: Foundations of Cryptography, vol. 2. Cambridge University Press, Cambridge (2004)
19. Goldreich, O., Micali, S., Wigderson, A.: How to play any mental game. In: Proc. 19th ACM Symp. Theory of Computing, pp. 218–229 (1987)
20. Gordon, D., Katz, J.: Rational secret sharing, revisited. In: De Prisco, R., Yung, M. (eds.) SCN 2006. LNCS, vol. 4116, pp. 229–241. Springer, Heidelberg (2006)
21. Halpern, J.Y., Teague, V.: Rational secret sharing and multiparty computation: extended abstract. In: Proc. 36th ACM Symp. Theory of Computing, pp. 623–632 (2004)
22. Heller, Y.: A minority-proof cheap-talk protocol (2005) (Unpublished manuscript)
23. Izmalkov, S., Micali, S., Lepinski, M.: Rational secure computation and ideal mechanism design. In: Proc. 46th IEEE Symp. Foundations of Computer Science, pp. 585–595 (2005)
24. Lamport, L.: The weak byzantine generals problem. J. ACM 30(3), 668–676 (1983)
25. Lepinksi, M., Micali, S., Shelat, A.: Collusion-free protocols. In: Proc. 37th ACM Symp. Theory of Computing, pp. 543–552 (2005)
26. Lepinski, M., Micali, S., Peikert, C., Shelat, A.: Completely fair SFE and coalition-safe cheap talk. In: Proc. 23rd ACM Symp. Principles of Distributed Computing, pp. 1–10 (2004)
27. Lysyanskaya, A., Triandopoulos, N.: Rationality and Adversarial Behavior in Multi-party Computation. In: Dwork, C. (ed.) CRYPTO 2006. LNCS, vol. 4117, pp. 180–197. Springer, Heidelberg (2006)
28. Myerson, R.B.: Game Theory: Analysis of Conflict. Harvard University Press (September 1997)
29. Rabin, T., Ben-Or, M.: Verifiable secret sharing and multiparty protocols with honest majority. In: Proc. 21st ACM Symp. Theory of Computing, pp. 73–85 (1989)
30. Shamir, A., Rivest, R.L., Adelman, L.: Mental poker. In: Klarner, D.A. (ed.) The Mathematical Gardner, Prindle, Weber, Schmidt, Boston, Mass, pp. 37–43 (1981)
31. Urbano, A., Vila, J.E.: Computational complexity and communication: Coordination in two-player games. Econometrica 70(5), 1893–1927 (2002)
32. Urbano, A., Vila, J.E.: Computationally restricted unmediated talk under incomplete information. Economic Theory 23(2), 283–320 (2004)
33. Yao, A.: Protocols for secure computation (extended abstract). In: Proc. 23rd IEEE Symp. Foundations of Computer Science, pp. 160–164 (1982)

Cryptography and Game Theory: Designing Protocols for Exchanging Information*

Gillat Kol and Moni Naor**

Department of Computer Science and Applied Mathematics
Weizmann Institute of Science, Rehovot 76100 Israel
{gillat.kol,moni.naor}@weizmann.ac.il

Abstract. The goal of this paper is finding *fair* protocols for the secret sharing and secure multiparty computation (SMPC) problems, when players are assumed to be rational.

It was observed by Halpern and Teague (STOC 2004) that protocols with bounded number of iterations are susceptible to backward induction and cannot be considered rational. Previously suggested cryptographic solutions all share the property of having an essential exponential upper bound on their running time, and hence they are also *susceptible to backward induction*.

Although it seems that this bound is an inherent property of every cryptography based solution, we show that this is not the case. We suggest *coalition-resilient* secret sharing and SMPC protocols with the property that after any sequence of iterations it is still a computational best response to follow them. Therefore, the protocols can be run any number of iterations, and are *immune to backward induction*.

The mean of communication assumed is a broadcast channel, and we consider both the *simultaneous* and *non-simultaneous* cases.

1 Introduction

1.1 Background and Related Work

The issue of fairness in multiparty computation has been actively investigated since the inception of the field. In fact, the goal of Yao's 1986 famous paper [33] (where Garbled Circuits were introduced) was to address this problem. In this work we consider the *rational*, game-theoretic version of the secure function evaluation problem, that is when the players are assumed to have utility functions they try to maximize.

Realizing the advantages of simulating an equilibrium without depending on an honest mediator, the Game Theory community began pursuing a similar goal to that of Yao's in Game Theoretic settings. The works [2,5,4,30,10,16] tried to

---

* Research supported by a grant from the Israel Science Foundation.
** Incumbent of the Judith Kleeman Professorial Chair.

R. Canetti (Ed.): TCC 2008, LNCS 4948, pp. 320–339, 2008.
© International Association for Cryptologic Research 2008

remove the mediator by allowing the players to have free communication (so-called "cheap talk") prior to playing the game. In [7] this problem was addressed using cryptographic tools.

Recently, the Cryptography community started exploring cryptographic information exchange problems, such as secret sharing and secure multiparty computation (SMPC), in Game Theoretic settings. Recall that in the classical problem of m-out-of-n secret sharing a dealer issues shares of a secret and privately assigns them to n players, such that any subset of m or more players can reconstruct the secret, but a subset of less than m players cannot learn anything about the secret. An SMPC protocol enables a group of players to evaluate a function on private inputs, but does not reveal *any* additional information about the players' inputs, over what is already disclosed by the function.

Since rational players will only participate in information exchange protocols when having an initial incentive to collaborate, we need to assume that players prefer getting the designated value (the secret or the function's value) to not getting it. In some papers it was further assumed that players prefer that as few as possible of the other players get the value. Although our protocols work without this last assumption, in the following discussions we always use this extreme case as an example.

The main difficulty in designing such fair protocols in rational settings is the players' desire to keep silent in the last round, if they can identify it (e.g., if the protocol is bounded), since they do no longer fear future punishment. Then, using a *backward induction* argument it can be shown that players prefer to keep silent in every round (see discussion in Section 1.3).

Several protocols overcoming this hurdle were offered by Halpern and Teague [15], Gordon and Katz [14], Abraham et al. [1], and Lysyanskaya and Triandopoulos [23]. All protocols require *simultaneous* channels (either a broadcast channel, or secure private channels) and use the key idea that in any given round players do not know whether the current round is going to be the last round, or whether this is a just a test round designed to catch cheaters. To prevent players from finding out the type of the round before it is carried out, the protocols in [1,23] used computational based cryptography.

We claim that those protocols have a weak point: they are still essentially bounded, since the cryptographic primitives used in the beginning of the protocols can surely be broken after an exponential number of rounds. Hence, they are also *susceptible to backward induction*. In a previous paper [19] we have offered a non-cryptographic protocol for rational secret sharing that is immune to backward induction. The protocol uses special formed shares taken from unbounded domains (we have shown that unbounded domains are necessary in this setting), and cannot be generalized to the case of rational SMPC.

In this work we show that new cryptographic tools can be used to get the best of all worlds. We start off by considering the case of a simultaneous broadcast channel (SBC), where all player broadcast messages at the same time (no rushing). We offer a fair, coalition-resilient rational secret sharing scheme that may use any set of shares (provided that they can be authenticated), and generalize our protocol to the case of rational SMPC. We then consider the case of a non-simultaneous broadcast channel (NSBC), where there is only a single sender per round. We show how to run the previous protocols using only an NSBC, at least when the function's range is small.[1] Unlike previously suggested cryptographic solutions, our protocols are *immune to backward induction*.

Another line of work was pursued by Lepinski et al. [20,21] and Izmalkov et al. [17] in their recent sequence of papers. Roughly speaking, they were able to obtain fair, rational SMPC protocols, prevent coalitions, and eliminate subliminal channels. However, the hardware requirements needed for these operations, including ideal envelopes and ballot boxes, are very strict; it is not clear how they can be implemented for distant participants, if at all.

1.2 Rationality Concepts

In Game Theoretic settings players are assumed to be *rational*. A great deal of effort was invested in trying to capture the nature of rational behavior, resulting in a long line of stability concepts. The best known concept is that of a Nash equilibrium: a vector of players strategies is a Nash equilibrium if given that all the other players are following their prescribed strategy, no player can gain from deviating from his strategy. In a Nash equilibrium, each player's strategy is a *best response* to the strategies of the others.

A natural generalization of a Nash equilibrium is a C-resilient equilibrium, where C is a collection of subsets of players (coalitions). In a C-resilient equilibrium, for any $C \in C$, *no* member of the coalition C can do better, even if the *whole* coalition C defects. A Nash equilibrium is a C-resilient equilibrium, where C is the set of all coalitions of size 1.

A cryptographic protocol cannot be expected to be the best response for all possible situations, since a relatively benign player may be very lucky and discover how to break a cryptographic primitive. Therefore, the previously suggested cryptographic protocols, as well the protocols suggested in this paper, are not exact Nash equilibria. However, they are computational Nash equilibria, i.e., they are "close" to being Nash in the sense that no player has an *efficient* (polynomial) deviating strategy that yields a *non-negligibly* greater payoff than the equilibrium strategy. A computational C-resilient equilibrium is defined similarly.

As pointed out by Halpern and Teague [15], when considering information exchange tasks, requiring protocols to induce a Nash equilibrium is not enough to ensure stability. For example, the famous m-out-of-n scheme due to Shamir [28], requiring players to broadcast their given shares, is a Nash equilibrium when $m < n$ and more than m players participate in the reconstruction, but is unstable since players prefer to keep silent rather than reveal their shares.

---

[1] Quite a lot of effort was invested into approximating an SBC via an NSBC and obtaining fair protocols using cryptographic techniques of gradual release (see [6,9,25] for recent work). Note, however, that such results do not take into account the rationality consideration that we use in this paper. Incorporating rationality considerations into such protocols is an interesting challenge.

This is due to the fact that silence strategy is never worse than the strategy of revealing the share, but it is sometimes strictly better. For example, if exactly $m - 1$ other players choose to reveal their shares.

To rule out such behaviors, two different strengthenings of the notion of Nash equilibrium were used in [15,14,23,1,19]: equilibrium surviving iterated deletion of weakly dominated strategies and strict equilibrium. Such notions are not discussed in this paper: we find the notion of surviving iterated dominance problematic (see [19] for discussion), and the notion of strict equilibrium unsuitable for the computational case since it demands a *unique* best response.

1.3   The Backward Induction Process

As observed by Halpern and Teague, no information exchange protocol with bounded number of rounds can be regarded as stable in the rational setting: suppose that the protocol is bounded by b rounds. When round b is reached players no longer fear future punishment and prefer to keep silent. As mentioned before, the silence strategy is always at least as good as cooperation strategy, but is sometimes strictly better. Consequently, round $b - 1$ is now essentially the last round, and players deviate from the same reason. The process continues in this way backwards in time, thus it is called backward induction, showing that players are better off keeping silent in rounds $b - 2, b - 3, ..., 1$ as well.

We sketch a basic version of the secret sharing schemes suggested in [23,1], and show that a similar problem arises. We start by describing a version of the scheme that requires an "on-line dealer" (i.e., the dealer is involved in the reconstruction process), and then show how the on-line dealer was removed.

The scheme with an on-line dealer proceeds in a sequence of iterations. At the beginning of each iteration the dealer distributes new (Shamir) shares: with probability β (whose value is discussed later) the distributed shares are of the original secret, and with probability $1 - \beta$ the shares are of a fictitious secret. Every player should then broadcast the last share given to him, as long as no player has deviated. If a deviation was detected, players abort the protocol.

When β is chosen to be small enough, as a function of the utility functions (the greater the ratio between the payoff for learning alone and learning with the others, the smaller β is), no player can improve his payoff by cheating. That is, the risk of deviating in a fake round and causing the others to abort overcomes the desire of getting a possibly higher payoff for deviating in a real round.

In order to remove the on-line dealer, players simulate the dealer using a (non-rational) SMPC protocol: the dealer only distributes initial shares of the secret, and in every iteration players run an SMPC protocol to compute the function that gets as input their initial shares and distributes new shares. It was shown in [1] that the described protocol is a computation $\mathcal{C}$-resilient equilibrium where $\mathcal{C}$ is the set of all coalitions of size smaller than the threshold.

We argue that a similar backward induction argument can be used to show the instability of the protocol without the on-line dealer, even in computational settings. To show our claim we first investigate the meaning of the phrase "*following a strategy*". We usually think of a strategy as a code of a program and say that player i follows the strategy σ_i if i runs the program σ_i line-by-line. However, the assumption that i runs the program σ_i, and not some other program σ_i' with the exact same "external functionality" (i.e., σ_i' broadcasts the same messages as σ_i), is not always realistic. *Therefore, we consider a strategy as satisfying the property X only if all possible* implementations *of it satisfy X.*[2] This approach of checking all possible "undetectable" deviating strategies resembles the "honest-but-curious" cryptographic approach.

Now suppose that players seem to be the running the protocol without the on-line dealer, but actually run an implementation of it for which each player works a polynomial "over time" in every iteration trying to crack information hidden about the shares from the SMPC used in the first iteration. This is done by checking one key in every iteration and storing the right key. Recall that in general an SMPC protocol only gives a computational protection, not information-theoretic one (this is certainly true when we want to be immune to arbitrary coalitions, or if we do not assume private lines). Therefore, after exponentially many iterations in the key size, even this new non-ambitious strategy will surely find the right key. This shows that there is an essential upper bound to the number of iterations this protocol can be run: if the Kth iteration is reached (where K is the number of possible keys), each player may be better off quitting and using his stored key to retrieve the secret and get a (non-negligible) extra payoff. From this point on, the same backward induction process can be applied.

The above example shows that the backward induction process in computational settings, where presumably we are not concerned with the protocol's stability in rare events, is as problematic as in the standard Game Theoretic settings, since it causes exponential events to be amplified: the instability of the protocol without the on-line dealer in the rare case that it runs for exponential number of iterations causes it to be unstable from round 1.

Although it seems that susceptibility to backward induction is an inherent property of every computational based cryptographic solution, this paper shows that this is not the case. Our protocols are not only computational $\mathcal{C}$-resilient equilibria, but satisfy the additional property that after *any sequence of iterations*, they still induce such equilibria. Thus, players will never have an incentive to deviate, and the backward induction argument cannot be used. We call such protocols computational $\mathcal{C}$-immune. Clearly, $\mathcal{C}$-immunity implies $\mathcal{C}$-resilience.[3]

---

[2] In classical Game Theory, where there are no computational limitations, the distinction between running σ_i and running σ_i' is insignificant: in both cases i's knowledge consists of his initial information and all previously selected actions. However, in settings such as ours, where resources are limited, the results of the calculations made by a player when running a *specific* program should also be considered as part of his knowledge, since it is not always possible for him to repeat them.

[3] We do not regard the $\mathcal{C}$-immunity property as a sufficient condition, ensuring the stability of information exchange protocols, as some unstable protocols satisfy it. For example, Shamir's m-out-of-n secret sharing scheme is $\mathcal{C}$-immune for the maximal possible set $\mathcal{C}$ (the set of all coalitions of size smaller than m), when $m < n$ and more than m players participate in the reconstruction, since its reconstruction protocol consists of a single communication round.

1.4   Organization and Summary

The main idea of our protocols is ensuring that no iteration until the last one contains *any* information, in the *information-theoretic sense*, about the players' private values. In order to so, we construct in Section 3 a new cryptographic tool called *meaningful/meaningless encryption* that has a special property: some public keys yield ciphertexts that cannot be decrypted (even with unbounded computational power). Such keys are called *meaningless*, while the other keys are called *meaningful* and provide semantic security. One can efficiently distinguish meaningful keys from meaningless ones only when given the private key.

In Section 4 we offer a rational secret sharing scheme for the SBC model that works for any kind of shares, provided that they can be authenticated. In every iteration of the scheme new private and public keys are created using a random seed via a (non-rational) SMPC. The public key is published and the seed is shared between the players. Players use the public key to encrypt their shares, and the ciphertexts are broadcasted. Then, the validity of the ciphertexts is verified by another SMPC. A key point is that the verification does not require knowledge of the original shares, thus leaks no information about the secret. After a successful verification the seed's shares are exchanged, allowing players regenerate the private key and check whether the public key is meaningful. If it is, the shares of the secret are retrieved from the ciphertexts, and the secret is regenerated. Otherwise, the protocol proceeds to the next iteration.

No information about the secret can be retrieved from the ciphertexts sent in iterations with meaningless keys, hence no coalition can benefit from deviating before the last iteration. Since players cannot efficiently identify this iteration before sending their encrypted share, they cannot prevent others from learning.

In Section 5 we offer a rational SMPC protocol, based on the secret sharing scheme. We first note that in a secret sharing scheme players are required to evaluate a "reconstruction function" on their shares in order to retrieve the secret. Since our secret sharing scheme works for any type of shares, it can be used to compute any reconstruction function. The main problem is that the computation is not secure, as players' shares are revealed during the last iteration. To protect players' inputs, the new rational SMPC protocol additionally creates a Garbled Circuit in each iteration, and requests players to encrypt their garbled strings instead of their original inputs.

Finally, in Section 6 we show how to get rid of the simultaneity assumption, at the price of causing the expected length of the protocol to depend (linearly) on the size of the function's range.

Our protocols are $\mathcal{C}$-immune for the maximal possible set of coalitions $\mathcal{C}$: the secret sharing scheme considers all coalitions of size smaller than the threshold. The SMPC protocols do not pose any new constraints on $\mathcal{C}$, over the ones already posed by the players ability to learn the function's value by colluding before the game starts. In general, we give no guarantee about the composability of our protocols with any other protocol.

Further details, as well as omitted proofs and definitions, can be found in the full version of this paper [18].

2 Definitions and Settings

2.1 Computing Games and Protocols

As discussed in Section 1.4, both rational secret sharing and rational SMPC require rational protocols allowing players to evaluate a function on their private values. Hence, we start off by describing a model for rational joint computation. This model is the computational analog of the one suggested in [19].

In rational joint computation a set of players $N = \{1, ..., n\}$ each holding an input are interested in evaluating an n-ary function $f : \mathbf{X} \to Y$ ($\mathbf{X} \subseteq \times_{i \in N} X_i$ for some sets X_i) with finite domain and range. Players are assumed to be rational, and try to maximize their utility function. Recall that utility functions associate numeric values to outcomes of the game, the value $u_i(o)$ is player i's payoff if outcome o was reached. In our case, an outcome consists of the players' inputs, and the sequence of actions taken by them.

Our input as protocol designers is the function f, the distribution over inputs $\mathcal{D}$, and players' utilities $(u_i)_{i \in N}$[4]. Actually, as discussed later, we only require partial information about the utilities and the distribution. We should then output a game and "rational" strategies allowing all players to "learn" $f(\mathbf{x})$.

We suggest a **computing game** for f (with respect to $(u_i)_{i \in N}$ and $\mathcal{D}$) that proceeds in a sequence of iterations, where each iteration may consist of multiple communication rounds. In every round players are allowed to broadcast *any* finite binary string of their choice and update their **state** (a private binary string). If an SBC is assumed, the broadcasts in every round are simultaneous. Otherwise, an NSBC is assumed, and only a single player may broadcast in every round. We make no assumptions regarding the NSBC's behavior when two or more players try to broadcast at the same time. In such cases, some players may get partial information about the messages. A player can leave the game in any round by broadcasting a **quit** sequence and **outputting** his guess of $f(\mathbf{x})$. Players observe the actions taken by the others in previous rounds, but do not view their guesses.

Throughout the paper we assume that players are computationally bounded and can only run **efficient** strategies to evaluate polynomial time computable functions. To define the computational power of the players, we introduce an external *initial security parameter* k into the game. The security parameter used in round t is $k + t$, and we require that the players' strategies can be computed in probabilistic polynomial time in the security parameter of the corresponding round. We assume that the parameters of the original game (like the payoff functions, the initial distribution over inputs, etc.) are all independent of the security parameter, and thus can always be computed "in constant time".

We say that strategy σ' implements strategy σ if they both choose the same action after witnessing the same transcript (sequence of messages broadcasted

---

[4] We regard the players' utility functions as given, and do not attempt to change them. Simpler solutions can be obtained by introducing a discounting factor to the utilities, causing them to decline over time. However, in such solutions when an advanced round is reached, the utilities assumed are very far from the original ones, thus do not properly reflect players preferences.

in previous rounds) when given the same input and random tape. Note that "implements" is a symmetric relation. A vector of strategies $\sigma = (\sigma_1, ..., \sigma_n)$ is called a protocol, and we say that σ computes the function f if it almost always ends, and in every finite run of it all players output $f(\mathbf{x})$.

2.2 The C-Immunity Property

In Game Theory, to show that an equilibrium σ is immune to backward induction, one needs to prove that it satisfies the following property: if players are running σ, then after *any* history, following σ is still an equilibrium. Such equilibria are called subgame perfect or sequential equilibria. Note that if this property holds, then no player will ever have an incentive to deviate from σ_i, and thus no backward induction process can be applied.

However, since our protocols involve cryptographic tools, there may be histories for which the cryptographic primitives are broken, and we cannot expect the protocol to induce an equilibrium in such cases. In particular, since we deal with protocols that proceed in a sequence of iterations, executing cryptographic primitives in each, we can only hope to satisfy a slightly weaker property. Namely, that following the protocols is still a (computational C-resilient) equilibrium after *any sequence of iterations*; i.e., after all histories that can be reached by σ, after which a new iteration begins. As discussed in Section 1.3, we need to require this property to also hold when players are running an implementation of σ, instead of σ. We call protocols satisfying this demand computational C-immune.

Definition 1 (computational C-immune). *Let σ be an efficient protocol for a computing game, and C be a set of coalitions (subsets of players). Let R^t be the set of sequences of random tapes for the first t iterations that do not cause σ to end. A sequence $\mathbf{r} \in R^t$ is of the form $\mathbf{r} = (\mathbf{r}^1, ..., \mathbf{r}^t)$ where $\mathbf{r}^s = (r_1^s, ..., r_n^s)$ and r_j^s is the random tape used by player j in iteration s.*

The protocol σ is computational C-immune *if for every coalition $C \in C$, and every sequence of tapes $\mathbf{r}_0 = (r_0^1, ..., r_0^t) \in R^t$ used by the players in the first t rounds, there exists a negligible function $\varepsilon(k)$ such that for every player $i \in C$, every efficient (deviating) joint strategy σ_C' for players in C, and every efficient joint strategy τ_{-C} for players in $N \backslash C$ implementing σ_{-C}, it hold that:*

$$\mathbf{E}\left[u_i(\tau_{-C}(k), \sigma_C(k))\right] + \varepsilon(k) \geq \mathbf{E}\left[u_i(\tau_{-C}(k), \sigma_C'(k))\right]$$

The expectation is taken over all sets of random tapes for the players assigning them the tapes $r_0^1, ..., r_0^t$ for the first t iterations.

2.3 Settings for Rational Secret Sharing and Rational SMPC

We review the models for rational secret sharing and rational SMPC assumed in this paper.

Definition 2 (computational rational secret sharing scheme). *A computational rational m-out-of-n secret sharing scheme for a set of secrets Y, with* respect to the distribution over secrets $\mathcal{D}$ and the utilities $(u_i)_{i \in N}$, consists of a dealer's algorithm for issuing shares, and a protocol allowing the players to reconstruct the secret. We require that:

- No subset C of less than m players can reveal any partial information about the secret before the game begins. I.e., the distribution over inputs given any shares of players in C is identical to the original distribution $\mathcal{D}$.

- The reconstruction protocol run by any group of at least m players is a computational $\mathcal{C}$-immune protocol for $\mathcal{C} = \{C \mid |C| \leq m - 1\}$ that computes the reconstruction function induced by the dealer's algorithm in the corresponding computing game.

Definition 3 (computational $\mathcal{C}$-rational SMPC protocol). *Let $\mathcal{C}$ be a set of coalitions. A computational $\mathcal{C}$-rational SMPC protocol for f, with respect to a distribution over inputs $\mathcal{D}$ and utilities $(u_i)_{i \in N}$, is:*

- *A secure protocol in the cryptographic sense for the one shot case (see [11], Definition 7.5.3).*

- *A computational $\mathcal{C}$-immune protocol that computes f in the corresponding computing game.*

2.4 Assumptions on the Utilities and the Distribution over Inputs

As mentioned in the Introduction, we must assume that players have initial motivation to participate in the computing games. As was done in previous papers, we assume that players prefer to learn the designated value. Formally, we say that a player learns the value when outcome o is reached, if according to o the player quits and outputs the right value. Our assumption is that for two possible outcomes o and o' it holds that $u_i(o) > u_i(o')$ whenever player i learns the value when o is reached, but does not learn when o' is reached.

In order to achieve $\mathcal{C}$-immune protocols, we additionally need to require that no coalition can guess the designated value or the last iteration of our protocol with a high enough probability. We denote by α an upper bound to the probability that a coalition $C \in \mathcal{C}$ can guesses the right value in advance, and by β' the probability (upto a negligible factor) that a coalition $C \in \mathcal{C}$ is able to identify the last iteration of the protocol before it is carried out. Note that in the protocol described in Section 1.3, as well as in our protocols, a value β determines the probability of proceeding to the next iteration and satisfies $\beta = \beta'$.

In the next sections we require $\alpha < \alpha_0$ and $\beta < \beta_0$, where α_0 and β_0 are functions of the utilities and of the set $\mathcal{C}$. The calculation of the functions is deferred to the full version of this paper [18]. As before, the greater the ratio between the payoff for learning the secret alone and learning with the others, the smaller α_0 and β_0 should be. Note that since players can always guess the value y with the highest probability according to $\mathcal{D}$, it holds that $\alpha \geq \mathcal{D}(y)$, and thus the requirement $\alpha < \alpha_0$ poses a condition on $\mathcal{D}$.

3   Cryptographic Tools

3.1   Standard Cryptographic Tools

Our protocols use several standard cryptographic tools:

A Commitment Scheme. We assume that $\mathsf{Commit}(x, r) = com$ generates a commitment for the value x using randomness r, and that the commitment is *perfectly binding*. We call (r, x) the *opening* of com.

A (Non-Rational) SMPC Protocol. We assume that the protocol allows the evaluation of randomized functions (in particular, we use it to select a random seed, and assume that the players cannot bias the result). In addition, we require that the SMPC protocol enables its participants to detect deviations with high probability. The protocol should work for an active adversary statically corrupting any number of parties ($\leq n - 1$). We do not consider premature suspension of execution a violation of security, and do not assume fairness. Our application of the SMPC ensures that players have an incentive to carry it out, allowing everybody to get the output.

A 1-Out-Of-2 OT Protocol. We assume that the OT protocol works for the active adversary model and provides computational security to the sender, and information-theoretic protection to the receiver. That is: (i) if the sender's values are (s_0, s_1) and the receiver's input is $b \in \{0, 1\}$, then the OT protocol is an SMPC (again, in the sense of Definition 7.5.3 in [11]) of the function $f((s_0, s_1), b) = s_b$, (ii) for every behavior of the sender, he witnesses the same distribution over transcripts when the receiver's input is 1 and when it is 0.

Such protection is possible under standard assumptions such as *enhanced trapdoor permutations* [8,11] and *Computational Diffie-Hellman* [3] for honest-but-curious players (the recent work [32] shows that OT is symmetric, thus a protocol that protects the sender information theoretically can be transformed to one that protects the receiver). In order to handle malicious behaviors, we use the compiler described in [11], with one change: the receiver uses a ZK *argument* with a *perfectly hiding* commitment ensuring information-theoretic security for its value in order to prove to the sender that he followed the protocol properly.

We assume that all the cryptographic primitives (the standard tools and the meaningful/meaningless encryption described next) are immune to non-uniform attacks. This assumption is needed in order to show that our protocols are stable after any number of iterations.

3.2   Meaningful/Meaningless Encryptions

In additional to the standard tools, we use a non-standard encryption scheme E called a meaningful/meaningless encryption. E has a special property: some public keys of it yield ciphertexts that cannot be decrypted (even with unbounded computational power). Such keys are called *meaningless*, while the other keys are called *meaningful*.

Definition 4 (meaningful/meaningless encryption). *An encryption scheme* $E(pub\_key, random, plain) = cipher$ *is a* β-Meaningful/Meaningless Encryption *if it satisfies the following properties:*

Key Generation: *Polynomial time generation of a private key, priv_key, and a public key, pub_key, on a given seed.*

Encryption: *Computing* $c = E(pub\_key, r, m)$ *can be done in polynomial time, given a public key pub_key, randomness r, and plaintext m.*

Meaningful and Meaningless Keys: *The public keys are partitioned into meaningful and meaningless sets. The probability, over the seeds, that the generated public key is 'meaningful' is β, and the probability of it being 'meaningless' is $1 - \beta$.*

If pub_key is meaningful, then given $c = E(pub\_key, r, m)$ and priv_key, the message m can be uniquely retrieved in polynomial time. Furthermore, for every ciphertext c there is only one plaintext m for which there exists a randomness r satisfying $c = E(pub\_key, r, m)$. The encryptions are computationally indistinguishable: for any two messages m and m', the distributions of $E(pub\_key, r, m)$ and $E(pub\_key, r, m')$ are computationally indistinguishable.

If pub_key is meaningless, then knowing c and priv_key yields no information about m. That is, for any two messages m and m', the distributions of $E(pub\_key, r, m)$ and $E(pub\_key, r, m')$ are identical.

Distinguishing Meaningful from Meaningless: *Given two public keys, one meaningful and one meaningless, guessing which is which cannot be done with a non-negligible advantage over $\frac{1}{2}$ by a probabilistic polynomial time tester. However, when supplied with the corresponding private key, the test is polynomial.*

Meaningful/meaningless encryption schemes can be constructed based on *Decisional Diffie Hellman*, using the construction in [24], on *Quadratic Residousity* [13], and on any *homomorphic encryption*[5].[6] For completeness we describe a construction of E that assumes the intractability of Quadratic Residousity, based on the scheme of Goldwasser and Micali [13].

Recall that in Goldwasser and Micali's scheme two distinct large prime numbers p and q are generated, and (p, q) is used as a private key. The public key generated is (N, x) where $N = pq$ and x is a quadratic *non*-residue of N ($x \neq z^2 \bmod N$) that has a Jacobi Symbol of $+1$. Each bit b_i of the message m is encrypted separately by choosing $y_i \in_R \mathbb{Z}_n^*$ and calculating $c_i = y_i^2 x^{b_i} \bmod N$.

---

[5] Homomorphic encryption is an encryption scheme with the additional special property: given two ciphertexts it is possible to generate a ciphertext for the sum (or multiplication) of the corresponding plaintexts.

[6] An interesting open problem is finding the minimal assumptions under which such a meaningful/meaningless encryption scheme can be constructed. The task requires non-trivial SZK: given a public key *pub_key* and two messages m and m' players should not be able to tell whether the two efficiently generated distributions $E(pub\_key, r, m)$ and $E(pub\_key, r, m')$ are identical or far apart. This problem was shown to be in SZK [27], and hence we must assume that there is a problem in SZK that is not in BPP.

The ciphertext is $(c_1, ..., c_n)$, and it can be decrypted using the private key (p, q): $b_i = 0$ iff c_i is a quadratic residue.

To construct a meaningful/meaningless encryption E we modify this scheme such that x is a random quadratic residue with probability $1 - \beta$, and a random quadratic non-residue with Jacobi Symbol of $+1$ with probability β. Note that if x is a quadratic residue, c_i is always a quadratic residue, and nothing can be learned about b_i, even when p and q are known.

Claim. The scheme E described above is a meaningful/meaningless encryption.

4   The Rational Secret Sharing Scheme

4.1   The Scheme

We describe an m-out-of-n rational secret sharing scheme for the SBC model.

The Dealer's Protocol. The scheme works for *any kind* of m-out-of-n shares the dealer may distribute (e.g. Shamir shares), provided that he additionally issues information-theoretic authentications for each share. For concreteness, it is assumed that the authentication information given to each player consists of a *tag* and a *hash function*. The hash function should allow the player to verify the authenticity of shares broadcasted by the others in probabilistic polynomial time and with error probability negligible in the security parameter. The tag should allow the player to prove the authenticity of the share he uses. The authentication information held by a group of players must not disclose any information about the other players' shares.[7]

The Players' Protocol. The reconstruction protocol is called `clean-slate` and it proceeds in a sequence of iterations. The protocol, like the one described in Section 1.3, uses a parameter β and has the property that after any sequence of iterations, the probability that the next iteration is the last one, revealing the secret, is β. Every iteration of the protocol consists of the following steps:

The *Key Generation* step. In each iteration new private and public keys for a β - meaningful/meaningless encryption are generated. This is done via a (non-rational) SMPC that takes no inputs, and generates the keys using a randomly chosen seed. The seed is shared between the players, and the public key, as well as a *perfectly binding* commitment to each of the seed's shares, are published.

If the public key generated is meaningful (which happens with probability β), we call the iteration *meaningful*, otherwise the iteration is *meaningless*. The protocol is designed not to reveal any information about the secret in meaningless iterations, and to allow the players to uncover the secret during the first meaningful iteration.

---

[7] For example, this can be done using the following method (see [31,26]): if player i's true information is $x \in \mathbb{F}$, then $s_i, b_i \in \mathbb{F}$, $b_i \neq 0$, are chosen at random and we set $c_i = b_i \cdot x + s_i \in \mathbb{F}$. The value s_i (the tag) is given to i. The other players each get b_i and c_i (the hash function). Player i is required to broadcast s_i in order to prove that x is his true information. The other players can then verify with high probability by checking that $c_i = b_i \cdot x + s_i$.

The *Encryption* and *Verification* steps. Players encrypt their share of the secret and authentication information (i.e., the tag and the hash function) using the meaningful/meaningless encryption with the public key generated in the last step. The ciphertexts are broadcasted and then validated by another SMPC.

The verification process takes as inputs the shares of the seed used to generate the keys, and additionally uses the broadcasted ciphertexts and the commitments published during the Key Generation step. It authenticates the seed's shares using the commitments, and uses them to regenerate the private key. Since the commitments are binding, the original private key is always the one generated, allowing the process to correctly determine whether the iteration is meaningful. If it is, the ciphertexts are decrypted and the retrieved authentication information is used to authenticate the retrieved shares of the secret, by verifying that all the tags and hash functions match.

The verification is considered to be successful if: (i) each seed share is a valid opening of the corresponding commitment, (ii) in case of a meaningful iteration, each ciphertext is valid encryption of a share of secret and a corresponding authentication.

A key point is that the verification process does not take the players' shares or authentication information as inputs, and when the public key is meaningless the ciphertexts it uses convey no information about the shares of the secret.

The *Exchange* step. If the verification process was successful, players *simultaneously* broadcast their shares of the seed. Each player then authenticates all seed's shares, regenerates the seed and determines by himself whether the iteration is meaningful. If it is, he decrypts the ciphertexts and uses the extracted shares of the secret to reconstruct the secret. Otherwise, the protocol proceeds to the next iteration.

Recall that players have only a small chance of discovering whether the key is meaningful before the seed's shares are revealed, since there is no efficient way of checking it. Thus, they are motivated to participate in the Exchange step. The complete protocol is described in Figure 1.

4.2   Scheme Analysis

We next argue that the suggested scheme is a computational rational secret sharing scheme. We first claim that `clean-slate` satisfies the following property, leading to its name: assuming that all players except (maybe) players in the coalition C are following the protocol, then no information about the secret is revealed before the last iteration (that is, every iteration "starts off with a clean slate"). The reason is that players' shares and authentications are only used by the protocol to create the encrypted messages. However, all iterations before the last one are meaningless, thus previous ciphertexts were created using meaningless keys and are simply random.

To show that no coalition C of size at most $m - 1$ has an incentive to deviate after any sequence of iterations, we note that for any joint strategy players in C may follow, they cannot be worse-off (up to an exponentially small factor) by

---

clean-slate$_i$(*share, authen*)

Let P be the set of players participating in the reconstruction, and denote $p = |P|$.

Repeat

If one of the following tests fail, or if a deviation was detected in one of the cryptographic schemes, quit.

Key Generation: Players run an SMPC of the function *GenarateKey*:

GenarateKey

- Choose p random strings, $(r_i)_{i \in P}$, of length $k + t$ where t is the iteration number and k is the initial security parameter.
- Generate public and private keys *pub_key*, *priv_key*, for E using $\oplus_{i \in P} r_i$ as a seed.
- Choose p random strings, $(rand\_r_i)_{i \in P}$, of length $k + t$ and set $com\_r_i = \mathsf{Commit}(r_i, rand\_r_i)$.
- **Public Output**: The public key *pub_key*, and the commitments $(com\_r_i)_{i \in P}$.
- **Private Output**: The values r_i and $rand\_r_i$ are given to player i.

Encryption: Encrypt *share* and *authen* using E with parameter β and with the public key *pub_key*, and broadcast the encrypted message C_i.

Verification: Players run an SMPC of the function $Verify$ that takes $(r_i, rand\_r_i)_{i \in P}$ as inputs:

$Verify$

- Check that each input pair is a valid opening of the corresponding commitment. That is, verify $com\_r_i = \mathsf{Commit}(r_i, rand\_r_i)$.
- Regenerate *priv_key* using $\oplus_{i \in P} r_i$ as a seed, and use it to check whether *pub_key* is meaningful.
- If so, decrypt each C_i using *priv_key*, and get the shares of the secret and authentication information of each player. Check that the shares are consistent with the authentications by verifying that all the tags and hash functions match.

Exchange:

- Broadcast r_i and $rand\_r_i$.
- Evaluate the first two stages of $Verify$ by yourself.
- If the *pub_key* is meaningful, reconstruct the secret using the retrieved shares (as done in the last step of $Verify$). Quit and Output the reconstructed secret.

Fig. 1. The rational secret sharing reconstruction protocol

always following the Key Generation, Encryption, and Verification steps: Key Generation and Verification are done via an SMPC, and therefore cannot be broken with a non-negligible probability. As to broadcasting a valid ciphertext - in a meaningless iteration no information can be gained anyway, and in a meaningful iteration the verification step detects invalid ciphertexts with high probability. Thus, we may assume that players only deviate during the Exchange step by broadcasting a seed share that does not open the commitment published in the Key Generation step. Such deviations are always detected, since the commitments to the shares are perfectly binding.

We argue that a coalition can only gain from deviating in the Exchange step of a meaningful iteration: if it deviates in a meaningless iteration, then no information about the secret is revealed due to the clean slate property, and thus the players are forced to guess the secret. Recall that a coalition cannot efficiently distinguish between meaningful and meaningless iterations before the Exchange step, if all its players have broadcasted valid encryptions (which is what we assume). Therefore, if the coalition deviates in meaningful iterations with a certain probability, it must deviate in meaningless ones with almost the same probability. As before, for a sufficiently small β, the risk of deviating in a meaningless iteration and causing the game to end is too great.

Theorem 1. *Let* $2 \leq m \leq n$, Y *be a finite set of secrets, and* `dealer` *be an algorithm assigning* m-*out-of-*n *information-theoretic authenticatable shares. Assume that* $\alpha < \alpha_0$ *and* $\beta < \beta_0$. *The scheme* (`dealer`, `clean-slate`) *is a computational rational* m-*out-of-*n *secret sharing scheme for* Y *with expected number of iterations* $O(1/\beta)$.

5   The Rational SMPC Protocol

5.1   The Protocol

We present the protocol `secure-clean-slate`, a rational SMPC protocol for the SBC model, based on protocol suggested in Section 4. The new protocol, like the previous one, ensures that no information is leaked until the final iteration (in an information theoretical sense). However, it additionally protects the inputs (in a computational sense) during the last iteration. This is done by composing the meaningful/meaningless technique with *Yao's Circuit Garbling* method.[8]

Recall that a Garbled Circuit is an encrypted form of an original circuit. It allows the circuit to be evaluated, but reveals no information except the result of the evaluation. A Garbled Circuit consists of: two random (garbled) strings assigned to each input wire (the first corresponds to a 0 value, and the other to a 1), gates tables, and translation tables for outputs. To evaluate the original circuit on a specific input, the Garbled Circuit is evaluated for the corresponding garbled strings using the gates tables. Then, the output is translated using the outputs translation tables. For a detailed description of Garbled Circuits see [22]. The `clean-slate` protocol in changed in the following way:

Adding the step of *Creating Garbled Circuit.* In every iteration the protocol constructs a new Garbled Circuit from the circuit representing f. The gates tables and translation tables are made public, and commitments to both garbled strings corresponding to each input wire are published *in an arbitrary order* (the reason for the arbitrary order will be made clear later). However, players are not

---

[8] General techniques for (non-rational) SMPC do not offer information-theoretic protection for both sides, thus cannot be used directly. In models in which such protocols can be constructed, we can use the secret sharing scheme from the last section in order to allow players to fairly exchange the last messages sent by the protocols.

given both garbled strings assigned to each of their input wires, since this will allow player i to learn $f(\mathbf{x}_{-i}, x_i')$ for every x_i'. Instead, a share of an n-out-of-n secret sharing of each garbled string assigned to an input wire is given to every player, and commitments to all shares are published.

Adding the step of *Obtaining Garbled Inputs*. Each player obtains *one* of the garbled strings chosen for each of his input wires according to the value assigned to the wire by his input. Player i gets all the shares of each such garbled string by engaging in a 1-out-of-2 OT protocol with every player j. When running the OT protocol, player j is the sender and his values are the shares of the two garbled strings chosen for i's input wire. Player i is the receiver, and his goal is to learn the value corresponding to his input bit. As discussed in Section 3, the OTs give information-theoretic protection to the receiver regarding the value he received, and computational security to the sender about the other value. This kind of protection is crucial, since we want to ensure that no information about i's input is leaked during meaningless iterations.

For ease of exposition we say that the sender (player j) sends encryptions of his two values to the receiver (player i) when the OT protocol is carried out. We require j to supply an additional ZK proof to convince i that both encryptions are valid. That is, after sending the encryptions, j must prove to i that each encryption contains a value that opens the corresponding commitment published during the Creating Garbled Circuit step.

Revising the steps of *Encryption* and *Verification*. Players encrypt their *garbled strings*, instead of their original inputs, using the β - meaningful/ meaningless encryption with the public key generated in the Key Generation step.

The verification process is changed: in a meaningful iteration it decrypts the ciphertexts and retrieves the garbled strings for each input bit. It then verifies that each extracted garbled string indeed opens *one* of the corresponding commitments. Note that since the commitments to the garbled strings corresponding to the same input wire were published in an arbitrary order when the Garbled Circuit was created, no information about the real value of this input wire is revealed to the other players.

During the Exchange step of a meaningful iteration the garbled strings are retrieved from the ciphertexts, allowing all players to learn the function's value, but protecting the original inputs. In a meaningless iteration, no information about the garbled strings encoding the real inputs is revealed, and hence no information about the real inputs is disclosed either. The complete protocol is described in Figure 2.

5.2 Protocol Analysis

We next argue that `secure-clean-slate` is a computational rational SMPC protocol. As discussed before, the protocol is secure (in the cryptographic sense), since no information about the inputs is revealed before the last iteration, and due to the fact that the Garbled Circuit created in the last iteration protects players' inputs computationally. To show that the protocol is also $\mathcal{C}$-immune, we

---

`secure-clean-slate`$_i$ $(input)$

Repeat

If one of the following tests fail, or if a deviation was detected in one of the cryptographic schemes, quit.

Key Generation: As in `clean-slate`$_i$ (with $P = N$).

Creating Garbled Circuit: Players run an SMPC of the function:
$CreateGarbledCiruit$

- Create a Garbled Circuit of the evaluated function f. The garbled string assigned to wire q and bit b is denoted W_q^b.
- Choose random strings $rand\_W_q^b$ of length $k + t$ where t is the iteration number and k is the initial security parameter. Denote $V_q^b = (W_q^b, rand\_W_q^b)$.
- Randomly select shares $V_q^{b,1}, ..., V_q^{b,n}$ such that $V_q^b = \oplus V_q^{b,i}$, and strings $rand\_V_q^{b,i}$ of length $k + t$.
- **Public Output**: (i) Tables for the garbled gates and translation tables for the outputs. (ii) The commitments $com\_W_q^b = \mathsf{Commit}(W_q^b, rand\_W_q^b)$. *For every input wire q, the commitments $com\_W_q^0, com\_W_q^1$ are output in an arbitrary order.* (iii) The commitments $com\_V_q^{b,i} = \mathsf{Commit}(V_q^{b,i}, rand\_V_q^{b,i})$.
- **Private Output**: The values $V_q^{b,i}$ and $rand\_V_q^{b,i}$ are given to player i.

Obtaining Garbled Inputs: If player i holds the qth input bit of f and its value is b, he engages in a 1-of-2 OTs (perfectly protecting player i) with every other player j, in order to get $V_q^{b,j}$ and $rand\_V_q^{b,j}$. When running an OT protocol, after player j sends encryptions of his two pair of values, $(V_q^{0,j}, rand\_V_q^{0,j})$ and $(V_q^{1,j}, rand\_V_q^{1,j})$, to player i, he supplies a ZK proof to convince i that each encryption contains a pair that is a valid opening the corresponding commitment $(comm\_V_q^{0,j}$ or $comm\_V_q^{1,j})$. Player i then reconstructs V_q^b using the received shares.

Encryption: Player i encrypts all V_q^b acquired during the previous step using E with parameter β and public key pub_key, and broadcasts the ciphertext C_i.

Verification: As in done in `clean-slate`$_i$, a $Verify$ procedure is run via an SMPC. The previous procedure is changed: if the key is meaningful, it decodes every C_i and checks that for every input bit q, the retrieved value $V_q^b = (W_q^b, rand\_W_q^b)$ is an opening of one of the commitments $com\_W_q^0$ or $com\_W_q^1$.

Exchange: As in `clean-slate`$_i$ with the exception that if the public key is meaningful, the function's value is obtained by evaluating the garbled circuit using the gates tables on the garbled strings extracted from the ciphertexts, and then translating the output using the outputs translation tables.

Fig. 2. The rational SMPC protocol

must first assume that players in every coalition $C \in \mathcal{C}$ have an initial incentive to use their true inputs when running a protocol that computes f. Note that although non-rational SMPC protocols allow players to change their inputs, we must rule out such behaviors since our utility functions only reward players for learning the value of f evaluated on the *original* inputs.

One way of ensuring such incentives is to assume that players in C would have reported their true inputs had a trusted mediator been running the computation.

That is, by using fictitious inputs, players in C are unlikely to be able to change the output of the calculation and still deduce the designated value (see discussion in [29]).[9] An alternative way is to assume the presence of an authenticator that produces authentication information for the inputs (as was done in the secret sharing scheme of Section 4). If one of the above options holds, we say that players in C *have an initial incentive to use their true shares*. When such an incentive is assumed, the described protocol can be shown to be C-immune using the arguments made for the `clean-slate` protocol.

Theorem 2. *Let f be a polynomial time computable function, and let C be a set of coalitions. Assume that players in every coalition $C \in C$ have an initial incentive to use their true shares, and that $\alpha < \alpha_0$ and $\beta < \beta_0$. The protocol* `secure-clean-slate` *is a computational rational SMPC protocol for f with expected number of iterations $O(1/\beta)$.*

6   The Rational SMPC Protocol for the NSBC Model

We describe the protocol `NSBC-secure-clean-slate`, a rational SMPC protocol for the NSBC model, based on the protocol suggested in Section 5. We first note that the trivial way of dividing every simultaneous round of the previous protocol into n non-simultaneous rounds fails: the last player to broadcast his share of the seed in the Exchange step of the meaningful iteration has already learned the value, and thus has no incentive to cooperate. We construct a new protocol in which players can retrieve the value even if the last player deviated, since the needed information is revealed by the number of the round he deviated in. The previous protocol is changed in the following way:

Revising the step of *Key Generation*. The new Key Generation step generates $|Y|$ pairs of keys, instead of just one. The set of public keys generated in every iteration has the property that at most one is meaningful. An iteration containing a meaningful key is called meaningful, and the others are called meaningless. As before, no information about the inputs is revealed in meaningless iterations, and players uncover the value during the first meaningful iteration.

Revising the steps of *Encryption* and *Verification*. In the Encryption step, players are required to encrypt their inputs $|Y|$ times using each of the public keys, and broadcast the ciphertexts one-by-one.

  The verifications process is changed: in addition to validating the ciphertexts, it also outputs a permutation of the public keys. In a meaningless iteration the published permutation is completely random. But, in a meaningful iteration the permutation places the (only) meaningful key in position y, where y is the designated value, and randomly orders the rest of the keys. Note that the verification

---

[9] For example, suppose that the players' inputs are bit strings and they wish to calculate the strings' XOR. A player benefits from using a fictitious input string, even if the computation is done by a trusted mediator: the other players will get a false value, but the deviating player will be able retrieve the real value by XORing the result with both his fictitious and real strings.

process can obtain y by evaluating the Garbled Circuit on the garbled strings retrieved from the ciphertexts, and then translating the output.

Revising the step of Exchange. The Exchange step is partitioned to $|Y| \cdot n$ non-simultaneous communication rounds in which shares of the seeds used to generate the keys are revealed *one by one*. First the shares of seed 1 are revealed in the first n rounds (call it cohort 1) with player j sending his share in round j, and so on for each of the $|Y|$ seeds. If a player deviates (e.g. refuses to reveal his share of the seed), and this is the last round of the yth cohort, the other players conclude that he already learned f's value, and hence it must be y.

Note 1. The described protocol is susceptible to existence of a malicious player: such a player can cause the others to output a wrong value by simply aborting prematurely. However, the deviating player will not be able to learn the secret himself. Since we assume that all players are rational individuals that prefer to learn above all else, there will never be an incentive to such behavior.

Theorem 3. *Let f be a polynomial time computable function, and let $\mathcal{C}$ be a set of coalitions. Assume that players in every coalition $C \in \mathcal{C}$ have an initial incentive to use their true shares, and that $\alpha < \alpha_0$ and $\beta < \beta_0$. The protocol* NSBC-secure-clean-slate *is a computational rational SMPC protocol for f with expected number of communication rounds $O\left(\frac{|Y|n}{\beta}\right)$.*

References

1. Abraham, I., Dolev, D., Gonen, R., Halpern, J.: Distributed Computing Meets Game Theory: Robust Mechanisms for Rational Secret Sharing and Multiparty Computation. In: PODC, pp. 53–62 (2006)
2. Barany, I.: Fair distribution protocols or how the players replace fortune. Mathematics of Operations Research 17, 327–340 (1992)
3. Bellare, M., Micali, S.: Non-Interactive Oblivious Transfer. In: Brassard, G. (ed.) CRYPTO 1989. LNCS, vol. 435, pp. 547–557. Springer, Heidelberg (1990)
4. Ben-Porath, E.: Cheap talk in games with incomplete information. Journal of Economic Theory 108, 45–71 (2003)
5. Ben-Porath, E.: Correlation Without Mediation: Expanding the Set of Equilibria Outcomes by "Cheap" Pre-Play Procedures. Journal of Economic Theory 80, 108–122 (1998)
6. Boneh, D., Naor, M.: Timed commitments. In: Bellare, M. (ed.) CRYPTO 2000. LNCS, vol. 1880, pp. 236–254. Springer, Heidelberg (2000)
7. Dodis, Y., Halevi, S., Rabin, T.: A Cryptographic Solution to a Game Theoretic Problem. In: Bellare, M. (ed.) CRYPTO 2000. LNCS, vol. 1880, pp. 112–130. Springer, Heidelberg (2000)
8. Even, S., Goldreich, O., Lempel, A.: A Randomized Protocol for Signing Contracts. Communications of the ACM 28(6), 637–647 (1985)
9. Garay, J., Jakobsson, M.: Timed Release of Standard Digital Signatures. In: Blaze, M. (ed.) FC 2002. LNCS, vol. 2357, pp. 168–182. Springer, Heidelberg (2003)
10. Gerardi, D.: Unmediated communication in games with complete and incomplete information. Journal of Economic Theory 114, 104–131 (2004)

11. Goldreich, O.: Foundations of Cryptography. Basic Applications, vol. 2. Cambridge University Press, Cambridge (2004)
12. Goldreich, O., Micali, S., Wigderson, A.: How to Play any Mental Game. In: STOC, pp. 218–229 (1987)
13. Goldwasser, S., Micali, S.: Probabilistic Encryption. Journal of Computer and System Sciences 28, 270–299 (1984)
14. Gordon, S.D., Katz, J.: Rational Secret Sharing, Revisited. In: De Prisco, R., Yung, M. (eds.) SCN 2006. LNCS, vol. 4116, pp. 229–241. Springer, Heidelberg (2006)
15. Halpern, J., Teague, V.: Rational Secret Sharing and Multiparty Computation. In: STOC, pp. 623–632 (2004)
16. Heller, Y.: A coalition-proof cheap-talk protocol (manuscript, 2005)
17. Izmalkov, S., Micali, S., Lepinski, M.: Rational Secure Computation and Ideal Mechanism Design. In: FOCS, pp. 585–595 (2005)
18. Kol, G., Naor, M.: Cryptography and Game Theory: Designing Protocols for Exchanging Information, full version:
    www.wisdom.weizmann.ac.il/%7enaor/PAPERS/crypto_games.html
19. Kol, G., Naor, M.: Games for Exchanging Information (manuscript, 2007)
20. Lepinski, M., Micali, S., Peikert, C., Shelat, A.: Completely Fair SFE and Coalition-Safe Cheap Talk. In: PODC, pp. 1–10 (2004)
21. Lepinski, M., Micali, S., Shelat, A.: Collusion-Free Protocols. In: STOC, pp. 543–552 (2005)
22. Lindell, Y., Pinkas, B.: A Proof of Yao's Protocol for Secure Two-Party Computation. In: ECCC, Report TR04-063 (2004)
23. Lysyanskaya, A., Triandopoulos, N.: Rationality and Adversarial Behavior in Multi-Party Computation. In: Dwork, C. (ed.) CRYPTO 2006. LNCS, vol. 4117, pp. 180–197. Springer, Heidelberg (2006)
24. Naor, M., Pinkas, B.: Efficient Oblivious Transfer Protocols. In: SODA, pp. 448–457 (2001)
25. Pinkas, B.: Fair Secure Two-Party Computation. In: Biham, E. (ed.) EUROCRYPT 2003. LNCS, vol. 2656, pp. 87–105. Springer, Heidelberg (2003)
26. Rabin, T., Ben-Or, M.: Verifiable Secret Sharing and Multiparty Protocols with Honest Majority. In: STOC, pp. 73–85 (1989)
27. Sahai, A., Vadhan, S.: A Complete Problem for Statistical Zero Knowledge. Journal of the ACM 50, 196–249 (2003)
28. Shamir, A.: How to share a secret. Communications of the ACM 22, 612–613 (1979)
29. Shoham, Y., Tennenholtz, M.: Non-Cooperative Computation: Boolean Functions with Correctness and Exclusivity. TCS 343(2), 97–113 (2005)
30. Urbano, A., Vila, J.: Computational Complexity and Communication: Coordination in Two-Player Games. Econometrica 70, 1893–1927 (1992)
31. Wegman, M., Carter, L.: New hash functions and their use in authentication and set equality. JCSS 22, 265–279 (1981)
32. Wolf, S., Wullschleger, J.: Oblivious Transfer is Symmetric. In: Vaudenay, S. (ed.) EUROCRYPT 2006. LNCS, vol. 4004, pp. 222–232. Springer, Heidelberg (2006)
33. Yao, A.: How to Generate and Exchange Secrets. In: FOCS, pp. 162–167 (1986)

Equivocal Blind Signatures and Adaptive UC-Security

Aggelos Kiayias* and Hong-Sheng Zhou*

Computer Science and Engineering
University of Connecticut
Storrs, CT, USA
{aggelos,hszhou}@cse.uconn.edu

Abstract. We study the design of adaptively secure blind signatures in the universal composability (UC) setting. First, we introduce a new property for blind signature schemes that is suitable for arguing security against adaptive adversaries: an *equivocal blind signature* is a blind signature where there exists a simulator that has the power of making signing transcripts correspond to any message signature pair. Second, we present a general construction methodology for building adaptively secure blind signatures: the starting point is a 2-move "equivocal lite blind signature", a lightweight 2-party signature protocol that we formalize and implement both generically as well as concretely; formalizing a primitive as "lite" means that the adversary is required to show all private tapes of adversarially controlled parties; this enables us to conveniently separate zero-knowledge (ZK) related security requirements from the remaining security properties in the blind signature design methodology. Next, we focus on the suitable ZK protocols for blind signatures. We formalize two special ZK ideal functionalities, single-verifier-ZK (SVZK) and single-prover-ZK (SPZK), both special cases of multi-session ZK that may be of independent interest, and we investigate the requirements for realizing them in a commit-and-prove fashion as building blocks for adaptively secure UC blind signatures. Regarding SPZK we find the rather surprising result that realizing it only against static adversaries is sufficient to obtain adaptive security for UC blind signatures.

We instantiate all the building blocks of our design methodology both generically based on the blind signature construction of Fischlin as well as concretely based on the 2SDH assumption of Okamoto, thus demonstrating the feasibility and practicality of our approach. The latter construction yields the first practical UC blind signature that is secure against adaptive adversaries. We also present a new more general modeling of the ideal blind signature functionality.

1 Introduction

A blind signature is a cryptographic primitive that was proposed by Chaum [12]; it is a digital signature scheme where the signing algorithm is split into a two-party protocol between a user (or client) and a signer (or server). The signing

---

* Research partly supported by NSF CAREER Award CNS-0447808.

R. Canetti (Ed.): TCC 2008, LNCS 4948, pp. 340–355, 2008.
© International Association for Cryptologic Research 2008

protocol's functionality is that the user can obtain a signature on a message that she selects in a blind fashion, i.e., without the signer being able to extract some useful information about the message from the protocol interaction. At the same time the existential unforgeability property of digital signatures should hold, i.e., after the successful termination of a number of n corrupted user instantiations, an adversary should be incapable of generating signatures for $(n + 1)$ distinct messages.

A blind signature is a very useful privacy primitive that has many applications in the design of electronic-cash schemes, the design of electronic voting schemes as well as in the design of anonymous credential systems. Since the initial introduction of the primitive, a number of constructions have been proposed [13,32,30,35,23,36,34,37,3,1,2,5,6,7,24,31,17,21,8]. The first formal treatment of the primitive in a stand-alone model and assuming random oracles (RO) was given by Pointcheval and Stern in [35].

Blind signatures is in fact one of the few complex cryptographic primitives (beyond digital signatures, public-key encryption, and key-exchange) that have been implemented in real world Internet settings (e.g., in the Votopia [27] voting system) and thus the investigation of more realistic attack models for blind signatures is of pressing importance. Juels, Luby and Ostrovsky [23] presented a formal treatment of blind signatures that included the possibility for an adversary to launch attacks that use arbitrary concurrent interleaving of either user or signer protocols. Still, the design of schemes that satisfied such stronger modeling proved somewhat elusive. In fact, Lindell [28] showed that unbounded concurrent security for blind signatures is impossible under a simulation-based security definition without any setup assumption; more recently in [21], the generic feasibility of blind signatures without setup assumptions was shown but using a game-based security formulation.

With respect to practical provably secure schemes, assuming random oracles or some setup assumption, various efficient constructions were proposed: for example, [5,6] presented efficient two-move constructions in the RO model, while [24,31] presented efficient constant-round constructions without random oracles employing a common reference string (CRS) model (i.e., when a trusted setup function initializes all parties' inputs) that withstand concurrent attacks. While achieving security under concurrent attacks is an important property for the design of useful blind signatures, a blind signature scheme may still be insecure for a certain deployment. Game-based security definitions [35,23,7,24,31,21,8] capture properties that are intuitively desirable. Nevertheless, the successive amendments of definitions in the literature and the differences between the various models exemplify the following: on the one hand capturing all desirable properties of a complex cryptographic primitive such as a blind signature is a difficult task, while on the other, even if such properties are attained, a "provably secure" blind signature may still be insecure if deployed within a larger system. For this reason, it is important to consider the realization of blind signatures under a general simulation-based security formulation such as the one provided in the Universal Composability (UC) framework of Canetti [9] that enables us to formulate cryptographic primitives so that they remain secure under arbitrary deployments and interleavings of protocol instantiations.

In the UC setting, against static adversaries, it was shown how to construct blind signatures in the CRS model [17] with two moves of interaction. Though the construction in [17] is round-optimal, it is unknown whether it can admit concrete practical instantiations. In addition, security is argued only against static adversaries; and while it should be feasible to extend the construction of [17] in the adaptive setting this can only exacerbate the difficulty of concretely realizing the basic design. Note that using the secure two party computation compiler of [11] one can derive adaptively secure blind signatures but this approach is also generic and does not suggest any concrete design.

1.1   Our Results

In this work we study the design of blind signatures in the UC framework against adaptive adversaries. Our approach is "practice-oriented" in the sense of minimizing communication complexity as well as entailing the following points: (i) a constant number of rounds, (ii) a choice of session scope that is consistent with how a blind signature would be implemented in practice, in particular a multitude of clients and one signer should be supported within a single session, (iii) a trusted setup string that is of constant length in the number of parties within a session, (iv) the avoidance, if possible, of cryptographic primitives that are "per-bit", such as bit-commitment, where one has to spend a communication length of $\Omega(l)$ where l is a security parameter per bit of private input. Our results are as follows:

Equivocal blind signatures. We introduce a new property for blind signatures, called equivocality that is suitable for arguing security against adaptive adversaries. In an equivocal blind signature there exists a simulator that has the power to construct the internal state of a client including all random tapes so that any simulated communication transcript can be mapped to any given valid message-signature pair. This capability should hold true even after a signature corresponding to the simulated transcript has been released to the adversary. Equivocality can be seen as a strengthening of the notion of blindness as typically defined in game-based security formulations of blind signatures: in an equivocal blind signature, signing transcripts can be simulated in an independent fashion to the message-signature pair they correspond to.

General methodology for building UC blind signatures. We present a general methodology for designing adaptively secure UC blind signatures. Our starting point is the notion of an *equivocal lite blind signature*: The idea behind "lite" blind signatures is that security properties should hold under the condition that the adversary deposits the private tapes of the parties he controls. This "open-all-private-tapes" approach simplifies the blind signature definitions substantially and allows one to separate security properties that relate to zero-knowledge compared to other necessary properties for blind signatures. Note that this is not an honest-but-curious type of adversarial formulation as the adversary is not required to be honestly simulating corrupted parties; in particular, the adversary may deviate from honest protocol specification (e.g., bias the random tapes) as long as he can present private tapes that match the communication transcripts.

We then demonstrate two instantiations of an equivocal lite blind signature, one that is based on generic cryptographic primitives that is inspired by the blind signature construction of [17] and one based on the design and the 2SDH assumption of [31].

Study of the ZK requirements for UC blind signatures. Having demonstrated equivocal lite blind-signatures as a feasible starting building block, we then focus on the formulation of the appropriate ZK-functionalities that are required for building blind signatures in the adaptive adversary setting. Interestingly, the user and the signer have different ZK "needs" in a blind signature. In particular the corresponding ZK-functionalities turn out to be simplifications of the standard multi-session ZK functionality $\mathcal{F}_{\mathrm{MZK}}$ that restrict the multi-sessions to occur either from many provers to a single verifier (we call this $\mathcal{F}_{\mathrm{SVZK}}$) or from a single prover to many verifiers (we call this $\mathcal{F}_{\mathrm{SPZK}}$). Note that this stems from our blind signature *session scope* that involves a multitude of users interacting with a single signer: this is consistent with the notion that a blind-signature signer is a server within a larger system and is expected that the number of such servers would be very small compared to a much larger population of users and verifiers.

First, regarding $\mathcal{F}_{\mathrm{SVZK}}$, the ZK protocol that users need to execute as provers, we show that it can be realized in a commit-and-prove fashion using a commitment scheme that, as it is restricted to the single-verifier setting, it does not require built-in non-malleability (while such property would be essential for general multi-session UC commitments). We thus proceed to realize $\mathcal{F}_{\mathrm{SVZK}}$ using mixed commitments [16,29] with only a constant length common reference string (as opposed to linear in the number of parties that is required in the multi-session setting). Second, regarding $\mathcal{F}_{\mathrm{SPZK}}$, the ZK protocol the signer needs to execute as a prover, we find the rather surprising result that it needs only be realized against static adversaries for the resulting blind signature scheme to satisfy adaptive security. This enables a much more efficient realization design for $\mathcal{F}_{\mathrm{SPZK}}$ as we can implement it using merely an extractable commitment and a Sigma protocol (alternatively, using an Ω-protocol [18]). The intuition behind this result is that in a blind signature the signer is not interested in hiding his input in the same way that the user is: this can be seen by the fact that the verification-key itself leaks a lot of information about the signing-key to the adversary/environment, thus, using a full-fledged zero-knowledge instantiation is an overkill from the signer's point of view; this phenomenon was studied in the context of zero-knowledge in [26]. We note that our $\mathcal{F}_{\mathrm{SPZK}}$ functionality can be seen as a special instance of client-server computation as considered in [38] (where the relaxed non-malleability requirement of such protocols was also noted); interestingly $\mathcal{F}_{\mathrm{SVZK}}$ falls outside that framework (despite its client-server nature).

Notations: $a \xleftarrow{r}$ RND denotes randomly selecting a in its domain; negl() denotes negligible function; poly() denotes polynomial function.

2 Equivocal Lite Blind Signatures

2.1 Building Block: Equivocal Lite Blind Signatures

A signature generation protocol is a tuple $\langle$CRSgen, gen, lbs$_1$, lbs$_2$, lbs$_3$, verify$\rangle$ where CRSgen is a common reference string generation algorithm, gen is a key-pair generation algorithm, lbs$_i$, $i = 1, 2, 3$, comprise a two-move signature generation protocol between the user U and the signer S as described in Figure 1 and verify is a signature verification algorithm. A *lite blind signature* is a signature generation protocol that satisfies completeness (see definition 1) as well as two security properties, *lite-unforgeability* and *lite-blindness*, defined below (consistency is another property [10] for signatures that will be trivially satisfied in our design and thus we omit it in this version).

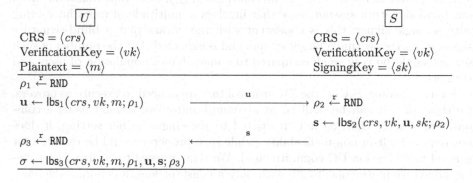

Fig. 1. Outline of a two-move signature generation protocol

Definition 1 (Completeness). *A signature generation protocol as in Figure 1 is complete if for all $(crs, \tau) \leftarrow$ CRSgen(1^λ), for all $(vk, sk) \leftarrow$ gen(crs), for all $\rho_1, \rho_2, \rho_3 \xleftarrow{r}$ RND, whenever $\mathbf{u} \leftarrow$ lbs$_1(crs, vk, m; \rho_1)$, $\mathbf{s} \leftarrow$ lbs$_2(crs, vk, \mathbf{u}, sk; \rho_2)$, and $\sigma \leftarrow$ lbs$_3(crs, vk, m, \rho_1, \mathbf{u}, \mathbf{s}; \rho_3)$, then verify$(crs, vk, m, \sigma) = 1$.*

Lite-unforgeability that we define below suggests informally that if we "collapse" the lbs$_1$, lbs$_2$ procedures into a single algorithm this will result to a procedure that combined with lbs$_3$ will be equivalent to the signing algorithm of an unforgeable digital signature sign in the sense of [19]. We note that lite-unforgeability is much weaker compared to regular unforgeability of blind signatures (as defined e.g., in [31,21]) since it requires from the adversary to open the internal tapes of each user instance (as opposed to hiding such internals in the usual unforgeability definition for blind signatures); note that this is not an honest-but-curious modeling as the adversary is not restricted to flip coins honestly.

Definition 2 (Lite-unforgeability). *A signature generation protocol as in Figure 1 is lite-unforgeable if for all PPT* $\mathcal{A} = (\mathcal{A}_1, \mathcal{A}_2)$ *and for any* $L = \mathsf{poly}(\lambda)$, *we have* $\mathsf{Adv}_{\mathrm{luf}}^{\mathcal{A},L}(\lambda) \leq \mathsf{negl}(\lambda)$, *where* $\mathsf{Adv}_{\mathrm{luf}}^{\mathcal{A},L}(\lambda) \overset{\mathrm{def}}{=} \Pr[\mathbf{Exp}_{\mathcal{A},L}^{\mathrm{LUF}}(\lambda) = 1]$ *and the experiment* $\mathbf{Exp}_{\mathcal{A},L}^{\mathrm{LUF}}(\lambda)$ *is defined below:*

Experiment $\mathbf{Exp}_{\mathcal{A},L}^{\mathrm{LUF}}(\lambda)$

$(crs, \tau) \leftarrow \mathtt{CRSgen}(1^{\lambda})$; $(vk, sk) \leftarrow \mathsf{gen}(crs)$; $state := \emptyset$; $k := 0$;

while $k < L$

$\quad (m_k, \rho_{1,k}, state) \leftarrow \mathcal{A}_1(state, crs, vk)$;

$\quad \mathsf{s}_k \leftarrow \mathsf{lbs}_2(crs, vk, \mathsf{lbs}_1(crs, vk, m_k; \rho_{1,k}), sk; \rho_{2,k})$; $\rho_{2,k} \xleftarrow{r} \mathtt{RND}$;

$\quad state \leftarrow state \| \mathsf{s}_k$; $k \leftarrow k + 1$;

$(m_1, \sigma_1, \ldots, m_\ell, \sigma_\ell) \leftarrow \mathcal{A}_2(state)$;

if $\ell > L$, *and* $\mathtt{verify}(crs, vk, m_i, \sigma_i) = 1$ *for all* $1 \leq i \leq \ell$,

$\quad$ *and* $m_i \neq m_j$ *for all* $1 \leq i \neq j \leq \ell$

then return 1 *else return* 0.

Similarly we can formulate blindness (as defined, e.g. in [8]) in the "lite" setting by requiring the adversary to open the private tape of the signer for each user interaction. Given that blindness is subsumed by our equivocality property (defined below), we will not explore this direction further here (the reader may refer to the full version [25] for more details). For simplicity we define equivocality only for two-move protocols following the skeleton of Figure 1. Informally an equivocal blind signature scheme is accompanied by a simulator procedure $\mathcal{I}$ which can produce signature generation transcripts without using the user input m and furthermore it can "explain" the transcripts to any adversarially selected m even after the signature σ for m has been generated. The property of equivocal blind signatures parallels the property of equivocal commitments [4] or zero-knowledge with state reconstruction, cf. [20]. We define the property formally below (cf. Figure 2). In the definition, we use the relation $\mathsf{KeyPair}$ defined as $(vk, sk) \in \mathsf{KeyPair}$ if and only if $(vk, sk) \leftarrow \mathsf{gen}(crs)$ (omitting crs to avoid cluttering the notation). Note that we require $(vk, sk), (vk, sk') \in \mathsf{KeyPair}$ to imply $sk = sk'$ (otherwise a blind signature *may* be susceptible to an attack due to [22]).

Definition 3 (Equivocality). *We say that a signature generation protocol is equivocal if there exists an interactive machine* $\mathcal{I} = (\mathcal{I}_1, \mathcal{I}_2)$, *such that for all PPT* $\mathcal{A}$, *we have* $\mathsf{Adv}_{\mathrm{eq}}^{\mathcal{A}}(\lambda) \leq \mathsf{negl}(\lambda)$,

$$\mathsf{Adv}_{\mathrm{eq}}^{\mathcal{A}}(\lambda) \overset{\mathrm{def}}{=} \left| \begin{array}{l} \Pr[(crs, \tau) \leftarrow \mathtt{CRSgen}(1^{\lambda}) : \mathcal{A}^{Users(crs, \cdot)}(crs) = 1] \\ - \Pr[(crs, \tau) \leftarrow \mathtt{CRSgen}(1^{\lambda}) : \mathcal{A}^{\mathcal{I}(crs, \tau, \cdot)}(crs) = 1] \end{array} \right|,$$

where oracle $Users(crs, \cdot)$ *operates as:*

- *Upon receiving message* (i, m, vk) *from* $\mathcal{A}$, *select* $\rho_1 \xleftarrow{r} \mathtt{RND}$ *and compute* $\mathbf{u} \leftarrow \mathsf{lbs}_1(crs, vk, m; \rho_1)$, *record* $\langle i, m, vk, \mathbf{u}, \rho_1 \rangle$ *into history$_i$, and return message* $(i, \mathbf{u})$ *to* $\mathcal{A}$.

- *Upon receiving message $(i, \mathbf{s}, \rho_2, sk)$ from $\mathcal{A}$, if there exists a record $\langle i, m, vk, \mathbf{u}, \rho_1 \rangle$ in $history_i$ and both $(vk, sk) \in$ KeyPair and $\mathbf{s} = \mathsf{lbs}_2(crs, vk, \mathbf{u}, sk; \rho_2)$ hold, then select $\rho_3 \xleftarrow{r}$ RND, compute $\sigma \leftarrow \mathsf{lbs}_3(crs, vk, m, \rho_1, \mathbf{u}, \mathbf{s}; \rho_3)$, update $\langle i, m, vk, \mathbf{u}, \rho_1 \rangle$ in $history_i$ into $\langle i, m, vk, \mathbf{u}, \sigma, \rho_1, \rho_3 \rangle$, and return to $\mathcal{A}$ the pair (i, σ); otherwise return to $\mathcal{A}$ the pair $(i, \perp)$.*
- *Upon receiving message (i, open), return to $\mathcal{A}$ the pair $(i, history_i)$.*

and oracle $\mathcal{I}(crs, \tau, \cdot)$ operates as:

- *Upon receiving message (i, m, vk) from $\mathcal{A}$, run $(\mathbf{u}, aux) \leftarrow \mathcal{I}_1(crs, \tau, vk)$, record $\langle i, m, vk, \mathbf{u}, aux \rangle$ into temp, and return message $(i, \mathbf{u})$ to $\mathcal{A}$.*
- *Upon receiving message $(i, \mathbf{s}, \rho_2, sk)$ from $\mathcal{A}$, if there exists a record $\langle i, m, vk, \mathbf{u}, aux \rangle$ in temp and both $(vk, sk) \in$ KeyPair and $\mathbf{s} = \mathsf{lbs}_2(crs, vk, \mathbf{u}, sk; \rho_2)$ hold, then select $\gamma \xleftarrow{r}$ RND, compute $\sigma \leftarrow \mathsf{sign}(crs, vk, sk, m, \gamma)$ (where sign is the "collapse" of lbs_i for $i = 1, 2, 3$), update $\langle i, m, vk, \mathbf{u} \rangle$ in temp into $\langle i, m, vk, \mathbf{u}, aux; \mathbf{s}, sk, \rho_2; \sigma, \gamma \rangle$, and return the pair (i, σ) to $\mathcal{A}$; otherwise return to $\mathcal{A}$ the pair $(i, \perp)$.*
- *Upon receiving message (i, open), if there exists a record $\langle i, m, vk, \mathbf{u}, aux \rangle$ in temp then run $\rho_1 \leftarrow \mathcal{I}_2(i, temp)$, record $\langle i, m, vk, \mathbf{u}, \rho_1 \rangle$ into $history_i$, and return to $\mathcal{A}$ the pair $(i, history_i)$; if there exists a record $\langle i, m, vk, \mathbf{u}, aux; \mathbf{s}, sk, \rho_2; \sigma, \gamma \rangle$ in temp, then run $(\rho_1, \rho_3) \leftarrow \mathcal{I}_2(i, temp)$, record $\langle i, m, vk, \mathbf{u}, \sigma, \rho_1, \rho_3 \rangle$ into $history_i$, and return message $(i, history_i)$ to $\mathcal{A}$.*

We call a signature generation protocol that satisfies completeness, lite-unforgeability as well as the equivocality property an *equivocal lite blind signature scheme*.

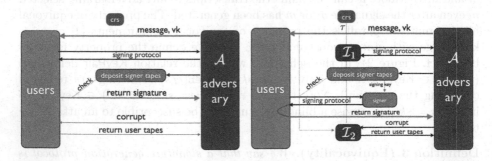

Fig. 2. The two worlds an equivocality adversary is asked to distinguish in Definition 3. In the left-hand the adversary is interacting with a set of users whereas in the right-hand side the users are interacting with an honest signer instantiation whereas the adversary is interacting with the simulator $\mathcal{I}$.

2.2   Constructions

In this subsection, we present two equivocal lite-blind signature constructions. The first construction is generic and is based on the blind signature design of

[17] whereas the second is a concrete construction that is based on [31]. In the full version of this work [25] we present additional constructions.

Generic equivocal lite blind signature. Our first construction is based on [17]; the main difference here is that we need the equivocality property (the original design employed two encryption steps for the user that are non-equivocal); in our setting, it is sufficient to have just one equivocal commitment (that is not extractable) in the first stage and then employ an extractable commitment in the second (that is not equivocal). Refer to the signature generation protocol in Figure 3: the CRSgen algorithm produces $crs = \langle pk_{\mathrm{eqc}}, pk_{\mathrm{exc}}, crs_{\mathrm{nizk}} \rangle$; EQC is a commitment scheme with committing key pk_{eqc} and EQCcom is its committing algorithm; EXC is a commitment scheme with committing key pk_{exc} and EXCcom is its committing algorithm; NIZK is an NIZK argument scheme with CRS crs_{nizk} where NIZKprove is the proof generation algorithm and NIZKverify is the proof verification algorithm. The **gen** algorithm produces a key-pair $\langle vk, sk \rangle$ for a signature scheme SIG where SIGsign is the signature generation algorithm and SIGverify is the corresponding verification algorithm. The language $\mathcal{L}_R \overset{\text{def}}{=} \{x | (x, w) \in R\}$ where $R \overset{\text{def}}{=} \{(crs, vk, E, m), (\mathbf{u}, \mathbf{s}, \rho_1, \rho_3) \,|\, \mathbf{u} = \mathrm{EQCcom}(pk_{\mathrm{eqc}}, m; \rho_1) \wedge \mathrm{SIGverify}(vk, \mathbf{u}, \mathbf{s}) = 1 \wedge E = \mathrm{EXCcom}(pk_{\mathrm{exc}}, \mathbf{u}, \mathbf{s}; \rho_3)\}$. The **verify** algorithm given a message m and signature σ operates as follows: parse σ into E and ϖ, and check that $\mathrm{NIZKverify}((crs, vk, E, m), \varpi) =^? 1$.

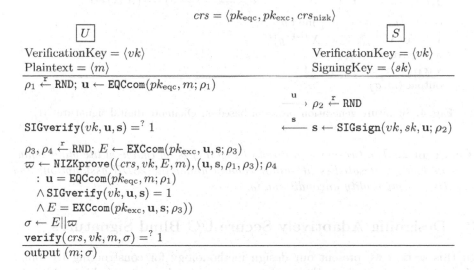

$$crs = \langle pk_{\mathrm{eqc}}, pk_{\mathrm{exc}}, crs_{\mathrm{nizk}} \rangle$$

U S

VerificationKey $= \langle vk \rangle$ VerificationKey $= \langle vk \rangle$

Plaintext $= \langle m \rangle$ SigningKey $= \langle sk \rangle$

$\rho_1 \overset{r}{\leftarrow} \mathrm{RND};\ \mathbf{u} \leftarrow \mathrm{EQCcom}(pk_{\mathrm{eqc}}, m; \rho_1)$

       $\overset{\mathbf{u}}{\longrightarrow}$ $\rho_2 \overset{r}{\leftarrow} \mathrm{RND}$

       $\overset{\mathbf{s}}{\longleftarrow}$ $\mathbf{s} \leftarrow \mathrm{SIGsign}(vk, sk, \mathbf{u}; \rho_2)$

SIGverify$(vk, \mathbf{u}, \mathbf{s}) =^? 1$

$\rho_3, \rho_4 \overset{r}{\leftarrow} \mathrm{RND};\ E \leftarrow \mathrm{EXCcom}(pk_{\mathrm{exc}}, \mathbf{u}, \mathbf{s}; \rho_3)$
$\varpi \leftarrow \mathrm{NIZKprove}((crs, vk, E, m), (\mathbf{u}, \mathbf{s}, \rho_1, \rho_3); \rho_4$
    $: \mathbf{u} = \mathrm{EQCcom}(pk_{\mathrm{eqc}}, m; \rho_1)$
    $\wedge \mathrm{SIGverify}(vk, \mathbf{u}, \mathbf{s}) = 1$
    $\wedge E = \mathrm{EXCcom}(pk_{\mathrm{exc}}, \mathbf{u}, \mathbf{s}; \rho_3))$
$\sigma \leftarrow E \| \varpi$
verify$(crs, vk, m, \sigma) =^? 1$

output $(m; \sigma)$

Fig. 3. A generic signature generation protocol

Theorem 1. *The two-move signature generation protocol in Figure 3 is an equivocal lite blind signature as follows: it satisfies lite-unforgeability provided that (i) SIG is EU-CMA secure, (ii) EQC is binding, (iii) EXC is extractable, and (iv) NIZK satisfies soundness; and it satisfies equivocality provided that (i) EQC is equivocal, (ii) EXC is hiding, and (iii) NIZK is non-erasure zero-knowledge.*

Concrete equivocal lite blind signature. In Figure 4 we present a lite blind signature $\langle \text{CRSgen}, \text{gen}, \text{lbs}_1, \text{lbs}_2, \text{lbs}_3, \text{verify} \rangle$ that uses the 2SDH assumption and is based on Okamoto's blind signature scheme [31]; the main contribution here is Theorem 2 that shows that the design is in fact equivocal (instead of merely blind as shown in [31]). In this scheme the CRSgen algorithm produces $crs = \langle p, g_1, g_2, \mathbb{G}_1, \mathbb{G}_2, \mathbb{G}_T, \hat{e}, \psi, u_2, v_2 \rangle$, where $\hat{e} : \mathbb{G}_1 \times \mathbb{G}_2 \to \mathbb{G}_T$ is a bilinear map, $\mathbb{G}_1, \mathbb{G}_2$ are groups of order p, the gen algorithm produces a key-pair $vk = \langle X \rangle$, $sk = \langle x \rangle$ such that $X = g_2^x$, and the verify algorithm given a message m and signature $\sigma = \langle \varsigma, \alpha, \beta, V_1, V_2 \rangle$, responds as follows: check that $m, \beta \in \mathbb{Z}_p$, $\varsigma, V_1 \in \mathbb{G}_1$, $\alpha, V_2 \in \mathbb{G}_2$, $\varsigma \neq 1$, $\alpha \neq 1$ and $\hat{e}(\varsigma, \alpha) = \hat{e}(g_1, g_2^m u_2 v_2^\beta)$, $\hat{e}(V_1, \alpha) = \hat{e}(\psi(X), X) \cdot \hat{e}(g_1, V_2)$.

$$crs = \langle p, g_1, g_2, \mathbb{G}_1, \mathbb{G}_2, \mathbb{G}_T, \hat{e}, \psi, u_2, v_2 \rangle$$

$\boxed{U}$		$\boxed{S}$
$vk = \langle X = g_2^x \rangle$		$vk = \langle X = g_2^x \rangle$
$msg = \langle m \rangle, m \in \mathbb{Z}_p$		$sk = \langle x \rangle$

$t, s \xleftarrow{\text{r}} \mathbb{Z}_p; W \leftarrow g_1^{mt} u_1^t v_1^{st}$	$\xrightarrow{\quad W \quad}$	
		$r, l \xleftarrow{\text{r}} \mathbb{Z}_p \text{ s.t. } x + r \neq 0$
$f, h \xleftarrow{\text{r}} \mathbb{Z}_p; \varsigma \leftarrow Y^{\frac{1}{ft} \bmod p}$	$\xleftarrow{\quad Y, l, r \quad}$	$Y \leftarrow (W v_1^l)^{\frac{1}{x+r}}$
$\alpha \leftarrow X^f g_2^{fr}; \beta \leftarrow s + \frac{l}{t} \bmod p$		
$V_1 \leftarrow \psi(X)^{\frac{1}{f}} g_1^h; V_2 \leftarrow X^{fh+r} g_2^{frh}$		
$\sigma \leftarrow \langle \varsigma, \alpha, \beta, V_1, V_2 \rangle$		
$\text{verify}(crs, vk, m, \sigma) =^? 1$		
output $(m; \sigma)$		

Fig. 4. Signature generation protocol based on Okamoto digital signature [31]

Theorem 2. *The two-move protocol of Figure 4 is an equivocal lite blind signature as follows: it satisfies lite-unforgeability under the 2SDH assumption; and it satisfies equivocality unconditionally.*

3 Designing Adaptively Secure UC Blind Signatures

In this section we present our design methodology for constructing UC-blind signatures secure against adaptive adversaries, i.e., the protocol obtained by our method can UC-realize the blind signature functionality $\mathcal{F}_{\text{BSIG}}$ (defined in Figure 5). A previous formalization of the blind signature primitive in the UC setting was given by [17]. In the full version, we include in the ideal functionality an explicit description of how corruptions are handled, and we justify our definition. Note that our $\mathcal{F}_{\text{BSIG}}$ does not require strong unforgeability from the underlying signing mechanism; this makes the presentation more general as strong unforgeability is not necessary for many applications of the blind signature primitive.

Functionality $\mathcal{F}_{\text{BSIG}}$

Key generation: Upon receiving (KEYGEN, sid) from party S, verify that $sid = (S, sid')$ for some sid'. If not, ignore the input. Else, forward (KEYGEN, sid) to the adversary $\mathcal{S}$.

Upon receiving (ALGORITHMS, sid, sig, ver) from the adversary $\mathcal{S}$, record $\langle \spadesuit, \text{sig}, \text{ver} \rangle$ in $history(S)$, and output (VERIFICATIONALG, sid, ver) to party S, where sig is a signing algorithm, and ver is a verification algorithm.

Signature generation: Upon receiving (SIGN, sid, m, ver') from party $U \neq S$, where $sid = (S, sid')$, record $\langle m, \text{ver}' \rangle$ in $history(U)$, and send (SIGN, sid, U, ver') to the adversary $\mathcal{S}$.

Upon receiving (SIGNSTATUS, sid, U, SignerComplete) from the adversary $\mathcal{S}$, where U is a user that has requested a signature, output (SIGNSTATUS, sid, U, complete) to party S, and record $\langle U, \text{complete} \rangle$ in $history(S)$.

Upon receiving (SIGNSTATUS, sid, U, SignerError) from the adversary $\mathcal{S}$, where U is a user that has requested a signature, output (SIGNSTATUS, sid, U, incomplete) to party S, and record $\langle U, \perp \rangle$ in $history(S)$.

Upon receiving (SIGNATURE, sid, U, UserComplete) from the adversary $\mathcal{S}$, where U is a user that has requested a signature,

- if S is not corrupted and $\langle U, \text{complete} \rangle$ is not recorded in $history(S)$, then halt.
- if S is not corrupted and $\langle U, \text{complete} \rangle$ has been recorded in $history(S)$ that also contains $\langle \spadesuit, \text{sig}, \text{ver} \rangle$, then compute $\sigma \leftarrow \text{sig}(m, rnd)$ flipping the required random coins rnd, and do the following: if $\text{ver}'(m, \sigma) \neq 1$, then halt; else if $\text{ver}'(m, \sigma) = 1$ but $\text{ver}(m, \sigma) \neq 1$, output (SIGNATURE, sid, σ) to party U, and update $history(U)$ into $\langle m, \sigma, rnd \rangle$; else if $\text{ver}'(m, \sigma) = \text{ver}(m, \sigma) = 1$, output (SIGNATURE, sid, σ) to party U, and update $history(U)$ into $\langle m, \sigma, rnd, \text{done} \rangle$.
- else if S is corrupted, then compute $\sigma \leftarrow \text{sig}'(m, rnd)$ flipping the required random coins rnd, where sig' is an algorithm that the adversary $\mathcal{S}$ has provided specifically for U (subject to the restriction that any sig' corresponds to a single ver'), and do the following: if $\text{ver}'(m, \sigma) = 1$, output (SIGNATURE, sid, σ) to party U, update $history(U)$ into $\langle m, \sigma, rnd \rangle$; else if $\text{ver}'(m, \sigma) \neq 1$, halt.

Upon receiving (SIGNATURE, sid, U, UserError) from the adversary $\mathcal{S}$, where U is a user that has requested a signature, output (SIGNATURE, sid, $\perp$) to party U and update $history(U)$ into $\langle m \rangle$.

Signature verification: Upon receiving (VERIFY, sid, m, σ, ver') from party V, where $sid = (S, sid')$, do: if, (i) the signer S is not corrupted, (ii) $history(S)$ contains $\langle \spadesuit, \text{sig}, \text{ver} \rangle$, (iii) $\text{ver}' = \text{ver}$, (iv) $\text{ver}(m, \sigma) = 1$, and (v) there is no U such that m is recorded with done in $history(U)$, then halt. Else, output (VERIFIED, sid, $\text{ver}'(m, \sigma)$) to party V.

Fig. 5. Blind signature functionality $\mathcal{F}_{\text{BSIG}}$. Each session contains a signer and unlimited number of users. Each user U obtains at most one signature.

Further, protocols realizing the functionality of [17] require a single "global trapdoor" that enables the functionality to produce a signature for a given message that will be valid for any given public-key; while this can be handy in the security proof, it is not a mandatory requirement for a UC-blind signature (which may allow for a different trapdoor to be used by the functionality in each signature generation); we reflect this in our ideal functionality by allowing the adversary in the corrupted signer setting to "patch" the ideal functionality with a different signing key for each user. In a blind signature session we allow for a single signer (whose identity is hard-coded into the session identifier sid) and a multitude of users. Our signer is active throughout the session and, after key-generation, is responsive to any user communicating with it via the network without waiting authorization by the environment.

Our design for UC-realizing $\mathcal{F}_{\mathrm{BSIG}}$ is modular and delineates the components required for designing UC blind signatures in the adaptive security setting. We present our methodology in two steps. First, we employ an equivocal lite blind signature scheme and we operate in a hybrid world where the following ideal functionalities exist: $\mathcal{F}_{\mathrm{CRS}}, \mathcal{F}_{\mathrm{SVZK}}^{R_U}, \mathcal{F}_{\mathrm{SPZK}}^{R_S}$. Here $\mathcal{F}_{\mathrm{CRS}}$ will be an appropriate common reference string functionality; on the other hand, $\mathcal{F}_{\mathrm{SVZK}}^{R_U}, \mathcal{F}_{\mathrm{SPZK}}^{R_S}$ will be two *different* zero-knowledge functionalities that are variations of the standard multi-session ZK functionality. This reflects the fact that the ZK "needs" of the user and the signer are different in a blind signature. (1) $\mathcal{F}_{\mathrm{SVZK}}^{R_U}$ is the "single-verifier zero-knowledge functionality for the relation R_U" where the user will be the prover and, (2) $\mathcal{F}_{\mathrm{SPZK}}^{R_S}$ is the "single-prover zero-knowledge functionality for the relation R_S" where the signer will be the prover. They differ from the multi-session ZK ideal functionality $\mathcal{F}_{\mathrm{MZK}}$ (e.g., see $\hat{\mathcal{F}}_{\mathrm{ZK}}$ in figure 7, page 49, in [11]) in the following manner: $\mathcal{F}_{\mathrm{SVZK}}$ assumes that there is only a single verifier that many provers wish to prove to it a certain type of statements; on the other hand, $\mathcal{F}_{\mathrm{SPZK}}$ assumes that only a single prover exists that wishes to convince many verifiers regarding a certain type of statement. Our setting is different from previous UC-formulations of ZK where multiple provers wish to convince multiple verifiers at the same time; while we could use such stronger primitives in our design, recall that we are interested in the simplest possible primitives that can instantiate our methodology as these highlight minimum sufficient requirements for blind signature design in the UC setting.

3.1   Construction in the $(\mathcal{F}_{\mathrm{CRS}}, \mathcal{F}_{\mathrm{SVZK}}, \mathcal{F}_{\mathrm{SPZK}})$-Hybrid World

In this section we describe our blind signature construction in the hybrid world. In Figure 6, we describe a UC blind signature protocol in the $(\mathcal{F}_{\mathrm{CRS}}, \mathcal{F}_{\mathrm{SVZK}}^{R_U}, \mathcal{F}_{\mathrm{SPZK}}^{R_S})$-hybrid world that is based on an equivocal lite blind signature protocol. The relations parameterized with the ZK functionalities are $R_U = \{((crs, vk, \mathbf{u}), (m, \rho_1)) \mid \mathbf{u} = \mathsf{lbs}_1(crs, vk, m; \rho_1)\}$ and $R_S = \{((crs, vk, \mathbf{u}, \mathbf{s}), (sk, \rho_2)) \mid \mathbf{s} = \mathsf{lbs}_2(crs, vk, \mathbf{u}, sk; \rho_2) \wedge (vk, sk) \in \mathsf{KeyPair}\}$. We prove the following theorem:

Theorem 3. *Given a signature generation protocol that is an equivocal lite blind signature, the protocol $\pi_{\Sigma(\mathrm{BSIG})}$ in Figure 6 securely realizes $\mathcal{F}_{\mathrm{BSIG}}$ in the* $(\mathcal{F}_{\mathrm{CRS}}, \mathcal{F}_{\mathrm{SVZK}}^{R_U}, \mathcal{F}_{\mathrm{SPZK}}^{R_S})$-*hybrid model.*

Protocol $\pi_{\Sigma(\mathrm{BSIG})}$ in the $(\mathcal{F}_{\mathrm{CRS}}, \mathcal{F}_{\mathrm{SVZK}}^{R_U}, \mathcal{F}_{\mathrm{SPZK}}^{R_S})$-Hybrid Model

CRS generation: $crs \leftarrow \mathsf{CRSgen}(1^\lambda)$ where λ is the security parameter.

Key generation: When party S is invoked with input (KEYGEN, sid) by $\mathcal{Z}$, it verifies that $sid = (S, sid')$ for some sid'; If not, it ignores the input; Otherwise, it runs $(vk, sk) \leftarrow \mathsf{gen}(crs)$, lets the verification algorithm ver $\stackrel{\mathrm{def}}{=}$ $\mathsf{verify}(crs, vk, \cdot, \cdot)$, and sends output $(\mathrm{VERIFICATIONALG}, sid, \mathsf{ver})$ to $\mathcal{Z}$.

Signature generation: On input $(\mathrm{SIGN}, sid, m, \mathsf{ver}')$ by $\mathcal{Z}$ where $sid = (S, sid')$, party U obtains vk' by parsing ver', selects random ρ_1, computes $\mathbf{u} \leftarrow \mathsf{lbs}_1(crs, vk', m; \rho_1)$ and sends $(\mathrm{PROVESVZK}, sid, U, (crs, vk', \mathbf{u}), (m, \rho_1))$ to $\mathcal{F}_{\mathrm{SVZK}}^{R_U}$.

Upon receiving $(\mathrm{VERIFIEDSVZK}, sid, U, (crs', vk', \mathbf{u}))$ from $\mathcal{F}_{\mathrm{SVZK}}^{R_U}$, party S verifies $crs' = crs$ and $vk' = vk$. If not, then party S outputs $(\mathrm{SIGNSTATUS}, sid, U, \mathsf{incomplete})$ to $\mathcal{Z}$. Else party S selects random ρ_2 and computes $\mathbf{s} \leftarrow \mathsf{lbs}_2(crs, vk, \mathbf{u}, sk; \rho_2)$ and sends $(\mathrm{PROVESPZK}, sid, U, (crs, vk, \mathbf{u}, \mathbf{s}), (sk, \rho_2))$ to $\mathcal{F}_{\mathrm{SPZK}}^{R_S}$, and outputs $(\mathrm{SIGNSTATUS}, sid, U, \mathsf{complete})$ to $\mathcal{Z}$.

Upon receiving $(\mathrm{VERIFIEDSVZK}, sid, U, \bot)$ from $\mathcal{F}_{\mathrm{SVZK}}^{R_U}$, party S outputs $(\mathrm{SIGNSTATUS}, sid, U, \mathsf{incomplete})$ to $\mathcal{Z}$.

Upon receiving $(\mathrm{VERIFIEDSPZK}, sid, U, (crs', vk'', \mathbf{u}', \mathbf{s}))$ from $\mathcal{F}_{\mathrm{SPZK}}^{R_S}$, party U verifies that $crs' = crs$ and $vk'' = vk'$ and $\mathbf{u}' = \mathbf{u}$. If not, then party U outputs $(\mathrm{SIGNATURE}, sid, \bot)$ to $\mathcal{Z}$. Else, party U selects random ρ_3 and computes $\sigma \leftarrow \mathsf{lbs}_3(crs, vk', m, \rho_1, \mathbf{u}, \mathbf{s}; \rho_3)$, and outputs $(\mathrm{SIGNATURE}, sid, \sigma)$ to $\mathcal{Z}$.

Upon receiving $(\mathrm{VERIFIEDSPZK}, sid, U, \bot)$ from $\mathcal{F}_{\mathrm{SPZK}}^{R_S}$, party U outputs $(\mathrm{SIGNATURE}, sid, \bot)$ to $\mathcal{Z}$.

Signature verification: When party V is invoked with input $(\mathrm{VERIFY}, sid, m, \sigma, \mathsf{ver}')$ by $\mathcal{Z}$ where $sid = (S, sid')$, it outputs $(\mathrm{VERIFIED}, sid, \mathsf{ver}'(m, \sigma))$ to $\mathcal{Z}$.

Fig. 6. Blind signature protocol $\pi_{\Sigma(\mathrm{BSIG})}$ in the $(\mathcal{F}_{\mathrm{CRS}}, \mathcal{F}_{\mathrm{SVZK}}^{R_U}, \mathcal{F}_{\mathrm{SPZK}}^{R_S})$-hybrid model based on a lite-blind signature scheme $\langle \mathsf{CRSgen}, \mathsf{gen}, \mathsf{lbs}_1, \mathsf{lbs}_2, \mathsf{lbs}_3, \mathsf{verify} \rangle$. Here functionalities $\mathcal{F}_{\mathrm{SVZK}}^{R_U}$ and $\mathcal{F}_{\mathrm{SPZK}}^{R_S}$ are parameterized with relations $R_U = \{((crs, vk, \mathbf{u}), (m, \rho_1)) \mid \mathbf{u} = \mathsf{lbs}_1(crs, m; \rho_1)\}$ and $R_S = \{((crs, vk, \mathbf{u}, \mathbf{s}), (sk, \rho_2)) \mid \mathbf{s} = \mathsf{lbs}_2(crs, vk, \mathbf{u}, sk; \rho_2) \wedge (vk, sk) \in \mathsf{KeyPair}\}$, respectively.

3.2   Realizing $\mathcal{F}_{\mathrm{SVZK}}$ and $\mathcal{F}_{\mathrm{SPZK}}$

In this subsection we focus on the requirements for the UC-realization of the two ZK functionalities $\mathcal{F}_{\mathrm{SVZK}}$ and $\mathcal{F}_{\mathrm{SPZK}}$. We note that they can be instantiated generically based on non-interactive zero-knowledge as in [11] or [20]. Nevertheless, by focusing on the exact requirements needed for the blind signature setting we manage to get more simplified concrete constructions; note that we will opt for minimizing the overall communication length as opposed to round complexity.

Realizing $\mathcal{F}_{\text{SVZK}}^{R_U}$**.** The functionality $\mathcal{F}_{\text{SVZK}}^{R_U}$ will be realized against adaptive adversaries. We proceed as follows: first given $(x, w) \in R_U$, we will have the prover commit the witness w into a value C; in order to obtain an efficient construction, we employ the mixed commitment primitive of [16,29]; a critical observation in our setting is that due to the fact that we have a single verifier (the signer) it is possible to maintain a constant size common reference string (independent in the number of committers). In contrast we note that in [16,29] it was necessary to rely on a linear length common reference string in the number of protocol participants; this was to suppress man-in-the-middle attacks that could be launched within their session scope (while such attacks are not possible within our session scope). Our construction also employs a non-erasure Sigma protocol based on which we show the consistency of the witness between the commitment C and the statement x by performing a proof of language membership; finally to defend against a dishonest verifier, our Sigma protocol will have to be strengthened so that it can be simulated without knowing the witness; this e.g., can be based on Damgård's trick [14].

Based on the above we obtain an efficient number-theoretic instantiation of the functionality that is secure under the Decisional Composite Residuosity assumption of Paillier [33]. The underlying mixed-commitment is based on Damgård-Jurik encryption [15]; it could be also based on other encryption schemes as well.

Realizing $\mathcal{F}_{\text{SPZK}}^{R_S}$**.** Regarding $\mathcal{F}_{\text{SPZK}}^{R_S}$ we find that, rather surprisingly, our task for attaining an adaptive secure UC blind signature is simpler since security against a static adversary suffices. The reason is that in the UC blind signature security proof, the simulator knows the signing secret which means the witness for $\mathcal{F}_{\text{SPZK}}^{R_S}$ is known by the simulator, and thus no equivocation of dishonestly simulated transcripts is ever necessary! This behavior was explored by the authors in the context of zero-knowledge in [26]; in the framework of that paper, we can say a blind signature protocol falls into the class of protocols where a leaking version of $\mathcal{F}_{\text{SPZK}}^{R_S}$ is sufficient for security and thus $\mathcal{F}_{\text{SPZK}}^{R_S}$ need be realized only against static adversaries.

Similarly to the realization of $\mathcal{F}_{\text{SVZK}}^{R_U}$, for $(x, w) \in R_S$, we have the prover commit to the witness w into the value C, but here we only need employ an extractable commitment considering we only need to realize $\mathcal{F}_{\text{SPZK}}^{R_S}$ against static adversaries; then we develop a Sigma protocol to show the consistency between the commitment C and the statement x by performing a proof of language membership; the first two steps together can be viewed as an Ω-protocol in [18]; further we need to wrap up such Ω-protocol by applying e.g., Damgård's trick to defend against dishonest verifiers.

3.3   Concrete Construction

In this section, we demonstrate how it is possible to derive an efficient UC blind signature instantiation based on Theorem 3 and the realization of its hybrid world with the related ZK-functionalities. Note that we opt for minimizing the overall communication complexity as opposed to round complexity. We need three ingredients: (1) an equivocal lite blind signature scheme, (2) a UC-realization of the ideal functionality $\mathcal{F}_{\text{SVZK}}^{R_U}$, (3) a UC-realization of the ideal functionality $\mathcal{F}_{\text{SPZK}}^{R_S}$. Regarding (1) we will employ the equivocal lite blind signature scheme of Figure 4. Regarding the two ZK functionalities we will follow the design strategy outlined in the previous subsection. Recall that in Figure 6, $R_U = \{((crs, vk, \mathbf{u}), (m, \rho_1)) \mid \mathbf{u} = \mathsf{lbs}_1(crs, m; \rho_1)\}$ and $R_S = \{((crs, vk, \mathbf{u}, \mathbf{s}), (sk, \rho_2)) \mid \mathbf{s} = \mathsf{lbs}_2(crs, vk, \mathbf{u}, sk; \rho_2) \wedge (vk, sk) \in \mathsf{KeyPair}\}$. Instantiating these relations for the protocol of Figure 4 we have that $R_U = \{((crs, X, W), (m, t, s)) \mid W = g_1^{mt} u_1^t v_1^{st}\}$ and $R_S = \{((crs, X, W, Y, l, r), x) \mid Y = (Wv_1^l)^{\frac{1}{x+r}} \wedge X = g_2^x\}$. Please refer to the full version for all the details [25] as well as the full description of the blind signature protocol.

Finally, we can obtain the corollary below:

Corollary 1. *Under the DCR assumption, the DLOG assumption, and the 2SDH assumption, and assuming existence of collision resistant hash function, there exists a blind signature protocol that securely realizes $\mathcal{F}_{\text{BSIG}}$ in the $\mathcal{F}_{\text{CRS}}$-hybrid model.*

Acknowledgements. We thank Jesper Nielsen for helpful clarifications on some of his zero-knowledge and commitment protocols and models. We also thank the anonymous referees for their constructive comments.

References

1. Abe, M.: A secure three-move blind signature scheme for polynomially many signatures. In: Pfitzmann, B. (ed.) EUROCRYPT 2001. LNCS, vol. 2045, pp. 136–151. Springer, Heidelberg (2001)
2. Abe, M., Ohkubo, M.: Provably secure fair blind signatures with tight revocation. In: Boyd, C. (ed.) ASIACRYPT 2001. LNCS, vol. 2248, pp. 583–602. Springer, Heidelberg (2001)
3. Abe, M., Okamoto, T.: Provably secure partially blind signatures. In: Bellare, M. (ed.) CRYPTO 2000. LNCS, vol. 1880, pp. 271–286. Springer, Heidelberg (2000)
4. Beaver, D.: Adaptive zero knowledge and computational equivocation (extended abstract). In: STOC 1996, pp. 629–638. ACM Press, New York (1996)
5. Bellare, M., Namprempre, C., Pointcheval, D., Semanko, M.: The one-more-RSA-inversion problems and the security of Chaum's blind signature scheme. J. Cryptology 16(3), 185–215 (2003); The preliminary version entitled as The power of RSA inversion oracles and the security of Chaum's RSA-based blind signature scheme appeared in Financial Cryptography 2001, LNCS 2339. Springer, Heidelberg (2001)
6. Boldyreva, A.: Threshold signatures, multisignatures and blind signatures based on the gap-Diffie-Hellman-group signature scheme. In: Desmedt, Y. (ed.) PKC 2003. LNCS, vol. 2567, pp. 31–46. Springer, Heidelberg (2002)
7. Camenisch, J., Koprowski, M., Warinschi, B.: Efficient blind signatures without random oracles. In: Blundo, C., Cimato, S. (eds.) SCN 2004. LNCS, vol. 3352, pp. 134–148. Springer, Heidelberg (2005)

8. Camenisch, J., Neven, G., Shelat, A.: Simulatable adaptive oblivious transfer. In: Naor, M. (ed.) EUROCRYPT 2007. LNCS, vol. 4515, pp. 573–590. Springer, Heidelberg (2007)
9. Canetti, R.: Universally composable security: A new paradigm for cryptographic protocols. In: FOCS 2001, pp. 136–145. IEEE Computer Society Press, Los Alamitos (2001)
10. Canetti, R.: Universally composable security: A new paradigm for cryptographic protocols. In: Cryptology ePrint Archive, Report 2000/067, Latest version at (December 2005), http://eprint.iacr.org/2000/067/
11. Canetti, R., Lindell, Y., Ostrovsky, R., Sahai, A.: Universally composable two-party and multi-party secure computation. In: STOC, pp. 494–503. ACM Press, Newyork (2002), Full version at , http://www.cs.biu.ac.il/ lindell/PAPERS/uc-comp.ps
12. Chaum, D.: Blind signatures for untraceable payments. In: Chaum, D., Rivest, R.L., Sherman, A.T. (eds.) Advances in Cryptology 1981 - 1997, pp. 199–203. Plemum Press (1982)
13. Damgård, I.: Payment systems and credential mechanisms with provable security against abuse by individuals. In: Goldwasser, S. (ed.) CRYPTO 1988. LNCS, vol. 403, pp. 328–335. Springer, Heidelberg (1990)
14. Damgård, I.: Efficient concurrent zero-knowledge in the auxiliary string model. In: Preneel, B. (ed.) EUROCRYPT 2000. LNCS, vol. 1807, pp. 418–430. Springer, Heidelberg (2000)
15. Damgård, I., Jurik, M.: A generalisation, a simplification and some applications of Paillier's probabilistic public-key system. In: Kim, K. (ed.) PKC 2001. LNCS, vol. 1992, pp. 119–136. Springer, Heidelberg (2001)
16. Damgård, I., Nielsen, J.B.: Perfect hiding and perfect binding universally composable commitment schemes with constant expansion factor. In: Yung, M. (ed.) CRYPTO 2002. LNCS, vol. 2442, pp. 581–596. Springer, Heidelberg (2002)
17. Fischlin, M.: Round-optimal composable blind signatures in the common reference string model. In: Dwork, C. (ed.) CRYPTO 2006. LNCS, vol. 4117, pp. 60–77. Springer, Heidelberg (2006)
18. Garay, J.A., MacKenzie, P.D., Yang, K.: Strengthening zero-knowledge protocols using signatures. J. Cryptology 19(2), 169–209 (2006) An extended abstract appeared in Eurocrypt 2003, Springer-Verlag (LNCS 2656), pp. 177–194, 2003.
19. Goldwasser, S., Micali, S., Rivest, R.L.: A digital signature scheme secure against adaptive chosen-message attacks. SIAM J. Comput. 17(2), 281–308 (1988)
20. Groth, J., Ostrovsky, R., Sahai, A.: Perfect non-interactive zero knowledge for NP. In: Vaudenay, S. (ed.) EUROCRYPT 2006. LNCS, vol. 4004, pp. 339–358. Springer, Heidelberg (2006)
21. Hazay, C., Katz, J., Koo, C.-Y., Lindell, Y.: Concurrently-secure blind signatures without random oracles or setup assumptions. In: Vadhan, S.P. (ed.) TCC 2007. LNCS, vol. 4392, pp. 323–341. Springer, Heidelberg (2007)
22. Horvitz, O., Katz, J.: Universally-composable two-party computation in two rounds. In: Menezes, A. (ed.) CRYPTO 2007. LNCS, vol. 4622, pp. 111–129. Springer, Heidelberg (2007)
23. Juels, A., Luby, M., Ostrovsky, R.: Security of blind digital signatures (extended abstract). In: Kaliski Jr., B.S. (ed.) CRYPTO 1997. LNCS, vol. 1294, pp. 150–164. Springer, Heidelberg (1997)
24. Kiayias, A., Zhou, H.-S.: Concurrent blind signatures without random oracles. In: De Prisco, R., Yung, M. (eds.) SCN 2006. LNCS, vol. 4116, pp. 49–62. Springer, Heidelberg (2006)

25. Kiayias, A., Zhou, H.-S.: Equivocal blind signatures and adaptive UC-security. In: Cryptology ePrint Archive: Report, /132, 2007. Full version (2007)
26. Kiayias, A., Zhou, H.-S.: Trading static for adaptive security in universally composable zero-knowledge. In: Arge, L., Cachin, C., Jurdziński, T., Tarlecki, A. (eds.) ICALP 2007. LNCS, vol. 4596, pp. 316–327. Springer, Heidelberg (2007)
27. Kim, K.: Lessons from Internet voting during 2002 FIFA WorldCup Korea/Japan(TM). In: DIMACS Workshop on Electronic Voting – Theory and Practice (2004)
28. Lindell, Y.: Bounded-concurrent secure two-party computation without setup assumptions. In: STOC 2003, pp. 683–692. ACM Press, New York (2003) Full version at, http://www.cs.biu.ac.il/~lindell/PAPERS/conc2party-upper.ps
29. Nielsen, J.B.: On protocol security in the cryptographic model. Dissertation Series DS-03-8, BRICS (2003), http://www.brics.dk/DS/03/8/BRICS-DS-03-8.pdf
30. Okamoto, T.: Provably secure and practical identification schemes and corresponding signature schemes. In: Brickell, E.F. (ed.) CRYPTO 1992. LNCS, vol. 740, pp. 31–53. Springer, Heidelberg (1993)
31. Okamoto, T.: Efficient Blind and Partially Blind Signatures Without Random Oracles. In: Halevi, S., Rabin, T. (eds.) TCC 2006. LNCS, vol. 3876, pp. 80–99. Springer, Heidelberg (2006)
32. Okamoto, T., Ohta, K.: Divertible zero knowledge interactive proofs and commutative random self-reducibility. In: Quisquater, J.-J., Vandewalle, J. (eds.) EUROCRYPT 1989. LNCS, vol. 434, pp. 134–148. Springer, Heidelberg (1990)
33. Paillier, P.: Public-key cryptosystems based on composite degree residuosity classes. In: Stern, J. (ed.) EUROCRYPT 1999. LNCS, vol. 1592, pp. 223–238. Springer, Heidelberg (1999)
34. Pointcheval, D.: Strengthened security for blind signatures. In: Nyberg, K. (ed.) EUROCRYPT 1998. LNCS, vol. 1403, pp. 391–405. Springer, Heidelberg (1998)
35. Pointcheval, D., Stern, J.: Provably secure blind signature schemes. In: Kim, K.-c., Matsumoto, T. (eds.) ASIACRYPT 1996. LNCS, vol. 1163, pp. 252–265. Springer, Heidelberg (1996)
36. Pointcheval, D., Stern, J.: New blind signatures equivalent to factorization (extended abstract). In: CCS 1997, pp. 92–99. ACM Press, New York (1997)
37. Pointcheval, D., Stern, J.: Security arguments for digital signatures and blind signatures. J. Cryptology 13(3), 361–396 (2000)
38. Prabhakaran, M., Sahai, A.: Relaxing environmental security: Monitored functionalities and client-server computation. In: Kilian, J. (ed.) TCC 2005. LNCS, vol. 3378, pp. 104–127. Springer, Heidelberg (2005)

P-signatures and Noninteractive Anonymous Credentials

Mira Belenkiy[1], Melissa Chase[1], Markulf Kohlweiss[2], and Anna Lysyanskaya[1]

[1] Brown University
{mira, melissa, anna}@cs.brown.edu
[2] KU Leuven
mkohlwei@esat.kuleuven.be

Abstract. In this paper, we introduce P-signatures. A P-signature scheme consists of a signature scheme, a commitment scheme, and (1) an interactive protocol for obtaining a signature on a committed value; (2) a *non-interactive* proof system for proving that the contents of a commitment has been signed; (3) a non-interactive proof system for proving that a pair of commitments are commitments to the same value. We give a definition of security for P-signatures and show how they can be realized under appropriate assumptions about groups with a bilinear map. We make extensive use of the powerful suite of non-interactive proof techniques due to Groth and Sahai. Our P-signatures enable, for the first time, the design of a practical non-interactive anonymous credential system whose security does not rely on the random oracle model. In addition, they may serve as a useful building block for other privacy-preserving authentication mechanisms.

1 Introduction

Anonymous credentials [Cha85, Dam90, Bra99, LRSW99, CL01, CL02, CL04] let Alice prove to Bob that Carol has given her a certificate. Anonymity means that Bob and Carol cannot link Alice's request for a certificate to Alice's proof that she possesses a certificate. In addition, if Alice proves possession of a certificate multiple times, these proofs cannot be linked to each other. Anonymous credentials are an example of a privacy-preserving authentication mechanism, which is an important theme in modern cryptographic research. Other examples are electronic cash [CFN90, CP93, Bra93, CHL05] and group signatures [CvH91, CS97, ACJT00, BBS04, BW06, BW07]. In a series of papers, Camenisch and Lysyanskaya [CL01, CL02, CL04] identified a key building block commonly called "a CL-signature" that is frequently used in these constructions. A CL-signature is a signature scheme with a pair of useful protocols.

The first protocol, called *Issue*, lets a user obtain a signature on a committed message without revealing the message. The user wishes to obtain a signature on a value x from a signer with public key pk. The user forms a commitment $comm$ to value x and gives $comm$ to the signer. After running the protocol, the user obtains a signature on x, and the signer learns no information about x other than the fact that he has signed the value that the user has committed to.

The second protocol, called *Prove*, is a zero-knowledge proof of knowledge of a signature on a committed value. The prover has a message-signature pair $(x, \sigma_{pk}(x))$.

R. Canetti (Ed.): TCC 2008, LNCS 4948, pp. 356–374, 2008.
© International Association for Cryptologic Research 2008

The prover has obtained it by either running the Issue protocol, or by querying the signer on x. The prover also has a commitment $comm$ to x. The verifier only knows $comm$. The prover proves in zero-knowledge that he knows a pair (x, σ) and a value $open$ such that $\mathsf{VerifySig}(pk, x, \sigma) = $ accept and $comm = \mathsf{Commit}(x, open)$.

It is clear that using general secure two-party computation [Yao86] and zero-knowledge proofs of knowledge of a witness for any NP statement [GMW86], we can construct the Issue and Prove protocols from any signature scheme and commitment scheme. Camenisch and Lysyanskaya's contribution was to construct specially designed signature schemes that, combined with Pedersen [Ped92] and Fujisaki-Okamoto [FO98] commitments, allowed them to construct Issue and Prove protocols that are efficient enough for use in practice. In turn, CL-signatures have been implemented and standardized [CVH02, BCC04]. They have also been used as a building block in many other constructions [JS04, BCL04, CHL05, DDP06, CHK+06, TS06].

A shortcoming of the CL signature schemes is that the Prove protocol is interactive. Rounds of interaction are a valuable resource. In certain contexts, proofs need to be verified by third parties who are not present during the interaction. For example, in off-line e-cash, a merchant accepts an e-coin from a buyer and later deposits the e-coin to the bank. The bank must be able to verify that the e-coin is valid.

There are two known techniques for making the CL Prove protocols non-interactive. We can use the Fiat-Shamir heuristic [FS87], which requires the random-oracle model. A series of papers [CGH04, DNRS03, GK03] show that proofs of security in the random-oracle model do not imply security. The other option is to use general techniques: [BFM88, DSMP88, BDMP91] show how any statement in NP can be proven in non-interactive zero-knowledge. This option is prohibitively expensive.

We give the first *practical* non-interactive zero-knowledge proof of knowledge of a signature on a committed message. We have two constructions using two different practical sigature schemes and a special class of commitments due to Groth and Sahai [GS07]. Our constructions are secure in the common reference string model.

Due to the fact that these protocols are so useful for a variety of applications, it is important to give a careful treatment of the security guarantees they should provide. In this paper, we introduce the concept of P-signatures — signatures with efficient Protocols, and give a definition of security. The main difference between P-signatures and CL-signatures is that P-signatures have non-interactive proof protocols. (Our definition can be extended to encompass CL signatures as well.)

OUR CONTRIBUTIONS. Our main contribution is the formal definition of a P-signature scheme and two efficient constructions.

Anonymous credentials are an immediate consequence of P-signatures (and of CL-signatures [Lys02]). Let us explain why (see full paper for an in-depth treatment). Suppose there is a public-key infrastructure that lets each user register a public key. Alice registers unlinkable pseudonyms A_B and A_C with Bob and Carol. A_B and A_C are commitments to her secret key, and so they are unlinkable by the security properties of the commitment scheme. Suppose Alice wishes to obtain a certificate from Carol and show it to Bob. Alice goes to Carol and identifies herself as the owner of pseudonym A_C. They run the P-signature Issue protocol as a result of which Alice gets

Carol's signature on her secret key. Now Alice uses the P-signature Prove protocol to construct a non-interactive proof that she has Carol's signature on the opening of A_B.

Our techniques may be of independent interest. Typically, a proof of knowledge π of a witness x to a statement s implies that there exists an efficient algorithm that can extract a value x' from π such that x' satisfies the statement s. Our work uses Groth-Sahai non-interactive proofs of knowledge [GS07] from which we can only extract $f(x)$ where f is a one-way function. We formalize the notion of an f-extractable proof of knowledge and develop useful notation for describing f-extractable proofs that committed values have certain properties. Our notation has helped us understand how to work with the GS proof system and it may encourage others to use the wealth of this powerful building block.

TECHNICAL ROADMAP. We use Groth and Sahai's f-extractable non-interactive proofs of knowledge [GS07] to build P-signatures. Groth and Sahai give three instantiations for their proof system, using SXDH, DLIN, and SDA assumptions. We can use either of the first two instantiations. The SDA-based instantiation does not give us the necessary extraction properties.

Another issue we confront is that Groth-Sahai proofs are f-extractable and not fully extractable. Suppose we construct a proof whose witness x contains $a \in Z_p$ and the opening of a commitment to a. For this commitment, we can only extract $b^a \in f(x)$ from the proof, for some base b. Note that the proof can be about multiple committed values. Thus, if we construct a proof of knowledge of (m, σ) where $m \in Z_p$ and VerifySig$(pk, m, \sigma) = $ accept, we can only extract some function $F(m)$ from the proof. However, even if it is impossible to forge (m, σ) pairs, it might be possible to forge $(F(m), \sigma)$ pairs. Therefore, for our proof system to be meaningful, we need to define F-unforgeable signature schemes, i.e. schemes where it is impossible for an adversary to compute a $(F(m), \sigma)$ pair on his own.

Our first construction uses the Weak Boneh-Boyen (WBB) signature scheme [BB04]. Using a rather strong assumption, we prove that WBB is F-unforgeable and our P-signature construction is secure. Our second construction uses a better assumption (because it is falsfiable [Nao03]) and Our construction is based on the Full Boneh-Boyen signature scheme [BB04]. We had to modify the Boneh-Boyen construction, however, because the GS proof system would not allow the knowledge extraction of the entire signature. Our first construction is much simpler, but, as it's security relies on an interactive and thus much stronger assumption, we have decided to focus here on our second construction. For details on the first construction, see the full version.

ORGANIZATION. Sections 2 and 3 define P-signatures and introduce complexity assumptions. Section 4 explains non-interactive proofs of knowledge, introduces our new notation, and reviews GS proofs. Finally, Section 5 contains our second construction.

2   Definition of a Secure P-signature Scheme

We say that a function $\nu : \mathbb{Z} \to \mathbb{R}$ is negligible if for all integers c there exists an integer K such that $\forall k > K, |\nu(k)| < 1/k^c$. We use the standard GMR [GMR88] notation to describe probability spaces.

Here we introduce P-signatures a primitive which lets a user (1) obtain a signature on a committed message without revealing the message, (2) construct a non-interactive *zero-knowledge proof of knowledge* of $(F(m), \sigma)$ such that VerifySig$(pk, m, \sigma) =$ accept and m is committed to in a commitment *comm*, and (3) a non-interactive method for proving that a pair of commitments are to the same value. In this section, we give the first formal definition of a non-interactive P-signature scheme. We begin by reviewing digital signatures and introducing the concept of F-unforgeability.

2.1 Digital Signatures

A signature scheme consists of four algorithms: SigSetup, Keygen, Sign, and VerifySig. SigSetup(1^k) generates public parameters $params_{Sig}$. Keygen$(params_{Sig})$ generates signing keys (pk, sk). Sign$(params_{Sig}, sk, m)$ computes a signature σ on m. VerifySig $(params_{Sig}, pk, m, \sigma)$ outputs accept if σ is a valid signature on m, reject if not.

The standard definition of a secure signature scheme [GMR88] states that no adversary can output (m, σ), where σ is a signature on m, without first previously obtaining a signature on m. This is insufficient for our purposes. Our P-Signature constructions prove that we know some value $y = F(m)$ (for an efficiently computable bijection F) and a signature σ such that VerifySig$(params_{Sig}, pk, m, \sigma) =$ accept. However, even if an adversary cannot output (m, σ) without first obtaining a signature on m, he might be able to output $(F(m), \sigma)$. Therefore, we introduce the notion of F-Unforgeability:

Definition 1 (F-Secure Signature Scheme). *We say that a signature scheme is F-secure (against adaptive chosen message attacks) if it is* Correct *and F-Unforgeable.*

Correct. VerifySig always accepts a signature obtained using the Sign algorithm.

F-Unforgeable. Let F be an efficiently computable bijection. No adversary should be able to output $(F(m), \sigma)$ unless he has previously obtained a signature on m. Formally, for every PPTM adversary $\mathcal{A}$, there exists a negligible function ν such that

$$\Pr[params_{Sig} \leftarrow \mathsf{SigSetup}(1^k); (pk, sk) \leftarrow \mathsf{Keygen}(params_{Sig});$$

$$(Q_{\mathsf{Sign}}, y, \sigma) \leftarrow \mathcal{A}(params_{Sig}, pk)^{\mathcal{O}_{\mathsf{Sign}}(params_{Sig}, sk, \cdot)} :$$

$$\mathsf{VerifySig}(params_{Sig}, pk, F^{-1}(y), \sigma) = 1 \wedge y \notin F(Q_{\mathsf{Sign}})] < \nu(k).$$

$\mathcal{O}_{\mathsf{Sign}}(params_{Sig}, sk, m)$ records m-queries on Q_{Sign} and returns Sign$(params_{Sig}, sk, m)$. $F(Q_{\mathsf{Sign}})$ evaluates F on all values on Q_{Sign}.

Lemma 1. *F-unforgeable signatures are secure in the standard [GMR88] sense.*

Proof sketch. Suppose an adversary can compute a forgery (m, σ). Now the reduction can use it to compute $(F(m), \sigma)$.

2.2 Commitment Schemes

Recall the standard definition of a non-interactive commitment scheme. It consists of algorithms ComSetup, Commit. ComSetup(1^k) outputs public parameters $params_{Com}$

for the commitment scheme. Commit($params_{Com}, x, open$) is a deterministic function that outputs $comm$, a commitment to x using auxiliary information $open$. We need commitment schemes that are *perfectly binding* and *strongly computationally hiding*:

Perfectly Binding. For every bitstring $comm$, there exists at most one value x such that there exists opening information $open$ so that $comm = $ Commit($params, x, open$). We also require that it be easy to identify the bitstrings $comm$ for which there exists such an x.

Strongly Computationally Hiding. There exists an alternate setup HidingSetup(1^k) that outputs parameters (computationally indistinguishable from the output of ComSetup(1^k)) so that the commitments become information-theoretically hiding.

2.3 Non-interactive P-signatures

A non-interactive P-signature scheme extends a signature scheme (Setup, Keygen, Sign, VerifySig) and a non-interactive commitment scheme (Setup, Commit). It consists of the following algorithms (Setup, Keygen, Sign, VerifySig, Commit, ObtainSig, IssueSig, Prove, VerifyProof, EqCommProve, VerEqComm).

Setup(1^k). Outputs public parameters $params$. These parameters include parameters for the signature scheme and the commitment scheme.

ObtainSig($params, pk, m, comm, open$) $\leftrightarrow$ IssueSig($params, sk, comm$). These two interactive algorithms execute a signature issuing protocol between a user and the issuer. The user takes as input ($params, pk, m, comm, open$) such that the value $comm = $ Commit($params, m, open$) and gets a signature σ as output. If this signature does not verify, the user sends "reject" to the issuer. The issuer gets ($params, sk, comm$) as input and gets nothing as output.

Prove($params, pk, m, \sigma$). Outputs the values ($comm, \pi, open$), such that $comm = $ Commit($params, m, open$) and π is a proof of knowledge of a signature σ on m.

VerifyProof($params, pk, comm, \pi$). Takes as input a commitment to a message m and a proof π that the message has been signed by owner of public key pk. Outputs accept if π is a valid proof of knowledge of $F(m)$ and a signature on m, and outputs reject otherwise.

EqCommProve($params, m, open, open'$). Takes as input a message and two commitment opening values. It outputs a proof π that $comm = $ Commit($m, open$) is a commitment to the same value as $comm' = $ Commit($m, open'$). This proof is used to bind the commitment of a P-signature proof to a more permanent commitment.

VerEqComm($params, comm, comm', \pi$). Takes as input two commitments and a proof and accepts if π is a proof that $comm, comm'$ are commitments to the same value.

Definition 2 (Secure P-Signature Scheme). *Let F be a efficiently computable bijection (possibly parameterized by public parameters). A P-signature scheme is secure if* (Setup, Keygen, Sign, VerifySig) *form an F-unforgeable signature scheme, if* (Setup, Commit) *is a perfectly binding, strongly computationally hiding commitment scheme, if* (Setup, EqCommProve, VerEqComm) *is a non-interactive proof system, and if the Signer privacy, User privacy, Correctness, Unforgeability, and Zero-knowledge properties hold:*

Correctness. An honest user who obtains a P-signature from an honest issuer will be able to prove to an honest verifier that he has a valid signature.

Signer privacy. No PPTM adversary can tell if it is running IssueSig with an honest issuer or with a simulator who merely has access to a signing oracle. Formally, there exists a simulator SimIssue such that for all PPTM adversaries $(\mathcal{A}_1, \mathcal{A}_2)$, there exists a negligible function ν so that:

$$\big| \Pr[params \leftarrow \mathsf{Setup}(1^k); (sk, pk) \leftarrow \mathsf{Keygen}(params);$$
$$(m, open, state) \leftarrow \mathcal{A}_1(params, sk);$$
$$comm \leftarrow \mathsf{Commit}(params, m, open);$$
$$b \leftarrow \mathcal{A}_2(state) \leftrightarrow \mathsf{IssueSig}(params, sk, comm) : b = 1]$$
$$- \Pr[params \leftarrow \mathsf{Setup}(1^k); (sk, pk) \leftarrow \mathsf{Keygen}(params);$$
$$(m, open, state) \leftarrow \mathcal{A}_1(params, sk);$$
$$comm \leftarrow \mathsf{Commit}(params, m, open); \sigma \leftarrow \mathsf{Sign}(params, sk, m);$$
$$b \leftarrow \mathcal{A}_2(state) \leftrightarrow \mathsf{SimIssue}(params, comm, \sigma) : b = 1] \big| < \nu(k)$$

Note that we ensure that IssueSig and SimIssue gets an honest commitment to whatever $m, open$ the adversary chooses.

Since the goal of signer privacy is to prevent the adversary from learning anything except a signature on the opening of the commitment, this is sufficient for our purposes. Note that our SimIssue will be allowed to rewind $\mathcal{A}$. to Also, we have defined Signer Privacy in terms of a single interaction between the adversary and the issuer. A simple hybrid argument can be used to show that this definition implies privacy over many sequential instances of the issue protocol.

User privacy. No PPTM adversary $(\mathcal{A}_1, \mathcal{A}_2)$ can tell if it is running ObtainSig with an honest user or with a simulator. Formally, there exists a simulator $\mathsf{Sim} = \mathsf{SimObtain}$ such that for all PPTM adversaries $\mathcal{A}_1, \mathcal{A}_2$, there exists negligible function ν so that:

$$\big| \Pr[params \leftarrow \mathsf{Setup}(1^k); (pk, m, open, state) \leftarrow \mathcal{A}_1(params);$$
$$comm = \mathsf{Commit}(params, m, open);$$
$$b \leftarrow \mathcal{A}_2(state) \leftrightarrow \mathsf{ObtainSig}(params, pk, m, comm, open) : b = 1]$$
$$- \Pr[(params, sim) \leftarrow \mathsf{Setup}(1^k); (pk, m, open, state) \leftarrow \mathcal{A}_1(params);$$
$$comm = \mathsf{Commit}(params, m, open);$$
$$b \leftarrow \mathcal{A}_2(state) \leftrightarrow \mathsf{SimObtain}(params, pk, comm) : b = 1] \big| < \nu(k)$$

Here again SimObtain is allowed to rewind the adversary.

Note that we require that only the user's input m is hidden from the issuer, but not necessarily the user's output σ. The reason that this is sufficient is that in actual applications (for example, in anonymous credentials), a user would never show σ in the clear; instead, he would just prove that he knows σ. An alternative, stronger way to define signer privacy and user privacy together, would be to require that the pair of algorithms ObtainSig and IssueSig carry out a secure two-party computation. This alternative definition would ensure that σ is hidden from the issuer as well. However, as explained above, this feature is not necessary for our application, so we preferred to give a special definition which captures the minimum properties required.

Unforgeability. We require that no PPTM adversary can create a proof for any message m for which he has not previously obtained a signature or proof from the oracle.

A P-signature scheme is unforgeable if an extractor (ExtractSetup, Extract) and a bijection F exist such that (1) the output of $\mathsf{ExtractSetup}(1^k)$ is indistinguishable from the output of $\mathsf{Setup}(1^k)$, and (2) no PPTM adversary can output a proof π that VerifyProof accepts, but from which we extract $F(m), \sigma$ such that either (a) σ is not valid signature on m, or (b) $comm$ is not a commitment to m or (c) the adversary has never previously queried the signing oracle on m. Formally, for all PPTM adversaries $\mathcal{A}$, there exists a negligible function ν such that:

$$\Pr[params_0 \leftarrow \mathsf{Setup}(1^k); (params_1, td) \leftarrow \mathsf{ExtractSetup}(1^k) : b \leftarrow \{0, 1\} :$$
$$\mathcal{A}(params_b) = b] < 1/2 + \nu(k), \text{ and}$$

$$\Pr[(params, td) \leftarrow \mathsf{ExtractSetup}(1^k); (pk, sk) \leftarrow \mathsf{Keygen}(params);$$
$$(Q_{\mathsf{Sign}}, comm, \pi) \leftarrow \mathcal{A}(params, pk)^{\mathcal{O}_{\mathsf{Sign}}(params, sk, \cdot)};$$
$$(y, \sigma) \leftarrow \mathsf{Extract}(params, td, \pi, comm) :$$
$$\mathsf{VerifyProof}(params, pk, comm, \pi) = \mathsf{accept}$$
$$\wedge \; (\mathsf{VerifySig}(params, pk, F^{-1}(y), \sigma) = \mathsf{reject}$$
$$\vee \; (\forall open, comm \neq \mathsf{Commit}(params, F^{-1}(y), open))$$
$$\vee \; (\mathsf{VerifySig}(params, pk, F^{-1}(y), \sigma) = \mathsf{accept} \wedge y \notin F(Q_{\mathsf{Sign}})))] < \nu(k).$$

Oracle $\mathcal{O}_{\mathsf{Sign}}(params, sk, m)$ runs the function $\mathsf{Sign}(params, sk, m)$ and returns the resulting signature σ to the adversary. It records the queried message on query tape Q_{Sign}. By $F(Q_{\mathsf{Sign}})$ we mean F applied to every message in Q_{Sign}.

Zero-knowledge. There exists a simulator $\mathsf{Sim} = (\mathsf{SimSetup}, \mathsf{SimProve}, \mathsf{SimEqComm})$, such that for all PPTM adversaries $\mathcal{A}_1, \mathcal{A}_2$, there exists a negligible function ν such that under parameters output by SimSetup, Commit is perfectly hiding and (1) the parameters output by SimSetup are indistinguishable from those output by Setup, but SimSetup also outputs a special auxiliary string sim; (2) when $params$ are generated by SimSetup, the output of $\mathsf{SimProve}(params, sim, pk)$ is indistinguishable from that of $\mathsf{Prove}(params, pk, m, \sigma)$ for all (pk, m, σ) where $\sigma \in \sigma_{pk}(m)$; and (3) when $params$ are generated by SimSetup, the output of $\mathsf{SimEqComm}(params, sim, comm, comm')$ is indistinguishable from that of $\mathsf{EqCommProve}(params, m, open, open')$ for all $(m, open, open')$ where $comm = \mathsf{Commit}(params, m, open)$ and $comm' = \mathsf{Commit}(params, m, open')$. In GMR notation, this is formally defined as follows:

$$|\Pr[params \leftarrow \mathsf{Setup}(1^k); b \leftarrow \mathcal{A}(params) : b = 1]$$
$$- \Pr[(params, sim) \leftarrow \mathsf{SimSetup}(1^k); b \leftarrow \mathcal{A}(params) : b = 1]| < \nu(k), \text{ and}$$
$$|\Pr[(params, sim) \leftarrow \mathsf{SimSetup}(1^k); (pk, m, \sigma, state) \leftarrow \mathcal{A}_1(params, sim);$$
$$(comm, \pi, open) \leftarrow \mathsf{Prove}(params, pk, m, \sigma); b \leftarrow \mathcal{A}_2(state, comm, \pi) : b = 1]$$
$$- \Pr[(params, sim) \leftarrow \mathsf{SimSetup}(1^k); (pk, m, \sigma, state) \leftarrow \mathcal{A}_1(params, sim);$$
$$(comm, \pi) \leftarrow \mathsf{SimProve}(params, sim, pk); b \leftarrow \mathcal{A}_2(state, comm, \pi)$$
$$: b = 1]| < \nu(k), \text{ and}$$

$| \Pr[(params, sim) \leftarrow \mathsf{SimSetup}(1^k); (m, open, open') \leftarrow \mathcal{A}_1(params, sim);$

$\quad \pi \leftarrow \mathsf{EqCommProve}(params, m, open, open'); b \leftarrow \mathcal{A}_2(state, \pi) : b = 1]$

$\quad - \Pr[(params, sim) \leftarrow \mathsf{SimSetup}(1^k); (m, open, open') \leftarrow \mathcal{A}_1(params, sim);$

$\quad\quad\quad \pi \leftarrow \mathsf{SimEqComm}(params, sim, \mathsf{Commit}(params, m, open),$

$\quad\quad\quad\quad\quad\quad\quad \mathsf{Commit}(params, m, open'));$

$\quad\quad b \leftarrow \mathcal{A}_2(state, \pi) : b = 1]| < \nu(k).$

3 Preliminaries

Let G_1, G_2, and G_T be groups. A function $e : G_1 \times G_2 \rightarrow G_T$ is called *a cryptographic bilinear map* if it has the following properties: *Bilinear.* $\forall a \in G_1, \forall b \in G_2, \forall x, y \in \mathbb{Z}$ the following equation holds: $e(a^x, b^y) = e(a, b)^{xy}$. *Non-Degenerate.* If a and b are generators of their respective groups, then $e(a, b)$ generates G_T. Let $\mathsf{BilinearSetup}(1^k)$ be an algorithm that generates the groups G_1, G_2 and G_T, together with algorithms for sampling from these groups, and the algorithm for computing the function e.

The function $\mathsf{BilinearSetup}(1^k)$ outputs $params_{BM} = (p, G_1, G_2, G_T, e, g, h)$, where p is a prime (of length k), G_1, G_2, G_T are groups of order p, g is a generator of G_1, h is a generator of G_2, and $e : G_1 \times G_2 \rightarrow G_T$ is a bilinear map.

We introduce a new assumption which we call TDH and review the HSDH assumption introduced by Boyen and Waters [BW07]. Groth-Sahai proofs use either the DLIN [BBS04] or SXDH [Sco02] assumption. For formal definitions, see the full version.

Definition 3 (Triple DH (TDH)). *On input $g, g^x, g^y, h, h^x, \{c_i, g^{1/(x+c_i)}\}_{i=1...q}$, it is computationally infeasible to output a tuple $(h^{\mu x}, g^{\mu y}, g^{\mu x y})$ for $\mu \neq 0$.*

Definition 4 (Hidden SDH [BW07]). *On input $g, g^x, u \in G_1, h, h^x \in G_2$ and $\{g^{1/(x+c_\ell)}, h^{c_\ell}, u^{c_\ell}\}_{\ell=1...q}$, it is computationally infeasible to output a new tuple $(g^{1/(x+c)}, h^c, u^c)$.*

Definition 5 (Decisional Linear Assumption (DLIN)). *On input $u, v, w, u^r, v^s \leftarrow G_1$ it is computationally infeasible to distinguish $z_0 \leftarrow w^{r+s}$ from $z_1 \leftarrow G_1$. The assumption is analogously defined for G_2.*

Definition 6 (Symmetric External Diffie-Hellman Assumption (SXDH)). *SXDH states that the Decisional Diffie Hellman problem is hard in both G_1 and G_2. This precludes efficient isomorphisms between these two groups.*

4 Non-interactive Proofs of Knowledge

Our P-signature constructions use the Groth and Sahai [GS07] non-interactive proof of knowledge (NIPK) system. De Santis et al. [DDP00] give the standard definition of NIPK systems. Their definition does not fully cover the Groth and Sahai proof system. In this section, we review the standard notion of NIPK. Then we give a useful generalization, which we call an f-extractable NIPK, where the extractor only extracts a function of the witness. We develop useful notation for expressing f-extractable NIPK systems, and explain how this notation applies to the Groth-Sahai construction. We then review Groth-Sahai commitments and pairing product equation proofs. Finally, we show how they can be used to prove statments about committed exponents, as this will be necessary later for our constructions.

4.1 Proofs of Knowledge: Notation and Definitions

In this subsection, we review the definition of NIPK, introduce the notion of f-extractability, and develop some useful notation. We review the De Santis et al. [DDP00] definition of NIPK. Let $L = \{s \ : \ \exists x \text{ s.t. } M_L(s, x) = \text{accept}\}$ be a language in NP and M_L a polynomial-time Turing Machine that verifies that x is a valid witness for the statement $s \in L$. A NIPK system consists of three algorithms: (1) PKSetup(1^k) sets up the common parameters $params_{PK}$; (2) PKProve($params_{PK}, s, x$) computes a proof π of the statement $s \in L$ using witness x; (3) PKVerify($params_{PK}, s, \pi$) verifies correctness of π. The system must be *complete* and *extractable*. Completeness means that for all values of $params_{PK}$ and for all s, x such that $M_L(s, x) = \text{accept}$, a proof π generated by PKProve($params_{PK}, s, x$) must be accepted by PKVerify($params_{PK}, s, \pi$). Extractability means that there exists a polynomial-time extractor (PKExtractSetup, PKExtract). PKExtractSetup(1^k) outputs $(td, params_{PK})$ where $params_{PK}$ is distributed identically to the output of PKSetup(1^k). For all PPT adversaries $\mathcal{A}$, the probability that $\mathcal{A}(1^k, params_{PK})$ outputs (s, π) such that PKVerify($params_{PK}, s, \pi$) = accept and PKExtract(td, s, π) fails to extract a witness x such that $M_L(s, x) = \text{accept}$ is negligible in k. We have *perfect* extractability if this probability is 0.

We first generalize the notion of NIPK for a language L to languages parameterized by $params_{PK}$ – we allow the Turing machine M_L to receive $params_{PK}$ as a separate input. Next, we generalize extractability to f-extractability. We say that a NIPK system is f-extractable if PKExtract outputs y, such that there $\exists x : M_L(params_{PK}, s, x) = \text{accept} \wedge y = f(params_{PK}, x)$. If $f(params_{PK}, \cdot)$ is the identity function, we get the usual notion of extractability. We denote an f-extractable proof π obtained by running PKProve($params_{PK}, s, x$) as

$$\pi \leftarrow \text{NIPK}\{params_{PK}, s, f(params_{PK}, x) : M_L(params_{PK}, s, x) = \text{accept}\}.$$

We omit the $params_{PK}$ where they are obvious. In our applications, s is a conditional statement about the witness x, so $M_L(s, x) = \text{accept}$ if Condition(x) = accept. Thus the statement $\pi \leftarrow \text{NIPK}\{f(x) : \text{Condition}(x)\}$ is well defined. Suppose s includes a list of commitments $c_n = \text{Commit}(x_n, open_n)$. The witness is $x = (x_1, \ldots, x_N, open_1, \ldots, open_N)$, however, we typically can only extract $x_1, \ldots, x_N$. We write

$$\pi \leftarrow \text{NIPK}\{(x_1, \ldots, x_n) : \text{Condition}(x)$$
$$\wedge \ \forall \ell \ \exists open_\ell : c_\ell = \text{Commit}(params_{Com}, x_\ell, open_\ell)\}.$$

We introduce shorthand notation for the above expression: $\pi \leftarrow \text{NIPK}\{((c_1 : x_1), \ldots, (c_n : x_n)) : \text{Condition}(x)\}$. For simplicity, we assume the proof π includes s.

4.2 Groth-Sahai Commitments [GS07]

We review the Groth-Sahai [GS07] commitment scheme. We use their scheme to commit to elements of a group G of prime order p. Technically, their constructions commit to elements of certain modules, but we can apply them to certain bilinear groups elements. Groth and Sahai also have a construction for composite order groups using the Subgroup Decision assumption; however it lacks the necessary extraction properties.

GSComSetup(p, G, g). Outputs a common reference string $params_{Com}$.

GSCommit($params_{Com}, x, open$). Takes as input $x \in G$ and some value $open$ and outputs a commitment $comm$. The extension GSExpCommit($params_{Com}, b, \theta, open$) takes as input $\theta \in Z_p$ and a base $b \in G$ and outputs $(b, comm)$, where $comm = $ GSCommit($params_{Com}, b^\theta, open$). (Groth and Sahai compute commitments to elements in Z_p slightly differently;

VerifyOpening($params_{Com}, comm, x, open$). Takes $x \in G$ and $open$ as input and outputs accept if $comm$ is a commitment to x. To verify that $(b, comm)$ is a commitment to exponent θ check VerifyOpening($params_{Com}, comm, b^\theta, open$).

For brevity, we write GSCommit(x) to indicate committing to $x \in G$ when the parameters are obvious and the value of $open$ is chosen appropriately at random. Similarly, GSExpCommit(b, θ) indicates committing to θ using $b \in G$ as the base.

GS commitments are *perfectly binding, strongly computationally hiding*, and *extractable*. Groth and Sahai [GS07] show how to instantiate commitments that meet these requirements using either the SXDH or DLIN assumptions. Commitments based on SXDH consist of 2 elements in G, while those based on DLIN setting require 3 elements in G. Note that in the Groth-Sahai proof system below, $G = G_1$ or $G = G_2$ for SXDH and $G = G_1 = G_2$ for DLIN.

4.3 Groth-Sahai Pairing Product Equation Proofs [GS07]

Groth and Sahai [GS07] construct an f-extractable NIPK system that lets us prove statements in the context of groups with bilinear maps.

GSSetup(1^k) outputs $(p, G_1, G_2, G_T, e, g, h)$, where G_1, G_2, G_T are groups of prime order p, with g a generator of G_1, h a generator of G_2, and $e : G_1 \times G_2 \to G_T$ a cryptographic bilinear map. GSSetup(1^k) also outputs $params_1$ and $params_2$ for constructing GS commitments in G_1 and G_2, respectively. (If the pairing is symmetric, $G_1 = G_2$ and $params_1 = params_2$.) The statement s to be proven consists of the following list of values: $\{a_q\}_{q=1...Q} \in G_1$, $\{b_q\}_{q=1...Q} \in G_2$, $t \in G_T$, and $\{\alpha_{q,m}\}_{m=1...M, q=1...Q}, \{\beta_{q,n}\}_{n=1...N, q=1...Q} \in Z_p$, as well as a list of commitments $\{c_m\}_{m=1...M}$ to values in G_1 and $\{d_n\}_{n=1...N}$ to values in G_2. Groth and Sahai show how to construct the following proof:

$$\text{NIPK}\{((c_1 : x_1), \ldots, (c_M : x_M), (d_1 : y_1), \ldots, (d_N : y_N)) :$$

$$\prod_{q=1}^{Q} e(a_q \prod_{m=1}^{M} x_m^{\alpha_{q,m}}, b_q \prod_{n=1}^{N} y_n^{\beta_{q,n}}) = t\}$$

The proof π includes the statement being proven; this includes the commitments $c_1, \ldots,$ c_M and $d_1, \ldots, d_N$. Groth and Sahai provide an efficient extractor that opens these commitments to values $x_1, \ldots, x_M, y_1, \ldots, y_N$ that satisfy the pairing product equation.

Recall the function $\mathsf{GSExpCommit}(params_1, b, \theta, open) = (b, \mathsf{GSCommit}$ $(params_1, b^\theta, open))$. We can replace any of the clauses $(c_m : x_m)$ with the clause $(c_m : b^\theta)$, and add b to the list of values included in the statement s (and therefore in the proof π). The same holds for commitments d_n. Groth-Sahai proofs also allow us to prove that the openings of $(c_1, \ldots, c_n, d_1, \ldots, d_n)$ satisfy several equations *simultaneously*.

We formally define the Groth-Sahai proof system. Let $params_{BM} \leftarrow \mathsf{BilinearSetup}(1^k)$.

$\mathsf{GSSetup}(params_{BM})$. Calls $\mathsf{GSComSetup}$ to generate $params_1$ and $params_2$ for constructing commitments in G_1 and G_2 respectively, and optional auxiliary values $params_\pi$. Outputs $params_{GS} = (params_{BM}, params_1, params_2, params_\pi)$.

$\mathsf{GSProve}(params_{GS}, s, (\{x_m\}_{1...M}, \{y_n\}_{1...N}, openings))$. Takes as input the parameters, the statement $s = \{(c_1, \ldots, c_M, d_1, \ldots, d_N), equations\}$ to be proven, (the statement s includes the commitments and the parameters of the pairing product equations), the witness consisting of the values $\{x_m\}_{1...M}, \{y_n\}_{1...N}$ and opening information $openings$. Outputs a proof π.

$\mathsf{GSVerify}(params_{GS}, \pi)$. Returns accept if π is valid, reject otherwise. (Note that it does not take the statement s as input because we have assumed that the statement is always included in the proof π.)

$\mathsf{GSExtractSetup}(params_{BM})$. Outputs $params_{GS}$ and auxiliary information (td_1, td_2). $params_{GS}$ are distributed identically to the output of $\mathsf{GSSetup}(params_{BM})$. (td_1, td_2) allow an extractor to discover the contents of all commitments.

$\mathsf{GSExtract}(params_{GS}, td_1, td_2, \pi)$. Outputs $x_1, \ldots, x_M \in G_1$ and $y_1, \ldots, y_N \in G_2$ that satisfy the *equations* and that correspond to the commitments (note that the commitments and the equations are included with the proof π).

Groth-Sahai proofs satisfy *correctness*, *extractability*, and *strong witness indistinguishability*. We explain these requirements in a manner compatible with our notation.

Correctness. An honest verifier always accepts a proof generated by an honest prover.

Extractability. If an honest verifier outputs accept, then the statement is true. This means that, given td_1, td_2 corresponding to $params_{GS}$, $\mathsf{GSExtract}$ extracts values from the commitments that satisfy the pairing product equations with probability 1.

Strong Witness Indistinguishability. A simulator $\mathsf{Sim} = (\mathsf{SimSetup}, \mathsf{SimProve})$ with the following two properties exists: (1) $\mathsf{SimSetup}(params_{BM})$ outputs $params_{GS}'$ such that they are computationally indistinguishable from the output of $\mathsf{GSSetup}(params_{BM})$. Let $params_1' \in params_{GS}'$ be the parameters for the commitment scheme in G_1. Using $params_1'$, commitments are perfectly hiding – this means that for all commitments $comm, \forall x \in G_1, \exists open : \mathsf{VerifyOpening}(params_1', comm, x, open) = \mathrm{accept}$ (analogous for G_2). (2) Using the $params_{GS}'$ generated by the challenger, GS proofs become perfectly witness indistinguishable. Suppose an unbounded adversary $\mathcal{A}$ generates a statement s consisting of the pairing product equations and a set of commitments $(c_1, \ldots, c_M, d_1, \ldots, d_N)$. The adversary opens the commitments in two different ways $W_0 = (x_1^{(0)}, \ldots, x_M^{(0)}, y_1^{(0)}, \ldots, y_N^{(0)}$, $openings_0)$ and $W_1 = (x_1^{(1)}, \ldots, x_M^{(1)}, y_1^{(1)}, \ldots, y_N^{(1)}, openings_1)$ (under the requirement that these witnesses must both satisfy s). The values $openings_b$ show how to open the commitments to $\{x_m^{(b)}, y_n^{(b)}\}$. (The adversary can do this because it is unbounded.) The challenger gets the statement s and the two witnesses W_0 and W_1. He chooses a bit $b \leftarrow \{0, 1\}$ and computes $\pi = \mathsf{GSProve}(params_{GS}', s, W_b)$. Strong witness indistinguishability means that π is distributed independently of b.

Composable Zero-Knowledge. Note that Groth and Sahai show that if in a given pairing product equation the constant t can be written as $t = e(t_1, t_2)$ for known t_1, t_2, then these proofs can be done in zero knowledge. However, their zero knowldge proof construction is significantly less efficient than the WI proofs. Thus, we choose to use only the WI construction as a building block. Then we can take advantage of special features of our P-signature construction to create much more efficient proofs that still have the desired zero knowledge properties. The only exception is our construction for EqCommProve, which does use the zero knowledge technique suggested by Groth and Sahai.

4.4 Proofs About Committed Exponents

We use the Groth-Sahai proof system to prove equality of committed exponents.

Equality of Committed Exponents in Different Groups. We want to prove the statement $\mathsf{NIPK}\{((c : g^\alpha), (d : h^\beta)) : \alpha = \beta\}$. We perform a Groth-Sahai pairing product equation proof $\mathsf{NIPK}\{((c : x), (d : y)) : e(x, h)e(1/g, y) = 1\}$. Security is straightforward due to the f-extractability property of the GS proof system.

Equality of Committed Exponents in the Same Group. We want to prove the statement $\mathsf{NIPK}\{((c_1 : g^\alpha), (c_2 : u^\beta)) : \alpha = \beta\}$, where $g, u \in G_1$. This is equivalent to proving $\mathsf{NIPK}\{((c_1 : g^\alpha), (c_2 : u^\beta), (d : h^\gamma) : \alpha = \gamma \wedge \beta = \gamma\}$.

Zero-Knowledge Proof of Equality of Committed Exponents. We want to prove the statement $\mathsf{NIZKPK}\{((c_1 : g^\alpha), (c_2 : g^\beta) : \alpha = \beta\}$ in zero-knowledge. We perform the Groth-Sahai *zero-knowledge* pairing product equation proof $\mathsf{NIPK}\{((c_1 : g^\alpha), (c_2 : g^\beta), (d : h^\theta) : e(a/b, h^\theta) = 1 \wedge e(g, h^\theta)e(1/g, h) = 1\}$. Proof of equality of committed exponents in group G_2 is done analogously. See full version for details.

Remark 1. We cannot directly use Groth-Sahai general arithmetic gates [GS07] to construct the above proofs because they assume that the commitments use the same base.

5 Efficient Construction of P-signature Scheme

In this section, we present a new signature scheme and then build a P-signature scheme from it. The new signature scheme is inspired by the full Boneh-Boyen signature scheme, and is as follows:

New-SigSetup(1^k) runs BilinearSetup(1^k) to get the pairing parameters $(p, G_1, G_2, G_T, e, g, h)$. In the sequel, by z we denote $z = e(g, h)$.

New-Keygen($params$) picks a random $\alpha, \beta \leftarrow Z_p$. The signer calculates $v = h^\alpha$, $w = h^\beta$, $\tilde{v} = g^\alpha$, $\tilde{w} = g^\beta$. The secret-key is $sk = (\alpha, \beta)$. The public-key is $pk = (v, w, \tilde{v}, \tilde{w})$. The public key can be verified by checking that $e(g, v) = e(\tilde{v}, h)$ and $e(g, w) = e(\tilde{w}, h)$.

New-Sign($params, (\alpha, \beta), m$) chooses $r \leftarrow Z_p - \{\frac{\alpha-m}{\beta}\}$ and calculates $C_1 = g^{1/(\alpha+m+\beta r)}$, $C_2 = w^r$, $C_3 = u^r$. The signature is (C_1, C_2, C_3).

New-VerifySig($params, (v, w, \tilde{v}, \tilde{w}), m, (C_1, C_2, C_3)$) outputs accept if $e(C_1, vh^m C_2) = z$, $e(u, C_2) = e(C_3, w)$, and if the public key is correctly formed, i.e., $e(g, v) = e(\tilde{v}, h)$, and $e(g, w) = e(\tilde{w}, h)$.[1]

Theorem 1. *Let $F(x) = (h^x, u^x)$, where $u \in G_1$ and $h \in G_2$ as in the HSDH and TDH assumptions. Our new signature scheme is F-secure given HSDH and TDH. (See full version for proof.)*

We extend the above signature scheme to obtain our second P-signature scheme (Setup, Keygen, Sign, VerifySig, Commit, ObtainSig, IssueSig, Prove, VerifyProof, EqCommProve, VerEqComm). The algorithms are as follows:

Setup(1^k) First, obtain $params_{BM} = (p, G_1, G_2, G_T, e, g, h) \leftarrow$ BilinearSetup(1^k). Next, obtain $params_{GS} = (params_{BM}, params_1, params_2, params_\pi) \leftarrow$ GSSetup($params_{BM}$). Pick $u \leftarrow G_1$. Let $params = (params_{GS}, u)$. As before, z is defined as $z = e(g, h)$.

Keygen($params$) Run the New-Keygen($params_{BM}$) and output $sk = (\alpha, \beta)$, $pk = (h^\alpha, h^\beta, g^\alpha, g^\beta) = (v, w, \tilde{v}, \tilde{w})$.

Sign($params, sk, m$) Run New-Sign($params_{BM}, sk, m$) to obtain $\sigma = (C_1, C_2, C_3)$ where $C_1 = g^{1/(\alpha+m+\beta r)}$, $C_2 = w^r$, $C_3 = u^r$, and $sk = (\alpha, \beta)$

VerifySig($params, pk, m, \sigma$) Run New-VerifySig($params_{BM}, pk, m, \sigma$).

Commit($params, m, open$) To commit to m, compute $C =$ GSExpCommit($params_2, h, m, open$). (Recall that GSExpCommit($params_2, h, m, open$) = GSCommit($params_2, h^m, open$), and $params_2$ is part of $params_{GS}$.)

ObtainSig($params, pk, m, comm, open$) $\leftrightarrow$ IssueSig($params, sk, comm$). The user and the issuer run the following protocol:

1. The user chooses $\rho_1, \rho_2 \leftarrow Z_p$.
2. The issuer chooses $r' \leftarrow Z_p$.
3. The user and the issuer run a secure two-party computation protocol where the user's private inputs are $(\rho_1, \rho_2, m, open)$, and the issuer's private inputs are $sk = (\alpha, \beta)$ and r'.
   The issuer's private output is $x = (\alpha + m + \beta\rho_1 r')\rho_2$ if $comm =$ Commit($params, m, open$), and $x = \bot$ otherwise.
4. If $x \neq \bot$, the issuer calculates $C_1' = g^{1/x}$, $C_2' = w^{r'}$ and $C_3' = u^{r'}$, and sends (C_1', C_2', C_3') to the user.
5. The user computes $C_1 = C_1'^{\rho_2}$, $C_2 = C_2'^{\rho_1}$, and $C_3 = C_3'^{\rho_1}$ and then verifies that the signature (C_1, C_2, C_3) is valid.

---

[1] The latter is needed only once per public key, and is meaningless in a symmetric pairing setting.

Prove($params, pk, m, \sigma$) Check if pk and σ are valid, and if they are not, output $\perp$. Then the user computes commitments $\Sigma = $ GSCommit($params_1, C_1, open_1$), $R_w = $ GSCommit($params_1, C_2, open_2$), $R_u = $ GSCommit($params_1, C_3, open_3$), $M_h = $ GSExpCommit($params_2, h, m, open_4$) $= $ GSCommit($params_2, h^m, open_4$) and $M_u = $ GSExpCommit($params_1, u, m, open_5$) $= $ GSCommit ($params_1, u^m, open_5$).

The user outputs the commitment $comm = M_h$ and the proof

$$\pi = \mathsf{NIPK}\{((\Sigma : C_1), (R_w : C_2), (R_u : C_3)(M_h : h^\alpha), (M_u : u^\beta)) :$$
$$e(C_1, vh^\alpha C_2) = z \wedge e(u, C_2) = e(C_3, w) \wedge \alpha = \beta\}.$$

VerifyProof($params, pk, comm, \pi$) Outputs accept if the proof π is a valid proof of the statement described above for $M_h = comm$ and for properly formed pk.

EqCommProve($params, m, open, open'$) Let commitment $comm = $ Commit ($params, m, open$) $= $ GSCommit($params_2, h^m, open$) and $comm' = $ Commit($params, m, open'$) $= $ GSCommit($params_2, h^m, open'$). Use the GS proof system as described in Section 4.4 to compute $\pi \leftarrow \mathsf{NIZKPK}\{((comm : h^\alpha), (comm' : h^\beta) : \alpha = \beta\}$.

VerEqComm($params, comm, comm', \pi$) Verify the proof π using the GS proof system as described in Section 4.4.

Theorem 2 (Efficiency). *Using SXDH GS proofs, each P-signature proof for our new signature scheme consists of 18 elements in G_1 and 16 elements in G_2. The prover performs 34 multi-exponentiation and the verifier 68 pairings. Using DLIN, each P-signature proof consists of 42 elements in $G_1 = G_2$. The prover has to do 42 multi-exponentiations and the verifier 84 pairings.*

Theorem 3 (Security). *Our second P-signature construction is secure given HSDH and TDH and the security of the GS commitments and proofs.*

Proof. Correctness. VerifyProof will always accept properly formed proofs.

Signer Privacy. We must construct the SimIssue algorithm that is given as input $params$, a commitment $comm$ and a signature $\sigma = (C_1, C_2, C_3)$ and must simulate the adversary's view. SimIssue will invoke the simulator for the two-party computation protocol. Recall that in two-party computation, the simulator can first extract the input of the adversary: in this case, some $(\rho_1, \rho_2, m, open)$. Then SimIssue checks that $comm = $ Commit($params, m, open$); if it isn't, it terminates. Otherwise, it sends to the adversary the values $(C_1' = C_1^{1/\rho_2}, C_2' = C_2^{1/\rho_1}, C_3' = C_3^{1/\rho_1})$. Suppose the adversary can determine that it is talking with a simulator. Then it must be the case that the adversary's input to the protocol was incorrect which breaks the security properties of the two-party computation.

User Privacy. The simulator will invoke the simulator for the two-party computation protocol. Recall that in two-party computation, the simulator can first extract the input of the adversary (in this case, some (α', β'), not necessarily the valid secret key). Then the simulator is given the target output of the computation (in this case, the value x

which is just a random value that the simulator can pick itself), and proceeds to interact with the adversary such that if the adversary completes the protocol, its output is x. Suppose the adversary can determine that it is talking with a simulator. Then it breaks the security of the two-party computation protocol.

Zero knowledge. Consider the following algorithms. SimSetup runs BilinearSetup to get $params_{BM} = (p, G_1, G_2, G_T, e, g, h)$. It then picks $t \leftarrow Z_p$ and sets up $u = g^a$. Next it calls GSSimSetup($params_{BM}$) to obtain $params_{GS}$ and sim. The final parameters are $params = (params_{GS}, u, z = e(g, h))$ and $sim = (a, sim)$. Note that the distribution of $params$ is indistinguishable from the distribution output by Setup. SimProve receives $params$, sim, and public key $(v, \tilde{v}, w, \tilde{w})$ and can use trapdoor sim to create a random P-signature forgery in SimProve as follows. Pick $s, r \leftarrow Z_p$ and compute $\sigma = g^{1/s}$. We implicitly set $m = s - \alpha - r\beta$. Note that the simulator does not know m and α. However, he can compute $h^m = h^s/(vw^r)$ and $u^m = u^s/(\tilde{v}^a \tilde{w}^{ar})$. Now he can use σ, h^m, u^m, w^r, u^r as a witness and construct the proof π in the same way as the real Prove protocol. By the witness indistinguishability of the GS proof system, a proof using the faked witnesses is indistinguishable from a proof using a real witness, thus SimProve is indistinguishable from Prove.

Finally, we need to show that we can simulate proofs of EqCommProve given the trapdoor sim_{GS}. This follows directly from composable zero knowledge of EqCommProve. See full version for details.

Unforgeability. Consider the following algorithms: ExtractSetup(1^k) outputs the usual $params$, except that it invokes GSExtractSetup to get alternative $params_{GS}$ and the trapdoor $td = (td_1, td_2)$ for extracting GS commitments in G_1 and G_2. The parameters generated by GSSetup are indistinguishable from those generated by GSExtractSetup, so we know that the parameters generated by ExtractSetup will be indistinguishable from those genrated by Setup.

Extract($params$, td, $comm$, π) extracts the values from commitment $comm$ and the commitments M_h, M_u contained in the proof π using the GS commitment extractor. If VerifyProof accepts then $comm = M_h$. Let $F(m) = (h^m, u^m)$.

Now suppose we have an adversary that can break the unforgeability of our P-signature scheme for this extractor and this bijection.

A P-signature forger outputs a proof from which we extract $(F(m), \sigma)$ such that either (1) VerifySig($params$, pk, m, σ) = reject, or (2) $comm$ is not a commitment to m, or (3) the adversary never queried us on m. Since VerifyProof checks a set of pairing product equations, f-extractability of the GS proof system trivially ensures that (1) never happens. Since VerifyProof checks that $M_h = comm$, this ensures that (2) never happens. Therefore, we consider the third possibility. The extractor calcualtes $F(m) = (h^m, u^m)$ where m is fresh. Due to the randomness element r in the signature scheme, we have two types of forgeries. In a Type 1 forgery, the extractor can extract from the proof a tuple of the form $(g^{1/(\alpha+m+\beta r)}, w^r, u^r, h^m, u^m)$, where $m + r\beta \neq m_\ell + r_\ell\beta$ for any (m_ℓ, r_ℓ) used in answering the adversary's signing or proof queries. The second type of forgery is one where $m + r\beta = m_\ell + r_\ell\beta$ for (m_ℓ, r_ℓ) used in one of these previous queries. We show that a Type 1 forger can be used to break the HSDH assumption, and a Type 2 forger can be used to break the TDH assumption.

Type 1 forgeries: $\beta r + m \neq \beta r_\ell + m_\ell$ for any r_ℓ, m_ℓ from a previous query. The reduction gets an instance of the HSDH problem $(p, G_1, G_2, G_T, e, g, X, \tilde{X}, h, u,$ $\{C_\ell, H_\ell, U_\ell\}_{\ell=1\ldots q})$, such that $X = h^x$ and $\tilde{X} = g^x$ for some unknown x, and for all ℓ, $C_\ell = g^{1/(x+c_\ell)}$, $H_\ell = h^{c_\ell}$, and $U_\ell = u^{c_\ell}$ for some unknown c_ℓ. The reduction sets up the parameters of the new signature scheme as $(p, G_1, G_2, e, g, h, u, z = e(g,h))$. Next, the reduction chooses $\beta \leftarrow Z_p$, sets $v = X, \tilde{v} = \tilde{X}$ and calculates $w = h^\beta, \tilde{w} = g^\beta$. The reduction gives the adversary the public parameters, the trapdoor, and the public-key $(v, w, \tilde{v}, \tilde{w})$.

Suppose the adversary's ℓth query is to Sign message m_ℓ. The reduction will implicitly set r_ℓ to be such that $c_\ell = m_\ell + \beta r_\ell$. This is an equation with two unknowns, so we do not know r_ℓ and c_ℓ. The reduction sets $C_1 = C_\ell$. It computes $C_2 = H_\ell/h^{m_\ell} = h^{c_\ell}/h^{m_\ell} = w^{r_\ell}$. Then it computes $C_3 = (U_\ell)^{1/\beta}/u^{m_\ell/\beta} = (u^{c_\ell})^{1/\beta}/u^{m_\ell/\beta} = u^{(c_\ell-m_\ell)/\beta} = u^{r_\ell}$ The reduction returns the signature (C_1, C_2, C_3).

Eventually, the adversary returns a proof π. Since π is f-extractable and perfectly sound, we extract $\sigma = g^{1/(x+m+\beta r)}, a = w^r, b = u^r, c = h^m$, and $d = u^m$. Since this is a P-signature forgery, $(c, d) = (h^m, u^m) \notin F(Q_{\mathsf{Sign}})$. Since this is a Type 1 forger, we also have that $m + \beta r \neq m_\ell + \beta r_\ell$ for any of the adversary's previous queries. Therefore, $(\sigma, ca, db^\beta) = (g^{1/(x+m+\beta r)}, h^{m+\beta r}, u^{m+\beta r})$ is a new HSDH tuple.

Type 2 forgeries: $\beta r + m = \beta r_\ell + m_\ell$ for some r_ℓ, m_ℓ from a previous query. The reduction receives $(p, G_1, G_2, G_T, e, g, h, X, Z, Y, \{\sigma_\ell, c_\ell\})$, where $X = h^x, Z = g^x,$ $Y = g^y$, and for all ℓ, $\sigma_\ell = g^{1/(x+c_\ell)}$. The reduction chooses $\gamma \leftarrow Z_p$ and sets $u = Y^\gamma$. The reduction sets up the parameters of the new signature scheme as $(p, G_1, G_2, e, g, h, u, z = e(g,h))$. Next the reduction chooses $\alpha \leftarrow Z_p$, and calculates $v = h^\alpha, w = X^\gamma, \tilde{v} = g^\alpha, \tilde{w} = Z^\gamma$. It gives the adversary the parameters, the trapdoor, and the public-key $(v, w, \tilde{v}, \tilde{w})$. Note that we set up our parameters and public-key so that β is implicitly defined as $\beta = x\gamma$, and $u = g^{\gamma y}$.

Suppose the adversary's ℓth query is to Sign message m_ℓ. The reduction sets $r_\ell = (\alpha + m_\ell)/(c_\ell\gamma)$ (which it can compute). The reduction computes $C_1 = \sigma_\ell^{1/(\gamma r_\ell)} = (g^{1/(x+c_\ell)})^{1/(\gamma r_\ell)} = g^{1/(\gamma r_\ell(x+c_\ell))} = g^{1/(\alpha+m_\ell+\beta r_\ell)}$. Since the reduction knows r_ℓ, it computes $C_2 = w^{r_\ell}, C_3 = u^{r_\ell}$ and send (C_1, C_2, C_3) to $\mathcal{A}$.

Eventually, the adversary returns a proof π. The proof π is f-extractable and perfectly sound, the reduction can extract $\sigma = g^{1/(x+m+\beta r)}, a = w^r, b = u^r, c = h^m$, and $d = u^m$. Therefore, VerifySig will always accept $m = F^{-1}(c, d), \sigma, a, b$. We also know that if this is a forgery, then VerifyProof accepts, which means that $comm = M_h$, which is a commitment to m. Thus, since this is a P-signature forgery, it must be the case that $(c, d) = (h^m, u^m) \notin F(Q_{\mathsf{Sign}})$. However, since this is a Type 2 forger, we also have that $\exists \ell : m + \beta r = m_\ell + \beta r_\ell$, where m_ℓ is one of the adversary's previous Sign or Prove queries. We implicitly define $\delta = m - m_\ell$. Since $m + \beta r = m_\ell + \beta r_\ell$, we also get that $\delta = \beta(r_\ell - r)$. Using $\beta = x\gamma$, we get that $\delta = x\gamma(r_\ell - r)$. We compute: $A = c/h^{m_\ell} = h^{m-m_\ell} = h^\delta, B = u^{r_\ell}/b = u^{r_\ell-r} = u^{\delta/(\gamma x)} = g^{y\delta/x}$ and $C = (d/u^{m_\ell})^{1/\gamma} = u^{(m-m_\ell)/\gamma} = u^{\delta/\gamma} = g^{\delta y}$. We implicitly set $\mu = \delta/x$, thus $(A, B, C) = (h^{\mu x}, g^{\mu y}, g^{\mu xy})$ is a valid TDH tuple.

Acknowledgments. Mira Belenkiy, Melissa Chase and Anna Lysyanskaya are supported by NSF grants CNS-0374661 CNS-0627553. Markulf Kohlweiss is supported by the European Commission's IST Program under Contracts IST-2002-507591 PRIME and IST-2002-507932 ECRYPT.

References

[ACJT00]   Ateniese, G., Camenisch, J., Joye, M., Tsudik, G.: A practical and provably secure coalition-resistant group signature scheme. In: Bellare, M. (ed.) CRYPTO 2000. LNCS, vol. 1880, pp. 255–270. Springer, Heidelberg (2000)

[BB04]   Boneh, D., Boyen, X.: Short signatures without random oracles. In: Cachin, C., Camenisch, J.L. (eds.) EUROCRYPT 2004. LNCS, vol. 3027, pp. 54–73. Springer, Heidelberg (2004)

[BBS04]   Boneh, D., Boyen, X., Shacham, H.: Short group signatures using strong Diffie-Hellman. In: Franklin, M. (ed.) CRYPTO 2004. LNCS, vol. 3152, pp. 41–55. Springer, Heidelberg (2004)

[BCC04]   Brickell, E., Camenisch, J., Chen, L.: Direct anonymous attestation. Technical Report Research Report RZ 3450, IBM Research Division (March 2004)

[BCL04]   Bangerter, E., Camenisch, J., Lysyanskaya, A.: A cryptographic framework for the controlled release of certified data. In: Cambridge Security Protocols Workshop (2004)

[BDMP91]   Blum, M., De Santis, A., Micali, S., Persiano, G.: Non-interactive zero-knowledge. SIAM J. of Computing 20(6), 1084–1118 (1991)

[BFM88]   Blum, M., Feldman, P., Micali, S.: Non-interactive zero-knowledge and its applications (extended abstract). In: STOC 1988, pp. 103–112 (1988)

[Bra93]   Brands, S.: An efficient off-line electronic cash system based on the representation problem. Technical Report CS-R9323, CWI (April 1993)

[Bra99]   Brands, S.: Rethinking Public Key Infrastructure and Digital Certificates— Building in Privacy. PhD thesis, Eindhoven Inst. of Tech. The Netherlands (1999)

[BW06]   Boyen, X., Waters, B.: Compact group signatures without random oracles. In: Vaudenay, S. (ed.) EUROCRYPT 2006. LNCS, vol. 4004, pp. 427–444. Springer, Heidelberg (2006)

[BW07]   Boyen, X., Waters, B.: Full-domain subgroup hiding and constant-size group signatures. In: Okamoto, T., Wang, X. (eds.) PKC 2007. LNCS, vol. 4450, pp. 1–15. Springer, Heidelberg (2007)

[CFN90]   Chaum, D., Fiat, A., Naor, M.: Untraceable electronic cash. In: Menezes, A., Vanstone, S.A. (eds.) CRYPTO 1990. LNCS, vol. 537, pp. 319–327. Springer, Heidelberg (1991)

[CGH04]   Canetti, R., Goldreich, O., Halevi, S.: The random oracle methodology, revisited. J. ACM 51(4), 557–594 (2004)

[Cha85]   Chaum, D.: Security without identification: Transaction systems to make big brother obsolete. Communications of the ACM 28(10), 1030–1044 (1985)

[CHK$^+$06] Camenisch, J., Hohenberger, S., Kohlweiss, M., Lysyanskaya, A., Meyerovich, M.: How to win the clonewars: efficient periodic n-times anonymous authentication. In: CCS 2006, pp. 201–210 (2006)

[CHL05]   Camenisch, J., Hohenberger, S., Lysyanskaya, A.: Compact E-Cash. In: Cramer, R.J.F. (ed.) EUROCRYPT 2005. LNCS, vol. 3494, pp. 302–321. Springer, Heidelberg (2005)

[CL01]   Camenisch, J., Lysyanskaya, A.: Efficient non-transferable anonymous multi-show credential system with optional anonymity revocation. In: Pfitzmann, B. (ed.) EUROCRYPT 2001. LNCS, vol. 2045, pp. 93–118. Springer, Heidelberg (2001)

[CL02]   Camenisch, J., Lysyanskaya, A.: A signature scheme with efficient protocols. In: Cimato, S., Galdi, C., Persiano, G. (eds.) SCN 2002. LNCS, vol. 2576, pp. 268–289. Springer, Heidelberg (2003)

[CL04]   Camenisch, J., Lysyanskaya, A.: Signature schemes and anonymous credentials from bilinear maps. In: Franklin, M. (ed.) CRYPTO 2004. LNCS, vol. 3152, pp. 56–72. Springer, Heidelberg (2004)

[CLM07]   Camenisch, J., Lysyanskaya, A., Meyerovich, M.: Endorsed e-cash. In: IEEE Symposium on Security and Privacy 2007, pp. 101–115 (2007)

[CP93]   Chaum, D., Pedersen, T.: Transferred cash grows in size. In: Rueppel, R.A. (ed.) EUROCRYPT 1992. LNCS, vol. 658, pp. 390–407. Springer, Heidelberg (1993)

[CS97]   Camenisch, J., Stadler, M.: Efficient group signature schemes for large groups. In: Kaliski Jr., B.S. (ed.) CRYPTO 1997. LNCS, vol. 1294, pp. 410–424. Springer, Heidelberg (1997)

[CvH91]   Chaum, D., van Heyst, E.: Group signatures. In: Davies, D.W. (ed.) EUROCRYPT 1991. LNCS, vol. 547, pp. 257–265. Springer, Heidelberg (1991)

[CVH02]   Camenisch, J., Van Herreweghen, E.: Design and implementation of the *idemix* anonymous credential system. In: Proc. 9th ACM CCS 2002, pp. 21–30 (2002)

[Dam90]   Damgård, I.: Payment systems and credential mechanism with provable security against abuse by individuals. In: Goldwasser, S. (ed.) CRYPTO 1988. LNCS, vol. 403, pp. 328–335. Springer, Heidelberg (1990)

[DDP00]   De Santis, A., Di Crescenzo, G., Persiano, G.: Necessary and sufficient assumptions for non-interactive zero-knowledge proofs of knowledge for all NP relations. In: ICALP 2000, pp. 451–462 (2000)

[DDP06]   Damgård, I., Dupont, K., Pedersen, M.: Unclonable group identification. In: Vaudenay, S. (ed.) EUROCRYPT 2006. LNCS, vol. 4004, pp. 555–572. Springer, Heidelberg (2006)

[DNRS03]   Dwork, C., Naor, M., Reingold, O., Stockmeyer, L.J.: Magic functions. J. ACM 50(6), 852–921 (2003)

[DSMP88]   De Santis, A., Micali, S., Persiano, G.: Non-interactive zero-knowledge proof systems. In: Pomerance, C. (ed.) CRYPTO 1987. LNCS, vol. 293, pp. 52–72. Springer, Heidelberg (1988)

[FO98]   Fujisaki, E., Okamoto, T.: A practical and provably secure scheme for publicly verifiable secret sharing and its applications. In: Nyberg, K. (ed.) EUROCRYPT 1998. LNCS, vol. 1403, pp. 32–46. Springer, Heidelberg (1998)

[FS87]   Fiat, A., Shamir, A.: How to prove yourself: Practical solutions to identification and signature problems. In: Odlyzko, A.M. (ed.) CRYPTO 1986. LNCS, vol. 263, pp. 186–194. Springer, Heidelberg (1987)

[GK03]   Goldwasser, S., Kalai, Y.: On the (in)security of the Fiat-Shamir paradigm. In: FOCS 2003, pp. 102–115 (2003)

[GMR88]   Goldwasser, S., Micali, S., Rivest, R.: A digital signature scheme secure against adaptive chosen-message attacks. SIAM J. on Computing 17(2), 281–308 (1988)

[GMW86]   Goldreich, O., Micali, S., Wigderson, A.: Proofs that yield nothing but their validity and a method of cryptographic protocol design. In: FOCS 1986, pp. 174–187 (1986)

[GS07]   Groth, J., Sahai, A.: Efficient non-interactive proof systems for bilinear groups, http://eprint.iacr.org/2007/155

[JS04]   Jarecki, S., Shmatikov, V.: Handcuffing big brother: an abuse-resilient transaction escrow scheme. In: Cachin, C., Camenisch, J.L. (eds.) EUROCRYPT 2004. LNCS, vol. 3027, pp. 590–608. Springer, Heidelberg (2004)

[LRSW99]   Lysyanskaya, A., Rivest, R., Sahai, A., Wolf, S.: Pseudonym systems. In: Emmerich, W., Tai, S. (eds.) EDO 2000. LNCS, vol. 1999, Springer, Heidelberg (2001)

[Lys02]   Lysyanskaya, A.: Signature Schemes and Applications to Cryptographic Protocol Design. PhD thesis, MIT, Cambridge, Massachusetts (September 2002)

[Nao03]   Naor, M.: On cryptographic assumptions and challenges. In: Boneh, D. (ed.) CRYPTO 2003. LNCS, vol. 2729, pp. 96–109. Springer, Heidelberg (2003)

[Ped92]   Pedersen, T.: Non-interactive and information-theoretic secure verifiable secret sharing. In: Brickell, E.F. (ed.) CRYPTO 1992. LNCS, vol. 740, pp. 129–140. Springer, Heidelberg (1993)

[Sco02]   Scott, M.: Authenticated id-based key exchange and remote log-in with insecure token and pin number, http://eprint.iacr.org/2002/164

[TFS04]   Teranishi, I., Furukawa, J., Sako, K.: k-times anonymous authentication (extended abstract). In: Lee, P.J. (ed.) ASIACRYPT 2004. LNCS, vol. 3329, pp. 308–322. Springer, Heidelberg (2004)

[TS06]   Teranishi, I., Sako, K.: k-times anonymous authentication with a constant proving cost. In: Yung, M., Dodis, Y., Kiayias, A., Malkin, T.G. (eds.) PKC 2006. LNCS, vol. 3958, pp. 525–542. Springer, Heidelberg (2006)

[Yao86]   Yao, A.: How to generate and exchange secrets. In: FOCS 1986, pp. 162–167 (1986)

Multi-property Preserving
Combiners for Hash Functions

Marc Fischlin and Anja Lehmann

Darmstadt University of Technology, Germany
www.minicrypt.de

Abstract. A robust combiner for hash functions takes two candidate
implementations and constructs a hash function which is secure as long
as at least one of the candidates is secure. So far, hash function combiners
only aim at preserving a single property such as collision-resistance or
pseudorandomness. However, when hash functions are used in protocols
like TLS they are often required to provide several properties simultaneously. We therefore put forward the notion of multi-property preserving combiners, clarify some aspects on different definitions for such
combiners, and propose a construction that provably preserves collision
resistance, pseudorandomness, "random-oracle-ness", target collision resistance and message authentication according to our strongest notion.

1 Introduction

Recent attacks on collision-resistant hash functions [17,19,18] have raised the
question how to achieve constructions that are more tolerant to cryptanalytic
results. One approach has been suggested by Herzberg in [11], where robust
combiners have been proposed as a viable strategy for designing less vulnerable
hash functions. The classical hash combiner takes two hash functions H_0, H_1 and
combines them into a failure-tolerant function by concatenating the outputs of
both functions, such that the combiner is collision resistant as long as at least
one of the two functions H_0 or H_1 obeys this property.

However, hash functions are currently used for various tasks that require numerous properties beyond collision resistance, e.g., the HMAC construction [2]
based on a keyed hash function is used (amongst others) in the IPSec and TLS
protocols as a pseudorandom function and as a MAC. In the standardized protocols RSA-OAEP [5] and RSA-PSS [6] even stronger properties are required for
the hash functions (cf. [3,4]), prompting Coron et al. [9] to give constructions
which propagate the random-oracle property from the compression function to
the hash function. A further example for the need of multiple properties is given
by Katz and Shin [13], where collision-resistant pseudorandom functions are required in order to protect authenticated group key exchange protocols against
insider attacks.[1]

---

[1] Technically, they require *statistical* collision-resistance for the keys of the pseudorandom function.

R. Canetti (Ed.): TCC 2008, LNCS 4948, pp. 375–392, 2008.
© International Association for Cryptologic Research 2008

Adhering to the usage of hash functions as "swiss army knives" Bellare and Ristenpart [7,8] have shown how to preserve multiple properties in the design of hash functions. In contrast to their approach, which starts with a compression function and aims at constructing a single multi-property preserving (MPP) hash function, a combiner takes two full-grown hash functions and tries to build a hash function which should preserve the properties, even if one of the underlying hash functions is already broken.

The Problem with Multiple Properties. Combiners which preserve a single property such as collision-resistance or pseudorandomness are quite well understood. Multi-property preserving combiners, on the other hand, are not covered by these strategies and require new techniques instead. As an example we discuss this issue for the case of collision-resistance and pseudorandomness.

Recall that the classical combiner for collision-resistance simply concatenates the outputs of both hash functions $\mathsf{Comb}(M) = H_0(M)\|H_1(M)$. Obviously, the combiner is collision-resistant as long es either H_0 or H_1 has this property. Yet, it does not guarantee for example pseudorandomness (assuming that the hash functions are keyed) if only one of the underlying hash functions is pseudorandom. An adversary can immediatly distinguish the concatenated output from a truly random value by simply examining the part of the insecure hash function.

An obvious approach to obtain a hash combiner that preserves pseudorandomness is to set $\mathsf{Comb}(M) = H_0(M) \oplus H_1(M)$. However, this combiner is not known to preserve collision-resistance anymore, since a collision for the combiner does not necessarily require collisions on both hash functions. In fact, this combiner also violates the conditions of [1,16] and [10], who have shown that the output of a (black-box) collision-resistant combiner cannot be significantly shorter than the concatenation of the outputs from all employed hash functions. Thus, already the attempt of combining only two properties in a robust manner indicates that finding a multi-property preserving combiner is far from trivial.

Our Construction. In this work we show how to build a combiner that provably preserves multiple properties, where we concentrate on the most common properties as proposed in [8], namely, collision resistance (CR), pseudorandomness (PRF), pseudorandom oracle (PRO), target collision resistance (TCR) and message authentication (MAC).

To explain the underlying idea of our construction it is instructive to recall the bit commitment scheme introduced by Naor [15]. There, the receiver sends a random $3n$-bit string t to the committing party who applies a pseudorandom generator to a random n-bit seed r and returns $G(r) \oplus t$ to commit to 1, or $G(r)$ to commit to 0. Due to the pseudorandomness of the generator's output, the receiver does not learn anything about the committed bit. An ambiguous opening of the commitment by the sender requires to find some $r' \neq r$ such that $G(r) = G(r')\oplus t$. Yet, since there are only 2^{2n} pairs of seeds for the pseudorandom generator but 2^{3n} random strings t, the probability that such a seed pair exists is at most 2^{-n}.

Adopting the approach of Naor we proceed as follows for each hash function H_b. First we hash the large message M with H_b into a short n-bit "seed" x_b. Then we expand this value into a $5n$-bit string (similar to the pseudorandom generator). Next, we xor the result with a subset of n random strings $t_i^b \in \{0,1\}^{5n}$, where the subset is determined by the bits $x_b[i] = 1$. We denote this output by $H_b^{\mathrm{prsrv}}(M)$. Only in the final step we combine the two resulting values for each function H_b into one output $\mathsf{Comb}(M) = H_0^{\mathrm{prsrv}}(M) \oplus H_1^{\mathrm{prsrv}}(M)$ by xor-ing them.

Due to the internal expansion of the short string x_b into five hash values, one can use a similar argumentation as in [15] together with the collision-resistance of one of the hash functions to prove that collision-resistance is preserved. At the same time, pseudorandomness is preserved by the final xor-combination of the results of the two hash functions. Moreover, we also show that this construction propagates several other properties, including PRO, TCR and MAC.

Weak vs. Strong Preservation. We prove our construction to be a *strongly* multi-property preserving combiner for $\{\mathsf{CR}, \mathsf{PRF}, \mathsf{PRO}, \mathsf{TCR}, \mathsf{MAC}\}$. That is, it suffices that each property is provided by at least one hash function, e.g., if H_0 or H_1 has property MAC, then so does the combiner, independently of the other properties. We also introduce further relaxations of MPP, denoted by weakly MPP and mildly MPP. In the weak case the combiner only inherits a set of multiple properties if they are all provided by at least one hash function (i.e., if there is a strong candidate which has all properties at the same time). Mildly MPP combiners are between strongly MPP and weakly MPP combiners, where all properties are granted, but different hash functions may cover different properties.

Our work then adresses several questions related to the different notions of multi-property preservation. Namely, we show that strongly MPP is indeed strictly stronger than mildly MPP which, in turn, implies weakly MPP (but not vice versa). We finally discuss the case of general tree-based combiners for more than two hash functions built out of combiners for two hash functions, as suggested in a more general setting by Harnik et al. [12]. As part of this result we show that such tree-combiners inherit the weakly and strongly MPP property of two-function combiners, whereas mildly MPP two-function combiners suprisingly do not propagate their security to trees.

Organization. We start by defining the three notions of multi-property preserving combiners and giving definitions of the desired properties in Section 2. In Section 3 we give the construction of our MPP combiner and prove that it achieves the strongest MPP notion. A brief discussion about variations of our construction, e.g., to reduce the key size, conclude this section. Section 4 deals again with the different notions of property preservation by showing the correlations between strongly, mildly and weakly MPP combiners. The issue of composing combiners resp. multi-hash combiners is then addressed in Section 5.

2   Preliminaries

2.1   Hash Function Properties

A hash function $\mathcal{H} = (\mathsf{HKGen}, \mathsf{H})$ is a pair of efficient algorithms such that HKGen for input 1^n returns (the description of) a hash function H, and H for input H and $M \in \{0,1\}^*$ deterministically outputs a digest $H(M) \in \{0,1\}^n$. Often, the hash function is also based on a public initial value IV and we therefore occassionally write $H(\mathrm{IV}, M)$ instead of $H(M)$. Similarly, we often identify the hash function with its digest values $H(\cdot)$ if the key generation algorithm is clear from the context.

A hash function may be attributed different properties $P_1, P_2, \ldots$, among which five important ones stand out (cf. [8]):

collision resistance (CR): The hash function is called *collision-resistant* if for any efficient algorithm $\mathcal{A}$ the probability that for $H \leftarrow \mathsf{HKGen}(1^n)$ and $(M, M') \leftarrow \mathcal{A}(H)$ we have $M \neq M'$ but $\mathsf{H}(H, M) = \mathsf{H}(H, M')$, is negligible (as a function of n).

pseudorandomness (PRF): A hash function can be used as a pseudorandom function if the inital value IV is replaced by a randomly chosen key K of the same size (i.e., the key generation algorithm outputs a public part (H, IV) and IV is replaced by a secret key K). Such a keyed hash function $H(K, \cdot)$ is called *pseudorandom* if for any efficient adversary $\mathcal{D}$ the advantage $\mathrm{Prob}\left[\mathcal{D}^{H(K, \cdot)}(H) = 1\right] - \mathrm{Prob}\left[\mathcal{D}^f(H) = 1\right]$ is negligible, where the probability in the first case is over $\mathcal{D}$'s coin tosses, the choice of $H \leftarrow \mathsf{HKGen}(1^n)$ and the key K, and in the second case over $\mathcal{D}$'s coin tosses, the choice of $H \leftarrow \mathsf{HKGen}(1^n)$, and the choice of the random function $f : \{0,1\}^* \to \{0,1\}^n$.

pseudorandom oracle (PRO): A hash function H^f based on a random oracle f is called a *pseudorandom oracle* if for any efficient adversary the construction H^f is indifferentiable from a random oracle $\mathcal{F}$, where indifferentiability [?] is a generalization of indistinguishability allowing to consider random oracles that are used as a public component. More formally, a hash function H^f is *indifferentiable* from a random oracle $\mathcal{F}$ if for any efficient adversary $\mathcal{D}$ there exists an efficient algorithm $\mathcal{S}$ such that the advantage $\mathrm{Prob}\left[\mathcal{D}^{H^f, f}(H) = 1\right] - \mathrm{Prob}\left[\mathcal{D}^{\mathcal{F}, \mathcal{S}^{\mathcal{F}}(H)}(H) = 1\right]$ is negligible in n, where the probability in the first case is over $\mathcal{D}$'s coin tosses, $H \leftarrow \mathsf{HKGen}(1^n)$ and the choice of the random function f, and in the second case over the coin tosses of $\mathcal{D}$ and $\mathcal{S}$, and $H \leftarrow \mathsf{HKGen}(1^n)$ and over the choice of $\mathcal{F}$.

target collision-resistance (TCR): Target collision-resistance is a weaker security notion than collision-resistance which obliges the adversary to first commit to a target message M before getting the description $H \leftarrow \mathsf{HKGen}(1^n)$ of the hash function. For the given H the adversary must then find a second message $M' \neq M$ such that $H(M) = H(M')$. More formally, an adversary $\mathcal{A}$ consists of two efficient algorithms $(\mathcal{A}^1, \mathcal{A}^2)$ where $\mathcal{A}^1(1^n)$ first generates the target message M and possibly some additional state information st. Then, a hash function $H \leftarrow \mathsf{HKGen}(1^n)$ is chosen and $\mathcal{A}^2$ has to compute on input (H, M, st) a colliding message $M' \neq M$. A hash function is called *target* *collision-resistant* if or any efficient adversary $\mathcal{A} = (\mathcal{A}^1, \mathcal{A}^2)$ the probability that for $(M, \mathsf{st}) \leftarrow \mathcal{A}^1(1^n)$, $H \leftarrow \mathsf{HKGen}(1^n)$ and $M' \leftarrow \mathcal{A}^2(H, M, \mathsf{st})$ we have $M \neq M'$ but $H(M) = H(M')$, is negligible.

message authentication (MAC): We assume again that the intial value is replaced by a secret random key K. We say that the hash function is a *secure MAC* if for any efficient adversary $\mathcal{A}$ the probability that for $H \leftarrow \mathsf{HKGen}(1^n)$ and random K and $(M, \tau) \leftarrow \mathcal{A}^{H(K, \cdot)}(H)$ we have $\tau = H(K, M)$ and M has never been queried to oracle $H(K, \cdot)$, is negligible.

For a set $\mathrm{PROP} = \{\mathsf{P}_1, \mathsf{P}_2, \dots, \mathsf{P}_N\}$ of properties we write $\mathrm{PROP}(\mathcal{H}) \subseteq \mathrm{PROP}$ for the properties which hash function $\mathcal{H}$ has.

2.2 Multi-property Preserving Combiners

A hash function combiner $\mathcal{C} = (\mathsf{CKGen}, \mathsf{Comb})$ for hash functions $\mathcal{H}_0, \mathcal{H}_1$ itself is also a hash function which combines the two functions $\mathcal{H}_0, \mathcal{H}_1$ such that, if at least one of the hash functions obeys property P, then so does the combiner. For multiple properties $\mathrm{PROP} = \{\mathsf{P}_1, \mathsf{P}_2, \dots, \mathsf{P}_N\}$ one can either demand that the combiner inherits the properties if one of the candidate hash functions is strong and has all the properties (weakly preserving), or that for each property at least one of the two hash functions has the property (strongly preserving). We also consider a notion in between but somewhat closer to the weak case, called mildly preserving, in which case all properties from PROP must hold, albeit different functions may cover different properties (instead of one function as in the case of weakly preserving combiners).[2] More formally,

Definition 1 (Multi-Property Preservation). *For a set* $\mathrm{PROP} = \{\mathsf{P}_1, \mathsf{P}_2, \dots, \mathsf{P}_N\}$ *of properties a hash function combiner* $\mathcal{C} = (\mathsf{CKGen}, \mathsf{Comb})$ *for hash functions* $\mathcal{H}_0, \mathcal{H}_1$ *is called* weakly multi-property preserving *(wMPP) for* PROP *iff*

$$\mathrm{PROP} = \mathrm{PROP}(\mathcal{H}_0) \text{ or } \mathrm{PROP} = \mathrm{PROP}(\mathcal{H}_1) \implies \mathrm{PROP} = \mathrm{PROP}(\mathcal{C}),$$

mildly multi-property preserving *(mMPP) for* PROP *iff*

$$\mathrm{PROP} = \mathrm{PROP}(\mathcal{H}_0) \cup \mathrm{PROP}(\mathcal{H}_1) \implies \mathrm{PROP} = \mathrm{PROP}(\mathcal{C}),$$

and strongly multi-property preserving *(sMPP) for* PROP *iff for all* $\mathsf{P}_i \in \mathrm{PROP}$,

$$\mathsf{P}_i \in \mathrm{PROP}(\mathcal{H}_0) \cup \mathrm{PROP}(\mathcal{H}_1) \implies \mathsf{P}_i \in \mathrm{PROP}(\mathcal{C}).$$

We remark that for weak and mild preservation all individual properties $\mathsf{P}_1, \mathsf{P}_2, \dots, \mathsf{P}_N$ from PROP are guaranteed to hold, either by a single function as in weak preservation, or possibly by different functions as in mild preservation. The combiner may therefore depend on some strong property $\mathsf{P}_i \in \mathrm{PROP}$ which

---

[2] One may also refine these notions further. We focus on these three "natural" cases.

one of the hash functions has, and which helps to implement some other property P_j in the combined hash function. But then, for a subset $PROP' \subseteq PROP$ which, for instance, misses this strong property P_i, the combiner may no longer preserve the properties $PROP'$. This is in contrast to strongly preserving combiners which support such subsets of properties by definition.

Note that for a singleton $PROP = \{P\}$ all notions coincide and we simply say that $\mathcal{C}$ is P-preserving in this case. However, for two or more properties the notions become strictly stronger from weak to mild to strong, as we show in Section 4. Finally, we note that our definition allows the case $\mathcal{H}_0 = \mathcal{H}_1$, which may require some care when designing combiners, especially if the hash functions are based on random oracles (see also the remark after Lemma 3).

3   Constructing Multi-property Preserving Combiners

In this section we propose our combiner for the properties CR, PRF, PRO, TCR and MAC. We then show it to be strongly multi-property preserving for these properties.

3.1   Our Construction

Our combiner for functions $\mathcal{H}_0, \mathcal{H}_1$ is a pair of efficient algorithms $\mathcal{C}_{sMPP} = (CKGen_{sMPP}, Comb_{sMPP})$. The key generation algorithm $CKGen_{sMPP}(1^n)$ generates a triple (H_0, H_1, T) consisting of two hash functions $H_0 \leftarrow HKGen_0(1^n)$, $H_1 \leftarrow HKGen_1(1^n)$ and a public string $T = (T_0, T_1)$ where $T_b = (t_1^b, \dots, t_n^b)$ consists of n uniformly chosen values $t_i^b \in \{0,1\}^{5n}$.

The evaluation algorithm $Comb_{sMPP}^{H_0,H_1,T}$ for parameters H_0, H_1, T and message M

$$Comb_{sMPP}^{H_0,H_1,T}(M) = H_0^{prsrv}(M) \oplus H_1^{prsrv}(M)$$

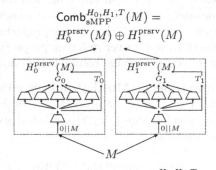

Fig. 1. Combiner $Comb_{sMPP}^{H_0,H_1,T}$

first computes two hash values $_b^{prsrv}(M)$ for $b = 0, 1$, each value based on hash function H_b and string T_b. For this it proceeds in three stages (see Figure 2):

- First hash the large message M into a short string $x_b \in \{0,1\}^n$ via the hash function H_b. For this step we prepend a 0-bit to M in order to make the hash function evaluation here somewhat independent from the subsequent stages.
- Then expand the short string x_b into five hash values $h_i^b = H_b(1||x_b|| \langle i \rangle_3)$ for $i = 0, 1, \dots, 4$, where $\langle i \rangle_3$ denotes the number i represented in binary with 3 bits. Concatenate these strings and denote the resulting $5n$-bit string by $G_b(x_b) = h_0^b||h_1^b|| \dots ||h_4^b$.
- Compute $T_b(x_b) = \oplus_{x_b[i]=1} t_i^b$ and add this value to $G_b(x_b)$. Denote the output by $H_b^{prsrv}(M) = G_b(x_b) \oplus T_b(x_b)$.

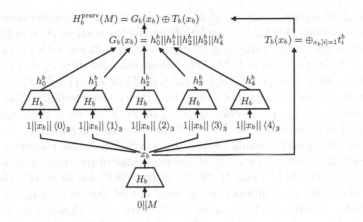

Fig. 2. Construction of H_b^{prsrv} based on hash function H_b

Our combiner now sets $\mathrm{Comb}_{\mathrm{sMPP}}^{H_0,H_1,T}(M) = H_0^{\mathrm{prsrv}}(M) \oplus H_1^{\mathrm{prsrv}}(M)$ as the final output.

3.2 Multi-property Preservation

We next show that the construction satisfies our strongest notion for combiners:

Theorem 1. *The combiner $\mathcal{C}_{sMPP}$ in Section 3.1 is a strongly multi-property preserving combiner for* PROP $= \{CR, PRF, PRO, TCR, MAC\}$.

The theorem is proven in five lemmas, each lemma showing that the combiner preserves one of the properties (as long as at least one hash functions guarantees this property). Since each lemma holds independently of further assumptions, the strong multi-property preservation follows.

Lemma 1. *The combiner $\mathcal{C}_{sMPP}$ is CR-preserving.*

Proof. The proof is by contradiction. Assume that an adversary $\mathcal{A}_{\mathrm{Comb}}$ on input H_0, H_1, T, with noticeable probability, outputs $M \neq M'$ with $\mathrm{Comb}_{\mathrm{sMPP}}^{H_0,H_1,T}(M) = \mathrm{Comb}_{\mathrm{sMPP}}^{H_0,H_1,T}(M')$. Then a collision

$$H_0^{\mathrm{prsrv}}(M) \oplus H_1^{\mathrm{prsrv}}(M) = H_0^{\mathrm{prsrv}}(M') \oplus H_1^{\mathrm{prsrv}}(M')$$
$$(G_0(x_0) \oplus T_0(x_0)) \oplus (G_1(x_1) \oplus T_1(x_1)) = (G_0(x_0') \oplus T_0(x_0')) \oplus (G_1(x_1') \oplus T_1(x_1'))$$

implies

$$G_0(x_0) \oplus G_0(x_0') \oplus G_1(x_1) \oplus G_1(x_1') = T_0(x_0) \oplus T_0(x_0') \oplus T_1(x_1) \oplus T_1(x_1'), \ (1)$$

where x_b denotes the hash value $H_b(0||M)$ of the first hash function evaluation and $G_b(x_b)$ the subsequent computation $h_0^b||h_1^b||h_2^b||h_3^b||h_4^b$ of the the hash values $h_i^b = H_b(1||x_b|| \langle i \rangle_3)$ for $i = 0, 1, \ldots, 4$.

The short inputs x_0, x_0', x_1, x_1' of n bits only give 2^{4n} possible values on the left side of equation (1), while the probability (over the random choice of the t_i's) that for such a fixed tuple with $x_0 \neq x_0'$ or $x_1 \neq x_1'$ a collision with $T_0(x_0) \oplus T_0(x_0') \oplus T_1(x_1) \oplus T_1(x_1')$ occurs, is 2^{-5n}. This follows since for $x_0 \neq x_0'$ or $x_1 \neq x_1'$ at least one of the sums $T_0(x_0) \oplus T_0(x_0') = T_0(x_0 \oplus x_0')$ or $T_1(x_1) \oplus T_1(x_1') = T_1(x_1 \oplus x_1')$ on the right hand side cannot cancel out. Hence the possibility that there exists some tuple x_0, x_0', x_1, x_1' with $x_0 \neq x_0'$ or $x_1 \neq x_1'$ such that equation (1) is statisfied, is at most $2^{4n} \cdot 2^{-5n} = 2^{-n}$ and therefore negligible.

Thus, with overwhelming probability a collision on the combiner only occurs if already the hash values x_b, x_b' at the first stage of the construction collide, i.e., $H_0(0||M) = H_0(0||M')$ and $H_1(0||M) = H_1(0||M')$ for $M \neq M'$. This, however, contradicts the assumption that at least one of the underlying hash functions is collision-resistant. This can be easily formalized through an adversary $\mathcal{A}_b$ for $b \in \{0,1\}$ which, on input $H_b \leftarrow \mathsf{HKGen}_b(1^n)$, samples the other public values $H_{\bar{b}} \leftarrow \mathsf{HKGen}_{\bar{b}}(1^n)$ and T and runs the adversary $\mathcal{A}_{\mathsf{Comb}}$ against the combiner on these data. Whenever $\mathcal{A}_{\mathsf{Comb}}$ outputs (M, M') adversary $\mathcal{A}_b$ returns $(0||M, 0||M')$. By assumption, both adversaries $\mathcal{A}_0, \mathcal{A}_1$ find collisions for H_0 and H_1, respectively, with noticeable probability then. $\square$

Lemma 2. *The combiner $\mathcal{C}_{sMPP}$ is PRF-preserving.*

Proof. The combiner $\mathsf{Comb}_{\mathsf{sMPP}}^{H_0, H_1, T}$ is pseudorandom if the distribution of the combiner's output cannot be distinguished from a truly random function by any polynomial-time adversary. Assume that one of the hash functions H_0 or H_1 is pseudorandom, yet the combiner is not pseudorandom, i.e., there is an adversary $\mathcal{D}_{\mathsf{Comb}}$ that can distinguish the function $\mathsf{Comb}_{\mathsf{sMPP}}^{H_0, H_1, T}(K_0, K_1, \cdot)$ from a random function F with non-negligible probability. We show that this allows to construct a successful distinguisher $\mathcal{D}_b$ for each underlying hash function H_b, which will contradict our initial assumption.

Recall that adversary $\mathcal{D}_{\mathsf{Comb}}$ has oracle access to a function that is either a random function $F : \{0,1\}^* \rightarrow \{0,1\}^{5n}$ or the keyed version of our construction $\mathsf{Comb}_{\mathsf{sMPP}}^{H_0, H_1, T}(K_0, K_1, \cdot)$, where the initial values $\mathrm{IV}_1, \mathrm{IV}_2$ in the applications of H_0 and H_1 are replaced by random strings K_0, K_1 of the same size. Then any efficient adversary $\mathcal{D}_{\mathsf{Comb}}$ can be transformed into an adversary $\mathcal{D}_b$ (for some $b \in \{0,1\}$) that distinguishes a random function $f : \{0,1\}^* \rightarrow \{0,1\}^n$ and a keyed hash function $H_b(K_b, \cdot) : \{0,1\}^* \rightarrow \{0,1\}^n$ for a randomly chosen key K_b with the same advantage.

First, the adversary $\mathcal{D}_b$ on input H_b samples $H_{\bar{b}} \leftarrow \mathsf{HKGen}_{\bar{b}}(1^n)$ and a key $K_{\bar{b}}$ and chooses random strings T. It then simulates $\mathcal{D}_{\mathsf{Comb}}$ on input (H_0, H_1, T). For each oracle query M of $\mathcal{D}_{\mathsf{Comb}}$, the adversary $\mathcal{D}_b$ computes a response by simulating the hash construction with the previously chosen key T, the function $H_{\bar{b}}(K_{\bar{b}}, \cdot)$ and its own oracle, i.e., each evaluation of the underlying hash function H_b in the computation of $H_b^{\mathrm{prsrv}}(K_b, M)$ is replaced by the response of $\mathcal{D}_b$'s oracle for the corresponding query. When $\mathcal{D}_{\mathsf{Comb}}$ eventually stops with output bit d algorithm $\mathcal{D}_b$, too, stops and returns d.

For the analysis recall that the underlying oracle of $\mathcal{D}_b$ is either a random function f or the hash function $H_b(K_b, \cdot)$. In the latter case, $\mathcal{D}_b$ perfectly simulates applications of our combiner and therefore generates outputs that are identically distributed to the hash values of the combiner. Hence, the output distribution of $\mathcal{D}_b$ in this case equals the one of $\mathcal{D}_{\mathsf{Comb}}$ with access to $\mathsf{Comb}_{\mathsf{sMPP}}^{H_0,H_1,T}$, i.e.,

$$\mathrm{Prob}\Big[\mathcal{D}_b^{H_b(K_b,\cdot)}(H_b) = 1\Big] = \mathrm{Prob}\Big[\mathcal{D}_{\mathsf{Comb}}^{\mathsf{Comb}_{\mathsf{sMPP}}^{H_0,H_1,T}(K_0,K_1,\cdot)}(H_0,H_1,T) = 1\Big].$$

If the oracle of $\mathcal{D}_b$ returns random values using a truly random function f, then the simulated response originates from a structured computation involving f. Yet we claim that the output still looks like a truly random function as long as no collision on the first stage of the construction occurs. With probability at most 2^{-n} any pair of queries $M \neq M'$ of $\mathcal{D}_{\mathsf{Comb}}$ yields a collision under f, i.e., such that $f(0\|M) = f(0\|M')$ which implies a collision on the final output of the simulation of H_b^{prsrv}. The probability that any collision among $q = q(n) = \mathrm{poly}(n)$ queries of $\mathcal{D}_{\mathsf{Comb}}$ occurs, is therefore at most $\binom{q}{2} \cdot 2^{-n}$. Given that this does not happen, each value $h_i^b = H_b(1\|x_b\|\langle i\rangle_3)$ for $i = 0,\ldots,4$ for the second stage is unique and the corresponding images under f are therefore independently and uniformly distributed. Hence $(h_0^b\|h_1^b\|h_2^b\|h_3^b\|h_4^b) \oplus T_b(x_b)$ is an independent random string, even when adding the value $H_{\bar{b}}^{\mathrm{prsrv}}(K_{\bar{b}}, M)$. This shows our claim.

Overall, the output distribution of $\mathcal{D}_b^f$ satisfies

$$\mathrm{Prob}\Big[\mathcal{D}_b^f(H_b) = 1\Big] \leq \mathrm{Prob}\Big[\mathcal{D}_b^f(H_b) = 1 \,\Big|\, \text{no Collision}\Big] + \mathrm{Prob}[\text{Collision}]$$

$$= \mathrm{Prob}\Big[\mathcal{D}_{\mathsf{Comb}}^F(H_0,H_1,T) = 1\Big] + \binom{q}{2} \cdot 2^{-n}.$$

Hence, the probability that $\mathcal{D}_b$ distinguishes H_b from f is

$$\mathrm{Prob}\Big[\mathcal{D}_b^{H_b(K_b,\cdot)}(H_b) = 1\Big] - \mathrm{Prob}\Big[\mathcal{D}_b^f(H_b) = 1\Big]$$

$$\geq \mathrm{Prob}\Big[\mathcal{D}_{\mathsf{Comb}}^{\mathsf{Comb}_{\mathsf{sMPP}}^{H_0,H_1,T}(K_0,K_1,\cdot)}(H_0,H_1,T) = 1\Big]$$

$$- \mathrm{Prob}\Big[\mathcal{D}_{\mathsf{Comb}}^F(H_0,H_1,T) = 1\Big] - \binom{q}{2} \cdot 2^{-n}$$

and thereby not negligible. This contradicts the assumption that either hash function H_0 or H_1 is a pseudorandom function.   □

Lemma 3. *The combiner $\mathcal{C}_{sMPP}$ is PRO-preserving.*

There is a small caveat here. Our definition of combiners allows to use the same hash function $\mathcal{H}_0 = \mathcal{H}_1$, albeit our combiner samples independent instances of the hash functions then. In this sense, it is understood that, if hash function $\mathcal{H}_0$ is given by a random oracle (as required for property PRO), then in case $\mathcal{H}_0 = \mathcal{H}_1$ the other hash function instance uses an independent random oracle.

Proof. We show that $\mathsf{Comb}_{\mathsf{sMPP}}^{H_0,H_1,T}$ is indifferentiable from a random oracle $\mathcal{F}$: $\{0,1\}^* \to \{0,1\}^{5n}$, when at least one underlying hash function H_0 or H_1 is a random oracle. By symmetry we can assume without loss of generality that $H_0 : \{0,1\}^* \rightarrow \{0,1\}^n$ is a random oracle. The (efficient) function H_1 can be arbitrary (but H_1 is sampled independently). It suffices that combiner and the simulator only have black-box access to H_1. The value T, required for the final output of the combiner, is chosen at random and given as input to all participating parties.

The adversary $\mathcal{D}$ has now oracle access either to the combiner $\mathrm{Comb}_{\mathrm{sMPP}}^{H_0,H_1,T}$ and the random oracle H_0 or to $\mathcal{F}$ and a simulator $\mathcal{S}^{\mathcal{F}}$. Our $\mathrm{Comb}_{\mathrm{sMPP}}^{H_0,H_1,T}$ is indifferentiable to $\mathcal{F}$ if there exists a simulator $\mathcal{S}^{\mathcal{F}}$, such that adversary $\mathcal{D}$ cannot have a significant advantage on deciding whether its interacting with $\mathrm{Comb}_{\mathrm{sMPP}}^{H_0,H_1,T}$ and H_0, or with $\mathcal{F}$ and $\mathcal{S}^{\mathcal{F}}$. We will use the simulator described below:

Simulator $\mathcal{S}_{H_0,H_1,T}^{\mathcal{F}}(X)$:   //use setEntry(), getEntry() to maintain list of queries/answers
on query X check if some entry $Y \leftarrow \mathrm{getEntry}(X)$ already exists
   if $Y = \perp$ //no entry so far
     if $X = 0||M$ for some M
       $\mathrm{setEntry}(X) = x_0$ where x_0 is randomly chosen from $\{0,1\}^n$
       get $U \leftarrow \mathcal{F}(M)$ for query M
       get $x_1 \leftarrow H_1(0||M)$ and subsequently $h_i^1 \leftarrow H_1(1||x_1||\langle i \rangle_3)$ for $i = 0, \ldots, 4$
       calculate $(h_0^0||h_1^0||h_2^0||h_3^0||h_4^0) = U \oplus (h_0^1||h_1^1||h_2^1||h_3^1||h_4^1) \oplus T_1(x_1) \oplus T_0(x_0)$
       save values $h_0^0, \ldots, h_4^0$ of potential queries $1||x_0||\langle 0 \rangle_3, \ldots, 1||x_0||\langle 4 \rangle_3$:
         $\mathrm{setEntry}(1||x_0||\langle i \rangle_3) = h_i^0$ for $i = 0, 1, \ldots, 4$

     if $X \neq 0||M$, choose a random $Y \in \{0,1\}^n$
       and save the value by $\mathrm{setEntry}(X) = Y$
  output $Y \leftarrow \mathrm{getEntry}(X)$

The simulator's goal is to mimic H_0, i.e., to produce an output that looks consistent to what the distinguisher can obtain from $\mathcal{F}$. To simulate H_0, the simulator $\mathcal{S}$ creates a database, where in addition to the previously processed queries and answers also answers to potential subsequent queries of $\mathcal{D}$ are stored. Those additional entries are generated if $\mathcal{S}$ receives a new query $X = 0||M$, that might be an attempt of $\mathcal{D}$ to simulate the construction of our combiner with the answers of $\mathcal{S}$. In this case, the simulator first chooses a random answer x_0. Then $\mathcal{S}$ invokes the random oracle $\mathcal{F}$ on input M and the black-box function H_1 on input X, where the answer $x_1 \leftarrow H_1(X)$ is used for further queries $1||x_1||\langle i \rangle_3$ to H_1. The responses to those queries correspond to the values $h_0^1, \ldots, h_4^1$ at the second stage of the H_1^{prsrv} evaluation. With the help of those values and the output $\mathcal{F}(M)$ of the random oracle, the simulator is able to compute the "missing" answers $h_0^0, \ldots, h_4^0$ that it has to return. Each h_i^0 for $i = 0, \ldots, 4$ is stored for the corresponding query $1||x_0||\langle i \rangle_3$ which $\mathcal{D}$ might submit later. For any new query X that is not of type $0||M$ the simulator responds with a random value from $\{0,1\}^n$ and stores the value.

Except for two events E_1, E_2 (defined below), the simulator will provide outputs that are consistent with $\mathcal{F}$, such that $\mathcal{D}$ cannot distinguish between $(\mathrm{Comb}_{\mathrm{sMPP}}^{H_0,H_1,T}, H_0)$ and $(\mathcal{F}, \mathcal{S}^{\mathcal{F}})$. The first event E_1 is a collision for $\mathcal{S}$ with $\mathcal{S}(0||M) = \mathcal{S}(0||M')$, $M \neq M'$ but $\mathcal{F}(M) \neq \mathcal{F}(M')$, that occurs with probability at most $\binom{q}{2} \cdot 2^{-n}$ where q denotes the number of queries by $\mathcal{D}$.

The second event E_2 occurs if $\mathcal{D}$ makes queries to $\mathcal{S}$ of the form $1||x_0|| \langle i \rangle_3$ where x_0 has not been an answer of the simulator before, but on a subsequent query $X = 0||M$ the simulator picks x_0 as its answer. In this case $\mathcal{S}$ has already fixed at least one value h_i^0 for $i = 0, \dots, 4$ and cannot later define this value after learning $\mathcal{F}(M)$. In particular, $\mathcal{S}$ is then unable to provide a consistent output. But, since $\mathcal{S}$ returns random values from $\{0, 1\}^n$ on new queries X, the probability for $\mathcal{S}(X) = x_0$ for any previous query x_0 is at most $q \cdot 2^{-n}$, where q is the maximal number of queries of type $1||x_0|| \langle i \rangle_3$ in $\mathcal{D}$'s execution. Overall, event E_2 happens with probability at most $q^2 \cdot 2^{-n}$.

Given that neither event occurs all replies by $\mathcal{S}$ are random (but consistent with the values provided by $\mathcal{F}$). Comparing the two games we note that, for a consistent run, the simulator's random choices and the replies of $\mathcal{F}$ to the simulator's queries implicitly define a random function f, where the only difference to the original construction of Π_0^{prsrv} and the "forward" usage of f is that f in the simulation here is defined "backwards" through $\mathcal{F}$. Still, the two experiments look identical from $\mathcal{D}$'s viewpoint.

The advantage of the adversary $\mathcal{D}$ is thus at most the probability that one of the events E_1 or E_2 happens, i.e., $\text{Prob}[E_2 \vee E_2] \leq (\binom{q}{2} + q^2) \cdot 2^{-n}$. Hence, the probability that $\mathcal{D}$ can distinguish whether it is communicating with $(\text{Comb}_{\text{sMPP}}^{H_0, H_1, T}, H_0)$ or with $(\mathcal{F}, \mathcal{S}^{\mathcal{F}})$, is negligible. $\square$

Lemma 4. *The combiner $\mathcal{C}_{sMPP}$ is TCR-preserving.*

The proof that our combiner is target collision-resistant follows the argument for collision-resistance closely and appears in the full version.

Lemma 5. *The combiner $\mathcal{C}_{sMPP}$ is MAC-preserving.*

Proof. Assume towards contradiction that our combiner is *not* a secure MAC. Then there exists an adversary $\mathcal{A}_{\text{Comb}}$ which, after learning several values $\tau_i = \text{Comb}_{\text{sMPP}}^{H_0, H_1, T}(K_0, K_1, M_i)$ for adaptively chosen M_i's, outputs $M \neq M_1, M_2, \dots, M_q$ and τ such that $\tau = \text{Comb}_{\text{sMPP}}^{H_0, H_1, T}(K_0, K_1, M)$ with noticeable probability.

Given $\mathcal{A}_{\text{Comb}}$ we construct a MAC-adversary $\mathcal{A}_b$ against hash function $\mathcal{H}_b$ for $b \in \{0, 1\}$. This adversary $\mathcal{A}_b$ is given H_b as input and oracle access to a function $H_b(K_b, \cdot)$ and uses the attacker $\mathcal{A}_{\text{Comb}}$ in a black-box way to produce a forgery. To this end, $\mathcal{A}_b$ first samples T and $H_{\bar{b}} \leftarrow \text{HKGen}_{\bar{b}}(1^n)$ and $K_{\bar{b}}$ as specified by the combiner, and then invokes $\mathcal{A}_{\text{Comb}}$ for input (H_0, H_1, T). For each query M_i of $\mathcal{A}_{\text{Comb}}$ our adversary computes the combiner's output with the help of its oracle $H_b(K_b, \cdot)$ and knowledge of the other parameters. In particular, for each query adversary $\mathcal{A}_b$ calls its oracle six times about $0||M_i$ and $1||x_{b,i}|| \langle 0 \rangle_3, \dots, 1||x_{b,i}|| \langle 4 \rangle_3$.

If, at the end, $\mathcal{A}_{\text{Comb}}$ returns M and τ such that M is not among the previous q queries M_i, then adversary $\mathcal{A}_b$ flips a coin $c \leftarrow \{0, 1\}$ and proceeds as follows:

- If $c = 0$ then $\mathcal{A}_b$ chooses an index i at random between 1 and q and looks up the answer $x_{b,i}$ it received in response to its query $0||M_i$. It stops with output $(0||M, x_{b,i})$.

- If $c = 1$ then $\mathcal{A}_b$ queries its oracle about $0||M$ to receive an answer x_b. It then uses its knowledge about the other parameters to compute $H_{\bar{b}}^{\text{prsrv}}(M)$ and calculates $y = \tau \oplus H_{\bar{b}}^{\text{prsrv}}(M) \oplus T_b(x_b)$. It outputs the message $1||x_b||000$ and the first n bits of y and stops.

If $\mathcal{A}_{\text{Comb}}$ fails to output a pair (M, τ) or returns a previously queried message $M = M_i$, then $\mathcal{A}_b$ reports failure and terminates.

For the analysis we consider the two exclusive cases of an successful $\mathcal{A}_{\text{Comb}}$. First, the adversary $\mathcal{A}_{\text{Comb}}$ manages to find a new M and a valid τ such that $H_b(K_b, 0||M)$ collides with some value $H_b(K_b, 0||M_i)$ for some query $0||M_i$. Given this, adversary $\mathcal{A}_b$ outputs $0||M$ and $H_b(K_b, 0||M)$ with probability $\frac{1}{2q}$, namely, if $c = 0$ and the guess for i is correct. But then $0||M$ is distinct from all of $\mathcal{A}_b$'s previous queries (because all $0||M_i$'s are distinct from $0||M$ and all other queries of $\mathcal{A}_b$ are prepended by a 1-bit). Hence, if $\mathcal{A}_{\text{Comb}}$ successfully forges such a MAC with noticeable probability, then so does $\mathcal{A}_b$. Put differently, the probability that $\mathcal{A}_{\text{Comb}}$ succeeds for such cases is negligible by the security of H_b.

The second case occurs if $\mathcal{A}_{\text{Comb}}$ outputs a fresh M and a valid tag τ such that $x_b = H_b(K_b, 0||M)$ is distinct from all values $x_{b,i} = H_b(K_b, 0||M_i)$ for the queries $0||M_i$. In this case, if $c = 1$, adversary $\mathcal{A}_b$ "unmasks" τ to recover $y = H_b(K_b, 1||x_b|| \langle 0 \rangle_3)|| \dots ||H_b(K_b, 1||x_b|| \langle 4 \rangle_3)$. Note that this requires $\mathcal{A}_b$ to make a further oracle query about value $0||M$. But this value (in addition to all other queries) is different from $1||x_b||000$, and $\mathcal{A}_b$ therefore returns a valid forgery with noticeable probability (if $\mathcal{A}_{\text{Comb}}$ would succeed with noticeable probability for this case).

In summary, it follows that any successful adversary on the combiner MAC immediately yields successful attacks on both hash functions, proving the claim.

□

3.3   Variations

In this section we briefly deal with some variations of our previous construction.

Reducing the Key Size. To reduce the key size in our construction we may assume that one of the hash functions is a random oracle and has property PRO, and move from strongly preserving combiners to mildly preserving ones. This also shows that such weaker combiners may come with a gain in efficiency.

If we assume that one hash function behaves like a random function then, instead of picking the t_i's at random and putting them into the key, we define $t_i^b := H_0(\ddagger||b||i) \oplus H_1(\ddagger||b||i)$ for a special symbol $\ddagger$ different from 0 and 1 (e.g., in practice encode 0 and 1 as 00 and 01, respectively, and set $\ddagger = 11$). The prefix $\ddagger$ makes the values independent of the intermediate values in the computation, and the values t_i^b can now be computed "on the fly" instead of storing them in the key.

Given that either hash function has property PRO the values t_i^b are pseudorandom and the proofs in the previous section carry over and we get a *mildly* multi-property preserving combiner for PROP = {CR, TCR, MAC, PRF, PRO}. The key size now equals the one for the two underlying hash functions.

Hash Functions with Different Output Sizes. Our construction utilizes the fact that both hash functions have the same output length n. This implies that the concatenation of 5 hash function values $H_b(1||x_b|| \langle i \rangle_3)$ for each function H_b^{prsrv} has the same length.

If we consider two hash functions with distinct output sizes n_0 and n_1, then we need to concatenate $5 \cdot \max\{n_0, n_1\}$ bits of output. For this we simply concatenate enough hash values $H_b(1||x_b|| \langle i \rangle_{\ell_b})$ (with increasing counter values i) for $\ell_b = \lceil \log_2(5 \cdot \max\{n_0, n_1\}/n_b) \rceil$, and truncate longer outputs to $5 \cdot \max\{n_0, n_1\}$ bits. At the same time the t_i^b's are also chosen to be of length $5 \cdot \max\{n_0, n_1\}$. With these modifications all the proofs carry over straightforwardly.

Combining More Hash Functions. To combine $h \geq 3$ hash functions, each with output size $n_0, n_1, \ldots, n_{h-1}$, we set again $n := \max\{n_0, n_1, \ldots, n_{h-1}\}$ and, this time, produce $(2h + 1) \cdot n$ output bits for each function H_b^{prsrv}. Accordingly, we let the t_i^b's be of length $(2h + 1) \cdot n$. As long as h is polynomial the proofs can be easily transferred to this case.

Alternatively, one can apply our general method to combine three or more hash functions as discussed in Section 5. Yet, this general construction yields a less efficient solution than the tailor-made solution above.

4   Weak vs. Mild vs. Strong Preservation

The first proposition shows that strong preservation implies mild preservation which, in turn, implies weak preservation. The proof is straightforward and given only for sake of completeness:

Proposition 1. *Let* PROP *be a set of properties. Then any strongly multi-property preserving combiner for* PROP *is also mildly preserving for* PROP, *and any mildly preserving combiner for* PROP *is also weakly preserving for* PROP.

Proof. Assume that the combiner is sMPP for PROP. Suppose further that $\mathrm{PROP}(\mathcal{C}) \not\subseteq \mathrm{PROP}$ such that there is some property $\mathsf{P}_i \in \mathrm{PROP} - \mathrm{PROP}(\mathcal{C})$. Then, since the combiner is sMPP, we must also have $\mathsf{P}_i \notin \mathrm{PROP}(\mathcal{H}_0) \cup \mathrm{PROP}(\mathcal{H}_1)$, else we derive a contradiction to the strong preservation. We therefore have $\mathrm{PROP} \not\subseteq \mathrm{PROP}(\mathcal{H}_0) \cup \mathrm{PROP}(\mathcal{H}_1)$, implying mild preservation via the contrapositive statement.

Now consider an mMPP combiner and assume $\mathrm{PROP} = \mathrm{PROP}(\mathcal{H}_0)$ or $\mathrm{PROP} = \mathrm{PROP}(\mathcal{H}_1)$. Then, in particular, $\mathrm{PROP} = \mathrm{PROP}(\mathcal{H}_0) \cup \mathrm{PROP}(\mathcal{H}_1)$ and the mMPP property says that also $\mathrm{PROP} = \mathrm{PROP}(\mathcal{C})$. This proves sMPP.   □

To separate the notions we consider the collision-resistance property CR and the property NZ (*non-zero output*) that the hash function should return $0 \cdots 0$ with small probability only. This may be for example required if the hash value should be inverted in a field:

non-zero output (NZ): A hash function $\mathcal{H}$ has property NZ if for any efficient adversary $\mathcal{A}$ the probability that for $H \leftarrow \mathsf{HKGen}(1^n)$ and $M \leftarrow \mathcal{A}(H)$ we have $H(M) = 0 \cdots 0$, is negligible.

Lemma 6. *Let* PROP = {*CR*, *NZ*} *and assume that collision-intractable hash functions exist. Then there is a hash function combiner which is weakly multi-property perserving for* PROP, *but not mildly multi-property preserving for* PROP.

Proof. Consider the following combiner (with standard key generation, (H_0, H_1) ← CKGen(1^n) for H_0 ← HKGen$_0$(1^n) and H_1 ← HKGen$_1$(1^n)):

> The combiner for input M first checks that the length of M is even, and if so, divides $M = L||R$ into halves L and R, and
> – checks that $H_0(L) \neq H_0(R)$ if $L \neq R$, and that $H_0(M) \neq 0 \cdots 0$,
> – verifies that $H_1(L) \neq H_1(R)$ if $L \neq R$, and that $H_1(M) \neq 0 \cdots 0$.
> If the length of M is odd or any of the two properties above holds, then the combiner outputs $H_0(M)||H_1(M)$. In any other case, it returns 0^{2n}.

We first show that the combiner is weakly preserving. For this assume that the hash function H_b for $b \in \{0, 1\}$ has both properties. Then the combiner returns the exceptional output 0^{2n} only with negligible probability, namely, if one finds an input with a non-trivial collision under H_b and which also refutes property NZ. In any other case, the combiner's output $H_0(M)||H_1(M)$ inherits the properties CR and NZ from hash function H_b.

Next we show that the combiner is not mMPP. Let H_1' be a collision-resistant hash function with $n - 1$ bits output (and let H_1 include a description of H_1'). Define the following hash functions:

$$H_0(M) = 1^n, \qquad H_1(M) = \begin{cases} 0^n & \text{if } M = 0^n 1^n \\ 1||H_1'(M) & \text{else} \end{cases}.$$

Clearly, H_0 has property NZ but is not collision-resistant. On the other hand, H_1 obeys CR but not NZ, as $0^n 1^n$ is mapped to zeros. But then we have PROP = {CR, NZ} = PROP(H_0)∪PROP(H_1) and mild preservation now demands that the combiner, too, has these two properties. Yet, for input $M = 0^n 1^n$ the combiner returns 0^{2n} since the length of M is even, but $L = 0^n$ and $R = 1^n$ collide under H_0, and M is thrown to 0^n under H_1. This means that the combiner does not obey property NZ.   □

Lemma 7. *Let* PROP = {*CR*, *NZ*}. *Then there exists a hash function combiner which is mildly multi-property perserving for* PROP, *but not strongly multi-property preserving for* PROP.

Proof. Consider the following combiner (again with standard key generation):

> The combiner for input M first checks that the length of M is even, and if so, divides $M = L||R$ into halves L and R and then verifies that $H_0(L) \neq H_1(R)$ or $H_1(L) \neq H_1(R)$ or $L = R$. If any of the latter conditions holds, or the length of M is odd, then the combiner outputs $H_0(M)||H_1(M)$. In any other case it returns 0^{2n}.

We first prove that the combiner above is mMPP. Given that $\text{PROP} \subseteq \text{PROP}(H_0) \cup \text{PROP}(H_1)$ at least one of the two hash functions is collision-resistant. Hence, even for $M = L\|R$ with even length and $L \neq R$, the hash values only collide with negligible probability. In other words, the combiner outputs $H_0(M)\|H_1(M)$ with overwhelming probability, implying that the combiner too has properties CR and NZ.

Now consider the constant hash functions $H_0(M) = H_1(M) = 1^n$ for all M. Clearly, both hash functions obey property $\text{NZ} \in \text{PROP}(H_0) \cup \text{PROP}(H_1)$. Yet, for input $0^n 1^n$ the combiner returns 0^{2n} such that $\text{NZ} \notin \text{PROP}(\mathcal{C})$, implying that the combiner is not strongly preserving. $\square$

The proof indicates how mildly (or weakly) preserving combiners may take advantage of further properties to implement other properties. It remains open if one can find similar separations for the popular properties like CR and PRF, or for CR and PRO.

5 Multiple Hash Functions and Tree-Based Composition of Combiners

So far we have considered combiners for two hash functions. The multi-property preservation definition extends to the case of more hash functions as follows:

Definition 2. *For a set* $\text{PROP} = \{P_1, P_2, \ldots, P_N\}$ *of properties an m-function combiner* $\mathcal{C} = (\text{CKGen}, \text{Comb})$ *for hash functions* $\mathcal{H}_0, \mathcal{H}_1, \ldots, \mathcal{H}_{m-1}$ *is called weakly multi-property preserving (wMPP) for* PROP *iff*

$$\exists j \in \{0, 1, \ldots, m-1\} \ s.t. \ \text{PROP} = \text{PROP}(\mathcal{H}_j) \implies \text{PROP} = \text{PROP}(\mathcal{C}),$$

mildly multi-property preserving (mMPP) for PROP *iff*

$$\text{PROP} = \bigcup_{j=0}^{m-1} \text{PROP}(\mathcal{H}_j) \implies \text{PROP} = \text{PROP}(\mathcal{C}),$$

and strongly multi-property preserving *(sMPP) for* PROP *iff for all* $P_i \in \text{PROP}$,

$$P_i \in \bigcup_{j=0}^{m-1} \text{PROP}(\mathcal{H}_j) \implies P_i \in \text{PROP}(\mathcal{C}).$$

For the above definitions we still have that sMPP implies mMPP and mMPP implies wMPP. The proof is a straightforward adaption of the case of two hash functions.

Given a combiner for two hash functions one can build a combiner for three or more hash functions by considering the two-function combiner itself as a hash function and applying it recursively. For instance, to combine three hash functions $\mathcal{H}_0, \mathcal{H}_1, \mathcal{H}_2$ one may define the "cascaded" combiner by $\mathcal{C}_2(\mathcal{C}_2(\mathcal{H}_0, \mathcal{H}_1), \mathcal{H}_2)$, where we assume that the output of C_2 allows to be used again as input to the combiner on the next level.

More generally, given m hash functions and a two-function combiner C_2 we define an m-function combiner C_{multi} as a binary tree, as suggested for general combiners by [12]. Each leaf is labeled by one of the m hash functions (different leaves may be labeled by the same hash function). Each inner node, including the root, with two descendants labeled by $\mathcal{F}_0$ and $\mathcal{F}_1$, is labeled by $C_2(\mathcal{F}_0, \mathcal{F}_1)$.

The key generation algorithm for this tree-based combiner now runs the key generation algorithm for the label at each node (each run independent of the others, even if two nodes contain the same label). To evaluate the multi-hash function combiner one inputs M into each leaf and computes the functions outputs recursively up to the root. The output of the root node is then the output of C_{multi}. We call this a *combiner tree for C_2 and $\mathcal{H}_0, \mathcal{H}_1, \ldots, \mathcal{H}_{m-1}$*.

For efficiency reasons we assume that there are at most polynomially many combiner evaluations in a combiner tree. Also, to make the output dependent on all hash functions we assume that each hash function appears in (at least) one of the leaves. If a combiner tree obeys these properties, we call it an *admissible combiner tree for C_2 and $\mathcal{H}_0, \mathcal{H}_1, \ldots, \mathcal{H}_{m-1}$*.

We first show that weak MPP and strong MPP preserve their properties for admissible combiner trees:

Proposition 2. *Let C_2 be a weakly (resp. strongly) multi-property preserving two-function combiner for* PROP. *Then any admissible combiner tree for C_2 and functions $\mathcal{H}_0, \mathcal{H}_1, \ldots, \mathcal{H}_{m-1}$ for $m \geq 2$ is also weakly (resp. strongly) multi-property preserving for* PROP.

Proof. We give the proof by induction for the depth of the tree. For depth $d = 1$ we have $m = 2$ and $C_{\text{multi}}(\mathcal{H}_0, \mathcal{H}_1) = C_2(\mathcal{H}_0, \mathcal{H}_1)$ or $C_{\text{multi}}(\mathcal{H}_0, \mathcal{H}_1) = C_2(\mathcal{H}_1, \mathcal{H}_0)$ and the claim follows straightforwardly for both cases.

Now assume $d > 1$ and that combiner Comb_2 is wMPP. Then the root node applies C_2 to two nodes N_0 and N_1, labeled by $\mathcal{F}_0$ and $\mathcal{F}_1$. Note that by the wMPP prerequisite we assume that there exists one hash function $\mathcal{H}_j$ which has all properties in PROP. Since this hash functions appears in at least one of the subtrees under N_0 or N_1, it follows by induction that at least one of the functions $\mathcal{F}_0$ and $\mathcal{F}_1$, too, has properties PROP. But then the combiner application in the root node also inherits these properties from its descendants.

Now consider $d > 1$ and the case of strong MPP. It follows analogoulsy to the previous case that for each property $\mathsf{P}_i \in$ PROP, one of the hash functions in the subtrees rooted at N_0 and N_1 must have property P_i as well. This carries over to the combiners at nodes N_0 or N_1 by induction, and therefore to the root combiner. □

Somewhat surprisingly, mild MPP in general does not propagate security for tree combiners, as we show by a counter-example described in the full version. Note that we still obtain, via the previous proposition, that the mMPP combiner is also wMPP and that the resulting tree combiner is thus also wMPP. Yet, it loses its mMPP property.

Proposition 3. *Let* PROP $= \{CR, NZ\}$ *and assume that there are collision-intractable hash functions. Then there exists a two-function weakly multi-property preserving combiner* C_2 *for* PROP, *and an admissible tree combiner for* C_2 *and hash functions* $\mathcal{H}_0, \mathcal{H}_1, \mathcal{H}_2$ *which is not mildly multi-property preserving for* PROP.

Note that the cascading combiner can also be applied to our combiner in Section 3 to compose three or more hash functions (with the adaption for hash functions with different output lengths discussed in Section 3.3). The derived combiner, however, is less efficient than the direct construction sketched there.

Acknowledgments

We thank the anonymous reviewers for valuable comments. Both authors are supported by the Emmy Noether Program Fi 940/2-1 of the German Research Foundation (DFG).

References

1. Boneh, D., Boyen, X.: On the Impossibility of Efficiently Combining Collision Resistant Hash Functions. In: Dwork, C. (ed.) CRYPTO 2006. LNCS, vol. 4117, pp. 570–583. Springer, Heidelberg (2006)
2. Bellare, M., Canetti, R., Krawczyk, H.: Keying hash functions for message authentication. In: Koblitz, N. (ed.) CRYPTO 1996. LNCS, vol. 1109, pp. 1–15. Springer, Heidelberg (1996)
3. Boldyreva, A., Fischlin, M.: Analysis of Random Oracle Instantiation Scenarios for OAEP and Other Practical Schemes. In: Shoup, V. (ed.) CRYPTO 2005. LNCS, vol. 3621, pp. 412–429. Springer, Heidelberg (2005)
4. Boldyreva, A., Fischlin, M.: On the Security of OAEP. In: Lai, X., Chen, K. (eds.) ASIACRYPT 2006. LNCS, vol. 4284, pp. 210–225. Springer, Heidelberg (2006)
5. Bellare, M., Rogaway, P.: Optimal Asymmetric Encryption — How to Encrypt with RSA. In: De Santis, A. (ed.) EUROCRYPT 1994. LNCS, vol. 950, pp. 92–111. Springer, Heidelberg (1995)
6. Bellare, M., Rogaway, P.: The exact security of digital signatures — How to sign with RSA and Rabin. In: Maurer, U.M. (ed.) EUROCRYPT 1996. LNCS, vol. 1070, pp. 399–416. Springer, Heidelberg (1996)
7. Bellare, M., Ristenpart, T.: Multi-Property Preserving Hash Domain Extensions and the EMD Transform. In: Lai, X., Chen, K. (eds.) ASIACRYPT 2006. LNCS, vol. 4284, pp. 299–314. Springer, Heidelberg (2006)
8. Bellare, M., Ristenpart, T.: Hash Functions in the Dedicated-Key Setting: Design Choices and MPP Transforms. In: Arge, L., Cachin, C., Jurdziński, T., Tarlecki, A. (eds.) ICALP 2007. LNCS, vol. 4596, pp. 399–410. Springer, Heidelberg (2007)
9. Coron, J.-S., Dodis, Y., Malinaud, C., Puniya, P.: Merkle-Damgard revisited: How to construct a hash function. In: Shoup, V. (ed.) CRYPTO 2005. LNCS, vol. 3621, Springer, Heidelberg (2005)
10. Canetti, R., Rivest, R.L., Sudan, M., Trevisan, L., Vadhan, S.P., Wee, H.: Amplifying Collision Resistance: A Complexity-Theoretic Treatment. In: Menezes, A. (ed.) CRYPTO 2007. LNCS, vol. 4622, pp. 264–283. Springer, Heidelberg (2007)

11. Herzberg, A.: On Tolerant Cryptographic Constructions. In: Menezes, A. (ed.) CT-RSA 2005. LNCS, vol. 3376, pp. 172–190. Springer, Heidelberg (2005)
12. Harnik, D., Kilian, J., Naor, M., Reingold, O., Rosen, A.: On Robust Combiners for Oblivious Transfer and other Primitives. In: Cramer, R.J.F. (ed.) EUROCRYPT 2005. LNCS, vol. 3494, pp. 96–113. Springer, Heidelberg (2005)
13. Katz, J., Shin, J.S.: Modeling Insider Attacks on Group Key-Exchange Protocols. In: Proceedings of the Annual Conference on Computer and Communications security (CCS), ACM Press, New York (2005)
14. Maurer, U., Renner, R., Holenstein, C.: Indifferentiability, Impossibility Results on Reductions, and Applications to the Random Oracle Methodology. In: Naor, M. (ed.) TCC 2004. LNCS, vol. 2951, pp. 21–39. Springer, Heidelberg (2004)
15. Naor, M.: Bit Commitment Using Pseudo-Randomness. Journal of Cryptology 4(2), 151–158 (1991)
16. Pietrzak, K.: Non-Trivial Black-Box Combiners for Collision-Resistant Hash-Functions don't Exist. In: Naor, M. (ed.) EUROCRYPT 2007. LNCS, vol. 4515, Springer, Heidelberg (2007)
17. Wang, X., Lai, X., Feng, D., Chen, H., Yu, X.: Cryptanalysis of the Hash Functions MD4 and RIPEMD. In: Cramer, R.J.F. (ed.) EUROCRYPT 2005. LNCS, vol. 3494, pp. 1–18. Springer, Heidelberg (2005)
18. Wang, X., Yu, H.: How to break MD5 and other hash functions. In: Cramer, R.J.F. (ed.) EUROCRYPT 2005. LNCS, vol. 3494, pp. 19–35. Springer, Heidelberg (2005)
19. Wang, X., Yin, Y.L., Yu, H.: Finding collisions in the full SHA-1. In: Shoup, V. (ed.) CRYPTO 2005. LNCS, vol. 3621, pp. 17–36. Springer, Heidelberg (2005)

OT-Combiners Via Secure Computation

Danny Harnik[1,*], Yuval Ishai[2,**], Eyal Kushilevitz[3,***],
and Jesper Buus Nielsen[4,†]

[1] IBM Research, Haifa, Israel
danny.harnik@gmail.com
[2] Technion, Israel and UCLA, USA
yuvali@cs.technion.ac.il
[3] Technion, Israel
eyalk@cs.technion.ac.il
[4] University of Aarhus, Denmark
buus@daimi.au.dk

Abstract. An *OT-combiner* implements a secure oblivious transfer (OT)
protocol using oracle access to n OT-candidates of which at most t may be
faulty. We introduce a new general approach for combining OTs by making
a simple and modular use of protocols for secure computation. Specifically,
we obtain an OT-combiner from any instantiation of the following two ingredients: (1) a t-secure n-party protocol for the OT functionality, in a
network consisting of secure point-to-point channels and a broadcast
primitive; and (2) a secure two-party protocol for a functionality determined by the former multiparty protocol, in a network consisting of a single OT-channel. Our approach applies both to the "semi-honest" and the
"malicious" models of secure computation, yielding the corresponding
types of OT-combiners.

Instantiating our general approach with secure computation protocols
from the literature, we conceptually simplify, strengthen the security, and
improve the efficiency of previous OT-combiners. In particular, we obtain
the first *constant-rate* OT-combiners in which the number of secure OTs
being produced is a constant fraction of the total number of calls to
the OT-candidates, while still tolerating a constant fraction of faulty
candidates ($t = \Omega(n)$). Previous OT-combiners required either $\omega(n)$ or
poly(k) calls to the n candidates, where k is a security parameter, and
produced only a single secure OT.

We demonstrate the usefulness of the latter result by presenting several applications that are of independent interest. These include:
Constant-rate OTs from a noisy channel. We implement n instances of a standard $\binom{2}{1}$-OT by communicating just $O(n)$ bits over
a noisy channel (binary symmetric channel). Our reduction provides

* Research conducted while at the Technion. Supported by grant 1310/06 from the
Israel Science Foundation and a fellowship from the Lady Davis Foundation.
** Supported by ISF grant 1310/06, BSF grant 2004361, and NSF grants 0205594,
0430254, 0456717, 0627781, 0716835, 0716389.
*** Supported by ISF grant 1310/06 and BSF grant 2002354.
† Funded by the Danish Agency for Science, Technology and Innovation.

R. Canetti (Ed.): TCC 2008, LNCS 4948, pp. 393–411, 2008.
© International Association for Cryptologic Research 2008

unconditional security in the semi-honest model. Previous reductions of this type required the use of $\Omega(kn)$ noisy bits.

Better amortized generation of OTs. We show that, following an initial "seed" of $O(k)$ OTs, each additional OT can be generated by only computing and communicating a *constant* number of outputs of a cryptographic hash function. This improves over a protocol of Ishai *et al.* (Crypto 2003), which obtained similar efficiency in the semi-honest model but required $\Omega(k)$ applications of the hash function for generating each OT in the malicious model.

1   Introduction

Secure Multiparty Computation (MPC) protocols allow a number of mutually distrusting parties to jointly evaluate functions over their local inputs without compromising the privacy of these inputs or the correctness of the output. (In the following we will also refer to functions over distributed inputs as "functionalities", capturing the general case where different parties may obtain distinct, and possibly randomized, outputs.) If a majority of the parties involved are honest, then there are "information-theoretic" solutions for this general task, requiring no computational assumptions [3,7]. On the other hand, if an honest majority is not guaranteed then, by [10,32], secure computation protocols for most functionalities imply the existence of *oblivious transfer* (OT) [37,19,39] — a secure two-party protocol for a simple functionality which allows a receiver to select one of two strings held by a sender. In an OT protocol the receiver learns the chosen string but no information about the other string, while the sender learns nothing about the receiver's selection. (By default, we use the term OT to refer to the basic *bit OT* primitive, where each string held by the sender consists of a single bit. OT of ℓ-bit strings can be implemented by making $O(\ell)$ calls to the basic OT primitive [4,5].) OT has proved to be a very useful building block in cryptographic protocols. Most notably, OT can serve as a building block for general secure two-party and multi-party protocols that tolerate an arbitrary number of corrupted parties [41,23,22,31,21,33].

1.1   Combiners and OT-Combiners

Often in cryptography there is uncertainty regarding the security of a construction (e.g., because of the reliance on unproven assumptions or placing too much trust in third parties). In such cases, it is handy to use a *combiner* for the underlying cryptographic task. An (m, n)-combiner (sometimes called a robust or tolerant combiner) is a method of taking n *candidates* for a cryptographic primitive and combining them into a single primitive that is secure as long as at least m of the n candidates were indeed secure. Combiners have been used implicitly in many cryptographic constructions as means of enhancing security and have recently been studied explicitly (initially in [26,25]).

In this paper, we focus on OT-combiners and their applications. The possibility of realizing combiners for OT or equivalent primitives has been investigated in [25,35,36,40]. Constructions of OT-combiners were given for the case that a *majority* of the OT-candidates are good [25,40]. These combiners are based on a technique of Damgård *et al.* [18] for reducing errors in weak versions of OT. On the other hand, there is a strong indication that there are no (black-box) OT-combiners if half of the candidates may be faulty [25]. We refer to $\frac{n-m}{n}$ as the *tolerance ratio* of the combiner. Thus, the results indicate that the tolerance ratio of an OT-combiner should be smaller than $\frac{1}{2}$.

A problem with OT-combiners as above is that, by definition, they are quite wasteful. One needs to invoke n OT-candidates (at least $m > \frac{n}{2}$ of which are good) in order to produce just a single instance of secure OT. Even worse, the known OT-combiners make a large number of *calls* to each candidate in order to produce this single secure OT. Thus, a desirable goal is to reduce the number of candidate calls made in order to produce each good OT. We refer to the latter quantity as the *production rate* of a combiner (or simply its *rate*).

The OT-combiners based on the technique of [18] have a production rate of $\Theta(k^2 n^4)$ (where k is a security parameter) for a majority of good candidates (see [40]). This rate can be improved to $\Theta(k^2)$ when assuming a constant tolerance ratio; i.e. $m > (\frac{1}{2} + \delta)n$, for a constant $\delta > 0$. Another downside of this construction is that it does not provide security when the identity of the faulty OTs can be determined adaptively.[1] A different approach to OT-combiners was taken by [36] (see also [2]) and requires $\Omega(n \log n)$ calls to the OT-candidates.

1.2 Our Results

We introduce a new general approach for constructing OT-combiners by making use of protocols for secure multiparty computation. Our approach follows a recent paradigm suggested by Ishai *et al.* [30] of employing secure *multiparty* protocols for the construction of secure *two-party* protocols. This allows us to benefit from the wide range of techniques that have been developed in the study of secure multiparty computation, obtaining conceptually simpler and more efficient OT-combiners.

More concretely, we show how to obtain an OT-combiner from any instantiation of the following two ingredients:

1. A t-secure n-party protocol for the OT functionality, in a network consisting of secure point-to-point channels and a broadcast primitive.[2] For instance, one could use here the (unconditionally secure) general-purpose protocols of [3,7,38,12].

---

[1] Adaptive security is not always required for combiners, however, at times it is crucial. For example, consider a setting where the OT-candidates are carried out simply by using third parties. A candidate is insecure if the corresponding third party is corrupted. In such a setting, an adversary can potentially corrupt a third party adaptively, during the execution of the combiner.

[2] This refers to a model in which the sender and the receiver are not considered to be among the n parties, but may each be corrupted by the adversary. Alternatively, one can use any $(t + 1)$-secure $(n + 2)$-party protocol in the standard MPC model.

2. A secure two-party protocol for a functionality determined by the former multiparty protocol, in a network consisting of a single OT-channel. For instance, one can use here the general-purpose unconditionally secure protocols of [23,31,22,21] or the computationally secure protocols of [41,33].

Our approach applies both to the "semi-honest" and to the "malicious" models of secure computation, yielding the corresponding types of OT-combiners. (In the semi-honest model the sender and the receiver follow the protocol as prescribed, while in the malicious model they may deviate from it.) In contrast to previous OT-combiners in the malicious model, the OT instances produced by our combiners are provably secure (in the malicious model) under standard simulation-based definitions, and can resist adaptive corruptions of candidates and parties.

By instantiating our general approach with efficient MPC protocols from the literature and by giving up just a constant fraction in the tolerance threshold, we get combiners with a constant production rate. In particular:

- There exists an OT-combiner in the *semi-honest* model with *constant production rate* and *constant tolerance ratio*. The combiner makes $O(1)$ calls to each of the n OT-candidates.
- There exists an OT-combiner in the *malicious* model with *constant production rate* and *constant tolerance ratio*. The combiner makes $s \leq \text{poly}(k)$ calls to each of the n OT-candidates (and generates $\Omega(ns)$ good OT calls). This combiner applies to *string OT* (rather than bit-OT) and requires additional calls to a one-way function.

Both results hold even if the bad candidates are chosen adaptively. Recall that previous OT-combiners required either $\omega(n)$ (with adaptive security) or $\text{poly}(k)$ (without adaptive security) calls to the OT-candidates in order to produce just a single instance of secure OT.

Techniques. The high-level idea behind our approach is to let the sender S and receiver R invoke the given MPC protocol between themselves and n additional "imaginary" parties called servers, where each server is jointly simulated by S and R using the given two-party protocol applied on top of a corresponding OT-candidate. Different instantiations of the underlying multiparty and two-party protocols yield different OT-combiners.

Our constant-rate combiners rely on MPC protocols in which the (amortized) communication complexity per gate of the circuit being evaluated is bounded by a constant, independently of the number of parties. For the type of functionalities we consider in this work, such protocols can be obtained by combining a protocol from [16] with secret-sharing schemes based on algebraic geometric codes or random linear codes [8,9] (see [30]). The protocol from [16], in turn, uses techniques from [3,20,27]. We also rely on OT-based secure two-party computation protocols in which the number of OT calls is a constant multiple of the input length. Such a protocol with a simulation-based proof of security was recently given in [33].

1.3   Applications

We demonstrate the usefulness of our constant-rate combiners by presenting several applications that are of independent interest.

Constant-rate OTs from a noisy channel. Crépeau and Kilian [14] demonstrated that two parties can implement an unconditionally secure OT by communicating over a binary symmetric channel (BSC) with some constant crossover probability $0 < p < \frac{1}{2}$. One can view such a channel as a secure implementation of a randomized functionality in which the receiver gets the sender's input bit with probability $1 - p$ and its negation with probability p. Thus, the result from [14] shows that this functionality is equivalent to OT. Unfortunately, the reduction from [14] is quite inefficient; its efficiency was later improved by Crépeau [13], but even this reduction requires $\Omega(k)$ noisy bits for producing a single OT, even in the semi-honest model.

Using our constant-rate combiners in the semi-honest model, we get n instances of $\binom{2}{1}$-OT by communicating just $O(n)$ bits over the noisy channel. Thus, the amortized cost of generating each OT call is just a *constant* number of calls to the noisy channel. Our reduction provides unconditional security in the semi-honest model and has error probability that vanishes exponentially with n. Combined with the OT-based secure computation protocol of [23,22,21], it implies that two parties can securely evaluate an arbitrary circuit of size s (with statistical security in the semi-honest model) by communicating only $O(s)$ bits over a noisy channel. It seems likely that our approach can be extended to yield similar results for the malicious model as well as for more general noise models and other probabilistic functionalities. We leave such extensions to future work.

Extending OTs efficiently in the malicious model. Current implementations of OT are quite expensive in practice, and thus form the efficiency bottleneck in protocols that make a heavy use of OTs. This state of affairs is backed up by the result of Impagliazzo and Rudich [28], which implies that there is no black-box construction of OT from one-way functions. As a next to best solution, Beaver [1] demonstrated how one can use just k OT calls (k being the security parameter) and extend them to polynomially many OT calls solely by adding calls to a one-way function. Beaver's protocol makes a non-black-box use of the underlying one-way function and is therefore considered inefficient in practice. Ishai *et al.* [29] gave an alternative construction that extends k OT calls to an essentially unbounded number of OT calls by making an additional black-box use of a cryptographic hash function. This protocol is highly efficient and has an amortized cost of computing and communicating just *two* outputs of the hash function for each produced OT. This approach can be viewed as the OT analogue of hybrid encryption, where an expensive asymmetric cryptosystem is used to encrypt a short secret key, allowing the bulk of the data to be encrypted efficiently using a symmetric encryption scheme.

Unfortunately, the efficient protocol of [29] applies only in the semi-honest model. In order to achieve security in the malicious model, a modified protocol is proposed based on a "cut-and-choose" approach. However, this approach increases the complexity of OT generation by a multiplicative factor of at least

$\Omega(k)$. In this paper, we utilize our constant-rate combiners to get OT extension in the malicious model that requires only a *constant* number of outputs of the hash function to be computed and communicated for each generated OT.

A first solution to this is by using a cut-and-choose approach similar to the one used in [29]. Namely, the semi-honest protocol of [29] is invoked $O(k)$ independent times on random inputs, and half of these invocations are "opened" to allow each party to verify that the other party followed the protocol's instructions. The unopened invocations have the guarantee that with overwhelming probability, a big majority of them generated secure OTs. This is exactly the setting required to apply our combiners. However, in this solution the seed of OT calls used grows substantially, which is undesirable.

Next, we develop a new solution that is not based on cut-and-choose. Instead, just a single instance of the semi-honest protocol is run, and a simple test is added communicating a constant number of hash values per produced OT. This test guarantees that with overwhelming probability, all but k of the produced OTs were secure. This allows to generate $(1 + \delta)k$ OT-candidates of which at most k are insecure, which again gives a big majority of secure OTs. As in [29], the reduction only makes a black-box use of the cryptographic hash function.

Reducing the number of OT channels in MPC. Consider an MPC protocol with security against a dishonest majority in a network of n parties. How many *OT channels* are required to allow such a non-trivial computation? An OT channel is a line over which two parties can carry out an unbounded number of OT calls. Do all pairs of parties require their own separate OT-channel? Harnik et al. [24] show that the answer is negative – if the number of corrupted parties is bounded by $t < (1 - \delta)n$ (for a constant δ) then $O(n)$ channels are sufficient. Namely, OT calls can be executed between every two parties using just OT calls on the existing $O(n)$ channels. However, in order to generate one OT call between a pair with no channel, the construction of [24] makes many OT calls over the existing channels: it first generates n candidates for OT (using just $O(1)$ OT calls to generate each candidate) and then it runs an OT-combiner on the n candidates. Using our new constant-rate combiners, we get the following result: the amortized cost of generating an OT call between two parties with no OT channel is $O(1)$ OT calls over the existing $O(n)$ channels.

Organization. In Section 2, we define OT-combiners. Section 3 describes our general approach for obtaining OT-combiners via secure computation and its instantiation for obtaining constant-rate combiners in the semi-honest model. Section 4 deals with the malicious model. The applications are described in Sections 5 (OT from noisy channels) and 6 (extending OTs efficiently). Due to space limitations, some of the details are deferred to the full version.

2   Definition of OT-Combiners

A combiner (see [26,25]) is given n implementation candidates for a cryptographic primitive and combines them into a single implementation that is secure if at least m of the original n candidates were indeed secure. Our study of combiners for OT follows this goal with an additional feature: we want the combiner to output many secure instances of OT rather than just one. This is desirable for efficiency reasons; indeed, invoking n OT-candidates and receiving just a single OT call in return seems quite wasteful. To accommodate this, we define the multi-OT functionality OT^ℓ in a straightforward manner (the sender holds ℓ pairs of secrets, the receiver holds ℓ choice bits and the receiver learns the secrets of his choice).

Our OT-combiners thus take several candidates for a secure OT protocol and produce a protocol for the OT^ℓ functionality. In general, the OT-candidates can be given in any representation, such as a code, or via a black-box access to a next message oracle. Our combiners work using such black-box access to the candidates. Accordingly, the definition we provide is that of a black-box combiner. For more comprehensive definitions of combiners, see [25].

We assume that the candidates are efficient (polynomial-time) algorithms. This guarantees that an efficient black-box combiner (counting each oracle call to a candidate as a single running step) remains efficient for any instantiation of the candidate.

When considering the functionality of the OT-candidates being combined, one should distinguish between two cases: (1) bad OT-candidates can compromise the privacy of the inputs but are guaranteed to have the correct functionality when executed honestly (namely, the receiver always ends up with the correct output); and (2) bad OT-candidates may produce arbitrary outputs. Combiners for the latter case are called error-tolerant combiners [36]. In the semi-honest model, we assume by default that bad candidates have the correct functionality, but our solutions can be easily extended to achieve error-tolerance with almost no loss of efficiency. In the malicious model, error-tolerance is always required; moreover, the functionality of each call to a bad candidate can be adaptively determined by the adversary during the execution of the combiner.

Definition 1 (OT-Combiner) *Let* $OT_1, \ldots, OT_n$ *be candidates for implementing* OT. *An* $(m, n; \ell, s)$-OT-Combiner *is an efficient two-party protocol with oracle access to the candidates such that: (1) If at least* m *of the* n *candidates securely compute the* OT^1 *functionality then the combiner securely computes the* OT^ℓ *functionality (where security is defined using a simulation-based definition, as in [6,21]); and (2) The combiner runs in polynomial time and makes a total of* s *calls to the candidates.*

The tolerance ratio *of the combiner is defined as* $\mu = \frac{m-n}{n}$. *The* production rate *(or simply the* rate*) of the combiner is defined as* $\rho = \frac{\ell}{s}$. *At times we omit the parameters* s *and* ℓ *and write "*(m, n)-OT-combiner*".*

The above definition views the number of candidates n as a constant. However, it is often useful to view the parameters of a combiner as functions of the security parameter k. (This is the case for the applications described in Sections 5,6.) The above definition can be naturally extended to this more general case.

We will sometimes refer to combiners that have *unconditional* (perfect or statistical) security. In such cases the above definition needs to be modified, since the candidates cannot be unconditionally secure. To this end, it is convenient to use the stronger notion of *third party black-box* combiners [25]. Such combiners are defined by viewing each candidate as a distinct external party that receives OT inputs from the sender and the receiver and sends the OT output to the receiver. A bad candidate is modelled by an external party that reveals both inputs to the adversary and (in the error-tolerant case) allows the adversary to control its output. All of our combiners satisfy this stronger definition with either perfect, statistical, or computational security.

3   OT-Combiners in the Semi-honest Model

In this section, we introduce our basic technique for obtaining combiners from protocols for secure computation. We begin by considering the semi-honest model and later (in Section 4) extend our results to the malicious model.

In the course of describing the OT-combiner, we define two intermediate models. The first is a tweak on the standard multiparty model which divides the parties into clients and servers, where the clients are the only parties to hold inputs and to receive outputs and the servers just assist in the computation. Such a variant of secure MPC setting was considered, e.g. in [11,16]. In our setting, there are only two clients, sender S and receiver R. In addition, there are n servers P_i that may aid the clients in the computation. However, up to t of the servers may be corrupted. Formally:

Definition 2 (Clients-Servers Model) *The network consists of $n+2$ parties: two clients, S and R, and n servers $P_1, \ldots, P_n$. There are secure channels between every two parties in the network. In the malicious model, we will also allow broadcast as an atomic primitive.*

FUNCTIONALITY: *f takes inputs from S and R and gives output to R.[3]*

ADVERSARIAL CORRUPTIONS: *The adversary may corrupt at most one of the clients S and R and at most t of the n servers. We refer to a protocol that is secure against such an adversary as a t-secure protocol in the clients-servers model. We consider adaptive adversaries by default; namely, we allow the adversary to decide which parties to corrupt during the execution of the protocol.*

The above model can be viewed as a refinement of the standard model for secure computation: every $(t + 1)$-secure $(n + 2)$-party protocol for f in the standard model is also a t-secure protocol for f in the clients-servers model.

In the second intermediate model we use, each server P_i is replaced by a pair of parties (S_i, R_i) that are connected by an OT channel. We call this the split-servers model. The intuition is that, at the end, we will have a two-party protocol where one party controls all R parties and the other controls all S parties.

---

[3] Our approach can be easily generalized to the case where f gives outputs to both S and R. However, in the malicious model it is impossible to guarantee *fairness* in this case.

Definition 3 (Split-Servers Model) *The network consists of $2n + 2$ parties: two clients, S and R, and n pairs of parties $(S_1, R_1), \ldots, (S_n, R_n)$. There is a secure channel between every two parties in the network (as well as a broadcast channel in the malicious model). In addition, there is a distinct OT channel between each pair (S_i, R_i).*

FUNCTIONALITY: *f takes inputs from S and R and gives output to R.*

ADVERSARIAL CORRUPTIONS: *The adversary has two possible corruption patterns: either (i) it corrupts the parties $S, S_1, \ldots, S_n$ and at most t of the R_i's; or (ii) it corrupts the parties $R, R_1, \ldots, R_n$ and at most t of the S_i's. We refer to a protocol that is secure against such an adversary as a t-secure protocol in the split-servers model. Again, we allow for adaptive adversaries.*

Our general construction employs two types of secure computation protocols: (1) Π_{MPC} is a t-secure *multi*party protocol in the clients-servers model. For $t < n/2$, every f admits such a protocol with perfect (resp., statistical) security against semi-honest (resp., malicious) adversaries [3,38]. (2) Π_{2party} is a secure 2-party protocol in the OT-hybrid model (i.e., using an ideal OT channel). Every f admits such a protocol with perfect (resp., statistical) security against semi-honest (resp., malicious) adversaries [22,31].

In our combiners, Π_{MPC} will always compute the functionality OT^ℓ.[4] On the other hand, we will need to employ protocols of type Π_{2party} for different functionalities. To simplify notation, we always use the notation Π_{2party} and make the actual functionality clear from the context.

Lemma 1. *Let f be a functionality taking inputs from S and R and returning output to R. There is a compiler that transforms any t-secure protocol Π_{MPC} for f in the semi-honest clients-servers model into a t-secure protocol Π_{split} for f in the semi-honest split-servers model. As a building block, the compiler requires a secure two-party protocol Π_{2party} for general functionalities in the semi-honest OT-hybrid model. If both Π_{MPC} and Π_{2party} are perfectly or statistically secure then so is Π_{split}.*

Proof: The idea is to distribute the local view of each server P_i in Π_{MPC} between the corresponding pair of parties S_i, R_i in Π_{split} using additive secret sharing. Thus, only an adversary corrupting both S_i and R_i can learn the view of P_i.

In the initialization stage of Π_{MPC}, the view of each client (S or R) contains its private input and its local randomness, while the view of each server P_i contains only its local randomness. To initialize the corresponding state in Π_{split}, let S and R remain as before and split the view of each P_i between S_i and R_i, by having each hold local random bits which together form an additive sharing of the randomness of P_i.

A typical intermediate step in the protocol Π_{MPC} is of the following form: Server P_i with local view v_{P_i} computes a function $m = \pi_{i,j}(v_{P_i})$ and sends the

---

[4] One can also consider cross-primitive combiners (see [34]), where the combiner implements a different functionality than the candidates. In such a case, the combiner's functionality will be computed by Π_{MPC}.

message m to server P_j. This step is simulated in Π_{split} by an interactive two-party protocol between S_i and R_i. Using the OT channel between them, S_i and R_i execute protocol Π_{2party} on the randomized functionality whose inputs are shares v_{S_i} and v_{R_i} of a view v_{P_i}, and whose outputs are random values m_{S_i} and m_{R_i} under the restriction that $m_{S_i} \oplus m_{R_i} = \pi_{i,j}(v_{S_i} \oplus v_{R_i})$. After the secure two-party protocol is executed, S_i sends m_{S_i} to S_j and R_i sends m_{R_i} to R_j. The recipients S_j and R_j append the new message to their share of the view v_{P_j}.

Another possible step in Π_{MPC} involves one or two of the clients, either as the party generating a new message m or as the receiving party. In case the recipient is one of the clients S or R, then both m_{S_i} and m_{R_i} are sent to this party. If a client (S or R) is generating the message m, then it computes m as in Π_{MPC} (no two-party protocol is needed in this case) and sends a random sharing of m to S_j and R_j, respectively. If both the sender and receiver of m are the clients then m is sent unchanged.

The new protocol produces a correct output since at each step the sum of the shares held by S_i and R_i is exactly the view of P_i, and thus the execution follows the protocol Π_{MPC} accurately. The proof of security hinges on the fact that if the ith pair is not corrupted (i.e., either S_i or R_i is uncorrupted), then the view of the simulated server P_i remains hidden from the adversary. Therefore, this view may be simulated in the same manner as in the original clients-servers model protocol Π_{MPC} (in the case that server P_i was not corrupted). On the other hand, if the ith pair is corrupted, then this corresponds to a corruption of P_i by the adversary. Further details are deferred to the full version. ∎

Lemma 2. *Given any protocol Π_{split} for the functionality OT^ℓ which is t-secure in the semi-honest split-servers model, one can construct (in a black-box way) an $(n-t,n)$-OT-combiner in the semi-honest model.*

Proof: We describe a two-party combiner with sender S' and receiver R' based on the split-server protocol Π_{split} with clients S and R and parties $S_1, \ldots, S_n$, $R_1, \ldots, R_n$. The combiner protocol is a straightforward simulation of Π_{split} where S' simulates the S-parties (i.e., $S, S_1, \ldots, S_n$) and R' simulates the R-parties $(R, R_1, \ldots, R_n)$. The simulation follows the protocol Π_{split} with the exception that, for every $i \in [n]$, all OT-calls between S_i and R_i are implemented using the candidate OT_i. Naturally, messages between the S-parties do not have to actually be sent as they are all simulated by S' (and similarly for the R-parties). Clearly, in a semi-honest environment, the combiner described is an execution of protocol Π_{split} and therefore it indeed implements the OT^ℓ functionality. Intuitively, security follows from the fact that for every bad candidate, the worst-case scenario is that the full view of the opposite party is revealed. But, as long as the adversary sees no more than t such views, security follows from the t-security of Π_{split}. A formal proof, deferred to the full version, uses a simulator for Π_{split} to obtain a simulator for the combiner. ∎

A first corollary of the above strategy is the existence of OT-combiners with a majority of good candidates. Such a result was already known (see [25,35,40]), based on a different approach stemming from the techniques of [18].

Corollary 4. *For any* m *and* n *such that* $m > n/2$, *there exists an* (m, n)-*OT-combiner in the semi-honest model. Furthermore, such a combiner can be perfectly secure.*

Proof: Use, for example, the protocol of [3] to implement Π_{MPC} with $t = n - m$ and the protocol of [23,21] to implement Π_{2party} in the semi-honest model. By Lemmas 1 and 2, this implies the desired OT-combiner. ∎

3.1 Constant-Rate OT-Combiners in the Semi-honest Model

We turn to optimizing the efficiency of the combiner described above. Its efficiency is inherited from the underlying protocols Π_{MPC} and Π_{2party}. For different purposes, one may choose different protocols Π_{MPC} and Π_{2party} with suitable properties. The key parameter that we investigate is the total number of calls to the OT-candidates. In our framework, calls to the candidates happen as part of executions of the protocol Π_{2party}, where each step of a server in the protocol Π_{MPC} requires an invocation of Π_{2party}. Therefore, the total number of calls is a function of the number of steps in Π_{MPC}, the complexity of the local computation in each such step, and the OT complexity of Π_{2party}.

A natural implementation of our combiner, using [3,23], has a polynomial rate and threshold $m = \lceil \frac{n+1}{2} \rceil$. More precisely, in order to compute ℓ OTs, the clients should compute a simple constant-size circuit on ℓ independent pairs of inputs. Using the BGW technique [3], the local computation required by each server for each OT is dominated by multiplying two elements from a field of size at least n. Simulating this action by a split server, requires a secure 2-party computation of such a functionality in the OT-hybrid model, which involves $\Omega(\log n)$ OT-calls. The overall OT complexity is therefore $s = \Omega(n\ell \log n)$.

In the following we show that, by a careful instantiation of Π_{MPC}, we can obtain a constant rate at the price of a slightly sub-optimal (yet still constant) tolerance ratio. The following two techniques allow this improvement:

- Using a generalization of Shamir's secret sharing scheme [20], which packs ℓ secrets into a single polynomial, one can run a joint computation for all ℓ inputs by sending just a constant number of field elements to each server. As a result of packing ℓ secrets into a single polynomial, the security threshold decreases from $t = \lfloor \frac{n-1}{2} \rfloor$ to $t' = t - \ell + 1$. Thus, letting ℓ be a sufficiently small constant fraction of n, the tolerance ratio remains constant.
- Using the techniques of [8,9], one can run all operations over a constant size field. This technique further deteriorates the security threshold to $t'' = t' - \delta n$, for some constant $\delta > 0$ that tends to 0 as the field size grows.

Combining the above two techniques, one gets a protocol Π_{MPC} in which each server performs a constant amount of work and ℓ, the number of OTs being computed, is a constant fraction of the number of servers. When compiling such a protocol to the split-servers model (and subsequently to the combiner) we get a constant number of OT-calls per split server. An appropriate choice of parameters (say, $\ell = 0.2n$ and $\delta = 0.2$) thus yields the following:

Theorem 5. *There is an OT-combiner with constant production rate and constant tolerance ratio in the semi-honest model. The combiner makes a constant number of calls to each OT-candidate.*

We end this section by noting that one can get an error-tolerant version of Theorem 5 by letting the clients in Π_{MPC} apply error-correction to the final n-tuple of field elements received from the n servers. This requires the underlying secret sharing scheme to be based on efficiently decodable codes (such as AG codes [8]). Furthermore, the (constant) fractional security threshold of Π_{MPC} should be further decreased in order to provide the redundancy required for error-correction. This yields the following:

Theorem 6. *There is an* error-tolerant *OT-combiner with constant production rate and constant tolerance ratio in the semi-honest model. The combiner makes a constant number of calls to each OT-candidate.*

4   OT-Combiners in the Malicious Model

Our constructions of combiners in the malicious model follow the same outline as in the semi-honest model. Namely, the combiner is a composition of two types of secure protocols, one in the two-party setting (with an OT channel) and one in the multiparty setting. Naturally, this time the components must be secure against malicious adversaries (and thus are inherently more complex). In addition, we must incorporate a mechanism to assure authenticity of intermediate shares supplied by the split servers.

Lemma 3. *Let f be a functionality taking inputs from S and R and returning output to R. There is a compiler that transforms any t-secure protocol Π_{MPC} for f in the malicious clients-servers model into a t-secure protocol Π_{split} for f in the malicious split-servers model. As a building block, the compiler requires a two-party protocol Π_{2party} for general functionalities in the malicious OT-hybrid model. If both Π_{MPC} and Π_{2party} are statistically secure then so is Π_{split}.*

The same general idea as in the semi-honest model applies here as well with one notable addition. Recall that in the general framework each server P_i in the protocol Π_{MPC} is simulated by two parties S_i, R_i that hold an additive secret sharing of P_i's view. The problem is that, in the split-servers model, the adversary may corrupt all of the parties on one side (either all of the R_i's or all of the S_i's). While the adversary has no information about the view of a server P_i unless both S_i and R_i are corrupted, it can still change the outgoing messages from this server. Namely, the adversary needs only to change the outgoing message of one side (say S_i) in order to change the effective outgoing message of server P_i in Π_{MPC} (recall that two messages in Π_{split} correspond to a single message in Π_{MPC}). To overcome this problem, we replace the use of standard additive secret sharing by *authenticated* secret sharing. Namely, each S_i gets, in addition to its additive share v_{S_i}, a signature on this share using a private key known to R_i (and vice versa). For the signature primitive it suffices to use a one-time MAC, which can be implemented with unconditional security using pairwise independent hash functions.

The two-party functionality realized by Π_{2party} takes the additive shares of the view together with the signatures and the keys as inputs. It then verifies that the signatures are valid; if this verification fails it sends an "abort" message to both parties. (Any party receiving an abort message broadcasts it to all parties and aborts; this cannot violate the fairness of Π_{split}, since there is only one party receiving an output.) The functionality returns a similar authenticated secret sharing of the next message sent from P_i to P_j in Π_{MPC}.

Note that, in the malicious model, Π_{2party} cannot achieve fairness. As before, if some honest party aborts it causes all honest parties to abort. Finally, if Π_{MPC} employs a broadcast primitive, then each message broadcasted by P_i can be naturally emulated in Π_{split} as follows. First, S_i and R_i broadcast their authenticated shares of the message. Then, each of them uses its secret key to verify the shares broadcasted by the other party, and broadcasts an abort message if the verification fails.

Lemma 4. *Given any protocol Π_{split} for the functionality OT^ℓ which is t-secure in the malicious split-servers model, one can construct (in a black-box way) an $(n-t, n)$-OT-combiner in the malicious model.*

The construction is essentially the same as in the semi-honest model. The only changes that need to be made involve broadcast messages and handling aborting parties. A broadcast message by S or S_i is emulated by simply having the sender S' of the combiner send this message to R' (and vice versa). In case some party S or S_i (resp., R or R_i) in Π_{split} aborts, then S' (resp., R') in the combiner aborts as well. A simulation-based security proof is deferred to the full version.

Corollary 7. *For any m and n such that $m > n/2$, there exists an (m, n)-OT-combiner in the malicious model. Furthermore, such a combiner can be statistically secure.*

Proof: For Π_{MPC}, we rely on the protocol of [38], which is statistically t-secure if $t < n/2$ (and employs a broadcast channel). For Π_{2party}, we can use the protocol of [31], which provides statistical security in the malicious OT-hybrid model. ∎

4.1 Constant-Rate OT-Combiners in the Malicious Model

The OT complexity of the combiner in the malicious model is higher than in the semi-honest model. This is because of the inherent complexity of the underlying protocols Π_{MPC} and Π_{2party} and because of the employment of authentication, which requires secure two-party computation of functionalities involving MACs. The underlying principles that allow for constant-rate combiners in the malicious model are the following:

- Use a (constant-round) protocol Π_{MPC} in which the overall communication of the servers is $O(\ell)$. This, in turn, translates to the size of inputs that the split-servers can run $\Pi_{2\text{party}}$ on. Such a protocol for arbitrary NC^0 functionalities (including OT^ℓ as a special case) is described in [30], building on [16]. In contrast to the semi-honest model, where each server can receive just a constant number of field elements, here we need ℓ to be sufficiently larger than n, say $\ell = nk$ for a security parameter k. In such a case, each server will receive $O(k)$ field elements. This, in turn, will translate into a larger (non-constant) number of invocations of each candidate.

- Use $\Pi_{2\text{party}}$ whose (amortized) OT complexity is a constant multiple of the input length I. Such a protocol was recently given in [33]. This protocol provides computational security and invokes $O(I+k)$ instances of *string-OT* with strings of length k (rather than bit-OT) along with a one-way function.

The above properties assure that the total number of calls to the (string-)OT-candidates is $O(\ell)$, provided that the MAC does not add a substantial overhead. The latter is guaranteed by the fact that the messages sent in the underlying protocol Π_{MPC} are long enough. Thus, the use of MACs does not increase the asymptotic length of the inputs.

A remaining caveat is that even though each of our candidates is a string-OT, the $O(\ell)$ calls to the candidates produce ℓ instances of *bit-OT*. Indeed, in the protocol Π_{split} obtained by our general compiler the length of the views of both S_i and R_i will be proportional to the total length of all strings (rather than the number of OTs). Thus, implementing ℓ good string-OTs would require $O(k\ell)$ calls to the candidates. To reduce the number of calls to $O(\ell)$, we observe that it is possible to modify our generic implementation of Π_{split} so that in all invocations of $\Pi_{2\text{party}}$ the inputs of the R_i's are short, namely of total size $O(\ell)$ (assuming $\ell = kn$), whereas the inputs of the S_i's are of total size $O(k\ell)$. Since the number of string OTs required by [33] is determined only by the length of the receiver's input, we will end up using $O(\ell)$ calls to string-OT candidates to produce ℓ good string-OTs. Overall we get:

Theorem 8. *There is a computationally secure combiner for string-OT with constant production rate and constant tolerance ratio in the* malicious model. *The combiner makes $O(k)$ calls to each OT-candidate, as well as black-box use of a one-way function.*

5 Application: Constant-Rate OTs from a Noisy Channel

In this section, we apply our constant-rate combiners for the semi-honest model in order to efficiently produce a reliable stream of bit-OTs from a noisy channel, namely a binary symmetric channel which flips each bit with probability p. (We will refer to the latter channel as a BSC with crossover probability p.) Known constructions for this task [14,13,17] require sending $\Omega(k)$ bits over the channel in order to generate just a single OT call, even in the semi-honest model. We show that, using our constant-rate combiners, one can achieve a number of OT calls that is a constant multiple of the number of bits sent over the channel. Formally:

Theorem 9 (Constant-rate OTs from a noisy channel). *For any constant* $0 < p < 1/2$, *there exists a two-party protocol that securely implements the* OT^ℓ *functionality in the semi-honest model by having the parties communicate* $O(\ell)$ *bits over a BSC with crossover probability* p. *The protocol has perfect privacy and statistical correctness, where the error probability is* $2^{-\Omega(\ell)}$.

Proof: The idea is that, instead of using a constant number of noisy bits to produce a single secure OT, we produce an instance of OT that has perfect privacy but a small constant error probability. Each such OT can in turn be viewed as an *OT-candidate*. These candidates are then combined, using our constant-rate error-tolerant combiner, to give a linear number of good OT calls (this time with exponentially small error). We use the combiner from Theorem 6, that has constant rate and constant tolerance ratio and makes just $O(1)$ calls to each candidate. Note that we can view each call to a candidate as a distinct candidate, at the cost of further reducing the tolerance ratio to some small constant $\epsilon > 0$.

We now present a variant of a protocol from [14] that can implement OT with perfect privacy and an arbitrarily small constant error $\epsilon > 0$ by communicating a constant number of bits over the BSC. By known reductions, it suffices to implement such OT on random inputs. Let w, z be sufficiently large constants. The sender, S, picks z random bits $r_1, \ldots, r_z$ and sends each one $2w$ times to R (we call each corresponding sequence of $2w$ bits received by R a "block"). A block is of "type I" if it has an equal number of 0's and 1's and is of "type II" if it has only 0's or only 1's. Since w is a constant, we expect a constant fraction of blocks of each type (this fraction is a function of w and the noise level p) and hence by increasing z we can guarantee that both types exist with high probability. Now the receiver assigns a block of type I to the sender's bit it should not learn and a block of type II to the sender's bit it should learn. As required, this gives perfect privacy (namely, R does not learn any information on the bit it should not learn), but has a small $(2^{-\Omega(w)})$ probability of error in the bits R should learn. We finally note that all the additional (reliable) communication required by the combiner can be implemented via standard error-correcting codes by communicating $O(\ell)$ bits over the noisy channel. ∎

6   Application: Extending OTs Efficiently

In this section, we present *efficient* black-box reductions of $OT^{p(k)}$ to $OT^{q(k)}$ in the malicious model, where $q(k)$ is a *fixed* polynomial and $p(k)$ is *any* polynomial. (Throughout this section, OT refers to *string-OT* of k-bit strings.) The reductions make an additional black-box use of a cryptographic hash function and build on a protocol from [29] for the semi-honest model.

We give two distinct reductions, each making just a constant number of calls to the hash function per each OT call generated. The first follows by applying our combiners after running a cut-and-choose procedure over the protocol of [29]. As in [29], the protocol uses a so-called *correlation-robust hash function (CorRH)*: an explicit function h such that for random strings $s, t_1, \ldots, t_m$ the distribution $(h(s \oplus t_1), \ldots, h(s \oplus t_m), t_1, \ldots, t_m)$ is pseudorandom (see [29] for further discussion). Alternatively, a non-programmable, non-extractable random oracle suffices to instantiate the CorRH. This solution requires a seed of k^3 OTs (more precisely, $O(k\sigma^2)$ OTs, where σ is a statistical security parameter) rather than the k OTs required in the semi-honest model.

Theorem 10 (Informal). *Let k be a security parameter. For any polynomial $p(k)$, there exists a black-box reduction of $OT^{p(k)}$ to OT^{k^3} in the malicious model, under the CorRH assumption. The construction requires only a constant number of calls to the hash function per each OT produced.*

Proof sketch: Consider the following ideal functionality, denoted IKNP$^\ell$ (as it captures a core idea of [29]): the sender S has input $\mathbf{a} = (a_1, \ldots, a_k) \in \{0,1\}^k$. For each $j \in [\ell]$, the receiver R has inputs $\mathbf{b}^j = (b_1^j, \ldots, b_k^j) \in \{0,1\}^k$ and $\mathbf{m}^j = (m_1^j, \ldots, m_k^j) \in \{0,1\}^k$. For $j \in [\ell]$, the sender has output $\mathbf{d}^j = \mathbf{a} \wedge \mathbf{b}^j \oplus \mathbf{m}^j$, where $\wedge$ and $\oplus$ denote bitwise operations. We also consider a committed version, called CIKNP$^\ell$, where parties are also committed to their inputs (if a party inputs a special symbol `reveal!`, its inputs will be leaked to the other party). Another stepping stone is a special version of OT$^\ell$, called SOT$^\ell$. It works as OT$^\ell$, except that a *malicious* R may give a special input `cheat!` before inputs are provided by S. In response to this, R will receive *all* inputs of S. Later, R can give another special input `open!` in response to which S is told whether R at some point input `cheat!`. As a side effect, `open!` leaks the choice bits of R.

The proof follows a series of reductions. First, CIKNP$^\ell$ is constructed from OT$^{k^2}$ (using just a single call to OT$^{k^2}$). This step follows the reduction from [29] of IKPN$^\ell$ to OT^k while the commitment property is achieved in the natural way by using k *committed OTs* [15] as the underlying primitive. As shown in [15], a committed OT can be implemented, in a black-box way, using $O(k)$ OTs and thus the overall $O(k^2)$ OTs. The second step, which is detailed below, builds SOT$^\ell$ from CIKNP$^\ell$. Finally, one constructs OT$^{p(k)}$ by making k calls to an instance of SOT$^\ell$. The last step calls k instances of SOT$^\ell$ and, using a simple cut-and-choose technique, one can produce $O(k)$ instances of SOT$^\ell$ of which a sufficiently small constant fraction is insecure. Then, one applies our constant-rate combiner to get an implementation of OT$^{O(k\ell)}$.

It remains to reduce SOT$^\ell$ to CIKNP$^\ell$ with an amortized constant number of hash function applications per produced OT. The protocol uses hash functions $H^j : \{0,1\}^k \rightarrow \{0,1\}^k$, for $j \in [\ell]$, where we let $H^j(x) = H(j\|x)$, for some fixed hash function H. Recall that SOT$^\ell$ takes as inputs secrets $(s_0^1, s_1^1), \ldots, (s_0^\ell, s_1^\ell)$ from the sender, and ℓ choice bits $c^1, \ldots, c^\ell$ from the receiver.

1. First, CIKNP$^\ell$ is called with S inputting a random $\mathbf{a}$ and R inputting a random $\mathbf{m}^j$ and $\mathbf{b}^j = (b^j, \ldots, b^j)$, where $b^j = c^j$ is the choice bit of R.
2. For $j \in [\ell]$, the sender S computes $\mathbf{d}_0^j \leftarrow \mathbf{d}^j$, $\mathbf{d}_1^j \leftarrow \mathbf{d}^j \oplus \mathbf{a}$, $r_0^j = H^j(\mathbf{d}_0^j)$ and $r_1^j = H^j(\mathbf{d}_1^j)$, and R computes $\mathbf{d}_{b^j}^j \leftarrow \mathbf{m}^j$ and $r_{b^j}^j = H^j(\mathbf{d}_{b^j}^j)$.

3. For $j \in [\ell]$, the sender S sends $e_0^j = r_0^j \oplus s_0^j$ and $e_1^j = r_1^j \oplus s_1^j$ to R, and R outputs $s_{b^j}^j = e_{b^j}^j \oplus r_{b^j}^j$.

To implement the **open!** command, the receiver will input **reveal!** to $CIKNP^\ell$ to show the values $\mathbf{m}^j$ and $\mathbf{b}^j$ to S. Sender S considers it a cheat if any $\mathbf{b}^j$ is not one of the monochromatic vectors 0^k or 1^k.

Correctness and security against a malicious sender are straightforward. The security against a malicious R is shown by a simulator with access to SOT^ℓ. At a high level, if any of the $\mathbf{b}^j$ input by R is polychromatic, then the simulator inputs **cheat!** to SOT^ℓ, learns the inputs of S and uses these to run the rest of the simulation as in the protocol. If, on the other hand, all $\mathbf{b}^j$ are monochromatic, then the simulator is reminiscent of that of [29] (including the use of the CorRH assumption). A complete proof appears in the full version.   ∎

The second result manages to work with a seed of just k OTs (rather than the k^3 OTs used in Theorem 10). For this result, we use a natural generalization of the CorRH assumption, called the *generalized correlation-robust hash function (GCorRH) assumption* and a more specialized variant called the *special xor correlation-robust hash function (S⊕CorRH) assumption*. As with the CorRH assumption, it holds that a random function satisfies the new assumption with overwhelming probability. This, in particular, implies the security of our protocol in the non-programmable, non-extractable random oracle model. We stress, however, that all assumptions are concrete computational assumptions, and our proofs are in the standard model.

Theorem 11. *Let k be a computational security parameter. For any polynomial $p(k)$, there exists a black-box reduction of $OT^{p(k)}$ to OT^k in the malicious model under the GCorRH and S⊕CorRH assumptions. The construction requires only a constant number of calls to the hash function per each OT produced.*

At a high level, we notice from the proof of the previous theorem that in order to gain an advantage, a cheating R must pick some $\mathbf{b}^j$ to be polychromatic. Note that $\mathbf{d}_0^j = \mathbf{a} \wedge \mathbf{b}^j \oplus \mathbf{m}^j$ and $\mathbf{d}_1^j = \mathbf{a} \wedge \bar{\mathbf{b}}^j \oplus \mathbf{m}^j$, where $\bar{\mathbf{b}}^j = 1^k \oplus \mathbf{b}^j$. This means that, when $\mathbf{b}^j$ is polychromatic, both $\mathbf{d}_0^j$ and $\mathbf{d}_1^j$ depend on some bits of $\mathbf{a}$. The honest R will always know $\mathbf{d}_{b^j}^j = \mathbf{m}^j$. We exploit this difference by introducing a test where R, for each j, shows that it knows $\mathbf{d}_0^j$ or $\mathbf{d}_1^j$ without revealing which. Essentially, we let S send the first k bits of each $H^j(\mathbf{d}_0^j) \oplus H^j(\mathbf{d}_1^j)$ to R (suppose H^j has $2k$-bit outputs). R must then return the first k bits of $H^j(\mathbf{d}_0^j)$. This is easy for an honest R, which knows $H^j(\mathbf{d}_{b^j}^j)$, but will catch a cheating R with some probability related to how many bits of $\mathbf{a}$ the receiver needs to guess one of $\mathbf{d}_0^j$ and $\mathbf{d}_1^j$. This test, however, introduces an opening for S to cheat, which requires an extra fix. After these tests, we have (with overwhelming probability) at most k bad OTs out of the ℓ OTs being produced, and we remove the bad OTs by using our combiner. The details and proof appear in the full version.

References

1. Beaver, D.: Correlated pseudorandomness and the complexity of private computations. In: 28th STOC, pp. 479–488 (1996)
2. Beaver, D.: Commodity-based cryptography. In: 29th STOC, pp. 446–455 (1997)
3. Ben-Or, M., Goldwasser, S., Wigderson, A.: Completeness theorems for noncryptographic fault-tolerant distributed computation. In: 20th STOC, pp. 1–10 (1988)
4. Brassard, G., Crépeau, C., Robert, J.-M.: All-or-nothing disclosure of secrets. In: Odlyzko, A.M. (ed.) CRYPTO 1986. LNCS, vol. 263, pp. 234–238. Springer, Heidelberg (1987)
5. Brassard, G., Crépeau, C., Wolf, S.: Oblivious Transfers and Privacy Amplification. J. Cryptology 16(4), 219–237 (2003)
6. Canetti, R.: Security and composition of multiparty cryptographic protocols. J. of Cryptology 13(1) (2000)
7. Chaum, D., Crépeau, C., Damgård, I.: Multiparty unconditionally secure protocols. In: 20th STOC, pp. 11–19 (1988)
8. Chen, H., Cramer, R.: Algebraic geometric secret sharing schemes and secure multiparty computations over small fields. In: Dwork, C. (ed.) CRYPTO 2006. LNCS, vol. 4117, pp. 521–536. Springer, Heidelberg (2006)
9. Chen, H., Cramer, R., Goldwasser, S., de Haan, R., Vaikuntanathan, V.: Secure computation from random error correcting codes. In: Naor, M. (ed.) EUROCRYPT 2007. LNCS, vol. 4515, pp. 291–310. Springer, Heidelberg (2007)
10. Chor, B., Kushilevitz, E.: A zero-one law for boolean privacy. SIAM Journal on Disc. Math. 4(1), 36–47 (1991); preliminary version in STOC 1989
11. Cramer, R., Damgård, I., Ishai, Y.: Share conversion, pseudorandom secret-sharing and applications to secure computation. In: Kilian, J. (ed.) TCC 2005. LNCS, vol. 3378, pp. 342–362. Springer, Heidelberg (2005)
12. Cramer, R., Damgård, I., Maurer, U.: General secure multi-party computation from any linear secret-sharing scheme. In: Preneel, B. (ed.) EUROCRYPT 2000. LNCS, vol. 1807, pp. 316–334. Springer, Heidelberg (2000)
13. Crépeau, C.: Efficient cryptographic protocols based on noisy channels. In: Nyberg, K. (ed.) EUROCRYPT 1998. LNCS, vol. 1403, pp. 306–317. Springer, Heidelberg (1998)
14. Crépeau, C., Kilian, J.: Achieving oblivious transfer using weakened security assumptions. In: 29th FOCS, pp. 42–52 (1988)
15. Crépeau, C., van de Graaf, J., Tapp, A.: Committed oblivious transfer and private multi-party computation. In: Coppersmith, D. (ed.) CRYPTO 1995. LNCS, vol. 963, pp. 110–123. Springer, Heidelberg (1995)
16. Damgård, I., Ishai, Y.: Scalable secure multiparty computation. In: Dwork, C. (ed.) CRYPTO 2006. LNCS, vol. 4117, pp. 501–520. Springer, Heidelberg (2006)
17. Damgaård, I., Fehr, S., Morozov, K., Salvail, L.: Unfair Noisy Channels and Oblivious Transfer. In: Proc. first TCC, pp. 355–373 (2004)
18. Damgård, I., Kilian, J., Salvail, L.: On the (im)possibility of basing oblivious transfer and bit commitment on weakened security assumptions. In: Stern, J. (ed.) EUROCRYPT 1999. LNCS, vol. 1592, pp. 56–73. Springer, Heidelberg (1999)
19. Even, S., Goldreich, O., Lempel, A.: A randomized protocol for signing contracts. Communications of the ACM 28(6), 637–647 (1985)
20. Franklin, M., Yung, M.: Communication complexity of secure computation (extended abstract). In: 24th STOC, pp. 699–710 (1992)

21. Goldreich, O.: Foundations of Cryptography, vol. 2. Cambridge University Press, Cambridge (2004)

22. Goldreich, O., Micali, S., Wigderson, A.: How to play any mental game - a completeness theorem for protocols with honest majority. In: 19th STOC, pp. 218–229 (1987)

23. Goldreich, O., Vainish, R.: How to solve any protocol problem - an efficiency improvement. In: Pomerance, C. (ed.) CRYPTO 1987. LNCS, vol. 293, pp. 73–86. Springer, Heidelberg (1988)

24. Harnik, D., Ishai, Y., Kushilevitz, E.: How many oblivious transfers are needed for secure multiparty computation? In: Menezes, A. (ed.) CRYPTO 2007. LNCS, vol. 4622, pp. 284–302. Springer, Heidelberg (2007)

25. Harnik, D., Kilian, J., Naor, M., Reingold, O., Rosen, A.: On tolerant combiners for oblivious transfer and other primitives. In: Cramer, R.J.F. (ed.) EUROCRYPT 2005. LNCS, vol. 3494, pp. 96–113. Springer, Heidelberg (2005)

26. Herzberg, A.: On tolerant cryptographic constructions. In: CT-RSA, pp. 172–190 (2005)

27. Hirt, M., Maurer, U.: Robustness for free in unconditional multi-party computation. In: Kilian, J. (ed.) CRYPTO 2001. LNCS, vol. 2139, pp. 101–118. Springer, Heidelberg (2001)

28. Impagliazzo, R., Rudich, S.: Limits on the provable consequences of one-way permutations. In: 21st STOC, pp. 44–61 (1989)

29. Ishai, Y., Kilian, J., Nissim, K., Petrank, E.: Extending oblivious transfers efficiently. In: Boneh, D. (ed.) CRYPTO 2003. LNCS, vol. 2729, pp. 145–161. Springer, Heidelberg (2003)

30. Ishai, Y., Kushilevitz, E., Ostrovsky, R., Sahai, A.: Zero-knowledge from secure multiparty computation. In: 39th STOC, pp. 21–30 (2007)

31. Kilian, J.: Founding cryptography on oblivious transfer. In: 20th STOC, pp. 20–31 (1988)

32. Kilian, J.: A general completeness theorem for two-party games. In: 23rd STOC, pp. 553–560 (1991)

33. Lindell, Y., Pinkas, B.: An efficient protocol for secure two-party computation in the presence of malicious adversaries. In: Naor, M. (ed.) EUROCRYPT 2007. LNCS, vol. 4515, pp. 52–78. Springer, Heidelberg (2007)

34. Meier, R., Przydatek, B.: On robust combiners for private information retrieval and other primitives. In: Dwork, C. (ed.) CRYPTO 2006. LNCS, vol. 4117, pp. 555–569. Springer, Heidelberg (2006)

35. Meier, R., Przydatek, B., Wullschleger, J.: Robuster combiners for oblivious transfer. In: Vadhan, S.P. (ed.) TCC 2007. LNCS, vol. 4392, pp. 404–418. Springer, Heidelberg (2007)

36. Przydatek, B., Wullschleger, J.: Error-tolerant combiners for oblivious primitives. In: Manuscript, Personal Communication (2006)

37. Rabin, M.O.: How to exchange secrets by oblivious transfer. TR-81, Harvard (1981)

38. Rabin, T., Ben-Or, M.: Verifiable secret sharing and multiparty protocols with honest majority (extended abstract). In: 21st STOC, pp. 73–85 (1989)

39. Wiesner, S.: Conjugate coding. SIGACT News 15(1), 78–88 (1983)

40. Wullschleger, J.: Oblivious transfer amplification. In: Naor, M. (ed.) EUROCRYPT 2007. LNCS, vol. 4515, pp. 555–572. Springer, Heidelberg (2007)

41. Yao, A.C.: How to generate and exchange secrets. In: 27th FOCS, pp. 162–167 (1986)

Semi-honest to Malicious Oblivious Transfer— The Black-Box Way

Iftach Haitner

Dept. of Computer Science and Applied Math., Weizmann Institute of Science,
Rehovot, Israel
iftach.haitner@weizmann.ac.il

Abstract. Until recently, all known constructions of oblivious transfer protocols based on general hardness assumptions had the following form. First, the hardness assumption is used in a black-box manner (i.e., the construction uses only the input/output behavior of the primitive guaranteed by the assumption) to construct a *semi-honest* oblivious transfer, a protocol whose security is guaranteed to hold only against adversaries that follow the prescribed protocol. Then, the latter protocol is "compiled" into a (malicious) oblivious transfer using non-black techniques (a Karp reduction is carried in order to prove an NP statement in zero-knowledge).

In their recent breakthrough result, Ishai, Kushilevitz, Lindel and Petrank (STOC '06) deviated from the above paradigm, presenting a black-box reduction from oblivious transfer to enhanced trapdoor permutations and to homomorphic encryption. Here we generalize their result, presenting a black-box reduction from oblivious transfer to semi-honest oblivious transfer. Consequently, oblivious transfer can be black-box reduced to each of the hardness assumptions known to imply a semi-honest oblivious transfer in a black-box manner. This list currently includes beside the hardness assumptions used by Ishai et al., also the existence of families of dense trapdoor permutations and of non trivial single-server private information retrieval.

1 Introduction

Since most cryptographic tasks are impossible to achieve with absolute, information-theoretic security, modern cryptography tries to design protocols that are *infeasible* to break. Namely, their security is based on computational hardness assumptions. These assumptions typically come in two flavors: ***specific hardness assumptions*** like discrete log, factoring and RSA, and ***general hardness assumptions*** like the existence of one-way functions. In this paper we refer to general hardness assumptions and how they are used. Primitives assumed to carry some hardness assumption can be used to construct a provably secure cryptographic tasks in two possible ways: "black-box usage", where the construction uses only the input/output behavior of the primitive, and "non-black-box usage", where the construction uses the internal structure

R. Canetti (Ed.): TCC 2008, LNCS 4948, pp. 412–426, 2008.
© International Association for Cryptologic Research 2008

of the primitive, e.g., its code. The above is formalized via the notion of **black-box reductions**. A black-box reduction from a primitive P to a primitive Q, is an efficient construction of P out of Q that ignores the internal structure of the implementation of Q and merely uses it as a "subroutine" (i.e., as a black-box). Such a reduction is **fully-black-box** [25] if the proof of security (showing that an adversary that breaks the implementation of P implies an efficient adversary that breaks the implementation of Q), is black-box as well. That is, the internal structure of the adversary that breaks the implementation of P is ignored. See Section 2.2 for more details.

Staring from the seminal paper of Impagliazzo and Rudich [16], a rich line of works tries to draw the border between possibility and impossibility for black-box reductions in cryptography. Currently, for most cryptographic tasks we either have a black-box reduction to a commonly believed hardness assumption, or have shown the impossibility of such a reduction. There are several important tasks, however, for which we have failed to apply the above black-box classification. Very interestingly, for most of those tasks we do have non-black-box reductions (typical examples are the reductions from oblivious transfer to semi-honest oblivious transfer [12], and from public-key encryption schemes secure against chosen cipher-text attack to semantically-secure encryption schemes [8,20,26]). In their recent breakthrough result, Ishai et al. [17] presented the first black-box reduction from oblivious transfer to "low-level" primitives (to homomorphic encryption and to enhanced trapdoor permutations). Yet, the question whether there exists a black-box reduction from oblivious transfer to semi-honest oblivious transfer, remained open.

A better understanding of the above might help up to resolve the intriguing question whether non-black-box techniques are superior to black-box ones also in the setting of reductions between cryptographic primitives. [1] On a more practical level, we mention that the non-black-box reductions of the above tasks are using Karp reductions for the purpose of using a (general) zero-knowledge proof/argument. Such reductions are highly inefficient and unlikely to be used in practice. Furthermore, in most cases the communication complexity in the resulting protocols depends on the complexity of computing the underlying primitive (i.e., of the trapdoor permutations), where black-box reductions, unaware of the inner structure of the underlying primitive, do not suffer from this phenomenon (see [17] for more details).

In this paper, we study the above issues w.r.t. oblivious transfer. Oblivious transfer, introduced by Rabin [24], is a fundamental primitive in cryptography and has several equivalent formulations [3,5,4,6,9,24]. The version we study here, defined by Even, Goldreich and Lempel [9], is that of **one-out-of-two oblivious transfer**. This version is an interactive protocol between a **sender** and a **receiver**. The sender gets as an input two secret bits: σ_0 and σ_1 and the receiver gets an index $i \in \{0, 1\}$. At the end of the protocol, R learns σ_i. Informally, the

---

[1] The superiority of non-black-box techniques was demonstrated by Barak [1] in the settings of zero-knowledge arguments for NP. In these settings, however, the black-box access is to the, possibly cheating, verifier and not to any underlying primitive.

security of the oblivious transfer states that the receiver does not learn σ_{1-i} and the sender does not learn i. Oblivious transfer is known to imply key-agreement signing contracts protocols [2,9,24] and, more generally, secure multiparty computation in the presence of malicious majority [12,18,28]. We sometimes add the term *malicious* to the above definition, to differentiate it from definitions that guarantee weaker security.

1.1 Defensible Privacy

The notion of defensible privacy, introduced by Ishai et al. [17], is a natural bridging step between semi-honest privacy and fully-fledged one. Informally, a two-party protocol (A, B) is *defensibly private* w.r.t. A and a function f defined over the parties' inputs (denoted as (A, f)-defensibly-private), if at the end of the interaction even a cheating A^* cannot simultaneously prove that it has acted honestly (i.e., as the honest party would) and learn the value of f. [2] [17] showed how to use enhanced trapdoor permutation (or homomorphic encryption) to construct *defensible oblivious transfer*. Where the latter is a protocol with the oblivious transfer functionally, which is defensibly-private w.r.t. to the sender and the input bit of the receiver, and w.r.t. to the receiver and the other secret of the sender. That is, it is (S, f_S) and (R, f_R) defensibly-private, where S and R stand for sender and the receiver respectively, $f_S(\sigma_0, \sigma_1, i) \stackrel{\text{def}}{=} i$ and $f_R(\sigma_0, \sigma_1, i) \stackrel{\text{def}}{=} \sigma_{1-i}$. [17] then show how to use such a defensible oblivious transfer to derive their main result.

1.2 Our Result

A two-party protocol (A, B) is (A, f)-*semi-honest-private*, if at the end of the interaction the semi-honest A does not learn the value of f. Our main technical contribution is the following theorem.

Theorem 1. *Let $\pi = (A, B)$ be a two-party protocol and let $f_A, f_B : \{0,1\}^k \times \{0,1\}^k \mapsto \{0,1\}^*$ be two functions defined over the parties' inputs. Assume that π is (A, f_A) and (B, f_B) semi-honest private. Then there exists a fully-black-box reduction from a protocol $\pi' = (\mathbb{A}, \mathbb{B})$ that has the same functionality as π and is $(\mathbb{A}, f_A)$ and $(\mathbb{B}, f_B)$ defensibly-private, to π and one-way functions.*

Since one-way functions can be black-box reduced to semi-honest oblivious transfer (see Theorem 4), we obtain the following corollary.

Corollary 1. *There exists a fully-black-box reduction from defensible oblivious transfer to semi-honest oblivious transfer.*

Combining the above with the reduction of [17] from malicious oblivious transfer to derisible one, we derive our main result.

---

[2] The above generalizes the definition of [17], which was only stated w.r.t. oblivious transfer protocols.

Theorem 2. *There exists a fully-black-box reduction from oblivious transfer to semi-honest oblivious transfer.*

As a corollary of Theorem 2, we have that there exists a fully-black-box reduction from oblivious transfer to each of the assumptions that known to imply semi-honest oblivious transfer in a fully-black-box manner. This list currently includes families of dense/enhanced trapdoor permutations [9,13], homomorphic encryption [19,27] and non-trivial single-server private-information retrieval [7]. In addition, Kilian [18] tells us that secure multiparty computation can be black-box reduced to oblivious transfer. Hence, we also have the following corollary.

Corollary 2. *There exist fully-black-box reductions from protocols for securely computing any multiparty functionality with an honest-minority and in the presence of static malicious adversaries, to semi-honest oblivious transfer.*

1.3   Our Technique - From Semi-Honest to Defensible Privacy

Given a protocol $\pi = (A, B)$ that is (A, f_A) and (B, f_B) semi-honest-private, and assuming that one-way functions exist, we create the protocol $\pi' = (\mathbb{A}, \mathbb{B})$ that is $(\mathbb{A}, f_A)$ and $(\mathbb{B}, f_B)$ defensibly-private. Our reduction is carried out in two steps. First, we create a protocol $(\mathbb{A}, \mathbb{B})$ with the same functionality as (A, B), which is $(\mathbb{A}, f_A)$-defensibly-private and $(\mathbb{B}, f_B)$-semi-honest-private. Then, we apply the same transformation on $(\mathbb{A}, \mathbb{B})$, to strengthen also the privacy w.r.t. f_B. In what follows we describe how to obtain the first step (the second step is analogous), but first let us describe what a commitment scheme is. In a commitment scheme the **sender** interacts with the **receiver** to commit to a private value; informally the commitment is **binding** if the sender cannot open the commitment into a different value than the one it had committed to, where the commitment is **hiding** if before the decommitment stage the receiver does not learn the committed value. Fully-black-box reductions from commitment schemes to one-way functions were given by [15,21] and [14,23].

In the new protocol $(\mathbb{A}, \mathbb{B})$, we embed an execution of (A, B) while using a commitment scheme in order to enforce the "defensible behavior" of $\mathbb{A}$. Let i_A, i_B and r_A, r_B be the inputs and random-coins of A and B respectively. We define $(\mathbb{A}(i_A, r_A), \mathbb{B}(i_B, r_B^1, r_B^2))$ as follows. First, $\mathbb{A}$ commits to (i_A, r_A) using a commitment scheme, followed by $\mathbb{B}$ sending r_B^1 over to $\mathbb{A}$. Then the two parties execute $(A(i_A, r_A \oplus r_B^1), B(i_B, r_B^2))$, where $\mathbb{A}$ and $\mathbb{B}$ act as A and B respectively. The hiding property of the commitment scheme yields that before the embedded execution of (A, B) starts, $\mathbb{B}$ does not learn any information about the input and random-coins that $\mathbb{A}$ uses in this execution. Thus, the semi-honest privacy of $(\mathbb{A}, \mathbb{B})$ w.r.t. $\mathbb{B}$ and f_B, follows by the semi-honest privacy of (A, B) w.r.t. B and f_B. In order to prove that $(\mathbb{A}, \mathbb{B})$ is $(\mathbb{A}, f_A)$-defensibly-private, we first note that a valid defense of $\mathbb{A}$ must include a valid opening of the commitment. Thus, the binding property of the commitment scheme yields that even a dishonest $\mathbb{A}^*$ can only provide a valid defense if it has acted in the embedded execution of (A, B) as A whose input and random-coins are set to i_A and $r_A \oplus r_B^1$ would. Namely, if it has acted as A whose input was decided *before* the execution has started, and its random-coins are chosen *at random* would. Hence, the defensible privacy of the protocol w.r.t. $\mathbb{A}$ and f_A, follows by the semi-honest privacy of (A, B) w.r.t. A and f_A. [3]

1.4 Paper Organization

Section 2 contains the notations and definitions used in this paper. In Section 3 we present our general transformation from semi-honest privacy to defensible one (Theorem 1) and in Section 4 we use this transformation to derive our main result (Theorem 2).

2 Preliminaries

2.1 Notation

We denote by U_n the random variable uniformly chosen in $\{0,1\}^n$. Given a distribution D, we denote its support by $\mathrm{Supp}(D)$. We adopt the convention that when the same random variable occurs several times in an expression, all occurrences refer to a single sample. For example, $\Pr[f(U_n) = U_n]$ is defined to be the probability that when $x \leftarrow U_n$, we have $f(x) = x$. Given a vector v of dimension n, we denote by $v[i_1, ..., i_k]$, where $i_1, ..., i_k \in [n]$, the vector $(v[i_1], ..., v[i_k])$. A function $\mu : \mathbb{N} \to [0, 1]$ is **negligible**, denoted $\mu = \mathrm{neg}$, if for every polynomial p we have that $\mu(n) < 1/p(n)$ for large enough n. Two distribution ensembles D_n and ξ_n are **computationally-indistinguishable** (denoted $D_n \approx_c \xi_n$), if no efficient algorithm distinguishes between them with more then negligible probability. Given a two-party protocol $\pi = (A, B)$, we denote the inputs and random-coins of A and B by i_A and i_B, and by r_A and r_B respectively. We denote by $\mathrm{View}_A^\pi((i_A, r_A), (i_B, r_B))$ the **view** of A after the execution of π on $((i_A, r_A), (i_B, r_B))$. This view consists on i_A, r_A and the messages A received thought the protocol. We denote by $\mathrm{View}_A^\pi(i_A, i_B)$, the random variable $\mathrm{View}_A^\pi((i_A, R_A), (i_B, R_B))$, where R_A and R_B are uniformly chosen among all strings of the right length.

2.2 Black-Box Reductions

A reduction from a primitive P to a primitive Q consists of showing that if there exists an implementation C of Q, then there exists an implementation M_C of P. This is equivalent to showing that for every adversary that breaks M_C, there exists an adversary that breaks C. Such a reduction is **semi-black-box** if it

---

[3] In their construction of defensible oblivious transfer from enhanced families of trapdoor permutations, [17] are using (perfectly-binding) commitment schemes for a similar purpose. More specifically, they employ the semi-honest oblivious transfer of [9] and use a commitment scheme for forcing the receiver to sample one of the two random elements it has to choose in the permutation domain honestly, i.e., choosing it as a random output of the domain sampler.

ignores the internal structure of Q's implementation, and it is ***fully-black-box*** if the proof of correctness is black-box as well, i.e., the adversary for breaking Q ignores the internal structure of both Q's implementation and of the (alleged) adversary breaking P. A taxonomy of black-box reductions was provided by [25], and the reader is referred to their paper for a more complete and formal view of these notions. All the reduction considered in this paper are fully-black-box ones.

2.3   Different Notions of Privacy

In the following we present the two privacy measures we use in this paper.

Semi-Honest Privacy
In the standard definitions of semi-honest privacy (c.f, [11]), it is required that the semi-honest party does not learn *any information* about the other party's input, save but the part it suppose to get according to the prescribed functionality. Here we present a natural relaxation to the above, defining the notion of ***semi-honest privacy w.r.t. a function***. Namely, we only require that the semi-honest party does not learn a predefined function of the parties' inputs. [4]

Definition 1 (semi-honest privacy w.r.t. a function). *Let $\pi = (A, B)$ be a two-party protocol getting security parameter 1^n and let $f : \{0,1\}^k \times \{0,1\}^k \mapsto \{0,1\}^*$ be a function defined over the parties' inputs. We say that π is (A, f)-semi-honest-private, if for every efficiently samplable input $i_A \in \{0,1\}^k$ it holds that*

$$(\text{View}_A^\pi(i_A, U_k), f(i_A, U_k)) \approx_c (\text{View}_A^\pi(i_A, U_k), f(i_A, U_k'))$$

Defensible Privacy

Definition 2 (defense). *Let $\pi = (A, B)$ be a two-party protocol and let t be a transcript of an interaction between some party A^* and B. We say that d is a* good defense *for t (w.r.t. A's role in π), if A whose input, including its random-coins, is set to d would have sent the same messages that A^* does in t. We use the following notations: given $v = \text{View}_{A^*}^{(A^*, B)}(\cdot)$, we let $\text{Defense}(v)$ be the defense that A^* locally output in the end of the interaction (set to $\perp$ is no such defense is given) and let the predicate $\text{IsGoodDef}^{\pi, A}(v)$ to be one if $\text{Defense}(v)$ is a good defense for (the transcript embedded in) v.*

Definition 3 (defensible privacy w.r.t. a function). *Let π and f be as in Definition 1. We say that π is (A, f)-defensibly-private, if the following holds for every* PPT *A^*:*

$$\Gamma(\text{View}_{A^*}^{(A^*, B)}(U_k), f(i_A^d, U_k)) \approx_c \Gamma(\text{View}_{A^*}^{(A^*, B)}(U_k), f(i_A^d, U_k')) ,$$

---

[4] We have chosen to work with this weaker form of semi-honest privacy, since we have found it simpler to handle and yet strong enough when considering semi-honest oblivious transfer protocols.

where $\Gamma(x,y)$ *equals* (x,y) *if* $\mathrm{IsGoodDef}^{\pi,\mathsf{A}}(x) = 1$ *and equals* $\perp$ *otherwise, and* i_{A}^{d} *is the value of* A*'s input in* $\mathrm{Defense}(\mathrm{View}_{\mathsf{A}^*}^{(\mathsf{A}^*,\mathsf{B})}(U_k))$. [5]

Remark 1. It seems natural to extend the above definition to a simulation based one. Namely, a protocol is ***defensibly private*** if a party that gives a valid defense learns nothing (in the simulation sense) other than the prescribed functionality. It is then seems tempting to try to reduce the above defensible privacy to semi-honest privacy (according to [11]). Namely, to prove that any semi-honest private protocol implies a defensibly private version of this protocol. We hope to address this issue in the full version.

2.4 Oblivious Transfer

Oblivious transfer is an interactive protocol between a sender, S, and a receiver, R. The sender gets as an input two secret bits: σ_0 and σ_1 and the receiver gets an index $i \in \{0, 1\}$, in the end of the protocol R locally outputs a single bit. We make the following correctness requirement: for all n and all valid values of σ_0, σ_1 and i, with save but negligible probability the output of R in the interaction $(\mathsf{S}(\sigma_0, \sigma_1), \mathsf{R}(i))$ is σ_i.

Let (S, R) be a protocol that computes the oblivious transfer functionality, let $f_{\mathsf{S}}(\sigma_0, \sigma_1, i) \stackrel{\mathrm{def}}{=} i$ and let $f_{\mathsf{R}}(\sigma_0, \sigma_1, i) \stackrel{\mathrm{def}}{=} \sigma_{1-i}$. We say that (S, R) is a ***semi-honest [resp. defensible]*** oblivious transfer if it is $(\mathsf{S}, f_{\mathsf{S}})$ and $(\mathsf{R}, f_{\mathsf{R}})$ semi-honest private [resp. defensibly private]. The protocol (S, R) is *(**malicious**) **oblivious transfer*** if its computation is secure according to the *real/ideal simulation paradigm* (see [11, Chapter 7] for formal definition). The following is implicit in [17].

Theorem 3 ([17]). *There exists a fully-black-box reduction from oblivious transfer to defensible oblivious transfer.*

2.5 Commitment Schemes

A commitment scheme is a two-stage protocol between a sender and a receiver. In the first stage, called the ***commit stage***, the sender commits to a private string σ. In the second stage, called the ***reveal stage***, the sender reveals σ and *proves* that it was the value to which she committed in the first stage. We require two properties of commitment schemes. The hiding property says that the receiver learns nothing about σ in the commit stage. The binding property says that after the commit stage, the sender is bound to a particular value of σ; that is, she cannot successfully open the commitment to two different values in the reveal stage. See [10] for a more formal definition. Fully-black-box reductions from commitment schemes to one-way functions were given by [15,21] and [14,23].

---

[5] It immediately follows that being (A, f)-defensibly-private implies being (A, f)-semi-honest-private. In Section 3, we show that the other direction is also true.

2.6 One-Way Functions

Definition 4. *Let* $f : \{0,1\}^* \mapsto \{0,1\}^*$ *be a polynomial-time computable function.* f *is* one-way *if the following is negligible for every* PPT A,

$$\Pr[A(1^n, U_n) \in f^{-1}(f(U_n))].$$

3 Reducing Semi-honest Protocols to Defensible Ones

Our transformation from semi-honest privacy to defensible privacy (Theorem 1) immediately follows by applying the next lemma twice. The lemma informally states that it is possible to "upgrade" the security of a protocol w.r.t. one of its parties while maintaining the initial security w.r.t. the other party.

Lemma 1. *Let* $\pi = (A, B)$ *be a two-party protocol and let* $f_A, f_B : \{0,1\}^k \times \{0,1\}^k \mapsto \{0,1\}^*$ *be two functions defined over the parties' inputs. Assume that* π *is* (A, f_A)-*semi-honest-private and* (B, f_B)-*x-private, where* x *stands for 'semi-honest' or 'defensibly'. Then there exists a fully-black-box reduction from a protocol* $\pi' = (\mathbb{A}, \mathbb{B})$ *that has the same functionality as* π *and is* $(\mathbb{A}, f_A)$-*defensibly-private and* $(\mathbb{B}, f_B)$-*x-private, to* π *and one-way functions.*

Proof. In the following definition of π' we are using a commitment scheme, Com. Recall that by [15,22] and by [14,23], there exists a fully-black-box reduction from Com to one-way functions.

Protocol 1 *[The defensible protocol* $\pi' = (\mathbb{A}, \mathbb{B})$*]*

Common input: 1^n.
A's inputs: $i_A \in \{0,1\}^k$ and $r_A = (r_A^1, r_A^2)$.
B's inputs: $i_B \in \{0,1\}^k$ and $r_B = (r_B^1, r_B^2, r_B^3)$.

1. $\mathbb{A}$ commits using Com to (i_A, r_A^2), where the security parameter of the commitment is set to 1^n and $\mathbb{A}$ and $\mathbb{B}$ are using the random-coins r_A^1 and r_B^1 respectively.
2. $\mathbb{B}$ sends r_B^3 to $\mathbb{A}$.
3. The two parties execute the protocol $(A(1^n, i_A, r_A^2 \oplus r_B^3), B(1^n, i_B, r_B^2))$, with $\mathbb{A}$ and $\mathbb{B}$ acting as A and B respectively.

Clearly π' has the same functionality as π. Lemma 2 states that π' maintains the *same* privacy w.r.t. $\mathbb{B}$ and f_B. The heart of our proof is in Lemma 3, where we show that π' has defensible privacy w.r.t. $\mathbb{A}$ and f_A.

Lemma 2. *Assume that* π *is* (B, f_B)-*x-private, then* π' *is* $(\mathbb{B}, f_B)$-*x-private.*

Proof. We assume that π is (B, f_B)-semi-honest-private and prove that π' is $(\mathbb{B}, f_B)$-semi-honest-private (the proof for the defensibly-private case is analogous). We first note that the hiding property of Com yields that for every $i_B \in$

$\{0,1\}^k$, the distribution $(\text{View}_{\mathsf{B}}^{\pi'}(U_k, i_{\mathsf{B}}), f_{\mathsf{B}}(U_k, i_{\mathsf{B}}))$ is computationally indistinguishable from $(\text{View}_{\mathsf{B}}^{\mathsf{Com}}(0^\ell), \text{View}_{\mathsf{B}}^{\pi}(U_k, i_{\mathsf{B}}), f_{\mathsf{B}}(U_k, i_{\mathsf{B}}))$. By the semi-honest privacy of π w.r.t. B and f_{B}, we have that $(\text{View}_{\mathsf{B}}^{\pi'}(U_k, i_{\mathsf{B}}), f_{\mathsf{B}}(U_k, i_{\mathsf{B}}))$ is computationally indistinguishable from $(\text{View}_{\mathsf{B}}^{\mathsf{Com}}(0^\ell), \text{View}_{\mathsf{B}}^{\pi}(U_k, i_{\mathsf{B}}), f_{\mathsf{B}}(U_k', i_{\mathsf{B}}))$. Using the hiding property of Com once more, we have that $(\text{View}_{\mathsf{B}}^{\pi'}(U_k, i_{\mathsf{B}}), f_{\mathsf{B}}(U_k, i_{\mathsf{B}}))$ is computationally indistinguishable from $(\text{View}_{\mathsf{B}}^{\pi'}(U_k, i_{\mathsf{B}}), f_{\mathsf{B}}(U_k', i_{\mathsf{B}}))$. Namely, we have proved that π' is $(\mathsf{B}, f_{\mathsf{B}})$-semi-honest-private.

Lemma 3. *Assume that π is $(\mathsf{A}, f_{\mathsf{A}})$-semi-honest-private, then π' is $(\mathsf{A}, f_{\mathsf{A}})$-defensibly-private.*

Proof. Assume toward a contradiction the existence of an efficient adversary A^* and a distinguisher $\mathbb{D}$ that violate the defensible privacy of π w.r.t. A and f_{A}. Namely, there exists a polynomial p such that for infinitely many n's $\mathbb{D}$ distinguishes with advantage at least $\frac{1}{p(n)}$ between $\Gamma(\text{View}_{\mathsf{A}^*}^{(\mathsf{A}^*, \mathsf{B})}(U_k), f_{\mathsf{A}}(i_{\mathsf{A}}^d, U_k))$ and $\Gamma(\text{View}_{\mathsf{A}^*}^{(\mathsf{A}^*, \mathsf{B})}(U_k), f_{\mathsf{A}}(i_{\mathsf{A}}^d, U_k'))$, where $\Gamma(x, y)$ equals (x, y) if $\text{IsGoodDef}^{\pi', \mathsf{A}}(x) = 1$ and equals $\perp$ otherwise, and i_{A}^d is the value of A's input in $\text{Defense}(\text{View}_{\mathsf{A}^*}^{(\mathsf{A}^*, \mathsf{B})}(U_k))$. In the following we use A^* and $\mathbb{D}$ to present an efficient distinguisher D with oracle access to A^* and $\mathbb{D}$ that violates the semi-honest privacy of (A, B) w.r.t. A and f_{A}. Recall that in order to violate the semi-honest privacy of (A, B), algorithm D should first sample an input element i_{A} for A. Then upon getting A's view from the execution of $(\mathsf{A}(i_{\mathsf{A}}), \mathsf{B}(U_k))$, algorithm D has to distinguish between $f_{\mathsf{A}}(i_{\mathsf{A}}, U_k)$ and $f_{\mathsf{A}}(i_{\mathsf{A}}, U_k')$. In order to make the dependencies between its two stages explicit, D uses the variable z to transfer information from its first stage to its second stage.

Algorithm 1 *[The distinguisher D]*

Sampling stage:
Input: 1^n

1. *Choose uniformly at random r_{A^*} and r_{B}^1 and fix A^*'s random-coins to r_{A^*}.*
2. *Simulate the first line of $(\mathsf{A}^*, \mathsf{B})$ (i.e., the execution of Com), where B uses r_{B}^1 as its random coins.*
3. *Do the following $np(n)$ times:*
   (a) *Simulate the last two lines of $(\mathsf{A}^*, \mathsf{B})$, choosing B's input and random-coins (i.e., i_{B}, r_{B}^2 and r_{B}^3) uniformly at random.*
   (b) *If A^* outputs a valid defense d, set $i_{\mathsf{A}} = i_{\mathsf{A}}$ and $z = (r_{\mathsf{A}^*}, r_{\mathsf{B}}^1, r_{\mathsf{A}}^2)$, where i_{A} and r_{A}^2 are the values of these inputs variables in d, and return.*
4. *Set $z = \perp$ and an arbitrary value for i_{A}.*

Predicting stage:
Input: z, v_{A}^{π} - *randomly chosen from* $\text{View}_{\mathsf{A}}^{\pi}(i_{\mathsf{A}}, U_k)$, *and* $c \in Im(f_{\mathsf{A}})$

1. *If $z = \perp$, output a random coin and return.*
2. *Fix the random-coins of A^* to $z[r_{\mathsf{A}^*}]$.*

3. *Simulate the first line of* $(\mathbb{A}^*, \mathbb{B})$ *(i.e., the execution of* Com*), where* $\mathbb{B}$ *uses* $z[r_{\mathbb{B}}^1]$ *as its random coins.*
4. *Simulate the second line of* $(\mathbb{A}^*, \mathbb{B})$, *where* $\mathbb{B}$ *sends* $r_{\mathbb{B}}^3 = v_{\mathbb{A}}^{\pi}[r_{\mathbb{A}}] \oplus z[r_{\mathbb{A}}^2]$ *to* $\mathbb{A}^*$.
5. *Simulate the last line of* $(\mathbb{A}^*, \mathbb{B})$, *where* $\mathbb{B}$ *sends the same messages that* $\mathbb{B}$ *sends in* $v_{\mathbb{A}}^{\pi}$.
6. *Let* $v_{\mathbb{A}^*}$ *be the view of* $\mathbb{A}^*$ *at the end of above simulation,*
   if $\text{IsGoodDef}^{\pi', \mathbb{A}}(v_{\mathbb{A}^*}) = 1$ *output* $\mathbb{D}(v_{\mathbb{A}^*}, c)$,
   otherwise output a random coin.

It is easy to verify that D is efficient given oracle access to $\mathbb{A}^*$ and $\mathbb{D}$, in the following we prove that D violates the semi-honest privacy of π w.r.t. A and f_{A}. We consider a random execution of $(\mathsf{A}, \mathsf{B}, \mathsf{D})$ with security parameter 1^n and define the random variable $Sim_n = (i_{\mathsf{A}}, i_{\mathsf{B}}, r_{\mathbb{A}^*}, r_{\mathbb{B}}, \text{trans})$ as A and B's inputs in the real execution of π, concatenated with $\mathbb{A}^*$ and $\mathbb{B}$'s views in the simulation of π' done in D's *predicting stage*. More precisely, $i_{\mathsf{A}} = v_{\mathsf{A}}^{\pi}[i_{\mathsf{A}}]$, $i_{\mathsf{B}} = v_{\mathsf{A}}^{\pi}[i_{\mathsf{B}}]$, $r_{\mathbb{A}^*} = z[r_{\mathbb{A}^*}]$, $r_{\mathbb{B}} = (z[r_{\mathbb{B}}^1], v_{\mathsf{A}}^{\pi}[r_{\mathsf{B}}], v_{\mathsf{A}}^{\pi}[r_{\mathsf{A}}] \oplus z[r_{\mathbb{A}}^2])$ (set to $\perp$ if $z = \perp$) and finally trans is the transcript of the simulation of π' done in the D's predicting stage (set to $\perp$ if no such simulation occurs).

Let Defense(x) and IsGoodDef(x) be Defense$(x[r_{\mathbb{A}^*}, \text{trans}])$ and IsGoodDef$^{\pi', \mathbb{A}}(x[r_{\mathbb{A}^*}, \text{trans}])$ respectively. For $c \in Im(f_{\mathsf{A}})$ let $\text{Out}_{\mathsf{D}}(x, c)$ be the output bit of D given x and c, note that $\text{Out}_{\mathsf{D}}(x, c)$ is a random variable that depends on the random-coins used by D to invoke $\mathbb{D}$. Finally, let $\text{Adv}_{\mathsf{D}}(x)$ be the advantage of D in predicting f_{A} given x. That is, $\text{Adv}_{\mathsf{D}}(x) \overset{\text{def}}{=} \Pr[\text{Out}_{\mathsf{D}}(x, f_{\mathsf{A}}(x[i_{\mathsf{A}}], x[i_{\mathsf{B}}])) = 1] - \Pr[\text{Out}_{\mathsf{D}}(x, f_{\mathsf{A}}(x[i_{\mathsf{A}}], U_k)) = 1]$. It is easy to verify that $|\text{Ex}_{x \leftarrow Sim_n}[\text{Adv}_{\mathsf{D}}(x)]|$ is exactly the advantage of D in breaking the semi-honest privacy of π w.r.t. A and f_{A}.

We would like the relate the above success probability to that of $\mathbb{D}$ in predicting f_{A} after a random execution of π'. We define the distribution $Real_n = (i_{\mathbb{A}}^d, i_{\mathbb{B}}, r_{\mathbb{A}^*}, r_{\mathbb{B}}, \text{trans})$ induced by a random execution of $(\mathbb{A}^*, \mathbb{B})$ with security parameter 1^n, where $i_{\mathbb{A}}^d$ is the value of this variable in the defense of $\mathbb{A}^*$ (set to $\perp$ is no good defense is given). Let $\text{Out}_{\mathbb{D}}(x, c)$ be the output bit of $\mathbb{D}$ given x and c, and let $\text{Adv}_{\mathbb{D}}(x)$ be the advantage of $\mathbb{D}$ in predicting f_{A} given x. It is easy to verify that $|\text{Ex}_{x \leftarrow Real_n}[\text{Adv}_{\mathbb{D}}(x)]|$ is exactly the advantage of $\mathbb{D}$ in breaking the defensible privacy of π' w.r.t. $\mathbb{A}$ and f_{A}. The following claim helps up to relate the advantage of D in breaking the semi-honest privacy of π to that of $\mathbb{D}$ in breaking the defensible privacy of π'.

Claim. The following hold:

1. For every $n \in \mathbb{N}$ and $x \in \text{Supp}(Real_n)$, it holds that $Sim_n(x) \leq Real_n(x)$
2. For large enough n there exists a set
   $L_n \subseteq \{x \in \text{Supp}(Real_n) : \text{IsGoodDef}(x) = 1\}$ for which the following hold:
   (a) $\Pr_{x \leftarrow Real_n}[\text{IsGoodDef}(x) \wedge x \notin L_n] \leq \frac{1}{4p(n)}$
   (b) For every $x \in L_n$ it holds that $Sim_n(x) \geq (1 - \frac{1}{4p(n)}) \cdot Real_n(x)$

Proof. When drawing a random $X_R = (i_A^d, i_B, r_{A^*}, r_B, \text{trans})$ from $Real_n$, its value is fully determined by the value of $X_R[i_B, r_{A^*}, r_B]$, where the latter value is uniformly distributed over all strings of the right length. On the other hand, when drawing a random X_S from Sim_n, the value of $X_S[i_B, r_{A^*}, r_B]$ is uniformly distributed over all strings, only when conditioning that $\text{IsGoodDef}(X_S) = 1$. Where otherwise, $X_S[i_B, r_{A^*}, r_B] = (*, *, \perp)$, a value that is never obtained by an element in $\text{Supp}(Real_n)$. In particular, for every $x \in \text{Supp}(Real_n)$ it holds that

$$Sim_n(x) = \Pr\big[X_S[i_A, i_B, r_{A^*}, r_B, \text{trans}] = x[i_A^d, i_B, r_{A^*}, r_B, \text{trans}]\big]$$
$$\le \Pr\big[X_S[r_{A^*}, i_B, r_B] = x[r_{A^*}, i_B, r_B]\big]$$
$$\le \Pr\big[X_R[r_{A^*}, i_B, r_B] = x[r_{A^*}, i_B, r_B]\big] = Real_n(x) \ ,$$

proving the first part of the claim. For $x \in \text{Supp}(Real_n)$, let $\text{Decom}(x)$ be the decommitment of Com given in $\text{Defense}(x)$ (we set it to $\perp$ if no valid defense is given). For $S \subseteq \{0,1\}^*$, we let $W_x(S)$ be the probability that the commitment embedded in x is decommitted to a value in S, conditioned *only* on the random-coins in x used for the commitment (and not on all x). That is, $W_x(S) = \Pr\big[\text{Decom}(X_R) \in S \mid X_R[r_B^1, r_{A^*}] = x[r_B^1, r_{A^*}]\big]$. Finally, let $\text{Heaviest}(x) = \text{argmax}_{\sigma \in \{0,1\}^*} W_x(\alpha)$, breaking ties arbitrarily (say, by choosing the lexicographic smallest α) and let $\text{Others}(x) = \{0,1\}^* \setminus \{\text{Heaviest}(x)\}$. We define $L_n = \{x \in \text{Supp}(Real_n) : \text{IsGoodDef}(x) = 1 \wedge W_x(\text{Others}(x)) < \frac{1}{8np(n)^2} \wedge W_x(\text{Heaviest}(x)) > \frac{1}{8p(n)} \wedge \text{Decom}(x) = \text{Heaviest}(x)\}$. In the following we prove the two properties of L_n.

Proving 2(a). We first observe that for every polynomial q, it holds that $\Pr[W_{X_R}(\text{Others}(X_R)) > \frac{1}{q(n)}] < \frac{1}{q(n)}$. Assume otherwise, then we can design an adversary for breaking the binding Com. In the commit stage, the adversary acts as A^* does in the first line of Protocol 1. Then it simulates the rest of the protocol twice (with the same prefix) and outputs the two decommitments implied by A^*'s defenses. Thus, whenever $\Pr[W_{X_R}(\text{Others}(X_R)) > \frac{1}{q(n)}] > \frac{1}{q(n)}$, our adversary breaks the binding of Com with probability $\Omega(\frac{1}{q(n)^3})$.

Since $\text{Decom}(x) \ne \perp$ only if x yields a good defense, it follows that $\Pr[\text{IsGoodDef}(X_R) \wedge (W_x(\text{Heaviest}(x)) + W_x(\text{Others}(x))) < \frac{1}{q(n)}] < \frac{1}{q(n)}$ for every polynomial q. We conclude that

$$\Pr\big[\text{IsGoodDef}(X_R) \wedge X_R \notin L_n\big]$$
$$\le \Pr\left[W_{X_R}(\text{Others}(X_R)) > \frac{1}{8np(n)^2}\right] + \Pr\bigg[\text{IsGoodDef}(X_R)$$
$$\wedge \, (W_x(\text{Heaviest}(x)) + W_x(\text{Others}(x))) < \left(\frac{1}{8p(n)} + \frac{1}{8np(n)^2}\right)\bigg]$$
$$+ \Pr\bigg[\text{Decom}(x) \ne \text{Heaviest}(x) \mid \text{IsGoodDef}(X_R) \wedge W_x(\text{Heaviest}(x)) > \frac{1}{8p(n)}$$

$$\wedge \; W_{X_R}(\text{Others}(X_R)) \le \frac{1}{8np(n)^2}\Big]$$

$$< \frac{1}{8np(n)^2} + \frac{1}{7p(n)} + \frac{8p(n)}{8np(n)^2} < \frac{1}{4p(n)}$$

Proving 2(*b*). Let $x \in L_n$, and let X be a random variable drawn from Sim_n conditioned that $X[r_{\mathsf{A}^*}, r_{\mathsf{B}}^1] = x[r_{\mathsf{A}^*}, r_{\mathsf{B}}^1]$. Recall that in order to sample X, algorithm D keeps sampling (up to $np(n)$ times) a random element x' in $Real_n$ conditioned that $x'[r_{\mathsf{A}^*}, r_{\mathsf{B}}^1] = x[r_{\mathsf{A}^*}, r_{\mathsf{B}}^1]$, until $\text{Decom}(x') \ne \perp$. It then set $(X[i_{\mathsf{A}}], z[r_{\mathsf{A}}^2])$ to $\text{Decom}(x')$, where z is the "state" that D transfers from its sampling stage to its predicting stage (the stage where the other parts of X are chosen). In order to keep notations simple, we define $X[r_{\mathsf{A}}^2]$ as $z[r_{\mathsf{A}}^2]$. By the above description it follows that

$$\Pr[X[i_{\mathsf{A}}, r_{\mathsf{A}}^2] \ne \text{Decom}(x)] \tag{1}$$

$$\le \Pr[\text{Decom}(X) = \perp] + \Pr[\text{Decom}(X) \notin \{\text{Decom}(x) \cup \perp\}]$$

$$\le \text{neg}(n) + \frac{np(n)}{8np(n)^2} < \frac{1}{4p(n)},$$

where the second inequality holds since $x \in L_n$. Since the value of $X[i_{\mathsf{B}}, r_{\mathsf{B}}^2, r_{\mathsf{B}}^3]$ is induced by a the parties' inputs and random-coins in a random execution of π, it follows that $X[i_{\mathsf{B}}, r_{\mathsf{B}}^2, r_{\mathsf{B}}^3]$ is uniformly distributed conditioned on $X[i_{\mathsf{A}}, r_{\mathsf{A}}^2] \ne \perp$ and every value of $X[i_{\mathsf{A}}, r_{\mathsf{A}^*}, r_{\mathsf{B}}^1, r_{\mathsf{A}}^2]$. Recall that the value of X_R is fully determined by the value of $X_R[i_{\mathsf{B}}, r_{\mathsf{A}^*}, r_{\mathsf{B}}]$ and that the latter is uniformly distributed over all possible strings. Hence,

$$\Pr[X_S[i_{\mathsf{B}}, r_{\mathsf{A}^*}, r_{\mathsf{B}}] = x[i_{\mathsf{B}}, r_{\mathsf{A}^*}, r_{\mathsf{B}}] \wedge X_S[i_{\mathsf{A}}, r_{\mathsf{A}}^2] = \text{Decom}(x)] \tag{2}$$

$$\ge (1 - \frac{1}{4p(n)}) \cdot \Pr[X_R[i_{\mathsf{B}}, r_{\mathsf{A}^*}, r_{\mathsf{B}}] = x[i_{\mathsf{B}}, r_{\mathsf{A}^*}, r_{\mathsf{B}}]]$$

$$= (1 - \frac{1}{4p(n)}) \cdot Real_n(x)$$

Let $X[r_{\mathsf{A}}]$ be the value of r_{A} in v_{A}^π as chosen in the sampling process of X and let $x[r_{\mathsf{A}}^2]$ be the value of r_{A}^2 in $\text{Defense}(x)$. Since $\text{IsGoodDef}(x) = 1$, A^* acts in the embedded execution of π in x, as $\mathsf{A}(x[i_{\mathsf{A}}^d], x[r_{\mathsf{A}}^2] \oplus x[r_{\mathsf{A}}^3])$ would. Thus, $X_S[r_{\mathsf{A}^*}, i_{\mathsf{B}}, r_{\mathsf{B}}] = x[r_{\mathsf{A}^*}, i_{\mathsf{B}}, r_{\mathsf{B}}]$ and $X_S[i_{\mathsf{A}}, r_{\mathsf{A}}^2] = \text{Decom}(x)$ implies that A^* acts in the embedded execution of π as $\mathsf{A}(X_S[i_{\mathsf{A}}], X_S[r_{\mathsf{A}}^2] \oplus X_S[r_{\mathsf{B}}^3])$ would, that is as $\mathsf{A}(X_S[i_{\mathsf{A}}], X_S[r_{\mathsf{A}}])$. Hence, $X_S[r_{\mathsf{A}^*}, i_{\mathsf{B}}, r_{\mathsf{B}}] = x[r_{\mathsf{A}^*}, i_{\mathsf{B}}, r_{\mathsf{B}}]$ and $X[i_{\mathsf{A}}, r_{\mathsf{A}}^2] = \text{Decom}(x)$ implies that $X_S[\text{trans}] = x[\text{trans}]$, and we conclude that

$$\Pr[X_S = x]$$

$$= \Pr[X_S[r_{\mathsf{A}^*}, i_{\mathsf{B}}, r_{\mathsf{B}}] = x[r_{\mathsf{A}^*}, i_{\mathsf{B}}, r_{\mathsf{B}}] \wedge X_S[i_{\mathsf{A}}, r_{\mathsf{A}}^2] = \text{Decom}(x)]$$

$$\ge (1 - \frac{1}{4p(n)}) \cdot Real_n(x)$$

$$\square$$

Back to the proof the lemma. Let n be large enough be large enough so that Claim 3 holds and assume w.l.o.g. that $\mathsf{Ex}[\mathsf{Adv_D}(X_R)] > \frac{1}{p(n)}$. Since $\mathbb{D}$ gains no advantage when $\mathsf{IsGoodDef}(X_R) = 0$, it follows that $\mathsf{Ex}[\mathsf{Adv_D}(X_R) \cdot \mathsf{IsGoodDef}(X_R)] > \frac{1}{p(n)}$ as well. We first observe that

$$
\begin{aligned}
\mathsf{Ex}[\mathsf{Adv_D}(X_S)] &= \mathsf{Ex}[\mathsf{Adv_D}(X_S) \cdot \mathsf{IsGoodDef}(X_S)] \\
&\geq \mathsf{Ex}[\mathsf{Adv_D}(X_S) \cdot 1_{X_S \in L_n}] - \Pr[\mathsf{IsGoodDef}(X_S) \wedge X_S \notin L_n] \\
&= \Pr\big[\mathsf{Out_D}(X_S, f_\mathsf{A}(X_S[i_\mathsf{A}, i_\mathsf{B}])) = 1) \wedge X_S \in L_n\big] \\
&\quad - \Pr\big[\mathsf{Out_D}(X_S, f_\mathsf{A}(X_S[i_\mathsf{A}], U_k)) = 1) \wedge X_S \in L_n\big] \\
&\quad - \Pr[\mathsf{IsGoodDef}(X_S) \wedge X_S \notin L_n] \ ,
\end{aligned}
$$

where $1_{x \in L_n}$ is one if $x \in L_n$ and zero otherwise, and the first equality holds since $\mathsf{Out_D}(x, c)$ is a random coin if $\mathsf{IsGoodDef}(x) = 0$. By Claim 3 we have that

$$
\Pr[\mathsf{IsGoodDef}(x) \wedge X_S \notin L_n] \tag{3}
$$
$$
\leq \Pr[\mathsf{IsGoodDef}(X_R) \wedge X_R \notin L_n] \leq \frac{1}{4p(n)}
$$

Since $\mathsf{Out_D}(x, c) = \mathsf{Out_{\mathbb{D}}}(x, c)$ for every $x \in \mathsf{Supp}(Real_n)$ such that $\mathsf{IsGoodDef}(x) = 1$, Claim 3 also yields that

$$
\Pr[\mathsf{Out_D}(X_S, f_\mathsf{A}(X_S[i_\mathsf{A}], U_k)) = 1 \wedge X_S \in L_n] \tag{4}
$$
$$
\leq \Pr[\mathsf{Out_{\mathbb{D}}}(X_R, f_\mathsf{A}(X_R[i_\mathsf{A}^d], U_k)) = 1 \wedge X_R \in L_n]
$$

and that

$$
\Pr[\mathsf{Out_D}(X_S, f_\mathsf{A}(X_S[i_\mathsf{A}, i_\mathsf{B}])) = 1 \wedge X_S \in L_n] \tag{5}
$$
$$
\geq (1 - \frac{1}{4p(n)}) \cdot \Pr[\mathsf{Out_{\mathbb{D}}}(X_R, f_\mathsf{A}(X_R[i_\mathsf{A}^d, i_\mathsf{B}])) = 1 \wedge X_R \in L_n]
$$

We conclude that

$$
\begin{aligned}
&\mathsf{Ex}[\mathsf{Adv_D}(X_S)] \\
&\geq (1 - \frac{1}{4p(n)}) \cdot \Pr[\mathsf{Out_{\mathbb{D}}}(X_R, f_\mathsf{A}(X_R[i_\mathsf{A}^d, i_\mathsf{B}])) = 1 \wedge X_R \in L_n] \\
&\quad - \Pr[\mathsf{Out_{\mathbb{D}}}(X_R, f_\mathsf{A}(X_R[i_\mathsf{A}^d], U_k)) = 1 \wedge X_R \in L_n] - \frac{1}{4p(n)} \\
&\geq (1 - \frac{1}{4p(n)}) \cdot \mathsf{Ex}[\mathsf{Adv_{\mathbb{D}}}(X_R) \cdot \mathsf{IsGoodDef}(X_R)] - \frac{1}{4p(n)} - \frac{1}{4p(n)} \\
&\geq (1 - \frac{1}{4p(n)}) \cdot \frac{1}{p(n)} - \frac{1}{2p(n)} > \frac{1}{4p(n)}
\end{aligned}
$$

Since the above holds for infinitely many n's, it concludes the proof of Lemma 3 and thus the proof of Theorem 1.

4 Achieving the Main Result

In the following we prove Theorem 2, the main result of this paper. As corollary of Theorem 1, we have that there exists a fully-black-box reduction from defensible oblivious transfer to semi-honest oblivious transfer and one-way functions. This corollary together with Theorem 3, yields the existence of a fully-black-box reduction from malicious oblivious transfer to semi-honest oblivious transfer and one-way functions. Thus, the proof of the Theorem 2 is concluded by the following folklore theorem (proof given in the full version).

Theorem 4. *There exists a fully-black-box reduction from one-way functions to semi-honest oblivious transfer.*

Acknowledgment

I am very grateful to Yuval Ishai and Omer Reingold for very helpful conversations and for reading early versions of this paper. I am also grateful to Oded Goldreich, Gil Segev and to the anonymous TCC 2008 referees for their useful comments regarding the write-up of the paper.

References

1. Barak, B.: How to go beyond the black-box simulation barrier. In: 42nd FOCS, pp. 106–115 (2001)
2. Blum, M.: How to exchange (secret) keys. ACM Transactions on Computer Systems (1983)
3. Brassard, G., Crépeau, C., Robert, J.-M.: Information theoretic reductions among disclosure problems. In: 27th FOCS (1986)
4. Crépeau, C.: Equivalence between two flavours of oblivious transfers. In: Pomerance, C. (ed.) CRYPTO 1987. LNCS, vol. 293, Springer, Heidelberg (1988)
5. Crépeau, C., Kilian, J.: Weakening security assumptions and oblivious transfer. In: Goldwasser, S. (ed.) CRYPTO 1988. LNCS, vol. 403, Springer, Heidelberg (1990)
6. Crépeau, C., Sántha, M.: On the reversibility of oblivious transfer. In: Davies, D.W. (ed.) EUROCRYPT 1991. LNCS, vol. 547, Springer, Heidelberg (1991)
7. Di Crescenzo, G., Malkin, T., Ostrovsky, R.: Single database private information retrieval implies oblivious transfer. In: Preneel, B. (ed.) EUROCRYPT 2000. LNCS, vol. 1807, Springer, Heidelberg (2000)
8. Dolev, D., Dwork, C., Naor, M.: Nonmalleable cryptography. JACM 30(2), 391–437 (2000)
9. Even, S., Goldreich, O., Lempel, A.: A randomized protocol for signing contracts. Communications of the ACM 28(6), 637–647 (1985)
10. Goldreich, O.: Foundations of Cryptography: Basic Tools. Cambridge University Press, Cambridge (2001)
11. Goldreich, O.: Foundations of Cryptography – vol. 2: Basic Applications. Cambridge University Press, Cambridge (2004)
12. Goldreich, O., Micali, S., Wigderson, A.: How to play any mental game or a completeness theorem for protocols with honest majority. In: 19th STOC, pp. 218–229 (1987)

13. Haitner, I.: Implementing oblivious transfer using collection of dense trapdoor permutations. In: 1st TCC, pp. 394–409 (2004)
14. Haitner, I., Reingold, O.: Statistically-hiding commitment from any one-way function. In: 39th STOC (2007)
15. Håstad, J., Impagliazzo, R., Levin, L.A., Luby, M.: A pseudorandom generator from any one-way function. SICOMP 28(4), 1364–1396 (1999)
16. Impagliazzo, R., Rudich, S.: Limits on the provable consequences of one-way permutations. In: 21st STOC, pp. 44–61. ACM Press, New York (1989)
17. Ishai, Y., Kushilevitz, E., Lindell, Y., Petrank, E.: Black-box constructions for secure computation. In: 38th STOC (2006)
18. Kilian, J.: Founding cryptography on oblivious transfer. In: pp. 20–31 (1988)
19. Kushilevitz, E., Ostrovsky, R.: Replication is NOT needed: SINGLE database, computationally-private information retrieval. In: 38th FOCS, pp. 364–373 (1997)
20. Lindell, Y.: A simpler construction of CCA2-secure public-key encryption under general assumptions. J. Cryptology 19(3), 359–377 (2006)
21. Naor, M.: Bit commitment using pseudorandomness. J. of Crypto. 4(2), 151–158 (1991)
22. Naor, M., Ostrovsky, R., Venkatesan, R., Yung, M.: Perfect zero-knowledge arguments for NP using any one-way permutation. J. of Crypto. 11(2), 87–108 (1998)
23. Nguyen, M.-H., Ong, S.J., Vadhan, S.: Statistical zero-knowledge arguments for NP from any one-way function. In: 47th FOCS, pp. 3–14 (2006)
24. Rabin, M.O.: How to exchange secrets by oblivious transfer. TR-81, Harvard (1981)
25. Reingold, O., Trevisan, L., Vadhan, S.P.: Notions of reducibility between cryptographic primitives. In: 1st TCC, pp. 1–20 (2004)
26. Sahai, A.: Non-malleable non-interactive zero knowledge and adaptive chosen-ciphertext security. In: 40th FOCS, pp. 543–553 (1999)
27. Stern, J.P.: A new and efficient all-or-nothing disclosure of secrets protocol. In: Ohta, K., Pei, D. (eds.) ASIACRYPT 1998. LNCS, vol. 1514, Springer, Heidelberg (1998)
28. Chi-Chih Yao, A.: How to generate and exchange secrets. In: 27th FOCS, pp. 162–167 (1986)

Black-Box Construction of a Non-malleable Encryption Scheme from Any Semantically Secure One

Seung Geol Choi, Dana Dachman-Soled, Tal Malkin*, and Hoeteck Wee

Columbia University
{sgchoi,dglasner,tal,hoeteck}@cs.columbia.edu

Abstract. We show how to transform any semantically secure encryption scheme into a non-malleable one, with a black-box construction that achieves a quasi-linear blow-up in the size of the ciphertext. This improves upon the previous non-black-box construction of Pass, Shelat and Vaikuntanathan (Crypto '06). Our construction also extends readily to guarantee non-malleability under a bounded-CCA2 attack, thereby simultaneously improving on both results in the work of Cramer et al. (Asiacrypt '07).

Our construction departs from the oft-used paradigm of re-encrypting the same message with different keys and then proving consistency of encryptions; instead, we encrypt an encoding of the message with certain locally testable and self-correcting properties. We exploit the fact that low-degree polynomials are simultaneously good error-correcting codes and a secret-sharing scheme.

Keywords: Public-key encryption, semantic security, non-malleability, black-box constructions.

1 Introduction

The most basic security guarantee we require of a public key encryption scheme is that of semantic security [GM84]: it is infeasible to learn anything about the plaintext from the ciphertext. In many cryptographic applications such as auctions, we would like an encryption scheme that satisfies the stronger guarantee of non-malleability [DDN00], namely that given some ciphertext c, it is also infeasible to generate ciphertexts of some message that is related to the decryption of c. Motivated by the importance of non-malleability, Pass, Shelat and Vaikuntanathan raised the following question [PSV06]:

> It is possible to *immunize* any semantically secure encryption scheme against malleability attacks?

Pass et al. gave a beautiful construction of a non-malleable encryption scheme from any semantically secure one (building on [DDN00]), thereby addressing the question in the affirmative. However, the PSV construction – as with previous constructions achieving non-malleability from general assumptions [DDN00, S99, L06] – suffers from the curse of inefficiency arising from the use of general NP-reductions. In this work, we show that we can in fact immunize any semantically secure encryption schemes against malleability attacks without paying the price of general NP-reductions:

---

* The work was partially supported by NSF grants CNS-0716245, CCF-0347839, and SBE-0245014.

R. Canetti (Ed.): TCC 2008, LNCS 4948, pp. 427–444, 2008.
ⓒ International Association for Cryptologic Research 2008

Main theorem (informal). There exists a (fully) black-box construction of a non-malleable encryption scheme from any semantically secure one.

That is, we provide a wrapper program (from programming language lingo) that given any subroutines for computing a semantically secure encryption scheme, computes a non-malleable encryption scheme, with a multiplicative overhead in the running time that is quasi-linear in the security parameter. Before providing further details, let us first provide some background and context for our result.

1.1 Relationships Amongst Cryptographic Primitives

Much of the modern work in foundations of cryptography rests on general cryptographic assumptions like the existence of one-way functions and trapdoor permutations. General assumptions provide an abstraction of the functionalities and hardness we exploit in specific assumptions such as hardness of factoring and discrete log without referring to any specific underlying algebraic structure. Constructions based on general assumptions may use the primitive guaranteed by the assumption in one of two ways:

Black-box usage: A construction is black-box if it refers only to the input/output behavior of the underlying primitive; we would typically also require that in the proof of security, we can use an adversary breaking the security of the construction as an oracle to break the underlying primitive. (See [RTV04] and references within for more details.). As emphasized earlier, our construction is black-box, using only oracle access to the key generation, encryption and decryption functionality of the underlying encryption scheme.

Non-black-box usage: A construction is non-black-box if it uses the code computing the functionality of the primitive. The PSV construction along with the work it builds on fall into this category: they use an NP reduction applied to the circuit computing the encryption functionality of the underlying encryption scheme in order to provide a non-interactive zero-knowledge proof of consistency.

Motivated by the fact that the vast majority of constructions in cryptography are black-box, a rich and fruitful body of work initiated in [IR89] seeks to understand the power and limitations of black-box constructions in cryptography, resulting in a fairly complete picture of the relations amongst most cryptographic primitives with respect to black-box constructions (we summarize several of the known relations pertaining to encryption in Figure 1). More recent work has turned to tasks for which the only constructions we have are non-black-box, yet the existence of a black-box construction is not ruled out. Two notable examples are general secure multi-party computation against a dishonest majority and encryption schemes secure against adaptive chosen-ciphertext (CCA2) attacks[1] (c.f. [GMW87, DDN00]).

---

[1] These are encryption schemes that remain semantically secure even under a CCA2 attack, wherein the adversary is allowed to query the decryption oracle except on the given challenge. A CCA1 attack is one wherein the adversary is allowed to query the decryption oracle before (but not after) seeing the challenge.

The general question of whether we can securely realize these tasks via black-box access to a general primitive is not merely of theoretical interest. A practical reason is related to efficiency, as non-black-box constructions tend to be less efficient due to the use of general NP reductions to order to prove statements in zero knowledge; this impacts both computational complexity as well as communication complexity (which we interpret broadly to mean message lengths for protocols and key size and ciphertext size for encryption schemes). Moreover, if resolved in the affirmative, we expect the solution to provide new insights and techniques for circumventing the use of NP reductions and zero knowledge in the known constructions. Finally, given that there has been no formal model that captures non-black-box constructions in a satisfactory manner, the pursuit of a positive result becomes all the more interesting.

Indeed, Ishai et al. [IKLP06] recently provided an affirmative answer for secure multi-party computation by exhibiting black-box constructions from some low-level primitive. Their techniques have since been used to yield secure multi-party computation via black-box access to an oblivious transfer protocol for semi-honest parties, which is complete (and thus necessary) for secure multi-party computation [H08]. This leaves the following open problem:

> Is it possible to realize CCA2-secure encryption via black-box access to a low-level primitive, e.g. enhanced trapdoor permutations or homomorphic encryption schemes?

Previous work pertaining to this problem is limited to non-black-box constructions of CCA2-secure encryption from enhanced trapdoor permutations [DDN00, s99, L06]; nothing is known assuming homomorphic encryption schemes. In work concurrent with ours, Peikert and Waters [PW07] made substantial progress towards the open problem – they constructed CCA2-secure encryption schemes via black-box access to a new primitive they introduced called lossy trapdoor functions, and in addition, gave constructions of this primitive from number-theoretic and worst-case lattice assumptions. Unfortunately, they do not provide a black-box construction of CCA2-secure encryption from enhanced trapdoor permutations.

Our work may also be viewed as a step towards closing this remaining gap (and a small step in the more general research agenda of understanding the power of black-box constructions). Specifically, the security guarantee provided by non-malleability lies between semantic security and CCA2 security, and we show how to derive non-malleability in a black-box manner from the minimal assumption possible, i.e., semantic security. In the process, we show how to enforce consistency of ciphertexts in a black-box manner. This issue arises in black-box constructions of both CCA2-secure and non-malleable encryptions. However, our consistency checks only satisfy a weaker notion of non-adaptive soundness, which is sufficient for non-malleability but not for CCA2-security (c.f. [PSV06]). As a special case of our result, we obtain a black-box construction of non-malleable encryptions from any (poly-to-1) trapdoor function. Our results are incomparable with those of Peikert and Waters since we start from weaker assumptions but derive a weaker security guarantee.

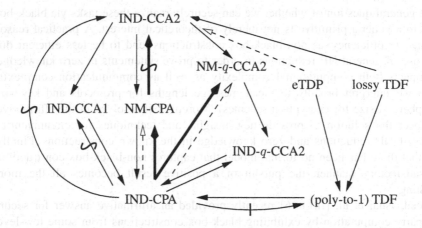

Fig. 1. Known relations among generic encryption primitives, and our results. Solid lines indicate black box constructions, and dotted lines indicate non-black-box constructions (c.f. [BHSV98, DDN00, PSV06, CHH⁺07, PW07]). The separations are with respect to black-box reductions, or black box shielding reductions (c.f. [GMR01, GMM07]). Our contributions are indicated with the thick arrows.

Related positive results. A different line of work focuses on (very) efficient constructions of CCA2-secure encryptions under specific number-theoretic assumptions [CS98, CS04, CHK04]. Apart from those based on identity-based encryption, these constructions together with previous ones based on general assumptions can be described under the following framework (c.f. [BFM88, NY90, RS91, ES02]). Start with some cryptographic hardness assumption that allows us to build a semantically secure encryption scheme, and then prove/verify that several ciphertexts satisfy certain relations in one of two ways:

- exploiting algebraic relations from the underlying assumption to deduce additional structure in the encryption scheme (e.g. homomorphic, reusing randomness) [CS98, CS04];
- apply a general NP reduction to prove in non-interactive zero knowledge (NIZK) statements that relate to the primitive [DDN00, S99, L06].

None of the previous approaches seems to yield black-box constructions under general assumptions. Indeed, our work (also [PW07]) does not use the above framework.

1.2   Our Results

As mentioned earlier, we exhibit a black-box construction of a non-malleable encryption scheme from any semantically secure one, the main novelty being that our construction is black-box. While this is interesting in and of itself, our construction also compares favorably with previous work in several regards:

- *Improved parameters.* We improve on the computational complexity of previous constructions based on general assumptions. In particular, we do not have to do an

NP-reduction in either encryption or decryption, although we do have to pay the price of the running time of Berlekamp-Welch for decryption. The running time incurs a multiplicative overhead that is quasi-linear in the security parameter, over the running time of the underlying CPA secure scheme. Moreover, the sizes of public keys and ciphertext are independent of the computational complexity of the underlying scheme.

- *Conceptual simplicity/clarity.* Our scheme (and the analysis) is arguably much simpler than many of the previous constructions, and like [PSV06], entirely self-contained (apart from the Berlekamp-Welch algorithm). We do not need to appeal to notions of zero-knowledge, nor do we touch upon subtle technicalities like adaptive vs non-adaptive NIZK. Our construction may be covered in an introductory graduate course on cryptography without requiring zero knowledge as a pre-requisite.

- *Ease of implementation.* Our scheme is easy to describe and can be easily implemented in a modular fashion.

We may also derive from our construction additional positive and negative results.

Bounded CCA2 non-malleability. Cramer et al. [CHH+07] introduced the bounded CCA2 attack, a relaxation of the CCA2 attack wherein the adversary is only allowed make an a-priori bounded number of queries q to the decryption oracle, where q is fixed prior to choosing the parameters of the encryption scheme. In addition, starting from any semantically secure encryption, they obtained[2]:

- an encryption scheme that is semantically secure under a bounded-CCA2 attack via a black-box construction, wherein the size of the public key and ciphertext are quadratic in q; and
- an encryption scheme that is non-malleable under a bounded-CCA2 attack via a non-black-box construction, wherein the size of the public key and ciphertext are linear in q.

Combining their approach for the latter construction with our main result, we obtain an encryption scheme that is non-malleable under a bounded-CCA2 attack via a black-box construction, wherein the size of the public key and ciphertext are linear in q.

Separation between CCA2 security and non-malleability. Our main construction has the additional property that the decryption algorithm does not query the encryption functionality of the underlying scheme. Gertner, Malkin and Myers [GMM07] referred to such constructions as shielding and they showed that there is no shielding black-box construction of CCA1-secure encryption schemes from semantically secure encryption. Combined with the fact that any shielding construction when composed with our construction is again shielding, this immediately yields the following:

> **Corollary (informal)** There exists no shielding black-box construction of CCA1-secure encryption schemes from non-malleable encryption schemes.

---

[2] While semantic security and non-malleability are equivalent under a CCA2 attack [DDN00], they are not equivalent under a bounded-CCA2 attack, as shown in [CHH+07].

Note that a CCA2-secure encryption scheme is trivially also CCA1-secure, so this also implies a separation between non-malleability and CCA2-security for shielding black-box constructions.

Our techniques. At a high level, we follow the cut-and-choose approach for consistency checks from [PSV06], wherein the randomness used for cut-and-choose is specified in the secret key. A crucial component of our construction is a message encoding scheme with certain locally testable and self-correcting properties, based on the fact that low-degree polynomials are simultaneously good error-correcting codes and a secret-sharing scheme; this has been exploited in the early work on secure multi-party computation with malicious adversaries [BGW88]. We think this technique may be useful in eliminating general NP-reductions in other constructions in cryptography (outside of public-key encryption).

Towards CCA2 Security? The main obstacle towards achieving full CCA2 security from either semantically secure encryptions or enhanced trapdoor permutations using our approach (and also the [PSV06] approach) lies in guaranteeing soundness of the consistency checks against an adversary that can adaptively determine its queries depending on the outcome of previous consistency checks. It seems conceivable that using a non-shielding construction that uses re-encryption may help overcome this obstacle.

1.3 Overview of Our Construction

Recall the DDN [DDN00] and PSV [PSV06] constructions: to encrypt a message, one (a) generates k encryptions of the same message under independent keys, (b) gives a non-interactive zero-knowledge proof that all resulting ciphertexts are encryptions of the same message, and (c) signs the entire bundle with a one-time signature. It is in step (b) that we use a general NP-reduction, which in return makes the construction non-black-box. In the proof of security, we exploit that fact that for a well-formed ciphertext, we can recover the message if we know the secret key for any of the k encryptions.

How do we guarantee that a tuple of k ciphertexts are encryptions of the same plaintext without using a zero-knowledge proof and without revealing any information about the underlying plaintext? Naively, one would like to use a cut-and-choose approach (as has been previously used in [LP07] to eliminate zero-knowledge proofs in the context of secure two-party computation), namely decrypt and verify that some constant fraction, say $k/2$ of the ciphertexts are indeed consistent. There are two issues with this approach:

- First, if only a constant number of ciphertexts are inconsistent, then we are unlikely to detect the inconsistency. To circumvent this problem, we could decrypt by outputting the majority of the remaining $k/2$ ciphertexts.
- The second issue is more fundamental: decrypting any of the ciphertexts will immediately reveal the underlying message, whereas it is crucial that we can enforce consistency while learning nothing about the underlying message.

We circumvent both issues by using a more sophisticated encoding of the message m based on low-degree polynomials instead of merely making k copies of the message as in the above schemes. Specifically, we pick a random degree k polynomial p such that $p(0) = m$ and we construct a $k \times 10k$ matrix such that the i'th column of the matrix comprises entirely of the value $p(i)$. To verify consistency, we will decrypt a random subset of k columns, and check that all the entries in each of these columns are the same.

- The issue that only a tiny number of ciphertexts are inconsistent is handled using the error-correcting properties of low-degree polynomials; specifically, each row of a valid encoding is a codeword for the Reed-Solomon code (and we output $\perp$ if it's far from any codeword).
- Low-degree polynomials are also good secret-sharing schemes, and learning a random subset of k columns in a valid encoding reveals nothing about the underlying message m. Encoding m using a secret-sharing scheme appears in the earlier work of Cramer et al. [CHH+07], but they do not consider redundancy or error-correction.

As before, we encrypt all the entries of the matrix using independent keys and then sign the entire bundle with a one-time signature. It is important that the encoding also provides a robustness guarantee similar to that of repeating the message k times: we are able to recover the message for a valid encryption if we can decrypt *any* row in the matrix. Indeed, this is essentially our entire scheme with two technical caveats:

- As with previous schemes, we will associate one pair of public/secret key pairs with each entry of the matrix, and we will select the public key for encryption based on the verification key of the one-time signature scheme.
- To enforce consistency, we will need a codeword check in addition to the column check outlined above. The reason for this is fairly subtle and we will highlight the issue in the formal exposition of our construction.

Decreasing ciphertext size. To encrypt an n-bit message with security parameter k, our construction yields $O(k^2)$ encryptions of n-bit messages in the underlying scheme. It is easy to see that this may be reduced to $O(k \log^2 k)$ encryptions by reducing the number of columns to $O(\log^2 k)$.

2 Preliminaries and Definitions

Notation. We adopt the notation used in [PSV06]. We use $[n]$ to denote $\{1, 2, \ldots, n\}$. If A is a probabilistic polynomial time (hereafter, ppt) algorithm that runs on input x, $A(x)$ denotes the random variable according to the distribution of the output of A on input x. We denote by $A(x; r)$ the output of A on input x and random coins r. Computational indistinguishability between two distributions A and B is denoted by $A \overset{c}{\approx} B$ and statistical indistinguishability by $A \overset{s}{\approx} B$.

2.1 Semantically Secure Encryption

Definition 1 (Encryption Scheme). *A triple* (Gen, Enc, Dec) *is an encryption scheme, if* Gen *and* Enc *are ppt algorithms and* Dec *is a deterministic polynomial-time algorithm which satisfies the following property:*

Correctness. *There exists a negligible function $\mu(\cdot)$ such that for all sufficiently large k, we have that with probability $1 - \mu(k)$ over* $(\text{PK}, \text{SK}) \leftarrow \text{Gen}(1^k)$: *for all m*, $\Pr[\text{Dec}_{\text{SK}}(\text{Enc}_{\text{PK}}(m)) = m] = 1$.

Definition 2 (Semantic Security). *Let* $\Pi = (\text{Gen}, \text{Enc}, \text{Dec})$ *be an encryption scheme and let the random variable* $\text{IND}_b(\Pi, A, k)$, *where $b \in \{0,1\}$, $A = (A_1, A_2)$ are ppt algorithms and $k \in \mathbb{N}$, denote the result of the following probabilistic experiment:*

$\text{IND}_b(\Pi, A, k)$:
   $(\text{PK}, \text{SK}) \leftarrow \text{Gen}(1^k)$
   $(m_0, m_1, \text{STATE}_A) \leftarrow A_1(\text{PK})$ *s.t.* $|m_0| = |m_1|$
   $y \leftarrow \text{Enc}_{\text{PK}}(m_b)$
   $D \leftarrow A_2(y, \text{STATE}_A)$
   Output D

(Gen, Enc, Dec) *is indistinguishable under a chosen-plaintext (CPA) attack, or semantically secure, if for any ppt algorithms $A = (A_1, A_2)$ the following two ensembles are computationally indistinguishable:*

$$\left\{ \text{IND}_0(\Pi, A, k) \right\}_{k \in \mathbb{N}} \stackrel{c}{\approx} \left\{ \text{IND}_1(\Pi, A, k) \right\}_{k \in \mathbb{N}}$$

It follows from a straight-forward hybrid argument that semantic security implies indistinguishability of multiple encryptions under independently chosen keys:

Proposition 1. *Let* $\Pi = (\text{Gen}, \text{Enc}, \text{Dec})$ *be a semantically secure encryption scheme and let the random variable* $\text{mIND}_b(\Pi, A, k, \ell)$, *where $b \in \{0,1\}$, $A = (A_1, A_2)$ are ppt algorithms and $k \in \mathbb{N}$, denote the result of the following probabilistic experiment:*

$\text{mIND}_b(\Pi, A, k, \ell)$:
   For $i = 1, \ldots, \ell$: $(\text{PK}_i, \text{SK}_i) \leftarrow \text{Gen}(1^k)$
   $(\langle m_0^1, \ldots, m_0^\ell \rangle, \langle m_1^1, \ldots, m_1^\ell \rangle, \text{STATE}_A) \leftarrow A_1(\langle \text{PK}_1, \ldots, \text{PK}_\ell \rangle)$
   s.t. $|m_0^1| = |m_1^1| = \cdots = |m_0^\ell| = |m_1^\ell|$
   For $i = 1, \ldots, \ell$: $y_i \leftarrow \text{Enc}_{\text{PK}_i}(m_b^i)$
   $D \leftarrow A_2(y_1, \ldots, y_\ell, \text{STATE}_A)$
   Output D

then for any ppt algorithms $A = (A_1, A_2)$ and for any polynomial $p(k)$ the following two ensembles are computationally indistinguishable:

$$\left\{ \text{mIND}_0(\Pi, A, k, p(k)) \right\}_{k \in N} \stackrel{c}{\approx} \left\{ \text{mIND}_1(\Pi, A, k, p(k)) \right\}_{k \in N}$$

2.2   Non-malleable Encryption

Definition 3 (Non-malleable Encryption [PSV06]). *Let* $\Pi = (\text{Gen}, \text{Enc}, \text{Dec})$ *be an encryption scheme and let the random variable* $\text{NME}_b(\Pi, A, k, \ell)$ *where* $b \in \{0, 1\}$, $A = (A_1, A_2)$ *are ppt algorithms and* $k, \ell \in \mathbb{N}$ *denote the result of the following probabilistic experiment:*

$$
\begin{aligned}
&\text{NME}_b(\Pi, A, k, \ell): \\
&\quad (\text{PK}, \text{SK}) \leftarrow \text{Gen}(1^k) \\
&\quad (m_0, m_1, \text{STATE}_A) \leftarrow A_1(\text{PK}) \text{ s.t. } |m_0| = |m_1| \\
&\quad y \leftarrow \text{Enc}_{\text{PK}}(m_b) \\
&\quad (\psi_1, \dots, \psi_\ell) \leftarrow A_2(y, \text{STATE}_A) \\
&\quad \textit{Output } (d_1, \dots, d_\ell) \textit{ where } d_i = \begin{cases} \perp & \textit{if } \psi_i = y \\ \text{Dec}_{\text{SK}}(\psi_i) & \textit{otherwise} \end{cases}
\end{aligned}
$$

$(\text{Gen}, \text{Enc}, \text{Dec})$ *is* non-malleable under a chosen plaintext (CPA) attack *if for any ppt algorithms* $A = (A_1, A_2)$ *and for any polynomial* $p(k)$, *the following two ensembles are computationally indistinguishable:*

$$
\left\{ \text{NME}_0(\Pi, A, k, p(k)) \right\}_{k \in \mathbb{N}} \stackrel{c}{\approx} \left\{ \text{NME}_1(\Pi, A, k, p(k)) \right\}_{k \in \mathbb{N}}
$$

It was shown in [PSV06] that an encryption that is non-malleable (under Definition 3) remains non-malleable even if the adversary A_2 receives several encryptions under many different public keys (the formal experiment is the analogue of mIND for non-malleability).

2.3   (Strong) One-Time Signature Schemes

Informally, a (strong) one-time signature scheme (GenSig, Sign, VerSig) is an existentially unforgeable signature scheme, with the restriction that the signer signs at most one message with any key. This means that an efficient adversary, upon seeing a signature on a message m of his choice, cannot generate a valid signature on a different message, or a different valid signature on the same message m. Such schemes can be constructed in a black-box way from one-way functions [R90, L79], and thus from any semantically secure encryption scheme (Gen, Enc, Dec) using black-box access only to Gen.

3   Construction

Given an encryption scheme $E = (\text{Gen}, \text{Enc}, \text{Dec})$, we construct a new encryption scheme $\Pi = (\text{NMGen}^{\text{Gen}}, \text{NMEnc}^{\text{Gen},\text{Enc}}, \text{NMDec}^{\text{Gen},\text{Dec}})$, summarized in Figure 2, and described as follows.

Polynomial encoding. We identify $\{0, 1\}^n$ with the field $\text{GF}(2^n)$. To encode a message $m \in \{0, 1\}^n$, we pick a random degree k polynomial p over $\text{GF}(2^n)$ such that $p(0) = m$ and construct a $k \times 10k$ matrix such that the i'th column of the matrix comprise entirely of the value $s_i = p(i)$ (where $0, 1, \dots, 10k$ are the lexicographically first

$10k + 1$ elements in $\mathrm{GF}(2^n)$ according to some canonical encoding). Note that $(s_1, \ldots, s_{10k})$ is both a $(k+1)$-out-of-$10k$ secret-sharing of m using Shamir's secret-sharing scheme and a codeword of the Reed-Solomon code $\mathcal{W}$, where

$$\mathcal{W} = \{ \, (p(1), \ldots, p(10k)) \mid p \text{ is a degree } k \text{ polynomial} \, \}.$$

Note that $\mathcal{W}$ is a code over the alphabet $\{0, 1\}^n$ with minimum relative distance 0.9, which means we may efficiently correct up to 0.45 fraction errors using the Berlekamp-Welch algorithm. [tm: add reference]

Encryption. The public key for Π comprises $20k^2$ public keys E indexed by a triplet $(i, j, b) \in [k] \times [10k] \times \{0, 1\}$; there are two keys corresponding to each entry of a $k \times 10k$ matrix. To encrypt a message m, we (a) compute $(s_1, \ldots, s_{10k})$ as in the above-mentioned polynomial encoding, (b) generate (SKSIG, VKSIG) for a one-time signature, (c) compute a $k \times 10k$ matrix $c = (c_{i,j})$ of ciphertexts where $c_{i,j} = \mathsf{Enc}_{\mathrm{PK}_{i,j}^{v_i}}(s_j)$, and (d) sign c using SKSIG.

$$
\begin{pmatrix}
\mathsf{Enc}_{\mathrm{PK}_{1,1}^{v_1}}(s_1) & \mathsf{Enc}_{\mathrm{PK}_{1,2}^{v_1}}(s_2) & \cdots & \mathsf{Enc}_{\mathrm{PK}_{1,10k}^{v_1}}(s_{10k}) \\
\mathsf{Enc}_{\mathrm{PK}_{2,1}^{v_2}}(s_1) & \mathsf{Enc}_{\mathrm{PK}_{2,2}^{v_2}}(s_2) & \cdots & \mathsf{Enc}_{\mathrm{PK}_{2,10k}^{v_2}}(s_{10k}) \\
\vdots & \vdots & \ddots & \vdots \\
\mathsf{Enc}_{\mathrm{PK}_{k,1}^{v_k}}(s_1) & \mathsf{Enc}_{\mathrm{PK}_{k,2}^{v_k}}(s_2) & \cdots & \mathsf{Enc}_{\mathrm{PK}_{k,10k}^{v_k}}(s_{10k})
\end{pmatrix}
$$

Consistency Checks. A valid ciphertext in Π satisfies two properties: (1) the first row is an encryption of a codeword in $\mathcal{W}$ and (2) every column comprises k encryptions of the same plaintext. We want to design consistency checks that reject ciphertexts that are "far" from being valid ciphertexts under Π. For simplicity, we will describe the consistency checks as applied to the underlying matrix of plaintexts. The checks depend on a random subset S of k columns chosen during key generation.

> COLUMN CHECK (column-check): We check that each of the k columns in
> S comprises entirely of the same value.
>
> CODEWORD CHECK (codeword-check): We find a codeword w that agrees
> with the first row of the matrix in at least $9k$ positions; the check fails if no
> such w exists. Then we check that the first row of the matrix agrees with
> w at the k positions indexed by S.

The codeword check ensures that with high probability, the first row of the matrix agrees with w in at least $10k - o(k)$ positions. We explain its significance after describing the alternative decryption algorithm in the analysis.

Decryption. To decrypt, we (a) verify the signature and run both consistency checks, and (b) if all three checks accept, decode the codeword w and output the result, otherwise output $\perp$. Note that to decrypt we only need the $20k$ secret keys corresponding to the first row of the matrix and $2k$ secret keys corresponding to each of the k columns in S.

NMGen(1^k):
1. For $i \in [k], j \in [10k], b \in \{0,1\}$, run Gen($1^k$) to generate key-pairs $(\text{PK}_{i,j}^b, \text{SK}_{i,j}^b)$.
2. Pick a random subset $S \subset [10k]$ of size k.

Set PK $= \left\{ (\text{PK}_{i,j}^0, \text{PK}_{i,j}^1) \mid i \in [k], j \in [10k] \right\}$ and SK $= \left\{ S, (\text{SK}_{i,j}^0, \text{SK}_{i,j}^1) \mid i \in [k], j \in [10k] \right\}$.

NMEnc$_{\text{PK}}(m)$:
1. Pick random $\alpha_1, \ldots, \alpha_k \in \text{GF}(2^n)$ and set $s_j = p(j), j \in [10k]$ where $p(x) = m_0 + \alpha_1 x + \ldots + \alpha_k x^k$.
2. Run GenSig(1^k) to generate (SKSIG, VKSIG). Let $(v_1, \ldots, v_k)$ be the binary representation of VKSIG.
3. Compute the ciphertext $c_{i,j} \leftarrow \text{Enc}_{\text{PK}_{i,j}^{v_i}}(s_j)$, for $i \in [k], j \in [10k]$.
4. Compute the signature $\sigma \leftarrow \text{Sign}_{\text{SKSIG}}(c)$ where $c = (c_{i,j})$.
Output the tuple $[c, \text{VKSIG}, \sigma]$.

NMDec$_{\text{SK}}([c, \text{VKSIG}, \sigma])$:
1. (sig-check) Verify the signature with $\text{VerSig}_{\text{VKSIG}}[c, \sigma]$.
2. Let $c = (c_{i,j})$ and VKSIG $= (v_1, \ldots, v_k)$. Compute $s_j = \text{Dec}_{\text{SK}_{1,j}^{v_1}}(c_{1,j})$, $j = 1, \ldots, 10k$ and the codeword $w = (w_1, \ldots, w_{10k}) \in \mathcal{W}$ that agrees with $(s_1, \ldots, s_{10k})$ in at least $9k$ positions. If no such codeword exists, output $\perp$.
3. (column-check) For all $j \in S$, check that $\text{Dec}_{\text{SK}_{1,j}^{v_1}}(c_{1,j}) = \text{Dec}_{\text{SK}_{2,j}^{v_2}}(c_{2,j}) = \cdots = \text{Dec}_{\text{SK}_{k,j}^{v_k}}(c_{k,j})$.
4. (codeword-check) For all $j \in S$, check that $s_j = w_j$.
If all three checks accept, output the message m corresponding to the codeword w; else, output $\perp$.

Fig. 2. THE NON-MALLEABLE ENCRYPTION SCHEME Π

Note that the decryption algorithm may be stream-lined, for instance, by running the codeword check only if the column check succeeds. We choose to present the algorithm as is in order to keep the analysis simple; in particular, we will run both consistency checks independent of the outcome of the other.

4 Analysis

Having presented our construction, we now formally state and prove our main result:

Theorem 1. (Main Theorem, restated).
Assume there exists an encryption scheme $E = $ (Gen, Enc, Dec) that is semantically secure under a CPA attack. Then there exists an encryption scheme $\Pi = $ (NMGen^{Gen}, NMEnc$^{\text{Gen,Enc}}$, NMDec$^{\text{Gen,Dec}}$) that is non-malleable under a CPA attack.

We establish the theorem (as in [DDN00, PSV06], etc) via a series of hybrid arguments and deduce indistinguishability of the intermediate hybrid experiments from the semantic security of the underlying scheme E under some set of public keys Γ. To do so, we will need to implement an alternative decryption algorithm NMDec* that is used in the intermediate experiments to simulate the actual decryption algorithm NMDec in the non-malleability experiment. We need NMDec* to achieve two conflicting requirements:

- NMDec* and NMDec must agree on essentially all inputs, including possibly malformed ciphertexts;
- We can implement NMDec* without having to know the secret keys corresponding to the public keys in Γ.

Of course, designing NMDec* is difficult precisely because NMDec uses the secret keys corresponding to the public keys in Γ.

Here is a high-level (but extremely inaccurate) description of how NMDec* works: Γ is the set of public keys corresponding to the first row of the $k \times 10k$ matrix. To implement NMDec*, we will decrypt the i'th row of the matrix of ciphertexts, for some $i > 1$, which the column check (if successful) guarantees to agree with the first row in most positions; error correction takes care of the tiny fraction of disagreements.

4.1   Alternative Decryption Algorithm NMDec*

Let $\text{VKSIG}^* = (v_1^*, \ldots, v_k^*)$ denote the verification key in the challenge ciphertext given to the adversary in the non-malleability experiment, and let $\text{VKSIG} = (v_1, \ldots, v_k)$ denote the verification key in (one of) the ciphertext(s) generated by the adversary. First, we modify the signature check to also output $\perp$ if there is a forgery, namely $\text{VKSIG} = \text{VKSIG}^*$. Next, we modify the consistency checks (again, as applied to the underlying matrix of plaintexts) as follows:

COLUMN CHECK (`column-check*`): This is exactly as before, we check that the each of the k columns in S comprises entirely of the same value.

CODEWORD CHECK (`codeword-check*`): Let i be the smallest value such that $v_i \neq v_i^*$ (which exists because $\text{VKSIG} \neq \text{VKSIG}^*$). We find a codeword w that agrees with the i'th row of the matrix in at least $8k$ positions (note agreement threshold is smaller than before); the check fails if so such w exists. Then we check that the first row of the matrix agrees with w at the k positions indexed by S.

To decrypt, run the modified signature and consistency checks, and if all three checks accept, decode the codeword w and output the result, otherwise output $\perp$. To implement the modified consistency checks and decryption algorithm, we only need the $10k$ secret keys indexed by $\overline{\text{VKSIG}^*}$ for each row of the matrix, and as before, the $2k$ secret keys corresponding to each of the k columns in S.

Remark on the Codeword Check. At first, the codeword check may seem superfluous. Suppose we omit the codeword check, and as before, define w to be a codeword that agrees with the first row in $9k$ positions and with the i'th row in $8k$ positions in the respective decryption algorithms; the gap is necessary to take into account inconsistencies not detected by the column check. Now, consider a malformed ciphertext ψ for Π where in the underlying matrix of plaintexts, each row is the same corrupted codeword that agrees with a valid codeword in exactly $8.5k$ positions. Without the codeword checks, ψ will be an invalid ciphertext according to NMDec and a valid ciphertext according to NMDec* and can be used to distinguish the intermediate hybrid distributions in the analysis; with the codeword checks, ψ is an invalid ciphertext according to both. It is also easy to construct a problematic malformed ciphertext for the case where both agreement thresholds are set to the same value (say $9k$).

4.2   A Promise Problem

Recall the guarantees we would like from NMDec and NMDec*:

- On input a ciphertext that is an encryption of a message m under Π, both NMDec and NMDec* will output m with probability 1.
- On input a ciphertext that is "close" to an encryption of a message m under Π, both NMDec and NMDec* will output m with the same probability (the exact probability is immaterial) and $\bot$ otherwise.
- On input a ciphertext that is "far" from any encryption, then both NMDec and NMDec* output $\bot$ with high probability.

To quantify and establish these guarantees, we consider the following promise problem (Π_Y, Π_N) that again refers to the underlying matrix of plaintexts. An instance is a matrix of k by $10k$ values in $\{0, 1\}^n \cup \bot$.

Π_Y (YES instances) — for some $w \in W$, every row equals w.

Π_N (NO instances) — either there exist two rows that are 0.1-far (i.e. disagree in at least k positions), or the first row is 0.1-far from every codeword in W (i.e. disagree with every codeword in at least k positions).

Valid encryptions correspond to the YES instances, while NO instances will correspond to "far" ciphertexts. To analyze the success probability of an adversary, we examine each ciphertext ψ it outputs with some underlying matrix M of plaintexts (which may be a YES or a NO instance or neither) and show that both NMDec and NMDec* agree on ψ with high probability. To facilitate the analysis, we consider two cases:

- If $M \in \Pi_N$, then it fails the column/codeword checks in both decryption algorithms with high probability, in which case both decryption algorithms output $\bot$. Specifically, if there are two rows that are 0.1-far, then column check rejects M with probability $1 - 0.9^k$. On the other hand, if the first row is 0.1-far from every codeword, then the codeword check in NMDec rejects M with probability 1 and that in NMDec* rejects M with probability at least $1 - 0.9^k$; that is, with probability $1 - 0.9^k$, both codeword checks in NMDec and NMDec* rejects M.

- If $M \notin \Pi_N$, then both decryption algorithms always output the same answer for all choices of the set S, provided there is no forgery. Fix $M \notin \Pi_N$ and a set S. The first row is 0.9-close to codeword $w \in \mathcal{W}$ and we know in addition that every other row is 0.9-close to the first row and thus 0.8-close to w. Therefore, we will recover the same codeword w and message m whether we decode the first row within distance 0.1, or any other row within distance 0.2. This means that the codeword checks in both decryption algorithms compare the first row with the same codeword w. As such, both decryption algorithms output $\perp$ with exactly the same probability, and whenever they do not output $\perp$, they output the same message m.

4.3 Proof of Main Theorem

In the hybrid argument, we consider the following variants of NME_b as applied to Π, where VKSIG^* denotes the verification key in the ciphertext $y = \mathsf{NMEnc}_{\mathrm{PK}}(m_b)$:

Experiment $\mathsf{NME}_b^{(1)}$ — $\mathsf{NME}_b^{(1)}$ proceeds exactly like NME_b, except we replace sig-check in NMDec with sig-check*:

> (sig-check*) Verify the signature with $\mathsf{VerSig}_{\mathrm{VKSIG}}[c, \sigma]$. Output $\perp$ if the signature fails to verify or if $\mathrm{VKSIG} = \mathrm{VKSIG}^*$.

Experiment $\mathsf{NME}_b^{(2)}$ — $\mathsf{NME}_b^{(2)}$ proceeds exactly like NME_b except we replace NMDec with NMDec*:

> $\mathsf{NMDec}^*_{\mathrm{SK}}([c, \mathrm{VKSIG}, \sigma])$:
> 1. (sig-check*) Verify the signature with $\mathsf{VerSig}_{\mathrm{VKSIG}}[c, \sigma]$. Output $\perp$ if the signature fails to verify or if $\mathrm{VKSIG} = \mathrm{VKSIG}^*$.
> 2. Let $c = (c_{i,j})$ and $\mathrm{VKSIG} = (v_1, \ldots, v_k)$. Let i be the smallest value such that $v_i \neq v_i^*$. Compute $s_j = \mathsf{Dec}_{\mathrm{SK}_{i,j}^{v_i}}(c_{i,j})$, $j = 1, \ldots, 10k$ and $w = (w_1, \ldots, w_{10k}) \in \mathcal{W}$ that agrees with $(s_1, \ldots, s_{10k})$ in at least $8k$ positions. If no such codeword exists, output $\perp$.
> 3. (column-check*) For all $j \in S$, check that $\mathsf{Dec}_{\mathrm{SK}_{1,j}^{v_1}}(c_{1,j}) = \mathsf{Dec}_{\mathrm{SK}_{2,j}^{v_2}}(c_{2,j})$
> $= \cdots = \mathsf{Dec}_{\mathrm{SK}_{k,j}^{v_k}}(c_{k,j})$.
> 4. (codeword-check*) For all $j \in S$, check that $\mathsf{Dec}_{\mathrm{SK}_{1,j}^{v_1}}(c_{1,j}) = w_j$.
> If all three checks accept, output the message m corresponding to the codeword w; else, output $\perp$.

Claim. For $b \in \{0, 1\}$, we have $\left\{ \mathsf{NME}_b(\Pi, A, k, p(k)) \right\} \overset{c}{\approx} \left\{ \mathsf{NME}_b^{(1)}(\Pi, A, k, p(k)) \right\}$

Proof. This follows readily from the security of the signature scheme.   □

Claim. For $b \in \{0, 1\}$, we have $\left\{ \mathsf{NME}_b^{(1)}(\Pi, A, k, p(k)) \right\} \overset{s}{\approx} \left\{ \mathsf{NME}_b^{(2)}(\Pi, A, k, p(k)) \right\}$

Proof. We will show that both distributions are statistically close for all possible coin tosses in both experiments (specifically, those of NMGen, A and NMEnc) except for the choice of S in NMGen. Once we fix all the coin tosses apart from the choice of S, the output $(\psi_1, \ldots, \psi_{p(k)})$ of A_2 are completely determined and identical in both experiments. We claim that with probability $1 - 2p(k) \cdot 0.9^k = 1 - \text{neg}(k)$ over the choice of S, the decryptions of $(\psi_1, \ldots, \psi_{p(k)})$ agree in both experiments. This follows from the analysis of the promise problem in Section 4.2. $\qquad\square$

Claim. For every ppt machine A, there exists a ppt machine B such that for $b \in \{0,1\}$,

$$\left\{ \text{NME}_b^{(2)}(\Pi, A, k, p(k)) \right\} \equiv \left\{ \text{mIND}_b(E, B, k, 9k^2) \right\}$$

Proof. The machine B is constructed as follows: B participates in the experiment mIND_b (the "outside") while internally simulating $A = (A_1, A_2)$ in the experiment $\text{NME}_b^{(2)}$.

- (pre-processing) Pick a random subset $S = \{u_1, \ldots, u_j\}$ of $[10k]$ and run $\text{GenSig}(1^k)$ to generate $(\text{SKSIG}^*, \text{VKSIG}^*)$ and set $(v_1^*, \ldots, v_k^*) = \text{VKSIG}^*$. Let ϕ be a bijection identifying $\{(i,j) \mid i \in [k], j \in [10k] \setminus S\}$ with $[9k^2]$.
- (key generation) B receives $\langle \text{PK}_1, \ldots, \text{PK}_{9k^2} \rangle$ from the outside and simulates NMGen as follows: for all $i \in [k], j \in [10k], \beta \in \{0,1\}$,

$$(\text{PK}_{i,j}^\beta, \text{SK}_{i,j}^\beta) = \begin{cases} (\text{PK}_{\phi(i,j)}, \bot) & \text{if } \beta = v_i^* \text{ and } j \notin S \\ \text{Gen}(1^k) & \text{otherwise} \end{cases}$$

- (message selection) Let (m_0, m_1) be the pair of messages A_1 returns. B then chooses k random values $(\gamma_{u_1}, \ldots, \gamma_{u_k}) \in \{0,1\}^n$ and computes two degree k polynomials p_0, p_1 where p_β interpolates the $k + 1$ points $(0, m_\beta), (u_1, \gamma_{u_1})$, $\ldots, (u_k, \gamma_{u_k})$ for $\beta \in \{0,1\}$. B sets $m_\beta^{\phi(i,j)} = p_\beta(j)$, for $i \in [k], j \in [10k] \setminus S$ and forwards $(\langle m_0^1, \ldots, m_0^{9k^2} \rangle, \langle m_1^1, \ldots, m_1^{9k^2} \rangle)$ to the outside.
- (ciphertext generation) B receives $\langle y_1, \ldots, y_{9k^2} \rangle$ from the outside (according to the distribution $\text{Enc}_{\text{PK}_1}(m_b^1), \ldots, \text{Enc}_{\text{PK}_{9k^2}}(m_b^{9k^2})$) and generates a ciphertext $[c, \text{VKSIG}^*, \sigma]$ as follows:

$$c_{i,j} = \begin{cases} y_{\phi(i,j)} & \text{if } j \notin S \\ \text{Enc}_{\text{PK}_{i,j}^{v_i^*}}(\gamma_j) & \text{otherwise} \end{cases}$$

B then computes the signature $\sigma \leftarrow \text{Sign}_{\text{SKSIG}^*}(c)$ and forwards $[c, \text{VKSIG}^*, \sigma]$ to A_2. It is straight-forward to verify that $[c, \text{VKSIG}^*, \sigma]$ is indeed a random encryption of m_b under Π.
- (decryption) Upon receiving a sequence of ciphertexts $(\psi_1, \ldots, \psi_{p(k)})$ from A_2, B decrypts these ciphertexts using NMDec^* as in $\text{NME}_b^{(2)}$. Note that to simulate NMDec^*, it suffices for B to possess the secret keys $\{\text{SK}_{i,j}^\beta \mid \beta = 1 - v_i^* \text{ or } j \in S\}$, which B generated by itself. $\qquad\square$

Combining the three claims, we conclude that for every ppt adversary A, there is a ppt adversary B such that for $b \in \{0, 1\}$,

$$\left\{ \mathsf{NME}_b(\Pi, A, k, p(k)) \right\} \stackrel{c}{\approx} \left\{ \mathsf{NME}_b^{(1)}(\Pi, A, k, p(k)) \right\}$$

$$\stackrel{s}{\approx} \left\{ \mathsf{NME}_b^{(2)}(\Pi, A, k, p(k)) \right\} \equiv \left\{ \mathsf{mIND}_b(E, B, k, 9k^2) \right\}$$

By Prop 1, $\mathsf{mIND}_0(E, B, k, 9k^2) \stackrel{c}{\approx} \mathsf{mIND}_1(E, B, k, 9k^2)$, which concludes the proof of Theorem 1.

5   Achieving Bounded-CCA2 Non-malleability

We sketch how our scheme may be modified to achieve non-malleability under a bounded-CCA2 attack. Here, we allow the adversary to query Dec at most q times in the non-malleability experiment (but it must not query Dec on y). The modification is the straight-forward analogue of the [CHH+07] modification of the [PSV06] scheme: we increase the number of columns in the matrix from $10k$ to $80(k+q)$, and the degree of the polynomial p and the size of S from k to $8(k+q)$, and propagate the changes accordingly. The analysis is basically as before, except for the following claim (where $\mathsf{NME}\text{-}q\text{-}\mathsf{CCA}_b^{(1)}$, $\mathsf{NME}\text{-}q\text{-}\mathsf{CCA}_b^{(2)}$ are the respective analogues of $\mathsf{NME}_b^{(1)}$, $\mathsf{NME}_b^{(1)}$):

Claim. For $b \in \{0, 1\}$, we have

$$\left\{ \mathsf{NME}\text{-}q\text{-}\mathsf{CCA}_b^{(1)}(\Pi, A, k, p(k)) \right\} \stackrel{s}{\approx} \left\{ \mathsf{NME}\text{-}q\text{-}\mathsf{CCA}_b^{(2)}(\Pi, A, k, p(k)) \right\}$$

Proof (sketch). As before, we will show that both distributions are statistically close for all possible coin tosses in both experiments (specifically, those of NMGen, A and NMEnc) except for the choice of S in NMGen. However, we cannot immediately deduce that the output of A_2 are completely determined and identical in both experiments, since they depend on the adaptively chosen queries to NMDec, and the answers depend on S. Instead, we will consider all 2^q possible computation paths of A which are determined based on the q query/answer pairs from NMDec. For each query, we consider the underlying matrix of plaintexts M:

– If $M \in \Pi_N$, then we assume NMDec returns $\perp$.
– If $M \notin \Pi_N$, then we consider two branches depending on the two possible outcomes of the consistency checks.

We claim that with probability $1 - 2^q \cdot p(k) \cdot 0.9^{8(k+q)} > 1 - \mathsf{neg}(k)$ over the choice of S, the decryptions of $(\psi_1, \ldots, \psi_{p(k)})$ agree in both experiments in all 2^q computation paths. $\square$

Remark on achieving (full) CCA2 security. It should be clear from the preceding analysis that the barrier to obtaining full CCA2 security lies in handling queries outside Π_N. Specifically, with even just a (full) CCA1 attack, an adversary could query NMDec on a series of adaptively chosen ciphertexts corresponding to matrices outside Π_N to learn the set S upon which it could readily break the security of our construction.

Acknowledgments. This work was initiated while the third and fourth authors were visiting IPAM. We would like to thank Vinod Vaikuntanathan for sharing his insights on non-malleability over the last two summers.

References

[BFM88]   Blum, M., Feldman, P., Micali, S.: Non-interactive zero-knowledge and its applications. In: STOC, pp. 103–112 (1988)

[BGW88]   Ben-Or, M., Goldwasser, S., Wigderson, A.: Completeness theorems for non-cryptographic fault-tolerant distributed computation. In: STOC, pp. 1–10 (1988)

[BHSV98] Bellare, M., Halevi, S., Sahai, A., Vadhan, S.P.: Many-to-one trapdoor functions and their relation to public-key cryptosystems. In: Goldwasser, S. (ed.) CRYPTO 1988. LNCS, vol. 403, pp. 283–298. Springer, Heidelberg (1990)

[CHH+07] Cramer, R., Hanaoka, G., Hofheinz, D., Imai, H., Kiltz, E., Pass, R., Shelat, A., Vaikuntanathan, V.: Bounded CCA2-secure encryption. In: Kurosawa, K. (ed.) ASIACRYPT 2007. LNCS, vol. 4833, Springer, Heidelberg (2007)

[CHK04]   Canetti, R., Halevi, S., Katz, J.: Chosen-ciphertext security from identity-based encryption. In: Cachin, C., Camenisch, J.L. (eds.) EUROCRYPT 2004. LNCS, vol. 3027, pp. 207–222. Springer, Heidelberg (2004)

[CS98]   Cramer, R., Shoup, V.: A practical public key cryptosystem provably secure against adaptive chosen ciphertext attack. In: Krawczyk, H. (ed.) CRYPTO 1998. LNCS, vol. 1462, pp. 13–25. Springer, Heidelberg (1998)

[CS04]   Cramer, R., Shoup, V.: Universal hash proofs and a paradigm for adaptive chosen ciphertext secure public-key encryption. In: Cachin, C., Camenisch, J.L. (eds.) EUROCRYPT 2004. LNCS, vol. 3027, pp. 45–64. Springer, Heidelberg (2004)

[DDN00]   Dolev, D., Dwork, C., Naor, M.: Nonmalleable cryptography. SIAM J. Comput. 30(2), 391–437 (2000)

[ES02]   Elkind, E., Sahai, A.: A unified methodology for constructing public-key encryption schemes secure against adaptive chosen ciphertext attack. Cryptology ePrint Archive, Report, /024, 2002. (2002), http://eprint.iacr.org/

[GM84]   Goldwasser, S., Micali, S.: Probabilistic encryption. J. Comput. Syst. Sci. 28(2), 270–299 (1984)

[GMM07]   Gertner, Y., Malkin, T., Myers, S.: Towards a separation of semantic and CCA security for public key encryption. In: TCC, pp. 434–455 (2007)

[GMR01]   Gertner, Y., Malkin, T., Reingold, O.: On the impossibility of basing trapdoor functions on trapdoor predicates. In: FOCS, pp. 126–135 (2001)

[GMW87]   Goldreich, O., Micali, S., Wigderson, A.: How to play any mental game or a completeness theorem for protocols with honest majority. In: STOC, pp. 218–229 (1987)

[H08]   Haitner, I.: Semi-Honest to Malicious Oblivious Transfer - The Black-Box Way. In: These proceedings (2008)

[IKLP06] Ishai, Y., Kushilevitz, E., Lindell, Y., Petrank, E.: Black-box constructions for secure computation. In: STOC, pp. 99–108 (2006)

[IR89]   Impagliazzo, R., Rudich, S.: Limits on the provable consequences of one-way permutations. In: STOC, pp. 44–61 (1989)

[L79]   Lamport, L.: Constructing digital signatures from a one-way function. Technical Report SRI-CSL-98, SRI International Computer Science Laboratory (1979)

[L06]   Lindell, Y.: A simpler construction of CCA2-secure public-key encryption under general assumptions. J. Cryptology 19(3), 359–377 (2006)

[LP07]   Lindell, Y., Pinkas, B.: An efficient protocol for secure two-party computation in the presence of malicious adversaries. In: Naor, M. (ed.) EUROCRYPT 2007. LNCS, vol. 4515, pp. 52–78. Springer, Heidelberg (2007)

[NY90]   Naor, M., Yung, M.: Public-key cryptosystems provably secure against chosen ciphertext attacks. In: STOC, pp. 427–437 (1990)

[PSV06]   Pass, R., Shelat, A., Vaikuntanathan, V.: Construction of a non-malleable encryption scheme from any semantically secure one. In: Dwork, C. (ed.) CRYPTO 2006. LNCS, vol. 4117, pp. 271–289. Springer, Heidelberg (2006)

[PW07]   Peikert, C., Waters, B.: Lossy trapdoor functions and their applications. Cryptology ePrint Archive, Report 2007/279 (2007), http://eprint.iacr.org/

[R90]   Rompel, J.: One-way functions are necessary and sufficient for secure signatures. In: STOC, pp. 387–394 (1990)

[RS91]   Rackoff, C., Simon, D.R.: Non-interactive zero-knowledge proof of knowledge and chosen ciphertext attack. In: Feigenbaum, J. (ed.) CRYPTO 1991. LNCS, vol. 576, pp. 433–444. Springer, Heidelberg (1992)

[RTV04]   Reingold, O., Trevisan, L., Vadhan, S.: Notions of reducibility between cryptographic primitives. In: TCC, pp. 1–20 (2004)

[S99]   Sahai, A.: Non-malleable non-interactive zero knowledge and adaptive chosen-ciphertext security. In: FOCS, pp. 543–553 (1999)

A Linear Lower Bound on the Communication Complexity of Single-Server Private Information Retrieval*

Iftach Haitner, Jonathan J. Hoch, and Gil Segev

Department of Computer Science and Applied Mathematics,
Weizmann Institute of Science, Rehovot 76100, Israel
{iftach.haitner,yaakov.hoch,gil.segev}@weizmann.ac.il

Abstract. We study the communication complexity of single-server Private Information Retrieval (PIR) protocols that are based on fundamental cryptographic primitives in a black-box manner. In this setting, we establish a tight lower bound on the number of bits communicated by the server in any polynomially-preserving construction that relies on trapdoor permutations. More specifically, our main result states that in such constructions $\Omega(n)$ bits must be communicated by the server, where n is the size of the server's database, and this improves the $\Omega(n/\log n)$ lower bound due to Haitner, Hoch, Reingold and Segev (FOCS '07). Therefore, in the setting under consideration, the naive solution in which the user downloads the entire database turns out to be optimal up to constant multiplicative factors. We note that the lower bound we establish holds for the most generic form of trapdoor permutations, including in particular enhanced trapdoor permutations.

Technically speaking, this paper consists of two main contributions from which our lower bound is obtained. First, we derive a tight lower bound on the number of bits communicated by the sender during the commit stage of any black-box construction of a statistically-hiding bit-commitment scheme from a family of trapdoor permutations. This lower bound asymptotically matches the upper bound provided by the scheme of Naor, Ostrovsky, Venkatesan and Yung (CRYPTO '92). Second, we improve the efficiency of the reduction of statistically-hiding commitment schemes to low-communication single-server PIR, due to Beimel, Ishai, Kushilevitz and Malkin (STOC '99). In particular, we present a reduction that essentially preserves the communication complexity of the underlying single-server PIR protocol.

1 Introduction

A single-server Private Information Retrieval (PIR) scheme is a protocol between a server and a user. The server holds a database $x \in \{0,1\}^n$ and the user holds an index $i \in [n]$ to an entry of the database. Informally, the user wishes to retrieve the ith entry of the database, without revealing the index i to the server. The

---

* Due to space limitations a more complete version is available as [19].

R. Canetti (Ed.): TCC 2008, LNCS 4948, pp. 445–464, 2008.
© International Association for Cryptologic Research 2008

notion of PIR was introduced by Chor, Goldreich, Kushilevitz and Sudan [4] to model applications that enable users to query public databases without revealing any information on the specific data that the users wish to retrieve. Chor et al. showed that in the information-theoretic setting any single-server PIR protocol has the server communicating at least n bits. Therefore in this setting the naive solution in which the user downloads the entire database is optimal.

Kushilevitz and Ostrovsky [26] were the first to construct a non-trivial single-server PIR protocol relying on computational assumptions. Their result initiated a sequence of papers showing that there exist single-server PIR protocols with poly-logarithmic communication complexity based on *specific* number-theoretic assumptions (see, for example, [2,3,12,26,28,40], and a recent survey by Ostrovsky and Skeith [35]). The only non-trivial construction based on *general* computational assumptions is due to Kushilevitz and Ostrovsky [27], and is based on enhanced trapdoor permutations. In their construction, however, the server is required to communicate $n - o(n)$ bits to the user.

Motivated by this ever-growing line of work, we study the communication complexity of single-server PIR protocols that are based on fundamental primitives. We establish a linear lower bound on the number of bits communicated by the server in constructions that rely on enhanced trapdoor permutations in a black-box manner. Therefore, in the setting under consideration in this paper, the naive solution in which the user downloads the entire database turns out to be optimal up to constant multiplicative factors. In the following paragraphs, we briefly describe the setting in which our lower bound is proved (a more formal description is provided in Section 2).

Black-box reductions. As previously mentioned, under widely believed specific number-theoretic assumptions, there are very efficient single-server PIR protocols. Therefore, if any of these assumptions holds, the existence of trapdoor permutations implies the existence of efficient single-server PIR protocols in a trivial sense. Faced with similar difficulties, Impagliazzo and Rudich [22] presented a paradigm for proving impossibility results under a restricted, yet very natural and important, subclass of reductions called *black-box reductions*. Informally, a black-box reduction of a primitive P to a primitive Q is a construction of P out of Q that ignores the internal structure of the implementation of Q and uses it as a "subroutine" (i.e., as a black-box). In addition, in the case of fully-black-box reductions (see, for example, [36]), the proof of security (showing that an adversary that breaks the implementation of P implies an adversary that breaks the implementation of Q), is black-box as well, that is, the internal structure of the adversary that breaks the implementation of P is ignored.

The strength of cryptographic reductions. Luby [30] provides a classification of the strength of cryptographic reductions into three classes: linearly-preserving, polynomially-preserving and weakly-preserving. In our setting, this classification comes into play when comparing the size of the server's database and the domain of the trapdoor permutations. Very informally, a reduction of single-server PIR for an n-bit database to a family of trapdoor permutations is linearly-preserving or polynomially-preserving if it uses trapdoor permutations over $\Omega(n)$ bits. Such a reduction is weakly-preserving if it uses trapdoor permutations over $\Omega(n^\epsilon)$ bits for some constant $0 < \epsilon \leq 1$. In linearly-preserving and polynomially-preserving reductions we are guaranteed that breaking the constructed primitive is essentially as hard as breaking the underlying primitive. However, in weakly-preserving reductions, we are only guaranteed that breaking the constructed primitive is as hard as breaking the underlying primitive for polynomially smaller security parameters. We refer the reader to [30] for a more comprehensive and complete discussion.

1.1 Related Work

Single-server PIR is one of the fundamental primitives in the foundations of cryptography. For example, non-trivial single-server PIR was shown to imply the existence of Oblivious Transfer protocols [5], and 2-move low-communication single-server PIR was shown to imply collision-resistant hash functions [23]. Single-server PIR was also shown to be tightly related to several other aspects of cryptography and complexity theory (see, for example, [6,20,24]). We note that it is far beyond the scope of this paper to present an exhaustive overview of the ever-growing line of work on single-server PIR, and we refer the reader to the recent survey of Ostrovsky and Skeith [35] for a more comprehensive discussion.

In the context of black-box reductions, Impagliazzo and Rudich [22] showed that there are no black-box reductions of key-agrement protocols to one-way permutations, and substantial additional work in this line followed (see, for example, [7,13,14,37,38]). Kim, Simon and Tetali [25] initiated a new line of impossibility results, by providing a lower bound on the *efficiency* of black-box reductions (rather than on their feasibility). They proved a lower bound on the efficiency, in terms of the number of calls to the underlying primitive, of any black-box reduction of universal one-way hash functions to one-way permutations. This result was later improved, to match the known upper bound, by Gennaro and Trevisan [11], which together with Gennaro et al. [8,9] provided tight lower bounds on the efficiency of several other black-box reductions. Building upon the technique developed by [11], Horvitz and Katz [21] provided lower bounds on the efficiency of black-box reductions of statistically-hiding and computationally-binding commitment schemes to one-way permutations. In the above results the measure of efficiency under consideration is the number of calls to the underlying primitives.

Di Crescenzo, Malkin and Ostrovsky [5] showed that any single-server PIR protocol in which the server communicates at most $n-1$ bits (where n is the size of the server's database) can be transformed in a fully-black-box manner to an Oblivious Transfer protocol. Gennaro, Lindell and Malkin [10] (refining Gertner et al. [13]) ruled out any black-box reduction of Oblivious Transfer to plain (i.e., non-enhanced) trapdoor permutations. The combination of these two results yields that there are no non-trivial black-box constructions of single-server PIR from non-enhanced trapdoor permutations. We note that although in this paper we rule out a more restricted class of constructions (that is, the class of fully-black-box constructions), our result holds for the most generic form of trapdoor permutations, including in particular enhanced trapdoor permutations.

Very recently, Haitner et al. [18], improving upon the work of Wee [41], proved that any polynomially-preserving fully-black-box reduction of a statistically-hiding bit-commitment scheme to trapdoor permutations has $\Omega(n/\log n)$ communication rounds (where n is the security parameter). As a corollary, they showed that any polynomially-preserving fully-black-box reduction of single-server PIR to trapdoor permutations has $\Omega(n/\log n)$ communication rounds, where n is the size of the server's database. In particular, the server is required to communicate $\Omega(n/\log n)$ bits to the user. Haitner et al. also established similar lower bounds on the communication complexity of Oblivious Transfer that guarantees statistical security for one of the parties and for Interactive Hashing.

In a slightly different setting, Ostrovsky and Skeith [34] proved a lower bound on the communication complexity of single-server PIR protocols with certain algebraic properties. For a class of PIR protocols, referred to as *abelian group algebraic PIR protocols*, with user-side communication complexity $g(n)$ and server-side communication complexity $h(n)$ they proved that $g(n)h(n) = \Omega(n)$.

1.2   Our Results

We study the class of black-box constructions of single-server PIR from trapdoor permutations, and establish a tight lower bound on the number of bits communicated by the server in such constructions. Our main result is the following:

Main Theorem (Informal). In any polynomially-preserving fully-black-box construction of a single-server PIR protocol from trapdoor permutations the server communicates $\Omega(n)$ bits, where n is the size of the server's database.

As mentioned above, the combination of the results of Di Crescenzo et al. [5] and of Gennaro et al. [10] rules out the more general class of black-box reductions of single-server PIR with $n - 1$ bits of communication to trapdoor permutations. This result, however, does not apply to enhanced trapdoor permutations. We note that our lower bound holds for the most generic form of trapdoor permutations, and in particular for enhanced trapdoor permutations.[1]

In addition, we note that our lower bound holds only for constructions which are polynomially-preserving. The construction of Kushilevitz and Ostrovsky [27], which is based on enhanced trapdoor permutations in a fully-black-box manner and in which the server communicates $n - o(n)$ bits, is only weakly-preserving (i.e., it is significantly easier to break their protocol than to break the security of the underlying family of trapdoor permutations [2]). Thus, the question of whether a tight linear lower bound can be established for weakly-preserving constructions as well remains open.

---

[1] Note that enhanced trapdoor permutations are, seemingly, stronger than plain trapdoor permutations. Therefore, although our result is weaker in terms of the class of reductions and the bound on the communication complexity, it provides the first evidence that enhanced trapdoor permutations are not sufficient to construct single-server PIR with sublinear communication (at least from a black-box perspective).

[2] Though the security guarantees of the two primitives are still polynomially-related.

The main technical contributions. This paper consists of two main contributions from which our lower bound is immediately obtained. First, we derive a tight lower bound on the communication complexity of black-box constructions of statistically-hiding bit-commitment schemes from trapdoor permutations. Very recently, Haitner et al. [18] proved that any polynomially-preserving fully-black-box construction of statistically-hiding bit-commitment scheme from a family of trapdoor permutations has $\Omega(n/\log n)$ communication rounds, where n is the security parameter of the scheme. In particular, this implies a lower bound on the number of bits communicated by the sender. In this paper we manage to improve their lower bound and prove the following theorem:

Theorem (Informal) 1.1. *In any polynomially-preserving fully-black-box construction of a statistically-hiding bit-commitment scheme from a family of trapdoor permutations the sender communicates $\Omega(n)$ bits during the commit stage, where n is the security parameter of the scheme.*

This lower bound asymptotically matches the upper bound given by the statistically-hiding commitment scheme of Naor et al. [31]. In addition, we improve the efficiency of the reduction of statistically-hiding commitment schemes to single-server PIR, presented by Beimel et al. [1]. Our reduction essentially uses the reduction of Beimel et al. instantiated with a better extractor, which enables us to preserve the communication complexity of the underlying single-server PIR protocol. As stating this result turns out to involve subtle technical details, here we only state a very informal statement:

Theorem (Informal) 1.2. *There is a linearly-preserving fully-black-box reduction of statistically-hiding commitment schemes to low-communication single-server PIR, which essentially preserves the communication complexity of the underlying single-server PIR protocol.*

Paper organization. In Section 2 we briefly present the notations and formal definitions used in this paper. In Section 3 we prove a tight lower bound on the number of bits communicated by the sender during the commit stage of statistically-hiding commitment schemes. In Section 4 we describe an improved reduction of statistically-hiding commitment schemes to single-server PIR. Finally, in Section 5 we provide some concluding remarks.

2   Preliminaries

We denote by Π_n the set of all permutations over $\{0,1\}^n$. For an integer n, we denote by U_n the uniform distribution over the set $\{0,1\}^n$. For a finite set X, we denote by $x \leftarrow X$ the experiment of choosing an element of X according to the uniform distribution. Similarly, for a distribution $\mathcal{D}$ over a set X, we denote by $x \leftarrow \mathcal{D}$ the experiment of choosing an element of X according to the distribution $\mathcal{D}$. The min-entropy of $\mathcal{D}$ is defined as $H_\infty(\mathcal{D}) = -\log(\max_x \Pr_\mathcal{D}[x])$. The statistical distance between two distributions X and Y over Ω is defined as $\mathrm{SD}(X,Y) = \frac{1}{2}\sum_{\omega \in \Omega}|\Pr_X[\omega] - \Pr_Y[\omega]|$.

Definition 2.1. *A function $E : \{0,1\}^n \times \{0,1\}^d \rightarrow \{0,1\}^m$ is a (k, ϵ)-extractor if for every distribution X over $\{0,1\}^n$ with $H_\infty(X) \geq k$, it holds that the distribution $E(X, U_d)$ is ϵ-close to uniform. Such a function E is a strong (k, ϵ)-extractor if the function $E'(x, y) = y \circ E(x, y)$ is a (k, ϵ)-extractor (where $\circ$ denotes concatenation).*

In our construction of a statistically-hiding commitment scheme from single-server PIR we will be using the following explicit construction of strong extractors, which is obtained as a corollary of [39, Corollary 3.4].

Proposition 2.1. *For any $k \in \omega(\log(n))$, there exists an explicit construction of a strong $(k, 2^{1-k})$-extractor* EXT $: \{0,1\}^n \times \{0,1\}^{3k} \rightarrow \{0,1\}^{k/2}$.

Trapdoor permutations. We briefly present the notion of trapdoor permutations, and refer the reader to [15] for a more comprehensive discussion. A collection of trapdoor permutations is represented by a triplet of the form $\tau = (G, F, F^{-1})$. Informally, G corresponds to a key generation procedure, which is queried on a string td (intended as the "trapdoor") and produces a corresponding public key pk. The procedure F is the actual collection of permutations, which is queried on a public key pk and an input x. Finally, the procedure F^{-1} is the inverse of F: If $G(td) = pk$ and $F(pk, x) = y$, then $F^{-1}(td, y) = x$. In this paper, since we are concerned with providing a lower bound, we do not consider the most general definition of a collection of trapdoor permutations. Instead, we denote by T_n the set of all triplets $\tau_n = (G_n, F_n, F_n^{-1})$ of the following form:

1. $G_n \in \Pi_n$.
2. $F_n : \{0,1\}^n \times \{0,1\}^n \rightarrow \{0,1\}^n$ is a function such that $F_n(pk, \cdot) \in \Pi_n$ for every $pk \in \{0,1\}^n$.
3. $F_n^{-1} : \{0,1\}^n \times \{0,1\}^n \rightarrow \{0,1\}^n$ is a function such that $F_n^{-1}(td, y)$ returns the unique $x \in \{0,1\}^n$ for which $F_n(G_n(td), x) = y$.

Our lower bound proof is based on analyzing random instances of such collections. A uniformly distributed $\tau_n \in T_n$ can be chosen as follows: G_n is chosen uniformly at random from Π_n, and for each $pk \in \{0,1\}^n$ a permutation $F_n(pk, \cdot)$ is chosen uniformly and independently at random from Π_n.

Definition 2.2. *A family $\tau = \{\tau_n = (G_n, F_n, F_n^{-1})\}_{n=1}^{\infty}$ of trapdoor permutations is $s(n)$-hard if for every probabilistic Turing-machine A that runs in time $s(n)$, and for all sufficiently large n,*

$$\Pr\left[A^\tau(1^n, G_n(td), y) = F_n^{-1}(td, y)\right] \leq \frac{1}{s(n)} ,$$

where the probability is taken uniformly over all the possible choices of $td \in \{0,1\}^n$ and $y \in \{0,1\}^n$, and over all the possible outcomes of the internal coin tosses of A.

Definition 2.2 refers to the difficulty of inverting a random permutation $F(pk, \cdot)$ on a uniformly distributed image y, when given only $pk = G(td)$ and y.

Some applications, however, require enhanced hardness conditions. For example, it may be required (cf. [16, Appendix C]) that it is hard to invert $F(pk, \cdot)$ on y even given the random coins used in the generation of y. Note that our formulation captures such hardness condition as well and therefore the impossibility results proved in this paper hold also for enhanced trapdoor permutations.[3]

Single-server Private Information Retrieval. A single-server Private Information Retrieval (PIR) scheme is a protocol between a server and a user. The server holds a database $x \in \{0,1\}^n$ and the user holds an index $i \in [n]$ to an entry of the database. Very informally, the user wishes to retrieve the ith entry of the database, without revealing the index i to the server. More formally, a single-server PIR scheme is defined via a pair of probabilistic polynomial-time Turing-machines (S, U) such that:

- S receives as input a string $x \in \{0,1\}^n$. Following its interaction it does not have any output.
- U receives as input an index $i \in [n]$. Following its interaction it outputs a value $b \in \{0, 1, \perp\}$.

Denote by $b \leftarrow \langle S(x), U(i) \rangle$ the experiment in which S and U interact (using the given inputs and uniformly chosen random coins), and then U outputs the value b. It is required that there exists a negligible function $\nu(n)$, such that for all sufficiently large n, and for every string $x = x_1 \circ \cdots \circ x_n \in \{0,1\}^n$, it holds that $x_i \leftarrow \langle S(x), U(i) \rangle$ with probability at least $1 - \nu(n)$ over the random coins of both S and R.

In order to define the security properties of such schemes, we first introduce the following notation. Given a single-server PIR scheme (S, U) and a Turing-machine S^* (a malicious server), we denote by $\text{view}_{\langle S^*, U(i) \rangle}(n)$ the distribution on the view of S^* when interacting with $U(i)$ where $i \in [n]$. This view consists of its random coins and of the sequence of messages it receives from U, where the distribution is taken over the random coins of both S^* and U.

Definition 2.3. *A single-server PIR scheme (S, U) is secure if for every probabilistic polynomial-time Turing-machines S^* and D, and for every two sequences of indices $\{i_n\}_{i=1}^{\infty}$ and $\{j_n\}_{i=1}^{\infty}$ where $i_n, j_n \in [n]$ for every n, it holds that*

$$\left| \Pr \left[v \leftarrow \text{view}_{\langle S^*, U(i_n) \rangle}(n) : D(v) = 1 \right] \right.$$
$$\left. - \Pr \left[v \leftarrow \text{view}_{\langle S^*, U(j_n) \rangle}(n) : D(v) = 1 \right] \right| \leq \nu(n) ,$$

for some negligible function $\nu(n)$ and for all sufficiently large n.

Commitment schemes. A commitment scheme is a two-stage interactive protocol between a sender and a receiver. Informally, after the first stage of the protocol, which is referred to as the *commit stage*, the sender is bound to at

---

[3] A different enhancement, used by [17], requires the permutations' domain to be polynomially dense in $\{0,1\}^n$. Clearly, our impossibility result holds for such an enhancement as well.

most one value, not yet revealed to the receiver. In the second stage, which is referred to as the *reveal stage*, the sender reveals its committed value to the receiver. More formally, a commitment scheme is defined via a triplet of probabilistic polynomial-time Turing-machines $(\mathcal{S}, \mathcal{R}, \mathcal{V})$ such that:

- $\mathcal{S}$ receives as input the security parameter 1^n and a string $x \in \{0,1\}^k$. Following its interaction, it outputs some information decom (the decommitment).
- $\mathcal{R}$ receives as input the security parameter 1^n. Following its interaction, it outputs a state information com (the commitment).
- $\mathcal{V}$ (acting as the receiver in the reveal stage[4]) receives as input the security parameter 1^n, a commitment com and a decommitment decom. It outputs either a string $x' \in \{0,1\}^k$ or $\perp$.

Denote by $(\text{decom}|\text{com}) \leftarrow \langle \mathcal{S}(1^n, x), \mathcal{R}(1^n) \rangle$ the experiment in which $\mathcal{S}$ and $\mathcal{R}$ interact (using the given inputs and uniformly chosen random coins), and then $\mathcal{S}$ outputs decom while $\mathcal{R}$ outputs com. It is required that for all n, every string $x \in \{0,1\}^k$, and every pair $(\text{decom}|\text{com})$ that may be output by $\langle \mathcal{S}(1^n, x), \mathcal{R}(1^n) \rangle$, it holds that $\mathcal{V}(\text{com}, \text{decom}) = x$.[5] In the remainder of the paper, it will often be convenient for us to identify $\mathcal{V}$ with $\mathcal{R}$, and refer to a commitment scheme as a pair $(\mathcal{S}, \mathcal{R})$.

The security of a commitment scheme can be defined in two complementary ways, protecting against either an all-powerful sender or an all-powerful receiver. In this paper, we deal with commitment schemes of the latter type, which are referred to as *statistically-hiding* commitment schemes. In order to define the security properties of such schemes, we first introduce the following notation. Given a commitment scheme $(\mathcal{S}, \mathcal{R})$ and a Turing-machine $\mathcal{R}^*$, we denote by $\text{view}_{\langle \mathcal{S}(x), \mathcal{R}^* \rangle}(n)$ the distribution on the view of $\mathcal{R}^*$ when interacting with $\mathcal{S}(1^n, x)$. This view consists of $\mathcal{R}^*$'s random coins and of the sequence of messages it receives from $\mathcal{S}$. The distribution is taken over the random coins of both $\mathcal{S}$ and $\mathcal{R}^*$. Note that whenever no computational restrictions are assumed on $\mathcal{R}^*$, without loss of generality we can assume that $\mathcal{R}^*$ is deterministic.

Definition 2.4. *A commitment scheme* $(\mathcal{S}, \mathcal{R})$ *is* $\rho(n)$*-hiding if for every deterministic Turing-machine* $\mathcal{R}^*$*, and for every two sequences of strings* $\{x_n\}_{i=1}^{\infty}$ *and* $\{x'_n\}_{i=1}^{\infty}$ *where* $x_n, x'_n \in \{0,1\}^{k(n)}$ *for every n the ensembles* $\{\text{view}_{\langle \mathcal{S}(x_n), \mathcal{R}^* \rangle}(n)\}$ *and* $\{\text{view}_{\langle \mathcal{S}(x'_n), \mathcal{R}^* \rangle}(n)\}$ *have statistical difference at most* $\rho(n)$ *for all sufficiently large n. Such a scheme is* statistically-hiding *if it is* $\rho(n)$*-hiding for some negligible function* $\rho(n)$.

Our lower bound for commitment schemes holds in fact under a weaker hiding requirement. We derive our results even for commitment schemes in which the

---

[4] Note that there is no loss of generality in assuming that the reveal stage is non-interactive. This is since any such interactive stage can be replaced with a non-interactive one as follows: The sender sends its internal state to the receiver, who then simulates the sender in the interactive stage.

[5] Although we assume perfect completeness, it is not essential for our results.

sender is statistically protected only against an honest receiver. Such schemes are referred to as *statistically-hiding honest-receiver* commitment schemes. Formally, it is only required that the statistical difference between the ensembles $\{\mathsf{view}_{\langle S(x_n), \mathcal{R} \rangle}(n)\}$ and $\{\mathsf{view}_{\langle S(x'_n), \mathcal{R} \rangle}(n)\}$ is some negligible function of n.

Definition 2.5. *A commitment scheme $(S, \mathcal{R}, \mathcal{V})$ is $\mu(n)$-binding if for every probabilistic polynomial-time Turing-machine S^* it holds that the probability that $((\mathsf{decom}, \mathsf{decom}') | \mathsf{com}) \leftarrow \langle S^*(1^n), \mathcal{R}(1^n) \rangle$ (where the probability is over the random coins of both S^* and $\mathcal{R}$) such that $\mathcal{V}(\mathsf{com}, \mathsf{decom}) \neq \mathcal{V}(\mathsf{com}, \mathsf{decom}')$ and $\mathcal{V}(\mathsf{com}, \mathsf{decom}), \mathcal{V}(\mathsf{com}, \mathsf{decom}') \neq \bot$ is negligible in n for all sufficiently large n. Such a scheme is* computationally-binding *if it is $\mu(n)$-binding for some negligible function $\mu(n)$, and is* weakly-binding *if it is $(1 - 1/p(n))$-binding for some polynomial $p(n)$.*

Black-box reductions. A reduction of a primitive P to a primitive Q is a construction of P out of Q. Such a construction consists of showing that if there exists an implementation C of Q, then there exists an implementation M_C of P. This is equivalent to showing that for every adversary that breaks M_C, there exists an adversary that breaks C. Such a reduction is *semi-black-box* if it ignores the internal structure of Q's implementation, and it is *fully-black-box* if the proof of correctness is black-box as well, i.e., the adversary for breaking Q ignores the internal structure of both Q's implementation and of the (alleged) adversary breaking P. Semi-black-box reductions are less restricted and thus more powerful than fully-black-box reductions. A taxonomy of black-box reductions was provided by Reingold, Trevisan and Vadhan [36], and the reader is referred to their paper for a more complete and formal view of these notions.

We now formally define the class of constructions considered in this paper. Our results in the current paper are concerned with the particular setting of fully-black-box constructions of single-server PIR and of statistically-hiding commitment schemes from trapdoor permutations. We focus here on specific definitions for these particular primitives and we refer the reader to [36] for a more general definition.

When examining efficiency measures of fully-black-box constructions, an essential parameter for such characterizations, as introduced by Haitner et al. [18], is the *security-parameter-expansion* of the construction. Consider, for example, a fully-black-construction of a commitment scheme from a family of trapdoor permutations. One ingredient of such a construction is a machine A that attempts to break the security of the trapdoor permutation family given oracle access to any malicious sender S^* that breaks the security of the commitment scheme. Then, A receives a security parameter 1^n (and possibly some additional inputs) and invokes S^* in a black-box manner. The standard definition does not restrict the range of security parameters that A is allowed to invoke S^* on. For example, A may invoke S^* on security parameter 1^{n^2}, or even on security parameter $1^{\Theta(s(n))}$, where $s(n)$ is the running time of A. In this paper, we will use the notion $\ell(n)$-expanding for short, and note that according to Luby's classification [30], any polynomially-preserving reduction is $O(n)$-expanding in our terminology.

Definition 2.6. *A fully-black-box $\ell(n)$-expanding construction of a single-server PIR scheme from an $s(n)$-hard family of trapdoor permutations is a triplet of probabilistic oracle Turing-machines $(\mathcal{S}, \mathcal{U}, A)$ for which the following hold:*

1. **Correctness:** *For every family τ of trapdoor permutations, $(\mathcal{S}^\tau, \mathcal{U}^\tau)$ is a single-server PIR scheme.*
2. **Black-box proof of security:** *For every family of trapdoor permutations $\tau = \left\{ \tau_n = \left(G_n, F_n, F_n^{-1}\right) \right\}_{n=1}^\infty$ and for every probabilistic polynomial-time Turing-machine $\mathcal{S}^*$, if $\mathcal{S}^*$ with oracle access to τ breaks the security of $(\mathcal{S}^\tau, \mathcal{U}^\tau)$, then*

$$\Pr\left[A^{\tau, \mathcal{S}^*}(1^n, G_n(td), y) = F_n^{-1}(td, y) \right] > \frac{1}{s(n)} \, ,$$

for infinitely many values of n, where A runs in time $s(n)$ and invokes $\mathcal{S}^*$ on security parameters which are at most $1^{\ell(n)}$. The probability is taken uniformly over all the possible choices of $td \in \{0,1\}^n$ and $y \in \{0,1\}^n$, and over all the possible outcomes of the internal coin tosses of A.

Definition 2.7. *A fully-black-box $\ell(n)$-expanding construction of a statistically-hiding (against an honest-receiver) and weakly-binding commitment scheme from an $s(n)$-hard family of trapdoor permutations is a triplet of probabilistic oracle Turing-machines $(\mathcal{S}, \mathcal{R}, A)$ for which the following hold:*

1. **Correctness:** *For every family τ of trapdoor permutations, $(\mathcal{S}^\tau, \mathcal{R}^\tau)$ is a statistically-hiding honest-receiver commitment scheme.*
2. **Black-box proof of binding:** *For every family of trapdoor permutations $\tau = \left\{ \tau_n = \left(G_n, F_n, F_n^{-1}\right) \right\}_{n=1}^\infty$ and for every probabilistic polynomial-time Turing-machine $\mathcal{S}^*$, if $\mathcal{S}^*$ with oracle access to τ breaks the binding of $(\mathcal{S}^\tau, \mathcal{R}^\tau)$, then*

$$\Pr\left[A^{\tau, \mathcal{S}^*}(1^n, G_n(td), y) = F_n^{-1}(td, y) \right] > \frac{1}{s(n)} \, ,$$

for infinitely many values of n, where A runs in time $s(n)$ and invokes $\mathcal{S}^*$ on security parameters which are at most $1^{\ell(n)}$. The probability is taken uniformly over all the possible choices of $td \in \{0,1\}^n$ and $y \in \{0,1\}^n$, and over all the possible outcomes of the internal coin tosses of A.

3   Communication Lower Bound for Statistically-Hiding Commitment Schemes

In this section we prove a lower bound on the communication complexity of fully-black-box constructions of statistically-hiding commitment schemes from trapdoor permutations. We establish a lower bound on the number of bits communicated by the sender during the commit stage of any such scheme. Since we are interested in proving an impossibility result for commitment schemes, it will be sufficient for us to deal with bit-commitment schemes. We prove the following theorem:

Theorem 3.1. *In any fully-black-box $O(n)$-expanding construction of a weakly-binding statistically-hiding honest-receiver bit-commitment scheme from a family of trapdoor permutations, the sender communicates $\Omega(n)$ bits during the commit stage.*

The proof of Theorem 3.1 follows the approach and technique of Haitner at el. [18] who constructed a "collision-finding" oracle in order to derive a lower bound on the round complexity of statistically-hiding commitment schemes. Given any fully-black-box $O(n)$-expanding construction $(\mathcal{S}, \mathcal{R}, A)$ of a weakly-binding statistically-hiding honest-receiver bit-commitment scheme from a family of trapdoor permutations τ, we show that relative to their oracle the following holds: (1) there exists a malicious sender $\mathcal{S}^*$ that breaks the binding of the scheme $(\mathcal{S}^\tau, \mathcal{R}^\tau)$, and (2) if the sender communicates $o(n)$ bits during the commit stage of $(\mathcal{S}^\tau, \mathcal{R}^\tau)$, then the machine A (with oracle access to $\mathcal{S}^*$) fails to break the security of τ.

3.1   The Oracle

We briefly describe the oracle constructed by Haitner et al. [18] and state its main property. The oracle is of the form $\mathcal{O} = (\tau, \mathsf{Sam}^\tau)$, where τ is a family of trapdoor permutations (i.e., $\tau = \{\tau_n\}_{n=1}^\infty$, where $\tau_n \in T_n$ for every n), and Sam^τ is an oracle that, very informally, receives as input a description of a circuit C (which may contain τ-gates) and a string z, and outputs a uniformly distributed preimage of z under the mapping defined by C. As discussed in [18], several essential restrictions are imposed on the querying of Sam that prevent it from assisting in inverting τ.

Description of Sam. The oracle Sam receives as input a query of the form $Q = (C_{\text{next}}^\tau, C^\tau, z)$, and outputs a pair (w', z') where w' is a uniformly distributed preimage of z under the mapping defined by the circuit C^τ, and $z' = C_{\text{next}}^\tau(w')$. We impose the following restrictions:

1. z was the result of a previous query with C^τ as the next-query circuit (note that this imposes a forest-like structure on the queries).
2. The circuit C_{next}^τ is a *refinement* of the circuit C^τ, where by a refinement we mean that $C_{\text{next}}^\tau(w) = (C^\tau(w), \widetilde{C}^\tau(w))$ for some circuit $\widetilde{C}^\tau$ and for every w. In particular, this implies that C^τ and C_{next}^τ have the same input length. Given a query Q, we denote this input length by $m(Q)$, and when the query Q is clear from the context we will write only m.
3. Each query contains a security parameter 1^n, and Sam answers queries only up to depth $\mathsf{depth}(n)$, for some "depth restriction" function $\mathsf{depth} : \mathbb{N} \to \mathbb{N}$ which is a part of the description of Sam. The security parameter is set such that a query with security parameter 1^n is allowed to contain circuits with queries to permutations on up to n bits. Note that although different queries may have different security parameters, we ask that in the same "query-tree", all queries will have the same security parameter (hence the depth of the tree is already determined by the root query).

In order to impose these restrictions, Sam is equipped with a family $\mathsf{sign} = \{\mathsf{sign}_k\}_{k=1}^{\infty}$ of (random) functions $\mathsf{sign}_k : \{0,1\}^k \to \{0,1\}^{2k}$ that will be used as "signatures" for identifying legal queries as follows: in addition to outputting (w', z'), Sam will also output the value $\mathsf{sign}(1^n, C_{\text{next}}^\tau, z', dep+1)$, where dep is the depth of the query, 1^n is the security parameter of the query, and by applying the "function" sign we actually mean that we apply the function sign_k for the correct input length. Each query of the form $Q = (1^n, C_{\text{next}}^\tau, C^\tau, z, dep, sig)$ is answered by Sam if and only if C_{next}^τ is a refinement of C^τ, $dep \leq \mathsf{depth}(n)$ and $sig = \mathsf{sign}(1^n, C^\tau, z, dep)$.

Finally, Sam is provided with a family of (random) permutations $\mathcal{F} = \{f_Q\}$, where for every possible query Q a permutation f_Q is chosen uniformly at random from $\Pi_{m(Q)}$. Given a query $Q = (1^n, C_{\text{next}}^\tau, C^\tau, z, dep, sig)$, the oracle Sam uses the permutation $f_Q \in \mathcal{F}$ in order to sample w' as follows: it outputs $w' = f_Q(t)$ for the lexicographically smallest $t \in \{0,1\}^m$ such that $C^\tau(f_Q(t)) = z$. Note that whenever the permutation f_Q is chosen from Π_m uniformly at random, and independently of all other permutations in $\mathcal{F}$, then w' is indeed a uniformly distributed preimage of z. In this paper, whenever we consider the probability of an event over the choice of the family $\mathcal{F}$, we mean that for each query Q a permutation f_Q is chosen uniformly at random from $\Pi_{m(Q)}$ and independently of all other permutations. A complete and formal description of the oracle is provided in Figure 3.1.

On input $Q = (1^n, C_{\text{next}}^\tau, C^\tau, z, dep, sig)$, $\mathsf{Sam}_{\text{depth}}^{\tau, \mathcal{F}, \mathsf{sign}}$ acts as follows:

1. If $C^\tau = \bot$, then output (w', z', sig') where $w' = f_Q(0^m)$, $z' = C_{\text{next}}^\tau(w')$, and $sig' = \mathsf{sign}(1^n, C_{\text{next}}^\tau, z', 1)$.

2. Else, if C_{next}^τ is a refinement of C^τ, $dep \leq \mathsf{depth}(n)$ and $sig = \mathsf{sign}(1^n, C^\tau, z, dep)$, then

   (a) Find the lexicographically smallest $t \in \{0,1\}^m$ such that $C^\tau(f_Q(t)) = z$.

   (b) Output (w', z', sig') where $w' = f_Q(t)$, $z' = C_{\text{next}}^\tau(w')$, and $sig' = \mathsf{sign}(1^n, C_{\text{next}}^\tau, z', dep+1)$.

3. Else, output $\bot$.

Fig. 1. The oracle Sam

Definition 3.1. *We say that a circuit A queries the oracle $\mathsf{Sam}_{\text{depth}}^{\tau, \mathcal{F}, \mathsf{sign}}$ up to depth d, if for every Sam-query $Q = (1^n, C_{\text{next}}^\pi, C^\pi, z, dep, sig)$ that A makes, it holds that $dep \leq d$.*

One of the main properties of the oracle Sam, as proved in [18], is the following: any circuit with oracle access to Sam that tries to invert a random trapdoor permutation, fails with high probability. More specifically, Haitner et al. managed to relate this success probability to the maximal depth of the Sam-queries made by the circuit, and to the size of the circuit. They proved the following theorem:

Theorem 3.2 ([18]). *For every circuit A of size $s(n)$ that queries* Sam *up to depth $d(n)$ such that $s(n)^{3d(n)+2} < 2^{n/8}$, for every depth restriction function* depth *and for all sufficiently large n, it holds that*

$$\Pr_{\substack{td \leftarrow \{0,1\}^n, \tau, \mathcal{F} \\ y \leftarrow \{0,1\}^n, \text{sign}}} \left[A^{\tau, \mathsf{Sam}_{\text{depth}}^{\tau, \mathcal{F}, \text{sign}}}(G_n(td), y) = F_n^{-1}(td, y) \right] \leq \frac{2}{s(n)} .$$

3.2 Breaking Low-Communication Statistically-Hiding Commitment Schemes

We show that a random instance of the oracle Sam can be used to break the binding of any statistically-hiding commitment scheme. Specifically, for every bit-commitment scheme $(\mathcal{S}, \mathcal{R})$ which is (1) weakly-biding, (2) statistically-hiding against an honest-receiver, and (3) has oracle access to a family τ of trapdoor permutations, we construct a malicious sender $\mathcal{S}^*$ which has oracle access to $\mathsf{Sam}_{\text{depth}}^{\tau, \mathcal{F}, \text{sign}}$, and breaks the binding of $(\mathcal{S}^\tau, \mathcal{R}^\tau)$ with sufficiently high probability over the choices of τ, $\mathcal{F}$ and sign. Formally, the following theorem is proved:

Theorem 3.3. *For any statistically-hiding bit-commitment scheme $(\mathcal{S}, \mathcal{R}, \mathcal{V})$ with oracle access to a family of trapdoor permutations in which the sender communicates at most $c(n)$ bits during the commit stage, and for any polynomial $p(n)$, there exists a polynomial-time malicious sender $\mathcal{S}^*$ such that*

$$\Pr_{\substack{\tau, \mathcal{F} \\ \text{sign}, r_{\mathcal{R}}}} \left[\begin{array}{c} ((\text{decom}, \text{decom}')|\text{com}) \leftarrow \left\langle \mathcal{S}^{* \ \mathsf{Sam}_{\text{depth}}^{\tau, \mathcal{F}, \text{sign}}}(1^n), \mathcal{R}^\tau(1^n, r_{\mathcal{R}}) \right\rangle : \\ \mathcal{V}^\tau(\text{com}, \text{decom}) = 0, \mathcal{V}^\tau(\text{com}, \text{decom}') = 1 \end{array} \right] > 1 - \frac{1}{p(n)}$$

for all sufficiently large n, where $\text{depth}(n) = \left\lceil \frac{c(n)}{\log n} \right\rceil + 1$.

We note that the above theorem holds even if the commitment scheme is statistically-hiding only against an honest receiver. In what follows we introduce the notation used in this section. We proceed with a brief presentation of the main ideas underlying the proof of Theorem 3.3, which is then followed by a formal description of the malicious sender $\mathcal{S}^*$.

Notations. Let $(\mathcal{S}, \mathcal{R})$ be a bit-commitment scheme with oracle access to a family of trapdoor permutations. We denote by $b \in \{0,1\}$ and $r_\mathcal{S}, r_\mathcal{R} \in \{0,1\}^*$ the input bit of the sender and the random coins of the sender and the receiver, respectively. We denote by $c(n)$ the maximal number of bits communicated from the sender to the receiver in the commit stage with security parameter 1^n. In addition we denote by $d(n)$ the number of communication rounds in the scheme with security parameter 1^n, and without loss of generality we assume that the receiver makes the first move. Each communication round consists of a message sent from the receiver to the sender followed by a message sent from the sender to the receiver. We denote by q_i and a_i the messages sent by the receiver and the sender in the i-th round, respectively, and denote by a_{d+1} the message sent by the sender in the reveal stage. Finally, we let $\bar{a}_i = (a_1, \ldots, a_i)$ and $\bar{q}_i = (q_1, \ldots, q_i)$.

Although the sender is a probabilistic polynomial-time *Turing-machine*, in order to interact with the oracle Sam we need to identify the sender with a sequence of polynomial-size *circuits* $S_1, \ldots, S_{d+1}$ as follows. In the first round, S sends a_1 by computing $a_1 = S_1(b, r_S, q_1)$. Similarly, in the following rounds, S sends a_i by computing $a_i = S_i(b, r_S, \bar{q}_i)$.

Finally, in order to simplify the notation regarding the input and output of the oracle Sam, in this section we ignore parts of the input and output of Sam: we ignore the security parameter and the "signatures" (since our malicious sender S^* will only ask legal queries), and consider queries of a simplified form $Q = (C_{\text{next}}^\tau, C^\tau, z)$, and answers that consist only of w' (i.e., an answer consists only of a uniformly distributed preimage of z under the mapping defined by C^τ). In addition, in what follows it will be more intuitive to replace z in the queries by its preimage w, but this is clearly not essential.

A brief overview. Informally, recall that the oracle Sam described in Section 3.1 acts as follows: Sam is given as input a query $Q = (C_{\text{next}}, C, z)$, and outputs a pair (w', z') where w' is a uniformly distributed preimage of z under the mapping defined by the circuit C, and $z' = C_{\text{next}}(w')$. In addition, we imposed the restriction that there was a previous query $(C, \cdot, \cdot)$ that was answered by (w, z) (note that this imposes a forest-like structure on the queries), and we only allow querying Sam up to depth $O(n/\log n)$.

Given a statistically-hiding bit-commitment scheme in which the sender communicates $c(n)$ bits during the commit stage, we assume without loss of generality that the commit stage of the scheme has $c(n)$ communication rounds, where in each round the sender communicates a single bit to the receiver. The malicious sender S^* operates as follows: it chooses a random input w (consisting of random coins and a random committed bit), and during the first $\log n$ rounds it simulates the honest sender. In these $\log n$ rounds, it receives $\log n$ messages $q_1, \ldots, q_{\log n}$ from the receiver. Then, S^* constructs the circuit $C_{q_1, \ldots, q_{\log n}}$ that receives as input the sender's input w and outputs the $\log n$ sender's messages corresponding to the receiver's messages $q_1, \ldots, q_{\log n}$. This circuit is used to query Sam for a random input w_1. It may be the case, however, that w_1 is not consistent with the actual messages $a_1, \ldots, a_{\log n}$ that S^* sent in the first $\log n$ rounds. In this case, S^* "rewinds" Sam for a polynomial number of times, and since the total length of the sender's messages in these $\log n$ rounds is only $\log n$ bits, then with sufficiently high probability S^* will obtain a consistent w_1. Now, in the next $\log n$ rounds the malicious sender S^* simulates the honest sender with input w_1, and at the end of these $\log n$ rounds it will query (and rewind) Sam again for another consistent input $w_{\log n+1}$, and so on. Finally, after completing the commit stage, S^* queries Sam to obtain two random inputs $w_{c(n)}$ and $w'_{c(n)}$ which are consistent with the transcript of the commit stage. Since the commitment scheme is statistically-hiding, with probability roughly half they can be used to break the binding of the protocol. A crucial point in this description, is that S^* queries Sam only up to depth $c(n)/\log n$ (S^* used Sam to obtain $c(n)/\log n$ values $w_1, w_{\log n+1}, \ldots, w_{c(n)}$). Therefore, if $c(n) = o(n)$, then an oracle Sam that answers queries only up to depth $c(n)/\log n$ cannot be used to invert a random trapdoor permutation, according to Theorem 3.2.

A formal description of $\mathcal{S}^*$. Given a bit-commitment scheme $(\mathcal{S}, \mathcal{R})$ in which the sender communicates $c(n)$ bits during the commit stage, we assume without loss of generality (and for simplicity of the presentation) that the scheme has $c(n)$ communication rounds (i.e., $d(n) = c(n)$) where in each round during the commit stage the sender communicates a single bit to the receiver (i.e., each of $a_1, \ldots, a_{d(n)}$ is one bit). Furthermore, in order to simplify the description of $\mathcal{S}^*$, we assume that $\log n$ is an integral value (where 1^n is the security parameter given as input to $\mathcal{S}^*$) and that $c(n) = M \cdot \log n + 1$ for some integer $M = M(n)$. We stress that these assumptions are not at all essential, but avoiding them will result in a more complicated description. On input 1^n, the malicious sender $\mathcal{S}^*$ with oracle access to $\mathsf{Sam}_{\mathsf{depth}}^{\tau, \mathcal{F}, \mathsf{sign}}$ interacts with the honest receiver $\mathcal{R}$ as follows.

1. **The commit stage:**

   (a) In the first round $\mathcal{S}^*$ receives $\mathcal{R}$'s message q_1, and computes the description of the circuit $C_1 = S_1(\cdot, \cdot, q_1)$ obtained from the circuit S_1 by fixing q_1 as its third input. Then, $\mathcal{S}^*$ queries $\mathsf{Sam}_{\mathsf{depth}}^{\tau, \mathcal{F}, \mathsf{sign}}$ with $(C_1, \perp, \perp)$, receives an answer $w_1 = (b_1, r_1)$ and sends $a_1 = S_1(b_1, r_1, q_1)$ to $\mathcal{R}$.

   (b) In every round $i \in \{2, \ldots, \log n\}$, $\mathcal{S}^*$ simulates the honest sender $\mathcal{S}$ with input w_1. That is, $\mathcal{S}^*$ receives $\mathcal{R}$'s message q_i and replies with $a_i = S_i(b_1, r_1, \bar{q}_i)$.

   (c) In round $\log n + 1$, $\mathcal{S}^*$ receives $\mathcal{R}$'s message $q_{\log n + 1}$, and computes the description of the circuit $C_{\log n + 1} = S_{\log n + 1}(\cdot, \cdot, \bar{q}_{\log n + 1})$ obtained from the circuit $S_{\log n + 1}$ by fixing $\bar{q}_{\log n + 1}$ as its third input. Then, $\mathcal{S}^*$ queries $\mathsf{Sam}_{\mathsf{depth}}^{\tau, \mathcal{F}, \mathsf{sign}}$ with $(C_{\log n + 1}, C_1, w_1)$ for $t = 2n^5 c(n) p(n)$ times and receives t answers. If one of these answers is consistent with the transcript of the protocol so far, then denote the first such answer by $w_{\log n + 1} = (b_{\log n + 1}, r_{\log n + 1})$, and in this case $\mathcal{S}^*$ sends the message $a_{\log n + 1} = S_{\log n + 1}(b_{\log n + 1}, r_{\log n + 1}, \bar{q}_{\log n + 1})$ to $\mathcal{R}$. Otherwise, $\mathcal{S}^*$ aborts the execution of the protocol.

   (d) In the remainder of the commit stage $\mathcal{S}^*$ acts as follows:

   i. For every k and in every round $i \in \{(k - 1) \log n + 2, \ldots, k \log n\}$, the malicious sender $\mathcal{S}^*$ simulates the honest sender $\mathcal{S}$ with input $w_{(k-1) \log n + 1}$.

   ii. For every integer k and in every round $k \log n + 1$ the malicious sender $\mathcal{S}^*$ receives $\mathcal{R}$'s message $q_{k \log n + 1}$, and computes the description of the circuit $C_{k \log n + 1} = S_{k \log n + 1}(\cdot, \cdot, \bar{q}_{k \log n + 1})$ obtained from the circuit $S_{k \log n + 1}$ by fixing $\bar{q}_{k \log n + 1}$ as its third input. Then, $\mathcal{S}^*$ queries $\mathsf{Sam}_{\mathsf{depth}}^{\tau, \mathcal{F}, \mathsf{sign}}$ with $(C_{k \log n + 1}, C_{(k-1) \log n + 1}, w_{(k-1) \log n + 1})$ for $t = 2n^5 c(n) p(n)$ times and receives t answers. If one of these answers is consistent with the transcript of the protocol so far, then denote the first such answer by $w_{k \log n + 1} = (b_{k \log n + 1}, r_{k \log n + 1})$, and in this case $\mathcal{S}^*$ sends $a_{k \log n + 1} = S_{k \log n + 1}(b_{k \log n + 1}, r_{k \log n + 1}, \bar{q}_{k \log n + 1})$ to $\mathcal{R}$. Otherwise, $\mathcal{S}^*$ aborts the execution of the protocol.

2. **The reveal stage:**

(a) $\mathcal{S}^*$ queries $\mathsf{Sam}_{\mathsf{depth}}^{\tau,\mathcal{F},\mathsf{sign}}$ with $(\bot, C_{d(n)}, w_{d(n)})$ for n times, and receives n pairs $\left\{ \left(b_{d(n)+1}^{(j)}, r_{d(n)+1}^{(j)} \right) \right\}_{j=1}^{n}$. If there exist $j_0, j_1 \in [n]$ such that $b_{d(n)+1}^{(j_0)} = 0$ and $b_{d(n)+1}^{(j_1)} = 1$, then $\mathcal{S}^*$ outputs the two values

$$\mathsf{decom} = S_{d(n)+1}\left(b_{d(n)+1}^{(j_0)}, r_{d(n)+1}^{(j_0)}, \bar{q}_{d(n)} \right)$$
$$\mathsf{decom}' = S_{d(n)+1}\left(b_{d(n)+1}^{(j_1)}, r_{d(n)+1}^{(j_1)}, \bar{q}_{d(n)} \right) \, .$$

Otherwise, $\mathcal{S}^*$ aborts the execution of the protocol.

Two minor technical details were omitted from the description. First, according to the description of Sam (Section 3.1), whenever Sam is queried multiple times with the same input, it returns the exact same answer. Thus, whenever $\mathcal{S}^*$ queries Sam more than once with the same input, $\mathcal{S}^*$ has to make sure that the queries are all different (for example, by artificially embedding the query number to one of the circuits in the query). Second, in order for $\mathcal{S}^*$'s queries to be legal, it should hold that the circuit $C_{k \log n+1}$ is a refinement of the circuit $C_{(k-1) \log n+1}$ for every integer k (as discussed in Section 3.1). This can be done very easily by embedding the description of each $C_{(k-1) \log n+1}$ inside each $C_{k \log n+1}$ (i.e., the output of C_i is the sequence of bits $\bar{a}_i$ instead of only the bit a_i).

The formal proof proceeds by arguing that $\mathcal{S}^*$ successfully completes the commit stage with high probability. Then, given that $\mathcal{S}^*$ has successfully completed the commit stage, we prove that the transcript of the commit stage is distributed identically to the transcript of the commit stage in an honest execution of the protocol. This enables us to use the fact that the commitment scheme is statistically-hiding, and therefore a random transcript can be revealed both as a commitment to $b = 0$ and as a commitment to $b = 1$, with almost equal probabilities. Due to space limitations we refer the reader to [19] for a formal proof, which then immediately implies the correctness of Theorem 3.1.

4 Refining the Relation Between Single-Server PIR and Commitment Schemes

The relation between single-server PIR and commitment schemes was first explored by Beimel et al. [1], who showed that any single-server PIR protocol in which the server communicates at most $n/2$ bits to the user (where n is the size of the server's database), can be used to construct a weakly-binding statistically-hiding bit-commitment scheme. In particular, this served as the first indication that the existence of low-communication PIR protocols implies the existence of one-way functions. In this section, we refine the relation between these two fundamental primitives by improving their reduction. Informally speaking, our reduction essentially uses the reduction of Beimel et al. instantiated with a better extractor. This enables the following improvements: (1) the communication complexity of the PIR protocol is essentially preserved, (2) given a single-server PIR protocol in which the server communicates $n - k$ bits, it is possible to commit to $\Omega(k)$ bits while executing the underlying single-server PIR protocol only once, and (3) whereas the construction of Beimel et al. was presented for single-server PIR protocols in which the server communicates at most $n/2$ bits, our construction can rely on single-server PIR in which the server communicates up to $n - \omega(\log n)$ bits.

In what follows we state our main theorem in the current section, and then turn to formally describe the construction and to provide intuition for its proof. Due to space limitations we refer the reader to [19] for the formal proof.

Theorem 4.1. *Let $d(n) \in \omega(\log n)$, $k(n) \geq 2d(n)$, and let $\mathcal{P}$ be a single-server PIR protocol in which the server communicates $n - k(n)$ bits, where n is the size of the server's database. Then, there exists a weakly-binding statistically-hiding commitment scheme $\mathcal{COM}^{\mathcal{P}}$ for $d(n)/6$ bits, in which the sender communicates less than $n - k(n) + 2d(n)$ bits during the commit stage. Moreover, the construction is fully-black-box and linearly-preserving.*

The construction. Fix $d(n)$, $k(n)$ and $\mathcal{P}$ as in Theorem 4.1. In the construction we use a strong $\left(d(n)/3, 2^{1-d(n)/3}\right)$-extractor $\text{EXT} : \{0,1\}^n \times \{0,1\}^{d(n)} \rightarrow \{0,1\}^{d(n)/6}$ whose existence is guaranteed by Proposition 2.1. Figure 4 describes our construction of the commitment scheme $\mathcal{COM}^{\mathcal{P}} = (\mathcal{S}, \mathcal{R})$. The correctness of $\mathcal{COM}^{\mathcal{P}}$ follows directly from the correctness of $\mathcal{P}$. In addition, notice that the total number of bits communicated by the sender in the commit stage is the total number of bits that the server communicates in $\mathcal{P}$ plus the seed length and the output length of the extractor EXT. Thus, the sender communicates less than $n - k(n) + 2d(n)$ bits during the commit stage.

Proof intuition. The commit stage consists of the sender and the receiver choosing random inputs $x \in \{0,1\}^n$ and $i \in [n]$, respectively, and executing the PIR protocol $\mathcal{P}$ on these inputs. As a consequence, the receiver obtains a bit x_i, which by the correctness of $\mathcal{P}$ is the ith bit of x. Now, notice that since the sender communicated only $n - \omega(\log n)$ bits, then the random variable corresponding to x still has $\omega(\log n)$ min-entropy from the receiver's point of view (with high probability). We take advantage of this fact, and exploit the remaining min-entropy of x in order to hide the committed string s in a statistical manner (note that since it is required to reveal the seed of the extractor during the commit stage, we need a *strong* extractor). The formal proof of the hiding property is similar to that of Lu [29] in the bounded storage model, which is in turn based on ideas that were used for constructing pseudorandom generators for space bounded computations [33]. We note that the proof of hiding does not rely on any computational properties of the underlying PIR protocol $\mathcal{P}$, but only on the assumed bound on the number of bits communicated by the server in $\mathcal{P}$. The binding property follows from the security of the PIR protocol: in the reveal stage, the sender must send a value x whose i^{th} bit is consistent with the bit obtained by the receiver during the commit stage – but this bit is not known to the sender.

Protocol $\mathcal{COM}^{\mathcal{P}} = (\mathcal{S}, \mathcal{R})$

Joint input: security parameter 1^n.
Sender's input: $s \in \{0,1\}^{d(n)/6}$.

Commit stage:
1. $\mathcal{S}$ chooses a uniformly distributed $x \in \{0,1\}^n$.
2. $\mathcal{R}$ chooses a uniformly distributed index $i \in [n]$.
3. $\mathcal{S}$ and $\mathcal{R}$ execute the single-server PIR protocol $\mathcal{P}$ for database of length n, where $\mathcal{S}$ acts as the server with input x and $\mathcal{R}$ acts as the user with input i. As a result, $\mathcal{R}$ obtains a bit $x_i \in \{0,1\}$.
4. $\mathcal{S}$ chooses a uniformly distributed seed $t \in \{0,1\}^{d(n)}$, computes $y = \mathrm{EXT}(x,t) \oplus s$, and sends (t,y) to $\mathcal{R}$.

Reveal stage:
1. $\mathcal{S}$ sends (s,x) to $\mathcal{R}$.
2. If the ith bit of x equals x_i and $y = \mathrm{EXT}(x,t) \oplus s$, then $\mathcal{R}$ outputs s. Otherwise, $\mathcal{R}$ outputs $\perp$.

Fig. 2. A construction of a commitment scheme from any low-communication single-server PIR protocol

5   Concluding Remarks

Our result does not rule out weakly-preserving (fully-black-box) constructions of single-server PIR from trapdoor permutations in which the sender communicates $o(n)$ bits to the user. We note that although weakly-preserving reductions guarantee much weaker security than polynomially-preserving reductions, investigating lower bounds for such reductions is still a very interesting research topic. Even more so as the sole construction to date of a single-server PIR protocol from trapdoor permutations uses such a reduction. A possible step towards tightening our bound is to first provide an improved lower bound on the communication complexity of statistically-hiding commitment schemes that allow the sender to commit to more than a single bit. Whereas in Section 4 we proved that any low-communication single-server PIR implies a statistically-hiding commitment scheme that allows the sender to commit to a relatively long string, our lower bound on the communication complexity of statistically-hiding commitment schemes in Section 3 serves as a bottleneck: it does not take into consideration the number of committed bits (the lower bound is only in terms of the security parameter).

It is quite possible that a much tighter lower bound can be proved for string-commitment schemes. Such a lower bound may extend the result of the current paper to the setting of weakly-preserving reductions, and prove the optimality of the single-server PIR protocol of Kushilevitz and Ostrovsky [27]. We note that the statistically-hiding commitment scheme of Naor et al. [31] (which is constructed from one-way permutations in a fully-black-box manner) can be used to commit to $O(\log n)$ bits while the sender communicates $O(n)$ bits (see, for example, [32]).

Acknowledgments. We are grateful to Yuval Ishai and Omer Reingold for many useful conversations and observations. We also thank the anonymous referees for their remarks and suggestions.

References

1. Beimel, A., Ishai, Y., Kushilevitz, E., Malkin, T.: One-way functions are essential for single-server private information retrieval. In: 31st STOC, pp. 89–98 (1999)
2. Cachin, C., Micali, S., Stadler, M.: Computationally private information retrieval with polylogarithmic communication. In: Stern, J. (ed.) EUROCRYPT 1999. LNCS, vol. 1592, pp. 402–414. Springer, Heidelberg (1999)
3. Chang, Y.: Single database private information retrieval with logarithmic communication. In: 9th ACISP, pp. 50–61 (2004)
4. Chor, B., Goldreich, O., Kushilevitz, E., Sudan, M.: Private information retrieval. In: 36th FOCS, pp. 41–50 (1995)
5. Di Crescenzo, G., Malkin, T., Ostrovsky, R.: Single database private information retrieval implies oblivious transfer. In: Preneel, B. (ed.) EUROCRYPT 2000. LNCS, vol. 1807, pp. 122–138. Springer, Heidelberg (2000)
6. Dziembowski, S., Maurer, U.M.: On generating the initial key in the bounded-storage model. In: Cachin, C., Camenisch, J.L. (eds.) EUROCRYPT 2004. LNCS, vol. 3027, pp. 126–137. Springer, Heidelberg (2004)
7. Fischlin, M.: On the impossibility of constructing non-interactive statistically-secret protocols from any trapdoor one-way function. In: CT-RSA, pp. 79–95 (2002)
8. Gennaro, R., Gertner, Y., Katz, J.: Lower bounds on the efficiency of encryption and digital signature schemes. In: 35th STOC, pp. 417–425 (2003)
9. Gennaro, R., Gertner, Y., Katz, J., Trevisan, L.: Bounds on the efficiency of generic cryptographic constructions. SIAM J. Comput. 35(1), 217–246 (2005)
10. Gennaro, R., Lindell, Y., Malkin, T.: Enhanced versus plain trapdoor permutations for non-interactive zero-knowledge and oblivious transfer. Manuscript (2006)
11. Gennaro, R., Trevisan, L.: Lower bounds on the efficiency of generic cryptographic constructions. In: 41st FOCS, pp. 305–313 (2000)
12. Gentry, C., Ramzan, Z.: Single-database private information retrieval with constant communication rate. In: 32nd ICALP, pp. 803–815 (2005)
13. Gertner, Y., Kannan, S., Malkin, T., Reingold, O., Viswanathan, M.: The relationship between public key encryption and oblivious transfer. In: 41st FOCS, pp. 325–335 (2000)
14. Gertner, Y., Malkin, T., Reingold, O.: On the impossibility of basing trapdoor functions on trapdoor predicates. In: 42nd FOCS, pp. 126–135 (2001)
15. Goldreich, O.: Foundations of Cryptography, Basic Tools, vol. 1. Cambridge University Press, Cambridge (2001)
16. Goldreich, O.: Foundations of Cryptography, Basic Applications, vol. 2. Cambridge University Press, Cambridge (2004)
17. Haitner, I.: Implementing oblivious transfer using collection of dense trapdoor permutations. In: 1st TCC, pp. 394–409 (2004)
18. Haitner, I., Hoch, J.J., Reingold, O., Segev, G.: Finding collisions in interactive protocols – A tight lower bound on the round complexity of statistically-hiding commitments. In: 48th FOCS, pp. 669–679 (2007)
19. Haitner, I., Hoch, J.J., Segev, G.: A linear lower bound on the communication complexity of single-server private information retrieval. Cryptology ePrint Archive, Report 2007/351 (2007)

20. Harnik, D., Naor, M.: On the compressibility of NP instances and cryptographic applications. In: 47th FOCS, pp. 719–728 (2006)
21. Horvitz, O., Katz, J.: Bounds on the efficiency of "black-box" commitment schemes. In: 32nd ICALP, pp. 128–139 (2005)
22. Impagliazzo, R., Rudich, S.: Limits on the provable consequences of one-way permutations. In: 21st STOC, pp. 44–61 (1989)
23. Ishai, Y., Kushilevitz, E., Ostrovsky, R.: Sufficient conditions for collision-resistant hashing. In: 2nd TCC, pp. 445–456 (2005)
24. Kalai, Y.T., Raz, R.: Succinct non-interactive zero-knowledge proofs with pre-processing for LOGSNP. In: 47th FOCS, pp. 355–366 (2006)
25. Kim, J.H., Simon, D.R., Tetali, P.: Limits on the efficiency of one-way permutation-based hash functions. In: 40th FOCS, pp. 535–542 (1999)
26. Kushilevitz, E., Ostrovsky, R.: Replication is NOT needed: SINGLE database, computationally-private information retrieval. In: 38th FOCS, pp. 364–373 (1997)
27. Kushilevitz, E., Ostrovsky, R.: One-way trapdoor permutations are sufficient for non-trivial single-server private information retrieval. In: Preneel, B. (ed.) EURO-CRYPT 2000. LNCS, vol. 1807, pp. 104–121. Springer, Heidelberg (2000)
28. Lipmaa, H.: An oblivious transfer protocol with log-squared communication. In: 8th ISC, pp. 314–328 (2005)
29. Lu, C.-J.: Encryption against storage-bounded adversaries from on-line strong extractors. J. Cryptology 17(1), 27–42 (2004)
30. Luby, M.: Pseudorandomness and Cryptographic Applications. Princeton University Press, Princeton (1996)
31. Naor, M., Ostrovsky, R., Venkatesan, R., Yung, M.: Perfect zero-knowledge arguments for NP using any one-way permutation. J. Cryptology 11(2), 87–108 (1998)
32. Nguyen, M.-H., Ong, S.J., Vadhan, S.P.: Statistical zero-knowledge arguments for NP from any one-way function. In: 47th FOCS, pp. 3–14 (2006)
33. Nisan, N., Zuckerman, D.: Randomness is linear in space. Journal of Computer and System Sciences 52(1), 43–52 (1996)
34. Ostrovsky, R., Skeith, W.E.: Algebraic lower bounds for computing on encrypted data. Cryptology ePrint Archive, Report 2007/064 (2007)
35. Ostrovsky, R., Skeith, W.E.: A survey of single database PIR: Techniques and applications. Cryptology ePrint Archive, Report 2007/059 (2007)
36. Reingold, O., Trevisan, L., Vadhan, S.P.: Notions of reducibility between cryptographic primitives. In: 1st TCC, pp. 1–20 (2004)
37. Rudich, S.: Limits on the provable consequences of one-way functions. PhD thesis, EECS Department, University of California, Berkeley (1988)
38. Simon, D.R.: Finding collisions on a one-way street: Can secure hash functions be based on general assumptions? In: Nyberg, K. (ed.) EUROCRYPT 1998. LNCS, vol. 1403, pp. 334–345. Springer, Heidelberg (1998)
39. Srinivasan, A., Zuckerman, D.: Computing with very weak random sources. SIAM J. Comput. 28(4), 1433–1459 (1999)
40. Stern, J.P.: A new efficient all-or-nothing disclosure of secrets protocol. In: Ohta, K., Pei, D. (eds.) ASIACRYPT 1998. LNCS, vol. 1514, pp. 357–371. Springer, Heidelberg (1998)
41. Wee, H.: One-way permutations, interactive hashing and statistically hiding commitments. In: 4th TCC, pp. 419–433 (2007)

Randomness Extraction Via δ-Biased Masking in the Presence of a Quantum Attacker

Serge Fehr* and Christian Schaffner**

CWI*** Amsterdam, The Netherlands
{S.Fehr,C.Schaffner}@cwi.nl

Abstract. Randomness extraction is of fundamental importance for information-theoretic cryptography. It allows to transform a raw key about which an attacker has some limited knowledge into a fully secure random key, on which the attacker has essentially no information. Up to date, only very few randomness-extraction techniques are known to work against an attacker holding quantum information on the raw key. This is very much in contrast to the classical (non-quantum) setting, which is much better understood and for which a vast amount of different techniques are known and proven to work.

We prove a new randomness-extraction technique, which is known to work in the classical setting, to be secure against a quantum attacker as well. Randomness extraction is done by xor'ing a so-called δ-biased mask to the raw key. Our result allows to extend the classical applications of this extractor to the quantum setting. We discuss the following two applications. We show how to encrypt a long message with a short key, information-theoretically secure against a quantum attacker, provided that the attacker has enough quantum uncertainty on the message. This generalizes the concept of entropically-secure encryption to the case of a quantum attacker. As second application, we show how to do error-correction without leaking partial information to a quantum attacker. Such a technique is useful in settings where the raw key may contain errors, since standard error-correction techniques may provide the attacker with information on, say, a secret key that was used to obtain the raw key.

1 Introduction

Randomness extraction allows to transform a raw key X about which an attacker has some limited knowledge into a fully secure random key S. It is required that the attacker has essentially no information on the resulting random key S, no

---

* Supported by a Veni grant from the Dutch Organization for Scientific Research (NWO).
** Supported by the EU projects SECOQC and QAP IST 015848 and a NWO Vici grant 2004-2009.
*** Centrum voor Wiskunde en Informatica, the national research institute for mathematics and computer science in the Netherlands.

R. Canetti (Ed.): TCC 2008, LNCS 4948, pp. 465–481, 2008.
© International Association for Cryptologic Research 2008

matter what kind of information he has about the raw key X, as long as his uncertainty on X is lower bounded in terms of a suitable entropy measure. One distinguishes between extractors which use a private seed (preferably as small as possible) [29], and, what is nowadays called *strong* extractors, which only use public coins [15,21]. In the context of cryptography, the latter kind of randomness extraction is also known as privacy amplification [5]. Randomness-extraction techniques play an important role in various areas of theoretical computer science. In cryptography, they are at the core of many constructions in information-theoretic cryptography, but they also proved to be useful in the computational setting. As such, there is a huge amount of literature on randomness extraction, and there exist various techniques which are optimized with respect to different needs; we refer to Shaltiel's survey [26] for an informative overview on classical and recent results.

Most of these techniques, however, are only guaranteed to work in a non-quantum setting, where information is formalized by means of classical information theory. In a quantum setting, where the attacker's information is given by a quantum state, our current understanding is much more deflating. Renner and König [23] have shown that privacy amplification via universal$_2$ hashing is secure against quantum adversaries. And, König and Terhal [18] showed security against quantum attackers for certain extractors, namely for one-bit-output strong extractors, as well as for strong extractors which work by extracting bit wise via one-bit-output strong extractors. Concurrent to our work, Smith has shown recently that Renner and König's result generalizes to *almost*-universal hashing, i.e., that Srinivasan-Zuckerman extractors remain secure against quantum adversaries [27]. On the negative side, Gavinsky *et al.* recently showed that there exist (strong) extractors that are secure against classical attackers, but which become completely insecure against quantum attackers [13]. Hence, it is not only a matter of lack of proof, but in fact classical extractors may turn insecure when considering *quantum* attackers.

We prove a new randomness-extraction technique to be secure against a quantum attacker. It is based on the concept of *small-biased spaces*, see e.g. [20]. Concretely, randomness extraction is done by xor'ing the raw key $X \in \{0,1\}^n$ with a δ-biased mask $A \in \{0,1\}^n$, chosen privately according to some specific distribution, where the distribution may be chosen publicly from some family of distributions. Roughly, A (or actually the family of distributions) is δ-biased, if any non-trivial parity of A can only be guessed with advantage δ. We prove that if A is δ-biased, then the bit-wise xor $X \oplus A$ is ε-close to random and independent of the attacker's quantum state with $\varepsilon = \delta \cdot 2^{(n-t)/2}$, where t is the attacker's quantum collision-entropy in X. Thus, writing $\delta = 2^{-\kappa}$, the extracted key $X \oplus A$ is essentially random as long as 2κ is significantly larger than $n - t$. Note that in its generic form, this randomness extractor uses public coins, namely the choice of the distribution, *and* a private seed, the sampling of A according to the chosen distribution. Specific instantiations though, may lead to standard extractors with no public coins (as in Section 5), or to a strong extractor with no private seed (as in Section 6). The proof of the new randomness-extraction result combines quantum-information-theoretic techniques developed by Renner [22,23] and techniques from Fourier analysis, similar to though slightly more involved than those used in [2].

We would like to point out that the particular extractor we consider, δ-biased masking, is well known to be secure against *non*-quantum attackers. Indeed, classical security was shown by Dodis and Smith, who also suggested useful applications [11,12]. Thus, our main contribution is the *security analysis* in the presence of a *quantum* attacker. Our positive result not only contributes to the general problem of the security of extractors against quantum attacks, but it is particularly useful in combination with the classical applications of δ-biased masking where it leads to interesting new results in the quantum setting. We discuss these applications and the arising new results below.

The first application is entropically secure encryption [25,12]. An encryption scheme is entropically secure if the ciphertext gives essentially no information away on the plaintext (in an information-theoretic sense), provided that the attacker's a priori information on the plaintext is limited. Entropic security allows to overcome Shannon's pessimistic result on the size of the key for information-theoretically secure encryption, in that a key of size essentially $\ell \approx n - t$ suffices to encrypt a plaintext of size n which has t bits of entropy given the attacker's a priori information. This key size was known to suffice for a non-quantum adversary [25,12]. By our analysis, this result carries over to the setting where we allow the attacker to store information as quantum states: a key of size essentially $\ell \approx n - t$ suffices to encrypt a plaintext of size n which has t bits of (min- or collision-) entropy given the attacker's quantum information about the plaintext.

Note that entropic security in a quantum setting was also considered explicitly in [8] and implicitly for the task of approximate quantum encryption [2,16,10]. However, all these results are on encrypting a *quantum* message into a quantum ciphertext on which the attacker has limited *classical* information (or none at all), whereas we consider encrypting a *classical* message into a classical ciphertext on which the attacker has limited *quantum* information. Thus, our result in quantum entropic security is in that sense orthogonal. As a matter of fact, the results in [2,16,10,8] about randomizing quantum states can also be appreciated as extracting "quantum randomness" from a quantum state on which the attacker has limited *classical* information. Again, this is orthogonal to our randomness-extraction result which allows to extract classical randomness from a *classical* string on which the attacker has limited *quantum* information. In independent recent work, Desrosiers and Dupuis showed that one can combine techniques to get the best out of both: they showed that δ-biased masking (as used in [2]) allows to extract "quantum randomness" from a *quantum* state on which the attacker has limited *quantum* information. This in particular implies our result.

The second application is in the context of private error-correction. Consider a situation where the raw key X is obtained by Alice and Bob with the help of some (short) common secret key K, and where the attacker Eve, who does not know K, has high entropy on X. Assume that, due to noise, Bob's version of the raw key X' is slightly different from Alice's version X. Such a situation may for instance occur in the bounded-storage model or in a quantum-key-distribution setting. Since Alice and Bob have different versions of the raw key, they first need to correct the errors before they can extract (by means of randomness extraction) a secure key S from X. However, since X and X' depend on K, standard techniques for correcting the errors between X and X' leak information not only on X but also on K to Eve, which prohibits that Alice and Bob can re-use K in a future session. In the case of a non-quantum attacker, Dodis and Smith showed how to do error-correction in such a setting without leaking information on K to Eve [11], and thus that K can be safely re-used an unlimited number of times. We show how our randomness-extraction result gives rise to a similar way of doing error correction without leaking information on K, even if Eve holds her partial information on X in a quantum state. Such a private-error-correction technique is a useful tool in various information-theoretic settings with a quantum adversary. Very specifically, this technique has already been used as essential ingredient to derive new results in the bounded-(quantum)-storage model and in quantum key distribution [7].

The paper is organized as follows. We start with some quantum-information-theoretic notation and definitions. The new randomness-extraction result is presented in Section 3 and proven in Section 4. The two applications discussed are given in Sections 5 and 6.

2 Preliminaries

2.1 Notation and Terminology

A *quantum system* is described by a complex Hilbert space $\mathcal{H}_A$ (in this paper always of finite dimension). The *state* of the system is given by a *density matrix*: a positive semi-definite operator ρ_A on $\mathcal{H}_A$ with trace $\mathrm{tr}(\rho_A) = 1$. We write $\mathcal{P}(\mathcal{H}_A)$ for the set of all positive semi-definite operators on $\mathcal{H}_A$, and we call $\rho_A \in \mathcal{P}(\mathcal{H}_A)$ *normalized* if it has trace 1, i.e., if it is a density matrix. For a density matrix $\rho_{AB} \in \mathcal{P}(\mathcal{H}_A \otimes \mathcal{H}_B)$ of a composite quantum system $\mathcal{H}_A \otimes \mathcal{H}_B$, we write $\rho_B = \mathrm{tr}_A(\rho_{AB})$ for the state obtained by tracing out system $\mathcal{H}_A$. A density matrix $\rho_{XB} \in \mathcal{P}(\mathcal{H}_X \otimes \mathcal{H}_B)$ is called *classical on* $\mathcal{H}_X$ *with* $X \in \mathcal{X}$, if it is of the form $\rho_{XB} = \sum_x P_X(x)|x\rangle\langle x| \otimes \rho_B^x$ with normalized $\rho_B^x \in \mathcal{P}(\mathcal{H}_B)$, where $\{|x\rangle\}_{x \in \mathcal{X}}$ forms an orthonormal basis of $\mathcal{H}_X$. Such a density matrix ρ_{XB} which is classical on $\mathcal{H}_X$ can be viewed as a random variable X with distribution P_X together with a family $\{\rho_B^x\}_{x \in \mathcal{X}}$ of *conditional density matrices*, such that the state of $\mathcal{H}_B$ is given by ρ_B^x if and only if X takes on the value x. We can introduce a new random variable Y which is obtained by "processing" X, i.e., by extending the distribution P_X to a consistent joint distribution P_{XY}. Doing so then naturally defines the density matrix $\rho_{XYB} = \sum_{x,y} P_{XY}(x,y)|x\rangle\langle x| \otimes |y\rangle\langle y| \otimes \rho_B^x$, and thus also the density matrix $\rho_{YB} = \mathrm{tr}_X(\rho_{XYB}) = \sum_y P_Y(y)|y\rangle\langle y| \otimes \left(\sum_x P_{X|Y}(x|y)\rho_B^x\right)$. If the meaning is clear from the context, we tend to slightly abuse notation and write the latter also as $\rho_{YB} = \sum_y P_Y(y)|y\rangle\langle y| \otimes \rho_B^y$, i.e., understand ρ_B^y as $\sum_x P_{X|Y}(x|y)\rho_B^x$. Throughout, we write $\mathbb{1}$ for the identity matrix of appropriate dimension.

2.2 Distance and Entropy Measures for Quantum States

We recall some definitions from [22]. Let $\rho_{XB} \in \mathcal{P}(\mathcal{H}_X \otimes \mathcal{H}_B)$. Although the following definitions make sense (and are defined in [22]) for arbitrary ρ_{XB}, we may assume ρ_{XB} to be normalized[1] and to be classical on $\mathcal{H}_X$.

Definition 2.1. *The L_1-distance from uniform of ρ_{XB} given B is defined by*

$$d(\rho_{XB}|B) := \|\rho_{XB} - \rho_U \otimes \rho_B\|_1 = \operatorname{tr}(|\rho_{XB} - \rho_U \otimes \rho_B|)$$

where $\rho_U := \frac{1}{\dim(\mathcal{H}_X)}\mathbb{1}$ is the fully mixed state on $\mathcal{H}_X$ and $|A| := \sqrt{A^\dagger A}$ is the positive square root of $A^\dagger A$ (where $A^\dagger$ is the complex-conjugate transpose of A).

If ρ_{XB} is classical on $\mathcal{H}_X$, then $d(\rho_{XB}|B) = 0$ if and only if X is uniformly distributed and ρ_B^x does not depend on x, which in particular implies that no information on X can be learned by observing system $\mathcal{H}_B$. Furthermore, if $d(\rho_{XB}|B) \leq \varepsilon$ then the real system ρ_{XB} "behaves" as the ideal system $\rho_U \otimes \rho_B$ except with probability ε in that for any evolution of the system no observer can distinguish the real from the ideal one with advantage greater than ε [23].

Definition 2.2. *The* collision-entropy *and the* min-entropy *of ρ_{XB} relative to a normalized and invertible $\sigma_B \in \mathcal{P}(\mathcal{H}_B)$ are defined by*

$$H_2(\rho_{XB}|\sigma_B) := -\log \operatorname{tr}\left(\left((\mathbb{1} \otimes \sigma_B^{-1/4}) \rho_{XB} (\mathbb{1} \otimes \sigma_B^{-1/4}) \right)^2 \right)$$

$$= -\log \sum_x P_X(x)^2 \operatorname{tr}\left(\left(\sigma_B^{-1/4} \rho_B^x \sigma_B^{-1/4} \right)^2 \right) \qquad and$$

$$H_\infty(\rho_{XB}|\sigma_B) := -\log \lambda_{max}\left((\mathbb{1} \otimes \sigma_B^{-1/2}) \rho_{XB} (\mathbb{1} \otimes \sigma_B^{-1/2}) \right)$$

$$= -\log \max_x \lambda_{max}\left(P_X(x) \sigma_B^{-1/2} \rho_B^x \sigma_B^{-1/2} \right),$$

respectively, where $\lambda_{max}(\cdot)$ denotes the largest eigenvalue of the argument. The collision-entropy *and the* min-entropy *of ρ_{XB} given $\mathcal{H}_B$ are defined by*

$$H_2(\rho_{XB}|B) := \sup_{\sigma_B} H_2(\rho_{XB}|\sigma_B) \qquad and \qquad H_\infty(\rho_{XB}|B) := \sup_{\sigma_B} H_\infty(\rho_{XB}|\sigma_B)$$

respectively, where the supremum ranges over all normalized $\sigma_B \in \mathcal{P}(\mathcal{H}_B)$.

Note that without loss of generality, the supremum over σ_B can be restricted to the set of normalized *and* invertible states σ_B which is dense in the set of

---

[1] For a non-normalized ρ_{XB}, there is a normalizing $1/\operatorname{tr}(\rho_{XB})$-factor in the definition of collision-entropy. Also note that $\operatorname{tr}(\sigma^{-1/2}\rho\sigma^{-1/2}) = \operatorname{tr}(\rho\sigma^{-1})$ for any invertible σ.

normalized states in $\mathcal{P}(\mathcal{H}_B)$. Note furthermore that it is not clear, neither in the classical nor in the quantum case, what the "right" way to define conditional collision- or min-entropy is, and as a matter of fact, it depends on the context which version serves best. An alternative way to define the collision- and min-entropy of ρ_{XB} given $\mathcal{H}_B$ would be as $\tilde{\mathrm{H}}_2(\rho_{XB}|B) := \mathrm{H}_2(\rho_{XB}|\rho_B)$ and $\tilde{\mathrm{H}}_\infty(\rho_{XB}|B) := \mathrm{H}_\infty(\rho_{XB}|\rho_B)$. For a density matrix ρ_{XY} that is classical on $\mathcal{H}_X$ and $\mathcal{H}_Y$, it is easy to see that $\tilde{\mathrm{H}}_2(\rho_{XY}|Y) = -\log \sum_y P_Y(y) \sum_x P_{X|Y}(x|y)^2$, i.e., the negative logarithm of the average conditional collision probability, and $\tilde{\mathrm{H}}_\infty(\rho_{XY}|Y) = -\log \max_{x,y} P_{X|Y}(x|y)$, i.e., the negative logarithm of the maximal conditional guessing probability. These notions of classical conditional collision- and min-entropy are commonly used in the literature, explicitly (see e.g. [24,6]) or implicitly (as e.g. in [5]). We stick to Definition 2.2 because it leads to stronger results, in that asking $\mathrm{H}_2(\rho_{XB}|B)$ to be large is a weaker requirement than asking $\tilde{\mathrm{H}}_2(\rho_{XB}|B)$ to be large, as obviously $\mathrm{H}_2(\rho_{XB}|B) \geq \tilde{\mathrm{H}}_2(\rho_{XB}|B)$, and similarly for the min-entropy.

3   The New Randomness-Extraction Result

We start by recalling the definition of a δ-biased random variable and of a δ-biased family of random variables [20,11].

Definition 3.1. *The* bias *of a random variable A, with respect to $\alpha \in \{0,1\}^n$, is defined as*

$$\mathrm{bias}_\alpha(A) := \sum_a P_A(a)(-1)^{\alpha \cdot a} = 2\big(P[\alpha \cdot A = 1] - \tfrac{1}{2}\big),$$

and A is called δ-biased if $\mathrm{bias}_\alpha(A) \leq \delta$ for all non-zero $\alpha \in \{0,1\}^n$. A family of random variables $\{A_i\}_{i \in \mathcal{I}}$ over $\{0,1\}^n$ is called δ-biased if, for all $\alpha \neq 0$,

$$\sqrt{\mathbb{E}_{i \leftarrow \mathcal{I}}\big[\mathrm{bias}_\alpha(A_i)^2\big]} \leq \delta$$

where the expectation is over a i chosen uniformly at random from $\mathcal{I}$.

Note that by Jensen's inequality, $\mathbb{E}_{i \leftarrow \mathcal{I}}[\mathrm{bias}_\alpha(A_i)] \leq \delta$ for all non-zero α is a necessary (but not sufficient) condition for $\{A_i\}_{i \in \mathcal{I}}$ to be δ-biased. In case though the family consists of only one member, then it is δ-biased if and only if its only member is.

Our main theorem states that if $\{A_i\}_{i \in \mathcal{I}}$ is δ-biased for a small δ, and if an adversary's conditional entropy $\mathrm{H}_2(\rho_{XB}|B)$ on a string $X \in \{0,1\}^n$ is large enough, then masking X with A_i for a random but known i gives an essentially random string.

Theorem 3.2. *Let the density matrix $\rho_{XB} \in \mathcal{P}(\mathcal{H}_X \otimes \mathcal{H}_B)$ be classical on $\mathcal{H}_X$ with $X \in \{0,1\}^n$. Let $\{A_i\}_{i \in \mathcal{I}}$ be a δ-biased family of random variables over $\{0,1\}^n$, and let I be uniformly and independently distributed over $\mathcal{I}$. Then*

$$d\big(\rho_{(A_I \oplus X)BI}\big|BI\big) \leq \delta \cdot 2^{-\frac{1}{2}(\mathrm{H}_2(\rho_{XB}|B) - n)}.$$

By the inequalities

$$H_\infty(X) - \log \dim(\mathcal{H}_B) \leq H_\infty(\rho_{XB}|B) \leq H_2(\rho_{XB}|B),$$

proven in [22], Theorem 3.2 may also be expressed in terms of conditional min-entropy $H_\infty(\rho_{XB}|B)$ or in terms of classical min-entropy of X minus the size of the quantum state (i.e. the number of qubits). If B is the "empty" quantum state, i.e., $\log \dim(\mathcal{H}_B) = 0$, then Theorem 3.2 coincides with Lemma 4 of [11]. Theorem 3.2 also holds, with a corresponding normalization factor, for non-normalized operators, from which it follows that it can also be expressed in terms of the *smooth* conditional min-entropy $H_\infty^\varepsilon(\rho_{XB}|B)$, as defined in [22], as

$$d(\rho_{(A_I \oplus X)BI}|BI) \leq 2\varepsilon + \delta \cdot 2^{-\frac{1}{2}(H_\infty^\varepsilon(\rho_{XB}|B)-n)}.$$

4 The Proof

We start by pointing out some elementary observations regarding the Fourier transform over the hypercube. In particular, we can extend the Convolution theorem and Parseval's identity to the case of matrix-valued functions. Further properties of the Fourier transform (with a different normalization) of matrix-valued functions over the hypercube have recently been established by Ben-Aron, Regev and de Wolf [4]. In Section 4.2, we introduce and recall a couple of properties of the L_2-distance from uniform. The actual proof of Theorem 3.2 is given in Section 4.3.

4.1 Fourier Transform and Convolution

For some fixed positive integer d, consider the complex vector space $\mathcal{MF}$ of all functions $M : \{0,1\}^n \to \mathbb{C}^{d \times d}$. The *convolution* of two such matrix-valued functions $M, N \in \mathcal{MF}$ is the matrix-valued function

$$M * N : x \mapsto \sum_y M(y)N(x-y)$$

and the *Fourier transform* of a matrix-valued function $M \in \mathcal{MF}$ is the matrix-valued function

$$\mathfrak{F}(M) : \alpha \mapsto 2^{-n/2} \sum_x (-1)^{\alpha \cdot x} M(x)$$

where $\alpha \cdot x$ denotes the standard inner product modulo 2. Note that if X is a random variable with distribution P_X and M is the matrix-valued function $x \mapsto P_X(x) \cdot \mathbb{1}$, then

$$\mathfrak{F}(M)(\alpha) = 2^{-n/2} \cdot \mathrm{bias}_\alpha(X) \cdot \mathbb{1}.$$

The *Euclidean* or L_2-norm of a matrix-valued function $M \in \mathcal{MF}$ is given by

$$\|M\|_2 := \sqrt{\mathrm{tr}\left(\sum_x M(x)^\dagger M(x)\right)}$$

where $M(x)^\dagger$ denotes the complex-conjugate transpose of the matrix $M(x)$.[2]

The following two properties known as Convolution Theorem and Parseval's Theorem are straightforward to prove (see Appendix A).

Lemma 4.1. *For all* $M, N \in \mathcal{MF}$:

$$\mathfrak{F}(M * N) = 2^{n/2} \cdot \mathfrak{F}(M) \cdot \mathfrak{F}(N) \qquad and \qquad \|\mathfrak{F}(M)\|_2 = \|M\|_2 .$$

4.2 The L_2-Distance from Uniform

The following lemmas together with their proofs can be found in [22]. Again, we restrict ourselves to the case where ρ_{XB} and σ_B are normalized and ρ_{XB} is classical on X, whereas the claims hold (partly) more generally.

Definition 4.2. *Let* $\rho_{XB} \in \mathcal{P}(\mathcal{H}_X \otimes \mathcal{H}_B)$ *and* $\sigma_B \in \mathcal{P}(\mathcal{H}_B)$. *Then the conditional* L_2-*distance from uniform of* ρ_{XB} *relative to* σ_B *is*

$$d_2(\rho_{XB}|\sigma_B) := \mathrm{tr}\left(\left((\mathbf{1} \otimes \sigma_B^{-1/4})(\rho_{XB} - \rho_U \otimes \rho_B)(\mathbf{1} \otimes \sigma_B^{-1/4}) \right)^2 \right),$$

where $\rho_U := \frac{1}{\dim(\mathcal{H}_X)}\mathbf{1}$ *is the fully mixed state on* $\mathcal{H}_X$.

Lemma 4.3. *Let* $\rho_{XB} \in \mathcal{P}(\mathcal{H}_X \otimes \mathcal{H}_B)$. *Then, for any normalized* $\sigma_B \in \mathcal{P}(\mathcal{H}_B)$,

$$d(\rho_{XB}|B) \le \sqrt{\dim(\mathcal{H}_X)}\sqrt{d_2(\rho_{XB}|\sigma_B)}.$$

Lemma 4.4. *Let* $\rho_{XB} \in \mathcal{P}(\mathcal{H}_X \otimes \mathcal{H}_B)$ *be classical on* $\mathcal{H}_X$ *with* $X \in \mathcal{X}$, *and let* ρ_B^x *be the corresponding normalized conditional operators. Then, for any* $\sigma_B \in \mathcal{P}(\mathcal{H}_B)$

$$d_2(\rho_{XB}|\sigma_B) = \sum_x \mathrm{tr}\left((\sigma_B^{-1/4} P_X(x)\rho_B^x \sigma_B^{-1/4})^2 \right) - \frac{1}{|\mathcal{X}|} \mathrm{tr}\left((\sigma_B^{-1/4} \rho_B \sigma_B^{-1/4})^2 \right).$$

4.3 Proof Theorem 3.2

Write $D_i = A_i \oplus X$ and $D_I = A_I \oplus X$. Since $\rho_{D_I BI} = \frac{1}{|\mathcal{I}|}\sum_i \rho_{D_i B}^i \otimes |i\rangle\langle i| = \frac{1}{|\mathcal{I}|}\sum_i \rho_{D_i B} \otimes |i\rangle\langle i|$, and similar for ρ_{BI}, it follows that the L_1-distance from uniform can be written as an expectation over the random choice of i from $\mathcal{I}$. Indeed

$$d(\rho_{D_I BI}|BI) = \frac{1}{|\mathcal{I}|}\mathrm{tr}\left(\left| \sum_i (\rho_{D_i B} - \rho_U \otimes \rho_B) \otimes |i\rangle\langle i| \right| \right)$$

$$= \frac{1}{|\mathcal{I}|}\sum_i \mathrm{tr}(|\rho_{D_i B} - \rho_U \otimes \rho_B|) = \frac{1}{|\mathcal{I}|}\sum_i d(\rho_{D_i B}|B) = \mathbb{E}_{i \leftarrow \mathcal{I}}[d(\rho_{D_i B}|B)] .$$

---

[2] We will only deal with Hermitian matrices $M(x)$ where $\|M\|_2 = \sqrt{\mathrm{tr}\left(\sum_x M(x)^2\right)}$.

where the second equality follows from the block-diagonal form of the matrix. With Lemma 4.3, the term in the expectation can be bounded in terms of the L_2-distance from uniform, that is, for any normalized $\sigma_B \in \mathcal{P}(\mathcal{H}_B)$,

$$d(\rho_{D_I BI}|BI) \le \sqrt{2^n} \, \mathbb{E}_{i \leftarrow \mathcal{I}}\left[\sqrt{d_2(\rho_{D_i B}|\sigma_B)}\right] \le 2^{n/2}\sqrt{\mathbb{E}_{i \leftarrow \mathcal{I}}\left[d_2(\rho_{D_i B}|\sigma_B)\right]}$$

where the second inequality is Jensen's inequality. By Lemma 4.4, we have for the L_2-distance

$$
\begin{aligned}
&d_2(\rho_{D_i B}|\sigma_B) \\
&= \mathrm{tr}\left(\sum_d (\sigma_B^{-1/4}\, P_{D_i}(d)\rho_B^d\, \sigma_B^{-1/4})^2\right) - \frac{1}{2^n}\,\mathrm{tr}\left((\sigma_B^{-1/4}\,\rho_B\,\sigma_B^{-1/4})^2\right) . \quad (1)
\end{aligned}
$$

Note that

$$
\begin{aligned}
P_{D_i}(d)\rho_B^d &= P_{D_i}(d)\sum_x P_{X|D_i}(x|d)\rho_B^x = \sum_x P_{XD_i}(x,d)\rho_B^x \\
&= \sum_x P_{XA_i}(x, d \oplus x)\rho_B^x = \sum_x P_X(x)P_{A_i}(d \oplus x)\rho_B^x
\end{aligned}
$$

so that the first term on the right-hand side of (1) can be written as

$$
\begin{aligned}
&\mathrm{tr}\left(\sum_d (\sigma_B^{-1/4}\, P_{D_i}(d)\rho_B^d\, \sigma_B^{-1/4})^2\right) \\
&= \mathrm{tr}\left(\sum_d \left(\sum_x P_X(x)\sigma_B^{-1/4}\,\rho_B^x\,\sigma_B^{-1/4}P_{A_i}(d \oplus x)\right)^2\right) .
\end{aligned}
$$

The crucial observation now is that the term that is squared on the right side is the convolution of the two matrix-valued functions $M : x \mapsto P_X(x)\sigma_B^{-1/4}\,\rho_B^x\,\sigma_B^{-1/4}$ and $N : x \mapsto P_{A_i}(x)\mathbf{1}$, and the whole expression equals $\|M * N\|_2^2$. Thus, using Lemma 4.1 we get

$$
\begin{aligned}
\mathrm{tr}\left(\sum_d (\sigma_B^{-1/4}\, P_{D_i}(d)\rho_B^d\, \sigma_B^{-1/4})^2\right) &= \|M * N\|_2^2 = \|\mathfrak{F}(M * N)\|_2^2 \\
&= \|2^{n/2} \cdot \mathfrak{F}(M) \cdot \mathfrak{F}(N)\|_2^2 = 2^n\,\mathrm{tr}\left(\sum_\alpha \left(\mathfrak{F}(M)(\alpha)\mathfrak{F}(N)(\alpha)\right)^2\right) \quad (2) \\
&= \frac{1}{2^n}\,\mathrm{tr}\left((\sigma_B^{-1/4}\,\rho_B\,\sigma_B^{-1/4})^2\right) + \mathrm{tr}\left(\sum_{\alpha \ne 0} \mathfrak{F}(M)(\alpha)^2\,\mathrm{bias}_\alpha(A_i)^2\right),
\end{aligned}
$$

where the last equality uses

$$\mathfrak{F}(M)(0) = 2^{-n/2}\sum_x P_X(x)\sigma_B^{-1/4}\,\rho_B^x\,\sigma_B^{-1/4} = 2^{-n/2}\sigma_B^{-1/4}\,\rho_B\,\sigma_B^{-1/4}$$

as well as

$$\mathfrak{F}(N)(0) = 2^{-n/2} \sum_x P_{A_i}(x)\mathbf{1} = 2^{-n/2}\mathbf{1} \quad \text{and} \quad \mathfrak{F}(N)(\alpha) = 2^{-n/2} \cdot \mathrm{bias}_\alpha(A_i)\mathbf{1}.$$

Substituting (2) into (1) gives

$$d_2(\rho_{D_iB}|\sigma_B) = \mathrm{tr}\left(\sum_{\alpha \neq 0} \mathfrak{F}(M)(\alpha)^2 \,\mathrm{bias}_\alpha(A_i)^2\right).$$

Using the linearity of the expectation and trace, and using the bound on the expected square-bias, we get

$$\mathbb{E}_{i \leftarrow \mathcal{I}}[d_2(\rho_{D_iB}|\sigma_B)] \leq \delta^2 \,\mathrm{tr}\left(\sum_{\alpha \neq 0} \mathfrak{F}(M)(\alpha)^2\right) \leq \delta^2 \,\mathrm{tr}\left(\sum_\alpha \mathfrak{F}(M)(\alpha)^2\right)$$

$$= \delta^2 \|\mathfrak{F}(M)\|_2^2 = \delta^2 \|M\|_2^2 = \delta^2 \sum_x \mathrm{tr}\left(P_X(x)^2(\sigma_B^{-1/4}\,\rho_B^x\,\sigma_B^{-1/4})^2\right)$$

$$= \delta^2 2^{-\,\mathrm{H}_2(\rho_{XB}|\sigma_B)},$$

where the second inequality follows because of

$$\mathrm{tr}\big(\mathfrak{F}(M)(0)^2\big) = 2^{-n}\,\mathrm{tr}\big((\sigma_B^{-1/4}\,\rho_B\,\sigma_B^{-1/4})^2\big) \geq 0.$$

Therefore,

$$d(\rho_{D_IBI}|BI) \leq 2^{n/2}\sqrt{\mathbb{E}_{i \leftarrow \mathcal{I}}[d_2(\rho_{D_iB}|\sigma_B)]} \leq \delta \cdot 2^{-\frac{1}{2}(\mathrm{H}_2(\rho_{XB}|\sigma_B)-n)}$$

and the assertion follows from the definition of $H_2(\rho_{XB}|B)$ because σ_B was arbitrary. $\square$

5   Application I: Entropic Security

Entropic security is a relaxed but still meaningful security definition for (information-theoretically secure) encryption that allows to circumvent Shannon's pessimistic result, which states that any perfectly secure encryption scheme requires a key at least as long as the message to be encrypted. Entropic security was introduced by Russell and Wang [25], and later more intensively investigated by Dodis and Smith [12]. Based on our result, and in combination with techniques from [12], we show how to achieve entropic security against quantum adversaries. We would like to stress that in contrast to perfect security e.g. when using the one-time-pad, entropic security does *not* a priori protect against a quantum adversary.

Informally, entropic security is defined as follows. An encryption scheme is entropically secure if no adversary can obtain any information on the message M from its ciphertext C (in addition to what she can learn from scratch), provided the message M has enough uncertainty from the adversary's point of view. The impossibility of obtaining any information on M is formalized by requiring that any adversary that can compute $f(M)$ for some function f when given C, can also compute $f(M)$ *without* C (with similar success probability). A different formulation, which is named *indistinguishability*, is to require that there exists a random variable C', independent of M, such that C and C' are essentially identically distributed. It is shown in [12], and in [8] for the case of a *quantum* message, that the two notions are equivalent if the adversary's information on M is classical. In recent work, Desrosiers and Dupuis proved this equivalence to hold also for an adversary with quantum information [9].

The adversary's uncertainty on M is formalized, for a *classical* (i.e. non-quantum) adversary, by the *min-entropy* $H_\infty(M|V = v)$ (or, alternatively, the collision-entropy) of M, conditioned on the value v the adversary's view V takes on. We formalize this uncertainty for a quantum adversary in terms of the quantum version of conditional min- or actually collision-entropy, as introduced in Section 2.2.

Definition 5.1. *We call a (possibly randomized) encryption scheme $E : \mathcal{K} \times \mathcal{M} \to \mathcal{C}$ (t,ε)-quantum-indistinguishable if there exists a random variable C' over $\mathcal{C}$ such that for any normalized $\rho_{MB} \in \mathcal{P}(\mathcal{H}_M \otimes \mathcal{H}_B)$ which is classical on $\mathcal{H}_M$ with $M \in \mathcal{M}$ and $H_2(\rho_{MB}|B) \geq t$, we have that*

$$\left\| \rho_{E(K,M)B} - \rho_{C'} \otimes \rho_B \right\|_1 \leq \varepsilon,$$

where K is uniformly and independently distributed over $\mathcal{K}$.

Note that in case of an "empty" state B, our definition coincides with the indistinguishability definition from [12] (except that we express it in collision- rather than min-entropy).

Theorem 3.2, with $\mathcal{I} = \{i_0\}$ and $A_{i_0} = K$, immediately gives a generic construction for a quantum-indistinguishable encryption scheme (with C' being uniformly distributed). Independently, this result was also obtained in [9].

Theorem 5.2. *Let $\mathcal{K} \subseteq \{0,1\}^n$ be such that the uniform distribution K over $\mathcal{K}$ is δ-biased. Then the encryption scheme $E : \mathcal{K} \times \{0,1\}^n \to \{0,1\}^n$ with $E(k,m) = k \oplus m$ is (t,ε)-quantum-indistinguishable with $\varepsilon = \delta \cdot 2^{\frac{n-t}{2}}$.*

Alon *et al.* [1] showed how to construct subsets $\mathcal{K} \subseteq \{0,1\}^n$ of size $|\mathcal{K}| = O(n^2/\delta^2)$ such that the uniform distribution K over $\mathcal{K}$ is δ-biased and elements in $\mathcal{K}$ can be efficiently sampled. With the help of this construction, we get the following result, which generalizes the bound on the key-size obtained in [12] to the quantum setting.

Corollary 5.3. *For any $\varepsilon \geq 0$ and $0 \leq t \leq n$, there exists a (t,ε)-quantum-indistinguishable encryption scheme encrypting n-bit messages with key length $\ell = \log |\mathcal{K}| = n - t + 2\log(n) + 2\log(\frac{1}{\varepsilon}) + O(1)$.*

In the language of extractors, defining a (t,ε)-*quantum extractor* in the natural way as follows, Corollary 5.3 translates to Corollary 5.5 below.

Definition 5.4. *A function $E : \mathcal{J} \times \mathcal{X} \to \{0,1\}^m$ is called a (t,ε)-weak quantum extractor if $d(\rho_{E(J,X)B}|B) \leq \varepsilon$, and a (t,ε)-strong quantum extractor if $d(\rho_{E(J,X)JB}|JB) \leq \varepsilon$ for any normalized $\rho_{XB} \in \mathcal{P}(\mathcal{H}_X \otimes \mathcal{H}_B)$ which is classical on $\mathcal{H}_X$ with $X \in \mathcal{X}$ and $\mathrm{H}_2(\rho_{XB}|B) \geq t$, and where J is uniformly and independently distributed over $\mathcal{J}$.*

Corollary 5.5. *For any $\varepsilon \geq 0$ and $0 \leq t \leq n$, there exists a (t,ε)-weak quantum extractor with n-bit output and seed length $\ell = \log|\mathcal{K}| = n - t + 2\log(n) + 2\log(\frac{1}{\varepsilon}) + O(1)$.*

6   Application II: Private Error Correction

Consider the following scenario. Two parties, Alice and Bob, share a common secret key K. Furthermore, we assume a "random source" which can be queried by Alice and Bob so that on identical queries it produces identical outputs. In particular, when Alice and Bob both query the source on input K, they both obtain the same "raw key" $X \in \{0,1\}^n$. We also give an adversary Eve access to the source. She can obtain some (partial) information on the source and store it possibly in a quantum state ρ_Z. However, we assume she has some uncertainty about X, because due to her ignorance of K, she is unable to extract "the right" information from the source. Such an assumption of course needs to be justified in a specific implementation. Specifically, we require that $\mathrm{H}_\infty(\rho_{XKZ}|KZ)$ is lower bounded, i.e., Eve has uncertainty in X even if at some later point she learns K but only the source has disappeared in the meantime.

Such a scenario for instance arises in the bounded-storage model [19,3] (though with classical Eve), when K is used to determine which bits of the long randomizer Alice and Bob should read to obtain X, or in a quantum setting when Alice sends n qubits to Bob and K influences the basis in which Alice prepares them respectively Bob measures them.

In this setting, it is well-known how to transform by public (authenticated) communication the weakly-secure raw key X into a fully secure key S: Alice and Bob do privacy amplification, as shown in [14,5] in case of a classical Eve, respectively as in [23,22] in case of a quantum Eve. Indeed, under the above assumptions on the entropy of X, privacy amplification guarantees that the resulting key S looks essentially random for Eve even given K. This guarantee implies that S can be used, say, as a one-time-pad encryption key, but it also implies that if Eve learns S, she still has essentially no information on K, and thus K can be safely re-used for the generation of a new key S.

Consider now a more realistic scenario, where due to noise or imperfect measurements Alice's string X and Bob's string X' are close but not exactly equal. There are standard techniques to do error correction (without giving Eve too much information on X): Alice and Bob agree on a suitable error-correcting code $\mathcal{C}$, Alice samples a random codeword C from $\mathcal{C}$ and sends $Y = X \oplus C$ to Bob, who can recover X by decoding $C' = Y \oplus X'$ to the nearest codeword C and compute $X = Y \oplus C$. Or equivalently, in case of a linear code, Alice can send the syndrome of X to Bob, which allows Bob to recover X in a similar manner. If Eve's entropy in X is significantly larger than the size of the syndrome, then one can argue that privacy amplification still works and the resulting key S is still (close to) random given Eve's information (including the syndrome) and K. Thus, S is still a secure key. However, since X depends on K, and the syndrome of X depends on X, the syndrome of X may give information on K to Eve, which makes it insecure to re-use K. A common approach to deal with this problem is to use part of S as the key K in the next session. Such an approach not only creates a lot of inconvenience for Alice and Bob in that they now have to be stateful and synchronized, but in many cases Eve can prevent Alice and Bob from agreeing on a secure key S (for instance by blocking the last message) while nevertheless learning information on K, and thus Eve can still cause Alice and Bob to run out of key material.

In [11], Dodis and Smith addressed this problem and proposed an elegant solution in case of a classical Eve. They constructed a family of codes which not only allow to efficiently correct errors, but at the same time also serve as randomness extractors. More precisely, they show that for every $0 < \lambda < 1$, there exists a family $\{C_j\}_{j \in J}$ of binary linear codes of length n, which allows to efficiently correct a constant fraction of errors, and which is δ-biased for $\delta < 2^{-\lambda n/2}$. The latter is to be understood that the family $\{C_j\}_{j \in J}$ of random variables, where C_j is uniformly distributed over C_j, is δ-biased for $\delta < 2^{-\lambda n/2}$. Applying Lemma 4 of [11] (the classical version of Theorem 3.2) implies that $C_j \oplus X$ is close to random for any X with large enough entropy, given j. Similarly, applying our Theorem 3.2 implies the following.

Theorem 6.1. *For every $0 < \lambda < 1$ there exists a family $\{C_j\}_{j \in J}$ of binary linear codes of length n which allows to efficiently correct a constant fraction of errors, and such that for any density matrix $\rho_{XB} \in \mathcal{P}(\mathcal{H}_X \otimes \mathcal{H}_B)$ which is classical on $\mathcal{H}_X$ with $X \in \{0,1\}^n$ and $H_2(\rho_{XB}|B) \geq t$, it holds that*

$$d\big(\rho_{(C_J \oplus X)BJ}\big|BJ\big) \leq 2^{-\frac{t-(1-\lambda)n}{2}} ,$$

where J is uniformly distributed over J and C_J is uniformly distributed over C_J.

Using a random code from such a family of codes allows to do error correction in the noisy setting described above without leaking information on K to Eve: By the chain rule [22, Sect. 3.1.3], the assumed lower bound on $H_\infty(\rho_{XKZ}|KZ)$ implies a lower bound on $H_\infty(\rho_{XSKZG}|SKZG)$ (essentially the original bound minus the bit length of S), where G is the randomly chosen universal hash function used to extract S from X. Combining systems S, K, Z and G into system B, Theorem 6.1 implies that $\rho_{(C_J \oplus X)SKZGJ} \approx \frac{1}{2^n} \mathbb{1} \otimes \rho_{SKZGJ}$. From standard privacy amplification follows that $\rho_{SKZGJ} \approx \frac{1}{2^\ell} \mathbb{1} \otimes \rho_{KZGJ}$. Using the independence of K, G, J (from Z and from each other), we obtain $\rho_{(C_J \oplus X)SKZGJ} \approx \frac{1}{2^n} \mathbb{1} \otimes \frac{1}{2^\ell} \mathbb{1} \otimes \rho_K \otimes \rho_Z \otimes \rho_G \otimes \rho_J$. This in particular implies that S is a secure key (even when K is given to Eve) and that K is still "fresh" and can be safely re-used (even when S is additionally given to Eve).

Specifically, our private-error-correction techniques allow to add robustness against noise to the bounded-storage model in the presence of a quantum attacker as considered in [17], without the need for updating the common secret key. The results of [17] guarantee that the min-entropy of the sampled substring is lower bounded given the attacker's quantum information and hence, security follows as outlined above. Furthermore, in [7] the above private-error-correction technique is an essential ingredient to add robustness against noise but also to protect against man-in-the-middle attacks in new quantum-identification and quantum-key-distribution schemes in the bounded-quantum-storage model.

In the language of extractors, we get the following result for arbitrary, not necessarily efficiently decodable, binary linear codes.

Corollary 6.2. *Let $\{C_j\}_{j \in J}$ be a δ-biased family of binary linear $[n, k, d]_2$-codes. For any $j \in J$, let G_j be a generator matrix for the code C_j and let H_j be a corresponding parity-check matrix. Then $E : J \times \{0,1\}^n \to \{0,1\}^{n-k}$, $(j, x) \mapsto H_j x$ is a (t, ε)-strong quantum extractor with $\varepsilon = \delta \cdot 2^{\frac{1}{2}(n-t)}$.*

This result gives rise to new privacy-amplification techniques, beyond using universal$_2$ hashing as in [23] or one-bit extractors as in [18]. Note that using arguments from [11], it is easy to see that the condition that $\{C_j\}_{j \in J}$ is δ-biased and thus the syndrome function H_j is a good strong extractor, is equivalent to requiring that $\{G_j\}_{j \in J}$ seen as family of (encoding) functions is δ^2-almost universal$_2$ [30,28].

For a family of binary linear codes $\{C_j\}_{j \in J}$, another equivalent condition for δ-bias of $\{C_j\}_{j \in J}$ is to require that for all non-zero α, $\Pr_{j \in J}[\alpha \in C_j^\perp] \leq \delta^2$, i.e. that the probability that α is in the dual code of C_j is upper bounded by δ^2 [11]. It follows that the family size $|J|$ has to be exponential in n to achieve an exponentially small bias δ and therefore, the seed length $\log |J|$ of the strong extractor will be linear in n as for the case of two-universal hashing.

7   Conclusion

We proposed a new technique for randomness extraction in the presence of a quantum attacker. This is interesting in its own right, as up to date only very few extractors are known to be secure against quantum adversaries, much in contrast to the classical non-quantum case. The new randomness-extraction technique has various cryptographic applications like entropically secure encryption, in the classical bounded-storage model and the bounded-quantum-storage model, and in quantum key distribution. Furthermore, because of the wide range of applications of classical extractors not only in cryptography but also in other areas of theoretical computer science, we feel that our new randomness-extraction technique will prove to be useful in other contexts as well.

Acknowledgments

We would like to thank Ivan Damgård, Renato Renner, and Louis Salvail for helpful discussions and the anonymous referees for useful comments.

References

1. Alon, N., Goldreich, O., Håstad, J., Peralta, R.: Simple constructions of almost k-wise independent random variables. In: 31st Annual IEEE Symposium on Foundations of Computer Science (FOCS), pp. 544–553 (1990)
2. Ambainis, A., Smith, A.: Small pseudo-random families of matrices: Derandomizing approximate quantum encryption. In: Jansen, K., Khanna, S., Rolim, J.D.P., Ron, D. (eds.) RANDOM 2004 and APPROX 2004. LNCS, vol. 3122, pp. 249–260. Springer, Heidelberg (2004)
3. Aumann, Y., Ding, Y.Z., Rabin, M.O.: Everlasting security in the bounded storage model. IEEE Transactions on Information Theory 48(6), 1668–1680 (2002)
4. Ben-Aroya, A., Regev, O., de Wolf, R.: A hypercontractive inequality for matrix-valued functions with applications to quantum computing (2007), http://arxiv.org/abs/0705.3806
5. Bennett, C.H., Brassard, G., Crépeau, C., Maurer, U.M.: Generalized privacy amplification. IEEE Transactions on Information Theory 41, 1915–1923 (1995)
6. Damgård, I.B., Fehr, S., Salvail, L., Schaffner, C.: Oblivious transfer and linear functions. In: Dwork, C. (ed.) CRYPTO 2006. LNCS, vol. 4117, pp. 427–444. Springer, Heidelberg (2006)
7. Damgård, I.B., Fehr, S., Salvail, L., Schaffner, C.: Secure identification and QKD in the bounded-quantum-storage model. In: Menezes, A. (ed.) CRYPTO 2007. LNCS, vol. 4622, pp. 342–359. Springer, Heidelberg (2007)
8. Desrosiers, S.P.: Entropic security in quantum cryptography (2007), http://arxiv.org/abs/quant-ph/0703046
9. Desrosiers, S.P., Dupuis, F.: Quantum entropic security and approximate quantum encryption, (July 5, 2007), http://arxiv.org/abs/0707.0691
10. Dickinson, P.A., Nayak, A.: Approximate randomization of quantum states with fewer bits of key. In: Quantum Computing: Back Action 2006, November 2006. American Institute of Physics Conference Series, vol. 864, pp. 18–36 (2006), http://arxiv.org/abs/quant-ph/0611033
11. Dodis, Y., Smith, A.: Correcting errors without leaking partial information. In: 37th Annual ACM Symposium on Theory of Computing (STOC), pp. 654–663 (2005)
12. Dodis, Y., Smith, A.: Entropic security and the encryption of high entropy messages. In: Kilian, J. (ed.) TCC 2005. LNCS, vol. 3378, pp. 556–577. Springer, Heidelberg (2005)
13. Gavinsky, D., Kerenidis, I., Kempe, J., Raz, R., de Wolf, R.: Exponential separations for one-way quantum communication complexity, with applications to cryptography. In: 39th Annual ACM Symposium on Theory of Computing (STOC), pp. 516–525 (2007), http://arxiv.org/abs/quant-ph/0611209
14. Håstad, J., Impagliazzo, R., Levin, L.A., Luby, M.: A pseudorandom generator from any one-way function. SIAM Journal on Computing 28(4) (1999)
15. Impagliazzo, R., Levin, L.A., Luby, M.: Pseudo-random generation from one-way functions. In: 21st Annual ACM Symposium on Theory of Computing (STOC), pp. 12–24 (1989)
16. Kerenidis, I., Nagaj, D.: On the optimality of quantum encryption schemes. Journal of Mathematical Physics 47, 92–102 (2006), http://arxiv.org/abs/quant-ph/0509169
17. König, R., Renner, R.: Sampling of min-entropy relative to quantum knowledge. In: Workshop on Quantum Information Processing (QIP 2008) (2007)

18. König, R., Terhal, B.M.: The bounded storage model in the presence of a quantum adversary (2006), http://arxiv.org/abs/quant-ph/0608101
19. Maurer, U.M.: A provably-secure strongly-randomized cipher. In: Damgård, I.B. (ed.) EUROCRYPT 1990. LNCS, vol. 473, pp. 361–373. Springer, Heidelberg (1991)
20. Naor, J., Naor, M.: Small-bias probability spaces: efficient constructions and applications. In: 22nd Annual ACM Symposium on Theory of Computing (STOC), pp. 213–223 (1990)
21. Nisan, N., Zuckerman, D.: More deterministic simulation in logspace. In: 25th Annual ACM Symposium on the Theory of Computing (STOC), pp. 235–244 (1993)
22. Renner, R.: Security of Quantum Key Distribution. PhD thesis, ETH Zürich (Switzerland) (September 2005), http://arxiv.org/abs/quant-ph/0512258
23. Renner, R., König, R.: Universally composable privacy amplification against quantum adversaries. In: Kilian, J. (ed.) TCC 2005. LNCS, vol. 3378, pp. 407–425. Springer, Heidelberg (2005)
24. Renner, R., Wolf, S.: Simple and tight bounds for information reconciliation and privacy amplification. In: Roy, B. (ed.) ASIACRYPT 2005. LNCS, vol. 3788, pp. 199–216. Springer, Heidelberg (2005)
25. Russell, A., Wang, H.: How to fool an unbounded adversary with a short key. In: Knudsen, L.R. (ed.) EUROCRYPT 2002. LNCS, vol. 2332, pp. 133–148. Springer, Heidelberg (2002)
26. Shaltiel, R.: Recent developments in explicit constructions of extractors. Bulletin of the EATCS 77, 67–95 (2002)
27. Smith, A.: Private communication (2007)
28. Stinson, D.R.: Universal hashing and authentication codes. In: Feigenbaum, J. (ed.) CRYPTO 1991. LNCS, vol. 576, pp. 74–85. Springer, Heidelberg (1992)
29. Ta-Shma, A.: On extracting randomness from weak random sources. In: 28th Annual ACM Symposium on the Theory of Computing (STOC), pp. 276–285 (1996)
30. Wegman, M.N., Carter, L.: New hash functions and their use in authentication and set equality. J. Comput. Syst. Sci. 22(3), 265–279 (1981)

A   Proof of Lemma 4.1

Concerning the first claim,

$$
\begin{aligned}
\mathfrak{F}(M * N)(\alpha) &= \frac{1}{2^{n/2}} \sum_x (-1)^{\alpha \cdot x} \sum_y M(y) N(x \oplus y) \\
&= 2^{-n/2} \sum_y (-1)^{\alpha \cdot y} M(y) \sum_x (-1)^{\alpha \cdot (x \oplus y)} N(x \oplus y) \\
&= 2^{-n/2} \sum_y (-1)^{\alpha \cdot y} M(y) \sum_z (-1)^{\alpha \cdot z} N(z) \\
&= 2^{n/2} \cdot \mathfrak{F}(M)(\alpha) \cdot \mathfrak{F}(N)(\alpha) .
\end{aligned}
$$

The second claim is argued as follows.

$$
\|\mathfrak{F}(M)\|_2^2 = \mathrm{tr}\left(\sum_\alpha \mathfrak{F}(M)(\alpha)^\dagger \mathfrak{F}(M)(\alpha) \right)
$$

$$= 2^{-n} \operatorname{tr} \left(\sum_\alpha \left(\sum_x (-1)^{\alpha \cdot x} M(x) \right)^* \left(\sum_{x'} (-1)^{\alpha \cdot x'} M(x') \right) \right)$$

$$= 2^{-n} \operatorname{tr} \left(\sum_{x,x'} M(x)^\dagger M(x') \sum_\alpha (-1)^{\alpha \cdot (x \oplus x')} \right)$$

$$= \operatorname{tr} \left(\sum_x M(x)^\dagger M(x) \right) = \|M\|_2^2$$

where the last equality follows from the fact that $\sum_\alpha (-1)^{\alpha \cdot y} = 2^n$ if $y = (0, \ldots, 0)$ and 0 otherwise. $\qquad\square$

An Equivalence Between Zero Knowledge and Commitments

Shien Jin Ong* and Salil Vadhan**

Harvard University, School of Engineering & Applied Sciences
33 Oxford Street, Cambridge, MA 02138, USA
{shienjin,salil}@eecs.harvard.edu

Abstract. We show that a language in NP has a zero-knowledge protocol if and only if the language has an "instance-dependent" commitment scheme. An instance-dependent commitment schemes for a given language is a commitment scheme that can depend on an instance of the language, and where the hiding and binding properties are required to hold only on the YES and NO instances of the language, respectively.

The novel direction is the *only if* direction. Thus, we confirm the widely held belief that commitments are not only sufficient for zero knowledge protocols, but *necessary* as well. Previous results of this type either held only for restricted types of protocols or languages, or used nonstandard relaxations of (instance-dependent) commitment schemes.

1 Introduction

From the early days in the study of zero knowledge, it has seemed that *commitment schemes* are the heart of *zero-knowledge protocols*. Indeed, the first construction of zero-knowledge proofs for all of NP, due to Goldreich, Micali, and Wigderson [GMW], shows that commitment schemes *suffice* for zero knowledge. Moreover, there have been a number of partial converses to this result, showing how to obtain certain kinds of commitments from certain kinds of zero-knowledge protocols for certain kinds of languages. In this paper, we present a complete equivalence between zero knowledge protocols and *instance-dependent* commitment schemes [BMO, IOS], in which the protocol depends on a given instance of a language (or promise problem). Specifically, we show that for every language $L \in$ NP, *L has a zero-knowledge protocol if and only if L has an instance-dependent commitment scheme*. Thus, we confirm the intuition that commitments are not only sufficient for zero knowledge, but necessary as well.

1.1 Review of Zero Knowledge and Commitments

In zero-knowledge protocols [GMR], a *prover* tries to convince a *verifier* that an assertion is true, namely that some string x is a YES instance of a (promise)

---

* Supported by NSF grant CNS-0430336.
** Supported by NSF grant CNS-0430336 and US-Israel BSF grant 2002246. Work done in part when visiting UC Berkeley, supported by a Guggenheim Fellowship and the Miller Institute for Basic Research in Science.

R. Canetti (Ed.): TCC 2008, LNCS 4948, pp. 482–500, 2008.
© International Association for Cryptologic Research 2008

problem Π,[1] without leaking any additional knowledge. Zero-knowledge protocols have two security requirements. Informally, *soundness* says that a cheating prover should not be able to convince the verifier of a false statement, and *zero knowledge* says that a cheating verifier should not be able to learn anything from the interaction other than the fact that the assertion being proven is true. Both security requirements come in two flavors — *statistical*, whereby we require security to hold even against computationally unbounded cheating strategies (except with negligible probability[2]), and *computational*, whereby we only require security against polynomial-time strategies (except with negligible probability). Protocols with statistical soundness are typically called interactive *proof systems* (which constitute the original model proposed by [GMR]), and those with computational soundness are typically called *argument systems* (which were introduced by [BCC]). Considering all combinations of computational and statistical versions of soundness and zero knowledge rise to four main flavors of zero knowledge protocols, and thus four complexity classes consisting of the problems Π having zero-knowledge protocols of a particular flavor. We denote these complexity classes SZKP, CZKP, SZKA, and CZKA, with the prefix of S or C denoting statistical or computational zero knowledge, and the suffix of P or A denoting proof systems (statistical soundness) or argument systems (computational soundness).

A commitment scheme is the cryptographic analogue of a locked box. It is a two-stage interactive protocol between a pair of probabilistic polynomial-time parties, the *sender* and the *receiver*. In the first stage, the sender "commits" to a string m, corresponding to locking an object in the box. In the second stage, the sender "reveals" m to the receiver, corresponding to opening the box. Like zero-knowledge protocols, commitment schemes have two security properties. Informally, *hiding* says that a cheating receiver should not be able to learn anything about m during the commit stage, and *binding* says that a cheating sender should not be able to reveal two different messages after the commit stage. Again, each of these properties can be statistical (holding against computationally unbounded cheating strategies, except with negligible probability) or computational (holding against polynomial-time cheating strategies, except with negligible probability). Thus we again get four flavors of commitment schemes, but it is easily seen to be impossible to simultaneously achieve statistical security for both hiding and binding. However, it is known that if one-way functions exist, then we can achieve statistical security for either one of the security properties [HILL, Nao, NOV, HR]. Conversely, commitment schemes, even with both properties computational, imply one-way functions [IL].

---

[1] A promise problem Π is a pair (Π_Y, Π_N) of disjoint sets of strings, corresponding to the YES instances and NO instances. Given a string x that is "promised" to be in $\Pi_Y \cup \Pi_N$, the task is to decide whether $x \in \Pi_Y$ or $x \in \Pi_N$.

[2] An even stronger notion that statistical security is *perfect security*, where the clause "except with negligible probability" is removed. We will not consider perfect security in this paper.

1.2 Previous Work

The classic construction of Goldreich, Micali, and Wigderson [GMW] shows how to construct zero-knowledge protocols for NP given any commitment scheme. Moreover, the security properties of the commitment scheme translate to the security properties of the zero-knowledge protocol: a statistically (resp., computationally) hiding commitment scheme yields statistical (resp., computational) zero knowledge, and a statistically (resp., computationally) binding commitment scheme yields a proof (resp. argument) system.[3] Thus, if one-way functions exist, CZKP, SZKA, and CZKA are very powerful in that they contain NP (and even the classes MA or IP [IY, BGG+], depending on whether or not we require the honest prover to be efficient).

Several papers, beginning with Damgård [Dam1], gave results of a converse nature, culminating in two theorems of Ostrovsky and Wigderson [OW]. The first theorem shows that a zero-knowledge protocol (of any type[4]) for a *hard-on-average* problem implies the existence of one-way functions. The second theorem shows that a zero-knowledge protocol for any problem that cannot be solved in probabilistic polynomial time (BPP) implies a "weak form" of one-way functions. (For problems in BPP, we do not expect to obtain any implication, since every problem in BPP has a trivial zero-knowledge proof in which the prover sends nothing and the verifier decides on its own.) These results suggest that the nontriviality of zero knowledge is equivalent to the existence of one-way functions, which in turn is equivalent to the existence of commitment schemes [HILL, Nao, IL]. However, they are only partial converses to [GMW], and do not provide an exact characterization of the power of zero knowledge. This is because for problems that are neither hard on average nor in BPP, a zero-knowledge protocol only implies the "weak form" of one-way functions in the second result, which seems too weak to construct commitment schemes and thus zero-knowledge protocols. Finally, note that first direction (one-way functions imply that zero knowledge is powerful) seems to say nothing about SZKP: to get an SZKP protocol out of [GMW], one would need a commitment scheme that is both statistically hiding and statistically binding, which is impossible.

The above difficulties no longer seem inherent, however, if one turns away from one-way functions and standard commitments to *instance-dependent* commitments [BMO, IOS]. These are commitment protocols where the sender and receiver both receive an instance x of some promise problem Π as an auxiliary input. We only require the commitment scheme to be hiding when x is a YES instance and binding when x is a NO instance. For example, GRAPH ISOMORPHISM has a simple instance-dependent commitment scheme: when the auxiliary input is $x = (G_0, G_1)$, the sender commits to a bit $b \in \{0, 1\}$ by sending a random isomorphic copy H of G_b, and reveals b by sending the isomorphism between H

---

[3] In [GMW], only computational zero-knowledge proof systems were considered. The original construction of statistical zero-knowledge arguments for NP [BCC] used stronger cryptographic primitives than commitment schemes.

[4] The results of [OW] are stated for CZKP, but are easily seen to hold even for the most general class CZKA.

and G_b. This protocol is perfectly hiding when $G_0 \cong G_1$, and perfectly binding when $G_0 \not\cong G_1$. It is possible to achieve both perfect hiding and perfect binding because we do not require the properties to hold at the same time.

As shown by Itoh, Ohta, and Shizuya [IOS], this relaxation of commitment schemes remains useful for constructing zero-knowledge protocols, because in many constructions, the hiding property is used for zero knowledge (which is required only when x is a YES instance) and the binding property is used for soundness (which is required ony when x is a NO instance). For example, using [GMW], we see that if a promise problem $\Pi \in$ NP has an instance-dependent commitment scheme, then Π has a zero-knowledge protocol, where the hiding property (statistical or computational) translates to the zero-knowledge property and the binding property translates to the soundness property.

In the last few years, there has been substantial progress on proving the converse: if a problem has a zero-knowledge protocol, then it has an instance-dependent commitment scheme. This progress started with SZKP, where both security properties are statistical.

- It was conjectured in [MV] that every problem in SZKP has an instance-dependent commitment scheme. As a first step, they constructed an instance-dependent commitment scheme for a *restricted version* of STATISTICAL DIFFERENCE, one of the complete problems for SZKP [SV].
- In [Vad], it was shown that SZKP consists exactly of the problems with instance-dependent commitment schemes in which the sender is computationally unbounded rather than polynomial time. The unbounded sender renders the result useless for the study of zero knowledge with efficient honest provers, which was the motivation of [MV]. But the result was useful for the study of CZKP; see below.
- In [NV], the sender was made efficient, at the price of working with a new variant of commitments, called *1-out-of-2-binding commitments* (denoted as $\binom{2}{1}$-binding). These $\binom{2}{1}$-binding commitments were shown to be sufficient for constructing zero-knowledge proofs for NP, but are otherwise cumbersome and of unclear value as cryptographic primitives on their own.
- In [Vad, OV], the classes involving computational security, namely CZKP, SZKA, and CZKA, were characterized in terms of SZKP and "instance-dependent one-way functions." Thus, combining the above types of instance-dependent commitments for SZKP with constructions of commitments from one-way functions [HILL, Nao, NOV, HR], the classes CZKP, SZKA, and CZKA could be characterized in terms of instance-dependent commitments, but inheriting the deficiencies of [Vad, OV] (namely, an unbounded sender or $\binom{2}{1}$-binding).[5] These instance-dependent commitments played a crucial role in the characterization of the classes CZKP, SZKA, and CZKA, and in proving various unconditional results about these classes (such as equiv-

---

[5] The proceedings version of [OV] actually quotes the main result (Theorem 1) of this present paper. However, this was done only to simplify the presentation there, and the main results of [OV] were actually obtained prior to Theorem 1.

alence of honest-verifier and cheating-verifier zero knowledge and closure under union).

- Instance-dependent commitments for a restricted class of zero-knowledge proofs, namely *3-round public-coin* zero-knowledge proofs, were implicit in the works of Damgård [Dam1, Dam2]. Indeed, Kapron, Malka, and Srinivasan [KMS] used Damgård's techniques to show that 3-round public-coin zero-knowledge proofs where the verifier just sends a single random bit — called *V-bit protocols* — are exactly characterized by *noninteractive* instance-dependent commitments.[6]

1.3 Our Results

In this paper, we show that zero knowledge proofs are equivalent to standard instance-dependent commitments, where the sender is efficient and there is no non-standard $\binom{2}{1}$-binding property. The main technical contribution is the construction for SZKP:

Theorem 1. *For every promise problem Π, $\Pi \in$ SZKP if and only if Π has an instance-dependent commitment scheme that is statistically hiding on the YES instances and statistically binding on the NO instances. Moreover, every $\Pi \in$ SZKP has an instance-dependent commitment scheme that is public coin and is constant round.*

As mentioned previously, a construction of instance-dependent commitments for SZKP implies ones for the other classes (by their characterizations in terms SZKP and instance-dependent one-way functions [Vad, OV] together with the constructions of commitments from one-way functions [HILL, Nao, NOV, HR]).

Corollary 1. *The following hold for every problem $\Pi \in$ NP:[7]*

1. *$\Pi \in$ CZKP if and only if Π has an instance-dependent commitment scheme that is computationally hiding on the YES instances and statistically binding on the NO instances. Moreover, this instance-dependent commitment scheme is public coin and is constant round.*
2. *$\Pi \in$ SZKA if and only if Π has an instance-dependent commitment scheme that is statistically hiding on the YES instances and computationally binding on the NO instances. Moreover, this instance-dependent commitment scheme is public coin.*

---

[6] *Noninteractive commitments* are commitments where the sender commits to a message in the commit stage by sending a *single* message to the receiver; hence, the receiver does not send any message, both in the commit and reveal stages.

[7] We state the result for problems in NP for simplicity. The direction stating that zero-knowledge implies instance-dependent commitments (which is our main contribution) actually holds without any constraint on Π other than being in the stated zero-knowledge class. The other direction actually generalizes to problems in MA when the honest prover is required to be efficient and IP when the honest prover is allowed to be computationally unbounded.

3. $\Pi \in$ CZKA *if and only if* Π *has an instance-dependent commitment scheme that is computationally hiding on the YES instances and computationally binding on the NO instances. Moreover, this instance-dependent commitment scheme is public coin.*

Note that for the case of *proof* systems (i.e., statistical binding), our instance-dependent commitment schemes are constant round. (For arguments, the polynomial round complexity is inherited from the statistically hiding commitments based on one-way functions [NOV, HR].) This enables us to resolve some open questions regarding the round complexity of zero-knowledge proofs. For example:

Corollary 2. *Every problem in* SKZP *(resp.,* CZKP $\cap$ NP[8]*) has a constant-round, public-coin statistical (resp., computational) zero-knowledge proof system with soundness error* $1/\operatorname{poly}(n)$ *and a black-box simulator.*[9]

It was known how to achieve constant rounds for CZKP under the assumption that one-way functions exist, but it was not known for SZKP under any assumption. Previously, it was only known that SZKP had constant-round *honest-verifier* statistical zero-knowledge proofs, and these were private coin [Oka].

Since SZKP is closed under complement [Oka], we can also obtain instance-dependent commitments in which the security properties are reversed (i.e., statistically binding on YES instances and statistically hiding on NO instances). Such commitments are useful for implementing commitments from the verifier. Using such commitments in the protocol of [GK1] (or, more easily, [Ros]), we obtain:

Corollary 3. *Every problem in* SZKP *has a constant-round (private-coin) zero-knowledge proof system with negligible soundness error.*

Following [MOSV], a potential application of our instance-dependent commitments is to show that every problem in SZKP has a *concurrent* statistical zero-knowledge proof system with $\omega(\log n)$ rounds. However, the analysis of [MOSV] is given only for noninteractive commitments, so it would need to be extended to handle our interactive commitments.

1.4 Overview of Our Techniques

Our proof of Theorem 1 uses techniques from Nguyen and Vadhan [NV] and Haitner and Reingold [HR]. Recall that [NV] constructed instance-dependent $\binom{2}{1}$-binding commitments for SZKP. In the standard, non-instance-dependent setting, [HR] showed how to convert $\binom{2}{1}$-binding commitments into standard

---

[8] This result actually generalizes to CZKP $\cap$ AM. No further restriction is needed for SZKP because SZKP $\subseteq$ AM [AH].

[9] Using [GMW] would yield a poor soundness error of $1 - 1/\operatorname{poly}(n)$. To obtain soundness error $1/\operatorname{poly}(n)$, we use an $O(\log n)$-fold parallel repetition of [Blu]. Negligible soundness error cannot be achieved with public coins and black-box simulation for problems outside BPP [GK2].

commitments using *universal one-way hash functions* [NY], whose existence is equivalent to that of one-way functions [Rom, KK]. Thus, we obtain our result by constructing an instance-dependent analogue of universal one-way hash functions for every problem in SZKP, and then applying the Haitner & Reingold transformation.

It is not immediately clear, however, how to define instance-dependent universal one-way hash functions in a way that allows for statistical security (as we need for Theorem 1). The standard definition of a universal one-way hash family is as a family $\mathcal{H}$ of *length-decreasing* functions $h\colon \{0,1\}^n \to \{0,1\}^m$, such that for every fixed $y \in \{0,1\}^n$, if we are given a random $h \xleftarrow{\text{R}} \mathcal{H}$, it is infeasible to find an $y' \neq y$ such that $h(y') = h(y)$. Note that the latter property is necessarily computational. Since the hash functions are length-decreasing, $h(y)$ will have many preimages y' with high probability over a random y and h, and thus an unbounded adversary could find a collision easily. We observe, however, that the Haitner & Reingold transformation [HR] does *not* really require a length-decreasing function. They only use the fact that $h(y)$ typically has many preimages, and they only use this to establish the hiding property of the resulting commitment scheme. For the binding property, they use infeasibility of finding collisions; for statistical security, this amounts to the functions being nearly injective. With these observations, our notion of an instance-dependent universal one-way hash family $\mathcal{H}_x$ is as a family of functions (typically not length decreasing) that also depend on an instance x of some promise problem Π. When x is a YES instance, a random hash function from the family has large preimages with high probability, and when x is a NO instance, a random hash function is nearly injective with high probability. We show that every problem in SZKP has an instance-dependent universal one-way hash family of this type, and thus are able to apply the Haitner & Reingold transformation to the $\binom{2}{1}$-binding commitments of [NV] to obtain our result.

1.5 Organization

In Section 2, we provide definitions to terminologies used in this paper. We prove our main result, Theorem 1, in Section 3. The proof of Corollary 1 can be found in [OV], and the proofs of the other corollaries will appear in the full version of this paper.

2 Preliminaries

If X is a random variable taking values in a finite set $\mathcal{U}$, then we write $x \xleftarrow{\text{R}} X$ to indicate that x is selected according to X. If S is a subset of $\mathcal{U}$, then $x \xleftarrow{\text{R}} S$ means that x is selected according to the uniform distribution on S. We adopt the convention that when the same random variable occurs several times in an expression, they refer to a single sample. For example, $\Pr[f(X) = X]$ is defined to be the probability that when $x \xleftarrow{\text{R}} X$, we have $f(x) = x$. We write U_n to denote the random variable distributed uniformly over $\{0,1\}^n$.

A function $\varepsilon : \mathbb{N} \mapsto [0,1]$ is called *negligible* if $\varepsilon(n) = n^{-\omega(1)}$. We let $\text{neg}(n)$ denote an arbitrary negligible function (i.e., when we say that $f(n) < \text{neg}(n)$ we mean that *there exists* a negligible function $\varepsilon(n)$ such that for every n, $f(n) < \varepsilon(n)$). Likewise, $\text{poly}(n)$ denotes an arbitrary function $f(n) = n^{O(1)}$.

PPT refers to probabilistic algorithms (i.e., Turing machines) that run in *strict* polynomial time. A *nonuniform* PPT algorithm is a pair $(A, \bar{z})$, where $\bar{z} = z_1, z_2, \ldots$ is an infinite sequence of strings where $|z_n| = \text{poly}(n)$, and A is a PPT algorithm that receives pairs of inputs of the form $(x, z_{|x|})$. (The string z_n is the called the *advice string* for A for inputs of length n.) Nonuniform PPT algorithms are equivalent to (nonuniform) families of polynomial-sized Boolean circuits.

Promise problems. Roughly speaking, a *promise problem* [ESY] is a decision problem where some inputs are excluded. Formally, a promise problem is specified by two disjoint sets of strings $\Pi = (\Pi_Y, \Pi_N)$, where we call Π_Y the set of *YES instances* and Π_N the set of *NO instances*. Such a promise problem is associated with the following computational problem: given an input that is "promised" to lie in $\Pi_Y \cup \Pi_N$, decide whether it is in Π_Y or in Π_N. Note that languages are a special case of promise problems (namely, a language L over alphabet Σ corresponds to the promise problem $(L, \Sigma^* \setminus L)$). Thus working with promise problems makes our results more general. Moreover, even to prove our results just for languages, it turns out to be extremely useful to work with promise problems along the way. All of the complexity classes in this paper are taken to be classes of promise problems. We refer the reader to the recent survey of Goldreich [Gol] for more on the utility and subtleties of promise problems.

2.1   Instance-Dependent Cryptographic Primitives

Instance-dependent functions. It will be very useful for us to work with cryptographic primitives that may depend on an instance x of a problem $\Pi = (\Pi_Y, \Pi_N)$, and where the security condition will hold only if x is in some particular set $I \subseteq \{0,1\}^*$. We begin our discussion of instance-dependent primitives with the following definition.

Definition 1. *An instance-dependent function is a family $\mathcal{F} = \{f_x \colon \{0,1\}^{n(|x|)} \to \{0,1\}^{m(|x|)}\}_{x \in \{0,1\}^*}$, where $n(\cdot)$ and $m(\cdot)$ are polynomials. We call $\mathcal{F}$ polynomial-time computable if there is a deterministic polynomial-time algorithm F such that for every $x \in \{0,1\}^*$ and $y \in \{0,1\}^{n(|x|)}$, we have $F(x, y) = f_x(y)$.*

To simplify notation, we often write $f_x \colon \{0,1\}^{n(|x|)} \to \{0,1\}^{m(|x|)}$ to mean the family $\{f_x \colon \{0,1\}^{n(|x|)} \to \{0,1\}^{m(|x|)}\}_{x \in \{0,1\}^*}$.

Indistinguishability of instance-dependent ensembles. The notions of statistical and computational indistinguishability have instance-dependent analogues. But first, we define an instance-dependent analogue of probability ensembles.

Definition 2. *An* instance-dependent probability ensemble *is a collection of random variables* $\{A_x\}_{x \in \{0,1\}^*}$, *where* A_x *takes values in* $\{0,1\}^{p(|x|)}$ *for some polynomial p. We call such an ensemble* samplable *if there is a probabilistic polynomial-time algorithm M such that for every x, the output M(x) is distributed according to* A_x.

Definition 3. *Two* instance-dependent probability ensembles $\{A_x\}_{x \in \{0,1\}^*}$ *and* $\{B_x\}_{x \in \{0,1\}^*}$ *are* computationally indistinguishable on $I \subseteq \{0,1\}^*$ *if for every nonuniform PPT D, there exists a negligible function* ε *such that for all* $x \in I$,

$$|\Pr[D(x, A_x) = 1] - \Pr[D(x, B_x) = 1]| \leq \varepsilon(|x|) \ .$$

Similarly, we say that $\{A_x\}_{x \in \{0,1\}^*}$ *and* $\{B_x\}_{x \in \{0,1\}^*}$ *are* statistically indistinguishable on $I \subseteq \{0,1\}^*$ *if the above is required for all functions D, instead of only nonuniform PPT ones. Equivalently,* $\{A_x\}_{x \in \{0,1\}^*}$ *and* $\{B_x\}_{x \in \{0,1\}^*}$ *are statistically indistinguishable on I iff* A_x *and* B_x *are have statistical distance at most* $\varepsilon(|x|)$ *for some negligible function* ε *and all* $x \in I$. *We write* $\approx_c$ *and* $\approx_s$ *to denote computational and statistical indistinguishability, respectively.*

Instance-dependent commitments. We give a definition of instance-dependent commitment schemes that extends the standard (that is, non-instance dependent) definition of commitment schemes in a natural way. Note that in our definition below, the reveal stage is *noninteractive* (that is, consisting of a single message from the sender to the receiver). This because in the reveal stage, without loss of generality, we can have the sender provide the receiver the random coin tosses it used in the commit stage, and have the receiver verify consistency.

Definition 4. *An* instance-dependent commitment scheme *is a family of protocols* $\{\mathsf{Com}_x\}_{x \in \{0,1\}^*}$ *with the following properties:*

1. *Scheme* Com_x *proceeds in two stages: a* commit stage *and a* reveal stage. *In both stages, the sender and receiver receive instance x as common input, and hence we denote the sender and receiver as* S_x *and* R_x, *respectively, and write* $\mathsf{Com}_x = (S_x, R_x)$.
2. *At the beginning of the commit stage, sender* S_x *receives a private input* $b \in \{0,1\}$, *which denotes the bit that S is supposed to commit to. At the end of the commit stage, both sender* S_x *and receiver* R_x *output a commitment* c.
3. *In the reveal stage, sender* S_x *sends a pair* (b, d), *where d is the* decommitment string for bit b. *Receiver* R_x *accepts or rejects based on x, b, d, and* c.
4. *The sender* S_x *and receiver* R_x *algorithms are computable in polynomial time (in* $|x|$), *given x as auxiliary input.*
5. *For every* $x \in \{0,1\}^*$, R_x *will always accept (with probability 1) if both sender* S_x *and receiver* R_x *follow their prescribed strategy.*

Instance-dependent commitment scheme $\{\mathsf{Com}_x = (S_x, R_x)\}_{x \in \{0,1\}^*}$ *is* public coin *if for every* $x \in \{0,1\}^*$, *all messages sent by* R_x *are independent random coins.*

To simplify notation, we write Com_x or (S_x, R_x) to denote instance-dependent commitment scheme $\{\mathsf{Com}_x = (S_x, R_x)\}_{x \in \{0,1\}^*}$.

The hiding and binding properties of standard commitments extend in a natural way to their instance-dependent analogues.

Definition 5. *Instance-dependent commitment scheme* $\mathsf{Com}_x = (S_x, R_x)$ *is statistically [resp., computationally] hiding on* $I \subseteq \{0,1\}^*$ *if for every [resp., nonuniform PPT]* R^*, *the ensembles* $\{\mathrm{view}_{R^*}(S_x(0), R^*)\}_{x \in I}$ *and* $\{\mathrm{view}_{R^*}(S_x(1), R^*)\}_{x \in I}$ *are statistically [resp., computationally] indistinguishable, where random variable* $\mathrm{view}_{R^*}(S_x(b), R^*)$ *denotes the view of* R^* *in the commit stage interacting with* $S_x(b)$. *For a problem* $\Pi = (\Pi_Y, \Pi_N)$, *an instance-dependent commitment scheme* Com_x *for* Π *is statistically [resp., computationally] hiding on the YES instances if* Com_x *is statistically [resp., computationally] hiding on* Π_Y.

Definition 6. *Instance-dependent commitment scheme* $\mathsf{Com}_x = (S_x, R_x)$ *is statistically [resp., computationally] binding on* $I \subseteq \{0,1\}^*$ *if for every [resp., nonuniform PPT]* S^*, *there exists a negligible function* ε *such that for all* $x \in I$, *the malicious sender* S^* *succeeds in the following game with probability at most* $\varepsilon(|x|)$.

> S^* *interacts with* R_x *in the commit stage obtaining commitment* c. *Then* S^* *outputs pairs* $(0, d_0)$ *and* $(1, d_1)$, *and succeeds if in the reveal stage,* $R_x(0, d_0, c) = R_x(1, d_1, c) = \mathsf{accept}$.

For a problem $\Pi = (\Pi_Y, \Pi_N)$, *an instance-dependent commitment scheme* Com_x *for* Π *is statistically [resp., computationally] binding on the NO instances if* Com_x *is statistically [resp., computationally] binding on* Π_N.

1-out-of-2-binding commitments. A *1-out-of-2-binding commitment scheme*—denoted as $\binom{2}{1}$-binding—is a commitment schemes with two *sequential* and *related* phases such that in each phase, the sender commits to and reveals a value. (They are related in the sense that the protocol for the second phase takes the transcript of the first phase as a common input to both the sender and receiver, and the sender may maintain private state between the two phases.) The hiding property of such commitments is strong: we require that at the end of each commit stage, the receiver has not learned anything about the value to which the sender is committing. The binding property, however, is relatively weak. It only says that it is infeasible for a cheating sender to break the commitment in *both* phases. That is, with high probability over the first commit stage, there is at most one value to which the sender can open that will result in the second phase being non-binding. A formal definition can be found in [NV, Sect. 2] (cf., [Ong, Sect. 3.4.1]).

3 Instance-Dependent Commitments for SZKP

Our goal in this section is to prove Theorem 1. We begin by recalling the result of Nguyen and Vadhan [NV], which is the starting point for our work. Their

construction started off from the SZKP-complete problem ENTROPY DIFFER-
ENCE [GV], ED = (ED$_Y$, ED$_N$), defined as:

$$ED_Y = \{(X, Y) : H(X) \geq H(Y) + 1\};$$
$$ED_N = \{(X, Y) : H(X) \leq H(Y) - 1\},$$

where X and Y are random variables specified by circuits that same from them
(by evaluating the circuit on a uniformly random input), and $H(\cdot)$ denotes the
(Shannon) entropy, i.e., $H(Z) = E_{z \leftarrow Z}[\log(1/\Pr[Z = z])]$. We assume, without
loss of generality, that the size of the circuits X and Y are upper bounded by
the square of their respective input lengths. (This can be guaranteed by padding
dummy input variables to circuits.)

The [NV] construction of instance-dependent schemes for ED does not provide
a commitment scheme with a standard binding property, but rather with the
weaker $\binom{2}{1}$ *binding* property (cf., Sect. 2.1). These commitments, even though
with a weaker binding property, suffice for getting efficient-prover statistical
zero-knowledge proofs for all of SZKP ∩ NP [NV].

Our construction of instance-dependent commitments for all of SZKP will
follow the same approach as [NV], except at the place where they get stuck with
$\binom{2}{1}$-binding commitments, we convert them into commitments with the standard
binding property using the ideas of Haitner and Reingold [HR]. Specifically, we
use an instance-dependent variant of the Haitner & Reingold transformation to
convert $\binom{2}{1}$-binding commitments into commitments with the standard binding
property.

The commitments of [NV] were not constructed directly from ED, but instead
utilized a Cook reduction from ED to a *restricted version* of the ENTROPY
APPROXIMATION [GSV] problem, denoted as EA' = (EA'$_Y$, EA'$_N$), and defined
below:[10]

$$EA'_Y = \{(X, t) : H(X) \geq t + 1, \text{ and } |X| \leq n^2\};$$
$$EA'_N = \{(X, t) : t - 1/n^{14} \leq H(X) \leq t, \text{ and } |X| \leq n^2\}.$$

Here n denotes the number of input gates to the circuit encoding X, and $|X|$ is
the size of that circuit. The condition $|X| \leq n^2$ simply allows us to use n as the
security parameter, even though the security properties of instance-dependent
commitment schemes are defined in terms of the size of the instance (X, t).

The problem EA' is considered a restricted version of ENTROPY APPROXIMA-
TION because (unrestricted version of) the ENTROPY APPROXIMATION problem
EA = (EA$_Y$, EA$_N$) does not lower-bound the entropy in the case of the NO
instances. EA is defined as follows:

---

[10] The definition of EA' in [NV] has an additional 'security parameter' k, which is
eventually set to $\max\{n^{14}, |X|\}$. For convenience, we have restricted to the case
that $|X| \leq n^2$; this is without loss of generality for ED and is preserved in the
reduction from ED to EA' in Proposition 1 below. Under this restriction, we can
simply set $k = n^14$, resulting in our definition of EA'.

$$EA_Y = \{(X,t) : H(X) \geq t + 1\};$$
$$EA_N = \{(X,t) : H(X) \leq t\}.$$

For instances of EA, we will assume, without loss of generality, that the size of the circuit X is upper bounded by the square of its input length (similar to what we assumed for instances of ED).

The Cook reduction from ED to EA' is established by the following proposition.

Proposition 1. *(Cook Reduction from* ED *to* EA'*; from [NV, Lem. 4.9], which builds on [GSV].) Let* (X,Y) *be an instance of the* ENTROPY DIFFERENCE *problem* ED $=$ (ED_Y, ED_N), *where the circuits encoding the random variables* X *and* Y *both have input length* n *and are of size at most* n^2 *(wlog). The Cook reduction from* ED *to* EA' *is as follows:*

$$(X,Y) \in ED_Y \Rightarrow \bigvee_{i=0}^{n \cdot k} \left((Y, i/k) \in \overline{EA'}_Y \wedge \bigwedge_{j=0}^{i} (X, j/k) \in EA'_Y \right) ;$$

$$(X,Y) \in ED_N \Rightarrow \bigwedge_{i=0}^{n \cdot k} \left((Y, i/k) \in \overline{EA'}_N \vee \bigvee_{j=0}^{i} (X, j/k) \in EA'_N \right) ,$$

where $k = n^{14}$.

Note that the reduction from ED to EA' in the above proposition does not alter the circuits; hence, the size of the circuits in both problems remain upper bounded by the square of their respective input lengths, which is what we require.

Using Proposition 1, [NV] noted that it suffices to construct instance-dependent commitments for both EA' and its complement $\overline{EA'}$ in order to obtain instance-dependent commitments for ED and hence all of SZKP. We capture that observation in the following lemma.

Lemma 1. *Suppose that both the special case of the* ENTROPY APPROXIMATION *problem* EA' *and its complement* $\overline{EA'}$ *have instance-dependent commitments. That is,*

- *there exist instance-dependent commitments that are statistically hiding on instances in* EA'_Y *and statistically binding on instances in* EA'_N, *and*
- *there exist instance-dependent commitments that are statistically hiding on instances in* EA'_N *and statistically binding on instances in* EA'_Y.

Then the ENTROPY DIFFERENCE *problem* ED *(and hence, every problem in SZKP) has an instance-dependent commitment scheme that is statistically hiding on the YES instances and statistically binding on the NO instances.*

Indeed, [NV] constructed instance-dependent schemes for both EA' and $\overline{EA'}$. Their scheme for EA' is a standard instance-dependent commitment scheme, but for $\overline{EA'}$, they only managed to only get a weaker $\binom{2}{1}$-binding commitment scheme (cf., Sect. 2.1).

Lemma 2. *(From [NV, Thm. 4.4].) The problem* EA' *has an instance-dependent commitment scheme that is statistically hiding on YES instances (namely, instances in* EA'$_Y$*) and statistically binding on NO instances (namely, instances in* EA'$_N$*). Moreover, this scheme is public coin and constant round.*

Lemma 3. *(From [NV, Thm. 4.5].) The problem* $\overline{\text{EA}'}$ *has an instance-dependent 2-phase commitment scheme that is statistically hiding on the YES instances (namely, instances in* EA'$_N$*) and statistically* $\binom{2}{1}$ *binding on NO instances (namely, instances in* EA'$_Y$*). Moreover, this scheme is public coin and constant round.*

3.1   The Haitner and Reingold Transformation

To obtain instance-dependent commitments (with the standard binding property) for $\overline{\text{EA}'}$, we use an instance-dependent variant of the Haitner & Reingold transformation [HR], which we informally describe now. (A detailed description can be found in [HNO+, Sect. 7].)

Overview of [HR]. The $\binom{2}{1}$ binding property of 2-phase commitment schemes states that it is infeasible for an adversarial sender S^* to break both phases of the commitment, but nonetheless it might be possible for S^* to break one of the two phases of its choice. With this in mind, suppose that after the first commitment phase, receiver R flips a coin *phase* $\leftarrow \{1, 2\}$. If *phase* $= 1$, the first commitment phase is used to do the commitment. On the other hand, if *phase* $= 2$, the second commitment phase is used to do the commitment (this is done by S^* revealing its first-phase commitment, and then proceeding to the second phase with R). Intuitively, this would make the scheme binding (with probability $1/2$) if S^* chooses which of the two phases it wants to break in advance. The problem, however, is that S^* could choose the phase that it wants to break after seeing the value of *phase*.

A way to overcome this problem is to force the adversary S^* to decide which of the two phases it wants to break before seeing the value of *phase*. Haitner and Reingold [HR] achieved this by having S^* send back a value $y = f(\sigma)$ before the value of *phase* is announced by the receiver R, where σ is the message committed to by S^* in the first phase, and f is a random hash function from a universal one-way hash family. A *universal one-way hash family* [NY] is a family of length-decreasing hash functions such that it is hard to find collisions with any particular value of x specified in advance. In other words, for a value of σ announced before a random hash function f is selected from that family, any efficient algorithm will not be able to find another σ' such that $f(\sigma') = f(\sigma)$. This property of a universal one-way hash family is termed *target collision resistance* by Bellare and Rogaway [BR].

We first argue the hiding property of this new scheme. Before y is sent, the value of σ, the message committed in the first phase, is hidden. If hash function f is *compressing enough*, then the value of $y = f(\sigma)$ leaks at most a few bits of *information* about σ, so the *entropy* of σ given y is still large. This means that we can apply a pairwise-independent hash on σ to get an almost uniform value (by the Leftover Hash Lemma [HILL]). Thus, this new scheme is hiding when $phase = 1$. When $phase = 2$, the sender reveals σ and proceeds on to the second phase, which is used for the commitment. In this case, the hiding property of this new scheme follows from the hiding property of the second commitment phase.

Next, we argue the binding property of this new scheme by making the following observation: the $\binom{2}{1}$ binding property says that after the first commitment phase, there exists at most one value of σ^* that allows an adversarial sender S^* to cheat in the second phase. In other words, if S^* reveals to a value other than σ^*, the second phase will be binding.

When it is the sender's turn to send y, after receiving a random hash function f from receiver R, sender S^* could decide to either send $y = f(\sigma^*)$ or send $y \neq f(\sigma^*)$. If it decides to send $y = f(\sigma^*)$, and if R selects $phase = 1$ following that, then S^* is bound to a single value, since to decommit to two different values it will have to reveal a $\sigma' \neq \sigma^*$ with $f(\sigma') = y = f(\sigma^*)$, and this is infeasible by the target collision resistance property of f. (The value of σ^* is determined by the first-phase commitment, which is completed before a random f is selected.) Instead if it decides to send $y \neq f(\sigma^*)$, and if R selects $phase = 2$ following that, then S^* will have to reveal to a value other than σ^* for its first-phase commitment. In this case, the commitments are done in the second phase, and by the $\binom{2}{1}$ binding property, this phase is guaranteed to be binding. Since the value of $phase$ is independent of y, both cases happen with probability $1/2$, which would make our scheme binding with probability close to $1/2$.

3.2 Instance-Dependent UOWHFs

Our approach to construction standard instance-dependent commitments for $\overline{EA}$' and hence all of SZKP is to carry out an instance-dependent analogue of the Haitner & Reingold transformation. To do this, we want to construct an instance-dependent analogue of universal one-way hash functions $\overline{EA}$' and apply it to the $\binom{2}{1}$-binding commitments of Lemma 3. Since we want instance-dependent commitments with statistical security (and are not making any complexity assumptions), we need to formulate the properties of universal one-way hash functions in a way that allows for statistical security. The properties used in the Haitner & Reingold transformation are that the functions should be compressing (used for hiding) and target collision-resistant (used for binding). Thus the first attempt would be to require that our instance-dependent universal one-way hash functions are compressing on YES instances and statistically target collision-resistant on NO instances. However, it seems unlikely that this is possible. Indeed, it would imply that $\overline{EA}$' is in BPP: statistical target collision resistance implies that the functions are not compressing, so we could distinguish YES and NO instances simply be checking whether the functions are compressing or not. To get around this difficulty, we observe that the hiding analysis sketched above only requires that σ retains a lot of entropy given $y = f(\sigma)$. This property can hold for non-compressing functions; it simply says that f has large preimage sizes.

This leads to the following definition.

Definition 7. *Problem* $\Pi = (\Pi_Y, \Pi_N)$ *has an* instance-dependent universal one-way hash family *if there exists a polynomial-time computable family* $\mathcal{F} = \bigcup_x \mathcal{F}_x = \{f \colon \{0,1\}^{n(|x|)} \to \{0,1\}^{m(|x|)}\}$, *where* $n(\cdot)$ *and* $m(\cdot)$ *are polynomials, such that the following two conditions hold.*

- *The family* $\mathcal{F}_Y = \bigcup_{x \in \Pi_Y} \mathcal{F}_x$ *has the* large preimages property*: there exists a function* $\alpha(\cdot) = \omega(1)$ *and a negligible function* ε, *such that the following holds for all* $x \in \Pi_Y$ *and every function* $f \in \mathcal{F}_x$:

$$\Pr_{y \leftarrow \{0,1\}^{n(|x|)}} \left[|f^{-1}(f(y))| \geq |x|^{\alpha(|x|)} \right] \geq 1 - \varepsilon(|x|) \ .$$

- *The family* $\mathcal{F}_N = \bigcup_{x \in \Pi_N} \mathcal{F}_x$ *has* statistical target collision resistance*: there exists a negligible function* ε *such that for every A, the following holds for all* $x \in \Pi_Y$ *and every* $y \in \{0,1\}^{n(|x|)}$:

$$\Pr_{f \leftarrow \mathcal{F}_x} \left[|f^{-1}(f(y))| = 1 \right] \geq 1 - \varepsilon(|x|) \ .$$

Following the discussion above, this definition allows $m(|x|) > n(|x|)$, and only insist that the family has the large preimages property on the YES instances. In fact, our construction of an instance-dependent universal one-way hash family for $\overline{\text{EA'}}$ will be such that $m(|x|)$ is much larger than $n(|x|)$.

With the above definition, we have the following instance-dependent analogue of the Haitner & Reingold transformation, obtained as corollary of Theorem 7.20 in [HNO+]:

Proposition 2. *(Corollary of [HNO+, Thm. 7.20].) Let* $\Pi = (\Pi_Y, \Pi_N)$ *be a promise problem, and suppose that the following two conditions hold:*

- *there exists an instance-dependent universal one-way hash family* $\mathcal{F} = \bigcup_x \mathcal{F}_x$ *for* Π, *and*
- *there is an instance-dependent 2-phase commitment scheme* $(\mathsf{S}_x, \mathsf{R}_x)$ *for* Π *that is statistically hiding on the YES instances, and statistically* $\binom{2}{1}$ *binding on NO instances.*

Then, there is an instance-dependent commitment scheme $(\mathsf{S}_x, \mathsf{R}_x)$ *for* Π *that is statistically hiding on the YES instances, and statistically binding on NO instances. Moreover,* $(\mathsf{S}_x, \mathsf{R}_x)$ *is public coin if* $(\mathsf{S}_x, \mathsf{R}_x)$ *is.*

Based on the above proposition, it suffices to construct an instance-dependent universal one-way hash family for $\overline{\text{EA'}}$ in order to get instance-dependent commitments for $\overline{\text{EA'}}$.

3.3   Instance-Dependent UOWHF for $\overline{\text{EA}}$

Although we just need an instance-dependent universal one-way hash family for $\overline{\text{EA'}}$, we will construct one for the slightly more general problem of $\overline{\text{EA}}$.

Working directly with $\overline{\text{EA}}$ on an instance (X, t) is difficult since we do not know any structure of the random variable X other than its entropy bound. To get more structure, we *flatten* X by taking multiple independent samples of it and outputting all of them. Let X' denote this new random variable. Doing this makes the probability masses of X' concentrated around $2^{-\text{H}(X')}$, and this is why we call it flattening the random variable. (This is also known as the Asymptotic Equipartition Property in the information theory literature; see [CT].) Following [GV], we give a quantitative definition of *flatness* as follows:

Definition 8. *Random variable X is Θ-flat if for every $r \geq 1$,*

$$\Pr_{x \leftarrow X} \left[2^{-r \cdot \Theta} < \frac{\Pr[X = x]}{2^{\text{H}(X)}} < 2^{r \cdot \Theta} \right] > 1 - 2^{-r^2} .$$

Consider the flattened version of the ENTROPY APPROXIMATION problem, denoted as FLATEA = (FLATEA$_\text{Y}$, FLATEA$_\text{N}$), and defined as follows:

$$\text{FLATEA}_\text{Y} = \{(X, t) : \text{H}(X) \geq t + n^{14/15}, X \text{ is } n^{8/15}\text{-flat, and } |X| \leq n^2\}$$

$$\text{FLATEA}_\text{N} = \{(X, t) : \text{H}(X) \leq t, X \text{ is } n^{8/15}\text{-flat, and } |X| \leq n^2\}$$

Here n denotes the number of input gates to the circuit encoding X, and $|X|$ is the size of that circuit. Recall that the condition $|X| \leq n^2$ simply allows us to use n as the security parameter, even though the security properties of instance-dependent commitment schemes are defined in terms of the size of the instance (X, t). Note that the entropy gap between the two cases is close to being linear in n, whereas the deviation from flatness is close to being $\sqrt{n}$.

It is clear that FLATEA is polynomial time reducible to EA, and the reverse reduction from EA to FLATEA is achieved by taking many (e.g. n^{28}) independent copies of X (cf., the Flattening Lemma of [GV, Lem. 3.5]). Hence, constructing an instance-dependent universal one-way hash family for $\overline{\text{EA}}$ is equivalent to constructing one for $\overline{\text{FLATEA}}$, and we do this next.

Theorem 2. *The complement of the flattened version of the* ENTROPY APPROXIMATION *problem, namely* $\overline{\text{FLATEA}}$ *has an instance-dependent universal one-way hash family.*

In the remaining of this section, we abuse notation by using $X : \{0, 1\}^n \rightarrow \{0, 1\}^m$ to denote the circuit that samples random variable X.

Proof Idea of Theorem 2

For the problem FLATEA, we will need to construct an instance-dependent (family of) functions that have statistical target collision resistance on the YES instances and large preimages property on the NO instances. These are reversed properties because we want to prove that the complement $\overline{\text{FLATEA}}$ has an instance-dependent universal one-way hash family.

For the YES instances of FLATEA, X has entropy at least $t + \gamma$, where $\gamma = n^{14/15}$. Since X is a *nearly flat* random variable, most of its preimages are small, i.e., their sizes are $\lesssim 2^{n-t-\gamma}$. So with high probability over a random $y \leftarrow \{0,1\}^n$, the preimage size of $X(y)$ is $\lesssim 2^{n-t-\gamma}$. By applying a pairwise-independent hash $h\colon \{0,1\}^n \to \{0,1\}^\beta$ to y, for $\beta \gtrsim n-t-\gamma$, it would make the function $g_h(y) = (X(y), h(y))$ *almost injective*, in that for almost every element in the range has a unique preimage. (An injective function is, by definition, collision resistant.)

The adversary, however, need not choose y uniformly at random; in particular, it could choose an element y such that $X^{-1}(X(y))$ is large, making $f(y) = (X(y), h(y))$ no longer injective. To prevent the adversary from gaining, we add a *shift* $s \in \{0,1\}^n$ to the circuit X. Specifically, let the new function be $f_{s,h}(y) = (X(y \oplus s), h(y))$. Since y is now randomly shifted by s, the preimage size of $X(y \oplus s)$ is small with high probability over a random $s \leftarrow \{0,1\}^n$. Thus, we can conclude that $f_{s,h}(y)$ is *almost injective* even for an adversarially chosen y. This will give us the desired target collision resistance property for $\beta \gtrsim n - t - \gamma$.

For the NO instances of FLATEA, X has entropy at most t. Since X is a *nearly-flat* random variable, most of its preimages are large, i.e., their sizes are $\gtrsim 2^{n-t}$. Restricting to a hash $h\colon \{0,1\}^n \to \{0,1\}^\beta$ will shrink the size of the preimages by a factor of approximately $2^{-\beta}$. So if $\beta \lesssim n - t$, the size of the preimages will still be large enough to satisfy the large preimages property.

The fact that the entropy gap $\gamma = n^{14/15}$ between the YES and NO instances is much greater than the deviation $\Theta = n^{8/15}$ from flatness is what allows us to find an appropriate value of β between $n - t - \gamma$ and $n - t$ that satisfies both cases. A complete proof will be given in the full version of this paper.

Acknowledgments

We are grateful to Iftach Haitner and Omer Reingold for a stimulating collaboration during the time that the results of this paper were obtained.

References

[AH]   Aiello, W., Håstad, J.: Statistical zero-knowledge languages can be recognized in two rounds. J. Comput. Syst. Sci. 42(3), 327–345 (1991)

[BCC]   Brassard, G., Chaum, D., Crépeau, C.: Minimum disclosure proofs of knowledge. J. Comput. Syst. Sci. 37(2), 156–189 (1988)

[BGG+]   Ben-Or, M., Goldreich, O., Goldwasser, S., Håstad, J., Kilian, J., Micali, S., Rogaway, P.: Everything provable is provable in zero-knowledge. In: Goldwasser, S. (ed.) CRYPTO 1988. LNCS, vol. 403, pp. 37–56. Springer, Heidelberg (1990)

[Blu]   Blum, M.: How to prove a theorem so no one else can claim it. In: Proc. International Congress of Mathematicians, pp. 1444–1451 (1987)

[BMO]   Bellare, M., Micali, S., Ostrovsky, R.: Perfect zero-knowledge in constant rounds. In: Proc. 22nd STOC, pp. 482–493 (1990)

[BR]   Bellare, M., Rogaway, P.: Collision-resistant hashing: towards making UOWHFs practical. In: Kaliski Jr., B.S. (ed.) CRYPTO 1997. LNCS, vol. 1294, pp. 470–484. Springer, Heidelberg (1997)

[CT] Cover, T.M., Thomas, J.A.: Elements of information theory, 2nd edn. Wiley-Interscience, New York (2006)

[Dam1] Damgård, I.: On the existence of bit commitment schemes and zero-knowledge proofs. In: Brassard, G. (ed.) CRYPTO 1989. LNCS, vol. 435, pp. 17–27. Springer, Heidelberg (1990)

[Dam2] Damgård, I.B.: Interactive hashing can simplify zero-knowledge protocol design without computational assumptions. In: Stinson, D.R. (ed.) CRYPTO 1993. LNCS, vol. 773, pp. 100–109. Springer, Heidelberg (1994)

[ESY] Even, S., Selman, A.L., Yacobi, Y.: The complexity of promise problems with applications to public-key cryptography. Inform. Control 61(2), 159–173 (1984)

[GK1] Goldreich, O., Kahan, A.: How to construct constant-round zero-knowledge proof systems for NP. J. Cryptol. 9(3), 167–190 (1996)

[GK2] Goldreich, O., Krawczyk, H.: On the composition of zero-knowledge proof systems. SIAM Journal on Computing 25(1), 169–192 (1996)

[GMR] Goldwasser, S., Micali, S., Rackoff, C.: The knowledge complexity of interactive proof systems. SIAM Journal on Computing 18(1), 186–208 (1989)

[GMW] Goldreich, O., Micali, S., Wigderson, A.: Proofs that yield nothing but their validity or all languages in NP have zero-knowledge proof systems. J. ACM 38(1), 691–729 (1991)

[Gol] Goldreich, O.: On promise problems (a survey in memory of Shimon Even [1935-2004]). Technical Report TR05-018, Electronic Colloquium on Computational Complexity (February 2005)

[GSV] Goldreich, O., Sahai, A., Vadhan, S.: Can statistical zero-knowledge be made non-interactive?, or On the relationship of SZK and NISZK. In: Wiener, M.J. (ed.) CRYPTO 1999. LNCS, vol. 1666, pp. 467–484. Springer, Heidelberg (1999)

[GV] Goldreich, O., Vadhan, S.P.: Comparing entropies in statistical zero knowledge with applications to the structure of SZK. In: Proc. 14th Computational Complexity, pp. 54–73 (1999)

[HILL] Håstad, J., Impagliazzo, R., Levin, L.A., Luby, M.: A pseudorandom generator from any one-way function. SIAM J. Comput. 28(4), 1364–1396 (1999)

[HNO+] Haitner, I., Nguyen, M.-H., Ong, S.J., Reingold, O., Vadhan, S.: Statistically hiding commitments and statistical zero-knowledge arguments from any one-way function. Preliminary versions appeared as [NOV] and [HR], (in submission, 2007), http://eecs.harvard.edu/~salil/papers/SHcommit-abs.html

[HR] Haitner, I., Reingold, O.: Statistically-hiding commitment from any one-way function. In: Proc. 39th STOC, pp. 1–10 (2007)

[IL] Impagliazzo, R., Luby, M.: One-way functions are essential for complexity based cryptography. In: Proc. 30th FOCS, pp. 230–235 (1989)

[IOS] Itoh, T., Ohta, Y., Shizuya, H.: A language-dependent cryptographic primitive. J. Cryptol. 10(1), 37–49 (1997)

[IY] Impagliazzo, R., Yung, M.: Direct minimum-knowledge computations (extended abstract). In: Pomerance, C. (ed.) CRYPTO 1987. LNCS, vol. 293, pp. 40–51. Springer, Heidelberg (1988)

[KK] Katz, J., Koo, C.-Y.: On constructing universal one-way hash functions from arbitrary one-way functions. Technical Report 2005/328, Cryptology ePrint Archive (2005)

[KMS]   Kapron, B., Malka, L., Srinivasan, V.: A characterization of non-interactive instance-dependent commitment-schemes (NIC). In: Arge, L., Cachin, C., Jurdziński, T., Tarlecki, A. (eds.) ICALP 2007. LNCS, vol. 4596, pp. 328–339. Springer, Heidelberg (2007)

[MOSV] Micciancio, D., Ong, S.J., Sahai, A., Vadhan, S.: Concurrent zero knowledge without complexity assumptions. In: Naor, M. (ed.) TCC 2004. LNCS, vol. 2951, pp. 1–20. Springer, Heidelberg (2004)

[MV]   Micciancio, D., Vadhan, S.: Statistical zero-knowledge proofs with efficient provers: lattice problems and more. In: Boneh, D. (ed.) CRYPTO 2003. LNCS, vol. 2729, pp. 282–298. Springer, Heidelberg (2003)

[Nao]   Naor, M.: Bit commitment using pseudorandomness. J. Cryptol. 4(2), 151–158 (1991)

[NOV]   Nguyen, M.-H., Ong, S.J., Vadhan, S.: Statistical zero-knowledge arguments for NP from any one-way function. In: Proc. 47th FOCS, pp. 3–14 (2006)

[NV]   Nguyen, M.-H., Vadhan, S.: Zero knowledge with efficient provers. In: Proc. 38th STOC, pp. 287–295 (2006)

[NY]   Naor, M., Yung, M.: Universal one-way hash functions and their cryptographic applications. In: Proc. 21st STOC, pp. 33–43 (1989)

[Oka]   Okamoto, T.: On relationships between statistical zero-knowledge proofs. J. Comput. Syst. Sci. 60(1), 47–108 (2000)

[Ong]   Ong, S.J.: Unconditional Relationships within Zero Knowledge. PhD thesis, Harvard University, Cambridge (May 2007)

[OV]   Ong, S.J., Vadhan, S.: Zero knowledge and soundness are symmetric. In: Naor, M. (ed.) EUROCRYPT 2007. LNCS, vol. 4515, pp. 187–209. Springer, Heidelberg (2007) Earlier version appeared as TR06-139 in the Electronic Colloquium on Computational Complexity

[OW]   Ostrovsky, R., Wigderson, A.: One-way functions are essential for non-trivial zero-knowledge. In: Proceedings of the 2nd Israel Symposium on Theory of Computing Systems, pp. 3–17. IEEE Computer Society, Los Alamitos (1993)

[Rom]   Rompel, J.: One-way functions are necessary and sufficient for secure signatures. In: Proc. 22nd STOC, pp. 387–394 (1990)

[Ros]   Rosen, A.: A note on constant-round zero-knowledge proofs for NP. In: Naor, M. (ed.) TCC 2004. LNCS, vol. 2951, pp. 191–202. Springer, Heidelberg (2004)

[SV]   Sahai, A., Vadhan, S.: A complete problem for statistical zero knowledge. J. ACM 50(2), 196–249 (2003)

[Vad]   Vadhan, S.P.: An unconditional study of computational zero knowledge. SIAM J. Comput. 36(4), 1160–1214 (2006)

Interactive and Noninteractive Zero Knowledge are Equivalent in the Help Model[*]

André Chailloux[1],[**], Dragos Florin Ciocan[2],[***], Iordanis Kerenidis[1],[**], and Salil Vadhan[2],[***]

[1] LRI, Université Paris-Sud, Orsay, France
{andre.chailloux,jkeren}@lri.fr
[2] School of Engineering and Applied Sciences, Harvard University, Cambridge, MA
{ciocan,salil}@eecs.harvard.edu

Abstract. We show that interactive and noninteractive zero-knowledge are equivalent in the 'help model' of Ben-Or and Gutfreund (*J. Cryptology*, 2003). In this model, the shared reference string is generated by a probabilistic polynomial-time dealer who is given access to the statement to be proven. Our results do not rely on any unproven complexity assumptions and hold for statistical zero knowledge, for computational zero knowledge restricted to AM, and for quantum zero knowledge when the help is a pure quantum state.

Keywords: cryptography, computational complexity, noninteractive zero-knowledge proofs, commitment schemes, Arthur–Merlin games, quantum zero knowledge.

1 Introduction

Zero-knowledge proofs [4] are protocols whereby a prover can convince a verifier that some assertion is true with the property that the verifier learns nothing else from the protocol. This remarkable property is easily seen to be impossible for the classical notion of a proof system, where the proof is a single string sent from the prover to the verifier, as the proof itself constitutes 'knowledge' that the verifier could not have feasibly generated on its own (assuming NP $\not\subseteq$ BPP). Thus zero-knowledge proofs require some augmentation to the classical model for proof systems.

The original proposal of Goldwasser, Micali, and Rackoff [4] augments the classical model with both randomization and multiple rounds of interaction between

---

[*] Preliminary versions of this work previously appeared on the Cryptology ePrint Archive [1,2], and in the second author's undergraduate thesis [3].
[**] Supported in part by ACI Securité Informatique SI/03 511 and ANR AlgoQP grants of the French Ministry and in part by the European Commission under the Integrated Project Qubit Applications (QAP) funded by the IST directorate as Contract Number 015848.
[***] Supported by NSF Grant CNS-0430336. Some of this work was done when the S. Vadhan was visiting U.C. Berkeley, supported by a Guggenheim Fellowship and the Miller Institute for Basic Research in Science.

R. Canetti (Ed.): TCC 2008, LNCS 4948, pp. 501–534, 2008.
© International Association for Cryptologic Research 2008

the prover and the verifier, leading to what are called *interactive zero-knowledge proofs*, or simply *zero-knowledge proofs*. An alternative model, proposed by Blum, Feldman, and Micali [5,6], augments the classical model with a set-up in which a trusted dealer randomly generates a *reference string* that is shared between the prover and verifier. After this reference string is generated, the proof consists of a single message from the prover to the verifier. Thus, these are referred to as *noninteractive zero-knowledge proofs*. Since their introduction, there have been many constructions of both interactive and noninteractive zero-knowledge proofs, and both models have found numerous applications in the construction of cryptographic protocols.

It is natural to ask what is the relation between these two models, that is:

> Can every assertion that can be proven with an interactive zero-knowledge proof also be proven with a noninteractive zero-knowledge proof?

Our main result is a positive answer to this question in the 'help model' of Ben-Or and Gutfreund [7], where the dealer is given access to the statement to be proven when generating the reference string. We hope that this will serve as a step towards answering the above question for more standard models of noninteractive zero knowledge, such as the common reference string model and the public parameter model.

1.1 Models of Zero Knowledge

Interactive Zero Knowledge. Recall that an *interactive proof system* [4] for a problem Π is an interactive protocol between a computationally unbounded prover P and a probabilistic polynomial-time verifier V that satisfies the following two properties:

- *Completeness:* if x is a YES instance of Π, then the V will accept with high probability after interacting with the P on common input x.
- *Soundness:* if x is a NO instance of Π, then for every (even computationally unbounded) prover strategy P^*, V will reject with high probability after interacting with P^* on common input x.

Here, we consider problems Π that are not only languages, but also ones that are *promise problems*, meaning that some inputs can be neither YES nor NO instances, and we require nothing of the protocol on such instances. (Put differently, we are 'promised' that the input x is either a YES or a NO instance.) We write IP for the class of promise problems possessing interactive proof systems.

As is common in complexity-theoretic studies of interactive proofs and zero knowledge, we allow the honest prover P to be computationally unbounded, and require soundness to hold against computationally unbounded provers. However, cryptographic applications of zero-knowledge proofs typically require an honest prover P that can be implemented in probabilistic polynomial-time given a witness of membership for x, and it often suffices for soundness to hold only for polynomial-time prover strategies P^* (leading to *interactive argument systems* [8]). It was recently shown how to extend the complexity-theoretic studies of interactive zero knowledge proofs to both polynomial-time honest provers [9], and to argument systems [10]; we hope that the same will eventually happen for noninteractive zero knowledge.

Intuitively, we say that an interactive proof system is *zero knowledge* if the verifier 'learns nothing' from the interaction other than the fact that the assertion being proven is true, even if the verifier deviates from the specified protocol. Formally, we require that there is an efficient algorithm, called the *simulator*, that can simulate the verifier's view of the interaction given only the YES instance x and no access to the prover P. The most general notion, *computational zero knowledge* or just *zero knowledge*, requires this to hold for all polynomial-time cheating verifier strategies (and the simulation should be computationally indistinguishable from the verifier's view). A stronger notion, *statistical zero knowledge*, requires security against even computationally unbounded verifier strategies (and the simulation should be statistically indistinguishable from the verifier's view). We write ZK (resp., SZK) to denote the class of promise problems possessing computational (resp., statistical) zero-knowledge proof systems.

Noninteractive Zero Knowledge. For noninteractive zero knowledge [5,6], we introduce a trusted third party, the *dealer*, who randomly generates a *reference string* that is provided to both the prover and verifier. After that, the prover sends a single message to the verifier, who decides whether to accept or reject. Completeness and soundness are defined analogously to interactive proofs, except that the probabilities are now also taken over the choice of the reference string. Computational and statistical zero knowledge are also defined analogously to the interactive case, except that now the reference string is also considered part of the verifier's view, and must also be simulated.

There are a number of variants of the noninteractive model, depending on the form of the trusted set-up performed by the dealer. In the original, *common random string (crs) model* proposed by Blum et al. [5,6], the reference string is simply a uniformly random string of polynomial length. This gives rise to the classes NIZK^{crs} and NISZK^{crs} of problems having noninteractive computational and statistical zero-knowledge proofs in the common random string model. A natural and widely used generalization is the *public parameter model*, where the reference string need not be uniform, but can be generated according to any polynomial-time samplable distribution. That is, we obtain the reference string by running a probabilistic polynomial-time *dealer* algorithm D on input 1^n, where n is the length of statements to be proven (or the security parameter). This model gives rise to the classes NIZK^{pub} and NISZK^{pub}.

A further generalization is the *help model* introduced by Ben-Or and Gutfreund [7]. In this model, the distribution of the reference string is allowed to depend on the statement x being proven. That is, the reference string is generated by running a probabilistic polynomial-time dealer algorithm D on input x. We denote the class of problems having computational (resp. statistical) zero-knowledge proofs in this model as NIZK^h (resp., NISZK^h). This model does not seem to suffice for most cryptographic applications, but its study may serve as a stepping stone towards a better understanding of the more standard models of noninteractive zero knowledge mentioned above. Indeed, any characterizations of noninteractive zero knowledge in the help model already serve as *upper bounds* on the power of noninteractive zero knowledge in the common random string and public parameter models.

We remark that one can also consider protocols in which we allow both a trusted dealer and many rounds of interaction. The most general model allows both help and interaction, yielding the classes ZK^h and SZK^h.

Quantum Interactive and Noninteractive Zero Knowledge. The definitions of interactive proofs and zero knowledge extend naturally to the quantum setting. A *quantum interactive proof system* ([11]) for a promise problem Π is an interactive protocol between a computationally unbounded prover P and a quantum polynomial-time verifier V that satisfies completeness and soundness properties as in the classical case and where the interaction is via quantum messages.

For quantum zero knowledge [12], we require that the verifier's view (which consists of qubits) can be simulated by a quantum polynomial-time machine. QSZK denotes the class of promise problems possessing quantum statistical zero-knowledge proof systems. Kobayashi [13] defined quantum noninteractive zero knowledge by having a dealer generate and share a maximally entangled quantum state between the prover and verifier. We write QNISZK to denote the class of promise problems possessing such quantum noninteractive statistical zero-knowledge proof systems.

In this paper, we define two more variants of the quantum noninteractive model, depending on the form of the trusted help created by the dealer. When the help is a *pure* quantum state that depends on the statement x being proven we have the class $QNISZK^h$. When the help is a *mixed* quantum state that depends on x, we have the class $QNISZK^{mh}$. Last, the class $QSZK^h$ refers to protocols where we allow both a pure quantum help and interaction.

1.2   Previous Work

Recall that we are interested in the relationship between the interactive zero-knowledge classes ZK and SZK and their various noninteractive counterparts, which we will denote by NIZK and NISZK when we do not wish to specify the model. That is, for a given model of noninteractive zero knowledge, we ask: Does ZK = NIZK and SZK = NISZK?

ZK vs. NIZK. A first obstacle to proving equality of ZK and NIZK is that NIZK is a subset of AM, the class of problems having constant-round interactive proof systems [14,15], whereas ZK may contain problems outside of AM. So, instead of asking whether ZK = NIZK, we should instead ask if ZK ∩ AM = NIZK.

Indeed, this equality is known to hold under complexity assumptions. If one-way permutations exist, then it is known that ZK = IP [16,17,18] and $NIZK^{crs}$ = AM [19], and thus ZK ∩ AM = $NIZK^{crs}$ = $NIZK^{pub}$ = $NIZK^h$. (In fact, if we replace NIZK^{crs} with NIZK^{pub}, these results hold assuming the existence of any one-way function [20,21,22,23].) Thus, for computational zero knowledge, the interesting question is whether we can prove that ZK $\cap$ AM = NIZK *unconditionally*, without assuming the existence of one-way functions. To our knowledge, there have been no previous results along these lines.

SZK vs. NISZK. For relating SZK and NISZK, the class AM no longer is a barrier, because it is known that SZK $\subseteq$ AM [24].

The relationship between SZK and NISZK was first addressed in the work of Goldreich et al. [25]. There it was shown that SZK and NISZK^{crs} have the 'same complexity' in the sense that SZK = BPP iff NISZK^{crs} = BPP. Moreover, it was proven that SZK = NISZK^{crs} iff NISZK^{crs} is closed under complement.

In addition to introducing the help model, Ben-Or and Gutfreund [7] studied the relationship between NISZK^{h} and SZK. They proved that NISZK$^{\mathrm{h}} \subseteq$ SZK (in fact that SZK^{h} = SZK), and posed as an open question whether SZK $\subseteq$ NISZK^{h}.[1]

1.3   Our Results

We show that interactive zero knowledge does in fact collapse to noninteractive zero knowledge in the help model, both for the computational case (restricted to AM) and the statistical case:

Theorem 1. ZK $\cap$ AM = NIZK^{h}.

Theorem 2. SZK = NISZK^{h}.

These results and their proofs yield new characterizations of the classes ZK and SZK. For example, we obtain a new complete problem for SZK, namely the NISZK^{h}-complete problem given in [7]. Similarly, we obtain a new characterization of ZK, which amounts to a computational analogue of the NISZK^{h}-complete problem. As suggested in [7], these results can also be viewed as first steps towards collapsing interactive zero knowledge to noninteractive zero knowledge in the public parameter or common reference string model. For example, to show SZK = NISZK^{crs} (the question posed in [26]), it now suffices to show that NISZK^{h} = NISZK^{crs}.

As mentioned above, one can consider even more general classes ZK^{h} and SZK^{h} that incorporate both help and interaction. Ben-Or and Gutfreund [7] showed that SZK^{h} = SZK. We prove an analogous result for computational zero knowledge:

Theorem 3. ZK^{h} = ZK.

In the quantum setting, very little is known about the relation of interactive and noninteractive quantum zero knowledge. Here, we start by providing two complete problems for the class QNISZK. Then, we define two variants of quantum

---

[1] In fact, their conference paper [22] claimed to prove that SZK = NISZK^{h}, but this was retracted in the journal version [7].

noninteractive zero knowledge depending on the 'help' created by the dealer. In the case where the help is a *pure* quantum state that depends on the input x, we prove an analogue of Theorem 2:

Theorem 4. $\mathrm{QNISZK}^h = \mathrm{QSZK} = \mathrm{QSZK}^h$.

In the case where the help is a *mixed* quantum state, we show that the class QNISZK^{mh} contains AM and hence is most probably larger than QSZK.

1.4 Techniques

Here we sketch the techniques underlying the forward inclusions in Theorems 1 and 2, showing that interactive zero knowledge is a subset of noninteractive zero knowledge in the help model.

We begin with the case of statistical zero knowledge. Our proof that SZK $\subseteq$ NISZK^h is similar to the approach suggested by Goldreich *et al.* [25] for showing that SZK $=$ NISZK^{crs}. They showed that this question boils down to proving that co-$\mathrm{NISZK}^{crs} = \mathrm{NISZK}^{crs}$ or in other words that the complement of the NISZK^{crs}-complete problem ENTROPY APPROXIMATION belongs to NISZK^{crs}. Similarly, the core part of our proof is showing that co-$\mathrm{NISZK}^{crs} \subseteq \mathrm{NISZK}^h$, which then we use to deduce that SZK $\subseteq \mathrm{NISZK}^h$.

More specifically, our goal is to reduce the SZK-complete problem ENTROPY DIFFERENCE (ED) to the NISZK-complete problem IMAGE INTERSECTION DENSITY (IID). Following [25], we start by reducing ED to *several* instances of ENTROPY APPROXIMATION (EA) and its complement ($\overline{\mathrm{EA}}$). We know that EA $\in$ NISZK^h since by definition $\mathrm{NISZK}^{crs} \subseteq \mathrm{NISZK}^h$. Next, inspired by Ben-Or and Gutfreund's attempt [22] to reduce ED to IID and relying on ideas from [27,28], we prove that $\overline{\mathrm{EA}}$ also belongs to NISZK^h. Thus we obtain a reduction from ED to several instances of IID. We then conclude our proof by showing that NISZK^h has enough boolean closure properties to combine these several instances into a *single* instance of IID. We establish these closure properties of NISZK^h and IID using techniques developed in [27,29] to show boolean closure properties for interactive SZK.

In the case of computational zero knowledge, we prove that ZK∩AM $\subseteq \mathrm{NIZK}^h$ by using certain variants of *commitment schemes*. Recall that a commitment scheme is a two-stage interactive protocol between a *sender* and a *receiver*. In the *commit stage*, the sender 'commits' to a secret message m. In the *reveal stage*, the sender 'reveals' m and tries to convince the verifier that it was the message committed to in the first stage. Commitments should be *hiding*, meaning that an adversarial receiver will learn nothing about m in the commit stage, and *binding*, meaning that after the commit stage, an adversarial sender should not be able to successfully reveal two different messages (except with negligible probability). Each of these security properties can be either *computational*, holding against polynomial-time adversaries, or *statistical*, holding even for computationally unbounded adversaries. Commitments are a basic building block for zero-knowledge protocols, e.g. they are the main cryptographic primitive used in the constructions of zero-knowledge proofs for all of NP [16] and IP [17,18].

A relaxed notion is that of *instance-dependent commitment schemes* [30,31,32]. Here the sender and receiver are given an instance x of some problem Π as auxiliary input. We only require the scheme to be hiding if x is a YES instance, and only require it to be binding if x is a NO instance. They are a relaxation of standard commitment schemes because we do not require hiding and binding to hold simultaneously. Still, as observed in [31], an instance-dependent commitment scheme for a problem $\Pi \in$ IP suffices to construct zero-knowledge proofs for Π because the constructions of [16,17,18] only use the hiding property for zero knowledge (which is only required on YES instances), and the binding property for soundness (which is only required on NO instances).

We show that a similar phenomenon holds for noninteractive zero knowledge in the help model: If a problem $\Pi \in$ AM has a certain kind of instance-dependent commitment scheme, then $\Pi \in$ NIZK^h. For this, the instance-dependent commitments naturally need to be *noninteractive*. On the other hand, they only need to be binding (on NO instances) in case the sender is *honest* during the commit phase. (Our observation is that such commitments can be used to implement the hidden bits model of [19].)

Thus our task is reduced to showing that every problem in ZK has a noninteractive instance-dependent commitment scheme that is computationally hiding on YES instances and statistically binding for honest senders on NO instances. To prove this, we begin by observing that a problem Π has such an instance-dependent commitment scheme with *statistical* hiding if and only if Π reduces to IID. Hence, the needed commitments already follow for all of SZK from our first result (SZK $\subseteq$ NISZK^h). To obtain commitments for all of ZK, we use a characterization of ZK in terms of SZK and 'instance-dependent one-way functions' [33], and combine the instance-dependent commitment schemes we obtain from both SZK and the instance-dependent one-way functions.

An alternative construction of the instance-dependent commitments we need can be obtained by using the concurrent work of Ong and Vadhan [34]. They showed that every problem in ZK (resp., SZK) has an instance-dependent commitment scheme that is computationally (resp., statistically) hiding on YES instances and statistically binding on NO instances. While their commitments are interactive, they can be made noninteractive if we assume that the sender is honest during the commit phase (by having the sender simulate both parties). Thus, our work can be viewed as a (substantial) simplification to their constructions for the case of honest senders.

2 Definitions and Preliminaries

2.1 Promise Problems

Promise problems are a more general variant of decision problems than languages. A promise problem Π is a pair of disjoint sets of strings (Π_Y, Π_N), where Π_Y is the set of YES instances and Π_N is the set of NO instances. The computational problem associated with any promise problem Π is: given a string that is "promised" to lie in $\Pi_Y \cup \Pi_N$, decide whether it is in Π_Y or Π_N. Reductions from one promise problem to another are natural extensions of reductions between languages. Namely, we say Π *reduces* to Γ (written $\Pi \preceq \Gamma$) if there exists a polynomial time computable function f such that $x \in \Pi_Y \Rightarrow f(x) \in \Gamma_Y$ and $x \in \Pi_N \Rightarrow f(x) \in \Gamma_N$. We can also naturally extend the definitions of complexity classes by letting the properties of the strings in the languages be conditions on the YES instances, and properties of strings outside of the language be conditions on NO instances.

2.2   Instance-Dependent Cryptographic Primitives

Many of the objects that we will be constructing for use in our zero knowledge constructions will be instance dependent. Hence, we will modify common cryptographic primitives such as one-way functions by allowing them to be parametrized by some string x, such that the cryptographic properties will only be guaranteed to hold if x is in some set I.

Definition 5. *An* instance-dependent function ensemble *is a collection of functions* $\mathcal{F} = \{f_x : \{0,1\}^{p(|x|)} \rightarrow \{0,1\}^{q(|x|)}\}_{x \in \{0,1\}^*}$, *where* $p(\cdot)$ *and* $q(\cdot)$ *are polynomials.* $\mathcal{F}$ *is* polynomial-time computable *if there exists a polynomial-time algorithm* F *such that for all* $x \in \{0,1\}^*$ *and* $y \in \{0,1\}^{p(|x|)}$, $F(x,y) = f_x(y)$.

Definition 6. *An* instance-dependent one-way function *on* I *is a polynomial-time instance-dependent function ensemble* $\mathcal{F} = \{f_x : \{0,1\}^{p(|x|)} \rightarrow \{0,1\}^{q(|x|)}\}_{x \in \{0,1\}^*}$, *such that for every nonuniform PPT* A, *there exists a negligible function* $\varepsilon(\cdot)$ *such that for all* $x \in I$,

$$\Pr\left[A(x, f_x(U_{p(|x|)})) \in f_x^{-1}(f_x(U_{p(|x|)}))\right] \leq \varepsilon(|x|)$$

Definition 7. *An* instance-dependent probability ensemble *on* I *is a collection of random variables* $\{X_x\}_{x \in \{0,1\}^*}$, *where* X_x *takes values in* $\{0,1\}^{p(|x|)}$ *for some polynomial* p. *We call such an ensemble* samplable *is there exists a probabilistic polynomial-time algorithm* M *such that for every input* x, $M(x)$ *is distributed according to* X_x.

Definition 8. *Two instance-dependent probabilistic ensembles* $\{X_x\}$ *and* $\{Y_x\}$ *are* computationally indistinguishable *on* $I \subset \{0,1\}^*$ *if for every nonuniform PPT* D, *there exists a negligible* $\varepsilon(\cdot)$ *such that for all* $x \in I$,

$$\Pr\left[D(x, X_x) = 1\right] - \Pr\left[D(x, Y_x) = 1\right]| \leq \varepsilon(|x|)$$

Similarly, we say $\{X_x\}$ *and* $\{Y_x\}$ *are* statistically indistinguishable *on* $I \subset \{0,1\}^*$ *if the above is required for all functions* D. *If* X_x *and* Y_x *are identically distributed for all* $x \in I$, *we say they are* perfectly indistinguishable .

We will sometimes use the informal notation $X \overset{c}{\equiv} Y$ to denote that ensembles X and Y are computationally indistinguishable.

Definition 9. *An* instance-dependent pseudorandom generator *on* I *is a polynomial-time instance-dependent function ensemble* $\mathcal{G} = \{G_x : \{0,1\}^{p(|x|)} \to \{0,1\}^{q(|x|)}\}$ *such that* $q(n) > p(n)$, *and the probability ensembles* $\{G_x(U_{p(|x|)})\}_x$ *and* $\{U_{q(|x|)}\}_x$ *are computationally indistinguishable on* I.

2.3 Probability Distributions

In this section, we define several tools that are useful for analysing properties of probability distributions.

Definition 10. *The* statistical difference *between two random variables* X *and* Y *taking values in some domain* $\mathcal{U}$ *is defined as:*

$$\Delta(X,Y) = \max_{S \subseteq \mathcal{U}} |\Pr[X \in S] - \Pr[Y \in S]| = \frac{1}{2} \sum_{x \in \mathcal{U}} |\Pr[X = x] - \Pr[Y = x]|$$

Definition 11. *For an ordered pair of random variables* (X,Y), *we define their* disjointness *to be:*

$$\mathrm{Disj}(X,Y) = \Pr_X[X \in \mathrm{Supp}(Y)]$$

and we define their mutual disjointness:

$$\mathrm{MutDisj}(X,Y) = \min(\mathrm{Disj}(X,Y), \mathrm{Disj}(Y,X)).$$

Note that disjointness is a more stringent measure of the disparity between two distributions than statistical difference. If two distributions have disjointness α, then their statistical difference is at least α. The converse, however, does not hold, since the two distributions could have statistical difference that is negligibly close to 1, yet have identical supports and mutual disjointness 0.

Moreover, we can go from disjoint to mutually-disjoint distributions by the following lemma:

Lemma 12. *[7,35] Given a pair of distributions* (X_0, X_1) *with* n *input gates, consider the following distributions:*

Y_0: *Choose* $r \xleftarrow{R} \{0,1\}^n$, $b \xleftarrow{R} \{0,1\}$, *output* $(X_b(r), b)$.
Y_1: *Choose* $r \xleftarrow{R} \{0,1\}^n$, $b \xleftarrow{R} \{0,1\}$, *output* $(X_b(r), \bar{b})$.
The following properties hold:

1. $\Delta(Y_0, Y_1) = \Delta(X_0, X_1)$
2. *If* (X_0, X_1) *is* α-*disjoint, then* (Y_0, Y_1) *is mutually* $\frac{\alpha}{2}$-*disjoint.*

Tensoring Distributions. For random variables X, Y, we let $X \otimes Y$ be the random variable consisting of a sample of X followed by an independent sample of Y. The $\otimes$ notation reflects the fact that the mass function of $X \otimes Y$ is the tensor product of the mass functions of X and Y. When the independence is clear from context, we sometimes write (X, Y) instead of $X \otimes Y$. $X^{\otimes k}$ is the random variable consisting of k independent copies of X.

Lemma 13 ([7,35]). *Given a parameter $k \in \mathbb{N}$ and the distributions $X_1, \ldots, X_k$ and $Y_1, \ldots, Y_k$, the pair $(X, Y) = X_1 \otimes \ldots \otimes X_k, Y_1 \otimes \ldots \otimes Y_k)$ will satisfy the following properties:*

1. *$1 - 2\exp(-k\delta^2/2) \leq \Delta(X, Y) \leq k\delta$ where $\delta = \sum_{i \in [k]} \Delta(X_i, Y_i)/k$.*
2. *$\mathrm{MutDisj}(X, Y) = 1 - \prod_{i \in [k]}(1 - \alpha_i)$, where $\alpha_i = \mathrm{MutDisj}(X_i, Y_i)$.*

XORing Distributions. We define the XOR operator which acts on pairs of distributions and returns a pair of distributions. Given two pairs (X_0, X_1) and (X'_0, X'_1), with n and n' input gates, respectively, $\mathrm{XOR}((X_0, X_1), (X'_0, X'_1))$ is defined by the circuits:

Y_0: Choose $b \overset{R}{\leftarrow} \{0,1\}, r \overset{R}{\leftarrow} \{0,1\}^n, r' \overset{R}{\leftarrow} \{0,1\}^{n'}$, output $(X_b(r), X'_b(r'))$.

Y_1: Choose $b \overset{R}{\leftarrow} \{0,1\}, r \overset{R}{\leftarrow} \{0,1\}^n, r' \overset{R}{\leftarrow} \{0,1\}^{n'}$, output $(X_b(r), X'_{\bar{b}}(r'))$.

Lemma 14 (XOR Lemma [7,35]). *If $(Y_0, Y_1) = \mathrm{XOR}((X_0, X_1), (X'_0, X'_1))$, then the following properties hold:*

1. *$\Delta(Y_0, Y_1) = \Delta(X_0, X_1) \cdot \Delta(X'_0, X'_1)$.*
2. *$\mathrm{MutDisj}(Y_0, Y_1) = \mathrm{MutDisj}(X_0, X_1) \cdot \mathrm{MutDisj}(X'_0, X'_1)$.*

By induction, the XOR Lemma implies the following method to decrease both statistical difference and mutual disjointness exponentially fast:

Lemma 15 ([7,35]). *Given circuits X_0, X_1 with n input gates and a parameter k, consider the following pair:*

Y_0: Choose $(b_1, \ldots, b_k) \overset{R}{\leftarrow} \{(c_1, \ldots, c_k) \in \{0,1\}^k : c_1 \oplus \ldots \oplus c_k = 0\}, (r_1, \ldots r_k) \overset{R}{\leftarrow} \{0,1\}^{kn}$, output $(X_{b_1}(r_1), \ldots, X_{b_k}(r_k))$.

Y_1: Choose $(b_1, \ldots, b_k) \overset{R}{\leftarrow} \{(c_1, \ldots, c_k) \in \{0,1\}^k : c_1 \oplus \ldots \oplus c_k = 1\}, (r_1, \ldots r_k) \overset{R}{\leftarrow} \{0,1\}^{kn}$, output $(X_{b_1}(r_1), \ldots, X_{b_k}(r_k))$.

The following properties hold:

1. *$\Delta(Y_0, Y_1) = \Delta(X_0, X_1)^k$.*
2. *$\mathrm{MutDisj}(Y_0, Y_1) = \mathrm{MutDisj}(X_0, X_1)^k$.*

Entropy and Hashing.

Definition 16. *The entropy of a random variable X is $\mathrm{H}(X) = \mathbb{E}_{x \leftarrow X}\left[\log \frac{1}{\Pr[X=x]}\right]$. The conditional entropy of X given Y is*

$$\mathrm{H}(X|Y) = \underset{y \leftarrow Y}{\mathbb{E}}[\mathrm{H}(X|_{Y=y})] = \underset{(x,y) \leftarrow (X,Y)}{\mathbb{E}}\left[\log \frac{1}{\Pr[X = x|Y = y]}\right] = \mathrm{H}(X, Y) - \mathrm{H}(Y).$$

For entropy, it holds that for every X, Y, $\mathrm{H}(X \otimes Y) = \mathrm{H}(X) + \mathrm{H}(Y)$. More generally, if $(X, Y)^{\otimes k} = ((X_1, Y_1), \ldots, (X_k, Y_k))$, then $\mathrm{H}((X_1, \ldots, X_k)|(Y_1, \ldots, Y_k)) = k \cdot \mathrm{H}(X|Y)$.

Definition 17. *The* relative entropy (Kullback-Liebler distance) *between two distributions* X, Y *is:*

$$\mathrm{KL}(X|Y) = \mathop{\mathrm{E}}_{x \leftarrow X} \left[\log \frac{\mathrm{Pr}\,[X = x]}{\mathrm{Pr}\,[Y = x]} \right]$$

We denote by $\mathrm{H}_2(p)$ the binary entropy function, which is the entropy of a $\{0,1\}$-valued random variable with expectation p. $\mathrm{KL}_2(p,q)$ denotes the relative entropy between two $\{0,1\}$-value random variables with expectations p and q.

Flat Distributions. Let X a distribution with entropy $\mathrm{H}(X)$. Elements x of X such that $|\log \mathrm{Pr}[X = x] - \mathrm{H}(X)| \leq k$ are called k-*typical*. We say that X is Δ-*flat* if for every $t > 0$ the probability that an element chosen from X is $t \cdot \Delta$-typical is at least $1 - 2^{-t^2+1}$.

Lemma 18 (Flattening Lemma [36]). *Let X be a distribution encoded by a circuit with n input gates. Then $X^{\otimes k}$ is $\sqrt{k} \cdot n$-flat.*

Definition 19. *A family $\mathcal{H}$ of functions from $A \to B$ is 2-universal if for every two elements $x \neq y \in A$ and $a, b \in B$, $\mathrm{Pr}_{h \in_R \mathcal{H}}[h(x) = a \text{ and } h(y) = b] = \frac{1}{|B|^2}$.*

We write $\mathcal{H}_{n,m}$ to denote the 2-universal family from $\{0,1\}^n$ to $\{0,1\}^m$.

Lemma 20 (Leftover Hash Lemma [37]). *Let $\mathcal{H}$ be a samplable family of 2-universal hashing functions from $A \to B$. Suppose X is a distribution on A such that with probability at least $1 - \delta$ over x selected from X, $\mathrm{Pr}[X = x] \leq \epsilon/|B|$. Consider the following distribution:*

$$Z : \text{Choose } h \leftarrow \mathcal{H} \text{ and } x \leftarrow X, \text{ return } (h, h(x)).$$

Then, $\Delta(Z, \mathcal{U}) \leq O(\delta + \epsilon^{1/3})$, where $\mathcal{U}$ is the uniform distribution on $\mathcal{H} \times B$.

3   Interactive Zero Knowledge

We consider a generalized version of interactive zero knowledge, introduced by Ben-Or and Gutfreund [7], in which the prover and the verifier have access to a help string output by a dealer algorithm that has access to the statement being proven. We will call this model of interactive zero knowledge the *help model*. Interactive zero-knowledge proofs are a special case of interactive zero-knowledge proofs in the help model.

We denote the three algorithms that make up an interactive zero-knowledge proof in the help model as D, P and V. All three receive as input x, the statement being proven. The dealer selects the help string $\sigma \leftarrow D(x)$ and sends it to P and V. P and V carry out an interactive protocol and, at the end of their interaction, they either output ACCEPT or REJECT. We call the *transcript* the sequence of messages which the triple (D, P, V) computes. $(D, P, V)(x)$ denotes the random variable of the possible outcomes of the protocol, while $\langle D, P, V \rangle(x)$ denotes the verifier's view of the transcripts (where the probability space is over the random coins of D, P and V).

Definition 21 (ZK^h, SZK^h [7]). *A zero-knowledge proof system in the help* model *for a promise problem Π is a triple of probabilistic algorithms (D, P, V) (where D and V are polynomial time bounded), satisfying the following conditions:*

1. *Completeness. For all $x \in \Pi_Y$, $\Pr[(D, P, V)(x) = 1] \geq \frac{2}{3}$, where the probability is taken over the coin tosses of D, P and V.*
2. *Soundness. For all $x \in \Pi_N$ and every prover strategy P^*, $\Pr[(D, P^*, V) = 1] \leq \frac{1}{3}$, where the probability is taken over the coin tosses of D, P^*, V.*
3. *Zero Knowledge. There exists a PPT S such that the ensembles $\{\langle D, P, V \rangle)(x)\}_x$ and $\{S(x)\}_x$ are computationally indistinguishable on Π_Y.*

If the ensembles are statistically indistinguishable, we call it a statistical zero knowledge proof system in the help model. *ZK^h (resp., SZK^h) is the class of promise problems possessing zero-knowledge (resp., statistical zero-knowledge) proof systems in the help model.*

If the help string σ is generated according to $D(1^{|x|})$, we call the proof system an interactive zero-knowledge proof system in the public parameter model. *The corresponding complexity class is ZK^{pub} (resp., SZK^{pub}). If the help string σ is generated from the uniform distribution on $\{0, 1\}^{|x|}$, we call the proof system an* interactive zero-knowledge proof system in the common random string model. *The corresponding complexity class is ZK^{crs} (resp., SZK^{crs}).*

If we remove the dealer's help, the resulting proof system is said to be an interactive zero-knowledge proof system. *The corresponding complexity class is* ZK *(resp., SZK).*

Note that, in the help model, the dealer is computable in polynomial time given only the instance, and not a witness (hence the notation $D(x)$).

It is simple to show (by having the verifier simulate the dealer's help) that ZK^h is contained in IP, the class of promise problems with interactive proofs:

Lemma 22. ZK$^h \subseteq$ IP.

3.1   Statistical Zero Knowledge

In this section, we state a few characterizations of statistical zero knowledge which will be related to the ones we will later obtain for the computational case. We begin by noting that, in the statistical case, Ben-Or and Gutfreund [7] showed that zero knowledge in the help model is equivalent to zero knowledge:

Theorem 23 ([7]). SZK^h = SZK.

The theorem above implies that all the characterizations of SZK will also hold for SZK^h. In particular, SZK^h shares the complete problems for SZK that are due to [36,35,33]:

Theorem 24 ([36,35,33]). *The following problems are SZK-complete:*

1. STATISTICAL DIFFERENCE:

$$\mathrm{SD}_Y = \{(X,Y) : \Delta(X,Y) \leq 1/3\}$$
$$\mathrm{SD}_N = \{(X,Y) : \Delta(X,Y) \geq 2/3\}$$

where X and Y are samplable distributions specified by circuits that sample from them.

2. ENTROPY DIFFERENCE:

$$\mathrm{ED}_Y = \{(X,Y) : \mathrm{H}(X) \geq \mathrm{H}(Y) + 1\}$$
$$\mathrm{ED}_N = \{(X,Y) : \mathrm{H}(Y) \geq \mathrm{H}(X) + 1\}$$

where X and Y are samplable distributions specified by circuits that sample from them.

3. CONDITIONAL ENTROPY APPROXIMATION:

$$\mathrm{CEA}_Y = \{(X,Y,r) : H(X|Y) \geq r\}$$
$$\mathrm{CEA}_N = \{(X,Y,r) : H(X|Y) \leq r-1\}$$

where (X,Y) is a joint samplable distribution specified by circuits that use the same coin tosses.

Note that we can change the thresholds of $1/3$ and $2/3$ in SD to other thresholds $\alpha < \beta$. We denote the resulting problem $\mathrm{SD}^{\alpha,\beta}$. It is known that $\mathrm{SD}^{\alpha,\beta}$ is SZK-complete for all constants α, β such that $0 \leq \alpha < \beta^2 \leq 1$ [35].

3.2 Computational Zero Knowledge

In the case of ZK, no natural complete problems are known (unless we assume that one-way functions exist, in which case ZK = IP = PSPACE [4,17,18,38,39,20,21]). However, characterizations that are analogous to the complete problems for SZK do exist in the form of the INDISTINGUISHABILITY CONDITION and the CONDITIONAL PSEUDOENTROPY CONDITION below. These conditions give 'if and only if' characterizations of ZK that provide essentially the same functionality that complete problems provide.

The first characterization is a natural computational analogue of STATISTICAL DIFFERENCE:

Definition 25. *A promise problem Π satisfies the* INDISTINGUISHABILITY CONDITION *if there is a polynomial-time computable function mapping strings x to pairs of samplable distributions (X,Y) such that:*

– *If $x \in \Pi_Y$, then X and Y are computationally indistinguishable.*
– *If $x \in \Pi_N$, then $\Delta(X,Y) \geq 2/3$.*

Theorem 26 ([33]). *$\Pi \in$ ZK if and only if $\Pi \in$ IP and Π satisfies the* INDISTINGUISHABILITY CONDITION.

The second characterization is based on the SZK-complete problem CEA:

Definition 27. *A promise problem Π satisfies the* CONDITIONAL PSEUDOEN-TROPY CONDITION *if there is a polynomial-time computable function mapping strings x to a samplable joint distribution (X, Y) such that:*

- *If $x \in \Pi_Y$, then there exists a (not necessarily samplable) joint distribution (X', Y') such that (X', Y') is computationally indistinguishable from (X, Y) and $H(X'|Y') \geq r$.*
- *If $x \in \Pi_N$, then $H(X|Y) \leq r - 1$.*

Theorem 28 ([33]). *$\Pi \in$ ZK if and only if $\Pi \in$ IP and Π satisfies the* CON-DITIONAL PSEUDOENTROPY CONDITION.

Another characterization that we will use is the SZK/OWF CONDITION of [33]. The SZK/OWF CONDITION states that any problem in ZK can be decomposed into a part with an SZK proof and another part on which instance-dependent one-way functions can be constructed:

Definition 29 (SZK/OWF CONDITION [33]). *A promise problem $\Pi = (\Pi_Y, \Pi_N)$ satisfies the* SZK/OWF CONDITION *if there exists a set $I \subseteq \Pi_Y$ of YES such that:*

1. *The promise problem $\Pi' = (\Pi_Y \backslash I, \Pi_N)$ is in SZK.*
2. *There exists an instance-dependent one-way function on I (in the sense of Definition 6).*

Theorem 30 ([33]). *$\Pi \in$ ZK if and only if $\Pi \in$ IP and Π satisfies the* SZK/OWF CONDITION.

4   Noninteractive Zero Knowledge

4.1   The Help Model

In this section, we define the noninteractive analogue of zero-knowledge proofs in the help model.

Definition 31 (NIZK^h, NISZK^h [7]). *A noninteractive zero-knowledge proof system in the help model for a promise problem Π is an interactive zero-knowledge proof in which there is only one message $\pi = P(x, \sigma)$ from prover to verifier.*

If the real transcripts are statistically indistinguishable from simulated ones, we call it a noninteractive statistical zero knowledge proof system. NIZK^h (resp., NISZK^h) is the class of promise problems possessing noninteractive zero-knowledge (resp., noninteractive statistical zero-knowledge) proof systems in the help model.

If the help string σ is generated according to $D(1^{|x|})$, we call the proof system a noninteractive zero-knowledge proof system in the public parameter model. The corresponding complexity class is NIZK^{pub} (resp., NISZK^{pub}). If the help string σ is generated from the uniform distribution on $\{0, 1\}^{|x|}$, we call the proof system an noninteractive zero-knowledge proof system in the common random string model. The corresponding complexity class is NIZK^{crs} (resp., NISZK^{crs}).

The main benefit of the public parameter model and the help model over the simpler CRS model is that they make it easier to construct NIZK proofs from simpler cryptographic primitives such as one-way functions ([7,23]), or, as we will show in this paper, from noninteractive, instance-dependent commitment schemes.

Like SZK, NISZK^{crs} and NISZK^{h} exhibit complete problems:

Theorem 32 ([25]). *The promise problem* ENTROPY APPROXIMATION, *defined as:*

$$\mathrm{EA}_Y = \{(X,t) : \mathrm{H}(X) \geq t + 1\}$$
$$\mathrm{EA}_N = \{(X,t) : \mathrm{H}(Y) \leq t - 1\}$$

is complete for NISZK^{crs}, *where X is a samplable distribution specified by a circuit that samples from it. We use the notation EA^t to specify an instance of EA with parameter t.*

Theorem 33 ([7]). *The promise problem* IMAGE INTERSECTION DENSITY, *defined as:*

$$\mathrm{IID}_Y = \{(X,Y) : \Delta(X,Y) \leq 1/3\}$$
$$\mathrm{IID}_N = \{(X,Y) : \mathrm{MutDisj}(X,Y) \geq 2/3\}$$

is complete for NISZK^{h}, *where X and Y are samplable distributions specified by circuits that sample from them.*

We note that our definition of IID is slightly different than the one used by [7]. In our definition, we are working with mutual disjointness, since it is easy to transform disjoint distributions to mutually disjoint ones (Lemma 12). Additionally, due to a stronger Polarization Lemma that we will describe in a subsequent section, we use constant thresholds of $1/3$ and $2/3$ rather than functions tending to 0 and 1.

We also recall the complexity class AM, which is is the class of promise problems possessing constant-round interactive proofs, or equivalently, 2-round public-coin interactive proofs [14,15]. Analogous to Lemma 22, AM proves to be a natural upper bound for NISZK^{h}, since we can just have the verifier replace the dealer in creating the reference string. Also, a lower bound for NISZK^{h} is NISZK^{crs}, which is definitionally a more restricted version of the help model.

5 Quantum Preliminaries and Definitions

5.1 The Quantum Formalism

Let $\mathcal{H}$ denote a 2-dimensional complex vector space, equipped with the standard inner product. We pick an orthonormal basis for this space, label the two basis vectors $|0\rangle$ and $|1\rangle$. A *qubit* is a unit length vector in this space, and so can be expressed as a linear combination of the basis states: $\alpha_0|0\rangle + \alpha_1|1\rangle$. Here α_0, α_1 are complex *amplitudes*, and $|\alpha_0|^2 + |\alpha_1|^2 = 1$.

An *m-qubit pure state* is a unit vector in the m-fold tensor space $H \otimes \cdots \otimes H$. The 2^m basis states of this space are the m-fold tensor products of the states $|0\rangle$ and $|1\rangle$. For example, the basis states of a 2-qubit system are the four 4-dimensional unit vectors $|0\rangle \otimes |0\rangle$, $|0\rangle \otimes |1\rangle$, $|1\rangle \otimes |0\rangle$, and $|1\rangle \otimes |1\rangle$. We abbreviate, e.g., $|1\rangle \otimes |0\rangle$ to $|0\rangle|1\rangle$, or $|1,0\rangle$, or $|10\rangle$, or even $|2\rangle$ (since 2 is 10 in binary). With these basis states, an m-qubit state $|\phi\rangle$ is a 2^m-dimensional complex unit vector $|\phi\rangle = \sum_{i \in \{0,1\}^m} \alpha_i|i\rangle$. We use $\langle\phi| = |\phi\rangle^*$ to denote the conjugate transpose of the vector $|\phi\rangle$, and $\langle\phi|\psi\rangle = \langle\phi| \cdot |\psi\rangle$ for the inner product between states $|\phi\rangle$ and $|\psi\rangle$. These two states are *orthogonal* if $\langle\phi|\psi\rangle = 0$. The *norm* of $|\phi\rangle$ is $\| \phi \| = \sqrt{\langle\phi|\phi\rangle}$.

A *mixed state* $\{p_i, |\phi_i\rangle\}$ is a classical distribution over pure quantum states, where the system is in state $|\phi_i\rangle$ with probability p_i. We can represent a mixed quantum state by the *density matrix* which is defined as $\rho = \sum_i p_i|\phi_i\rangle\langle\phi_i|$. Note that ρ is a positive semidefinite operator with trace (sum of diagonal entries) equal to 1. The density matrix of a pure state $|\phi\rangle$ is $\rho = |\phi\rangle\langle\phi|$.

A quantum system is called *bipartite* if it consists of two subsystems. We can describe the state of each of these subsystems separately with the *reduced density matrix*. For example, if the joint quantum state of two subsystems A, B has the form $|\phi\rangle = \sum_i \sqrt{p_i}|i\rangle_A|\phi_i\rangle_B$, then the state of the subsystem B, i.e., the subsystem which contains only the second part of $|\phi\rangle$ is described by the (reduced) density matrix $\sum_i p_i|\phi_i\rangle\langle\phi_i|$.

A quantum state evolves by a unitary operation or by a measurement. A *unitary* transformation U is a linear mapping that preserves the complex ℓ_2 norm. If we apply U to a state $|\phi\rangle$, it evolves to $U|\phi\rangle$. A mixed state ρ evolves to $U\rho U^\dagger$.

The most general measurement allowed by quantum mechanics is specified by a family of positive semidefinite operators $E_i = M_i^* M_i$, $1 \le i \le k$, subject to the condition that $\sum_i E_i = I$. Given a density matrix ρ, the probability of observing the ith outcome under this measurement is given by the trace $p_i = \text{Tr}(E_i\rho) = \text{Tr}(M_i\rho M_i^*)$. These p_i are nonnegative because E_i and ρ are positive semidefinite and they also sum to 1. If the measurement yields outcome i, then the resulting mixed quantum state is $M_i\rho M_i^*/\text{Tr}(M_i\rho M_i^*)$. In particular, if $\rho = |\phi\rangle\langle\phi|$, then $p_i = \langle\phi|E_i|\phi\rangle = \| M_i|\phi\rangle \|^2$, and the resulting state is $M_i|\phi\rangle/\| M_i|\phi\rangle \|$. A special case is where $k = 2^m$ and $B = \{|\psi_i\rangle\}$ forms an orthonormal basis of the m-qubit space. 'Measuring in the B-basis' means that we apply the measurement given by $E_i = M_i = |\psi_i\rangle\langle\psi_i|$. Applying this to a pure state $|\phi\rangle$ gives resulting state $|\psi_i\rangle$ with probability $p_i = |\langle\phi|\psi_i\rangle|^2$.

The trace norm of a matrix A is denoted by $\|A\|$ and is equal to the trace of $|A|$, where $|A| = \sqrt{A^\dagger A}$ is the positive square root of $A^\dagger A$. For two density matrices ρ_1, ρ_2 we define their trace distance as the trace norm of the matrix $\rho_1 - \rho_2$, i.e., $\|\rho_1 - \rho_2\|$.

The von Neumann Entropy of a mixed quantum state ρ with eigenvalues λ_i is defined as $S(\rho) = -\sum_i \lambda_i \log \lambda_i$.

5.2   Quantum Interactive and Noninteractive Statistical Zero-Knowledge

Quantum statistical zero knowledge proofs are a special case of quantum interactive proofs. We can think of a *quantum interactive protocol* $\langle P, V \rangle(x)$ as a series of circuits $(V_1(x), P_1(x), \ldots, V_k(x), P_k(x))$ on the space $\mathcal{V} \otimes \mathcal{M} \otimes \mathcal{P}$. $\mathcal{V}$ are the verifier's private qubits, $\mathcal{M}$ are the message qubits and $\mathcal{P}$ are the prover's private qubits. $V_i(x)$ (resp. $P_i(x)$) represents the ith action of the verifier (resp. the prover) during the protocol and acts on $\mathcal{V} \otimes \mathcal{M}$ (resp. $\mathcal{M} \otimes \mathcal{P}$). β_i corresponds to the state that appears after the ith action of the protocol. We define completeness and soundness exactly the same way as in the case of classical protocols. We say that a protocol $\langle P, V \rangle$ *solves* Π if it has completeness greater than $2/3$ and soundness less than $1/3$.

In the zero knowledge setting, we also want that the verifier learns nothing from the interaction other than the fact that $x \in \Pi_Y$ when it is the case. The way it is formalized is that for $x \in \Pi_Y$, the verifier can simulate his view of the protocol. We are interested only in honest verifier protocols where the verifier and the prover use unitary operations, since by Watrous [40] we know that honest verifier with unitary operations is equivalent to cheating verifier (that is allowed to use any permissible operation).

Let $\langle P, V \rangle$ a quantum protocol and β_j defined as before. The verifier's *view* of the protocol is his private qubits and the message qubits, $\text{view}_{\langle P,V \rangle}(j) = Tr_{\mathcal{P}}(\beta_j)$. We also want to separate the verifier's view based on whether the last action was made by the verifier or the prover. We note ρ_0 the input state, ρ_i the verifier's view of the protocol after P_i and ξ_i the verifier's view of the protocol after V_i.

Definition 34. *A quantum protocol $\langle P, V \rangle$ has the zero knowledge property for Π if there exists a quantum polynomial-time simulator σ and a negligible function μ such that for every input $x \in \Pi_Y$ and $\forall j \ \|\sigma_j(x) - \rho_j\| \le \mu(|x|)$.*

Note that for a state σ such that $\|\sigma - \rho_i\| \le \mu(|x|)$ it is easy to see that $\sigma' = V_{i+1}\sigma V_{i+1}^\dagger$ is close to $\xi_{i+1} = V_{i+1}\rho_i V_{i+1}^\dagger$ in this sense that $\|\sigma' - \xi_{i+1}\| \le \mu(|x|)$. Therefore, in the definition we just need to simulate the ρ_i's. Also note that the simulation in the quantum case is done round by round which seems to be a weaker definition than in the classical case. However, since the message qubits are reused in every round, the notion of a transcript can not be defined in the quantum case.

Definition 35. *$\Pi \in QSZK$ iff there exists a quantum protocol $\langle P, V \rangle$ that solves Π and that has the zero-knowledge property for Π.*

In the setting of quantum noninteractive statistical zero knowledge, first defined by Kobayashi [13], the prover and verifier share a maximally entangled state $\sum_i |i\rangle_P |i\rangle_V$ created by a trusted third party: the dealer D. Then the prover sends a single quantum message to the verifier. We can assume that the message from the dealer to the verifier goes into his private space $\mathcal{V}$. Hence, after the prover's message, the verifier's view ρ_1 also contains the message from the dealer.

In this setting, we define the zero knowledge property as follows:

Definition 36. *A quantum noninteractive protocol* $\langle D, P, V \rangle$ *has the zero knowledge property for* Π *if there exists a quantum polynomial-time simulator* σ *and a negligible function* μ *such that for every input* $x \in \Pi_Y$ $\|\sigma(x) - \rho_1\| \leq \mu(|x|)$.

Definition 37. $\Pi \in \text{QNISZK}$ *iff, when the prover and verifier share the maximally entangled state* $\sum_i |i\rangle_P |i\rangle_V$ *created by the dealer* D, *there exists a quantum noninteractive protocol* $\langle D, P, V \rangle$ *that solves* Π *and that has the zero-knowledge property for* Π.

6   Statistical Zero Knowledge

6.1   The Polarization Lemma

Zero knowledge protocols usually require from promise problems some parameters that are exponentially close to 0 or 1. Polarizations are reductions from promise problems with weak parameters to promise problems that can be solved by the protocols. For example, there is a polarization for the promise problem SD that transforms $\text{SD}^{a,b}$ with $a^2 > b$ to $\text{SD}^{1-2^{-k},2^{-k}}$ for any $k = \text{poly}(n)$ [35].

The best polarization that was known for IID was that $\text{IID}^{1/n^2,1-1/n^2}$ reduces to $\text{IID}^{2^{-k},1-2^{-k}}$ and henceforth $\text{IID}^{1/n^2,1-1/n^2}$ is complete for NISZK^h [7]. We will show here that $\text{IID}^{a,b}$ is complete for NISZK^h with $b > a$ (where a and b are constants).

Lemma 38 (Polarization Lemma [7,35]). *There exists an algorithm that takes a pair of distributions* (X_0, X_1) *and parameters* $n \in \mathbb{N}, 0 \leq \alpha < \beta \leq 1$, *and outputs a pair of distributions* (Y_0, Y_1) *such that:*

1. $\Delta(X_0, X_1) \leq \alpha \Rightarrow \Delta(X_0, X_1) \leq 2^{-n}$.
2. $\text{MutDisj}(X_0, X_1) \geq \beta \Rightarrow \text{MutDisj}(Y_0, Y_1) \geq 1 - 2^{-n}$.

The algorithm runs in time $\text{poly}\left(|(X_0, X_1)|, n, \exp\left(\frac{\alpha \log(1/\beta)}{\beta - \alpha}\right)\right)$.

Proof. Let $\lambda = \min\{\beta/\alpha, 2\} > 1$.

We first apply Lemma 15 with $k = \log_\lambda 2n$, obtaining two distributions which are either statistically α^k close, or have β^k mutual disjointness.

Then, we apply Lemma 13 with $m = \lambda^k/(2\beta^k) \leq 1/(2\alpha^k)$. This gives two distributions with either statistical difference at most $m\alpha^k \leq 1/2$, or mutual disjointness of at most $1 - (1 - \beta^k)^m \geq 1 - e^{-\beta^k m} = 1 - e^{-\beta^k \cdot \lambda^k/(2\beta^k)} = 1 - e^{-\lambda^k/2} = 1 - e^{-n}$.

Finally, we apply again Lemma 15 with parameter n to get either statistical difference at most 2^{-n}, or mutual disjointness at most $(1 - e^{-n})^n \geq 1 - ne^{-n} \geq 1 - 2^{-n}$, for sufficiently large n.

The running time of the algorithm is $\text{poly}(|(X_0, X_1)|, n, k)$, where $k = O(\log n/(\lambda - 1)) = O(\alpha \log n/(\beta - \alpha))$ and $m \leq 1/2 \cdot (2/\beta)^k = \exp\left(O\left(\frac{\alpha \log n \log(2/\beta)}{\beta - \alpha}\right)\right)$. This gives the claimed running time if either $n = O(1)$

or if $\beta - \alpha = \Omega(1)$. Thus we can obtain the lemma by applying the transformation in two steps, first with $n' = 2$ to polarize to thresholds $\alpha' = 1/4$ and $\beta' = 3/4$, and then once more with the desired value of n.

This can be compared to the original Polarization Lemma of [35], which refers to statistical difference in Item 2 (rather than mutual disjointness), but only achieves polarization from thresholds such that $0 \le \alpha < \beta^2 \le 1$, and for which it is known that the gap between thresholds is inherent for a natural class of transformations [41].

6.2   SZK and NISZK^h are Equivalent

We show in this section that help and interaction are equivalent in the statistical zero knowledge setting.

Theorem 39. SZK = NISZK^h

The inclusion NISZK$^h \subseteq$ SZK was proven by Ben-Or and Gutfreund [7], since the NISZK^h-complete problem IMAGE INTERSECTION DENSITY (IID) trivially reduces to STATISTICAL DIFFERENCE (SD), the SZK-complete problem. In what follows, we prove the opposite inclusion by reducing the SZK-complete problem ENTROPY DIFFERENCE (ED) to IID. Ben-Or and Gutfreund claimed to have proven this reduction in [22] but due to a flaw they retracted it in [7]. Their reduction from ED to IID was in fact only a reduction to SD. Still, part of our proof is inspired by their method.

In order to prove that SZK $\subseteq$ NISZK^h, we follow [25] and reduce the SZK-complete problem ED to several instances of ENTROPY APPROXIMATION and its complement (EA and $\overline{\text{EA}}$) using the following fact:

Fact 40 ([25]) Let $X' = X^{\otimes 3}$ and $Y' = Y^{\otimes 3}$. Let n the output size of X' and Y'. It holds that:

$$(X, Y) \in \text{ED}_Y \Leftrightarrow \forall t \in \{1, \ldots, n\} \left[((X', t) \in \text{EA}_Y) \vee ((Y', t) \in \overline{\text{EA}}_Y) \right]$$
$$(X, Y) \in \text{ED}_N \Leftrightarrow \exists t \in \{1, \ldots, n\} \left[((X', t) \in \text{EA}_N) \wedge ((Y', t) \in \overline{\text{EA}}_N) \right]$$

We know that EA $\in$ NISZK^h (since by definition NISZK$^{crs} \subseteq$ NISZK^h), so it remains to show the following two things:

1. $\overline{\text{EA}} \in$ NISZK^h: in order to this, we reduce $\overline{\text{EA}}$ to IID, inspired by Ben-Or and Gutfreund's attempt [22] to reduce ED to IID. This reduction relies on ideas from [27,28].
2. NISZK^h has certain boolean closure properties: this will allow us to reduce ED to a single instance of IID. Since IID and SD are closely related, we use similar techniques to the ones used in [27,29].

Note that our proof's structure is similar to the approach suggested by Goldreich et al. [25] for showing that NISZK^{crs} = SZK. They proved that if NISZK^{crs} = co-NISZK^{crs} then NISZK^{crs} = SZK. We show here that co-NISZK$^{crs} \subseteq$ NISZK^h, and using the closure properties, conclude that NISZK^h = SZK.

6.3 $\overline{\text{EA}}$ Belongs to NISZK^{h}

In this section, we prove the following lemma:

Lemma 41. $\overline{\text{EA}} \in \text{NISZK}^{\text{h}}$.

Proof. We will reduce $\overline{\text{EA}}$ to IID, which is complete for NISZK^{h}.

Let (X, t) an instance of $\overline{\text{EA}}$. By artificially adding input gates or output gates to X, we can assume that X has m input and output gates. Let k a large constant that will be specified later on and $X' = X^{\otimes s}$ with $s = 4km^2$. Note that X' has $m' = s \cdot m$ input and output gates and $\text{H}(X') = s \cdot \text{H}(X)$. We have:

Fact 42

1. X' is Δ-flat with $\Delta = 2\sqrt{k}m^2$, where s was chosen such that $s = 2\sqrt{k}\Delta$.
2. $\Pr[X'$ is $\sqrt{k}\Delta$-typical $] \geq 1 - 2^{-\Omega(k)}$.

Given (X, t), we can create two distributions Z as Z' as following

Z: Choose $r \stackrel{R}{\leftarrow} \{0,1\}^{m'}$, $x = X'(r)$, $h \stackrel{R}{\leftarrow} \mathcal{H}_{m'+st,m'}$, $z \stackrel{R}{\leftarrow} \{0,1\}^{m'}$. Return (x, h, z).

Z': Choose $r \stackrel{R}{\leftarrow} \{0,1\}^{m'}$, $x = X'(r)$, $h \stackrel{R}{\leftarrow} \mathcal{H}_{m'+st,m'}$, $u \stackrel{R}{\leftarrow} \{0,1\}^{st}$. Return $(x, (h, h(r, u)))$.

Note that Z' is of the form $Z' = (X', A)$. We write A_x to denote the distribution of A conditioned on $X' = x$. Note that we can describe A_x as follows :

A_x : Choose $r \stackrel{R}{\leftarrow} (X')^{-1}(x)$, $h \stackrel{R}{\leftarrow} \mathcal{H}_{m'+st,m'}$, $u \stackrel{R}{\leftarrow} \{0,1\}^{st}$ and return $(h, h(r, u))$.

Hence, we need to show that, when conditioning on $X' = x$, we have either $\Delta(\mathcal{U}, A_x)$ small (on the YES instances) or $\text{Disj}(\mathcal{U}, A_x)$ large (on the NO instances).

For $x \in \text{Supp}(X')$, let $\text{wt}(x) = \log|(X')^{-1}(x)| = m' - \log(\frac{1}{\Pr[X'=x]})$. The number of different possible inputs (r, u) that are hashed in A_x is $2^{\text{wt}(x)+st}$. Using Fact 42, it is easy to see that, if $\text{H}(X) \leq t - 1$, then $\text{wt}(x)$ will be large with high probability, whereas, if $\text{H}(X) \geq t + 1$, then $\text{wt}(x)$ will be small with high probability. We can now show the following two claims which will allow us to conclude the proof.

Claim. $(X, t) \in \overline{\text{EA}}_Y \Rightarrow \Delta(Z, Z') = 2^{-\Omega(k)}$.

Proof. For all $x \in \text{Supp}(X')$ that are $\sqrt{k}\Delta$-typical, $\left|\log(\frac{1}{\Pr[X'=x]}) - \text{H}(X')\right| \leq \sqrt{k}\Delta$. Hence,

$$\text{wt}(x) \geq m' - s \cdot \text{H}(X) - \sqrt{k}\Delta \geq m' - st + s - \sqrt{k}\Delta \geq m' - st + \sqrt{k}\Delta.$$

Therefore, the number of inputs (r, u) such that $X'(r) = x$ and $u \in \{0,1\}^{st}$ is greater than $2^{m'+\sqrt{k}\Delta} \geq 2^{m'+k}$. By the Leftover Hash Lemma (Lemma 20), $\Delta(\mathcal{U}, A_x) = 2^{-\Omega(k)}$. By Fact 42, the probability of a $\sqrt{k}\Delta$-typical x is $1 - 2^{-\Omega(k)}$ and hence we can conclude that $\Delta(Z, Z') = 2^{-\Omega(k)}$.

Claim. $(X, t) \in \overline{\mathrm{EA}}_N \Rightarrow \mathrm{Disj}(Z, Z') = 1 - 2^{-\Omega(k)}$.

Proof. For all $x \in \mathrm{Supp}(X')$ that are $\sqrt{k}\Delta$-typical, we have:

$$\mathrm{wt}(x) \le m' - s \cdot \mathrm{H}(X) + \sqrt{k}\Delta \le m' - st - s + \sqrt{k}\Delta \le m' - st - \sqrt{k}\Delta.$$

Therefore, the number of inputs (r, u) such that $X'(r) = x$ and $u \in \{0, 1\}^{st}$ is smaller than $2^{m' - \sqrt{k}\Delta} \le 2^{m' - k}$. Since we hash at most $2^{m' - k}$ values into $\{0, 1\}^{m'}$, we get only a 2^{-k} fraction of the total support and hence $\mathrm{Disj}(\mathcal{U}, A_x) = 1 - 2^{-\Omega(k)}$. By Fact 42, the probability of a $\sqrt{k}\Delta$-typical x is $1 - 2^{-\Omega(k)}$ and hence we can conclude that $\mathrm{Disj}(Z, Z') = 1 - 2^{-\Omega(k)}$.

By taking k a large enough constant, we can ensure that $(X, t) \in \overline{\mathrm{EA}}_Y \Rightarrow \Delta(Z, Z') \le 1/4$ and also $(X, t) \in \overline{\mathrm{EA}}_N \Rightarrow \mathrm{Disj}(Z, Z') \ge 3/4$.

The only thing that remains is to transform the disjointness in the NO instances to mutual disjointness. We first apply Lemma 12 to create distributions (A, B) such that $\Delta(A, B) \le 1/4$ or $\mathrm{Disj}(A, B) \ge 3/8$. Then, by the polarization Lemma shown in Subsection 6.1, we create distributions (A', B') such that $(X, t) \in \overline{\mathrm{EA}}_Y \Rightarrow \Delta(A', B') \le 1/3$ and $(X, t) \in \overline{\mathrm{EA}}_N \Rightarrow \mathrm{Disj}(A', B') \ge 2/3$.

In conclusion, we see that from (X, t), we have created distributions A', B' in polynomial time such that :

- $(X, t) \in \overline{\mathrm{EA}}_Y \Rightarrow (A', B') \in \mathrm{IID}_Y$.
- $(X, t) \in \overline{\mathrm{EA}}_N \Rightarrow (A', B') \in \mathrm{IID}_N$.

Hence, $\overline{\mathrm{EA}}$ reduces to IID and from the completeness of IID for NISZK^h, we have $\overline{\mathrm{EA}} \in \mathrm{NISZK}^h$.

6.4   Closure Properties for NISZK^h

We now prove some closure properties of NISZK^h that we will use to complete the proof of Theorem 39. Every promise problem $\Pi \in \mathrm{NISZK}^h$ reduces to IID and hence, we just have to concentrate on this problem. Note that this problem is very similar to the SZK-complete promise problem SD and hence we use similar techniques to those developed in [29,27] to show closure properties for SZK. In our case, we just need to show some limited closure properties that will be enough to prove that $\overline{\mathrm{ED}} \in \mathrm{NISZK}^h$.

Definition 43. *Let Π some promise problem. We define $\mathrm{AND}(\Pi)$ to be the following promise problem:*

- $\mathrm{AND}(\Pi)_Y = \{(x_1, \ldots, x_k) : \forall i \in \{1, \ldots, k\} \; x^i \in \Pi_Y\}$.
- $\mathrm{AND}(\Pi)_N = \{(x_1, \ldots, x_k) : \exists i \in \{1, \ldots, k\} \; x^i \in \Pi_N\}$.

Similarly, we define $\mathrm{OR}(\Pi)$ for a pair of instances of Π.

Definition 44. *Let Π a promise problem. We define $\mathrm{OR}(\Pi)$ to be the following promise problem:*

- $\mathrm{OR}(\Pi)_Y = \{(x_1, x_2) : \exists i \in \{1, 2\} \ x^i \in \Pi_Y\}$.
- $\mathrm{OR}(\Pi)_N = \{(x_1, x_2) : \forall i \in \{1, 2\} \ x^i \in \Pi_N\}$.

We show that NISZK^h is closed under AND and OR.

Lemma 45. NISZK^h *is closed under* AND.

Proof. Let Π be in NISZK^h and $(x_1, \ldots, x_k)$ be an instance of $\mathrm{AND}(\Pi)$. We reduce Π to the IID problem which means that we transform each x_i into a pair of distributions (X^i, Y^i) such that $x_i \in \Pi_Y \Rightarrow (X^i, Y^i) \in \mathrm{IID}_Y$ and $x_i \in \Pi_N \Rightarrow (X^i, Y^i) \in \mathrm{IID}_N$. Let $X = X^1 \otimes \cdots \otimes X^k$ and $Y = Y^1 \otimes \cdots \otimes Y^k$. We first polarize each pair (X^i, Y^i) to have statistical difference at most $1/3k$ or mutual disjointness at least $2/3$. From Lemma 13, we can easily see that $(x_1, \ldots, x_k) \in \mathrm{AND}(\Pi)_Y \Rightarrow (X, Y) \in \mathrm{IID}_Y$ and that $(x_1, \ldots, x_k) \in \mathrm{AND}(\Pi)_N \Rightarrow (X, Y) \in \mathrm{IID}_N$, which concludes our proof.

Lemma 46. NISZK^h *is closed under* OR.

Proof. Let Π be in NISZK^h. Let (x_1, x_2) be an instance of $\mathrm{OR}(\Pi)$. We reduce Π to the IID problem which means that we transform each x_i into a pair of distributions (X^i, Y^i) such that $x_i \in \Pi_Y \Rightarrow (X^i, Y^i) \in \mathrm{IID}_Y$ and $x_i \in \Pi_N \Rightarrow (X^i, Y^i) \in \mathrm{IID}_N$. We first polarize each pair (X^i, Y^i) to have statistical difference at most $1/3$ or mutual disjointness at least $\sqrt{2/3}$. Now, consider the pair (A, B) obtained by XORing (X_1, Y_1) and (X_2, Y_2) (in the sense of Lemma 14). Using this Lemma, we conclude that $(x_1, x_2) \in \mathrm{OR}(\Pi)_Y \Rightarrow (A, B) \in \mathrm{IID}_Y$ and that $(x_1, x_2) \in \mathrm{OR}(\Pi)_N \Rightarrow (A, B) \in \mathrm{IID}_N$.

6.5 Putting It Together

We can now prove that $\mathrm{SZK} \subseteq \mathrm{NISZK}^h$ and hence conclude the proof of Theorem 39. In the language of the previous section, Fact 40 says that the SZK-complete problem ED reduces to $\mathrm{AND}(\mathrm{OR}(\overline{\mathrm{EA}}, \mathrm{EA}))$ via a standard Karp (*i.e.*, many-one) reduction. Since EA and $\overline{\mathrm{EA}}$ are in NISZK^h (Lemma 41) and NISZK^h is closed under AND and OR (Lemma 45 and 46), we conclude that $\mathrm{ED} \in \mathrm{NISZK}^h$ and that $\mathrm{SZK} \subseteq \mathrm{NISZK}^h$.

An interesting corollary is the following new complete problem for SZK.

Corollary 47. IID *is complete for* SZK.

7   Computational Zero Knowledge

In this section, we extend the results presented in the previous section to computational zero knowledge. However, the techniques that we have used in the statistical case cannot be applied directly here, so we take a more indirect route to proving an equivalence for the computational case. We define the COMPUTATIONAL IMAGE INTERSECTION DENSITY CONDITION (CIIDC), a natural computational analogue of IID in the style of the INDISTINGUISHABILITY CONDITION and the

CONDITIONAL PSEUDOENTROPY CONDITION used in [33] (see Section 3.2), and prove that all problems in ZK satisfy the CIIDC, building on our proof that every problem in SZK reduces to IID. Next we want to show that every problem in AM satisfying the CIIDC is in NISZK^h. However, as the approach used in [7] to show IID is in NISZK^h does not generalize to the computational case, following [33], we get around this difficulty by interpreting the COMPUTATIONAL IMAGE INTERSECTION DENSITY CONDITION as a special type of commitment scheme that is sufficient for constructing NIZK^h proofs. Hence, we show that any promise problem in ZK∩AM has a NIZK^h proof. For the other direction, we prove that ZK equals ZK^h, a class which contains NIZK^h, concluding that NIZK^h = ZK ∩ AM.

7.1 The COMPUTATIONAL IMAGE INTERSECTION DENSITY CONDITION

We define the COMPUTATIONAL IMAGE INTERSECTION DENSITY CONDITION, and show that any promise problem with a ZK proof satisfies this condition.

Definition 48 (COMPUTATIONAL IMAGE INTERSECTION DENSITY CONDITION (CIIDC)). *A promise problem Π satisfies CIIDC if there is a polynomial time mapping from strings $x \in \Pi$ to two distributions (X, Y) specified by circuits sampling from them such that*

1. *If $x \in \Pi_Y$, then X and Y are computationally indistinguishable.*
2. *If $x \in \Pi_N$, then (X, Y) have mutual disjointness at least $1/3$.*

Lemma 49. *Every promise problem $\Pi \in$ ZK satisfies CIIDC.*

Proof. Since every problem $\Pi \in$ ZK satisfies the SZK/OWF CONDITION, it follows that Π can be decomposed into two promise problems, Γ and Θ, such that $\Pi = \Gamma \cup \Theta$, $\Gamma \in$ SZK = NISZK^h and for $x \in \Theta$, instance dependent one-way functions can be constructed.

On the instances x in Γ, a reduction to IID gives a pair (X_0, X_1) such that on $x \in \Gamma_Y$, $\Delta(X_0, Y_0)$ is close to 0, and, on $x \in \Gamma_N$, MutDisj(X_0, X_1) is close to 1. Informally, on the instances in Θ, we apply [20] to the instance-dependent one-way function to obtain an instance-dependent pseudorandom generator $G_x(\cdot)$, and consider the pair (Y_0, Y_1) obtained by comparing the output of $G_x(\cdot)$ to the uniform distribution. Note that on $x \in \Theta_Y$, (Y_0, Y_1) will be computationally indistinguishable, while on $x \in \Theta_N$, it will be disjoint (since $G_x(\cdot)$ has a small support), and hence mutually disjoint by Lemma 12.

Since it might not be possible to efficiently distinguish between instances in Γ and those in Θ, it is not sufficient to simply map x to (X_0, X_1) when $x \in \Gamma$, and to (Y_0, Y_1) when $x \in \Theta$. Rather, we map x to $(X, Y) = \text{XOR}((X_0, X_1), (Y_0, Y_1))$, which satisfies the CIIDC (by a computational analogue of Lemma 14).

7.2 Noninteractive, Instance-Dependent Commitments

We begin by reviewing Ben-Or and Gutfreund's [7] proof that IID is in NISZK^h and note that this proof cannot be replicated in the computational case to show that every Π satisfying the CIIDC is in NISZK^h. Ben-Or and Gutfreund show that IID is in NISZK^h by polarizing $(X_0, X_1) \in$ IID to the distributions (Y_0, Y_1), setting the help string to $\sigma = Y_0(r)$ and having P prove to V that $\sigma \in \mathrm{Supp}(Y_1)$ by sending a random preimage in $Y_1^{-1}(\sigma)$. However, this protocol may fail to even have completeness for promise problems satisfying CIIDC, since the images of Y_0 and Y_1 might even be disjoint, although they are computationally indistinguishable. Indeed, we do not expect to show that every problem satisfying CIIDC is in NIZK^h, since NIZK$^h \subseteq$ AM but problems outside AM may satisfy CIIDC (indeed, if one-way functions exist, *every* promise problem satisfies the CIIDC). Thus, in showing an equivalence between interactive and noninteractive zero knowledge in the computational case, it is necessary to use a different approach. Following [33], we view IID/CIIDC as a kind of instance-dependent commitment scheme, and use it to implement the general construction of noninteractive zero-knowledge proofs for AM [19].

We show that promise problems that reduce to IID or that satisfy CIIDC have a natural form of noninteractive, instance-dependent commitment schemes. In particular, for a promise problem Π which reduces to IID (resp., satisfies the CIIDC), the sender and the receiver can use the Polarization Lemma to obtain a pair of distributions (Y_0, Y_1) that are statistically close on YES instances, and mutually disjoint on NO instances. To commit to a bit b, the sender draws c from Y_b and outputs c as the commitment. To reveal b, the sender only needs to prove that c is drawn from Y_b by presenting to the receiver the randomness used in sampling from Y_b. Note that this binding property requires that the *sender generates the commitments honestly*. (Otherwise, it could always generate the commitment from the intersection of the supports, even if it negligibly small.) While assuming an honest sender is usually not suitable in applications of commitments, it turns out to be fine for constructing NIZK^h proofs, because the dealer generates the commitments.

We note that this commitment-based approach can also be used as an alternate, more circuitous proof of NISZK^h = SZK, since our results regarding commitments apply to both IID and CIIDC. Hence, the definitions and theorems presented below will deal with both the statistical and computational variants.

We now give a formal definition of the noninteractive, instance-dependent commitment schemes we will be using:

Definition 50. *A noninteractive, instance-dependent commitment scheme is a family* $\{\mathrm{Com}_x\}_{x \in \{0,1\}^*}$ *with the following properties:*

1. *The scheme* Com_x *proceeds in the stages: the* commit stage *and the* reveal stage. *In both stages, both the* sender *and the* receiver *share as common input the instance* x. *Hence we denote the sender and receiver as* S_x *and, respectively,* R_x, *and we write* $\mathrm{Com}_x = (S_x, R_x)$.
2. *At the beginning of the commit stage, the sender* S_x *receives as private input the bit* $b \in \{0, 1\}$ *to commit to. The sender then sends a single message* $c = S(x, b)$ *to the receiver.*

3. In the reveal stage, S_x sends a pair (b, d), where d is the decommitment *string* for bit b. Receiver R_x either accepts or rejects based on inputs x, b, d and c.

4. The sender S_x and receiver R_x algorithms are computable in time $\text{poly}(|x|)$, given the instance x.

5. For every $x \in \{0, 1\}^*$, R_x will always accept (with probability 1) if both S_x and R_x follow their prescribed strategy.

Security Properties. We now define the security properties of noninteractive, instance-dependent commitment schemes. These properties will be natural extensions of the hiding and binding requirements of standard commitments:

Definition 51. *A noninteractive, instance-dependent commitment scheme* $\text{Com}_x = (S_x, R_x)$ *is statistically (resp., computationally) hiding on* $I \subseteq \{0, 1\}^*$ *if for every (resp., nonuniform PPT) R^*, the ensembles* $\{S_x(0))\}_{x \in I}$ *and* $\{(S_x(1))\}_{x \in I}$ *are statistically (resp., computationally) indistinguishable.*

For a promise problem $\Pi = (\Pi_Y, \Pi_N)$, *a noninteractive, instance-dependent commitment scheme* Com_x *is statistically (resp., computationally) hiding on the* YES *instances if* Com_x *is statistically (resp., computationally) hiding on* Π_Y.

Definition 52. *A noninteractive instance-dependent commitment scheme* $\text{Com}_x = (S_x, R_x)$ *is statistically (resp., computationally) binding for honest senders on* $I \subseteq \{0, 1\}^*$ *if there exists a negligible function* ε *such that for all* $x \in I$, *a computationally unbounded (resp., nonuniform PPT) algorithm* S^* *succeeds in the following game with probability at most* $\varepsilon(|x|)$:

> S outputs a commitment c. Then, given the coin tosses of S, S^* outputs pairs $(0, d_0)$ and $(1, d_1)$ and succeeds if in the reveal stage, $R_x(0, d_0, c) = R_x(1, d_1, c) - \text{ACCEPT}$.

For a promise problem $\Pi = (\Pi_Y, \Pi_N)$, *a noninteractive, instance-dependent commitment scheme* Com_x *is statistically (resp., computationally) binding for honest senders on the* YES *instances if* Com_x *is statistically (resp., computationally) binding on* Π_Y.

Having defined noninteractive, instance-dependent commitment schemes, we proceed to show that they are equivalent to IID (resp., CIIDC), and consequently, SZK (resp., ZK).

Lemma 53. *A promise problem* Π *has a noninteractive, instance-dependent commitment scheme that is statistically (resp., computationally) hiding on* YES *instances and statistically binding for honest senders on* NO *instances if and only if* Π *reduces to IID (resp., if and only if* Π *satisfies the CIIDC).*

Proof. For the backwards direction, consider a problem Π that reduces to IID (the computational case will be similar). We construct the following protocol:

Commitment protocol for Π:

1. **Preprocessing:**
   First, reduce $x \in \Pi$ to an instance (X_0, X_1) of IID. Use the Polarization Lemma on (X_0, X_1) to obtain (Y_0, Y_1) such that, if $x \in \Pi_Y$, $\Delta(Y_0, Y_1) \leq 2^{-n}$, and, if $x \in \Pi_N$, (Y_0, Y_1) have mutual disjointness $(1 - 2^{-n})$, where $n = |x|$.
2. **Commit Stage:**
   $S_x(x, b)$: To commit to bit $b \in \{0, 1\}$, choose $d \xleftarrow{R} \{0, 1\}^m$, where m is the input length of Y_b, set $c = Y_b(d)$ and output (c, d).
3. **Reveal Stage:**
   $R_x(x, c, b, d)$: Accept if and only if $Y_b(d) = c$.

On $x \in \Pi_Y$, we know that Y_0 and Y_1 have negligible statistical difference. Hence, a commitment to 1 is statistically indistinguishable from a commitment to 0. Hence, the scheme is computationally hiding on YES instances (actually, the scheme is statistically hiding.)

When $x \in \Pi_N$, the pair (Y_0, Y_1) has mutual disjointness $(1 - 2^{-n})$. It directly follows that only a negligible fraction of commitments can be opened in two ways.

In the case that we are working with a problem which satisfies the CIIDC, we use the same scheme. However, instead of polarizing, we will simply take direct products to amplify the mutual disjointness on NO instances while preserving computational indistinguishability on YES instances (Lemma 13).

For the forward direction, let $\text{Com}_x = (S_x, R_x)$ be a noninteractive, instance-dependent commitment scheme that is statistically hiding on YES instances and statistically binding for honest senders on NO instances, and consider $X = S_x(0)$ and $Y = S_x(1)$:

- If $x \in \Pi_Y$, we know that $\Delta(\text{view}_R(S_x(0), R), \text{view}_R(S_x(1), R)) \leq \varepsilon(|x|)$, and hence, $\Delta(S_x(0), S_x(1)) \leq \varepsilon(|x|)$.
- If $x \in \Pi_N$, assume that there exists no negligible function $\mu(|x|)$ such that $\text{MutDisj}(S_x(0), S_x(1)) = (1 - \mu(|x|))$. Hence for all negligible functions $\mu(|x|)$ and $c \leftarrow S_x(b)$, $\Pr\left[c \in S_x(\bar{b})\right] > \mu(|x|)$. But then, S can always succeed with probability greater than $\mu(|x|)$ at the game described in Definition 52. So, for some negligible μ, $(S_x(0), S_x(1))$ have mutual disjointness $(1 - \mu(|x|))$, and Π reduces to IID.

The proof for the computational case is analogous.

By combining our previous results concerning IID and CIIDC with Lemma 53, we obtain the following theorem:

Theorem 54. *If a promise problem Π is in SZK (resp., ZK), then Π also has a noninteractive instance-dependent commitment scheme that is statistically (resp., computationally) hiding on YES instances and statistically binding for honest senders on NO instances.*

Proof. This follows from the fact that any $\Pi \in$ SZK (resp., ZK) reduces to IID (resp., satisfies CIIDC) (Lemma 49). By Lemma 53, Π has a noninteractive, instance-dependent commitment scheme.

7.3   From Noninteractive, Instance-Dependent Commitments to NIZK^h

In section, we will show that noninteractive, instance-dependent commitment schemes are sufficient to obtain NIZK^h. We start from the *hidden bits model*, a fictitious construction that implements noninteractive zero knowledge unconditionally for all promise problems in AM. Then, we show how our commitments can be employed in conjunction with this model to construct NIZK^h proofs.

The Hidden Bits Model. The hidden bits model is a model due to Feige, Lapidot and Shamir [19] that allows for an unconditional construction of NIZK. It assumes that both the prover P and the verifier V share a common reference string σ, which we will call the hidden random string (HRS). However, only the prover can see the HRS. We can imagine that the individual bits of σ are locked in boxes, and only the prover has the keys to unlock them. The prover can selectively unlock boxes and reveal bits of the hidden random string. However, without the prover's help, the verifier has no information about any of the bits in the HRS.

Definition 55 (NIZK in the Hidden Bits Model [19]). *A noninteractive zero knowledge proof system in the hidden-bits model for a promise problem Π is a pair of probabilistic algorithms (P, V) (where P and V polynomial-time bounded) and a polynomial $l(|x|) = |\sigma|$, satisfying the following conditions:*

1. *Completeness. For all $x \in \Pi_Y$, $\Pr\left[\exists (I, \pi)s.t.\ V(x, \sigma_I, I, \pi) = 1\right] > \frac{2}{3}$, where $(I, \pi) = P(x, \sigma)$, I is a set of indices in $\{0, \dots, l(k)\}$, and σ_I is the sequence of opened bits of σ, $(\sigma_i : i \in I)$, and where the probability is taken over $\sigma \xleftarrow{R} \{0,1\}^{l(|x|)}$ and the coin tosses of P and V.*
2. *Soundness. For all $x \in \Pi_N$ and all P^*, $\Pr\left[\exists (I, \pi)s.t.\ V(x, \sigma_I, I, \pi) = 1\right] \leq \frac{1}{3}$, where $(I, \pi) = P^*(x, \sigma)$, where the probability is taken over $\sigma \xleftarrow{R} \{0,1\}^{l(|x|)}$ and the coin tosses of P^* and V.*
3. *Zero Knowledge. There exists a PPT S such that the ensembles of transcripts $\{(x, \sigma, P(x, \sigma))\}_x$ and $\{S(x)\}_x$ are statistically indistinguishable on Π_Y, where $\sigma \xleftarrow{R} \{0,1\}^{l(|x|)}$.*

Note that we have defined the zero-knowledge condition in this model to be statistical rather than computational. Indeed, the known construction of hidden bits NIZK proof systems is unconditional and yields statistically indistinguishable proof systems.

Theorem 56 ([19]). *Every promise problem $\Pi \in$ NP has a hidden bits zero knowledge proof system (P, V).*

As has been observed before (e.g. [23]), this construction for NP automatically implies one for all of AM.

Corollary 57 ([19]). *Every promise problem $\Pi \in$ AM has a hidden bits zero knowledge proof system (P, V).*

Proof. Informally, this result can be obtained by transforming an AM proof into a statement that there exists some message from the prover that the verifier accepts. Since this statement is an NP statement, it can be proven in the hidden bits NIZK model.

The corollary above shows that there exists an unconditional construction of NIZK for all problems in AM. However, this construction holds only in the impractical hidden bits model. In proving our results, we show how to implement this construction in the help model by exploiting a novel connection to noninteractive, instance-dependent commitment schemes:

Theorem 58. *If $\Pi \in$ AM and Π has a noninteractive, honest-sender, instance-dependent commitment scheme that is statistically (resp., computationally) hiding on* YES *instances and statistically binding for honest senders on* NO *instances, then $\Pi \in$ NISZK^{h} (resp., $\Pi \in$ NIZK^{h}).*

Proof. Our general strategy will be to exploit the correspondence between the algorithms in our definition of an instance-dependent commitment scheme, and the three algorithms in a NIZK^{h} proof system. More specifically, we will have the dealer D use the sender algorithm to commit to a hidden bits string (this is why we can afford to assume the sender is honest). Since the prover P is allowed to be unbounded, we will use it to exhaustively search for openings to D's commitments. Finally, the verifier V will use the receiver algorithm to check P's openings.

Let $(P^{\mathrm{HB}}, V^{\mathrm{HB}})$ be a hidden bits proof system for Π and let $(\mathrm{Sen}, \mathrm{Rec})$ be the noninteractive, honest-sender bit commitment scheme for Π. Then, the following proof system (D, P, V) is NIZK^{h}:

1. $D(x, 1^k)$: Select $\sigma^D \overset{R}{\leftarrow} \{0,1\}^m$, and run $\mathrm{Sen}(x, \sigma_i^D)$ to generate a commitment c_i, for all i. Output $c = (c_1, \ldots, c_m)$ as the public help parameter.
2. $P(x, c)$: Exhaustively find a random opening o_i^P for each c_i (and, implicitly, each σ_i^D). If one commitment c_i can be opened as both 0 or 1, P outputs o_i^P according to the distribution $O|_{C=c_i}$, where (O, C) is the output of S on a random bit b. Let σ^P be the secret string obtained by P opening D's help string. P runs $P^{\mathrm{HB}}(x, \sigma^P)$ to obtain (I, π). Send $(I, \sigma_I^P, o_I^P, \pi)$ to V.
3. $V(x, I, o_I^P, \pi)$: Compute $\sigma_j^P, \forall j \in I$. Use Rec to check that the commitments are consistent. Run $V^{\mathrm{HB}}(x, I, \sigma_I^P, \pi)$ and accept if and only if V^{HB} accepts.

In the full version of the paper, we show that the construction above satisfies the completeness, soundness and zero knowledge properties, concluding that Π is in NIZK^{h}.

7.4 From ZK^{h} to ZK

In this section, we generalize the results of Ben-Or and Gutfreund [7] that SZK^{h} = SZK (Theorem 23) to show that adding help to ZK proofs does not confer any additional power:

Theorem 59 (Theorem 3, restated). $\mathrm{ZK}^{\mathrm{h}} = \mathrm{ZK}$.

To prove Theorem 23, Ben-Or and Gutfreund employ the techniques of [24,42,36], by considering the output of the simulator S for a zero-knowledge proof for Π as the moves of a *virtual prover* and a *virtual verifier*. The simulated transcripts are compared to the transcripts output by a cheating strategy for a real prover P_S (called the *simulation-based prover*), which tries to imitate the behavior of the virtual prover. Intuitively, on YES instances, the output of the simulator should be statistically close to the output of the simulation-based prover interacting with the real verifier. On NO instances, however, if we modify the simulator to accept with high probability (we can easily modify it to do that), the difference between the two transcripts must be significant. [7] exploit this to show that any problem in $\mathrm{SZK}^{\mathrm{h}}$ can be reduced to the intersection of the SZK-complete problems STATISTICAL DIFFERENCE([35]) and ENTROPY DIFFERENCE([36]). Since the other direction (SZK $\subseteq$ $\mathrm{SZK}^{\mathrm{h}}$) follows from the definitions, the conclusion that SZK $=$ $\mathrm{SZK}^{\mathrm{h}}$ follows immediately. We will use the same strategy with ZK^{h}, replacing statistical measures of closeness with computational ones. To do this, we replace the SZK-complete problems SD and ED with the INDISTINGUISHABILITY CONDITION and the CONDITIONAL PSEUDOENTROPY CONDITION, which characterize the class ZK, and show that for every $\Pi \in \mathrm{ZK}^{\mathrm{h}}$, Π can be reduced to the intersection of a problem which satisfies INDISTINGUISHABILITY CONDITION and a problem which satisfies CONDITIONAL PSEUDOENTROPY CONDITION, and is thus in ZK.

7.5 Putting It Together

We can now use the previous sections' results to prove our main theorems regarding computational zero knowledge:

Theorem 60 (Theorem 1, restated). $\mathrm{ZK}^{\mathrm{h}} \cap \mathrm{AM} = \mathrm{ZK} \cap \mathrm{AM} = \mathrm{NIZK}^{\mathrm{h}}$.

Proof. By definition, $\mathrm{NIZK}^{\mathrm{h}} \subseteq \mathrm{ZK}^{\mathrm{h}} \cap \mathrm{AM}$. For the other direction, we know any $\Pi \in \mathrm{ZK}$ has a noninteractive, instance-dependent commitment scheme (Theorem 54), so a $\mathrm{NIZK}^{\mathrm{h}}$ proof can built for Π (Theorem 58). Hence, $\mathrm{ZK}^{\mathrm{h}} \cap \mathrm{AM} \subseteq \mathrm{NIZK}^{\mathrm{h}}$, which completes the proof of our theorem.

Theorem 61. $\Pi \in \mathrm{ZK} = \mathrm{ZK}^{\mathrm{h}}$ *if and only if* $\Pi \in \mathrm{IP}$ *and* Π *satisfies the* CIIDC.

Proof. Since a promise problem that satisfies the CIIDC also satisfies the INDISTINGUISHABILITY CONDITION (this follows from the fact that of two distributions have disjointness α, they must have statistical difference at least α), the promise problem must have a ZK proof system by Theorem 26. Conversely, any problem in $\mathrm{ZK}^{\mathrm{h}} = \mathrm{ZK}$ satisfies CIIDC by Lemma 49.

8   Quantum Statistical Zero Knowledge

In this section, we study different variants of help for quantum noninteractive statistical zero knowledge. We start by providing complete problems for the class QNISZK defined by Kobayashi [13] and proceed to define the following two types of help: *pure quantum help* and *mixed quantum help*.

8.1   Complete Problems for QNISZK

Kobayashi [13] gave a complete problem for the class of quantum noninteractive perfect zero-knowledge, but not for statistical zero-knowledge. We continue this line of work and give two complete problems for QNISZK, QUANTUM ENTROPY APPROXIMATION (QEA) and QUANTUM STATISTICAL CLOSENESS TO UNIFORM (QSCU).

Let ρ be a quantum mixed state of n qubits which can be created in time polynomial in n by a quantum machine and t a positive integer. Then,

$$\text{QEA}_Y = \{(\rho, t) : S(\rho) \geq t + 1\} \qquad \text{QSCU}_Y = \{\rho : \|\rho - \mathcal{U}\| \leq 1/n\}$$
$$\text{QEA}_N = \{(\rho, t) : S(\rho) \leq t - 1\} \qquad \text{QSCU}_N = \{\rho : \|\rho - \mathcal{U}\| \geq 1 - 1/n\}$$

Note that these problems are the quantum equivalents of EA and SCU where the statistical difference is replaced by the trace distance and the Shannon entropy by the von Neumann entropy.

Theorem 62. QEA *and* QSCU *are complete for* QNISZK.

Proof Sketch: We start by showing that QEA belongs to QNISZK by using results of Ben-Aroya and Ta-Shma ([43]) on quantum expanders. Then, similarly to the classical case we reduce QSCU to QEA and last by Kobayashi's results ([13]) we know that QSCU is hard for QNISZK. This concludes the proof.   □

8.2   Help in Quantum Noninteractive Zero-Knowledge

In quantum noninteractive zero knowledge, the only model we defined so far is the model where the prover and the verifier share the maximally entangled state $\sum_i |i\rangle_P |i\rangle_V$ which can be created by a dealer with quantum polynomial power ([13]). In the previous section, we provided two complete problems for this class. Here, we extend this definition to allow the dealer to create as help a quantum state that depends on the input.

We define two types of help and study the resulting classes:

- *Pure Help*: In the usual framework of quantum zero-knowledge protocols, the prover and the verifier use only unitaries. We define QNISZK^{h} as the class where the prover and the verifier share a pure state (*i.e.*, the outcome of a unitary operation) created by the dealer in quantum polynomial time. This state can depend on the input. Note that since the maximally entangled state is a pure state QNISZK $\subseteq$ QNISZK^{h}. In fact, we show that QNISZK^{h} = QSZK = QSZK^{h}.

– *Mixed Help*: The previous definition does not allow the dealer to have some private coins and hence does not fully correspond to NISZK^h. We suppose now that the prover and verifier share a *mixed* quantum state created by the dealer. As before, the dealer has quantum polynomial power and the state depends on the input. We call the resulting class QNISZK^{mh} and show that this kind of help is most probably stronger than quantum interaction.

For these classes, the definition of the zero knowledge property remains the same as in the case of QNISZK (Section 5).

Pure Help. We suppose here that there is a trusted dealer with quantum polynomial power. On input x, he performs a unitary D_x and creates a pure state $D_x(|0\rangle) = |h_{PV}\rangle$ in the space $\mathcal{P} \times \mathcal{V}$. The prover gets $h_P = \mathrm{Tr}_\mathcal{V}(h_{PV})$ and the verifier gets $h_V = \mathrm{Tr}_\mathcal{P}(h_{PV})$. Note that the state h_{PV} is a pure state and depends on the input.

Definition 63. *We say that $\Pi \in$ QSZK^h (resp. $\Pi \in$ QNISZK^h) if there is an interactive (resp. noninteractive) protocol $\langle D, P, V \rangle$ that solves Π, has the zero knowledge property and where the verifier and the prover share a pure state h_{PV} created by a dealer D that has quantum polynomial power and access to the input. They also start with an arbitrary polynomial number of qubits initialized at $|0\rangle$.*

Next, we prove a quantum analogue of Theorem 39, *i.e.*, interactive and noninteractive zero knowledge are equivalent in the pure help model. We remark that the proof of this statement is much more straightforward than in the classical case.

Theorem 64. QNISZK^h = QSZK = QSZK^h

Proof. We start by showing that QSZK$^h \subseteq$ QSZK (and hence by definition QNISZK$^h \subseteq$ QSZK). Let $\Pi \in$ QSZK^h and $\langle D, P, V \rangle$ denote the protocol. Since h_{PV} is a pure state, we can create another protocol $\langle \widetilde{P}, \widetilde{V} \rangle$ where the verifier takes the place of the dealer. That is, V generates for his first message the state $|h_{PV}\rangle$ and sends the h_P part to the dealer while keeping the h_V part for himself. At this point, note that the verifier and prover have exactly the same states then when the dealer generates the state $|h_{PV}\rangle$ and sends it to them.

The protocol is the same so soundness and completeness are preserved. The first message in $\langle \widetilde{P}, \widetilde{V} \rangle$ can be simulated because the circuit of the dealer is public and computable in quantum polynomial time. The remaining messages in $\langle \widetilde{P}, \widetilde{V} \rangle$ can be simulated because of the zero-knowledge property of the protocol $\langle D, P, V \rangle$.

The inclusion QSZK $\subseteq$ QNISZK^h (and hence by definition QSZK $\subseteq$ QSZK^h) follows immediately from Watrous' two-message protocol for the QSZK-complete problem QSD [12]. The first message of the verifier can be replaced by the dealer's help.

532   A. Chailloux et al.

Mixed help. In the most general case, the dealer can create as help a mixed quantum state, *i.e.*, a state that can depend on some private coins or measurements as well as the input.

Definition 65. *We say that $\Pi \in$ QNISZK^{mh} if there is a noninteractive protocol $\langle D, P, V \rangle$ that solves Π with the zero-knowledge property, where the verifier and the prover share a mixed state h_{PV} created by a dealer D that has quantum polynomial power and access to the input. They also start with $|0\rangle$ qubits.*

Note that the only difference between QNISZK^h and QNISZK^{mh} is that the verifier and the prover share a mixed state instead of a pure state; however, we show that this difference is significant. In the classical case, a model was studied where the dealer flips some coins r and sends correlated messages $m_P(r)$ and $m_V(r)$ to the prover and the verifier. The resulting class was called NISZK^{sec} and it was shown by Pass and shelat in [23] that NISZK^{sec} = AM. To create the secret correlated messages $m_P(r)$ and $m_V(r)$ in our quantum setting, we just have to create the following state : $|\phi\rangle = \sum_r |r\rangle|m_P(r)\rangle|m_V(r)\rangle$. This state can be created in polynomial time because $m_P(r)$ and $m_V(r)$ can be created with a classical circuit. The dealer keeps the r part, sends the m_P part to the prover and the m_V part to the verifier. From this construction, we can easily see that AM = NISZK^{sec} $\subseteq$ QNISZK^{mh}. Note that it is not known that NP $\subseteq$ QSZK = QNISZK^h so this may be interpreted as evidence that QNISZK^h is a strict subset of QNISZK^{mh}.

Last, when we also allow the verifier to use non-unitary operations (*i.e.*, private coins and measurements), we don't know if help and interaction are equivalent. The case of quantum zero knowledge protocols with non-unitary players is indeed very interesting and we refer the reader to [44] for more results.

Acknowledgements. We thank the anonymous referees for their helpful comments.

References

1. Chailloux, A., Kerenidis, I.: The role of help in classical and quantum zero-knowledge. Cryptology ePrint Archive, Report 2007/421 (2007), http://eprint.iacr.org/
2. Ciocan, D.F., Vadhan, S.: Interactive and noninteractive zero knowledge coincide in the help model. Cryptology ePrint Archive, Report 2007/389 (2007), http://eprint.iacr.org/
3. Ciocan, D.: Constructions and characterizations of non-interactive zero-knowledge. Undergradute thesis, Harvard University (2007)
4. Goldwasser, S., Micali, S., Rackoff, C.: The knowledge complexity of interactive proof systems. SIAM Journal on Computing 18(1), 186–208 (1989)
5. Blum, M., Feldman, P., Micali, S.: Non-interactive zero-knowledge and its applications (extended abstract). In: STOC 1988: Proceedings of the twentieth annual ACM symposium on Theory of computing, pp. 103–112 (1988)
6. Blum, M., De Santis, A., Micali, S., Persiano, G.: Noninteractive zero-knowledge. SIAM Journal on Computing 20(6), 1084–1118 (1991)

7. Ben-Or, M., Gutfreund, D.: Trading help for interaction in statistical zero-knowledge proofs. Journal of Cryptology 16(2) (2003) (Preliminary version appeared as [22])
8. Brassard, G., Chaum, D., Crépeau, C.: Minimum disclosure proofs of knowledge. Journal of Computer and System Sciences 37(2), 156–189 (1988)
9. Nguyen, M.-H., Vadhan, S.: Zero knowledge with efficient provers. In: STOC 2006: Proceedings of the thirty-eighth annual ACM symposium on Theory of computing, pp. 287–295. ACM Press, New York (2006)
10. Ong, S.J., Vadhan, S.: Zero knowledge and soundness are symmetric. In: Naor, M. (ed.) EUROCRYPT 2007. LNCS, vol. 4515, Springer, Heidelberg (2007)
11. Kitaev, A., Watrous, J.: Parallelization, amplification, and exponential time simulation of quantum interactive proof systems. In: Proceedings of the 32nd ACM Symposium on Theory of computing, pp. 608–617 (2000)
12. Watrous, J.: Limits on the power of quantum statistical zero-knowledge. In: FOCS 2002: Proceedings of the 43rd Symposium on Foundations of Computer Science, Washington, DC, USA, pp. 459–468. IEEE Computer Society Press, Los Alamitos (2002)
13. Kobayashi, H.: Non-interactive quantum perfect and statistical zero-knowledge. In: Ibaraki, T., Katoh, N., Ono, H. (eds.) ISAAC 2003. LNCS, vol. 2906, pp. 178–188. Springer, Heidelberg (2003)
14. Babai, L., Moran, S.: Arthur-Merlin games: A randomized proof system and a hierarchy of complexity classes. Journal of Computer and System Sciences 36, 254–276 (1988)
15. Goldwasser, S., Sipser, M.: Private coins versus public coins in interactive proof systems. In: Micali, S. (ed.) Advances in Computing Research, JAC Press, Inc., vol. 5, pp. 73–90 (1989)
16. Goldreich, O., Micali, S., Wigderson, A.: Proofs that yield nothing but their validity, or All languages in NP have zero-knowledge proof systems. Journal of the Association for Computing Machinery 38(3), 691–729 (1991)
17. Impagliazzo, R., Yung, M.: Direct Minimum Knowledge Computations. In: Pomerance, C. (ed.) CRYPTO 1987. LNCS, vol. 293, pp. 40–51. Springer, Heidelberg (1988)
18. Goldreich, O., Håstad, J., Goldwasser, S., Micali, S., Rogaway, P., Kilian, J., Ben-Or, M.: Everything Provable Is Provable in Zero-Knowledge. In: Goldwasser, S. (ed.) CRYPTO 1988. LNCS, vol. 403, pp. 37–56. Springer, Heidelberg (1990)
19. Feige, U., Lapidot, D., Shamir, A.: Multiple non-interactive zero knowledge proofs under general assumptions. SIAM Journal on Computing 29(1), 1–28 (1999)
20. Håstad, J., Impagliazzo, R., Levin, L.A., Luby, M.: A pseudorandom generator from any one-way function. SIAM Journal on Computing 28(4), 1364–1396 (1999)
21. Naor, M.: Bit commitment using pseudorandomness. Journal of Cryptology 4(2), 151–158 (1991)
22. Gutfreund, D., Ben-Or, M.: Increasing the power of the dealer in non-interactive zero-knowledge proof systems. In: Okamoto, T. (ed.) ASIACRYPT 2000. LNCS, vol. 1976, pp. 429–442. Springer, Heidelberg (2000), (Journal version appeared as [7])
23. Pass, R., Shelat, A.: Unconditional characterizations of non-interactive zero-knowledge. In: Shoup, V. (ed.) CRYPTO 2005. LNCS, vol. 3621, pp. 118–134. Springer, Heidelberg (2005)
24. Aiello, W., Håstad, J.: Statistical zero-knowledge languages can be recognized in two rounds. Journal of Computer and System Sciences 42(3), 327–345 (1991)

25. Goldreich, O., Sahai, A., Vadhan, S.: Can statistical zero-knowledge be made non-interactive?, or On the relationship of SZK and NISZK. In: Wiener, M.J. (ed.) CRYPTO 1999. LNCS, vol. 1666, pp. 467–484. Springer, Heidelberg (1999)
26. Goldreich, O., Sahai, A., Vadhan, S.: Honest verifier statistical zero-knowledge equals general statistical zero-knowledge. In: Proceedings of the 30th Annual ACM Symposium on Theory of Computing, pp. 399–408 (1998)
27. Sahai, A., Vadhan, S.: Manipulating statistical difference. In: Pardalos, P., Rajasekaran, S., Rolim, J. (eds.) Randomization Methods in Algorithm Design (DIMACS Workshop, December 1997. DIMACS Series in Discrete Mathematics and Theoretical Computer Science, vol. 43, pp. 251–270. American Mathematical Society (1999)
28. Okamoto, T.: On relationships between statistical zero-knowledge proofs. Journal of Computer and System Sciences 60(1), 47–108 (2000)
29. De Santis, A., De Crescenzo, G., Persiano, G., Yung, M.: On monotone formula closure of SZK. In: Proc. 26th ACM Symp. on Theory of Computing, Montreal, Canada, pp. 454–465. ACM, New York (1994)
30. Bellare, M., Micali, S., Ostrovsky, R.: Perfect zero-knowledge in constant rounds. In: STOC 1990: Proceedings of the twenty-second annual ACM symposium on Theory of computing, pp. 482–493 (1990)
31. Itoh, T., Ohta, Y., Shizuya, H.: A language-dependent cryptographic primitive. Journal of Cryptology 10(1), 37–49 (1997)
32. Micciancio, D., Vadhan, S.: Statistical zero-knowledge proofs with efficient provers: Lattice problems and more. In: Boneh, D. (ed.) CRYPTO 2003. LNCS, vol. 2729, pp. 282–298. Springer, Heidelberg (2003)
33. Vadhan, S.: An unconditional study of computational zero knowledge. SIAM Journal on Computing 36(4), 1160–1214 (2006) (Special Issue on Randomness and Complexity)
34. Ong, S.J., Vadhan, S.: An equivalence between zero knowledge and commitments, These proceedings (2008)
35. Sahai, A., Vadhan, S.: A complete problem for statistical zero knowledge. Journal of the ACM 50(2), 196–249 (2003)
36. Goldreich, O., Vadhan, S.: Comparing entropies in statistical zero-knowledge with applications to the structure of SZK. In: Proceedings of the Fourteenth Annual IEEE Conference on Computational Complexity, Atlanta, GA, pp. 54–73 (1999)
37. Impagliazzo, R., Levin, L.A., Luby, M. (Pseudo-random generation from one-way functions (extended abstracts)) 12–24
38. Shamir, A.: IP = PSPACE. Journal of the ACM 39(4), 869–877 (1992)
39. Lund, C., Fortnow, L., Karloff, H., Nisan, N.: Algebraic methods for interactive proof systems. Journal of the ACM 39(4), 859–868 (1992)
40. Watrous, J.: Zero-knowledge against quantum attacks. In: STOC 2006: Proceedings of the thirty-eighth annual ACM Symposium on Theory of Computing, pp. 296–305. ACM Press, New York (2006)
41. Holenstein, T., Renner, R.: One-way secret-key agreement and applications to circuit polarization and immunization of public-key encryption. In: CRYPTO 2005, pp. 478–493. ACM Press, New York (2005)
42. Petrank, E., Tardos, G.: On the knowledge complexity of NP. In: IEEE Symposium on Foundations of Computer Science, pp. 494–503 (1996)
43. Ben-Aroya, A., Ta-Shma, A.: Quantum expanders and the quantum entropy difference problem. ArXiv Quantum Physics e-prints, quant-ph/0702129 (2007)
44. Chailloux, A., Kerenidis, I.: Increasing the power of the verifier in quantum zero knowledge. Arxiv Quantum Physics e-prints, quant-ph/07114032 (2007)

The Round-Complexity of Black-Box Zero-Knowledge: A Combinatorial Characterization

Daniele Micciancio and Scott Yilek

Dept. of Computer Science & Engineering, University of California, San Diego
9500 Gilman Drive, La Jolla, CA 92093-0404, USA
{daniele,syilek}@cs.ucsd.edu
http://www-cse.ucsd.edu/users/{daniele,syilek}

Abstract. The round-complexity of black-box zero-knowledge has for years been a topic of much interest. Results in this area generally focus on either proving lower bounds in various settings (e.g., Canetti, Kilian, Petrank, and Rosen [3] prove concurrent zero-knowledge (cZK) requires $\Omega(\log n / \log \log n)$ rounds and Barak and Lindell [2] show no constant-round single-session protocol can be zero-knowledge with strict poly-time simulators), or giving upper bounds (e.g., Prabhakaran, Rosen, and Sahai [15] give a cZK protocol with $\omega(\log n)$ rounds). In this paper we show that though proving upper bounds seems to be quite different from demonstrating lower bounds, underlying both tasks there is a single, simple combinatorial game between two players: a rewinder and a scheduler. We give two theorems relating the success of rewinders in the game to both upper and lower bounds for black-box zero-knowledge in various settings (sequential composition, concurrent composition, etc). Our game and theorems unify the previous results in the area, simplify the task of proving upper and lower bounds, and should be useful in showing future results in the area.

1 Introduction

Zero-Knowledge proofs, introduced by Goldwasser, Micali, and Rackoff [9], have been the focus of much research since their invention, both in cryptography and complexity theory. Interest in these proofs sparks from the fact that they provide both a useful tool for the construction of higher level security protocols, and a test-bed to explore new security issues, e.g., security under various forms of protocol composition. As a consequence, numerous variants of zero-knowledge have been considered and studied, resulting in many general possibility and impossibility results.

Informally, a zero-knowledge proof system is a two party protocol between a prover P and a verifier V that allows P to prove some assertion (e.g., membership of a string x in an $\mathcal{NP}$-language L) to V without leaking *any* information other than the truth of the assertion. This is typically proven by exhibiting a *black-box* simulator, i.e., an efficient procedure S that, given oracle access to the

R. Canetti (Ed.): TCC 2008, LNCS 4948, pp. 535–552, 2008.
© International Association for Cryptologic Research 2008

program of any (possibly misbehaving) verifier V^*, produces (without interacting with the prover) an output which is essentially identical to that of V^* when interacting with P. Non-black-box simulation methods are also possible [1], but most theoretical work, as well as essentially all protocols of practical interest, fall in the black-box model, making black-box simulation an interesting area of research on its own.

Much of the work on black-box zero-knowledge focuses on *round-complexity*, as interaction is at the same time essential (to achieve the zero-knowledge property) and expensive (from a practical performance point of view). In the last decade, many general upper and lower bounds on the round complexity of black-box zero-knowledge proof systems have been established [8,12,18,16,11,15].

Negative results (i.e., round complexity lower bounds) typically show that no non-trivial[1] language admits a black-box zero-knowledge proof system with less than a given number of rounds. For example, Barak and Lindell have shown that if the black-box simulator is constrained to run in *strict* polynomial time, then only trivial languages admit constant round zero-knowledge proofs [2]. Similarly, [3] has shown that only trivial languages admit black-box concurrent zero-knowledge proofs with $o(\log n / \log \log n)$ rounds. Such negative results are proved by exhibiting a carefully crafted hard-to-simulate verifier V^* such that any efficient black-box algorithm S that simulates the interaction between V^* and P can be transformed into an efficient decision procedure for the language L being proven.

Positive results (i.e., round complexity upper bounds) typically assert that under general cryptographic assumptions (e.g., the existence of commitment schemes) every language in $\mathcal{NP}$ admits a black-box zero-knowledge proof system with low round complexity. Such positive results are usually proved by giving an explicit proof system for a single $\mathcal{NP}$-complete problem (e.g., 3-coloring or graph hamiltonicity), and proof systems for all other $\mathcal{NP}$ languages follow by reduction. For example, Goldreich and Kahan [7] give a constant-round zero-knowledge proof for 3-coloring, with an expected polynomial-time black-box simulator.

It would appear that the tasks of proving lower and upper bounds for black-box zero-knowledge are quite different: the former needs to be completely general and hold for any language and associated proof system; the latter considers a specific ($\mathcal{NP}$-complete) language and provides an explicit prover and simulator strategy for that language.

1.1   Our Results

We describe a simple combinatorial game (parameterized by an integer r) that closely characterizes (up to a small constant additive term) the round complexity of black-box zero-knowledge, in the sense that (under standard cryptographic

---

[1] In the context of zero-knowledge proofs, "non-trivial" typically refers to languages that are not known to be decidable in probabilistic polynomial time, since any such language admits a trivial zero-knowledge proof system where the verifier checks the validity of the assertion on its own, without the help of the prover.

assumptions) any non-trivial language (in $\mathcal{NP} \setminus \mathcal{BPP}$) admits a black-box zero-knowledge proof system with $r + O(1)$ rounds *if and only if* the combinatorial game admits a solution. (See Sect. 3 for a formal statement of the results.)

The game is simple, yet general enough to study, in a unified way, black-box zero-knowledge in many important settings, including

- zero-knowledge with sequential composition, where only a single prover interacts with a single verifier at any time (no interleavings of sessions),
- concurrent zero-knowledge, where many colluding cheating verifiers can interact with independent provers while interleaving their actions in the most adversarially possible way,
- a form of parallel zero knowledge, where n provers send all their first messages in order, before moving to the next round, and so on,
- zero-knowledge under various forms of bounded concurrency, where different sessions can be interleaved, provided not too many sessions are active at the same time.

All of these settings are treated in a uniform way simply by parameterizing the combinatorial game with a set Γ of "forbidden patterns", i.e., sequences of interleavings that are guaranteed not to occur. For example, in the case of concurrent zero-knowledge, $\Gamma = \emptyset$ and all interleavings are allowed, while in sequential zero-knowledge Γ is the set of all sequences in which labels are interleaved (i.e., between the first and last labels in some session, there is a node with a different label). For simplicity of exposition, we focus on the case of *concurrent* zero-knowledge, where $\Gamma = \emptyset$ and can be omitted. The reader can easily check that all our results and proofs immediately extend to arbitrary Γ.

Our game (described below) is similar, though not identical, to games implicitly defined in previous papers providing upper and lower bounds on the round complexity of black-box concurrent zero-knowledge [3,15]. We remark that the combinatorial games implicitly used in previous proofs differed from each other, leaving a (perhaps small, but interesting) gap between upper and lower bounds. A technical contribution of our paper is to identify (and precisely define) a variant of the game that simultaneously yields both upper and lower bounds. Moreover, our characterization result holds in a variety of settings (basically, any kind of protocol composition that can be described by a set of forbidden patterns). This can be useful not only to unify and simplify many previous results for the sequential and concurrent composition setting (e.g., [2,3,7,15,17],) but also as a starting point to study the round complexity under other forms of composition (e.g., parallel composition, bounded concurrency)

1.2 The Game

The game (parameterized by three integers d, h and r) is played by two players, called the scheduler and the rewinder. The scheduler strategy is described by a labeled tree, where each internal node v has exactly d children $v_1, \ldots, v_d$, and all leaves are at level h. Each node v is represented by a sequence $\{1, \ldots, d\}^*$ in the standard way, where nodes at level l of the tree are represented by sequences of length l, and the parent of a node is obtained by removing the last element in the sequence. Each edge $v \to v_m$ (from a node v to its mth child) carries the label m, and every internal node carries a label $\pi(v)$ which equals either the node itself $\pi(v) = v$, or a previous label $\pi(v) = \pi(w)$, where w is an ancestor of v.

For every node $v \in \{1, \ldots, d\}^*$ in the tree, let $\mu(v)$ be the subsequence of v corresponding to the edges immediately following a node w with label $\pi(w) = \pi(v)$. (See Fig. 2. The reader is referred to Sect. 3 for a formal definition.) In an execution of the game, the rewinder explores the labeled tree defined by the scheduler, learning the label $\pi(v)$ when a node v is visited. Starting from the root node, the rewinder selects at every step a child of some previously visited node, subject to the following restriction:

- At any point during the game, if $|\mu(v)| = r$, then the rewinder is allowed to select a child of v only if he has already visited some other node w with the same label $\pi(v) = \pi(w)$ such that $\mu(w)$ is not a prefix of $\mu(v)$.

The rewinder is allowed to perform random choices during the game, with the goal of reaching a random leaf of the tree, while visiting the smallest possible number of nodes. Formally, we measure the success of the rewinder in the game by the number of visited nodes (which should be small) and the distance from random of the distribution over leaves defined by an execution of the game.

While the above game can be conveniently studied in a purely combinatorial setting, in order to establish a link between the game and computational zero-knowledge proof and argument systems one has to consider a natural computational version of the game where,

- the scheduler and rewinder strategies are required to be efficiently computable, and
- the leaf reached by the rewinder is only required to be pseudo-randomly distributed.

We remark that these computational restrictions are mostly a technicality, as they do not seem to affect the combinatorial complexity of solving the game. In particular, in all settings that we are aware of, the best known solution to the game is also efficiently computable and it generates distributions over leaves that are statistically close to (rather than simply indistinguishable from) random. We give examples in the full version of the paper [14].

2   Preliminaries

2.1   Notation

We denote by Σ^* the (infinite) set of all strings. We sometimes write $\{0, 1\}^{\leq n}$ to denote the set of all bit-strings with length at most n. We will often use bold letters or overbars to represent (possibly empty) sequences of messages (e.g., $\mathbf{q}$, $\bar{\beta}$).

For interactive machines P and V, let $\langle P, V \rangle(x)$ be the local output of V after interacting with P on common input x. We denote by $\mathsf{view}_V^P(x)$ the random tape of V followed by the sequence of messages V receives while interacting with P

on common input x. We will generally use α to denote a message from V, while β will represent a message from P. We will often use next-message functions when dealing with interactive machines. Let $V(x, z, \bar{\beta}; r)$ denote the *next-message function* which takes as input a sequence of messages $\bar{\beta}$ and a random tape r and outputs the result of running interactive machine V with random tape r on common input x, auxiliary input z, and sequence of incoming message $\bar{\beta}$. When dealing with a joint computation between two interactive machines, we define a *round* to be two messages, one from each machine. So if we say a protocol is four rounds, for example, there are actually eight messages exchanged. It is sometimes convenient to run an interactive machine for a number of rounds which is higher than the number specified by its program. We use the convention that if an interactive machine implementing an r-round protocol is sent more than r messages, it replies to the messages beyond round r with an empty dummy message. For Turing machine M, we let $\mathsf{desc}(M)$ denote the description of M.

A function $\nu : \mathbb{N} \to [0,1]$ is called *negligible* if $\nu(n) = n^{-\omega(1)}$, and *non-negligible* (or *noticeable*) if $\nu(n) = n^{-O(1)}$. We say two ensembles $\{X_w\}_{w \in S}$ and $\{Y_w\}_{w \in S}$ indexed by strings in some infinite set S, are *computationally indistinguishable* (denoted $\{X_w\}_{w \in S} \equiv_c \{Y_w\}_{w \in S}$) if for every probabilistic (strict) polynomial time (PPT) algorithm D (called the distinguisher) we have $|\mathbf{Pr}[D(X_w, w) = 1] - \mathbf{Pr}[D(Y_w, w) = 1]| < \nu(|w|)$ for some negligible function ν. Conversely, we say that the same ensembles are *computationally distinguishable* if there is some PPT distinguisher D such that $|\mathbf{Pr}[D(X_w, w) = 1] - \mathbf{Pr}[D(Y_w, w) = 1]| \geq \nu(|w|)$ for some non-negligible function ν.

2.2 Interactive Proofs and Black-Box Zero Knowledge

We use the standard definition of interactive proofs with negligible soundness error:

Definition 1. *A pair of interactive machines (P, V) is an interactive proof system for language L if machine V runs in polynomial time and there exists a negligible function $\nu : \mathbb{N} \to [0,1]$ such that*

- *Completeness: For every $x \in L$, $\mathbf{Pr}[\langle P, V \rangle(x) = 1] \geq 1 - \nu(|x|)$*
- *Soundness: For every $x \notin L$ and every interactive machine B,*
  $\mathbf{Pr}[\langle B, V \rangle(x) = 1] \leq \nu(|x|)$

We say that such an interactive protocol has almost-perfect completeness or negligible completeness error, and negligible soundness error. If the soundness condition holds only against PPT B, then (P, V) is called an *argument* system (also known as a computationally sound proof). The results in this paper hold for both proof and argument systems, but for simplicity we focus on proof systems.

Definition 2. *Let (P, V) be an interactive proof system for some language L. We say that (P, V) is **black-box zero-knowledge** if there exists a PPT oracle machine S such that for every PPT interactive machine V^*, the ensembles*

$$\{\mathsf{view}_{V^*(z)}^{P(w_x)}(x)\}_{x \in L, z \in \{0,1\}^*} \text{ and } \{S^{V^*(x,z,\cdot)}(x)\}_{x \in L, z \in \{0,1\}^*}$$

are computationally indistinguishable.

Note that in this definition the simulator S is given oracle access to the next-message function of the adversarial verifier.

2.3 Composition of Interactive Proofs

We will consider a setting in which a verifier may interact with multiple, independent copies of the prover, each of which is attempting to prove the same theorem. Specifically, the prover's reply in some session should only depend on previous messages *from that session* and not messages from some other session. On the other hand, the verifier may make decisions based on messages it has seen from any session. To model this, we will consider a single adversarial verifier V^* which sends messages of the form (α, s) to the prover, where s is just some arbitrary string identifying the session. The prover's next message must be a reply for this session. The last message of the interaction is then considered the output of V^*, which we require to have the form (α^*, end) for some α^*. So, the transcript of an interaction between P and V^* will be of the form $((\alpha_1, s_1), \beta_1, (\alpha_2, s_2), \beta_2, \ldots, (\alpha_v, s_v), \beta_v, (\alpha^*, \mathsf{end}))$.

In this paper, we only consider adversarial verifiers which never abort, where by abort we mean send messages that are malformed in some session (or deviate in some other detectable way). This is without loss in the upper bound because a simulator can easily modify its verifier to be non-aborting. In the case of the lower bound, the adversarial verifiers we construct do not send abort messages at any time. Previous lower bounds (e.g., [3]) rely on aborting sessions to force the simulator to do extra work. Nevertheless, the act of the verifier *informing* the simulator that it is going to abort some session simply makes the task of simulating easier, since the simulator knows that on the current path it no longer needs to worry about the aborted session. In our paper, the verifier can still implicitly "abort" a session by never again sending a message for that session; it just doesn't tell the simulator it is doing this.

2.4 Black-Box Concurrent Zero Knowledge

In the most general form of concurrent composition, the verifier is allowed to interact with a polynomial number of independent provers while maintaining complete control over the scheduling of messages. Specifically, the verifier may interleave messages from different sessions any way it chooses. This situation was first explored for witness indistinguishability (a weaker notion than zero-knowledge) in [6,5] and later for zero-knowledge in [4].

As observed in [3], the standard definition of black-box zero-knowledge is insufficient in the concurrent setting, since the running time of the simulator must be a fixed polynomial and independent of its oracle. Thus, a verifier could initiate more sessions than the simulator has time to handle (yet still a polynomial number). To overcome this subtlety, we give the simulator an additional input representing an upper bound on the length of the interaction with the verifier (how many messages the verifier will send in all sessions before halting with some local output). A similar approach was used in [3][2]. We then require the simulator to run in polynomial time in both the length of the common input and the maximum length of an interaction.

Definition 3. *Let (P, V) be an interactive proof system for some language L. We say that (P, V) is **black-box concurrent zero-knowledge** if there exists a PPT oracle machine S such that for every polynomial $h(\cdot)$ and every PPT interactive machine V^* sending at most $h(|x|)$ messages in any interaction, the ensembles*

$$\{\text{view}^P_{V^*(z)}(x)\}_{x \in L, z \in \{0,1\}^*} \text{ and } \{S^{V^*(x,z,\cdot)}(x, h(|x|))\}_{x \in L, z \in \{0,1\}^*}$$

are computationally indistinguishable.

3   A Simple Combinatorial Game

In this section we formally define our combinatorial game. We first describe a purely combinatorial version, and then modify the game to satisfy some natural computational restrictions. The game is parameterized by positive integers h, r, and d and has two players: a rewinder S and a scheduler V.

3.1   The Scheduler

The *scheduler* V uses a private random input to specify a labeled tree where each internal node v has exactly d children $v_1, \ldots, v_d$ and all leaves are at level h (see Fig. 1). Each edge $v \to v_m$ (from a node v to its mth child) carries the label m. Each internal node v (represented by a sequence in $D^{\leq h} = \{1, \ldots, d\}^{\leq h}$ of edges) carries a label $\pi(v)$ which equals either the node itself ($\pi(v) = v$), or a previous label $\pi(v) = \pi(w)$ for some ancestor w (i.e. prefix) of v. The labeling function $V : D^{\leq h} \to D^{\leq h}$ takes as input some node v and returns its label $\pi(v)$.

We let $U_{d,h}$ denote the uniform distribution on leaves in a tree of degree d and height h. Leaves are sequences in D^h.

For any $v = (m_1, \ldots, m_j)$, let $\mu(v) = v[I]$, where I is the set of positions i in v satisfying $\pi(v[1, \ldots, i-1]) = \pi(v)$. This means that $\mu(v)$ is the sequence of edges coming out of nodes with the same label as v, on the path from the root to v. See Fig. 2 for an illustration.

3.2   The Rewinder

The *rewinder* is a probabilistic oracle algorithm S^V where $V : D^{\leq h} \to D^{\leq h}$. The rewinder S uses the oracle to explore the scheduler tree, and it may query the oracle on a child of some previously visited node v subject to one restriction:

– If $|\mu(v)| = r$, then the rewinder must have already visited some other node w with the same label $\pi(w) = \pi(v)$ such that $\mu(w)$ is not a prefix of $\mu(v)$.

---

[2] In [3], the simulator takes an additional input representing an upper bound on the number of sessions the verifier will initiate during the interaction.

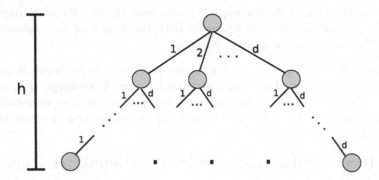

Fig. 1. The scheduler tree. Each internal node has degree d, with edges labeled 1 to d. The height of the tree is always h and the leaves are sequences in $\{1, \ldots, d\}^h$.

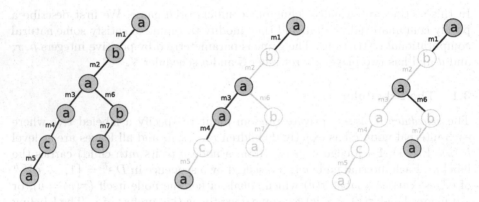

Fig. 2. The above depicts the explored portion of some tree. For brevity we let $a = \epsilon$, $b = m_1$, and $c = (m_1, m_2, m_3, m_4)$. The middle picture depicts $\mu(m_1, m_2, m_3, m_4, m_5) = (m_1, m_3, m_4)$. The right picture shows how the node (m_1, m_2, m_6, m_7) has the same label $a = \epsilon$ as the node $(m_1, m_2, m_3, m_4, m_5)$, but $\mu(m_1, m_2, m_6, m_7) = (m_1, m_6)$ is not a prefix of $\mu(m_1, m_2, m_3, m_4, m_5) = (m_1, m_3, m_4)$.

The restriction, which we often call the μ restriction, is illustrated in Fig. 2. It is easy to see that the condition can be efficiently checked by $\mathbb{S}$ (without making any additional queries) because it only depends on the previously issued queries and respective answers. We consider rewinders $\mathbb{S}$ that always terminate (or, at least terminate with probability 1), and always make at least one query $v_0 = \epsilon$.

We denote by $\mathbb{S}^{\mathbb{V}}$ the probability distribution over leaves D^h, specified by the last query asked by $\mathbb{S}$ to the oracle $\mathbb{V}$. The goal of the rewinder is to produce a distribution over leaves as close as possible to uniform $U_{d,h}$.

3.3 A Computational Version

To make the game useful in asymptotic settings (like black-box computational zero-knowledge proof systems), we need to impose some computational restrictions.

Instead of considering a single scheduler $\mathbb{V}$, we consider a family of schedulers $\{\mathbb{V}_x\}_{x \in \Sigma^*}$ indexed by a parameter string x. Each $\mathbb{V}_x$ is a defined as usual as a function $\mathbb{V}_x : D_x^{h_x} \to D_x^{h_x}$ where $D_x = \{1, \ldots, d_x\}$, i.e., $\mathbb{V}_x$ is a labeled tree of degree d_x and height h_x. We say that $\mathbb{V}$ is an *efficient* scheduler if $h_x \leq |x|^{O(1)}$, $d_x \leq 2^{|x|^{O(1)}}$ and the labeling function $(x, v) \mapsto \mathbb{V}_x(v)$ is computable in polynomial time $|x|^{O(1)}$. (Notice that by the bounds on h_x and d_x, the size of the second argument $|v| \leq h_x \cdot \log_2 d_x \leq |x|^{O(1)}$ is always polynomial in $|x|$.)

Similarly, we consider families of oracle algorithms $\mathbb{S}_x^{\mathbb{V}}(h)$ indexed by the string x. We say that $\mathbb{S}$ is an *efficient rewinder* if the strategy $\mathbb{S}_x^{(\cdot)}(h)$ can be implemented in PPT in $|x|$ and h. The goal of $\mathbb{S}$ is to produce a distribution over leaves $D_x^{h_x}$ which is computationally indistinguishable (in $|x|$) from the uniform distribution.

Finally, the round number r_x defining the rules of the game, can also depend on x. Typically, h_x, d_x and r_x are just functions of the parameter length $|x|$.

4 Two Theorems

In this section we give two theorems (for upper and lower bounds) which relate the success of rewinders in our combinatorial game to the success of black-box simulators in the more complicated concurrent zero-knowledge setting. For simplicity, in both of the following theorems, fix d_x to be some super-polynomial function of $|x|$.

Theorem 1 (Lower Bound). *For any round parameter r_x and some language L, if the following two conditions hold:*

1. *The game does not have a solution: For every efficient rewinder $\mathbb{S}$ there exists an efficient scheduler $(h_x, \mathbb{V}_x)$ such that the ensembles $\{\mathbb{S}_x^{\mathbb{V}_x}(h_x)\}_{x \in \Sigma^*}$ and $\{U_{d_x, h_x}\}_{x \in \Sigma^*}$ are distinguishable in polynomial time*
2. *There exists an $(r_x + 1)$-round black-box concurrent zero-knowledge proof system for L*

Then $L \in \mathcal{BPP}$.

Theorem 2 (Upper Bound). *For any game parameter r_x, if the following two conditions hold:*

1. *The game has a solution: There exists an efficient rewinder $\mathbb{S}$ such that for all efficient schedulers $(h_x, \mathbb{V}_x)$, the ensembles $\{\mathbb{S}_x^{\mathbb{V}_x}(h_x)\}_{x \in \Sigma^*}$ and $\{U_{d_x, h_x}\}_{x \in \Sigma^*}$ are computationally indistinguishable*
2. *Perfectly-hiding commitment schemes exist*

Then there exists an $(r_x + O(1))$-round black-box concurrent zero-knowledge proof system for all $L \in \mathcal{NP}$.

We prove Thm. 1 in Sect. 5 and Thm. 2 in Sect. 6. Combining the two theorems gives us our main result: Any non-trivial language (in $\mathcal{NP} \setminus \mathcal{BPP}$) has an $(r + O(1))$-round black-box concurrent zero-knowledge proof system if and only if our combinatorial game with parameter r admits a solution.

5   Proof of Theorem 1 (Lower Bound)

In this section we show the relationship between the combinatorial game and black-box zero-knowledge lower bounds. Before getting into the details of the proof we will first provide some intuition for our results.

Like in the lower bound proof of [3], we run the simulator S for language L inside of a $\mathcal{BPP}$ decision procedure while simulating its oracle (an adversarial concurrent verifier) using sufficiently independent copies of the single-session honest verifier. However, the way we use the single-session honest verifier differs, as does our accepting criteria. We force the decision procedure to accept if and only if at some point S causes a single-session verifier to accept without successfully rewinding the session. Otherwise, the decision procedure rejects.

It is fairly straightforward to show that if x is not in L, then our decision procedure will reject with overwhelming probability, due to the soundness of the proof system. More difficult is showing that if x is in the language, then the decision procedure accepts with noticeable probability. To accomplish this, we want to make it difficult for the simulator to successfully rewind every session. This is where we use the combinatorial game and our assumption that for any rewinder there is a difficult scheduler.

Intuitively, a simulator making oracle queries to an adversarial verifier (and possibly rewinding) is similar to a rewinder exploring a scheduler tree. Following this intuition, we show that given S we can build a rewinder $\mathbb{S}$. The rewinder runs S internally and uses its scheduler oracle to help simulate a concurrent adversarial verifier oracle for S. Given such a rewinder $\mathbb{S}$, our assumption guarantees a corresponding difficult scheduler $\mathbb{V}^*$. The hope is that this difficult scheduler will make the corresponding adversarial verifier hard-to-simulate. We then simply make our $\mathcal{BPP}$ decision procedure run S and simulate its oracle in the same way as the rewinder $\mathbb{S}$, and using the difficult scheduler $\mathbb{V}^*$.

There is still one issue to overcome. Our assumption tells us that it is difficult for $\mathbb{S}$ to reach a random leaf in $\mathbb{V}^*$, so we need to make sure the simulator's task of properly simulating and the rewinder's task of reaching a random leaf are related. To do this, we have the simulated adversarial concurrent verifier randomize queries before querying the difficult scheduler $\mathbb{V}^*$. As long as the independence used to randomize is greater than the maximum length of an interaction between the honest prover and the verifier, the prover's interaction with this verifier will correspond to reaching a random leaf in the scheduler tree. Since S must simulate this interaction properly (because of the zero-knowledge property), we have our desired relationship between the game and the proof system. It should then be the case that if x is in L, S will have difficult time rewinding every session, and thus it will be forced to make a single-session honest verifier accept without rewinding. This, as we said above, causes the decision procedure to accept x.

5.1   Details of the Proof

We will now expand on the ideas explained in the previous section. Fix some function r_x, fix function d_x to be super-polynomial in $|x|$, and let L be some language. Assume the following:

1. For every probabilistic oracle machine $\mathbb{S}_x(h)$ running in polynomial time in $|x|$ and h, there exists a polynomial time (in $|x|$) computable $\mathbb{V}_x(\cdot)$, such that the ensembles $\{\mathbb{S}_x^{\mathbb{V}_x}(h_x)\}_{x \in \Sigma^*}$ and $\{U_{d_x,h_x}\}_{x \in \Sigma^*}$ are distinguishable in polynomial time.
2. There exists an $(r_x + 1)$-round black-box concurrent zero-knowledge proof system for L.

We aim to show that $L \in \mathcal{BPP}$. Since we are assuming L has a black-box zero-knowledge proof, let S be its simulator. As we explained above, we will define a decision procedure D which runs S and simulates for it a verifier oracle V, with the help of some scheduler $\mathbb{V}$. We would like our decision procedure to simulate a verifier oracle which makes it difficult for S to properly rewind every session. To do this we first show how to construct a difficult verifier after placing a few restrictions on S.

Restrictions on S. We make some assumptions about S which are without loss of generality. First, we assume that S only makes queries for which it has already queried all shorter prefixes. Second, we assume that before S outputs a view with prover messages β, it queries its oracle one last time with $\bar{\beta}$ (and all shorter prefixes). Clearly, any PPT S can be transformed into a simulator S' that satisfies the above properties and still runs in polynomial time.

Let $t(|x|, h)$ and $m(|x|)$ be polynomial bounds on the number of queries and the message size for all queries made by S.

A Difficult Verifier. As with past lower bound proofs, we wish to describe a verifier which will be difficult for the simulator S to simulate inside of the decision procedure. We first describe an oracle verifier $\hat{V}$ which is given oracle access to a scheduler $\mathbb{V}$ which uses p bits of its private random input.

We will rely extensively on two hash functions, which we say are given to $\hat{V}$ as auxiliary input. Let $\mathcal{F} = \{F_{n,h}\}_{n,h \in \mathbb{N}}$ be a family of $t(n, h)$-wise independent hash functions. Let $\mathcal{G} = \{G_{n,h}\}_{n,h \in \mathbb{N}}$ be a family of $t(n + h)$-wise independent hash functions. Each function $f \in F_{n,h}$ will be from $\{0,1\}^{\leq h \cdot m}$ to $\{0,1\}^\rho$, where ρ is an upper bound on the number of bits the single-session honest verifier reads from its random tape. Functions from $\mathcal{F}$ will be used to hash all of the messages leading up to the start of a session in order to generate randomness for the single-session honest verifier used in that session. Each function $g \in G_{n,h}$

will be from $\{0,1\}^{\leq h \cdot m} \times \{0,1\}^{\leq h \cdot m} \times m$ to $\{0,1\}^{\log d}$. Functions from $\mathcal{G}$ are used to randomize messages for use as edges in a scheduler tree. They contain independence greater than the running time of S and therefore also greater than the maximum length of an interaction. In particular, for a randomly chosen function from the family, queries of length h will lead to a uniformly random leaf as long as the inputs are unique.

Given any function g from the family just described, for simplicity we define another function $\hat{g}_{\mathbb{V}}$ which is with respect to some scheduler $\mathbb{V}$. Now, define $\mathbb{V} \circ g(\bar{\beta}) = \mathbb{V}(\hat{g}_{\mathbb{V}}(\bar{\beta}))$ and recursively define $\hat{g}$ as

$$\hat{g}_{\mathbb{V}}(\epsilon) = \epsilon$$
$$\hat{g}_{\mathbb{V}}(\bar{\beta}, \beta) = (\hat{g}_{\mathbb{V}}(\bar{\beta}), g(\pi_{\mathbb{V} \circ g}(\bar{\beta}), \mu_{\mathbb{V} \circ g}(\bar{\beta}), \beta)),$$

where π and μ are defined as they were in Sect. 3. We are ready to describe the verifier $\hat{V}$ with oracle access to some scheduler $\mathbb{V}$. We describe the next-message function, which takes as input a sequence of messages $\bar{\beta}, \beta$. The verifier is given as auxiliary input two hash functions f and g from the families we described above.

Algorithm $\hat{V}^{\mathbb{V}_x}(x, (f,g), (\bar{\beta}, \beta))$
1. If $|\bar{\beta}, \beta| = h$ or $V(x, (\mu_{\mathbb{V} \circ g}(\bar{\beta}), \beta); f(\pi_{\mathbb{V} \circ g}(\bar{\beta}))) = \mathsf{reject}$ then
   return $((\bar{\beta}, \beta), \mathsf{end})$
2. Else return $(V(x, \mu_{\mathbb{V} \circ g}(\bar{\beta}, \beta); f(\pi_{\mathbb{V} \circ g}(\bar{\beta}, \beta))), \mathbb{V} \circ g(\bar{\beta}, \beta))$

The adversarial concurrent verifier intuitively executes multiple sessions and interleaves them based on queries to its scheduler oracle (the queries are first randomized using g before querying the oracle). The content of the replies from $\hat{V}$ comes from a single-session honest verifier V which $\hat{V}$ runs internally. Each session's messages are determined by a sufficiently independent copy of V (the independence is due to the hash function f).

The adversarial verifier $\hat{V}$ sends final output (a sequence of messages) in two cases. One is if at any time a single-session honest verifier rejects. The other is if the end of the interaction is reached (the sequence of messages is length h). We should mention that it is possible a valid rewinding of some session by S might not yield a valid rewinding in $\mathbb{V}$. This is because each message in the queries from S is at most size m (a polynomial in $|x|$), but when g randomizes it maps these messages into a message space of size d (which is super-polynomial in $|x|$). However, since S makes at most polynomially many queries, the probability of a collision is negligible, so we ignore this event for the rest of the proof.

Now, as we said earlier, we want to run the simulator inside of a decision procedure, giving it oracle access to a difficult-to-simulate verifier. To make the verifier just described "difficult", we want it to have oracle access to a difficult scheduler. To get such a scheduler, we need to define a rewinder and then use our assumption about the combinatorial game not admitting a solution. The rewinder is as follows.

Algorithm $\mathbb{S}_x^{V_x}(h)$
1. Randomly choose f and g from their respective families.
2. Run $[S^{\hat{V}}(x, h)]^{V_x}$
3. Watch for queries to V_x which violate the condition of the combinatorial game.
4. If such a query is made then halt and output $\perp$.

The rewinder $\mathbb{S}$ internally runs S, giving it oracle access to $\hat{V}$. Since $\hat{V}$ expects a scheduler oracle, $\mathbb{S}$ will simulate it using its own scheduler oracle V. However, $\mathbb{S}$ will monitor these queries to V and if at any time there is a query which violates the condition of the game (the scheduler was not properly "rewound"; see Sect. 3) $\mathbb{S}$ will halt and fail. Otherwise $\mathbb{S}$ will continue until S halts with some final output.

Given the efficient rewinder above, our assumption about the game not admitting a solution guarantees there is some difficult efficient scheduler V^*. Our adversarial concurrent verifier $\hat{V}$ should now, when given oracle access to V^*, be hard-to-simulate for S. Let p be an upper bound on the number of private random bits that V^* uses.

The Decision Procedure. We are ready to give the $\mathcal{BPP}$ decision procedure D for language L.

Algorithm $D(x)$
1. Choose f, g randomly from the required families.
2. Choose private random input $R \xleftarrow{\$} \{0,1\}^p$.
3. Run $[S^{\hat{V}(w,(f,g),\cdot)}(x, h_x)]^{V_x^*}$
4. If there is ever an attempted query to V_x^* which violates the conditions of the game, then halt and ACCEPT

The decision procedure D, on input x, runs the simulator S on input x and with interaction length equal to the height h_x of the difficult scheduler tree. The simulator expects an oracle so to simulate it D uses $\hat{V}$ with auxiliary input two randomly chosen hash function f and g. Scheduler oracle queries from $\hat{V}$ are answered using the difficult scheduler V^*. Like in the rewinder defined above, D monitors the oracle queries to V^*. If there is ever a query which would violate the game condition, D halts and accepts. Otherwise, S will eventually halt with some output $(\bar{\beta}, \beta)$, in which case D rejects. The correctness of our decision procedure follows from two lemmas.

Lemma 1. *For all but finitely many $x \in L$, $\mathbf{Pr}[D(x)$ accepts$] \geq 1/q(|x|)$ for some polynomial $q(\cdot)$.*

Lemma 2. *For all but finitely many $x \notin L$, $\mathbf{Pr}[D(x)$ accepts$] < \nu(|x|)$ for some negligible function $\nu(\cdot)$.*

Proof of Lemma 1. The idea of the proof is simple. The decision procedure only accepts if there is some query which violates the game condition. Notice that this is the only difference between the execution of the rewinder $\mathbb{S}_x^{V_x^*}(h_x)$ and the execution of $S^{\hat{V}^{V^*}}(x, h_x)$. We show that this latter execution (without any check for game conditions) results in a random leaf when x is in the language. The inability of the rewinder to reach a random leaf then allows us to argue that the difference between the two executions above is noticeable.

We now give details. Consider some adversarial verifier V^* which is identical to $\hat{V}$ except that it does not make oracle calls to a scheduler. Instead, it is given the code of a scheduler as auxiliary input (as well as hash function f and g and private random input for the scheduler).

We first claim that for randomly chosen f and g, the ensembles

$$\{\hat{g}_{V^*}(\langle P, V^*_{f,g,\text{desc}(V^*)},R\rangle(x))\}_{x\in L} \text{ and } \{U_{d_x,h_x}\}_{x\in L}$$

are computationally indistinguishable. To see this, notice first that the output of V^* when interacting with the honest prover on input $x \in L$ will be a prover query of length h. This is because the proof system has negligible completeness error, so any invocation of the honest verifier will accept with overwhelming probability. Thus, with high probability V^* will only reply with final output when the length of the query is h. This means that with high probability P will cause V^* to query its scheduler at a leaf. Now, because the independence of g is greater than h, g was randomly chosen from the family, and because each input to g (inside the definition of $\hat{g}$) is distinct (appending π and μ ensures this), then g, when applied to the query of length h will result in a uniformly random leaf.

The zero-knowledge property then ensures us that ensembles

$$\{\hat{g}(S^{V^*(x,(f,g),\text{desc}(V^*),R,\cdot)}(x, h_x))\}_{x\in L} \text{ and } \{\hat{g}(\langle P, V^*_{f,g,\text{desc}(V^*)},R\rangle(x))\}_{x\in L}$$

must be computationally indistinguishable as well (again for randomly chosen f and g as above). Now, we know by our assumption that there must be some Δ which can distinguish between U_{d_x,h_x} and $\mathbb{S}^{V^*}_x(h_x)$ with noticeable probability for all sufficiently large strings x. So there must be some $\tilde{\Delta}$ which can distinguish between the ensembles $\mathbb{S}^{V^*}_x(h_x)$ and $\hat{g}(S^{V^*(x,(f,g),\text{desc}(V^*),R,\cdot)}(x, h_x))$ for all sufficiently large $x \in L$.

Recall that the random variable $\mathbb{S}^{V^*}_x(h_x)$ is considered to be the function $\hat{g}$ applied to the last query from S, which we have said must be the same as the sequence of messages S outputs. Yet for any x the statistical distance between $\hat{g}(S^{V^*(x,(f,g),\text{desc}(V^*),R,\cdot)}(x))$ and $\mathbb{S}^{V^*}_x(h_x)$ is at most the probability that some invalid query from S causes an honest-verifier to accept plus the probability of a collision, and the probability of collision is negligible (since d is a fixed super-polynomial) so it must be the case that the former probability is noticeable. The lemma immediately follows.

Proof Sketch of Lemma 2. The proof uses the standard technique from past lower bound papers (cf. [3]), which relies on the soundness of the (single-session) proof system. Since D accepts if and only if S is able to make the single-session honest verifier accept without rewinding, we can use S to build a single-session "cheating prover" which, while interacting with some single-session honest verifier, runs S

internally. The cheating prover uses the actual verifier it is interacting with to simulate one session with S, while for all other sessions the cheating prover follows the strategy of D and runs copies of the single-session honest verifier internally. The negligible soundness error of the proof system ensures that S must only be able to make an unrewound session accept with negligible probability, and the lemma follows.

6 Proof of Theorem 2 (Upper Bound)

In this section we show how our combinatorial game relates to black-box zero-knowledge upper bounds. We follow the usual procedure for proving there is a black-box zero-knowledge proof system for all languages in $\mathcal{NP}$ by giving such a proof system for an $\mathcal{NP}$-complete language (this was first done in [10]).

We want this proof system to closely resemble our combinatorial game. Specifically, we desire a proof system in which

- for r rounds, an honest prover sends uniformly random messages, and
- if a simulator executes one successful rewind during these r rounds, it will be able to successfully simulate that session.

Intuitively, we want the first property since in our combinatorial game the rewinder must reach a uniformly random leaf. The second property matches our μ restriction on the rewinder.

The Proof System. As a starting point, we will focus on $\mathcal{NP}$-complete languages with 3-message, public-coin, committed-verifier zero-knowledge (CVZK) proof systems (formalized in [13]). This means the proof system (P, V) has three messages, the verifier simply outputs bits from its random tape, and there exists a simulator S such that for all challenges c the ensembles $\{S(x, c)\}_{x \in L}$ and $\{\text{view}_{V_c}^P(x)\}_{x \in L}$ are computationally indistinguishable. Hamiltonicity is one such $\mathcal{NP}$-complete language.

We then follow the technique of Rosen [17] (which also appeared in [15]) and augment the above proof system with a preamble which proceeds as follows. The verifier initially commits (using a perfectly-hiding commitment scheme) to a challenge σ and shares $\sigma_{i,j}^0, \sigma_{i,j}^1$ for $1 \le i \le l$ and $1 \le j \le r$ of σ such that $\sigma_{i,j}^0 \oplus \sigma_{i,j}^1 = \sigma$ for all i and j. Numerous rounds of challenge-response follow. In each round, the prover sends a random bit-string and the verifier opens the shares corresponding to this bit-string. The prover then executes the 3-message, public-coin, committed verifier zero-knowledge proof system (which we now call the second stage). In this second stage, the verifier sends σ as its challenge, as well as openings for the rest of the shares to show that it did not cheat in the preamble. See Fig. 3 for details.

The Simulator. For an $\mathcal{NP}$-complete language L with the property mentioned above, we wish to show that there is an $(r(\cdot) + O(1))$-round black-box concurrent zero-knowledge proof system, assuming that there is an efficient rewinder $\mathbb{S}$ that

Preamble:
 $V \to P$: commit to σ and $\sigma_{i,j}^0, \sigma_{i,j}^1$ for $1 \le i \le l$ and $1 \le j \le r$.
 For $j = 1, \ldots, r$:
   $P \to V$: Send $b_{1,j}, \ldots, b_{l,j} \xleftarrow{\$} \{0,1\}^l$.
   $V \to P$: Decommit to $\sigma_{1,j}^{b_{1,j}}, \ldots, \sigma_{l,j}^{b_{l,j}}$.
Second Stage:
   $P \to V$: Send first message of three-round protocol.
   $V \to P$: Decommit to σ and $\sigma_{i,j}^{1-b_{i,j}}$ for $1 \le i \le l$ and $1 \le j \le r$.
   $P \to V$: Answer according to the value of σ.

Fig. 3. The Rosen protocol

can win our combinatorial game with parameter $r(\cdot)$, and that perfectly-hiding commitment schemes exist.

To accomplish this task, we will need to build a successful simulator S with an output distribution computationally indistinguishable from the view of any adversarial concurrent verifier V^* interacting with the real prover P. The simulator is given oracle access to V^* to aid it in this task. We assume without loss that V^* always sends exactly h messages in any concurrent interaction.

From a high level, our simulator S will make queries to V^* and use its one advantage over a prover (the ability to rewind V^*) to help it achieve its goal. At some point S will make one final query to V^* and output a view corresponding to this query.

The rewinding strategy employed by S will be dictated by the efficient, successful rewinder $\mathbb{S}$, which S will run internally. Since $\mathbb{S}$ expects a scheduler oracle which it uses to explore a labeled tree, we make S simulate the oracle using an internal tree data structure which it labels with the help of its own oracle, V^*.

Each node $v = (m_1, \ldots, m_j)$ (denoted by the sequences of edges leading to it) in the tree will correspond to some query made by $\mathbb{S}$, and will contain three pieces of information. The first is the label of the node $\pi(v)$, which is either v or $\pi(w)$, the label of some ancestor. The other information will be a sequence of prover message $\bar{\beta}$ and the corresponding reply (α, s) from V^*. Intuitively, s is the session label which S uses to determine the node label $\pi(v)$. If s is the same as in some ancestor node w, then $\pi(v) = \pi(w)$. If s is a new label, then $\pi(v) = v$.

When $\mathbb{S}$ asks its oracle for the label of v, some previously unexplored node, the simulator S adds the node v to the tree (recall nodes are specified by a sequence of edges), and looks up the information $(\pi(v'), \bar{\beta}, (\alpha, s))$ stored at the parent node $v' = (m_1, \ldots, m_{j-1})$. There are now two cases.

In the first case, there are less than $r + 1$ nodes on the path to v labeled $\pi(v')$,i.e., $|\mu(v')| < r$, in which case the simulator appends m_j to $\bar{\beta}$ and queries it to V^*. This corresponds to taking some previous query $\bar{\beta}$ and making a longer query with the addition of preamble message m_j.

In the second case, there are either $r + 1$ or $r + 2$ nodes with label $\pi(v')$ on the path, so the preamble is finished. In this case, S searches the tree for a

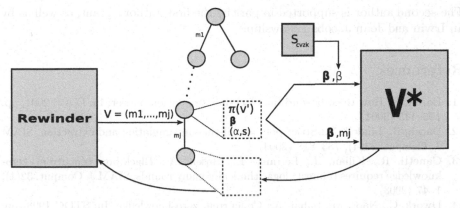

Fig. 4. Illustration of how S uses $\mathbb{S}$ to make queries to V^*

node y with label $\pi(y) = \pi(v')$ and $\mu(y)$ not a prefix of $\mu(v')$. This constitutes a proper rewinding, so S should have V^*'s challenge revealed to it by comparing the messages stored at each node. The simulator, armed with the challenge, uses the committed-verifier simulator to generate a message to send to V^*. The simulator will find the required information in the tree, since we are assuming verifiers that do not abort and rewinders that do not make invalid queries. In either case, the simulator will make some query (call it $\bar{\beta}_j$) to V^*.

We still need to give $\mathbb{S}$ an answer to its query v. To do so we need to find out which information to store at node v in the tree data structure. We will use V^* to accomplish this task. Let (α_j, s) be the reply from V^* on query $\bar{\beta}_j$. If there exists some ancestor w of v with a message (α_i, s) stored at it, then at node v store $\pi(w), \bar{\beta}_j, (\alpha_j, s)$ and otherwise store $v, \bar{\beta}_j, (\alpha_j, s)$. Finally, once the tree has been updated, S returns the label $\pi(v)$ to $\mathbb{S}$. The sequences of messages S ultimately outputs will be the $\bar{\beta}$ stored at the last node $\mathbb{S}$ queries. The entire process is illustrated in Fig. 4.

Showing zero-knowledge is straightforward. To do so we gradually modify the simulator until it acts identically to the honest prover, arguing indistinguishability for each modification. We first use a hybrid argument to argue that we can replace the invocations of the CVZK simulator with the honest prover given a real witness. Then we use the fact that $\mathbb{S}$ is a good rewinder, meaning it can reach a random leaf given any scheduler, to replace the $\mathbb{S}$ inside of S with a single, random path down the tree. It then is the case that S follows a single path down the tree, preamble messages are randomly chosen, and second-stage messages are simulated using the honest prover with an actual witness. This is the same as the behavior of the honest prover, so the zero-knowledge property follows.

Acknowledgements

The authors would like to thank the anonymous referees for their comments. The first author is supported in part by NSF Cybertrust grant CNS–0430595.

The second author is supported in part by the first author's grant, as well as by an Irwin and Joan Jacobs Fellowship.

References

1. Barak, B.: How to go beyond the black-box simulation barrier. In: FOCS 2001, pp. 106–115 (2001)
2. Barak, B., Lindell, Y.: Strict polynomial-time in simulation and extraction. SIAM J. Comput. 33(4), 783–818 (2004)
3. Canetti, R., Kilian, J., Petrank, E., Rosen, A.: Black-box concurrent zero-knowledge requires (almost) logarithmically many rounds. SIAM J. Comput. 32(1), 1–47 (2003)
4. Dwork, C., Naor, M., Sahai, A.: Concurrent zero-knowledge. In: STOC 1998, pp. 409–418 (1998)
5. Feige, U.: Alternative Models for Zero-Knowledge Interactive Proofs. PhD thesis, Weizmann Institute of Science (1990)
6. Feige, U., Shamir, A.: Witness indistinguishable and witness hiding protocols. In: STOC 1990, pp. 416–426. ACM Press, New York (1990)
7. Goldreich, O., Kahan, A.: How to construct constant-round zero-knowledge proof systems for NP. Jour. of Cryptology 9(2), 167–189 (1996)
8. Goldreich, O., Krawczyk, H.: On the composition of zero-knowledge proof systems. SIAM J. Comput. 25(1), 169–192 (1996)
9. Goldwasser, S., Micali, S., Rackoff, C.: The knowledge complexity of interactive proof systems. SIAM J. Comput. 18(1), 186–208 (1989)
10. Goldreich, O., Micali, S., Wigderson, A.: Proofs that yield nothing but their validity or all languages in NP have zero-knowledge proof systems. J. ACM 38(3), 690–728 (1991)
11. Kilian, J., Petrank, E.: Concurrent and resettable zero-knowledge in poly-logorithmic rounds. In: STOC 2001, pp. 560–569. ACM, New York (2001)
12. Kilian, J., Petrank, E., Rackoff, C.: Lower bounds for zero knowledge on the internet. In: FOCS 1998, pp. 484–492 (1998)
13. Micciancio, D., Ong, S.J., Sahai, A., Vadhan, S.: Concurrent zero knowledge without complexity assumptions. In: Halevi, S., Rabin, T. (eds.) TCC 2006. LNCS, vol. 3876, pp. 1–20. Springer, Heidelberg (2006)
14. Micciancio, D., Yilek, S.: The round-complexity of black-box zero-knowledge: A combinatorial characterization. Full Version of this paper, http://www-cse.ucsd.edu/users/syilek/
15. Prabhakaran, M., Rosen, A., Sahai, A.: Concurrent zero knowledge with logarithmic round-complexity. In: FOCS 2002, pp. 366–375. IEEE, Los Alamitos (2002)
16. Richardson, R., Kilian, J.: On the concurrent composition of zero-knowledge proofs. In: Stern, J. (ed.) EUROCRYPT 1999. LNCS, vol. 1592, pp. 415–431. Springer, Heidelberg (1999)
17. Rosen, A.: A note on constant-round zero-knowledge proofs for NP. In: Naor, M. (ed.) TCC 2004. LNCS, vol. 2951, pp. 191–202. Springer, Heidelberg (2004)
18. Rosen, A.: A note on the round-complexity of concurrent zero-knowledge. In: Bellare, M. (ed.) CRYPTO 2000. LNCS, vol. 1880, pp. 451–468. Springer, Heidelberg (2000)

On Constant-Round Concurrent Zero-Knowledge

Rafael Pass and Muthuramakrishnan Venkitasubramaniam

Cornell University
{rafael,vmuthu}@cs.cornell.edu

Abstract. Loosely speaking, an interactive proof is said to be *zero-knowledge* if the view of every "efficient" verifier can be "efficiently" simulated. An outstanding open question regarding zero-knowledge is whether constant-round concurrent zero-knowledge proofs exists for non-trivial languages. We answer this question to the affirmative when modeling "efficient adversaries" as probabilistic quasi-polynomial time machines (instead of the traditional notion of probabilistic polynomial-time machines).

1 Introduction

Zero-knowledge interactive proofs [14] are paradoxical constructs that allow one player (called the Prover) to convince another player (called the Verifier) of the validity of a mathematical statement $x \in L$, while providing *zero additional knowledge* to the Verifier. This is formalized by requiring that the view of every "efficient" adversary verifier V^* interacting with the honest prover P be simulated by an "efficient" machine S (a.k.a. the *simulator*). The idea behind this definition is that whatever V^* might have learned from interacting with P, he could have actually learned by himself (by running the simulator S). As "efficient" adversaries normally are modelled as probabilistic polynomial-time machines ($\mathcal{PPT}$), the traditional definition of $\mathcal{ZK}$ models both the verifier and the simulator as $\mathcal{PPT}$ machines. In this paper, we investigate alternative models of efficient adversaries—in particular, as in [21], we model adversaries as probabilistic quasi-polynomial time machines ($\mathcal{PQT}$).

Concurrency and $\mathcal{ZK}$. The notion of concurrent $\mathcal{ZK}$, first introduced and achieved, by Dwork, Naor and Sahai [8] considers the execution of zero-knowledge proofs in an asynchronous setting and concurrent setting. More precisely, we consider a single adversary mounting a coordinated attack by acting as a verifier in many concurrent executions. Concurrent zero-knowledge proofs are significantly harder to construct (and analyze).

Since the original protocols by Dwork, Naor and Sahai (which relied on so called "timing assumptions"), various other protocols have been obtained based on different set-up assumptions (e.g., [9] [6] [4]). On the other hand, in the "plain" model without any set-up Canetti, Kilian, Petrank and Rosen [5] (building on earlier works by [17] [26]) show that concurrent $\mathcal{ZK}$ proofs for non-trivial

R. Canetti (Ed.): TCC 2008, LNCS 4948, pp. 553–570, 2008.
© International Association for Cryptologic Research 2008

languages, with so called "black-box" simulators, require at least $\Omega(\frac{\log n}{\log\log n})$ number of communication rounds. Richardson and Kilian [25] constructed the first concurrent zero-knowledge argument in the standard model. Their protocol which uses a black-box simulator requires $O(n^\epsilon)$ number of rounds. Kilian and Petrank [16] later obtained a round complexity of $\tilde{O}(\log^2 n)$, and finally Prabhakaran, Rosen and Sahai [23] essentially closed the gap by obtaining a round complexity of $\tilde{O}(\log n)$.

All of the above results rely on the traditional modeling of adversaries as $\mathcal{PPT}$ machines. Thus, it is feasible that there exists some super-polynomial, but "well-behaved", model of adversaries that admits constant-round concurrent $\mathcal{ZK}$ proofs.

Concurrent $\mathcal{ZK}$ w.r.t super-polynomial adversaries. The lower bound of [17] shows that only languages decidable in probabilistic subexponential-time have 4-round concurrent black-box zero-knowledge arguments w.r.t to probabilistic subexponential-time adversaries. On the other hand, [21] constructs constant-round concurrent zero-knowledge arguments w.r.t $\mathcal{PQT}$ verifiers (and consequently also simulators); however the soundness condition of those argument systems only holds w.r.t. $\mathcal{PPT}$ adversaries—in fact, the simulator succeeds in its simulation by breaking the soundness condition of the argument system. Additionally, it is noted in [21] that there exist 3-round concurrent $\mathcal{ZK}$ proofs w.r.t. exponential-time adversaries (as any witness indistinguishable proof is also zero-knowledge with respect to exponential-time verifiers). Finally, [25] claimed that a constant-round version of their protocol remains secure w.r.t $\mathcal{PQT}$ adversaries, when considering a "benign" type of concurrent adversary (which never sends any invalid messages and has a fixed—i.e., non-adaptively chosen—scheduling), but as far as we know a proof of this has never appeared.

Thus, the above results leave open the question of whether there exist $r(n)$-round concurrent black-box zero-knowledge proofs w.r.t super-polynomial, but sub-exponential, adversaries, as long as $4 < r(n) < \log n$. In particular,

> *Does there exists constant-round concurrent zero-knowledge arguments w.r.t. $\mathcal{PQT}$ (or even sub-exponential time) adversaries?*

1.1 Our Results

Our main result answers the above question in the affirmative. Let $\mathcal{PQT}$ denote the class of probabilistic quasi-polynomial time machines, i.e., randomized machines that run in time $n^{poly(\log(n))}$. Let $\omega(\mathcal{PQT})$ denote the class of *probabilistic super quasi-polynomial time* machines, i.e. randomized machines that run in time $n^{\omega(poly(\log(n)))}$.

Theorem 1 (Main Theorem). *Assume the existence of claw-free permutations w.r.t $\mathcal{PQT}$. Then, every language in $\mathcal{NP}$ has an $O(1)$-round perfect concurrent black-box $\mathcal{ZK}$ argument w.r.t $\mathcal{PQT}$.*

In addition, we show:

Theorem 2. *Assume the existence of one-way functions that are secure w.r.t* $\omega(\mathcal{PQT})$ *and collision-resistant hash function that are secure w.r.t* $\mathcal{PQT}$. *Then, every language in* $\mathcal{NP}$ *has an* $O(1)$-*round concurrent computational black-box* $\mathcal{ZK}$ *proof w.r.t* $\mathcal{PQT}$.

Theorem 3. *Assume the existence of one-way function that are secure w.r.t* $\omega(\mathcal{PQT})$. *Then, every language in* $\mathcal{NP}$ *has an* $O(1)$-*round concurrent computational black-box* $\mathcal{ZK}$ *arguments w.r.t* $\mathcal{PQT}$.

Theorem 4. *There exists an* $O(1)$-*round concurrent perfect* $\mathcal{ZK}$ *proof w.r.t* $\mathcal{PQT}$ *for* Graph Non-Isomorphism *and* Quadratic Non-Residuosity

We emphasize that in the above theorems, "$\mathcal{ZK}$ proofs and arguments w.r.t $\mathcal{PQT}$" refer to proofs/ arguments where both the soundness condition and the $\mathcal{ZK}$ condition holds w.r.t to $\mathcal{PQT}$ adversaries; in particular, for the $\mathcal{ZK}$ property we also require that the distinguishability gap is smaller than the inverse of any quasi-polynomial function.

A note on expected running-time. In contrast to earlier work on concurrent zero-knowledge (e.g. [25,16,23]), our simulators run in *expected* $\mathcal{PQT}$. This is inherent: by the work of Barak-Lindell [1] it follows that only languages decidable in $\mathcal{PQT}$ have constant-round $\mathcal{ZK}$ protocols w.r.t $\mathcal{PQT}$ if requiring a strict $\mathcal{PQT}$ simulator (let alone the question of concurrency). In particular, this shows that none of the previous simulation techniques can be extended to get constant-round protocols w.r.t $\mathcal{PQT}$ (at least when requiring that the output of the simulation is also indistinguishable for $\mathcal{PQT}$).[1]

Additional results. Finally, we mention that our techniques apply also to concurrent $\mathcal{ZK}$ proofs w.r.t $\mathcal{PPT}$. As a result we obtain the first concurrent perfect $\mathcal{ZK}$ arguments/proofs w.r.t $\mathcal{PPT}$.

Theorem 5. *Assume the existence of claw-free permutations (w.r.t* $\mathcal{PPT}$). *Then, every language in* $\mathcal{NP}$ *has an* $O(n^\epsilon)$-*round perfect concurrent black-box* $\mathcal{ZK}$ *argument w.r.t* $\mathcal{PPT}$, *for every* $\epsilon > 0$.

Theorem 6. *For every* $\epsilon > 0$, *there exists a* $O(n^\epsilon)$-*round concurrent perfect* $\mathcal{ZK}$ *proof for* Graph Non-Isomorphism *and* Quadratic Non-Residuosity.

As an additional contribution, we believe that both our protocols and their analysis provides the simplest proof of the existence of concurrent $\mathcal{ZK}$ proofs (w.r.t $\mathcal{PPT}$).[2]

PQT v.s. PPT: What is right model for adversarial computation? Recall that to show that $\mathcal{ZK}$ is closed under sequential composition, the original definition of $\mathcal{ZK}$ was extended to consider *non-uniform* $\mathcal{PPT}$ adversaries

---

[1] On the other hand, it might still be plausible that the technique of [25] can be extended to give constant-round protocols w.r.t $\mathcal{PQT}$, when allowing the indistinguishability gap to be a polynomial (or even some *fixed* quasi-polynomial) function.

[2] In a related work [24], joint with Dustin Tseng we provide a simple proof for existence of concurrent $\mathcal{ZK}$ proofs with logarithmic round complexity.

[13]—in other words, in the context of $\mathcal{ZK}$ the notion of non-uniform $\mathcal{PPT}$ (for modeling adversaries) is more robust than simply $\mathcal{PPT}$. Additionally, security is guaranteed w.r.t a stronger class of adversaries. Of course, the extra price to pay is that all hardness assumptions now must hold also with respect to non-uniform $\mathcal{PPT}$.

In this paper we show that by considering an even stronger class of adversaries —namely $\mathcal{PQT}$—we get a notion that is even more robust; in particular, it is now possible to get constant-round concurrent $\mathcal{ZK}$ protocols. Again, this requires us to rely on hardness assumptions against $\mathcal{PQT}$, but this seems like a weak strengthening of traditional hardness assumptions (especially since the known attacks on traditional conjectured hard functions require subexponential time).

A note on plausible deniability. The notion of $\mathcal{ZK}$ is traditionally associated with *plausible deniability*—i.e., that the interaction leaves "no trace" which the verifier can use later to convince that the interaction took place. Intuitively, this holds since the verifier could have executed the simulator (on its self) to generate its view of the interaction. We mention, however, that since the traditional definition of $\mathcal{ZK}$ allows the simulator to have an arbitrary (polynomial) overhead with respect to the verifier (who's view it is supposed to simulate), the deniability guarantee offered by traditional $\mathcal{ZK}$ proofs is weak: consider for instance a verifier with a running-time of $t = 2^{40}$ computational steps, and a simulator with running-time, say, t^3; although 2^{40} is very feasible, 2^{120} seems like a stretch! The example is not hypothetic—the "tightest" concurrent $\mathcal{ZK}$ protocols [16,23] indeed have a running-time of t^2 not counting the time need to emulate the verifier. Additionally, as demonstrated in [18], the traditional notion of $\mathcal{ZK}$ does not guarantee that the running-time of the simulator is (even polynomially) related to the running-time of the verifier in the view it is outputting, but rather the *worst-case* running-time of the verifier; this makes deniability even harder to argue.[3]

Nevertheless, in this respect, $\mathcal{ZK}$ w.r.t $\mathcal{PQT}$ provides even worse guarantees (as the overhead is now allowed to be quasi-polynomial).

1.2   Our Techniques

The concurrent $\mathcal{ZK}$ protocols of Richardson and Kilian (RK) [25], Kilian and Petrank (KP) [16] and Prabhakaran, Rosen and Sahai(PRS) [23] rely on the same principal idea: provide the simulator with multiple possibilities (called "slots") to rewind the verifier. If a rewinding is successful, the simulator obtains a trapdoor that allows it to complete the execution that has been rewound. The RK simulator is "adaptive" and dynamically decides when and where to rewind, while making sure there are not too many recursive rewinding (which would result in a large running-time). On a high-level this is done by recursively

---

[3] In a recent work [19], joint with Pandey, Sahai and Tseng we also show how to obtain precise concurrent $\mathcal{ZK}$ proofs. Precise zero knowledge guarantees that the view of any verifier V can be simulated in time closely related to the *actual* (as opposed to the worst-case) time spent by V in the generated view.

invoking the simulator, but ensuring that the number of levels of the recursion stays small (in fact, constant). On the other hand the KP (and PRS) simulator is "oblivious"; the simulator has a fixed rewinding scheduling, thereby ensuring a fixed (and bounded) running-time. The core of the argument is then to show that every execution has a slot that is rewound at least once.

Our approach is based on the approach taken by RK. As RK, we consider an adaptive simulator that makes recursive calls to itself, while ensuring that the depth of the recursion stays small. Our actual simulation procedure is, however, quite different. On a high-level, our approach will perform a straight-line simulation until a "good" slot has been found, and then continue rewinding that slot until a trapdoor has been found. Thus, in contrast to the previous approach, we can not bound the worst-case running-time of our simulator, instead we are forced to bound the expected running-time of the simulator.

The benefit of our approach is that 1) it enables us to achieve perfect simulation, and 2) our analysis works no matter how many slots we have and what the depth of recursion is. In fact, we can achieve both of these properties while still guaranteeing the same expected running-time as RK—namely $O(m^{O(\log_r m)})$, where r is the number of slots. As a consequence, when applied to constant-round protocols (and considering a logarithmic recursive depth) we get a quasi-polynomial running time. As already mentioned, for this application, it is inherent to have an *expected* quasi-polynomial running-time.

1.3 Open Questions

We have demonstrated that constant-round concurrent $\mathcal{ZK}$ is possible w.r.t $\mathcal{PQT}$ adversaries. Our protocol currently uses 10 communication rounds[4]. A natural open question is to either improve the round-complexity or to strengthen the 4-round lower bound of [17]. Another question is to investigate the possibility of using an even weaker (but still super-polynomial) model of computation. Rosen [26] shows that only languages in probabilistic sub quasi-polynomial time have 7-round concurrent black-box zero-knowledge arguments when adversaries are modelled as probabilistic sub quasi-polynomial time machines; thus, such protocols would require more than 7-rounds.

1.4 Organization

Definitions are found in Section 2. The proof of main theorem is contained in Sections 3 and 4. We give proof sketches for the remaining theorems in Section 5.

2   Definitions and Notations

We assume familiarity with the basic notions of an Interactive Turing Machine (ITM for brevity) and a protocol (in essence a pair of ITMs. Briefly, a protocol

---

[4] To obtain a 10 round protocol, we require non-interactive commitment schemes, which can be constructed from one-way-permutations. If we assume only existence of one-way functions, we get a 11-round protocol.

is pair of ITMs computing in turns. A round ends with the active machine either halting - in which case the protocol halts - or by sending a message m to the other machine, which becomes active with m as a special input.

We let $\mathcal{C}$ denote any class of functions.

2.1   Interactive Proofs and Arguments

Given a pair of interactive Turing machines, P and V, we denote by $\langle P, V \rangle(x)$ the random variable representing the (local) output of V when interacting with machine P on common input x, when the random input to each machine is uniformly and independently chosen.

Definition 1 ($T(\cdot)$-sound Interactive Proof System). *A pair of interactive machines $\langle P, V \rangle$ is called $T(\cdot)$-sound interactive proof system for a language L if machine V is polynomial-time and the following two conditions hold :*

- Completeness: *For every $x \in L$, $\Pr[\langle P, V \rangle(x) = 1] = 1$*
- Soundness: *For every $x \notin L$, and every interactive machine B,*
  $\Pr[\langle B, V \rangle(x) = 1] \leq \frac{1}{T(|x|)}$

In case that the soundness condition holds only with respect to a $T(n)$-bounded prover, the pair $\langle P, V \rangle$ is called an $T(\cdot)$-sound interactive argument.

$\langle P, V \rangle$ is an interactive proofs (interactive argument) w.r.t. $\mathcal{C}$ if for all $T(\cdot) \in \mathcal{C}$ the protocol is a $T(\cdot)$-sound interactive proof ($T(\cdot)$-sound interactive argument).

2.2   Indistinguishability

We rely on a generalization of the notion of indistinguishability [27], which considers $T(n)$-bounded distinguishers and require the indistinguishability gap to be smaller than $\frac{1}{\text{poly}(T(n))}$.

Definition 2 (Strong $T(\cdot)$-indistinguishability[21]). *Let X and Y be countable sets. Two ensembles $\{A_{x,y}\}_{x \in X, y \in Y}$ and $\{B_{x,y}\}_{x \in X, y \in Y}$ are said to be indistinguishable in time $T(\cdot)$ over $x \in X$, if for every probabilistic "distinguishing" algorithm D with running time $T(\cdot)$ in its first input, and every $x \in X, y \in Y$ it holds that:*

$$|\Pr[a \leftarrow A_{x,y} : D(x, y, a) = 1] - \Pr[b \leftarrow B_{x,y} : D(x, y, b) = 1]| < \frac{1}{\text{poly}(T(|x|))}$$

Definition 3 (Computational indistinguishability w.r.t $\mathcal{C}$). *Let X and Y be countable sets. Two ensembles $\{A_{x,y}\}_{x \in X, y \in Y}$ and $\{B_{x,y}\}_{x \in X, y \in Y}$ are said to be indistinguishable w.r.t $\mathcal{C}$ over $x \in X$, if A, B are $q(\cdot)$-indistinguishable for every function $q(\cdot) \in \mathcal{C}$.*

2.3   Witness Indistinguishability

An interactive proof is said to be *witness indistinguishable* ($\mathcal{WI}$) if the verifier's view is "computationally independent" of the witness used by the prover for proving the statement—i.e. the view of the Verifier in the interaction with a prover using witness w_1 or w_2 for two different witnesses are indistinguishable.

Definition 4 (Witness-indistinguishability w.r.t C). *Let $\langle P, V \rangle$ be an interactive proof system for a language $L \in \mathcal{NP}$. We say that $\langle P, V \rangle$ is C-witness-indistinguishable for R_L, if for every probabilistic polynomial-time interactive machine V^* and for every two sequences $\{w_x^1\}_{x \in L}$ and $\{w_x^2\}_{x \in L}$, such that $w_x^1, w_x^2 \in R_L(x)$ for every $x \in L$, the probability ensembles $\{\text{VIEW}_2[P(x, w_x^1) \leftrightarrow V^*(x, z)]\}_{x \in L, z \in \{0,1\}^*}$ and $\{\text{VIEW}_2[P(x, w_x^2) \leftrightarrow V^*(x, z)]\}_{x \in L, z \in \{0,1\}^*}$ are computationally indistinguishable w.r.t C over $x \in L$.*

We say that the proof system is perfectly witness indistinguishable (Perfect-$\mathcal{WI}$) if the corresponding views are identically distributed.

2.4   Black-Box Concurrent Zero-Knowledge

Let $\langle P, V \rangle$ be an interactive proof for a language L. Consider a concurrent adversary verifier V^* that, given an input instance $x \in L$ interacts with m independent copies of P concurrently, without any restrictions over the scheduling of the messages in the different interactions with P. Let $\Big\{ \text{VIEW}_2[P(x, y) \leftrightarrow V^*(x, z)] \Big\}_{x \in L, w \in R_L(x), z \in \{0,1\}^*}$ denote the random variable describing the view of the adversary V^* on common input x and auxiliary input z, in an interaction with P.

Definition 5 (Black-box concurrent zero-knowledge w.r.t C:). *Let $\langle P, V \rangle$ be an interactive proof system for a language L. We say that $\langle P, V \rangle$ is black-box concurrent zero-knowledge w.r.t C if for every functions $q, m \in C$, there exists a probabilistic algorithm $S_{q,m}$, such that for every concurrent non-uniform adversary V^* that on common input x and auxiliary input z has a running-time bounded by $q(|x|)$ and opens up $m(|x|)$ executions, $S_{q,m}(x, z)$ runs in time polynomial in $|x|$. Furthermore, the ensembles $\Big\{ S_{q,m}(x, z) \Big\}_{x \in L, w \in R_L(x), z \in \{0,1\}^*}$ and $\Big\{ \text{VIEW}_2[P(x, y) \leftrightarrow V^*(x, z)] \Big\}_{x \in L, w \in R_L(x), z \in \{0,1\}^*}$ are computationally indistinguishable w.r.t C over $x \in L$.*

2.5   Other Primitives

We informally define the other primitives we use in the construction of our protocols.

Special-sound proofs: A 3-round public-coin interactive proof for the language $L \in \mathcal{NP}$ with witness relation R_L is special-sound with respect to R_L, if for any two transcripts (α, β, γ) and $(\alpha', \beta', \gamma')$ such that the initial messages α, α' are the same but the challenges β, β' are different, there is a deterministic procedure to extract the witness from the two transcripts that runs in polynomial time. Special-sound $\mathcal{WI}$ proofs for languages in $\mathcal{NP}$ can be based on the existence of non-interactive commitment schemes, which in turn can be based on one-way permutations. Assuming only one-way functions, 4-round special-sound $\mathcal{WI}$ proofs for NP exists[5]. For simplicity, we use 3-round special-sound proofs in our protocol though our proof works also with 4-round proofs.

Proofs of knowledge: Informally an interactive proof is a proof of knowledge if the prover convinces the verifier not only of the validity of a statement, but also that it possesses a witness for the statement. If we consider computationally bounded provers, we only get a "computationally convincing" notion of a proof of knowledge (a.k.a *arguments of knowledge*)

3   Our Protocol and Simulator

3.1   Description of the Protocol

Our concurrent $\mathcal{ZK}$ protocol (also used in [24]) is a slight variant of the precise $\mathcal{ZK}$ protocol of [20], which in turn is a modification of the Feige-Shamir protocol [10]. The protocol proceeds in the following two stages, on a common input statement $x \in \{0,1\}^*$ and security parameter n,

1. In Stage 1, the Verifier picks two random strings $s_1, s_2 \in \{0,1\}^n$, and sends their image $c_1 = f(r_1)$, $c_2 = f(r_2)$ through a one-way function f to the Prover. The Verifier sends $\alpha_1, \ldots, \alpha_r$, the first messages of r invocations of a $\mathcal{WI}$ special-sound proof of the fact that c_1 and c_2 have been constructed properly (i.e., that they are in the image set of f). This is followed by r iterations so that in the jth iteration, the Prover sends $\beta_j \leftarrow \{0,1\}^{n^2}$, a random second message for the jth proof and the Verifier sends the third message γ_j for the jth proof.

2. In Stage 2, the Prover provides a WI proof of knowledge of the fact that either x is in the language, or (at least) one of c_1 and c_2 are in the image set of f.

More precisely, let $f : \{0,1\}^n \rightarrow \{0,1\}^n$ be a one-way function and let the witness relation $R_{L'}$, where $((x_1, x_2), (y_1, y_2)) \in R_{L'}$ if $f(x_1) = y_1$ or $f(x_2) = y_2$, characterize the language L'. Let the language $L \in \mathcal{NP}$. Protocol ConcZKArg for proving that $x \in L$ is depicted in Figure 1.

The soundness and the completeness of the protocol follows directly from the proof of Feige and Shamir [10]; in fact, the protocol is an instantiation of theirs. (Intuitively, to cheat in the protocol a prover must "know" an inverse to either c_1 or c_2, which requires inverting the one-way function f.).

---

[5] A 4-round protocol is special sound if a witness can be extracted from any two transcripts $(\tau, \alpha, \beta, \gamma)$ and $(\tau', \alpha', \beta', \gamma')$ such that $\tau = \tau, \alpha = \alpha'$ and $\beta \neq \beta'$.

PROTOCOL CONCZKARG

Common Input: an instance x of a language L with witness relation R_L.
Auxiliary Input for Prover: a witness w, such that $(x, w) \in R_L(x)$.
Stage 1:
    V uniformly chooses $r_1, r_2 \in \{0, 1\}^n$.
    V $\rightarrow$ P: $c_1 = f(r_1), c_2 = f(r_2)$. r first messages $\alpha_1, \ldots, \alpha_r$ for $\mathcal{WI}$ special-sound
        proof of the statement. (called the **start** message)
        either there exists a value r_1 s.t $c_1 = f(r_1)$
        or there exists a value r_2 s.t $c_2 = f(r_2)$
        The proof of knowledge is with respect to the witness relation R'_L
    For $j = 1$ to r do
        P $\rightarrow$ V: Second message $\beta_j \leftarrow \{0, 1\}^{n^2}$ for j^{th} $\mathcal{WI}$ special-sound proof.
        (called the **opening** of slot j)
        V $\rightarrow$ P: Third message γ_j for j^{th} $\mathcal{WI}$ special-sound proof. (called the **closing**
        of slot j)
Stage 2:
    P $\leftrightarrow$ V: a perfect-$\mathcal{WI}$ argument of knowledge of the statement
        either there exists values r'_1, r'_2 s.t either $c_1 = f(r'_1)$ or $c_2 = f(r'_2)$.
        or $x \in L$
    The argument of knowledge is with respect to the witness relation
    $R_{L \lor L'}(c_1, c_2, x) = \{(r'_1, r'_2, w) | (r'_1, r'_2) \in R_{L'}(c_1, c_2) \lor w \in R_L(x)\}$.

Fig. 1. Concurrent Perfect $\mathcal{ZK}$ argument for $\mathcal{NP}$

3.2 Description of the Simulator

On a very high-level the simulation follows that of Feige and Shamir [10]: the simulator will attempt to rewind one of the special-sound proofs—each such proof, i.e. the challenge(β) and the response(γ) is called a **slot**. If the simulator gets two accepting proof transcripts, the special-soundness property allows the simulator to extract a "fake" witness r_i such that $c_i = f(r_i)$. This witness can later be used in the second phase of the protocol. We call an execution "solved" if a witness is extracted. More precisely, our simulation is defined recursively in the following manner.

On the recursive level ℓ, the simulator feeds random Stage 1 messages to V^* (Step 3). Whenever a **slot** s closes, S decides whether or not to rewind s depending on the number of new executions that started between the **opening** and the **closing** of s. If the number of executions is "small" (where small is defined based on the level ℓ), S begins rewinding the slot, i.e. S sends a new challenge β for slot s and recursively invokes itself on recursive level $\ell + 1$, and continues executing until one of the following happens:

1. *S is "stuck" at Stage 2 of an unsolved execution that started at level $\ell + 1$:* S halts and outputs fail.
2. *The* **closing** *message* γ *for slot s occurs:* S extracts a "fake" witness using the special-sound property and continues its simulation (on level ℓ).

3. V^* *aborts or starts "too many" executions:* S restarts its rewinding using a new challenge β for s. We show that S in expectation restarts $O(1)$ times because of this. (Intuitively this follows since during the execution at level ℓ, S only starts rewinding if V^* did not abort and only opened a "small" number of executions).

4. S *gets "stuck" at Stage 2 of an unsolved execution that started at level ℓ:* Again, S restarts its rewinding. We show that this case can happen at most $m - 1$ times, where m is the total number of executions.

5. S *gets "stuck" at Stage 2 of an unsolved execution that started at level $\ell' < \ell$:* S returns the view to level ℓ'.

In the unlikely event that S asks the same challenge β twice, S performs a brute-force search for the witness. Furthermore, to simplify the analysis of the running-time, the simulation is cut-off if it runs "too long" and S extracts witnesses for each execution using brute-force search.

The basic idea behind the simulation is similar to [25]: if we define "small" appropriately we can ensure that some slot of every execution is rewound and the expected running time is bounded. A first approach would be to ensure that at recursive level l at most $\frac{m}{r^l}$ executions start, and define "small" to be $\frac{m}{r^{l+1}}$, where m is the number of executions and r is the number of slots. Then, for every execution that started at level ℓ and completed r slots, S is guaranteed to rewind at least one slot. Furthermore, if we show that the expected number of rewindings of each slot is $O(m)$, then the expected running time of the simulator is at most $poly(m^{\log_r m})$; letting $r = 2$, the running time becomes $poly(m^{\log_2 m})$. However, to make sure that the simulator does not output fail, our analysis requires the simulator to be able to rewind at least two slots—in fact, we require that once the simulator reaches the last slot, it has already performed one rewinding. To ensure this, we make sure that at level ℓ, there are at most $\frac{m}{(r-1)^\ell}$ executions and define "small" to be $\frac{m}{(r-1)^{\ell+1}}$; now letting $r = 3$ we get a running-time of $poly(m^{\log_r m})$.

A formal description of our simulator can be found in Figure 4.2. We rely on the following notation.

- $d = \lceil \log_{r-1} m \rceil$ will denote the maximum depth of recursion.
- slot (i, j) will denote slot j of execution i.
- A partial view h is defined to be good w.r.t. (s, l), if in h, V^* does not abort on s and does not open more that $(r-1)^{d-l}$ new executions after the opening of the s.
- W is a repository that stores the witness for each execution. The update W command extracts a witness from two transcripts of a slot (using the special-sound property). If the two transcripts are identical (i.e. the openings of the slot are the same), the simulator performs a brute-force search to extract a "fake" witness r_i s.t. $c_i = f(r_i)$ for $i \in \{1, 2\}$.
- R is a repository that stores the transcripts of slots of unsolved executions. Transcripts are stored in R when the simulator gets stuck in a rewinding (cases 4 and 5 mentioned in the high-level description).

4   Analysis of the Simulator

To prove correctness of the simulator, we show that the output of the simulator is correctly distributed and its expected running-time is bounded. We first prove in Claim 1 that the simulator never outputs fail. Using Claim 1, we show in Proposition 1 that the output distribution of the simulator is correct. In Proposition 2, we show that the expected running time of the simulation is at most $poly(m^d r^d)$. Throughout this proof we assume without loss of generality the adversary verifier V^* is deterministic (as it can always get its random coins as part of the auxiliary input).

4.1   Simulation Never Fails

Claim 1. *For every* $x \in L$, $S^{V^*}(x, z)$ *never outputs* fail.

Proof: Recall that $S^{V^*}(x, z)$ outputs fail only if $\text{SOLVE}_d^{V^*}(x, 0, , ,)$ outputs fail. Furthermore, SOLVE outputs fail at recursive level ℓ only if it reaches Stage 2 of an unsolved execution that started at level ℓ (i.e. only in Step 3 of SOLVE). Note that at recursive level ℓ, at most $(r-1)^{d-\ell}$ executions are opened up. Hence, for all executions that start and complete $r-1$ slots at level ℓ, there is some slot, inside which have fewer than $(r-1)^{d-(\ell+1)}$ executions opened; SOLVE must have rewound that slot "completely"—i.e. executed Step 5.d to obtain m good views without returning to a lower recursive level. Below, we show that whenever SOLVE rewinds a slot completely a witness is extracted and thus the proof of the claim follows.

Assume for contradiction that SOLVE fails to extract a witness after rewinding a particular slot. Let level ℓ and slot j of execution i be the first time this happens. This means at the end of Step 5.d, m good views are obtained and none of them contained a second transcript for slot j. Furthermore, in each such view, SOLVE got stuck only on unsolved executions that started at level ℓ (since otherwise SOLVE would have returned the view to the lower level). We now show that SOLVE can get stuck on the (at most $m-1$) other executions that started on level ℓ at most once; this contradicts the fact that m good views were obtained.

For every execution i' that SOLVE gets stuck on, both the opening and the closing of the last slot occurs inside the rewinding of slot (i, j); otherwise, SOLVE would have rewound one of the $r-1$ slots that occurred before the opening of slot (i, j) and by our assumption that l, i, j was the first "failed" slot, extracted a witness. Furthermore, the transcript of this slot enables SOLVE to never get stuck on execution i' again, since next time the last slot of execution i' closes a witness for that execution will be extracted.   ∎

4.2   Indistinguishability of the Simulation

Proposition 1. *The ensembles* $\{\text{VIEW}_2[P(x, w) \leftrightarrow V^*(x, z)]\}_{x \in L, w \in R_L(x), z \in \{0,1\}^*}$ *and* $\{S^{V^*}(x, z)\}_{x \in L, w \in R_L(x), z \in \{0,1\}^*}$ *are identical.*

$\text{SOLVE}_d^{V^*}(x, \ell, h_{initial}, s, W, R)$:

Let $h \leftarrow h_{initial}$.

Repeat forever:

1. If v is a Stage 2 verifier message of some execution, continue.
2. If V^* aborts or the number of executions that started after $h_{initial}$ in h exceeds $(r-1)^{d-\ell}$, return h.
3. If the next scheduled message is a Stage 2 prover message for execution i and $W(i) \neq \bot$, then use $W(i)$ to complete the $\mathcal{WI}$ proof of knowledge; if $W(i) = \bot$ and start message of execution i is in $h_{initial}$ return h, otherwise halt with output **fail**.
4. If the next scheduled message is a Stage 1 prover message for slot s', pick a random message $\beta \leftarrow \{0,1\}^{n^2}$. Append β to h. Let $v \leftarrow V^*(h)$.
5. Otherwise, if v is the closing message for $s' = \text{slot}\ (i', j')$, then update W with v (using R) and proceed as follows.
   (a) If $s = s'$, then return h.
   (b) Otherwise, if execution i' starts in $h_{initial}$, then return h.
   (c) Otherwise, if $W(i') \neq \bot$ or the number of executions started inside s' exceeds $(r-1)^{d-(\ell+1)}$, then continue.
   (d) Otherwise, let h' be the prefix of the history h where the prover message for s' is generated. Set $R' \leftarrow \phi$.
   Repeat m times:
       i. Repeat $h^* \leftarrow \text{SOLVE}_d^{V^*}(x, \ell + 1, h', s', W, R')$ until h^* is "good" w.r.t $(s', \ell + 1)$.
       ii. If h^* contains an accepting proof transcript for slot s', extract witness for execution i' from h and h^* and update W.
       iii. Otherwise, if the last message in h^* is the closing message for the last slot of an execution that started in $h_{initial}$ return h^*.
       iv. Otherwise, add h^* to R'.

$S^{V^*}(x, z)$:

1. Let $d \leftarrow \lceil \log_{r-1} m \rceil$. Run $\text{SOLVE}_d^{V^*}(x, 0, , , ,)$ and output whatever SOLVE outputs, with the following exception. If in the execution of $\text{SOLVE}_d^{V^*}(x, 0, , , ,)$, it queries V^* more that 2^n times, proceed as follows: Let h denote the view reached in the "main-line" simulation (i.e., in the top-level of the recursion). Continue the simulation in a "straight-line" fashion from h by using a brute-force search to find a "fake" witness each time Stage 2 of an execution i is reached.

Fig. 2. Description of Simulator

Proof: Consider the following hybrid simulator $\tilde{S}^{V^*}$ that receives the real witness w to the statement x. $\tilde{S}^{V^*}$ on input x,w, and z proceeds just like S^{V^*} in order to generate the prover messages in Stage 1, but proceeds as the honest prover using the witness w in order to generate messages in Stage 2 (instead of using the "fake" witness as S^{V^*} would have). Using the same proof as in Claim 1, we can show that $\tilde{S}^{V^*}(x, (w, z))$ never outputs fail. Furthermore, as the prover messages in Stage 1 are chosen uniformly and $\tilde{S}^{V^*}$ behaves like an honest prover in Stage 2. Therefore, we get:

Claim 2. *The ensembles* $\{\mathsf{VIEW}_2[P(x, w) \leftrightarrow V^*(x, z)]\}_{x \in L, w \in R_L(x), z \in \{0,1\}^*}$ *and* $\{\tilde{S}^{V^*}(x, (w, z))\}_{x \in L, w \in R_L(x), z \in \{0,1\}^*}$ *are identical.*

To show the proposition, it suffices to show that output distributions of $\tilde{S}^{V^*}$ and S^{V^*} are identical. This follows from the perfect-$\mathcal{WI}$ property of Stage 2 of the protocols, since the only difference between the simulators $\tilde{S}^{V^*}$ and S^{V^*} is the choice of witness used. For completeness, we provide a proof below.

Claim 3. *The ensemble* $\{\tilde{S}^{V^*}(x, (w, z))\}_{x \in L, w \in R_L(x), z \in \{0,1\}^*}$ *is identical to* $\{S^{V^*}(x, z)\}_{x \in L, w \in R_L(x), z \in \{0,1\}^*}$

Proof: To prove the claim we will rely on the fact that the running time of the simulator is bounded. This holds since S stops executing SOLVE whenever it performs more than 2^n queries and continues the simulation in a straight-line fashion, extracting "fake" witnesses using brute-force search. Assume, for contradiction, that the claim is false, i.e. there exists a deterministic verifier V^* (we assume w.l.o.g that V^* is deterministic, as its random-tape can be fixed) such that the ensembles are not identical.

We consider several hybrid simulators, S_i for $i = 0$ to N, where N is an upper-bound on the running time of the simulator. S_i receives the real witness w to the statement x and behaves exactly like S, with the exception that Stage 2 messages in the first i proofs are generated using the honest prover strategy (and the witness w). By construction, $S_0 = \tilde{S}$ and $S_N = S$. Since, by assumption, the outputs of S_1 and S_N are not identically distributed, there must exist some j such that the output of S_j and S_{j+1} are different. Furthermore, since S_j proceeds exactly as S_{j+1} in the first j executions, and also the same in Stage 1 of the $j+1$'th execution, there exists a partial view v—which defines an instance $x' \in L \vee L'$ for Stage 2 of the $j+1$'th execution—such that outputs of S_j and S_{j+1} are not identical also conditioned on the event that S_j and S_{j+1} feed V^* the view v. Since the only only difference between the view of V^* in S_j and S_{j+1} is the choice of the witness used for the statement x' used in Stage 2 of the $j + 1$'the execution, we contradict the perfect-$\mathcal{WI}$ property of Stage 2. ∎

∎

4.3 Running-Time of S

We consider the hybrid simulator $\tilde{S}^{V^*}$ constructed in proof of Proposition 1. It follows by the same proof as in Claim 3 that the running time distributions of $\tilde{S}$ and S are identical. Therefore, it suffices to analyze the expected running time of $\tilde{S}$.

Proposition 2. *For all* $x \in L, z \in \{0,1\}^*$, *and all* V^* *such that* $V^*(x, z)$ *opens up at most* m *executions,* $E[\mathrm{time}_{\tilde{S}^{V^*}(x,z)}] \leq poly(m^d r^d)$

Proof: Recall that $\tilde{S}^{V^*}(x, z)$ starts running SOLVE, but in the event that SOLVE uses more than 2^n queries to V^*, it instead continues in a straight-line simulation using a brute-force search. By linearity of expectation, the expected running time of S is

$poly(E[\#$ queries made to V^* by SOLVE $])$

$$+ E[\text{time spent in straight-line simulation}]$$

In Claim 4 below, we show that expected time spent in straight-line simulation is negligible. In Claim 5 we show that the expected number of queries made by $SOLVE$ to V^* is at most $m^{2(d+1-\ell)}(2r)^{d+1-\ell}$. The proof of the proposition follows.

Claim 4. *The expected time spent by $\tilde{S}^{V^*}$ in straight-line simulation is negligible.*

Proof: The straight-line simulation takes at most $poly(2^n)$ steps since it takes $O(2^n)$ steps to extract a "fake" witness. Recall that, SOLVE runs the brute-force search only if it picks the same challenge (β) twice. Since, SOLVE is cut-off after 2^n steps, it can pick at most 2^n challenges. Therefore, by the union bound, the probability that it obtains the same challenge twice is at most $\frac{2^n}{2^{n^2}}$. Thus, the expected time spent by S^{V^*} in straight-line simulation is at most $\frac{2^n}{2^{n^2}}poly(2^n)$, which is negligible.   ∎

Claim 5. *For all $x \in L, h, s, \mathsf{W}, \mathsf{R},\ \ell \leq d$ such that $SOLVE_d^{V^*}(x, \ell, h, s, \mathsf{W}, \mathsf{R})$ never outputs* fails, $E[\#$ *queries by* $SOLVE_d^{V^*}(x, \ell, h, s, \mathsf{W}, \mathsf{R})] \leq m^{2(d+1-\ell)}$ $(2r)^{d+1-\ell}$.

Proof: We prove the claim by induction on ℓ. To simplify notation let $\alpha(\ell) = m^{2(d+1-\ell)}(2r)^{d+1-\ell}$. When $\ell = d$ the claim follows since SOLVE does not perform any recursive calls and the number of queries made by SOLVE can be at most the total number of messages, which is mr.

Assume the claim is true for $\ell = \ell' + 1$. We show that it holds also for $\ell = \ell'$. Consider some fixed $x \in L$, h, s, W, R such that $SOLVE_d^{V^*}(x, \ell', h, s, \mathsf{W}, \mathsf{R})$ never outputs fails. We show that

$$E[\#\text{ queries by SOLVE}_d^{V^*}(x, \ell', h, s, \mathsf{W}, \mathsf{R})] \leq m^{2(d+1-\ell')}r^{d+1-\ell'}$$
$$= \alpha(\ell') = m^2(2r)\alpha(\ell' + 1)$$

Towards this goal we introduce some additional notation. Given a view $\hat{h}$ extending the view h,

- Let $q_{\hat{s}}^{\ell'}(\hat{h})$ denote the probability that the view $\hat{h}$ occurs in the "main-line" execution of
  $\text{SOLVE}_d^{V^*}(x, \ell', h, s, \mathsf{W}, \mathsf{R})$ (i.e., starting on level ℓ) and that slot $\hat{s}$ opens immediately after $\hat{h}$.
- Let $\Gamma_{\hat{s}}$ denote the set of views such that $q_{\hat{s}}^{\ell'}(\hat{h}) > 0$.

We bound the number of queries made by $\text{SOLVE}_d^{V^*}(x, \ell', h, s, \mathsf{W}, \mathsf{R})$ as the sum of the queries SOLVE makes on level ℓ', and the queries made by recursive calls. The number of queries made by SOLVE on level ℓ' is at most the total number of messages in an execution, i.e. mr. The number of queries made on recursive calls is computed by summing the queries made by recursive calls on over every slot $\hat{s}$ and taking expectation over every view $\hat{h}$ (such that $q_{\hat{s}}^{\ell'}(\hat{h}) > 0$).

More precisely,

$$E[\# \text{ queries by SOLVE}_d^{V^*}(x, \ell', h, s, \mathsf{W}, \mathsf{R})] \leq mr + \sum_{\hat{s}} \sum_{\hat{h} \in \Gamma_{\hat{s}}} q_{\hat{s}}^{\ell'}(\hat{h}) E_{\hat{s}}(\hat{h})$$

where $E_{\hat{s}}(\hat{h})$ denotes the expected number of queries made by SOLVE from the view $\hat{h}$ on $\hat{s}$. There are two steps involved in computing $E_{\hat{s}}(\hat{h})$. The first step involves finding the expected number of times SOLVE is run on a slot and the second step using the induction hypothesis computing a bound for $E_{\hat{s}}(\hat{h})$.

Step 1: Given a view $\hat{h}$ from where slot $\hat{s}$ opens, let p^ℓ denote the probability that SOLVE rewinds slot $\hat{s}$ from $\hat{h}$, i.e. p^ℓ is the probability that in the simulation from $\hat{h}$ at level ℓ, V^* completes $\hat{s}$ with an accepting proof while opening fewer than $(r-1)^{d-\ell'}$ new executions within the slot $\hat{s}$. Let y^ℓ denote the probability that when executing SOLVE at level ℓ from $\hat{h}$, V^* either aborts or opens more than $(r-1)^{d-\ell'}$ new executions in slot $\hat{s}$. We clearly have that $p^\ell \leq 1 - y^\ell$ (note that equality does not necessarily hold since SOLVE might also return to a lower recursive level). Furthermore, it holds that $y^\ell = y^{\ell+1}$. This follows since SOLVE generates random Stage 1 messages, and uses the same (real) witness to generate Stage 2 messages, independent of the level of the recursion; additionally, since by Claim 4.1, SOLVE never halts outputting fail, we conclude that the view of V^* in the "main-line" simulation by SOLVE on level l is identically distributed to its view on level $l + 1$.

Therefore, the expected number of times SOLVE recursively executes $\hat{s}$ at level $\ell + 1$, before obtaining a good view, is at most $\frac{1}{1 - y^{\ell+1}} = \frac{1}{1 - y^\ell} \leq \frac{1}{p^\ell}$. Using linearity of expectation, the expected number of times SOLVE executes $\hat{s}$ before obtaining m good views is at most $\frac{m}{p^\ell}$. Since, SOLVE rewinds $\hat{s}$ from $\hat{h}$ only with probability p^ℓ, the expected number of recursive calls to level $\ell + 1$ from $\hat{h}$ is at most $p^\ell \frac{m}{p^\ell} = m$.

Step 2: From the induction hypothesis, we know that the expected number of queries made by SOLVE at level $\ell' + 1$ is at most $\alpha(\ell' + 1)$. Therefore, if SOLVE is run u times on a slot, the expected total number of queries made by SOLVE is bounded by $u\alpha(\ell' + 1)$. We conclude that

$$E_{\hat{s}}(\hat{h}) \leq \sum_{u \in \mathbf{N}} Pr[u \text{ recursive calls are made by SOLVE from } \hat{h}] u\alpha(\ell' + 1)$$

$$= \alpha(\ell' + 1) \sum_{u \in \mathbf{N}} u \cdot Pr[u \text{ recursive calls are made by SOLVE from } \hat{h}]$$

$$\leq m\alpha(\ell' + 1)$$

Therefore, $E[\# \text{ queries by SOLVE}_d^{V^*}(x, \ell', h, s, \mathsf{W}, \mathsf{R})] \leq$

$$mr + \sum_{\hat{s}} \sum_{\hat{h} \in \Gamma_{\hat{s}}} q_{\hat{s}}^{\ell'}(\hat{h}) E_{\hat{s}}(\hat{h}) \leq mr + \sum_{\hat{s}} m\alpha(\ell' + 1) \sum_{\hat{h} \in \Gamma_{\hat{s}}} q_{\hat{s}}^{\ell'}(\hat{h})$$

$$\leq mr + \sum_{\hat{s}} m\alpha(\ell' + 1) \leq mr + (mr)m\alpha(\ell' + 1) \leq \alpha(\ell')$$

This completes the induction step and concludes the proof of Claim 2. ∎

4.4   Concluding the Proof of Theorem 1 (and Theorem 4)

Using $r = 3$, we get by Proposition 2 that the expected running-time of S is $poly(m^{log_2 m})$, and by Proposition 1 that its output is correctly distributed. This concludes the proof of Theorem 1. We also remark that the proof of Theorem 4 is directly obtained by instead relying on an n^ϵ-rounds version of the protocol.

5   Proving the Other Theorems

Due to lack of space, we provide only proof ideas for the remaining theorems. The complete proofs will be contained in the full version.

Proof idea of Theorem 2: To prove the theorem, we rely on a slight variant of the $\mathcal{ZK}$ proof of [18,20] (which is an instantiation of the protocol of [23]); the protocol is described in Figure 3. We assume the existence of honest-verifier $\mathcal{ZK}$ proofs that are secure w.r.t $\omega(\mathcal{PQT})$. Such proofs exists if one-way functions that are secure w.r.t $\omega(\mathcal{PQT})$ exists. Furthermore, we require constant round

---

PROTOCOL COMPZKPROOF

Common Input: an instance x of a language L with witness relation R_L.
Auxiliary Input for Prover: a witness w, such that $(x, w) \in R_L(x)$.
Stage 1:
  V uniformly chooses $\bar{r} = r_1, r_2, ..., r_n \in \{0, 1\}^n$, $s \in \{0, 1\}^{poly(n)}$.
  V → P: $c = $ COM$(\bar{r}; s)$, where COM is a statistically hiding commitment, which has the property that the commiter must communicate at least m bits in order to commit to m strings.
  V → P: r first messages $\alpha_1, \ldots, \alpha_r$ for $\mathcal{WI}$ special-sound proofs of the statement. (called the start message)
      there exists values $\bar{r}', s'$ s.t $c = $ COM$(\bar{r}'; s')$
  The proof of knowledge is with respect to the witness relation $R'_L(c) = \{(v, s) | c = $ COM$(v; s)\}$.
  For $j = 1$ to r do
      P → V: Second message $\beta_j \leftarrow \{0, 1\}^{n^2}$ for j^{th} $\mathcal{WI}$ special-sound proof. (called the opening of slot j)
      V → P: Third message γ_j for j^{th} $\mathcal{WI}$ special-sound proof. (called the closing of slot j)
Stage 2:
  P ↔ V: P and V engage in n parallel executions of the GMW's (3-round) Graph 3-Coloring protocol, where V uses the strings $r_1, .., r_n$ as its challenges:
      1. P → V: n (random) first messages of the GMW proof system for the statement x.
      2. V ← P: V decommits to $\bar{r} = r_1, .., r_n$.
      3. P → V: For $i = 1..n$, P computes the answer (i.e., the 3rd message of the GMW proof system) to the challenge r_i and sends all the answers to V.

Fig. 3. Computational $\mathcal{ZK}$ Proof for $\mathcal{NP}$

statistically hiding commitments that are computationally binding w.r.t $\mathcal{PQT}$ adversaries. Such commitment schemes can be constructed from collision resistant hash functions that are secure w.r.t $\mathcal{PQT}$ [7,15]. The simulator and the proof of indistinguishability is essentially similar to Section 3.2. However, to bound the running-time of the simulator we require the Stage 2 of the protocol to satisfy the honest-verifier $\mathcal{ZK}$ property w.r.t. $\omega(\mathcal{PQT})$.

Proof idea of Theorem 3: The protocol is obtained by using a computational $\mathcal{WI}$ protocol w.r.t $\mathcal{PQT}$ instead of the perfect $\mathcal{WI}$ protocol in Stage 2 described in Section 3.1, which can be constructed based on the existence of OWF secure for $\mathcal{PQT}$. The simulator and the analysis from Section 3.2 essentially works for this protocol too, except that to show indistinguishability we use the computational $\mathcal{WI}$ property of the protocol in Stage 2.

Proof idea of Theorems 5 and 6: Our constructions are essentially identical to the protocols in [18,20]. On a high level, the protocols show how to recast the $\mathcal{ZK}$ protocols for Graph Non-Isomorphism and Quadratic Non-Residuosity into the Feige-Shamir paradigm, after which we can rely on the same proof as in the previous section. Finally, Theorem 6 is directly obtained by relying on an $r = n^\epsilon$-rounds version of the protocol.

Acknowledgements

We are very grateful to both Joe Kilian and Alon Rosen for insightful and helpful conversations.

References

1. Barak, B., Lindell, Y.: Strict Polynomial-Time in Simulation and Extraction. In: 34th STOC, pp. 484–493 (2002)
2. Benaloh, J.D.: Cryptographic Capsules: A disjunctive primitive for interactive protocols. In: Odlyzko, A.M. (ed.) CRYPTO 1986. LNCS, vol. 263, pp. 213–222. Springer, Heidelberg (1987)
3. Brassard, G., Chaum, D., Crépeau, C.: Minimum Disclosure Proofs of Knowledge. JCSS 37(2), 156–189 (1988); Preliminary version by Brassard and Crépeau. In: 27th FOCS (1986)
4. Canetti, R., Goldreich, O., Goldwasser, S., Micali, S.: Resettable Zero-Knowledge. In: 32nd STOC, pp. 235–244 (2000)
5. Canetti, R., Kilian, J., Petrank, E., Rosen, A.: Black-Box Concurrent Zero-Knowledge Requires (almost) Logarithmically Many Rounds. SIAM Jour. on Computing 32(1), 1–47 (2002)
6. Damgård, I.: Efficient Concurrent Zero-Knowledge in the Auxiliary String Model. In: Preneel, B. (ed.) EUROCRYPT 2000. LNCS, vol. 1807, pp. 418–430. Springer, Heidelberg (2000)
7. Damgård, I., Pedersen, T., Pfitzmann, B.: On the Existence of Statistically Hiding Bit Commitment Schemes and Fail-Stop Signatures. In: Stinson, D.R. (ed.) CRYPTO 1993. LNCS, vol. 773, pp. 250–265. Springer, Heidelberg (1994)

8. Dwork, C., Naor, M., Sahai, A.: Concurrent Zero-Knowledge. In: 30th STOC, pp. 409–418 (1998)
9. Dwork, C., Sahai, A.: Concurrent Zero-Knowledge: Reducing the Need for Timing Constraints. In: Krawczyk, H. (ed.) CRYPTO 1998. LNCS, vol. 1462, pp. 442–457. Springer, Heidelberg (1998)
10. Fiat, A., Shamir, A.: How to Prove Yourself: Practical Solutions to Identification and Signature Problems. In: Odlyzko, A.M. (ed.) CRYPTO 1986. LNCS, vol. 263, pp. 181–187. Springer, Heidelberg (1987)
11. Goldreich, O., Kahan, A.: How to Construct Constant-Round Zero-Knowledge Proof Systems for NP. Jour. of Cryptology 9(2), 167–189 (1996)
12. Goldreich, O., Micali, S., Wigderson, A.: Proofs that Yield Nothing But Their Validity or All Languages in NP Have Zero-Knowledge Proof Systems. JACM 38(1), 691–729 (1991)
13. Goldreich, O., Oren, Y.: Definitions and Properties of Zero-Knowledge Proof Systems. Jour. of Cryptology 7(1), 1–32 (1994)
14. Goldwasser, S., Micali, S., Rackoff, C.: The Knowledge Complexity of Interactive Proof Systems. SIAM Jour. on Computing 18(1), 186–208 (1989)
15. Halevi, S., Micali, S.: Practical and Provably-Secure Commitment Schemes from Collision-Free Hashing. In: Koblitz, N. (ed.) CRYPTO 1996. LNCS, vol. 1109, pp. 201–215. Springer, Heidelberg (1996)
16. Kilian, J., Petrank, E.: Concurrent and Resettable Zero-Knowledge in Polylogarithmic Rounds. In: 33rd STOC, pp. 560–569 (2001)
17. Kilian, J., Petrank, E., Rackoff, C.: Lower Bounds for Zero-Knowledge on the Internet. In: 39th FOCS, pp. 484–492 (1998)
18. Micali, S., Pass, R.: Local Zero Knowledge. In: STOC 2006 (2006)
19. Pandey, O., Pass, R., Sahai, A., Tseng, D., Venkitasubramaniam, M.: Precise Concurrent Zero-Knowledge (manuscript)
20. Pass, R.: A Precise Computational Approach to Knowledge. PhD thesis, MIT (2006)
21. Pass, R.: Simulation in Quasi-Polynomial Time and Its Application to Protocol Composition. In: Biham, E. (ed.) EUROCRYPT 2003. LNCS, vol. 2656, pp. 160–176. Springer, Heidelberg (2003)
22. Pass, R., Rosen, A.: Bounded-Concurrent Two-Party Computation in Constant Number of Rounds. In: 44th FOCS, pp. 404–413 (2003)
23. Prabhakaran, M., Rosen, A., Sahai, A.: Concurrent Zero-Knowledge with Logarithmic Round Complexity. In: 43rd FOCS, pp. 366–375 (2002)
24. Pass, R., Tseng, D., Venkitasubramaniam, M.: Concurrent Zero Knowledge: A Simplified Proof (submission)
25. Richardson, R., Kilian, J.: On the Concurrent Composition of Zero-Knowledge Proofs. In: Stern, J. (ed.) EUROCRYPT 1999. LNCS, vol. 1592, pp. 415–431. Springer, Heidelberg (1999)
26. Rosen, A.: A note on the round-complexity of Concurrent Zero-Knowledge. In: Bellare, M. (ed.) CRYPTO 2000. LNCS, vol. 1880, pp. 451–468. Springer, Heidelberg (2000)
27. Goldwasser, S., Micali, S.: Probabilistic Encryption. JCSS 28(2), 270–299 (1984)

Concurrent Non-malleable Commitments from Any One-Way Function

Huijia Lin, Rafael Pass,
and Muthuramakrishnan Venkitasubramaniam

Cornell University
{huijia,rafael,vmuthu}@cs.cornell.edu

Abstract. We show the existence of concurrent non-malleable commitments based on the existence of one-way functions. Our proof of security only requires the use of black-box techniques, and additionally provides an arguably simplified proof of the existence of even stand-alone secure non-malleable commitments.

1 Introduction

Often described as the "digital" analogue of sealed envelopes, commitment schemes enable a *sender* to commit itself to a value while keeping it secret from the *receiver*. For some applications, however, the most basic security guarantees of commitments are not sufficient. For instance, the basic definition of commitments does not rule out an attack where an adversary, upon seeing a commitment to a specific value v, is able to commit to a related value (say, $v - 1$), even though it does not know the actual value of v. This kind of attack might have devastating consequences if the underlying application relies on the *independence* of committed values (e.g., consider a case in which the commitment scheme is used for securely implementing a contract bidding mechanism). The state of affairs is even worsened by the fact that many of the known commitment schemes are actually susceptible to this kind of attack. In order to address the above concerns, Dolev, Dwork and Naor (DDN) introduced the concept of *non-malleable commitments* [6]. Loosely speaking, a commitment scheme is said to be non-malleable if it is infeasible for an adversary to "maul" a commitment to a value v into a commitment of a related value $\tilde{v}$.

The first non-malleable commitment protocol was constructed by Dolev, Dwork and Naor [6]. The security of their protocol relies on the existence of one-way functions and requires $O(\log n)$ rounds of interaction, where $n \in N$ is the length of party identifiers (or alternatively, a security parameter). A more recent result by Barak presents a constant-round protocol for non-malleable commitments whose security relies on the existence of trapdoor permutations and hash functions that are collision-resistant against sub-exponential sized circuits [2]. Even more recently, Pass and Rosen present a constant-round protocol, assuming only collision resistant hash function secure against polynomial sized circuits [12].

R. Canetti (Ed.): TCC 2008, LNCS 4948, pp. 571–588, 2008.
© International Association for Cryptologic Research 2008

1.1   Concurrent Non-malleable Commitments

The basic definition of non-malleable commitments only considers a scenario in which two executions take place at the same time. A natural extension of this scenario (already suggested in [6]) is one in which more than two invocations of the commitment protocol take place concurrently. In the concurrent scenario, the adversary is receiving commitments to multiple values $v_1, \ldots, v_m$, while attempting to commit to related values $\tilde{v}_1, \ldots, \tilde{v}_m$. As argued in [6], non-malleability with respect to two executions can be shown to guarantee *individual* independence of any $\tilde{v}_i$ from any v_j. However, it does not rule out the possibility of an adversary creating *joint* dependencies between more than a single individual pair (see [6], Section 3.4.1 for an example in the context of non-malleable encryption). Resolving this issue has been stated as a major open problem in [6].

Partially addressing this issue, Pass demonstrated the existence of commitment schemes that remain non-malleable under *bounded concurrent* composition [10]. That is, for any (predetermined) polynomial $p(\cdot)$, there exists a non-malleable commitment that remains secure as long as it is not executed more than $p(n)$ times, where $n \in N$ is a security parameter. More recently, Pass and Rosen [12] constructed a commitment scheme that remains non-malleable also under an unbounded number of concurrent executions. Their construction uses only a constant number of rounds and is based on the existence of (certified) claw-free permutations. The protocol—which is a variant of the protocol of [11]—relies on the message-length technique of [10], which in turn relies on the non-black box zero-knowledge protocol of Barak [1]. As such, it seems that practical implementations of this approach currently are not within reach.

In contrast, the original construction of Dolev, Dwork and Naor (which is only stand-alone secure) relied on the minimal assumption of one-way functions and had a black-box security proof. Natural questions left open are thus:

> *Can concurrent non-malleable commitments be based solely on the existence of one-way functions?*

> *Does there exist concurrent non-malleable commitments with black-box proofs of security?*

A partial answer to the second question was provided by Pass and Vaikuntanathan [13], demonstrating the existence of concurrent non-malleable commitments with black-box security proofs; their construction, however, relies on a new (and non-standard) hardness assumption with a strong non-malleability flavor.[1]

1.2   Our Results

In this work, we fully resolve both of the above questions. Namely, we show the following theorem using only black-box techniques.

---

[1] More precisely, they assume the existence of, so called, *adaptive one-way permutations*—namely permutations which remain one-way even when the adversary has access to an inversion oracle.

Main Theorem If one-way functions exist, then there exists a statistically-binding commitment scheme that is concurrent non-malleable.

Our protocol, which is a variant of the protocol of [6] (and in particular relies on the same scheduling techniques as in [6]), uses $O(n)$ number of communication rounds. Moreover, it seems that by relying on specific (number theoretic) hardness assumptions (and appropriate Σ-protocols [4]), one can obtain an "implementable" instantiation of our protocol (without going through Cook's reductions).

Additional results. All previous constructions of non-malleable commitments require complex and subtle proofs. As an additional contribution, our protocol and its proof provide the arguably simplest proof of existence of non-malleable commitments (let alone the question of concurrency); more precisely, it provides a new (and arguably simpler) proof of the feasibility result of [6].

Furthermore, by relying on the concurrent security of our protocol, we also obtain a simple (and self-contained) proof of the existence of $\log n$-round (stand-alone secure) non-malleable commitment schemes based on only the existence of one-way functions. As far as we know, a complete proof of this statement (which appeared only with a proof sketch in [6]) has never appeared before.

Finally, we mention that our protocols satisfy a notion of non-malleability called *strict* (as opposed to *liberal*) non-malleability—this notion, which was defined (but not achieved) in [6], requires simulation to be performed by a strict polynomial-time machine (as opposed to an expected polynomial-time machine). Our results provide the first construction of strictly non-malleable commitments based on one-way functions, or using a black-box security proof.

1.3 Overview

Section 2 contains basic notation and definitions of commitment schemes and concurrent non-malleability. In Section 3, we present our $O(n)$-round commitment scheme, and in Section 4, we prove that the commitment scheme is concurrent non-malleable. In Section 5, we additionally provide the construction of a $O(\log n)$-round (stand-alone secure) non-malleable commitment scheme based on any $O(n)$-round concurrent non-malleable commitment scheme.

2 Definitions and Notations

We let N denote the set of all integers. For any integer $m \in N$, denote by $[m]$ the set $\{1, 2, \ldots, m\}$. For any $x \in \{0, 1\}^*$, we let $|x|$ denote the size of x (i.e., the number of bits used in order to write it). For two machines M, A, we let $M^A(x)$ denote the output of machine M on input x and given oracle access to A. The term negligible is used for denoting functions that are (asymptotically) smaller than one over any polynomial. More precisely, a function $\nu(\cdot)$ from non-negative integers to reals is called negligible if for every constant $c > 0$ and all sufficiently large n, it holds that $\nu(n) < n^{-c}$.

2.1 Witness Relations

We recall the definition of a witness relation for an $\mathcal{NP}$ language [8].

Definition 1 (Witness relation). *A witness relation for a language $L \in \mathcal{NP}$ is a binary relation R_L that is polynomially bounded, polynomial time recognizable and characterizes L by $L = \{x : \exists y \, s.t. \, (x,y) \in R_L\}$*

We say that y is a witness for the membership $x \in L$ if $(x,y) \in R_L$. We will also let $R_L(x)$ denote the set of witnesses for the membership $x \in L$, i.e., $R_L(x) = \{y : (x,y) \in L\}$. In the following, we assume a fixed witness relation R_L for each language $L \in \mathcal{NP}$.

2.2 Interactive Proofs

We use the standard definitions of interactive proofs (and interactive Turing machines) [9] and arguments (a.k.a computationally-sound proofs) [3]. Given a pair of interactive Turing machines, P and V, we denote by $\langle P, V \rangle(x)$ the random variable representing the (local) output of V when interacting with machine P on common input x, when the random input to each machine is uniformly and independently chosen.

Definition 2 (Interactive Proof System). *A pair of interactive machines $\langle P, V \rangle$ is called an interactive proof system for a language L if for every probabilistic polynomial time machine (PPT) V there is a negligible function $\nu(\cdot)$ such that the following two conditions hold :*

- *Completeness: For every $x \in L$, $\Pr[\langle P, V \rangle(x) = 1] = 1$*
- *Soundness: For every $x \notin L$, and every interactive machine B,*
  $\Pr[\langle B, V \rangle(x) = 1] \leq \frac{1}{\nu(|x|)}$

In case that the soundness condition is required to hold only with respect to a computationally bounded prover, the pair $\langle P, V \rangle$ is called an interactive argument system.

Special-sound proofs. A 3-round public-coin interactive proof for the language $L \in \mathcal{NP}$ with witness relation R_L is special-sound with respect to R_L, if for any two transcripts (α, β, γ) and $(\alpha', \beta', \gamma')$ such that the initial messages α, α' are the same but the challenges β, β' are different, there is a deterministic procedure to extract the witness from the two transcripts and runs in polynomial time. Special-sound $\mathcal{WI}$ proofs for languages in $\mathcal{NP}$ can be based on the existence of non-interactive commitment schemes, which in turn can be based on one-way permutations. Assuming only one-way functions, 4-round special-sound $\mathcal{WI}$ proofs for $\mathcal{NP}$ exist.[2] For simplicity, we use 3-round special-sound proofs in our protocol though our proof works also with 4-round proofs.

---

[2] A 4-round protocol is special sound if a witness can be extracted from any two transcripts $(\tau, \alpha, \beta, \gamma)$ and $(\tau', \alpha', \beta', \gamma')$ such that $\tau = \tau, \alpha = \alpha'$ and $\beta \neq \beta'$.

2.3 Indistinguishability

Definition 3 ((Computational) Indistinguishability). *Let X and Y be countable sets. Two ensembles $\{A_{x,y}\}_{x \in X, y \in Y}$ and $\{B_{x,y}\}_{x \in X, y \in Y}$ are said to be* computationally indistinguishable over $x \in X$, *if for every probabilistic "distinguishing" machine D whose running time is polynomial in its first input, there exists a negligible function $\nu(\cdot)$ so that for every $x \in X, y \in Y$:*

$$|\Pr[a \leftarrow A_{x,y} \; : \; D(x,y,a) = 1] - \Pr[b \leftarrow B_{x,y} \; : \; D(x,y,b) = 1]| < \nu(|x|)$$

2.4 Witness Indistinguishability

An interactive proof is said to be *witness indistinguishable* ($\mathcal{WI}$) if the verifier's output is "computationally independent" of the witness used by the prover for proving the statement. In this context, we focus on languages $L \in \mathcal{NP}$ with a corresponding witness relation R_L. Namely, we consider interactions in which on common input x the prover is given a witness in $R_L(x)$. By saying that the output is computationally independent of the witness, we mean that for any two possible $\mathcal{NP}$-witnesses that could be used by the prover to prove the statement $x \in L$, the corresponding outputs are computationally indistinguishable.

Definition 4 (Witness-indistinguishability). *Let $\langle P, V \rangle$ be an interactive proof system for a language $L \in \mathcal{NP}$. We say that $\langle P, V \rangle$ is* witness-indistinguishable *for R_L, if for every probabilistic polynomial-time interactive machine V^* and for every two sequences $\{w_x^1\}_{x \in L}$ and $\{w_x^2\}_{x \in L}$, such that $w_x^1, w_x^2 \in R_L(x)$ for every $x \in L$, the probability ensembles $\{\langle P(w_x^1), V^*(z) \rangle(x)\}_{x \in L, z \in \{0,1\}^*}$ and $\{\langle P(w_x^2), V^*(z) \rangle(x)\}_{x \in L, z \in \{0,1\}^*}$ are computationally indistinguishable over $x \in L$.*

2.5 Commitment Schemes

Commitment schemes are used to enable a party, known as the *sender*, to commit itself to a value while keeping it secret from the *receiver* (this property is called hiding). Furthermore, the commitment is binding, and thus in a later stage when the commitment is opened, it is guaranteed that the "opening" can yield only a single value determined in the committing phase. In this work, we consider commitment schemes that are statistically-binding, namely while the hiding property only holds against computationally bounded (non-uniform) adversaries, the binding property is required to hold against unbounded adversaries. More precisely, a pair of PPT machines $\langle C, R \rangle$ is said to be a commitment scheme if the following two properties hold.

Computational hiding: For every (expected) PPT machine R^*, it holds that, the following ensembles are computationally indistinguishable over $\{0,1\}^n$.

- $\{\mathsf{sta}_{\langle C,R \rangle}^{R^*}(v_1, z)\}_{v_1, v_2 \in \{0,1\}^n, n \in N, z \in \{0,1\}^*}$
- $\{\mathsf{sta}_{\langle C,R \rangle}^{R^*}(v_2, z)\}_{v_1, v_2 \in \{0,1\}^n, n \in N, z \in \{0,1\}^*}$

where $\text{sta}^{R^*}_{\langle C,R\rangle}(v,z)$ denotes the random variable describing the output of R^* after receiving a commitment to v using $\langle C,R\rangle$.

Statistical binding: Informally, the statistical-binding property asserts that, with overwhelming probability over the coin-tosses of the receiver R, the transcript of the interaction fully determines the value committed to by the sender. We refer to [8] for more details.

2.6   Concurrent Non-Malleable Commitments

Our definition of concurrent non-malleable commitments is very similar to that of [11], but different in two aspects: first, our definition of non-malleability is w.r.t identities (in analogy with DDN [6])[3]; second, our definition considers not only the values the adversary commits to, but also the view of the adversary.[4] Let $\langle C,R\rangle$ be a commitment scheme, and let $n \in N$ be a security parameter. Consider man-in-the-middle adversaries that are participating in left and right interactions in which $m = \text{poly}(n)$ commitments take place. We compare between a *man-in-the-middle* and a *simulated* execution. In the man-in-the-middle execution, the adversary A is simultaneously participating in m left and right interactions. In the left interactions the man-in-the-middle adversary A interacts with C receiving commitments to values $v_1,\ldots,v_m$, using identities $\text{id}_1,\ldots,\text{id}_m$ of its choice. In the right interaction A interacts with R attempting to commit to a sequence of related values $\tilde{v}_1,\ldots,\tilde{v}_m$, again using identities of its choice $\tilde{\text{id}}_1,\ldots,\tilde{\text{id}}_m$. If any of the right commitments are invalid, or undefined, its value is set to $\perp$. For any i such that $\tilde{\text{id}}_i = \text{id}_j$ for some j, set $\tilde{v}_i = \perp$—i.e., any commitment where the adversary uses the same identity as one of the honest committers is considered invalid. Let $\text{mim}^A_{\langle C,R\rangle}(v_1,\ldots,v_m,z)$ denote a random variable that describes the values $\tilde{v}_1,\ldots,\tilde{v}_m$ and the view of A, in the above experiment.

In the simulated execution, a simulator S directly interacts with R. Let $\text{sim}^S_{\langle C,R\rangle}(1^n,z)$ denote the random variable describing the values $\tilde{v}_1,\ldots,\tilde{v}_m$ committed to by S, and the output *view* of S; again, whenever *view* contains a right interaction i where the identity is the same as any of the left interactions, $\tilde{v}_i$ is set to $\perp$.

Definition 5. *A commitment scheme $\langle C,R\rangle$ is said to be concurrent non-malleable (with respect to commitment) if for every polynomial $p(\cdot)$, and every probabilistic polynomial-time man-in-the-middle adversary A that participates in at most $m = p(n)$ concurrent executions, there exists a probabilistic polynomial time simulator S such that the following ensembles are computationally indistinguishable over $\{0,1\}^n$:*

---

[3] That is, we disallow even copying of commitment as long as the adversary uses a different identity (than all the committers he receives commitments from). In contrast, [11] defined non-malleability w.r.t content; i.e., the adversary allowed copy commitments. This difference is inconsequential as any commitment non-malleable w.r.t content can be turned into one that is non-malleable w.r.t identities, and vice versa.

[4] This point is particularly important when considering our definition w.r.t composability; see Proposition 1 and Section 5.

$$\left\{ \mathsf{mim}^{A}_{\langle C,R \rangle}(v_1,\ldots,v_m,z) \right\}_{v_1,\ldots,v_m \in \{0,1\}^n, n \in N, z \in \{0,1\}^*}$$

$$\left\{ \mathsf{sim}^{S}_{\langle C,R \rangle}(1^n,z) \right\}_{v_1,\ldots,v_m \in \{0,1\}^n, n \in N, z \in \{0,1\}^*}$$

We also consider relaxed notions of concurrent non-malleability: one-many, many-one and one-one secure non-malleable commitments. In a one-one (i.e., a stand-alone secure) non-malleable commitment, we consider only adversaries A that participate in one left and one right interaction; in one-many, A participates in one left and many right, and in many-one, A participates in many left and one right.

Dolev, Dwork and Naor [6] argued that one-one commitments are also many-one secure. Pass and Rosen [11] additionally showed that one-many non-malleability implies (many-many) concurrent non-malleability if the commitment protocol is "natural". Given our stronger definition, which also considers the view of the adversary, we prove that *any* protocol that is one-many non-malleable is also concurrent non-malleable. Namely,

Proposition 1. *Let $\langle C, R \rangle$ be a one-many concurrent non-malleable commitment. Then, $\langle C, R \rangle$ is also a concurrent non-malleable commitment.*

Proof. Let A be a man-in-the-middle adversary that participates in at most $m = p(n)$ concurrent executions. Below, we provide a simulator S for A. S proceeds as follows on input 1^n and z. S incorporates $A(z)$ and internally emulates all the left interactions for A by simply honestly committing to the string 0^n. Messages from the right interactions are instead forwarded externally. Finally S outputs the view of A.

We show that the values that S commits to are indistinguishable from the values that A commits to. Suppose, for contradiction, that this is not the case. Then, there exists a polynomial-time distinguisher D and a polynomial $p(n)$ such that for infinitely many n, there exist strings $v_1,\ldots,v_m \in \{0,1\}^n, z \in \{0,1\}^*$ such that D distinguishes $\mathsf{mim}^{A}_{\langle C,R \rangle}(v_1,\ldots,v_m,z)$ and $\mathsf{sta}^{S}_{\langle C,R \rangle}(1^n,z)$ with probability $\frac{1}{p(n)}$. Fix a generic n for which this happens. Consider the hybrid simulator S_i that on input $1^n, z' = v_1,\ldots,v_m,z$, proceeds just as S, with the exception that in left interactions $j \leq i$, it instead commits to v_j. It directly follows that $\mathsf{mim}^{A}_{\langle C,R \rangle}(v_1,\ldots,v_m,z) = \mathsf{sta}^{S_m}_{\langle C,R \rangle}(1^n,z')$ and $\mathsf{sta}^{S}_{\langle C,R \rangle}(1^n,z) = \mathsf{sta}^{S_0}_{\langle C,R \rangle}(1^n,z')$. By a standard hybrid argument there exists an $i \in [m]$ such that

$$\left| \Pr\left[a \leftarrow \mathsf{sta}^{S_{i-1}}_{\langle C,R \rangle}(1^n,z') : D(1^n,z',a) = 1 \right] \right.$$
$$\left. - \Pr\left[b \leftarrow \mathsf{sta}^{S_i}_{\langle C,R \rangle}(1^n,z') : D(1^n,z',b) = 1 \right] \right| \geq \frac{1}{p(n)m}$$

Note that the only difference between the executions by $S_{i-1}(1^n,z')$ and $S_i(1^n,z')$ is that in the former A receives a commitment to 0^n in session i, whereas in the latter it receives a commitment to v_i. Consider the one-many adversary $\tilde{A}$ that on input $\tilde{z} = z',n,i$ executes $S_{i-1}(1^n,z')$ with the exception that the i'th left interaction is forwarded externally. Consider, the function reconstruct that on input $\text{mim}^{\tilde{A}}_{\text{com}}(0^n, \tilde{z})$, i.e. values $v'_1, \ldots, v'_m$, and the view of $\tilde{A}$, reconstructs the view $view$ of A in the emulation by $\tilde{A}$, and sets $\tilde{v}_i = v'_1$ if A did not copy the identity of any of the left interactions, and $\bot$ otherwise, and finally outputs $\tilde{v}_1, \ldots, \tilde{v}_m, view$. By construction, it follows that

$$\text{reconstruct}(\text{mim}^{\tilde{A}}_{\langle C,R \rangle}(0^n, \tilde{z})) = \text{sta}^{S_{i-1}}_{\langle C,R \rangle}(1^n, z')$$

$$\text{reconstruct}(\text{mim}^{\tilde{A}}_{\langle C,R \rangle}(v_i, \tilde{z})) = \text{sta}^{S_i}_{\langle C,R \rangle}(1^n, z')$$

Since reconstruct is polynomial-time computable, this contradicts the one-many non-malleability of $\langle C, R \rangle$.

3   The Protocol

Our protocol is based on Feige-Shamir's zero-knowledge protocol [7] while relying on the *message scheduling technique* of Dolev, Dwork and Naor[6]. For simplicity of exposition, our description below relies on the existence of one-way functions with efficiently recognizable range, but the protocol can be easily modified to work with any arbitrary one-way function (by simply providing a witness hiding proof that an element is in the range of the one-way function). The protocol proceeds in the following three stages on common input the identity id $\in \{0,1\}^l$ of the committer, and security parameter n.

1. In Stage 1, the Receiver picks a random string $r \in \{0,1\}^n$, and sends its image $s = f(r)$ through a one-way function f with an efficiently recognizable range to the Committer. The Committer checks that s is in the range of f and aborts otherwise.
2. In Stage 2, the Committer sends $c = \text{com}(v)$, where $\text{com}(\cdot)$ is any commitment scheme that is statistically-binding.
3. In Stage 3, the Committer proves that c is a valid commitment for v or s is in the image set of f. This is proved by $4l$ invocations of a special-sound $\mathcal{WI}$ proof where the messages are scheduled based on the id (very similar to the scheduling presented in [6]). More precisely, there are l rounds, where in round i, the schedule $\text{design}_{\text{id}_i}$ is followed by $\text{design}_{1-\text{id}_i}$ (See Figure 1).

We remark that the scheduling (essentially identical to [6]) in Stage 3 of the protocol is the key in achieving concurrent non-malleability. Loosely speaking, the purpose of the scheduling is to guarantee that for each of the commitments that a man-in-the-middle adversary gives, there exists a point at which the adversary cannot answer the challenge from the receiver simply by "mauling" the commitments on the left (provided that the identity of the commitment is different from any of the commitments on the left).

One important difference between our protocol and the protocol of [6] is that the designs we use consist of two *three-round* protocols, whereas the protocol in [6] uses more rounds; this makes the analysis clearer. An additional simplification is the use of only $\mathcal{WI}$ proofs (instead of zero-knowledge proofs as in [6]).

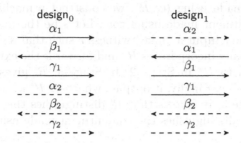

Fig. 1. Description of the schedules used in Stage 3 of the protocol

Claim 1. $\langle C, R \rangle$ *is a statistically-binding commitment scheme.*

Proof. We show that the $\langle C, R \rangle$ scheme satisfies the binding and hiding properties.

Protocol ConcNMCom

Common Input: An identifier id $\in \{0,1\}^l$.
Auxiliary Input for Committer: A string $v \in \{0,1\}^n$.
Stage 1:
    R uniformly chooses $r \in \{0,1\}^n$.
    R $\rightarrow$ C: $s = f(r)$.
    C aborts if s not in the range of f.
Stage 2:
    C uniformly chooses $r' \in \{0,1\}^{poly(n)}$.
    C $\rightarrow$ R: $c = \mathsf{com}(v, r')$.
Stage 3:
    C $\rightarrow$ R: $4l$ special-sound $\mathcal{WI}$ proofs of the statement
      either there exists values v, r' s.t $c = \mathsf{com}(v, r')$
      or there exists a value r s.t $s = f(r)$
    with verifier query of length $2n$, in the following schedule:
    For $j = 1$ to l do: Execute design_{id_j} followed by Execute design_{1-id_j}

Fig. 2. Non-Malleable String Commitment Scheme $\langle C, R \rangle$

Binding: The binding property follows directly from the binding property of com.

Hiding: The hiding property essentially follows from the hiding property of com and the fact that Stage 3 of the protocol is $\mathcal{WI}$ (since $\mathcal{WI}$ proofs are closed under concurrent composition [7]). For completeness, we provide the proof. We show that any adversary R^* that violates the hiding property of $\langle C, R \rangle$ can be used to violate the hiding property of com. More precisely, given any adversary R^* (without loss of generality, deterministic) that distinguishes a commitment made using $\langle C, R \rangle$, we construct a machine R' that distinguishes a commitment made using com. Let s be the first message sent by R^*. R' on auxiliary-input a "fake" witness r such that $s = f(r)$, proceeds as follows. It internally incorporates R^* and forwards the external commitment made using com to R^* in Stage 2. In Stage 3, R' gives $\mathcal{WI}$ proofs using the "fake witness" r. Finally, it outputs whatever R^* outputs. From the $\mathcal{WI}$ property of Stage 3, it follows that R' distinguishes the commitment made using com, if R^* distinguishes the commitment made using $\langle C, R \rangle$.

4   Proof of Security

Theorem 1. $\langle C, R \rangle$ *is one-many concurrent non-malleable.*

Proof: Let A be a man-in-the-middle adversary that participates in one execution in the left and many executions in the right. We construct a simulator S such that the following ensembles are computationally indistinguishable over $\{0, 1\}^*$.

$$\left\{ \mathsf{mim}^A_{\langle C,R \rangle}(v, z) \right\}_{v \in \{0,1\}^n, n \in N, z \in \{0,1\}^*}$$
$$\left\{ \mathsf{sim}^S_{\langle C,R \rangle}(1^n, z) \right\}_{v \in \{0,1\}^n, n \in N, z \in \{0,1\}^*}$$

The simulator S on input $(1^n, z)$ proceeds as follows. S incorporates $A(z)$ and internally emulates the left interaction by *honestly* committing to the string 0^n. Messages in the right interactions are instead forwarded externally. Finally, S outputs the view of A. We show that the values that S commits to combined with the output view are indistinguishable from the values that A commits to combined with its view. Since S emulates the left interaction by *honestly* committing to 0^n, this is equivalent to showing that

$$\left\{ \mathsf{mim}^A_{\langle C,R \rangle}(v, z) \right\}_{v \in \{0,1\}^n, n \in N, z \in \{0,1\}^*} \approx \left\{ \mathsf{mim}^A_{\langle C,R \rangle}(0^n, z) \right\}_{v \in \{0,1\}^n, n \in N, z \in \{0,1\}^*}$$

Towards this goal, we define a new commitment scheme $\langle \hat{C}, \hat{R} \rangle$ (much like the adaptor scheme in DDN [6]), which is a variant of $\langle C, R \rangle$ where the receiver can ask for an arbitrary number of special-sound $\mathcal{WI}$ designs in Stage 3. Furthermore, $\langle \hat{C}, \hat{R} \rangle$ does not have a fixed scheduling in Stage 3; the receiver instead gets to choose which design to execute in each iteration (by sending bit i to select design$_i$). Note that, clearly, any execution of $\langle C, R \rangle$ can be emulated by an execution of $\langle \hat{C}, \hat{R} \rangle$ by simply requesting the appropriate designs.

Using the same proof as in Claim 1, it follows that $\langle \hat{C}, \hat{R} \rangle$ is hiding, i.e.

Lemma 1. *For every (expected) PPT machine M,*

$$\left\{ \mathsf{sta}^M_{\langle \hat{C},\hat{R} \rangle}(v, z) \right\}_{v \in \{0,1\}^n, n \in N, z \in \{0,1\}^*} \approx \left\{ \mathsf{sta}^M_{\langle \hat{C},\hat{R} \rangle}(0^n, z) \right\}_{v \in \{0,1\}^n, n \in N, z \in \{0,1\}^*}$$

Below, in Lemma 2, we show that for every adversary A, there exists an expected *non-uniform* PPT machine R^* whose output, upon receiving a commitment using $\langle \hat{C}, \hat{R} \rangle$ to v, is indistinguishable from the view and the *values committed to* by $A(z)$ when receiving a commitment to v using $\langle C, R \rangle$; by the hiding property of $\langle \hat{C}, \hat{R} \rangle$ we then conclude that $\text{mim}^A_{\langle C, R \rangle}(v, z)$ and $\text{mim}^A_{\langle C, R \rangle}(0^n, z)$ are indistinguishable. On a high-level, R^* will emulate an execution of $\langle C, R \rangle$ for A (by requesting the appropriate design in $\langle \hat{C}, \hat{R} \rangle$) and then will attempt to extract the values committed to by A. In fact, it suffices for R^* to extract only the values committed to *after* the left execution starts (as all values committed to before-hand can be non-uniformly given to R^*).

Let $\Gamma(A, z)$ denote the distribution of all joint views τ of A and the receivers in the right, such that A sends its first message in the left interaction directly after receiving the messages in τ. Let the function $\mathcal{Z} : \{0, 1\}^* \times \{0, 1\}^* \to \{0, 1\}^*$ be such that, $\mathcal{Z}(z, \tau) = z \| \tau \| \tilde{v}_1 \| \dots \| \tilde{v}_\ell$ where $\tilde{v}_1 \dots \tilde{v}_\ell$, $\ell \in [m]$ are the values committed to by $A(z)$ in τ (using com).

The main technical content of Theorem 1 is in proving the following lemma.

Lemma 2. *For every PPT adversary A, there exists an expected PPT adversary R^* such that the following ensembles are indistinguishable over $\{0, 1\}^*$.*

$$- \left\{ \tau \leftarrow \Gamma(A, z), \; z' \leftarrow \mathcal{Z}(z, \tau) : \text{sta}^{R^*}_{\langle \hat{C}, \hat{R} \rangle}(v, z') \right\}_{v \in \{0,1\}^n, n \in N, z \in \{0,1\}^*}$$

$$- \left\{ \text{mim}^A_{\langle C, R \rangle}(v, z) \right\}_{v \in \{0,1\}^n, n \in N, z \in \{0,1\}^*}$$

Before proceeding to the proof of lemma 2, note that by lemma 1, it holds that the following ensembles are indistinguishable

$$- \left\{ \text{sta}^{R^*}_{\langle \hat{C}, \hat{R} \rangle}(v, z') \right\}_{v \in \{0,1\}^n, n \in N, z, \tau, z' \in \{0,1\}^*}$$

$$- \left\{ \text{sta}^{R^*}_{\langle \hat{C}, \hat{R} \rangle}(0^n, z') \right\}_{v \in \{0,1\}^n, n \in N, z, \tau, z' \in \{0,1\}^*}$$

It thus follows that the following ensembles also are indistinguishable

$$- \left\{ \tau \leftarrow \Gamma(A, z), \; z' \leftarrow \mathcal{Z}(z, \tau) : \text{sta}^{R^*}_{\langle \hat{C}, \hat{R} \rangle}(v, z') \right\}_{v \in \{0,1\}^n, n \in N, z \in \{0,1\}^*}$$

$$- \left\{ \tau \leftarrow \Gamma(A, z), \; z' \leftarrow \mathcal{Z}(z, \tau) : \text{sta}^{R^*}_{\langle \hat{C}, \hat{R} \rangle}(0^n, z') \right\}_{v \in \{0,1\}^n, n \in N, z \in \{0,1\}^*}$$

By lemma 2, we thus conclude that the following ensembles are indistinguishable,

$$- \left\{ \text{mim}^A_{\langle C, R \rangle}(v, z) \right\}_{v \in \{0,1\}^n, n \in N, z \in \{0,1\}^*}$$

$$- \left\{ \text{mim}^A_{\langle C, R \rangle}(0^n, z) \right\}_{v \in \{0,1\}^n, n \in N, z \in \{0,1\}^*}$$

which concludes the proof of theorem 1.

Proof (of lemma 2). Recall that by the definition of $\mathcal{Z}$ it holds that $z' = z \| \tau \| \tilde{v}_1 \| \dots \| \tilde{v}_\ell$ where $\tilde{v}_1 \dots \tilde{v}_\ell$, $\ell \in [m]$, are the values committed to by $A(z)$ using com in the view τ. On a high-level, R^* on auxiliary input z', internally incorporates $A(z)$ and emulates the left and the right executions for A. First, however, it starts by feeding A its part of the joint view τ. It, then, emulates the left interaction for A by externally forwarding messages using $\langle \hat{C}, \hat{R} \rangle$ (by

---

Description of R^*

Input: R^* receives auxiliary input $z' = z\|\tau\|\tilde{v}_1\|\ldots\|\tilde{v}_\ell$.

Procedure: R^* interacts externally as a receiver using $\langle \hat{C}, \hat{R} \rangle$. Internally it incorporates $A(z)$ and emulates a one-many man-in-the-middle execution by simulating all right receivers and emulating the left $\langle C, R \rangle$ interaction by requesting the appropriate designs expected by $A(z)$ using $\langle \hat{C}, \hat{R} \rangle$ from outside.

Main Execution Phase: Feed the view in τ to A and all right receivers. Emulate all the interactions from τ and complete the execution with A. Let Δ be the transcript of messages obtained.

Rewinding Phase: For $k = \ell+1$ to m, if interaction k is convincing and its identity is different from the left interaction, do:

- In Δ, find the first point ρ that is a safe-point for interaction k; let the associated proof be $(\alpha_\rho, \beta_\rho, \gamma_\rho)$.
- Repeat until a second-proof transcript $(\alpha_\rho, \beta'_\rho, \gamma'_\rho)$ is obtained:
  Emulate the left interaction as in the Main-Execution Phase. For the left interaction:
  - If A expects to get a new proof from the external committer (case (i) in Figure 5): Emulate the proof, by requesting for design_0 from outside committer. Forward one of the two proofs internally.
  - If A sends a challenge for a proof whose first message occurs in ρ: Cancel the execution, rewind to ρ and continue.
- If $\beta_\rho \neq \beta'_\rho$ extract witness w from $(\alpha_\rho, \beta_\rho, \gamma_\rho)$ and $(\alpha_\rho, \beta'_\rho, \gamma'_\rho)$. Otherwise halt and output fail.
- If $w = (v, r)$ is valid commitment for interaction k, i.e. $\mathsf{com}(v, r) = c_k$, where c_k is the Stage 2 message in interaction k, then set $\hat{v}_k = v$. Otherwise halt and output fail.

Note that, since right interactions $\ell+1$ to n all have their Stage 2 and 3 occurring after τ, none of the rewinding can make A request a new commitment from the external committer.

Output Phase: For every interaction k that is not convincing or if the identity of the right interaction is the same as the left interaction, set $\hat{v}_k = \perp$. Output $(\hat{v}_1, \ldots, \hat{v}_m)$ and the view from the Main Execution Phase.

Finally, if it runs for more than 2^n steps, halt and output fail.

Fig. 3. The construction of R^*

appropriately choosing the "right" designs); the right interactions are instead dealt with internally by first honestly emulating the receivers on the right, from the view in τ—this is called the *main execution*. In a second phase, it then attempts to extract all the values committed to on the right—this is called the *rewinding phase*. Finally, in the *output phase*, it outputs the view of A and all the values extracted, including the ones received as auxiliary input (additionally, if A fails in completing one of the commitments that started in τ, or if it uses the same identity as the left interaction, that value is replaced by $\perp$). The core of the proof is to show that extraction during the rewinding phase is successful. Towards this goal, we need to ensure that there exist some point where we can rewind A on the right interaction, *without rewinding on the left*; this is possible in two cases: (1) if rewinding on the right does not cause A to request any new messages on the left, or (2) if rewinding on the right causes A to only request a new special-sound proof—in this case R^* can perfectly emulate this new proof by simply requesting another design from $\langle \hat{C}, \hat{R} \rangle$.

We show below that there exist certain points—called safe-points—in each execution, from which it will be possible to perform extraction by simply rewinding until we obtain a second proof transcript, without rewinding on the left (and aborting all rewindings where A requests a message on the left which would require us to rewind also the left execution). (Actually, to simplify our analysis this extraction procedure is cut-off if it runs "too long" (2^n steps) in which case R^* halts and outputs fail.)

Below we provide a definition of safe-points. A formal description of R^* (which relies on the notion of safe-points) is found in Figure 3.

Intuitively, a safe point ρ is a prefix of some transcript Δ which has the property that if during a rewinding from ρ, A uses the same "scheduling" of messages as in Δ, then the left execution can be perfectly emulated without rewinding (on the left). As we show later, if we rewind only from such points we ensure that the expected running time is polynomially bounded (even if A adaptively schedule the messages on the left).

Definition 6. *A prefix ρ of a transcript Δ is called a* safe-point *for interaction k, if there exists an accepting proof $(\alpha_r, \beta_r, \gamma_r)$ in the kth right interaction, such that:*

1. *α_r occurs in ρ, but not β_r (and γ_r).*
2. *for any proof $(\alpha_l, \beta_l, \gamma_l)$ in the left interaction, if α_l occurs in ρ, then β_l occurs after γ_r.*

If ρ is a safe-point, let $(\alpha_\rho, \beta_\rho, \gamma_\rho)$ denote the canonical "safe" right proof.[5]

Note that the only case a right-interaction proof does not have a safe-point is if it is "aligned" with a left execution proof (such that A can forward messages between the left and the right interactions); see Figure 4. In contrast, in all other cases, a right-interaction proof has a safe-point. In Figure 5, we present the three characteristic types of safe-point. Note that in the first case (see Figure 5 (i)), when rewinding from ρ, R^* can emulate the left proof by requesting a new design from $\langle \hat{C}, \hat{R} \rangle$; in the second case (Figure 5 (ii)), R^* can simply re-send the third message of the left proof (since it is determined by the first two messages in the proof); and in the last case (Figure 5 (ii)), no new message is requested by A, so the left interaction can be "trivially" emulated (by doing nothing).

---

[5] We remark that our definition of safe-points is analogous to the "safe" rewinding points inside *exposed triplets* defined in DDN [6]. Loosely speaking, for every *exposed triplet*, there is a "safe" rewinding point that one can rewind to extract the committed value on the right without "affecting" the left interaction. Defining safe-point this way, avoids the complication of finding the "safe" rewinding point in each type of the *exposed triplet*.

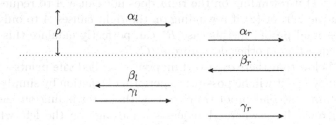

Fig. 4. Prefix ρ that is not a safe point

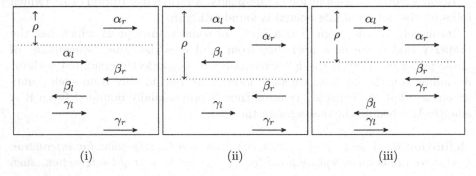

Fig. 5. Three characteristic safe-points

Running-time analysis of R^*. We show that R^* is expected PPT. Note that the time spent by R^* in the Main Execution Phase is $poly(n)$ (where n is the security parameter), since A is a strict polynomial time machine. We show below that the expected time spent by R^* in the Rewinding Phase is $poly(n)$. To bound the expected running time, we assume for simplicity that R^* does not check the fail conditions and may run for more than 2^n steps (since this only increases the running time).

Recall that in the Rewinding Phase, R^* rewinds A from all safe points. Let $T_k(i)$ be the random variable that describes the time spent in rewinding a proof in interaction k after i messages have been exchanged. We show that $E[T_k(i)] \le poly(n)$ and then by linearity of expectation, we conclude that the expected time spent by R^* in the Rewinding phase is

$$\sum_{k=1}^{m}\sum_{i} E[T_k(i)] \le \sum_{k=1}^{m}\sum_{i} poly(n) \le poly(n),$$

where the total number of messages exchanged and m is $poly(n)$.

Bounding $E[T_k(i)]$. Given a (partial) transcript of messages ρ, let $\Pr[\rho]$ denote the probability that ρ occurs as a prefix of the execution emulated in the Main Execution phase. Furthermore, let p_ρ denote the probability that ρ is a safe-point and is rewound—i.e. p_ρ is the probability that, conditioned on the prefix ρ occurring, the right interaction k is convincing and ρ is a safe-point for interaction k. Recall that R^* rewinds until it finds another transcript for the proof $(\alpha_\rho, \beta_\rho, \gamma_\rho)$ associated with ρ, cancelling each rewinding for which A requests the second message of a proof in the left-interaction whose first message occurs in ρ. We claim that the probability of cancelling a rewinding from ρ, is at most $1 - p_\rho$ since ρ is not a safe-point for every rewinding that is cancelled, and conditioned on ρ, the probability of a view occurring in a rewinding from ρ is same as occurring in the Main Execution phase (as the emulated receiver picks uniformly random messages in Stage 3 of the protocol). Thus, the expected number of rewindings is at most $\frac{1}{p_\rho}$ Therefore, the expected number of rewindings from ρ is at most $p_\rho \cdot \frac{1}{p_\rho} = 1$ and each rewinding takes at most $poly(n)$ steps, i.e.

$$E[T_k(i)|\rho] \le poly(n)$$

Thus,

$$E[T_k(i)] = \sum_{\rho \text{ of length } i} E[T_k(i)|\rho] \Pr[\rho] \le poly(n) \times \sum_{\rho \text{ of length } i} \Pr[\rho] \le poly(n)$$

Output distribution of R^* is correct. We proceed to show that the output distribution of R^* is correct. This follows from the following two claims:

Claim 2. *Assume that R^* does not output* fail, *then except with negligible probability, its output is identical to the values committed to by A in the right interactions combined with its view.*

Proof. We first note that since in the Main Execution Phase, R^* feeds A messages according to the correct distribution, the view of A in the simulation by R^* is identical to the view of A in a real interaction. We show in Lemma 3 that there is a safe point for every right interaction that has an identity different from the left interaction. Hence, for every convincing right interaction $k > \ell$ that has a different identity, R^* rewinds that interaction and eventually will either output fail or a witness is extracted from the rewinding phase of R^*. Conditioned on R^* not outputting fail, by the statistical-binding property of com, except with negligible probability the witnesses extracted by R^* are the values committed to by A.

Lemma 3 (Safe-point Lemma). *In any one-many man-in-the-middle execution with m right interactions, for any right interaction k, $k \in [m]$, such that it has a different identity from the identity of the left interaction, there exists a* safe-point *for interaction k.*

Proof. Consider a one-many man-in-the-middle execution Δ, where the identities in the left and right interaction are different. Assume for contradiction, that there is some right interaction k which does not have a safe-point, i.e. every prefix of Δ is not a safe-point for interaction k.

Consider any proof $(\alpha_r, \beta_r, \gamma_r)$ in the right interaction k. Let ρ be the prefix after which β_r is sent immediately. By assumption, ρ is not a safe-point. This means there exists a proof $(\alpha_l, \beta_l, \gamma_l)$ in the left interaction, such that α_l occurs before ρ, β_l occurs after ρ and before γ_r, as depicted in Figure 4. That is, β_l occurs in between β_r and γ_r; we say a left proof is associated with a right proof in this case. Note that each left proof can be associated with at most one right proof. For the interaction k to not have a safe-point, the proofs in the left and right interactions must match up each other one by one: the ith proof in the left is associated with the ith proof in the right.

Since the identities in the left and right interactions are different, there must be a position j they differ at. Let the jth bit in the left be b and that in the right be $1 - b$. Recall that, in the j^{th}round of Stage 3 of the protocol, the left interaction has design_b followed by design_{1-b}; and the right interaction has design_{1-b} followed by design_b. Since all the proofs are "matched up", it must be the case that there is a design_0 on the left that is matched with a design_1 on the right, as depicted in Figure 6. Let $(\alpha_i^l, \beta_i^l, \gamma_i^l)$, $i = 1, 2$, be the two proofs in design_0, and $(\alpha_i^r, \beta_i^r, \gamma_i^r)$, $i = 1, 2$, be the ones on the right in design_1. In this case, consider ρ to be the prefix that includes all the message up until β_1^l. Consider the second proof $(\alpha_2^r, \beta_2^r, \gamma_2^r)$; there is no proof on the left having its first message before ρ and its challenge before γ_2^r at the same time. Hence, we arrive at a contradiction to our assumption that there is no safe-point for that right interaction.

Claim 3. R^* *outputs* fail *with negligible probability.*

Proof. R^* outputs fail only in the following cases:

R^* **runs for more than** 2^n **steps:** We know that the expected running time of R^* is $poly(n)$. Using Markov inequality, we conclude that the probability that R^* runs more than 2^n steps is at most $\frac{poly(n)}{2^n}$.

The same proof transcript is obtained from some safe-point**:** This case occurs if R^* picks some challenge β in the Rewinding Phase that appeared as a challenge in the Main Execution Phase. As R^* runs for at most 2^n steps, it picks at most 2^n challenges. Furthermore, the length of each challenge is

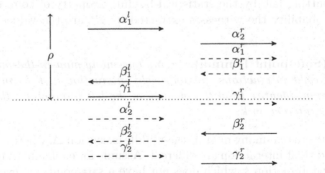

Fig. 6. A design_0 matches up with design_1

$2n$. By applying the union bound, we obtain that the probability that a β is picked twice is at most $\frac{2^n}{2^{2n}}$. Since there are at most polynomially many challenges picked in the Main Execution Phase, using the union bound again, we conclude that the probability that it outputs fail in this case is negligible.

The witness extracted is not a valid decommitment: Suppose, the witness extracted is not the decommitment information, then by the special-sound property it follows that it must be a value r such that $f(r) = s$. We show that if this happens with non-negligible probability, then we can invert the one-way function f. More precisely, given A, z and v, we construct A^* that inverts f; A^* on input y, picks τ uniformly at random from $\Gamma(A, z)$ (by emulating an execution of $A(z)$ internally) and proceeds identically as R^* with inputs τ, z' where $z' = z\|\tau\|\bot\|\bot\|\ldots$ with the exception that it picks a random right interaction, say k, and feeds y as the Stage 1 message in that interaction. On the left interaction it honestly commits to the string v using $\langle \hat{C}, \hat{R} \rangle$. Finally, if the value r' output for interaction k is the inverse image of y w.r.t f (*i.e.* $f(r') = y$), then A^* outputs r'. (Notice that it is not necessary to compute z' according to the definition of $\mathcal{Z}$, since R^* uses the values in z' only in the output phase and not in its extraction procedure). Therefore, the probability that A^* inverts f is identical to the probability that R^* inverts f which is non-negligible; this contradicts the one-wayness of f.

Since each of the above cases occur with negligible probability, using the union bound, we conclude that R^* outputs fail with negligible probability.

5   A $\log n$-Round Non-malleable Commitment Scheme

In this section, we show how to construct a $O(\log n)$-round commitment scheme that is stand-alone non-malleable using any $O(n)$-round commitment scheme that is one-many non-malleable. In particular, using the scheme $\langle C, R \rangle$ described in the previous section, we obtain a $O(\log n)$-round commitment scheme that is stand-alone non-malleable. The idea for this construction is almost identical to the $O(\log n)$-round protocol constructed in [6], except that our construction is more general, as it can be applied to *any* commitment scheme that satisfies our notion of one-many non-malleability; we here rely on the fact that our definition considers not only the values committed to by the adversary but also its view.

Description of the Protocol $\langle \tilde{C}, \tilde{R} \rangle$: To commit to value $v \in \{0,1\}^n$, choose random shares $r_1, \ldots, r_n \in \{0,1\}^n$, such that $v = r_1 \oplus \ldots \oplus r_l$. If id $\in \{0,1\}^l$ is the identity of the $\langle \tilde{C}, \tilde{R} \rangle$ interaction, then for each i, commit to r_i (in parallel) using $\langle C, R \rangle$ with identity (i, id_i), where id_i is the ith bit of id.

In the full version of the paper, we show the following claim.

Claim 4. $\langle \tilde{C}, \tilde{R} \rangle$ *is stand-alone non-malleable.*

Acknowledgement

We are very grateful to Danny Dolev for helpful conversations and for his contagious enthusiasm. We are also grateful to Cynthia Dwork and Moni Naor for helpful clarifications.

References

1. Barak, B.: How to go Beyond the Black-Box Simulation Barrier. In: 42nd FOCS, pp. 106–115 (2001)
2. Barak, B.: Constant-Round Coin-Tossing or Realizing the Shared-Random String Model. In: 43rd FOCS, pp. 345–355 (2002)
3. Brassard, G., Chaum, D., Crépeau, C.: Minimum Disclosure Proofs of Knowledge. JCSS 37(2), 156–189 (1988); Preliminary version by Brassard and Crépeau. In: 27th FOCS (1986)
4. Cramer, R., Damgård, I., Schoenmakers, B.: Proofs of Partial Knowledge and Simplified Design of Witness Hiding Protocols. In: Desmedt, Y.G. (ed.) CRYPTO 1994. LNCS, vol. 839, pp. 174–187. Springer, Heidelberg (1994)
5. di Crescenzo, G., Persiano, G., Visconti, I.: Constant-Round Resettable Zero Knowledge with Concurrent Soundness in the Bare Public-Key Model. In: Franklin, M. (ed.) CRYPTO 2004. LNCS, vol. 3152, pp. 237–253. Springer, Heidelberg (2004)
6. Dolev, D., Dwork, C., Naor, M.: Non-Malleable Cryptography. SIAM Journal on Computing 30(2), 391–437 (2000)
7. Feige, A., Shamir, A.: How to Prove Yourself: Practical Solutions to Identification and Signature Problems. In: Odlyzko, A.M. (ed.) CRYPTO 1986. LNCS, vol. 263, pp. 181–187. Springer, Heidelberg (1987)
8. Goldreich, O.: Foundations of Cryptography – Basic Tools. Cambridge University Press, Cambridge (2001)
9. Goldwasser, S., Micali, S., Rackoff, C.: The Knowledge Complexity of Interactive Proof Systems. SIAM Jour. on Computing 18(1), 186–208 (1989)
10. Pass, R.: Bounded-Concurrent Secure Multi-Party Computation with a Dishonest Majority. In: 36th STOC, pp. 232–241 (2004)
11. Pass, R., Rosen, A.: Bounded-Concurrent Secure Two-Party Computation in a Constant Number of Rounds. In: 44th FOCS (2003)
12. Pass, R., Rosen, A.: New and Improved Constructions of Non-Malleable Cryptographic Protocols. In: 37th STOC, pp. 533–542 (2005)
13. Pass, R., Vaikuntanathan, V.: New-Age Cryptography (manuscript)

Faster and Shorter
Password-Authenticated Key Exchange

Rosario Gennaro

IBM T.J. Watson Research Center
19 Skyline Drive,
Hawthorne 10532, USA
rosario@us.ibm.com

Abstract. This paper presents an improved password-based authenticated key exchange protocol in the common reference string model. Its
security proof requires no idealized assumption (such as random oracles).

The protocol is based on the GL framework introduced by Gennaro
and Lindell, which generalizes the KOY key exchange protocol of Katz
et al. Both the KOY and the GL protocols use (one-time) signatures as
a non-malleability tool in order to prevent a man-in-the-middle attack
against the protocol. The efficiency of the resulting protocol is negatively
affected, since if we use regular signatures, they require a large amount
of computation (almost as much as the rest of the protocol) and further computational assumptions. If one-time signatures are used, they
substantially increase the bandwidth requirement.

Our improvement avoids using digital signatures altogether, replacing
them with faster and shorter message authentication codes. The crucial
idea is to leverage as much as possible the non-malleability of the encryption scheme used in the protocol, by including various values into
the ciphertexts as *labels*. As in the case of the GL framework, our protocol can be efficiently instantiated using either the DDH, Quadratic
Residuosity or N-Residuosity Assumptions.

For typical security parameters our solution saves as much as 12
Kbytes of bandwidth if one-time signatures are implemented in GL with
fast symmetric primitives. If we use number-theoretic signatures in the
GL framework, our solution saves several large exponentiations (almost a
third of the exponentiations computed in the GL protocol). The end result
is that we bring provable security in the realm of password-authenticated
key exchange one step closer to practical.

1 Introduction

The central problem of cryptography is to enable reliable and secure communication among parties in the presence of an adversary. In order to do this,
parties must share a common secret key to secure communication using known
techniques (e.g., applying encryption and message authentication codes to all
messages).

A protocol that allows two parties to establish such a secret key is called a
key exchange protocol. The key exchange problem was initially studied by Diffie

R. Canetti (Ed.): TCC 2008, LNCS 4948, pp. 589–606, 2008.
© International Association for Cryptologic Research 2008

and Hellman [15] who considered a passive adversary that can eavesdrop on the honest parties' communication, but cannot actively modify it. In other words, parties are assumed to be connected by reliable, albeit non-private, channels. Many efficient and secure protocols are known for this scenario. The more realistic scenario however is that of a far more powerful adversary who can modify and delete messages sent between the parties, as well as insert messages of its own choice. This is the scenario we consider in this paper.

Once we allow such a powerful adversary it becomes clear that in order to securely exchange a key, any two parties (call them Alice and Bob) must hold some secret information. Otherwise, there is nothing preventing an adversary from pretending to be Bob while communicating with Alice (and vice versa). The most common type of secret information considered are (i) parties already share a high entropy secret key; (ii) each party holds a secret key matching an authenticated public key (i.e. a public key securely associated with his identity) and (iii) the case we consider in this paper: parties share only a low entropy *password* that can be remembered and typed in by human users.

Cryptography has long been concerned with cases (i) and (ii), while the scenario of low entropy passwords (arguably the most commonly used case) has only recently received attention. This paper proposes a new and improved family of protocols (a framework) for password-based key exchange in the face of a powerful, active adversary.

Password-based authenticated key-exchange. Our model consists of a group of parties, with each pair of them sharing a password chosen uniformly at random from some small dictionary (the assumption of uniformity is made for simplicity only). The parties communicate over a network in the presence of an active adversary who has full control over the communication lines. In other words all communication between parties is basically carried out through the adversary. Nevertheless, the goal of the parties is to generate session keys in order to secretly and reliably communicate with each other.

An immediate observation is that the small size of the dictionary implies a non-negligible probability that the attacker will succeed in impersonating one of the parties, since the adversary can always guess Bob's password and pretend to be him while communicating with Alice. This type of attack is called an on-line guessing attack and is inherent whenever security depends on low entropy passwords. The severity of on-line guessing attacks can be limited with other mechanisms (such as locking an account after a number of failed attempts). A more dangerous attack is the *off-line guessing attack*, in which the adversary obtains a transcript of an execution of the key exchange protocol and is then able to check guesses for Bob's password against this transcript *off-line*. The aim of password-based authenticated key exchange is to limit the adversary only to on-line guessing attacks, and rule out possible off-line ones.

Prior related work. Bellovin and Merritt [3] proposed the first protocol for password-based session-key generation. Although the specific protocol of [3] can be attacked (see [28]), small modifications to the protocol prevent these attacks [28], and even allow one to prove it secure under the ideal cipher and random oracle models [1]. Bellovin and Merritt's work was very influential and was followed by many protocols (e.g. [4,29,23,27,28,30]) which, however, have not been proven secure and their conjectured security is based on heuristic arguments.

A first rigorous treatment of the problem was provided by Halevi and Krawczyk [22]. They consider an *asymmetric* model in which some parties (called servers) hold certified public keys available to the all parties, including the clients who instead hold only passwords. In this model (which requires a public-key infrastructure) Halevi and Krawczyk provide a secure password-based key exchange. The first (and only currently known) protocol to achieve security without *any* additional setup is that of Goldreich and Lindell [20]. Their protocol is based on general assumptions (i.e., the existence of trapdoor permutations) and constitutes a proof that password-based authenticated key exchange can actually be obtained. Unfortunately, the protocol of [20] is not very efficient and thus cannot be used in practice.

Katz, Ostrovsky and Yung (KOY) [24] present an efficient and practical protocol for the problem of password-authenticated key-exchange in the common reference string model. In this model, the extra setup assumption is that all parties have access to some public parameters, chosen by some trusted third party. This assumption is clearly weaker than assuming a public-key infrastructure, and there are settings in which it can be implemented safely and efficiently (such as a corporation wanting to provide secure password login for its employees, and thus can be trusted to choose and distribute the common reference string). The KOY protocol is based on the security against chosen-ciphertext attack [16] of the original Cramer-Shoup encryption scheme [11]. This in turn can be reduced to the Decisional Diffie-Hellman (DDH) assumption. The complexity of the KOY protocol is only 5–8 times the complexity of a Diffie-Hellman unauthenticated key-exchange protocol.

The KOY protocol was generalized by Gennaro and Lindell [19], using generic building blocks instead of specific number-theoretic assumptions. More specifically, they use the notion of *projective hash functions* and the CCA-secure encryption schemes defined in [12]. The resulting protocol GL has a much more intuitive proof of security and can be proven secure under a variety of computational assumptions (such as Quadratic Residuosity and N-Residuosity).

We note that there are password-authenticated key-exchange protocols which are more efficient than KOY and GL , but whose proof holds in an idealized model of computation such as the ideal cipher and random oracle models [1,6]. The common interpretation of such results is that security is *likely* to hold even if the random oracle is replaced by a ("reasonable") concrete function known explicitly to all parties (e.g., SHA-1). However, it has been shown that it is impossible to replace the random oracle in a generic manner with any concrete function [7]. Thus, the proofs of security of these protocols are actually heuristic in nature.

1.1 Our Contributions

We improve on the both the GL and KOY protocols, in particular by reducing the communication bandwidth required by the protocol.

Both the KOY and the GL protocol use (one-time) signatures as a non-malleability tool in order to prevent a man-in-the-middle attack against the protocol. This negatively affects the efficiency of the resulting protocol. Indeed in order to preserve provable security without use of the random oracle an implementation of the KOY or GL protocol is presented with two choices.

One-time signature schemes (i.e. signature schemes which are secure if the key is used to sign only one message) can be implemented from fast symmetric key primitives (such as one-way functions). However the length of the resulting keys and signatures is problematic and causes a substantial increase in the required bandwidth.

One could use "regular" signature schemes (i.e. secure for many messages) but then, if we require provable security in the standard model, the amount of computation would substantially increase. Moreover if we want to use the most efficient provably secure signature schemes in the literature (e.g. [18,14,25]) we would introduce new computational assumptions such as the Strong RSA assumption, on top of the ones required by the GL protocol.

Our improvement avoids using digital signatures altogether, replacing them with faster and shorter message authentication codes. The crucial idea is to leverage as much as possible the non-malleability of the encryption scheme used in the protocol, by including various values into the encryption as *labels*. For typical security parameters our improvement saves as much as 12 Kbytes of bandwidth in a protocol execution.

As in the case of the GL framework, our protocol can be efficiently instantiated using either the DDH, Quadratic Residuosity or N-Residuosity Assumption.

1.2   Our Construction in a Nutshell

Let us describe our construction informally. We start by first describing the tools that we are going to use and then describing the protocol.

Chosen-Ciphertext Secure Public-Key Encryption [16]: We use an encryption scheme $\mathcal{E}$ which is secure against chosen-ciphertext attack. The common reference string for our password protocol is simply the public key PK for such an encryption scheme. We stress that the corresponding secret key does not have to be known by any party[1].

Smooth projective hashing [12]: Let X be a set and $L \subset X$ a language. Loosely speaking, a hash function H_k that maps X to some set is projective if there exists a projection key that defines the action of H_k over the subset L of the domain X. That is, there exists a projection function $\alpha(\cdot)$ that maps

---

[1] In the GL protocol the requirement is actually weaker, as all is needed is a non-interactive non-malleable (with respect to many commitments) commitment which in the common reference string can be built out of CCA-Secure Encryption. For simplicity we describe our protocol using encryption, but in the final version we show that we can also use commitments. In practice this does not make much difference since CCA encryption is the most efficient known implementation of for this type of non-malleable commitments.

keys k into their projections $s = \alpha(k)$. The projection key s is such that for every $x \in L$ it holds that the value of $H_k(x)$ is uniquely determined by s and x. In contrast, nothing is guaranteed for $x \notin L$, and it may not be possible to compute $H_k(x)$ from s and x. A smooth projective hash function has the additional property that for $x \notin L$, the projection key s actually says *nothing* about the value of $H_k(x)$. More specifically, given x and $s = \alpha(k)$, the value $H_k(x)$ is uniformly distributed (or statistically close) to a random element in the range of H_k.

What makes smooth projective hashing a powerful tool (in both our application and the original one in [12]) is that if L is an NP-language, then for every $x \in L$ it is possible to efficiently compute $H_k(x)$ using the projection key $s = \alpha(k)$ and a witness of the fact that $x \in L$. Alternatively, given k itself, it is possible to efficiently compute $H_k(x)$ even without knowing a witness. Gennaro and Lindell [19] also prove another important property of smooth projective hash functions that holds when L is a hard-on-the-average NP-language. For a random $x \in_R L$, given x and $s = \alpha(k)$ the value $H_k(x)$ is *computationally indistinguishable* from a random value in the range of $H_k(x)$. Thus, even if $x \in L$, the value $H_k(x)$ is pseudorandom, unless a witness is known.

The basic idea behind the KOY and GL protocols is to have the parties exchange non-malleable encryptions of the joint password. The session key is then computed as the result of applying smooth projective hash functions to these encryptions (in this case the hard-on-the-average NP language consists of correct ciphertext/message pairs). Figure 1 shows the basic layout of the protocol.

The basic problem with the protocol described in Figure 1 is that the projective hash function themselves can be malleable, and an adversary could manage to get information about the session key by playing man-in-the-middle. In order to avoid this attack, the GL and KOY protocols add a signature step. A verification key is chosen by party A in the first message and *bound* together with the first encryption, by including it as a *label*[2]. Then A signs the whole transcript in the third message. Party B accepts only if the signature is correct. Since the verification key cannot be changed (being protected by the non-malleability of the encryption in the first step), the adversary cannot modify the projection keys, unless it is able to produce a forgery.

In our protocol we expand the use of encryption labels. We protect the first projection s, by including it as a label in the second ciphertext[3] c'. Now the adversary is left with the possibility of manipulating the second projection key s'. But at this point the master key computed by party A is already pseudorandom

---

[2] A *label* is a public string that accompanies a ciphertext and is an integral part of it. It must be submitted together with the ciphertext in order to obtain a decryption and the adversary should not be able to modify it. See Section 2.1 for details.

[3] Interestingly this is already done in the KOY protocol, but it is not used in the proof in any significant way. Indeed the GL proof shows that the use of digital signatures make this step unnecessary. We reinstate it exactly because we want to avoid using signatures.

Insecure Password-Authenticated Key Exchange

- **Common reference string:** The public key PK for a chosen-ciphertext secure encryption scheme $\mathcal{E}$. A description of a smooth projective hashing family H_k over the set X of ciphertext/password pairs (c, w).

- **Common input:** a shared (low-entropy) password w.

- **The protocol:**
  1. Party A computes an encryption $c = \mathcal{E}_{PK}(w)$ and sends it to party B.

  2. Party B chooses a key k for the smooth projective hash function, and computes its projection $s = \alpha(k)$. Also B computes the projective hash over (c, w), i.e. $sk_B = H_k(c, w)$.
     Finally B computes another encryption of the password i.e., $c' = \mathcal{E}_{PK}(w)$.
     B sends s, c' to party A.

  3. Party A chooses another key k' for the smooth projective hash function, and computes its projection $s' = \alpha(k')$. Also A computes the projective hash over (c', w), i.e. $sk_A = H_{k'}(c', w)$.
     A sends s', t to party B.

- **Session Key Definition:**
  1. Party B computes sk_A using the projection s' and its knowledge of a witness for the fact that c' is an encryption to the password w (it knows a witness because it generated c') and outputs the session key $sk = sk_A \oplus sk_B$.

  2. Party A also computes sk_B using the projection s and its knowledge of a witness for the fact that c is an encryption to the string w (it knows a witness because it generated c) and outputs $sk = sk_A \oplus sk_B$.

Fig. 1. Common skeleton of KOY, GL and our protocol

for the adversary and thus it can be used as a key to MAC the projection key s' in order to prevent $\mathcal{A}$ from changing it.

A technical issue arises here, as party B has to use the same key that party A uses to compute the MAC, but party B has to yet finish the protocol and compute such key. Moreover the adversary can make B compute a different key from A, by modifying the projection s'. This issue can be solved by using sk_B as a MAC key since B already knows it. However the explicit use of only one component of the session key would allow an off-line attack from $\mathcal{A}$ (see Section 4).

The final solution is to MAC the transcript with sk_B and then use sk_A to "mask" the value of the MAC from the adversary. In the proof if B accepts after $\mathcal{A}$ modified s' he will be able to retrieve a forgery on the MAC keyed with sk_B. The protocol in full details is shown in Figure 2.

1.3 Efficiency Gains

Using symmetric primitives. In terms of computation, the most efficient implementation of the KOY or GL protocols uses one-time signatures based on symmetric primitives, such as one-way functions. One example of such a signature is the Lamport signature [26]: to sign a single bit b the public key consists of two values y_0, y_1 and the secret key is x_0, x_1 where $y_i = F(x_i)$ for a one-way function F. To sign bit b the signer reveals x_b.

Assuming a security parameter of 128 (e.g. a one-way function applied to 128 bits input, and messages hashed to 256 bits using a collision resistant function), we have that transmitting the key and the signature requires about 12 KBytes. Other solutions exist that create shorter signatures at the expense of an increase in computation time (see a survey of possible one-time signatures in e.g. [10]). In contrast our solution requires only 256 bits for the MAC.

Number-Theoretic Signatures. Of course one could implement the signature step in the KOY or GL protocol using provably secure signature schemes such as Gennaro-Halevi-Rabin [18] or Cramer-Shoup [14] which are based on the Strong RSA Assumption: they not only introduce another computational assumption for the security of the scheme, but require several modular exponentiations and about 4 Kbit of bandwidth to transmit keys and signatures. A shorter alternative would be the Boneh-Boyen [5] which requires only 160 bit for the signature, but it would still require 2 Kbits to send the verification key. Moreover signature verification in the Boneh-Boyen scheme is particularly expensive since it requires the computation of a bilinear map.

The above signatures are secure against many messages. There are more efficient number-theoretic one-time signatures such as the one obtained by a chain of length two in the GMR scheme [21], or the one recently proposed in [10] based on chameleon hashing with two trapdoors. Still because of the computation of modular exponentiations and the transmission of verification key and signature, these options are much more expensive then sending a simple MAC. It is not hard to see that for each one of these options the reduction in the number of exponentiations is at least a third.

1.4 Organization

We first recall the cryptographic tools that we need in Section 2: chosen-ciphertext secure public-key encryption, and message authentication codes. In Section 3 we review the notions of smooth projective hash functions (mostly lifted *verbatim*, with permission, from [19]). The protocol is then presented in Section 4 with an intuitive informal proof. Some concluding remarks are presented in Section 5.

For lack of space we refer the reader to [19] for the formal definition of password-authenticated key exchange. Also the formal proof of our protocol can be found in the expanded version of this paper [17].

2   Cryptographic Tools

We denote by n the security parameter.

If S is a set, with $|S|$ we denote its cardinality. $|m|$ denotes the bit length of m, if m is a string or a number.

If $A(\cdot, \cdot, \cdots)$ is a probabilistic algorithm, then $x \in_R A(x_1, x_2, \cdots)$ denotes the experiment of running A on input $x_1, x_2, \cdots$ with x being the outcome. If S is a set, $x \in_R S$ denotes the experiment of choosing $x \in S$ uniformly at random. If X is a probability distribution over S then $x \in_R X$ denotes the experiment of choosing $x \in S$ according to the distribution X.

Finally, we denote statistical closeness of probability ensembles by $\overset{s}{\equiv}$, and computational indistinguishability (with respect to non-uniform polynomial-time machines[4]) by $\overset{c}{\equiv}$.

We say that a real-valued function $\epsilon(\cdot)$ defined over the integers is *negligible* if for every constant $c \geq 0$ there exists an integer n_c such that for all $n > n_c$ $\epsilon(n) < n^{-c}$.

2.1   Chosen-Ciphertext Secure Public-Key Encryption

A public key encryption scheme is a tuple of three algorithms $\mathsf{PKE} = (\mathcal{K}, \mathcal{E}, \mathcal{D})$. The key generation algorithm $\mathcal{K}$ generates a pair $(PK, SK) \in_R \mathcal{K}(1^n)$, where PK is a public key and SK is a secret key.

We use *labeled* encryption, which means that the encryption algorithm $\mathcal{E}$ takes a public key PK a plaintext m, and a label ℓ and returns a ciphertext $c \in_R \mathcal{E}_{PK}(m, \ell)$. The decryption algorithm $\mathcal{D}$ takes a secret key SK, a ciphertext c and a label ℓ, and returns $\mathcal{D}_{SK}(c, \ell)$ which is either a message m or *reject*. If $c \in_R \mathcal{E}_{PK}(m, \ell)$ then $m = \mathcal{D}_{SK}(c, \ell)$.

The adaptive chosen ciphertext attack (IND-CCA) game is defined as follows. A key pair is generated by the key generation algorithm: $(PK, SK) \in_R \mathcal{K}(1^n)$. Then a PPT adversary $\mathcal{A}$, on input the public key PK, queries a pair of equal length messages m_0 and m_1 and a label ℓ^* to an encryption oracle. The encryption oracle chooses $b \in_R \{0, 1\}$ and computes a challenge ciphertext $c^* \in_R \mathcal{E}_{PK}(m_b, \ell^*)$, which is given to $\mathcal{A}$. In the course of the game the adversary A is given access to a decryption oracle, $\mathcal{D}_{SK}(\cdot, \cdot)$ which A can query on any ciphertext/label pair except the challenge ciphertext/label pair c^*, ℓ^*. The game ends with the adversary outputting a bit $\tilde{b}$.

We say that the encryption scheme is secure against (adaptive) chosen-ciphertext attack if for any adversary $\mathcal{A}$, the probability that $b = \tilde{b}$ is negligible (in the security parameter n).

Notice that the adversary is allowed to query the decryption oracle on *any* ciphertext/label pair which is not the target pair. In particular this definition guarantees that the adversary will not get any information from querying a ciphertext with a label different from the one used when the ciphertext was created.

---

[4] All of our results also hold with respect to uniform adversaries.

Notice that in our notation the first argument of the encryption algorithm is always the message, the second argument is the label, and the random coins are implicit.

2.2 Message Authentication Codes

A message authentication code MAC is a function

$$MAC : \{0,1\}^n \times \{0,1\}^* \longrightarrow \{0,1\}^n.$$

The first input is the key $k \in \{0,1\}^n$, and the second input is the message $m \in \{0,1\}^*$. The output is called a "tag" $t = MAC_k(m)$.

The chosen message attack (CMA) game is defined as follows. A key is selected uniformly at random $k \in_R \{0,1\}^n$. The adversary $\mathcal{A}$ is given $t^* = MAC_k(m^*)$ for many adaptively adversarially chosen m^*, after which the adversary outputs a pair (m,t). We say that (m,t) is a *forgery* if $m \neq m^*$ for all the queried m^* and $t = MAC_k(m)$.

We say that a MAC is secure if for every adversary $\mathcal{A}$ the probability of computing a forgery is negligible. We say that a MAC is *1-time secure* if the adversary in the above game is restricted to querying a single message. We note that 1-time secure MACs can be constructed unconditionally.

3 Smooth Projective Hash Functions

Following Gennaro and Lindell [19] we use a modified version of the notion of *smooth projective hashing* introduced by Cramer and Shoup [12]. We recall the definition from [19] which is needed here and refer the reader to [19] for a description of the differences between this definition and the original one from [12].

Subset membership problems. Intuitively, a hard subset membership problem is a problem for which "hard instances" can be efficiently sampled. More formally, a subset membership problem $\mathcal{I}$ specifies a collection $\{I_n\}_{n\in\mathbb{N}}$ such that for every n, I_n is a probability distribution over *problem instance descriptions* Λ. A problem instance description defines a set and a hard language for that set. Formally, each instance description Λ specifies the following:

1. Finite, non-empty sets $X_n, L_n \subseteq \{0,1\}^{\mathrm{poly}(n)}$ such that $L_n \subset X_n$, and distributions $D(L_n)$ over L_n and $D(X_n \setminus L_n)$ over $X_n \setminus L_n$.
2. A witness set $W_n \subseteq \{0,1\}^{\mathrm{poly}(n)}$ and an NP-relation $R_n \subseteq X_n \times W_n$. R_n and W_n must have the property that $x \in L_n$ if and only if there exists $w \in W_n$ such that $(x,w) \in R_n$.

We are interested in subset membership problems $\mathcal{I}$ which are efficiently samplable. That is, the following algorithms must exist:

1. *Problem instance samplability:* a probabilistic polynomial-time algorithm that upon input 1^n, samples an instance $\Lambda = (X_n, D(X_n \setminus L_n), L_n, D(L_n), W_n, R_n)$ from I_n.
2. *Instance member samplability:* a probabilistic polynomial-time algorithm that upon input 1^n and an instance $(X_n, D(X_n \setminus L_n), L_n, D(L_n), W_n, R_n)$, samples $x \in L_n$ according to distribution $D(L_n)$, together with a witness w for which $(x, w) \in R_n$.
3. *Instance non-member samplability:* a probabilistic polynomial-time algorithm that upon input 1^n and an instance $(X_n, D(X_n \setminus L_n), L_n, D(L_n), W_n, R_n)$, samples $x \in X_n \setminus L_n$ according to distribution $D(X_n \setminus L_n)$.

We are now ready to define hard subset membership problems:

Definition 1. (hard subset membership problems): *Let $V(L_n)$ be the following random variable: Choose a problem instance Λ according to I_n, a value $x \in L_n$ according to $D(L_n)$ (as specified in Λ), and then output (Λ, x). Similarly, define $V(X_n \setminus L_n)$ as follows: Choose a problem instance Λ according to I_n, a value $x \in X_n \setminus L_n$ according to $D(X_n \setminus L_n)$ (as specified in Λ) and then output (Λ, x). Then, we say that a subset membership problem $\mathcal{I}$ is* hard *if*

$$\left\{ V(L_n) \right\}_{n \in \mathbb{N}} \stackrel{c}{\equiv} \left\{ V(X_n \setminus L_n) \right\}_{n \in \mathbb{N}}.$$

In other words, $\mathcal{I}$ is hard if random members of L_n cannot be distinguished from random non-members. In order to simplify notation, from here on we drop the subscript of n from all sets. However, all mention of sets X and L etc., should be understood as having been sampled according to the security parameter n.

Smooth projective hash functions. Loosely speaking a smooth projective hash function is a function with two keys. The first key maps the entire set X to some set G. The second key (called the projection key) is such that it can be used to correctly compute the mapping of L to G. However, it gives no information about the mapping of $X \setminus L$ to G. In fact, given the projection key, the distribution over the mapping of $X \setminus L$ to G is statistically close to uniform (or "smooth"). We now present the formal definition.

Let X and G be finite, non-empty sets and let $\mathcal{H} = \{H_k\}_{k \in K}$ be a collection of hash functions from X to G. We call K the key space of the family. Now, let L be a non-empty, proper subset of X (i.e., L is a language). Then, we define a *key projection* function $\alpha : K \times X \to S$, where S is the space of key projections. Informally, the above system defines a projective hash system if for $x \in L$, the projection key $s_x = \alpha(k, x)$ uniquely determines $H_k(x)$. (Ignoring issues of efficiency, this means that $H_k(x)$ can be computed given only s_x and $x \in L$.) We stress that the projection key $s_x = \alpha(k, x)$ is only guaranteed to determine $H_k(x)$ for $x \in L$, and nothing is guaranteed for $x' \neq x$. Formally,

Definition 2. (projective hash functions): *The family $(\mathcal{H}, K, X, L, G, S, \alpha)$ is a* projective hash family *if for all $k \in K$ and $x \in L$, it holds that the value of $H_k(x)$ is uniquely determined by $\alpha(k, x)$ and x.*

Of course, projective hash functions can always be defined by taking $\alpha(\cdot, \cdot)$ to be the identity function. However, we will be interested in *smooth* projective hash functions which have the property that for every $x \notin L$, the projection key $s_x = \alpha(k, x)$ reveals (almost) nothing about $H_k(x)$. More exactly, for every $x \notin L$, the distribution of $H_k(x)$ given $\alpha(k, x)$ should be statistically close to uniform. Formally,

Definition 3. (smooth projective hash functions [12]): *Let* $(\mathcal{H}, K, X, L, G, S, \alpha)$ *be a projective hash family. Then, for every* $x \in X \backslash L$ *define the random variable* $V(x, \alpha(k), H_k(x))$ *by choosing* $k \in_R K$ *and output* $(x, \alpha(k, x), H_k(x))$. *Similarly, define* $V(x, \alpha(k, x), g)$ *as follows: choose* $k \in_R K$, $g \in_R G$ *and output* $(x, \alpha(k, x), g)$. *Then, the projective hash family* $(\mathcal{H}, K, X, L, G, S, \alpha)$ *is* smooth *if for every* $x \in X \backslash L$:

$$\left\{ V(x, \alpha(k), H_k(x)) \right\}_{n \in \mathsf{N}} \overset{\mathsf{s}}{\equiv} \left\{ V(x, \alpha(k), g) \right\}_{n \in \mathsf{N}}.$$

Efficient smooth projective hash functions. We say that a smooth projective hash family is efficient if the following algorithms exist:

1. *Key sampling:* a probabilistic polynomial-time algorithm that upon input 1^n samples $k \in K$ uniformly at random.
2. *Projection computation:* a deterministic polynomial-time algorithm that upon input 1^n, $k \in K$ and $x \in X$ outputs $s = \alpha(k, x)$.
3. *Efficient hashing from key:* a deterministic polynomial-time algorithm that upon input 1^n, $k \in K$ and $x \in X$, outputs $H_k(x)$.
4. *Efficient hashing from projection key and witness:* a deterministic polynomial-time algorithm that upon input 1^n, $x \in L$ with a witness w such that $(x, w) \in R$, and $\alpha(k, x)$ (for some $k \in K$), computes $H_k(x)$.

We note an interesting and important property of such hash functions. For $x \in L$, it is possible to compute $H_k(x)$ in two ways: either by knowing the key k (as in item 3 above) or by knowing the projection s_x of the key, and a witness for x (as in item 4 above). This property plays a central role in our password-based protocol.

Another interesting property formalized by Gennaro and Lindell in [19] is that these are the only ways to compute $H_k(x)$. Specifically, for $x \in_R D(L)$ (where an appropriate witness w is not known), the value $H_k(x)$ is *computationally* indistinguishable from random, given the projection s_x.

Since we use smooth projective hashing in our password protocol, it is necessary to prove the above statement even when the adversary sees many tuples $(x, s_x, H_k(x))$ with $x \in_R D(L)$. Let M be a (non-uniform) polynomial-time oracle machine. Define the following two experiments.

Expt-Hash(M): An instance $\Lambda = (X, D(X \backslash L), L, D(L), W, R)$ of a hard subset membership problem is chosen from I_n. Then, the machine M is given access to two oracles: Ω_L and Hash($\cdot$). The Ω_L oracle receives an empty input and returns $x \in L$ chosen according to the distribution $D(L)$. The Hash oracle receives an input x. It first checks that x was previously output by the Ω_L oracle. If no, then it returns nothing. Otherwise, it chooses a key $k \in_R K$ and returns the pair $(\alpha(k, x), H_k(x))$. We stress that the Hash oracle *only* answers for inputs x that were generated by Ω_L. The output of the experiment is whatever machine M outputs.

Expt-Unif(M): This experiment is defined exactly as above except that the Hash oracle is replaced by the following Unif oracle. On input x, Unif first checks that x was previously output by the Ω_L oracle. If no, it returns nothing. Otherwise, it chooses a key $k \in_R K$ and a random element $g \in_R G$, and returns the pair $(\alpha(k, x), g)$. As above, the output of the experiment is whatever M outputs.

In [19] it is proven that no efficient M can distinguish between the experiments. In other words, when $x \in_R D(L)$, the value $H_k(x)$ is pseudorandom in G, even given $\alpha(k, x)$. This lemma is used a number of times in the proof of our password protocol.

Lemma 1. *Assume that $\mathcal{I}$ is a hard subset membership problem. Then, for every (non-uniform) polynomial-time oracle machine M it holds that,*

$$\left| \Pr[\text{Expt-Hash}(M) = 1] - \Pr[\text{Expt-Unif}(M) = 1] \right| < \text{negl}(n).$$

Hard partitioned subset membership problems. We now consider a variant of hard subset membership problems, where the set X can be *partitioned* into disjoint subsets of hard problems. That is, assume that the set X contains pairs of the form (i, x), where $i \in \{1, \ldots, \ell\}$ is an index. We denote by $X(i)$ the subset of pairs in X of the form (i, x). Furthermore, we denote by $L(i)$ the subset of pairs in the language L of the form (i, x). (We also associate sampling distributions $D(L(i))$ and $D(X(i) \backslash L(i))$ to each partition.) Then, such a problem constitutes a hard partitioned subset membership problem if for *every* i, it is hard to distinguish $x \in_R D(L(i))$ from $x \in_R D(X(i) \backslash L(i))$. (In the notation of Definition 1, we require that for every i, the ensembles $\{V(L(i))\}$ and $\{V(X(i) \backslash L(i))\}$ are computationally indistinguishable.) We stress that the definition of smooth projective hashing is unchanged when considered in the context of hard partitioned subset problems. That is, the smoothness is required to hold with respect to the entire sets X and L, and not with respect to individual partitions.

Lemma 1 also holds for hard partitioned subset membership problems (see [19]). Specifically, the definitions of the oracles in the experiments are modified as follows. The Ω_L oracle is modified so that instead of receiving the empty input, it is queried with an index i, and returns $x \in_R D(L(i))$. Likewise, $\Omega_{X \backslash L}$ receives an index i and returns an element $x \in_R D(X(i) \backslash L(i))$. Notice that in this scenario, the distinguishing machine M is given some control over the choice of x. Specifically, M can choose the index i that determines from which partition an element x is sampled.

Corollary 1. *Assume that* $\mathcal{I}$ *is a family of hard* partitioned *subset membership problem. Then, for every (non-uniform) polynomial-time oracle machine* M

$$|\Pr[\mathsf{Expt\text{-}Hash}(M) = 1] - \Pr[\mathsf{Expt\text{-}Unif}(M) = 1]| < \mathsf{negl}(n).$$

4   The Protocol

Our protocol uses a chosen-ciphertext secure public-key labeled encryption scheme $\mathcal{E}$. The common reference string for the protocol is a public key PK for $\mathcal{E}$.

We then use a family of smooth projective functions $\mathcal{H} = \{H_k\}$ such that for every k in the key space K, $H_k : C_{PK} \times M \to \{0,1\}^{2n}$, where M is the message space, C_{PK} is an efficiently recognizable superset of the ciphertext space. Notice that we are assuming that the projective hash function outputs $2n$-bit strings[5]. If sk is a $2n$-bit string we denote with $sk^{(1)}$ and $sk^{(2)}$ the first and second half of it respectively.

Finally, we assume that there is a mechanism that enables the parties to differentiate between different concurrent executions and to identify who they are interacting with. This can easily be implemented by having P_i choose a sufficiently long random string r and send the pair (i, r) to P_j along with its first message. P_i and P_j will then include r in any future messages of the protocol. We stress that the security of the protocol does not rest on the fact that these values are not modified by the adversary. Rather, this just ensures correct communication for protocols that are not under attack. The protocol appears in Figure 2.

Intuitive Security Proof. First notice that both A and B can compute the session key as instructed. Specifically, A can compute $H_k(c, w, A \circ B)$ because it has the projection key s and the witness (coins) for c. Furthermore, it can compute $H_{k'}(c', w, c \circ s)$ because it has the key k' (and therefore does not need the witness for c'). Likewise, B can also correctly compute both the hash values (and thus the session key). Second, when both parties A and B see the same messages (c, s, c', s', t) the session keys that they compute are the same. This is because the same hash value is obtained when using the hash keys (k and k') and when using the projection keys (s and s'). This implies that the correctness property holds for the protocol.

We now proceed to motivate why the adversary cannot distinguish a session key from a random key with probability greater than $Q_{\mathsf{send}}/|\mathcal{D}|$, where Q_{send} equals the number of Send oracle calls made by the adversary to different protocol instances and $\mathcal{D}$ is the password dictionary. In order to see this, notice that if A, for example, receives c' that is not an encryption to w with label $c \circ s$ under PK, then A's component of the session key sk_A will be statistically close to uniform.

---

[5] We note that the constructions in [19] output values in a large algebraic group G. It is a standard application of randomness extraction (e.g. via universal hashing) to map such elements into $2n$-bit strings, assuming the group G is large enough.

RG-PaKE

- **Common reference string:** The public key PK for a chosen-ciphertext secure encryption scheme $\mathcal{E}$. A description of a smooth projective hashing family H_k over the set X of ciphertext/password pairs (c, w). The NP language L is composed of the tuples (c, m, ℓ) where $c = \mathcal{E}_{PK}(m, \ell)$ i.e. c is an encryption of m with label ℓ under PK. A message authentication code MAC.

- **Common input:** a shared (low-entropy) password w.

- **The protocol:**
  1. Party A computes an encryption $c = \mathcal{E}_{PK}(w, A \circ B)$ and sends it to party B.

  2. Party B chooses a key k for the smooth projective hash function (for the language L described above), and computes its projection $s = \alpha(k, c)$. Also B computes the projective hash over $(c, w, A \circ B)$, i.e. $sk_B = H_k(c, w, A \circ B)$.
     Finally B computes another encryption of the password with label $c \circ s$ i.e., $c' = \mathcal{E}_{PK}(w, c \circ s)$.
     B sends s, c' to party A.

  3. Party A chooses another key k' for the smooth projective hash function (for the language L described above), and computes its projection $s' = \alpha(k', c')$. Also A computes the projective hash over $(c', w, c \circ s)$, i.e. $sk_A = H_{k'}(c', w, c \circ s)$.
     A also computes sk_B using the projection s and its knowledge of a witness for the fact that c is an encryption to the string w with label $A \circ B$ (it knows a witness because it generated c).
     Set $t = MAC_{sk_B^{(1)}}(c, s, c', s') \oplus sk_A^{(1)}$.
     A sends s' to party B.

- **Session Key Definition:**
  1. Party B computes sk_A using the projection s' and its knowledge of a witness for the fact that c' is an encryption to the password w with label $c \circ s$ (it knows a witness because it generated c')
     It tests if $t = MAC_{sk_B^{(1)}}(c, s, c', s') \oplus sk_A^{(1)}$. If the test fails it outputs an error message, otherwise it outputs $sk = sk_A^{(2)} \oplus sk_B^{(2)}$.

  2. Party A outputs $sk = sk_A^{(2)} \oplus sk_B^{(2)}$.

Session-Identifier Definition: Both parties take the series of messages (c, s, c', s') to be their session identifiers.

Fig. 2. Improved Password-Based Session-Key Exchange

This is because A computes $H_{k'}(c', w, c \circ s)$ for $c' \notin \mathcal{E}_{PK}(w, c \circ s)$ i.e. on an input outside the language. Therefore, by the definition of smooth projective hashing, $\{c', w, \alpha(k, c'), H_k(c', w, c \circ s)\}$ is statistically close to $\{c', w, \alpha(k, c'), r\}$, where r is a random $2n$-bit string.

The same argument holds if B receives c that is not a encryption of w with label $A \circ B$. It therefore follows that if the adversary is to distinguish the session key from a random element, it must hand the parties encryptions of the valid messages (and in particular containing the correct passwords). One way for the adversary to do this is to copy (valid) commitments that are sent by the honest parties in the protocol executions. However, in this case, the adversary does not know the random coins used in generating the commitment, and once again the result of the projective hash function is a pseudorandom $2n$-bit string (see Lemma 1). This means that the only option left to the adversary is to come up with valid commitments that were not previously sent by honest parties. However, by the non-malleability of the encryption scheme, the adversary cannot succeed in doing this with probability non-negligibly greater than just a priori guessing the password. Thus, its success probability is limited to $Q_{\text{send}}/|\mathcal{D}| + \mathsf{negl}(n)$.

This intuitive explanation of the security of the protocol is not complete. Indeed it does not address the use of message authentication codes in the protocol. The MAC is needed to prevent further malleability attacks. Indeed while the ciphertexts containing the password are not malleable (because of the strong security of the encryption scheme used), the computation of the projective hash function could be malleable, and by recycling messages from previous executions the adversary could gain some knowledge about a session key. For example, it is possible that for some smooth projective hash family it holds that for every k, $H_{2k}(x) = 2H_k(x)$, and that by seeing $s = \alpha(k, c)$, the value $\hat{s} = \alpha(2k, c)$ is efficiently computable.

If this were the case, in the basic protocol described in Figure 1 an adversary could cause two instances to accept with *different* session identifiers and *related* session-keys. For example, given the message $s = \alpha(k, c), c'$ by party B in Round 2, the adversary could forward to A the message $\hat{s} = \alpha(2k, c), c'$. By requesting a Reveal for one of the instances, it could then distinguish the other instance's session-key from random, in contradiction to the security requirements.

Notice that the projection s is protected by malleability attacks because is incorporated as a label in the ciphertext c'. The surprising thing is that at this point the value sk_B is already pseudo-random to the eyes of the adversary, and known to both parties A and B. The most intuitive thing would be to use it to MAC the other projection s'.

But if A were to send $t = MAC_{sk_B}(c, s, c', s')$ the adversary could perform the following off-line attack. The adversary would start a session with A pretending to be B and obtain the commitment c. Next, the adversary $\mathcal{A}$ chooses k and returns $s = \alpha(k, c)$ to A together with an incorrect encryption c'. The response from A is s' and t computed as above with $sk_B = H_k(c, w, A \circ B)$. Now the adversary can traverse the entire dictionary $\mathcal{D}$ and for all possible w's compute $sk_B = H_k(c, w, A \circ B)$ (it can do this because it knows k and so can compute $H_k(c, w)$ without a witness for c). The right password is the one for which sk_B verifies the above MAC t.

The final solution is then to "mask" the MAC, using sk_A which is not known to the adversary, and cannot even be computed off-line by traversing the dictionary (because the adversary does not know the coins used to produce c'). If the adversary modifies s' and makes B accept then it must produce a MAC forgery with key sk_B. We note that a specific MAC key is used to MAC a single message, so it is sufficient to assume 1-time security for the MAC algorithm.

Of course this is just an intuition and the proof presented in [17] works out all the details[6].

Theorem 1. *Assume that $\mathcal{E}$ is a public-key encryption secure against adaptive chosen ciphertext attack, MAC is a 1-time secure message authentication code and $\mathcal{H}$ is a family of smooth projective hash functions. Then, Protocol* RG-PaKE *in Figure 2 is a secure password-based session-key generation protocol.*

5   Extension and Conclusions

The reader is referred to [19] to see examples of efficient chosen-ciphertext secure encryption schemes that admit the type of projective hash functions needed in this protocol. They are based on the encryption schemes proposed by Cramer and Shoup in [11,12], and can be based on the DDH, Quadratic Residuosity and N-Residuosity Assumptions.

For simplicity we have presented the protocol using chosen-ciphertext secure encryption. It is possible using techniques used in [19] to prove the protocol assuming that $\mathcal{E}$ is non-malleable commitment scheme, which admits a projective hash function. The protocol needs to be modified and the proof is more complicated. However in practice it does not make much of a difference, as the only known efficient implementations of such commitment schemes are the ones mentioned above (i.e. based on chosen-ciphertext secure encryption and described by [19]).

Canetti et al. extend the KOY and GL protocol to the Universal Composability framework in [8]. Their protocol also uses one-time signatures to prevent malleability attacks and our modification is applicable to their protocol as well.

Conclusions. We have shown an improvement of the KOY and GL protocols, which does not require one-time signatures. Our protocol works in the common reference string and its proof does not require idealized assumptions such as the random oracle. For typical security parameters our protocol saves about 12 Kbytes of bandwidth, thus bringing provable security in the realm of password-authenticated key exchange one step closer to practical.

References

1. Bellare, M., Pointcheval, D., Rogaway, P.: Authenticated Key Exchange Secure Against Dictionary Attacks. In: Preneel, B. (ed.) EUROCRYPT 2000. LNCS, vol. 1807, pp. 139–155. Springer, Heidelberg (2000)
2. Bellare, M., Rogaway, P.: Entity Authentication and Key Distribution. In: Stinson, D.R. (ed.) CRYPTO 1993. LNCS, vol. 773, pp. 232–249. Springer, Heidelberg (1994)

---

[6] One final technicality: in in order to protect the semantic security of the final session key we use the fist half of sk_A and sk_B to perform the MAC test during the protocol and the second half to compute the actual session key.

3. Bellovin, S.M., Merritt, M.: Encrypted Key Exchange: Password Based Protocols Secure Against Dictionary Attacks. In: Proceedings 1992 IEEE Symposium on Research in Security and Privacy, pp. 72–84. IEEE Computer Society, Los Alamitos (1992)
4. Bellovin, S.M., Merritt, M.: Augmented Encrypted Key Exchange: A Password-Based Protocol Secure Against Dictionary Attacks and Password File Compromise. In: Proceedings of the 1st ACM Conference on Computer and Communication Security, pp. 244–250 (1993)
5. Boneh, D., Boyen, X.: Short Signatures Without Random Oracles. In: Cachin, C., Camenisch, J.L. (eds.) EUROCRYPT 2004. LNCS, vol. 3027, pp. 56–73. Springer, Heidelberg (2004)
6. Boyko, V., MacKenzie, P., Patel, S.: Provably Secure Password-Authenticated Key Exchange Using Diffie-Hellman. In: Preneel, B. (ed.) EUROCRYPT 2000. LNCS, vol. 1807, pp. 156–171. Springer, Heidelberg (2000)
7. Canetti, R., Goldreich, O., Halevi, S.: The Random Oracle Methodology, Revisited. Journal of the ACM 51(4), 557–594 (2004)
8. Canetti, R., Halevi, S., Katz, J., Lindell, Y., MacKenzie, P.: Universally Composable Password-Based Key Exchange. In: Cramer, R.J.F. (ed.) EUROCRYPT 2005. LNCS, vol. 3494, pp. 404–421. Springer, Heidelberg (2005)
9. Canetti, R., Krawczyk, H.: Analysis of Key-Exchange Protocols and Their Use for Building Secure Channels. In: Pfitzmann, B. (ed.) EUROCRYPT 2001. LNCS, vol. 2045, pp. 453–474. Springer, Heidelberg (2001)
10. Catalano, D., Di Raimondo, M., Fiore, D., Gennaro, R.: Some Theoretical and Experimental Results about Off-Line/On-Line Signatures. In : PKC 2008 (to appear, 2008)
11. Cramer, R., Shoup, V.: A Practical Public-Key Cryptosystem Secure Against Adaptive Chosen Ciphertexts Attacks. In: Krawczyk, H. (ed.) CRYPTO 1998. LNCS, vol. 1462, pp. 13–25. Springer, Heidelberg (1998)
12. Cramer, R., Shoup, V.: Universal Hash Proofs and a Paradigm for Adaptive Chosen Ciphertext Secure Public-Key Encryption. In: Knudsen, L.R. (ed.) EUROCRYPT 2002. LNCS, vol. 2332, pp. 45–64. Springer, Heidelberg (2002)
13. Cramer, R., Shoup, V.: Design and Analysis of Practical Public-Key Encryption Schemes Secure Against Adaptive Chosen Ciphertext Attack. SIAM Journal of Computing 33, 167–226 (2003)
14. Cramer, R., Shoup, V.: Signature schemes based on the strong RSA assumption. ACM Trans. Inf. Syst. Secur. 3(3), 161–185 (2000)
15. Diffie, W., Hellman, M.E.: New Directions in Cryptography. IEEE Trans. on Inf. Theory IT-22, 644–654 (1976)
16. Dolev, D., Dwork, C., Naor, M.: Non-Malleable Cryptography. SIAM Journal of Computing 30(2), 391–437 (2000)
17. Gennaro, R.: Faster and Shorter Password-Authenticated Key Exchange, http://eprint.iacr.org/2007/325
18. Gennaro, R., Halevi, S., Rabin, T.: Secure Hash-and-Sign Signatures Without the Random Oracle. In: Stern, J. (ed.) EUROCRYPT 1999. LNCS, vol. 1592, pp. 123–139. Springer, Heidelberg (1999)
19. Gennaro, R., Lindell, Y.: A framework for password-based authenticated key exchange. ACM Transactions on Information and System Security (TISSEC) 9(2), 181–234 (2006)
20. Goldreich, O., Lindell, Y.: Session Key Generation using Human Passwords Only. In: Kilian, J. (ed.) CRYPTO 2001. LNCS, vol. 2139, pp. 408–432. Springer, Heidelberg (2001)

21. Goldwasser, S., Micali, S., Rivest, R.L.: A Digital Signature Scheme Secure Against Adaptive Chosen-Message Attacks. SIAM Journal on Computing 17(2), 281–308 (1988)
22. Halevi, S., Krawczyk, H.: Public-Key Cryptography and Password Protocols. ACM Transactions on Information and System Security (TISSEC) 2(3), 230–268 (1999)
23. Jablon, D.P.: Strong Password-Only Authenticated Key Exchange. SIGCOMM Computer Communication Review 26(5), 5–26 (1996)
24. Katz, J., Ostrovsky, R., Yung, M.: Practical Password-Authenticated Key Exchange Provably Secure under Standard Assumptions. In: Pfitzmann, B. (ed.) EUROCRYPT 2001. LNCS, vol. 2045, pp. 475–494. Springer, Heidelberg (2001)
25. Kurosawa, K., Schmidt-Samoa, K.: New Online/Offline Signature Schemes Without Random Oracles. In: Yung, M., Dodis, Y., Kiayias, A., Malkin, T.G. (eds.) PKC 2006. LNCS, vol. 3958, pp. 330–346. Springer, Heidelberg (2006)
26. Lamport, L.: Constructing digital signatures from a one-way function. Technical Report CSL-98, SRI International (October 1979)
27. Lucks, S.: Open Key Exchange: How to Defeat Dictionary Attacks Without Encrypting Public Keys. In: Christianson, B., Lomas, M. (eds.) Security Protocols 1997. LNCS, vol. 1361, pp. 79–90. Springer, Heidelberg (1998)
28. Patel, S.: Number Theoretic Attacks on Secure Password Schemes. In: Proceedings of the 1997 IEEE Symposium on Security and Privacy, pp. 236–247 (1997)
29. Steiner, M., Tsudik, G., Waidner, M.: Refinement and Extension of Encrypted Key Exchange. ACM SIGOPS Oper. Syst. Rev. 29(3), 22–30 (1995)
30. Wu, T.: The Secure Remote Password Protocol. In: 1998 Internet Society Symposium on Network and Distributed System Security, pp. 97–111 (1998)

Saving Private Randomness
in One-Way Functions
and Pseudorandom Generators

Nenad Dedić[1,2,4], Danny Harnik[3,4], and Leonid Reyzin[1,4]

[1] Boston University, Department of Computer Science, 111 Cummington St.,
Boston, MA 02215
reyzin@cs.bu.edu
[2] Google, Inc., 76 9th Ave., 6th Floor,
New York, NY 10011
nenad.dedic@gmail.com
[3] IBM Research, Haifa, Israel
danny.harnik@gmail.com
Research conducted while at the Technion, Haifa, Israel
[4] Research conducted, in part, at the Institute for Pure and Applied Mathematics at
UCLA, whose hospitality the authors gratefully acknowledge

Abstract. Can a one-way function f on n input bits be used with fewer
than n bits while retaining comparable hardness of inversion? We show
that the answer to this fundamental question is negative, if one is limited
black-box reductions.

Instead, we ask whether one can save on *secret* random bits at the
expense of more *public* random bits. Using a shorter secret input is highly
desirable, not only because it saves resources, but also because it can yield
tighter reductions from higher-level primitives to one-way functions. Our
first main result shows that if the number of output elements of f is at
most 2^k, then a simple construction using pairwise-independent hash
functions results in a new one-way function that uses only k secret bits.
We also demonstrate that it is not the knowledge of *security* of f, but
rather of its *structure*, that enables the savings: a black-box reduction
cannot, for a general f, reduce the secret-input length, even given the
knowledge that security of f is only 2^{-k}; nor can a black-box reduction
use fewer than k secret input bits when f has 2^k distinct outputs.

Our second main result is an application of the public-randomness
approach: we show a construction of a pseudorandom generator based
on any *regular* one-way function with output range of *known* size 2^k.
The construction requires a seed of only $2n + \mathcal{O}(k \log k)$ bits (as opposed to $\mathcal{O}(n \log n)$ in previous constructions); the savings come from
the reusability of public randomness. The secret part of the seed is of
length only k (as opposed to n in previous constructions), less than the
length of the one-way function input.

1   Introduction

PRG Seed Length It is important to keep the seed required for a pseudorandom
generator (PRG) as short as possible, lest the amount of true random bits needed

R. Canetti (Ed.): TCC 2008, LNCS 4948, pp. 607–625, 2008.
© International Association for Cryptologic Research 2008

to run it exceed the amount of pseudorandom bits its application requires, thus rendering it pointless. Moreover, in reductions from PRGs (or other constructs) to one-way functions, the blowup in the input length turns out to be the most central parameter in determining the security of the construct. It is therefore a major goal to reduce this parameter (as was addressed in [GIL$^+$90, HL92, HHR06b, Hol06, HHR06a]). The ultimate goal is a linear blowup, a necessary, although not a sufficient, condition to achieve a reduction with tight security preservation, i.e. a linear preserving one [HL92, HILL99].

Consider, therefore, the following problem: when is it possible to build a pseudorandom generator out of a one-way function f while keeping the generator seed length linear in the one-way function input length n? Certainly this is possible if f is a permutation—in fact, in the original PRG construction of [BM82, Yao82] the seed length is equal to the one-way function input length. However, no broader class of one-way functions satisfying this condition is currently known: even one-way bijections, if their output range is not easily mapped to $\{0,1\}^n$, are not known to satisfy this condition (the best constructions for them are the same as for other regular one-way functions, discussed below).

In this paper we demonstrate constructions of PRGs with the linear input length condition for a large class of *known regular* one-way functions. Specifically, if every output of f has α preimages (thus f has 2^k distinct outputs where $k = n - \log \alpha$) and (a lowerbound on) α is known, then we can build a PRG with seed length $2n + \mathcal{O}(k \log k)$. Thus, for functions with high enough degeneracy, where $k = \mathcal{O}(n/\log n)$, our PRG has a linear-length seed, like the Blum-Micali-Yao PRG built from one-way permutations. The construction, described in Section 4, builds upon the techniques of Haitner, Harnik and Reingold [HHR06b], which require longer seed length of $\mathcal{O}(n \log n)$, but assume only regularity rather than *known* regularity.

New Tool: One-Way Functions with Short Secret Inputs. We arrive at our pseudorandom generator as part of a study of a more fundamental problem: when is it possible to reduce the input length of a one-way function while maintaining some of its security? In other words, given a one-way function f with input length n, when is it possible to build another function g of input length $\ell(n) < n$ with comparable security? Indeed, if this were possible, then one could, for example, build a pseudorandom generator from g rather than from f, and maintain a reasonable seed length even if the PRG construction blows up the input size. However, we show that in general it is impossible to significantly reduce the input length of one-way function in a black-box manner, even for regular one-way functions (Theorem 5). That is, one must invest essentially the full n random bits when calling a one-way function.

This result, however, does not doom all efforts of using the one-way function with a shorter input. The insight is to use the paradigm introduced by Herzberg and Luby [HL92], which separates *public* randomness from *secret* randomness. It turns out to be possible to reduce the amount of *secret* randomness at the cost of additional *public* randomness. In Theorem 1 we show how to convert any one-way function f with 2^k distinct outputs into a *collection* of one-way functions f_h with inputs of length k, where the index h into the collection is the public randomness. The simple construction uses a pairwise independent family of *expanding* hash functions. The choice of the function from the collection is a choice of a hash function h, and we define $f_h(x) = f(h(x))$. This choice is made using $2n$ *public* random coins, which are available to any potential inverter.

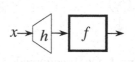

One way to achieve such a result is by using a technical Lemma of Dodis and Smith [DS05, Lemma 12], which shows the same construction secure if it uses $k + 2\log\frac{1}{\varepsilon} + 1$ secret input bits, where ε is the additive security loss. In particular, even if one needs to ensure that extra security loss is exponentially small, the result of [DS05] requires only linearly more input bits. However, the linear improvement we achieve over [DS05] is crucial for building our pseudorandom generator, as we explain shortly. To achieve this improvement, we take a different path from [DS05]: instead of showing that the distributions $(f(x), h)$ and $(f(h(x)), h)$ are statistically close, we show they have polynomially related subset weights, a relation between distributions that we call *g-domination*.

The secret input to our one-way function need not consist of k uniform independent bits: inputs from any distribution of entropy[1] k suffice (the same is true for our pseudorandom generator construction). This is beneficial, because uniform random bits may be harder to obtain that simply strings of high entropy.[2] Moreover, this enables our pseudorandom generator construction.

Application: The PRG Construction We construct our pseudorandom generator by applying the randomized iterate construction of [HHR06b] (henceforth called "the HHR construction") to f_h for a known regular f. Because f_h is secure even when h is public, the coins for h can be given only once and used for all iterations, resulting in a shorter seed. As compared to the HHR construction, we replace the need for many large hash functions with one large hash function (the $\hat{h}$ used for $f_{\hat{h}}$), and many small ones ($h^1, \ldots, h^k$ used in the randomized iterate construction). Our construction is illustrated in Figure 1.

To get some intuition for the construction, observe that if f is regular, then the number of secret random input bits we require for f_h is the entropy of the output of f_h. This enables iteration, because the output of f_h has enough entropy to be used (after an appropriate transformation) as an input to the next f_h. We could not use the result of [DS05], because it requires more input entropy than is output; nor could we use functions that are not regular, because they produce less output entropy than the input requires. The proof of pseudorandomness is

---

[1] Specifically, Renyi entropy of order 2, i.e., negative logarithm of collision probability.

[2] Of course, almost uniform independent bits can be obtained from a distribution of high entropy through the use of a strong extractor (whose seed can be public), but extractors necessarily lose entropy, so this approach would require a secret input with entropy higher than k, which, as we already pointed out, would create difficulties for our PRG construction.

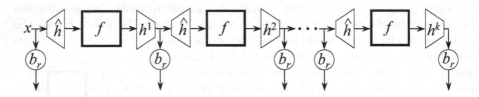

Fig. 1. Our pseudorandom generator on seed x. $\hat{h}$ is a pairwise-independent hash function from k bits to n bits; $h^1, h^2, \ldots, h^k$ are almost-pairwise independent hash functions from the output space of f to k bits, generated by a bounded space generator from a common seed s of length $\mathcal{O}(k \log k)$; b_r is the Goldreich-Levin hardcore bit (the same r is used throughout). $\hat{h}$, s and r are included in the output or, equivalently, are public.

not as simple as applying the HHR result to $f_{\hat{h}}$, because the HHR construction needs to start with a regular one-way function, and $f_{\hat{h}}$ is not necessarily regular even if f is.

In Appendix A we show how one can further exploit the knowledge of the regularity and further shorten the seed of our PRG to $2n + \mathcal{O}(k \log \log k)$, albeit at the cost of lowering its security.

In addition to considering the overall PRG seed length, it is also important to consider how much of the generator seed must be secret, because secret random bits tend to be much harder to obtain than nonsecret ones (again, this was already observed in [HL92]). Our PRG is the first to require a *sublinear* number of secret bits, namely, just k (the HHR generator, like the generators of [BM82, GKL93], requires n secret bits). Moreover, just like for our one-way function, the secret input to our PRG need not consist of uniform independent bits, but can come from any distribution of entropy k.

Example: One-way Function and PRG Based on Factoring. Consider the problem of building a one-way function based on the hardness of factoring products of two b-bit randomly chosen primes. If one is willing to assume a trusted party with secret coins, then it is easy: the trusted party chooses two secret random b-bit primes p and q, publishes $N = pq$, and the function can be, for example, squaring modulo N.

However, without trusted setup, there is no such easy construction. In order to work on the domain $\{0, 1\}^n$, the one-way function needs to include the process of generating the two random primes. A natural way to do this is to test some number of random integers for primality. To guarantee that two primes are found with probability 2^{-s} for some security parameter s, the number of integers tested should be $\Theta(sb)$ (because the probability that a random b-bit integer is prime is $\Theta(1/b)$). The natural function therefore gets $n = \Theta(sb^2)$ bits as input, splits them into $\Theta(sb)$ integers of length b each, finds the first two such integers p, q that are prime (if they do not exist, output 0), and outputs their product $N = pq$. We call this function f_{mult} (observe that, for sufficiently large s, it is one-way under the assumption that factoring is hard).

For reasonably secure values for b (e.g., 2048) and s (e.g., 64), the input length n of f_{mult} will be on the order of tens of megabytes. To come up with such a long secret input is, naturally, quite costly. Because the output of f_{mult} is short, however, we can apply our result on converting one-way functions to families with shorter secret inputs. Setting $k = 2b = o(\sqrt{n})$, we obtain a family of one-way functions with secret inputs of length only $2b$—as short as the description of the two primes p and q. To sample a function from this family, one still needs $\Theta(n)$ random bits, but they can be public, and are therefore much less expensive to obtain (e.g., from adversarially observable sources such as user behavior or ambient noise). Finally we note that using our techniques, one can generate a product $N = pq$ of two secret b-bit primes p,q using private randomness of entropy $2b$ (and the appropriate amount of public randomness). This can be used, for example, for generating public/secret key pairs for RSA or Paillier functions, from a modest amount of private randomness.

Consider now trying to make a PRG out of f_{mult}. The prior most efficient way (in terms of seed length) to achieve this is to notice that f_{mult} is a regular one-way function (except the negligible 2^{-s} portion that leads to the 0 output) and use the HHR construction, which takes a seed of $\mathcal{O}(n \log n)$ bits with $\mathcal{O}(n)$ of the bits being secret.[3] For reasonable parameter settings, it would be useful only in applications that can afford to gather tens of megabytes of secret randomness and gigabytes of public randomness before invoking the PRG.

Instead, observe that f_{mult} is also a known[4] regular one-way function, with $k < 2b$. Applying our PRG construction, we get a pseudorandom generator with just $2b = o(\sqrt{n})$ secret seed bits (which is roughly what's required to describe the two primes, anyway) and $\mathcal{O}(n)$ seed bits total (which is linear in what's anyway required as an input to f_{mult}).

Impossibility Results. As already mentioned, Theorem 5 shows that the total input length of a one-way function cannot be reduced in a black-box manner, thus leading us to use public randomness in order to reduce the amount of secret randomness. It is natural to ask if this approach can also work for one-way functions with a large number of outputs. On the positive side, we show in Theorem 2 that if a sufficiently large portion of the inputs goes to a sufficiently small portion of the outputs, then the answer is yes. In general, however, this appears unlikely to be the case, for the following reasons. In Theorem 6 we show that the number of *secret* random bits used when calling a one-way *permutation* f cannot be reduced to be substantially smaller than n by use of black-box

---

[3] It seems fruitless to try to turn f_{mult} into a permutation to order to apply the efficient construction of [BM82, Yao82]. Indeed, a natural way to build a bijection from f_{mult} is to include in the output all the unused bits as well as information on where p and q were in the sequence. However, this does not make it a permutation, because the output range (which includes the product of two primes) is not easily mapped back to the input domain of bit strings. Unfortunately, known solutions for bijections are not any better than those for regular functions.

[4] Our results apply to a weaker notion of "known": α can be a lower bound on the regularity of f, rather than its exact value.

reductions. This theorem is actually more general, and shows that our positive result is indeed tight for regular one-way functions, and the number of secret bits cannot be reduced any further in a black-box manner. Moreover, Theorem 7 shows that there is no black-box reduction that takes a one-way function f with hardness 2^s on n input bits and produce a collection of one-way functions on $n - s + \mathcal{O}(\log n)$ input bits. Thus, unless f has hardness very close to 2^n, in general the number of secret inputs bits must remain linear if one wants to have *any* hardness at all.

Discussion. Ideally, one would like to use only as many secret bits as the security one gets from the one-way function (it is clear that at least that many bits are necessary: a one-way function with n secret input bits can be easily inverted with probability 2^{-n}). Indeed, typical conjectured one-way functions, for example, RSA or discrete logarithm, are known to provide less security than 2^n (for the above examples, at most roughly $2^{n^{1/3}}$). Our negative results show that this is not possible in general with a black-box reduction (although we do not rule it out for specific functions such as discrete logarithm, of course). Our positive result, however, shows that if this weaker than optimal security manifests itself in a "structural" way, i.e., with the function having fewer outputs (a one-way function with k output bits can be easily inverted with probability 2^{-k}), then reduction in the number of inputs bits is possible.

It is natural to ask, of course, if one can not simply use the same one-way function f on a shorter input. It should be noted that our negative results do not consider such constructions, and hence do not rule them out. However, this option is unavailable when f is a fixed-length function secure in a concrete sense, such as a 128-bit block cipher or a hardware device implementing modular exponentiation for a 2,048-bit modulus. In this case, our impossibility results indicate that if we are given a hardware implementation of a one-way function we should use it with its full input length (unless we can look inside the box and learn something from there). This last observation adds motivation to results that take as input an exponentially hard one-way function and construct from it a pseudorandom generator with *weaker* security (of $n^{\log n}$) (e.g., some of the results stated in [Hol06, HHR06a] and the one in Appendix A in this paper). These results would be less interesting if there was a direct method of trading input length for security.

Even when the one-way function has variable input length, using it on a shorter input will reduce security. Of course, our construction also reduces security, but the security loss (i.e., security of f_h with n-bit f as compared to security of f on n bits) is polynomial. In contrast, simply using f on a shorter input can reduce security more than polynomially when the reduction in input length is superlinear.

Security comparison of the original f and our construction f_h depends on what parameters are set to equal each other. For example, we can compare the security of f on n bits to the security of f_h with a n-bit f (thus equating the input length to f, and hence the output length and likely most of the computational cost). In that case, f_h incurs a polynomial deterioration in security. Herzberg and Luby [HL92] advocate equating the secret input length. In that comparison, our constructions can actually be *more* secure that f, because f needs all n bits to be secret, while f_h and our PRG need only $k < n$ secret bits.

2 Definitions and Notation

If Y is a set, we denote by Y also the uniform distribution over that set, unless another distribution on Y is specified. We denote by U_n the uniform distribution over $\{0,1\}^n$. Given a distribution X and a function $f : X \to Y$, we denote by $f(X)$ the induced distribution on Y.

Let P and Q be distributions over some finite domain X. The collision-probability of P is $CP(P) = \sum_{x \in X} P(x)^2$. P and Q ε-close (or have statistical distance ε) if for every $A \subseteq X$ it holds that $|\Pr_{x \leftarrow P}(A) - \Pr_{x \leftarrow Q}(A)| \leq \varepsilon$ (equivalently, $\frac{1}{2}\sum_{x \in X} |\Pr_P[x] - \Pr_Q[x]| \leq \varepsilon$).

We assume familiarity with the standard notions of computational indistinguishability, one-way functions and pseudorandom generators (with public inputs, or equivalently, as public-coin collections), which are given in the full version of this paper [DHR07].

Definition 1 (Regular functions). *A function $f : \{0,1\}^* \to \{0,1\}^*$ is* regular *if for any $x,y \in \{0,1\}^n$, $|f^{-1}(f(x))| = |f^{-1}(f(y))|$. If $k(n) = -\log(|\{f(x) \mid x \in \{0,1\}^n\}|)$ then f is said to be* regular with output entropy k. *When k is also polynomial-time computable on input 1^n, f is* known-regular.

It is also customary to say that f *is an α-regular function* (for some $\alpha : \mathbb{N} \to \mathbb{N}$) — this means that f is a regular function with output entropy $k(n) = n - \log \alpha(n)$, i.e. preimage sizes are equal to $\alpha(n)$.

Definition 2 (Family of almost pairwise-independent hash functions). *Let $\{X_n\}_{n \in \mathbb{N}}, \{Y_n\}_{n \in \mathbb{N}}$ be two families of subsets of $\{0,1\}^*$. For any $n \in \mathbb{N}$ let $\mathcal{H}_n$ be a collection of functions where each $h \in \mathcal{H}_n$ is from X_n to Y_n. $\{\mathcal{H}_n\}_{n \in \mathbb{N}}$ is an* (efficient) family of δ-almost pairwise-independent hash functions *if: 1. there is a polynomial-time sampler which on $n \in \mathbb{N}$ outputs a description of randomly chosen $h \in \mathcal{H}_n$, 2. for any $h \in \mathcal{H}_n$, $|h|$ (i.e., the description length of h) is polynomial in $\log |X_n|$, 3. each $h \in \mathcal{H}_n$ is a polynomially-computable function, and 4. for all $x \neq x' \in X_n$ and all $y, y' \in Y_n$,*

$$\left| \Pr_{h \leftarrow \mathcal{H}_n} [h(x) = y \bigwedge h(x') = y'] - \frac{1}{|Y_n|^2} \right| \leq \delta(n).$$

A 0-almost pairwise independent family is called simply pairwise independent.

There are various constructions of efficient families of pairwise-independent hash functions (i.e. $\delta = 0$) for any $X_n = \{0,1\}^n$ and $Y_n = \{0,1\}^{\ell(n)}$ whose description length (i.e., $|h|$) is linear in $max\{n, \ell(n)\}$ (e.g., [CW77]). It is possible to construct δ-almost pairwise independent families for $\delta > 0$ whose description size depends very mildly on the input size. In particular, using [CW77], [WC81] and

[NN93] one gets constructions of efficient families of almost pairwise-independent hash functions for $X_n = \{0,1\}^n$ and $Y_n = \{0,1\}^{\ell(n)}$ whose description length is $\mathcal{O}(\log(n) + \ell(n) + \log(1/\delta))$.

Proposition 1. *Let $\{\mathcal{H}_n\}$ be a family of δ-almost pairwise independent hash functions from X_n to Y_n. Then for any n, and any distinct $x_1, x_2 \in X_n$ the following distributions have statistical distance at most $\delta|Y_n|^2/2$: 1. uniform on $Y_n \times Y_n$, 2. $(h(x_1), h(x_2))$ for uniformly random $h \in \mathcal{H}_n$.*

Proof: For any $y_1, y_2 \in Y_n$,
$\left| \Pr_h[(h(x_1), h(x_2)) = (y_1, y_2)] - \Pr_{(z_1, z_2) \in Y_n \times Y_n}[(z_1, z_2) = (y_1, y_2)] \right| \le \delta$ by definition. Summing over all $y_1, y_2 \in Y_n$ and dividing by 2, we get the desired result.   □

To simplify exposition, we will often work with (almost) pairwise independent hash functions on some fixed domain and range X and Y (rather than consider families $\{X_n\}$, $\{Y_n\}$).

Definition 3 (g-Domination). *Let B and C be distributions on the same set Π, and g a real-valued function. We will say that C g-dominates B if $\forall S \subseteq \Pi$, $\Pr_C[S] \ge g(\Pr_B[S])$ (this is a generalization of the notion of "dominates" from [Lev86], which contemplated linear g).*

Lemma 1. *If C g-dominates B for a convex function g, then for any distribution D on a set Φ, $D \times C$ g-dominates $D \times B$.*

Proof: Let $E \subset \Phi \times \Pi$. Let $p(\pi)$, for $\pi \in \Pi$, be $\Pr_{\phi \leftarrow D}[(\phi, \pi) \in E]$.

$$
\begin{aligned}
\Pr_{D \times C}[E] &= \underset{\pi \leftarrow C}{\mathbb{E}}\, p(\pi) \\
&= \int_0^1 \Pr_{\pi \leftarrow C}[p(\pi) > \alpha]\, d\alpha \quad \text{(using } \mathbb{E}(x) = \int \Pr[x > \alpha]\, d\alpha) \\
&\ge \int_0^1 g\left(\Pr_{\pi \leftarrow B}[p(\pi) > \alpha] \right) d\alpha \\
&\ge g\left(\int_0^1 \Pr_{\pi \leftarrow B}[p(\pi) > \alpha]\, d\alpha \right) \quad \text{(Jensen's inequality, since } g \text{ is convex)} \\
&= g\left(\underset{\pi \leftarrow B}{\mathbb{E}}\, p(\pi) \right) = g\left(\Pr_{D \times B}[E] \right).
\end{aligned}
$$
    □

A common approach in cryptographic reduction is to focus only on the subset of B for which $p(\pi)$ is large, and use Markov's inequality to obtain g'-domination of $D \times B$ by $D \times C$, for $g' \in \omega(g)$. Instead, this lemma, which takes all subsets into account, saves the increase in g and the corresponding loss of tightness in reductions.

3   One-Way Functions and Public Randomness

Here we show that a one-way function needs only as many secret input bits as the number of output bits it produces. We state our theorem in terms of bits in order to get a more concise statement; neither the domain nor the range need to be restricted to bit strings of a particular length, as shown in Lemma 2.

Theorem 1. *Let* $f : \{0,1\}^* \to \{0,1\}^*$ *be a one-way function that on n-bit inputs has at most* 2^k *distinct outputs. Let* $\mathcal{H}_{k,n}$ *be a family of pairwise-independent functions from* $\{0,1\}^k$ *to* $\{0,1\}^n$. *Define the domain-sampled* f *as* $f_h(x) \stackrel{def}{=} f(h(x))$ *for* $h \in \mathcal{H}_{k,n}$ *and* $x \in \{0,1\}^k$. *Then* $\{f_h\}_{h \in \mathcal{H}_{k,n}}$ *is a public-coin collection of one-way functions.*

The theorem is immediate from the following lemma.

Lemma 2. *Let* $f : Y \to Z$ *be a function, where* $|Z| = K$. *Let* X *be a distribution with collision probability at most* $1/K$, *and let* $\mathcal{H}_{X,Y}$ *be a family of pairwise-independent functions from the elements of* X *to* Y. *For every* $h \in \mathcal{H}_{X,Y}$ *define* $f_h : X \to Z$ *as* $f_h(x) \stackrel{def}{=} f(h(x))$. *Then any adversary* A *that inverts* f_h *with probability at least* ε *over* $x \in X$ *and* $h \in \mathcal{H}_{X,Y}$ *can be used to invert* f *on uniformly random inputs from* Y *with probability at least* $\varepsilon^4/21 - 1/(4K^2)$ *(* $\varepsilon^2/2$ *if* f *is regular) in the* *same running time as* A *(plus the time required to pick and evaluate a random hash function from* $\mathcal{H}_{X,Y}$*).*

Proof: Suppose that an algorithm A, when given $(f_h(x), h)$ computes x' such that $f_h(x') = f_h(x)$ with probability ε. That is,

$$\Pr_{(x,h) \leftarrow (X, \mathcal{H}_{X,Y})} [f_h(A(f_h(x), h)) = f_h(x)] \geq \varepsilon$$

Consider the following procedure M^A for inverting f: on input $z \in Z$, choose a random $h' \in \mathcal{H}_{X,Y}$, let $x' = A(z, h')$, and output $h'(x')$. Note that the notation h' in M^A, rather than h, emphasizes that the h' does not necessarily have to be consistent with z. While there exist many h with x such that $z = f_h(x)$, the chosen h' might not be one of them.

We will analyze the success probability of M^A as follows. The success of A (and therefore M^A) is determined by its internal coin flips and its input (z, h'). We will show that the distribution of (coinflips, input) pairs that A sees when run within M g-dominates the distribution for which A is designed, for a polynomial g; therefore, the probability of the event that M^A succeeds in inverting f is polynomially related to the probability of the event that A inverts the domain-sampled f. We will first show g-domination for inputs only, ignoring the coinflips, and take care of the coinflips later.

It is worth comparing the following proposition, about g-domination of inputs, to the aforementioned lemma by Dodis and Smith [DS05, Lemma 12], which analyzes the same construction but with longer inputs to h, showing that $(f(y), h')$ is close to $(f(h(x)), h)$. Our proof technique is entirely different and builds on the technique of [HHR06b].

Proposition 2. *For any (not necessarily one-way)* $f : Y \to Z$ *with* K *distinct outputs, distribution* X *with* $CP(X) \leq 1/K$, *and pairwise-independent hash family* $\mathcal{H}_{X,Y}$, *the distribution* $(f(y), h')$ *(where* $y \leftarrow Y, h' \leftarrow \mathcal{H}_{X,Y}$*) g-dominates*

$(f(h(x)), h)$ (where $x \leftarrow X, h \leftarrow \mathcal{H}_{X,Y}$), for $g(\delta) = \delta^4/21 - 1/(4K^2)$, or $g(\delta) = \delta^2/2$ if f is regular.

Proof: We need show that for any $S \subseteq Z \times \mathcal{H}_{X,Y}$,

$$\Pr_{(x,h) \leftarrow X \times \mathcal{H}_{X,Y}} [(f_h(x), h) \in S] \geq \delta \Rightarrow \Pr_{(y,h') \leftarrow Y \times \mathcal{H}_{X,Y}} [(f(y), h') \in S] \geq \frac{\delta^4}{21} - \frac{1}{4K^2}$$

(replace the right-hand-side with $\delta^2/2$ if f is regular).

First we give a one-paragraph outline of the proof of this proposition. Call the points in S *good*. Let $(y, h) \in Y \times \mathcal{H}_{X,Y}$ be called *good* if and only if $(f(y), h)$ is good. We will divide the space Y of inputs to f into K equal-size chunks, producing a set of chunks called C. Call $(c, h) \in C \times \mathcal{H}_{X,Y}$ *good* if $\exists y \in c$ such that (y, h) is good (i.e., a chunk is good if contains a preimage of a good point in Z). We will show, simply using properties of $\mathcal{H}_{X,Y}$, that the fraction of good chunks (under the uniform distribution) is at least $\delta^2/2.125$. This will imply that A works on some portion of sufficiently many chunks. Then, using the fact that f has only K outputs, we will show that A works on a sufficiently large portion of most of these chunks. The actual proof is in in the full version of this paper [DHR07]. □

M^A succeeds whenever A succeeds; in turn, the success or failure of A depends on the point (z, h') chosen, and on the coin flips of A. Let Φ, with probability distribution D, be the space of all coin flips of A. Let $\Pi = Z \times \mathcal{H}_{X,Y}$, let B be the distribution on Π obtained by choosing $x \leftarrow X, h \in \mathcal{H}_{X,Y}$, and $z = f_h(x)$, and let C be the distribution on Π obtained by choosing a uniform $y \in Y$, $h' \in \mathcal{H}_{X,Y}$, and $z = f(y)$. Applying Lemma 1 below to the event E that that A succeeds (here $g(\delta) = \delta^4/21 - 1/4K^2$, or $\delta^2/2$ in the case of regular functions), we obtain the desired statement. □

3.1   The Case of Many Outputs

Theorem 1 can be used to reduce the number of secret input bits to a one-way function provided the function has a sufficiently small output range. As we show in this section, the same technique is useful even if the function has large output range, as long as an appreciable fraction of the inputs falls into a rather small subset of the output range. Namely, suppose there is a set of outputs O_H of size 2^k such that $\Pr_{y \in \{0,1\}^n}[f(y) \in O_H] \geq p_H$. If $k < \sqrt{p_H n}$, then it is possible to reduce the number of secret input bits from n to k^2/p_H, as follows.

Let X be a distribution of collision probability $1/2^k$, and $\mathcal{H}_{X,Y}$ and $f_h(x)$ as above. In the full version [DHR07], we show that $f_h(x)$ is a collection of *weak* one-way functions, i.e., is not invertible with probability appreciably more than $1 - p_H$.

We can then use the standard hardness amplification technique of Yao [Yao82] in order to convert the weak one-way function collection into a strong one. The technique simply concatenates many independent copies of the weak one-way function. The number of repetitions needed to reduce the easily invertible fraction of inputs to (negligibly more than) $1/2^k$ from $1 - p_H$ is k/p_H (thus requiring k^2/p_H secret bits) This gives the following result, whose proof is similar to the proof of Theorem 1 and is outlined in the full version of this paper.

Theorem 2. *Let* $f : \{0,1\}^* \to \{0,1\}^*$ *be a one-way function and suppose for every* n *there exists a set* $O_H(n)$ *of size* $k(n)$ *such that* $\Pr_{y \in \{0,1\}^n}[f(y) \in O_H(n)] \geq p_H(n)$. *For every* $n \in \mathbb{N}$ *let* $\mathcal{H}_{k,n}$ *be a family of pairwise-independent functions from* k *bits to* n *bits. Denote* $\ell = k/p_H$ *and define* $\overline{f_{\overline{h}}}(x_1, \ldots, x_\ell) \stackrel{def}{=} (f_{h_1}(x_1), \ldots, f_{h_\ell}(x_\ell))$ *for* $\overline{h} = (h_1, \ldots, h_\ell) \in \mathcal{H}_{k,n}^\ell$ *and* $x_1, \ldots, x_\ell \in \{0,1\}^k$. *Then* $\{\overline{f_{\overline{h}}}\}_{\overline{h} \in \mathcal{H}_{k,n}^\ell}$ *is a public-coin collection of one-way functions.*

4   Pseudorandom Generator Collection from Any Known Regular OWF

In this section we show a construction of a pseudorandom generator collection from any regular one-way function. Unlike in the randomized iterate constructions of [GKL93, HHR06b], here the underlying function f has *known* (i.c. efficiently computable) regularity. We use this knowledge to get a PRG collection with particularly short secret input and little security loss.

Namely, suppose f is a regular OWF with output entropy $k(n)$, and that $(t(n), \epsilon(n))$ is the security of f. On secret seed of length $s_S(n) = k(n)$, our PRG collection attains the security of $(\mathsf{poly}(n) + t(n), \mathsf{poly}(\epsilon(n)))$ (Theorem 3). For example, if $k(n) = n^{1/3}$, then we get security comparable to $(t(n), \epsilon(n))$ using only $n^{1/3}$ secret bits. And since, for sufficiently small k, the public index of our PRG collection is of linear size $\mathcal{O}(n)$, one can also view it as a PRG, rather than collection, with good security preservation: on seed length $\mathcal{O}(n)$ it attains security $(\mathsf{poly}(n) + t(n), \mathsf{poly}(\epsilon(n)))$.

Our construction in fact requires a somewhat weaker condition on f than known regularity: f still must be regular, but it is sufficient to have an efficiently computable upper bound $k(n)$ on the output entropy of f. Note that a more accurate bound leads to greater savings in the number of secret seed bits.

Theorem 3. *Let* f *be a regular one-way function with security* $(t(n), \epsilon(n))$ *and output entropy at most* $k(n)$ *(for* k *computable in time polynomial in* n*). Then there is a public-coin PRG collection* G, *which is* $(\mathsf{poly}(n) + t(n), \mathsf{poly}(\epsilon(n)))$-*indistinguishable on secret seeds of length* $s_S(n) = k(n)$ *and public seeds of length* $s_P(n) = 2n + \mathcal{O}(k(n) \log k(n))$. *(In particular* $s_P(n) = \mathcal{O}(n)$ *if* $k = \mathcal{O}(n/\log n)$.*)*

Before the actual construction we present the basic tool of the randomized iterate [GKL93, HHR06b]. We define it slightly differently than [GKL93, HHR06b]: theirs outputs a value in $Im(f)$, and ours outputs a hash function image.

Definition 4 (The mth Randomized Iterate of f). *Let* $f : \{0,1\}^k \to \{0,1\}^\ell$ *and let* $\mathcal{H}$ *be a family of functions from* $\{0,1\}^\ell$ *to* $\{0,1\}^k$. *For input* $x \in \{0,1\}^k$

and $\overline{h} = (h^1, \ldots, h^t) \in \mathcal{H}^t$ define the mth Randomized Iterate $f^m : \{0,1\}^k \times \mathcal{H}^t \to Im(f)$ for every $m \in [t]$ recursively as:

$$f^m(x, \overline{h}) = h^m(f(f^{m-1}(x, \overline{h})))$$

where $f^0(x, \overline{h}) = x$.

We first show a construction with public seed length $2n + \mathcal{O}(k^2)$ and then describe how it may be reduced to as low as $2n + \mathcal{O}(k \log k)$, following the same technique as in the HHR construction.

Construction 1. *The generator takes the following as inputs:*

1. *A secret random $x \in \{0,1\}^k$*
2. *A (public) description of one hash function $\hat{h}$ from a family $\mathcal{H}_{k,n}$ of pairwise independent hash functions from k bits to n bits (requires $2n$ bits).*
3. *(Public) descriptions of k hash functions $\overline{h} = (h^1, \ldots, h^k)$ from a family $\mathcal{H}_{\ell,k}$ of 2^{-3k}-almost pairwise independent hash functions from ℓ bits to k bits (requires $\mathcal{O}(k)$ bits each).*
4. *A (public) random string $r \in \{0,1\}^k$ for the Goldreich-Levin [GL89] hardcore bit b_r (requires k bits).*

The generator is defined as follows:

$$G_{\hat{h}, \overline{h}, r}(x) = b_r(x), b_r(f_{\hat{h}}^1(x, \overline{h})), \ldots, b_r(f_{\hat{h}}^k(x, \overline{h})),$$

where $f_{\hat{h}}^i$ denotes the ith randomized iterate of the function $f_{\hat{h}} = \hat{h} \circ f$ (see Figure 1).

Theorem 4. *Suppose f is regular one-way with output entropy at most $k(n)$ and security $(t(n), \epsilon(n))$. Then G in Construction 1 is a public-coin pseudorandom generator collection. It is $(\mathsf{poly}(n) + t(n), \mathsf{poly}(\epsilon(n)))$-indistinguishable on secret seeds of length $s_S(n) = k(n)$ (and public seeds of length $s_P(n) = 2n + \mathcal{O}(k^2)$). (In particular, $s_P(n) = \mathcal{O}(n)$ if $k(n) = \mathcal{O}(\sqrt{n})$.)*

Proof: G takes k bits and outputs $k + 1$ bits. Thus it is expanding. We must now prove that it is indistinguishable. It is tempting to first fix $\hat{h}$ and since by Theorem 1 $f_{\hat{h}}$ is a one-way function, simply plug $f_{\hat{h}}$ in the HHR construction. However, the HHR construction relies heavily on the fact that the underlying function is regular or at least very close to regular. The function $f_{\hat{h}}$ on the other hand is *not* guaranteed to be regular once $\hat{h}$ is fixed, even if f is regular to begin with. If $\hat{h}$ were from a k-wise independent family (rather than a pairwise independent one) then one can prove that with overwhelming probability $f_{\hat{h}}$ is close to regular. This is not the case with pairwise independent $\hat{h}$ and on the contrary, it is likely that with noticeable probability $f_{\hat{h}}$ will deviate too much from a regular function. Our proof follows the basic structure of the proof of the HHR construction, so we give a sketch, detailing the parts which differ from [HHR06b].

As in the previous iterative constructions (such as [BM82, Yao82, Lev87], [GKL93, HHR06b]), the key to the proof is the unpredictability of the sequence

$$\left(f_{\hat{h}}^k(x, \overline{h}), f_{\hat{h}}^{k-1}(x, \overline{h}), \ldots, f_{\hat{h}}^1(x, \overline{h}), x \right),$$

even for an adversary who is given $(\overline{h}, \hat{h})$. Once this is shown (Lemma 3), it follows from the stronger Goldreich-Levin theorem [Lev93], that the output of the PRG is next-bit unpredictable with essentially the same security. Next-bit unpredictability is equivalent to indistinguishability with a security loss $1/k^{\mathcal{O}(1)}$ (see [Gol01], Theorem 3.3.7). Thus the output of G is indeed pseudorandom, with security essentially the same as of the above sequence. We now turn to the proof of unpredictability.

Let $\mathsf{Supp}(n) = \mathcal{H}_{\ell,k}^k \times \mathcal{H}_{k,n} \times \{0,1\}^k$, and call an element of Supp an *instance*. Let $\Phi = \{0,1\}^{\mathbb{N}}$ denote the set of all coin toss sequences. We say that an algorithm A *inverts i-th iteration (on random coins ω and instance $(\overline{h}, \hat{h}, f_{\hat{h}}^i(x, \overline{h}))$)* if

$$A(\omega, \overline{h}, \hat{h}, f_{\hat{h}}^i(x, \overline{h})) = f_{\hat{h}}^{i-1}(x, \overline{h}).$$

Let $D(n)$ be the distribution of instances produced by the generator, i.e. $(\overline{h}, \hat{h}, f_{\hat{h}}^i(x, \overline{h}))$ for uniform $(\overline{h}, \hat{h}, x)$. Let $Z(n)$ be the uniform distribution of instances, i.e. uniform $(\overline{h}, \hat{h}, z)$.

Lemma 3. *Let A be an algorithm with running time $\leq t(n)$. Suppose that*

$$\Pr[A \text{ inverts } i\text{-th iteration on } (\omega, \overline{h}, \hat{h}, f_{\hat{h}}^i(x, \overline{h}))] \geq \epsilon(n),$$

where ω is uniform and $(\overline{h}, \hat{h}, f_{\hat{h}}^i(x, \overline{h}))$ is distributed according to $D(n)$. Then there is an algorithm B which runs in time $\leq \mathsf{poly}(n) + t(n)$ and inverts $f(x)$ with probability $\geq \epsilon^{2.5}(n)/(16(k+1))$ (for $|x| = n$).

Proof: On input y, the algorithm B generates random $(\overline{h}, \hat{h})$, sets $u \leftarrow A(\overline{h}, \hat{h}, h^i(y))$, and outputs $\hat{h}(u)$.

Fix some n and then we can omit it from the notation. B chooses the hash functions independently of y, i.e. it produces instances distributed according to Z. However, A is guaranteed to invert with probability ϵ on a different distribution D. The bulk of the proof is devoted to proving that A inverts with comparable probability $\approx \epsilon^2$ also on distribution Z. The basic idea of the proof is similar to [HHR06b]: we show that collision probabilities of Z and D are closely related $CP(Z) \geq \mathcal{O}(k) \cdot CP(D)$, and from that we conclude that event probabilities are closely related as well $\Pr_Z[S] \geq (\Pr_D[S])^2/\mathcal{O}(k)$. In particular, the inversion event happens with probability $\epsilon^2/\mathcal{O}(k)$ under Z. The actual proof is more involved than this simple outline, the main complications being: 1. there is a single expanding hash function $\hat{h}$ which is used in every iteration, so the technique of [HHR06b] is not directly applicable, 2. contracting hash functions h^i cause collisions, so an inverse of i-th iteration may be unrelated to y. The details of the proof are given in the full version of this paper [DHR07]. $\square$

Reducing the public seed length. To reduce the public seed length of the above construction from $2n + \mathcal{O}(k^2)$ to $2n + \mathcal{O}(k \log k)$, we follow exactly the same derandomization technique as in the HHR construction. The idea is to not use independent choices of hash functions for $\overline{h} = (h_1, \ldots, h_k)$ but rather choose functions that are correlated yet satisfy the proof of the previous section. The central observation is that the collision probability of a randomized iterate can be computed by a *bounded space* program. More precisely, there is a simple bounded space branching program such that its input tape consists of the choice of $\overline{h}$ and its acceptance probability is precisely the collision probability of $f_{\hat{h}}^k$ (the probability is over inputs $x, \overline{h}$) for every fixed $\hat{h}$. Thus replacing the hash functions in the input tape by the output of a generator that fools bounded space programs (such as the generators of [Nis92, INW94]) changes the collision probability only by a small additive error. This is sufficient to make the proof of the previous section go through. Loosely speaking, the bounded space program takes two initial inputs x_1 and x_2.[5] At the first step the program reads the randomizing hash function h^1 and computes $f_{\hat{h}}^1(x_1, h^1)$ and $f_{\hat{h}}^1(x_2, h^1)$ and stores only these two intermediate values (not storing x_1 and x_2). At each iteration the program reads a new randomizing hash and computes the next randomized iterate of the two values, while not storing the previous one. At the end the program simply compares the two values and outputs 1 only if they are the same value. An accurate account of such a program, bounded space generators and the revisions needed in the proof appears in [HHR06b].

Construction 2. *The generator takes the following as inputs:*

1. *A secret random* $x \in \{0, 1\}^k$
2. *Description of one hash function* $\hat{h}$ *from a family* $\mathcal{H}_{k,n}$ *of pairwise independent hash functions from k bits to n bits (requires 2n bits).*
3. *Seed* $s \in \{0, 1\}^{\mathcal{O}(k \log k)}$ *to a bounded space generator BSG with space bound $2k$ and error 2^{-k}. The output $BSG(s) = (h^1, \ldots, h^k)$ of the generator consists of the descriptions of k hash functions from a family $\mathcal{H}_{\ell,k}$ of almost pairwise independent hash functions from ℓ bits to k bits.*
4. *A random string* $r \in \{0, 1\}^k$ *for the Goldreich-Levin hardcore bit b_r (requires k bits).*

The generator is defined as follows:

$$G'(x, \hat{h}, s, r) = b_r(x), b_r(f_{\hat{h}}^1(x, BSG(s))), \ldots, b_r(f_{\hat{h}}^k(x, BSG(s))), \hat{h}, s, r$$

Where $f_{\hat{h}}^i$ denotes the ith randomized iterate of the function $f_{\hat{h}} = \hat{h} \circ f$.

The seed length of the aforementioned generators is $\mathcal{O}(\log |\mathcal{H}_{\ell,k}| \cdot \log k)$ (which equals $\mathcal{O}(k \log k)$ with our choice of parameters) and thus the overall construction takes seed length $2n + \mathcal{O}(k \log k)$.

---

[5] The program actually computes the collision probability for one fixed pair of inputs x_1, x_2. The actual collision probability is the average over all possible input fixings. But since the generator fools each program separately, it will also fool the average.

On using secret seeds from non-uniform distributions. A simple modification makes our PRG secure even when used with secret seed drawn from any distribution X as long as $CP(X) \leq 2^{-k}$. The modification can be applied to either Construction 2 or Construction 1. The public seed then increases by only $\mathcal{O}(k)$ bits, therefore it remains unchanged asymptotically. Please see Appendix B for a brief description of the modification.

5   Black-Box Separations

As discussed in the introduction, it is natural to ask under which conditions one can reduce the input length to a one-way function below its "native" length n. More abstractly, we want to know: *Is there a generic way of securely using a OWF on n-bit inputs, if we are given only $\ell < n$ random bits? How small can ℓ be?*

We formalize these questions using circuits, where it is easy to talk about security on fixed-length input. (It is possible to formulate them in the uniform context, but they become too cumbersome.) We then give some indications that improving upon our results requires non-black-box reductions. Roughly, by "no black-box reduction of P to Q" we mean that the security proof *"if Q is secure then P is too"* is necessarily non-black-box (the construction of P from Q, however, may be black-box). Before elaborating, let us informally summarize the optimality results:

1. For any $l < n$, there is no black-box reduction of l-bit input OWF to regular n-bit-input OWF (and, as a corollary, no black-box reduction to either OWF of known hardness, or arbitrary OWF).
2. For any $l < n - \log \alpha$, there is no black-box reduction of l-bit input one-way-collection to α-regular n-bit-input OWF (and, as a corollary, no black-box reduction to either OWF of known hardness $< 2^n/\alpha$, or arbitrary OWF).
3. For any $s < n$ and $l < n - s$, there is no black-box reduction of l-bit input one-way-collection to an n-bit input OWF of hardness at most s.

5.1   Formal Statements

Let $\mathcal{F}^n$ denote the set of all $f : \{0,1\}^n \to \{0,1\}^n$. Let $\nu(n)$ denote a negligible function (one decaying faster than any inverse polynomial). Note that $1/\nu(n)$ is then a superpolynomial function.

Circuits, oracle circuits. Let $|A|$ denote the size of the circuit A. For an oracle circuit A and a function $f : \{0,1\}^n \to \{0,1\}^m$, A^f denotes the oracle circuit in which each oracle gate with input x outputs $f(x)$. If $\mathcal{G} = \{g_i\}_{i \in \{0,1\}^n}$ is a collection of functions $g_i : \{0,1\}^n \to \{0,1\}^m$ then $A^{\mathcal{G}}$ denotes the oracle circuit in which each oracle gate, on input (i, x) outputs $g_i(x)$.

Inverter. A circuit $A : \{0,1\}^l \to \{0,1\}^n$ is a *p-inverter* for $f : \{0,1\}^n \to \{0,1\}^l$ if $\Pr_{x \in \{0,1\}^n}[A(f(x)) \in f^{-1}(f(x))] \geq p$. A 1-inverter is called *perfect*.

Black-box reduction. Let $\mathcal{F} \subseteq \mathcal{F}^n$. A pair of circuits (R, g) is an (l, p)-*reduction to* $\mathcal{F}$ if for any $f \in \mathcal{F}$:

1. g has l input wires.
2. If V is a perfect inverter for g^f, then $R^{V,f}$ is a p-inverter for f.

A sequence (R_n, g_n) of (l_n, p_n)-reductions to $\mathcal{H}_n \subseteq \mathcal{F}^n$ is called $d(n)$-*saving* if:
1. $(|R_n| + |g_n|)/p_n$ is polynomial in n, 2. $n - l_n = d(n)$.

Let $\mathcal{F}_{\text{REG}}^{n,\alpha} \subseteq \mathcal{F}_{\text{ALL}}$ denote its subset of all α-regular functions. Let $\mathcal{F}_{\text{LOW}}^{n,s} \subseteq \mathcal{F}_{\text{ALL}}$ denote the subset of all *at most s-hard permutations* (permutations which have a $1/2$-inverter of size $< s$).

Black-box collection reduction. A pair of circuits (R, g) is a (l, m, p)-*collection-reduction to* $\mathcal{F}$ if:

1. For any $f \in \mathcal{F}$, and any $(i, x) \in \{0, 1\}^m \times \{0, 1\}^l$, $g^f(i, x)$ is of the form (i, y).
2. If V is a perfect inverter for g^f, then $R^{V,f}$ is a p-inverter for f.

A sequence (R_n, g_n) of (l_n, m_n, p_n)-reductions to $\mathcal{H}_n \subseteq \mathcal{F}^n$ is called $d(n)$-*saving* if: 1. $m_n(|R_n| + |g_n|)/p_n$ is polynomial in n, 2. $n - l_n = d(n)$.

The following two technical lemmas are at the heart of our separations. Their proofs can be found in the full version of the present article [DHR07].

Lemma 4. *Let $l = n - c$ and $p \geq 2^{-c/2+1}$. If (R, g) is an (l, p)-reduction to $\mathcal{F}_{\text{REG}}^{n,\alpha}$ then $|g| > 2^{c/2}$ or $|R| > p2^{n-a+3}$.*

Lemma 5. *Let $l = n - \log \alpha - d$. If (R, g) is a (l, m, p)-collection-reduction from $\mathcal{F}_{\text{REG}}^{n,\alpha}$, then $|R| > p2^{d-4}/m$.*

Theorem 5. *Let $\alpha(n) = \nu(n)2^n$. There is no $\omega(\log n)$-saving reduction to $\mathcal{F}_{\text{REG}}^{n,\alpha(n)}$.*

Proof: Suppose to the contrary that (R, g) is a $\omega(\log n)$-saving reduction to $\mathcal{F}_{\text{REG}}^{n,\alpha(n)}$. Consider some particular f, and let D be the set of all possible oracle queries that g^f can ask, on any input. Then $|S| \leq |g|2^l$, because on each of the 2^l distinct inputs, g asks at most $|g|$ queries. The basic idea of the lower bound proof is that, for $l < n - \omega(\log n)$, and polynomial-sized g, S occupies a negligible fraction of f's domain. But the one-way f can be easy on S, and g^f is then not one-way.

Formally: apply Lemma 4 to (R_n, g_n) with $c = \omega(\log(n))$ and $p = p(n)$. Since $2^{c/2} = 1/\nu(n)$ and $2^{n-\log \alpha(n)} = 2^n/\alpha(n) = 1/\nu(n)$ we conclude that $|R_n| + |g_n|$ is superpolynomial. $\square$

Theorem 6. *Let $\alpha(n) = \nu(n)2^n$. There is no $(\omega(\log(n)) + \log \alpha(n))$-saving collection-reduction to $\mathcal{F}_{\text{REG}}^{n,\alpha(n)}$.*

Proof: Suppose that (R, g) is the collection-reduction which contradicts the theorem statement, and let l be the number of g's input wires. We show that it is possible to build from (R, g) a circuit B of size about 2^l which inverts any $f \in \mathcal{F}_{\text{REG}}^{n,\alpha(n)}$. To do this, note that R^V inverts any $f \in \mathcal{F}_{\text{REG}}^{n,\alpha(n)}$ as long as it is given an inverter V for g^f. But V can be implemented as a circuit of size $2^l/\nu(n)$. Therefore R^V can be implemented (without any oracle) as a circuit of size about $|R|2^l/\nu(n)$. But this is too small to invert any function $f \in \mathcal{F}_{\mathrm{REG}}^{n,\alpha(n)}$. The formal argument follows.

If $|g_n|$ is superpolynomial we are done. Else suppose $|g_n|$ grows polynomially fast. Apply Lemma 5 with $d = \omega(\log(n))$ (and $\log|I| < |g_n|$ since $\log|I|$ is at most the number of input wires of g_n), to get that $|R_n| > p(n)2^{\omega(\log(n))}/|g_n|$ which is superpolynomial. $\qquad\square$

Theorem 7. *Let $s(n) < n$. There is no $(\omega(\log(n)) + s(n))$-saving collection-reduction to $\mathcal{F}_{\mathrm{LOW}}^{n,s(n)}$.*

Proof Sketch: Let f be a random permutation and let $h(p, y)$ output $x = f^{-1}(y)$ if p is an s-bit prefix of x. This ensures that f is "exactly" s-hard. For any construction g^f with input size $l = n - s - d$ (and description of family index m polynomial in n), we can show an oracle V which inverts it, but such that V does not significantly reduce the hardness of f. Some minor modifications are needed to ensure that (f, h) is a permutation.

V, on input (i, y), simply outputs a random x for which $g_i^f(x) = y$. To see that f is still s-hard, suppose there is a poly-size inverter $A^{(f,h),V}$ for f. From it one can build a circuit B^f which perfectly simulates $A^{(f,h),V}$. Each call to h can be simulated using 2^{n-s} queries to f, and each call to V using $\approx 2^l$ queries to f. So B^f calls f about $|B|(2^l + 2^{n-s}) < |B|(2 \cdot 2^{n-s})$ times. With this many queries, the probability of inverting f cannot exceed $\approx 2^{-s}$, so f is still s-hard. $\qquad\square$

Corollary 1 (To Theorem 5). *There is no $\omega(\log n)$-saving reduction to $\mathcal{F}^n$.*

Corollary 2 (To Theorem 6). *There is no $(\omega(\log n) + \log\alpha(n))$-saving reduction to $\mathcal{F}^n$.*

Acknowledgements

We thank anonymous referees for their many helpful suggestions. Research of N.D. and L.R. was supported, in part, by IPAM at UCLA, and by US NSF grants CCF-0515100, CNS-0546614 and CNS-0202067. Research of D.H. was supported by a Lady Davis Fellowship and by a grant from the Israeli Science Foundation.

References

[BM82]   Blum, M., Micali, S.: How to generate cryptographically strong sequences of pseudo random bits. In: 23rd FOCS, pp. 112–117 (1982)

[CW77]   Carter, I., Wegman, M.: Universal classes of hash functions. In: 9th ACM Symposium on Theory of Computing, pp. 106–112 (1977)

[DHR07]   Dedić, N., Harnik, D., Reyzin, L.: Saving private randomness in one-way functions and pseudorandom generators. Technical Report 2007/458, Cryptology e-print archive (2007), http://eprint.iacr.org

[DS05]   Dodis, Y., Smith, A.: Correcting errors without leaking partial information. In: 37th STOC, pp. 654–663 (2005)

[GIL+90]   Goldreich, O., Impagliazzo, R., Levin, L., Venkatesan, R., Zuckerman, D.: Security preserving amplification of hardness. In: 31st IEEE Symposium on Foundations of Computer Science, pp. 318–326 (1990)

[GKL93]   Goldreich, O., Krawczyk, H., Luby, M.: On the existence of pseudorandom generators. SIAM Journal of Computing 22(6), 1163–1175 (1993)

[GL89]   Goldreich, O., Levin, L.A.: A hard-core predicate for all one-way functions. In: 21st ACM Symposium on the Theory of Computing, pp. 25–32 (1989)

[Gol01]   Goldreich, O.: Foundations of Cryptography. Cambridge University Press, Cambridge (2001)

[HHR06a]   Haitner, I., Harnik, D., Reingold, O.: Efficient pseudorandom generators from exponentially hard one-way functions. In: Bugliesi, M., Preneel, B., Sassone, V., Wegener, I. (eds.) ICALP 2006. LNCS, vol. 4052, pp. 228–239. Springer, Heidelberg (2006)

[HHR06b]   Haitner, I., Harnik, D., Reingold, O.: On the power of the randomized iterate. In: Dwork, C. (ed.) CRYPTO 2006. LNCS, vol. 4117, pp. 22–40. Springer, Heidelberg (2006)

[HILL99]   Håstad, J., Impagliazzo, R., Levin, L.A., Luby, M.: A pseudorandom generator from any one-way function. SIAM Journal of Computing 29(4), 1364–1396 (1999)

[HL92]   Herzberg, A., Luby, M.: Pubic randomness in cryptography. In: Brickell, E.F. (ed.) CRYPTO 1992. LNCS, vol. 740, pp. 421–432. Springer, Heidelberg (1993)

[Hol06]   Holenstein, T.: Pseudorandom generators from one-way functions: A simple construction for any hardness. In: Halevi, S., Rabin, T. (eds.) TCC 2006. LNCS, vol. 3876, pp. 443–461. Springer, Heidelberg (2006)

[INW94]   Impagliazzo, R., Nisan, N., Wigderson, A.: Pseudorandomness for network algorithms. In: 26th STOC, pp. 356–364 (1994)

[Lev86]   Levin, L.A.: Average case complete problems. SIAM Journal on Computing 15(1), 285–286 (1986)

[Lev87]   Levin, L.A.: One-way functions and pseudorandom generators. Combinatorica 7, 357–363 (1987)

[Lev93]   Levin, L.A.: Randomness and nondeterminism. The Journal of Symbolic Logic 58(3), 1102–1103 (1993)

[Nis92]   Nisan, N.: Pseudorandom generators for space-bounded computation. Combinatorica 12(4), 449–461 (1992)

[NN93]   Naor, J., Naor, M.: Small-bias probability spaces: Efficient constructions and applications. SIAM Journal on Computing 22(4), 838–856 (1993)

[WC81]   Wegman, M., Carter, J.: New hash functions and their use in authentication and set equality. Journal of Computer and System Sciences (1981)

[Yao82]   Yao, A.C.: Theory and application of trapdoor functions. In: 23rd IEEE Symposium on Foundations of Computer Science, pp. 80–91 (1982)

A   Further Shortening the PRG Seed

In our pseudorandom generator, the output of the last hash function has, intuitively, almost k bits of entropy. It entropy can be converted to pseudorandomness using an extractor with a public seed (of length k). To get this pseudorandomness to be, e.g., $n^{\log^c n}$-close to uniform for some c, one will lose $\Theta(\log^{c+1} n)$

bits. If we take this approach, then the we need to run the randomized iterate construction not k times, but $\Theta(\log^{c+1} n)$ times; thus, we need the space-bounded generator to produce not k, but $\Theta(\log^{c+1} n)$ hash functions, which can be done in space $\mathcal{O}(k \log(\log^{c+1} n)) = \mathcal{O}(k \log \log k)$. The result is a PRG with seed length $2n + \mathcal{O}(k \log \log k)$ of which only k bits needs to be secret, but security reduced to the bare minimum $n^{\log^c n}$.

B   On Using Secret Seeds from Non-uniform Distributions

Suppose X is a distribution with the only guarantee that $CP(X) \leq 2^{-k}$. We outline the modification which makes our PRG secure even when its seed x is drawn from X. Namely, suppose that the support of X is $\{0,1\}^m$, and let $\mathcal{H}_{m,k}$ be a family of 2^{-3k}-almost pairwise independent hash functions from $\{0,1\}^m$ to $\{0,1\}^k$. The modified generator first pre-processes its seed x by applying a random $h^0 \in \mathcal{H}_{m,k}$ to x, and then uses our PRG (either of Construction 2 or of Construction 1) on secret seed $h^0(x)$. The hash function h^0 need not be secret. As explained in Section 2, h^0 can be specified using $\mathcal{O}(k)$ bits, therefore the public seed length remains essentially unchanged ($\mathcal{O}(k \log k)$ for Construction 2, or $\mathcal{O}(k^2)$ for Construction 1).

Degradation and Amplification of Computational Hardness

Shai Halevi and Tal Rabin

IBM T.J. Watson Research Center, Hawthorne, NY USA

Abstract. What happens when you use a partially defective bit-commitment protocol to commit to the same bit many times? For example, suppose that the protocol allows the receiver to guess the committed bit with advantage ε, and that you used that protocol to commit to the same bit more than $1/\varepsilon$ times. Or suppose that you encrypted some message many times (to many people), only to discover later that the encryption scheme that you were using is partially defective, and an eavesdropper has some noticeable advantage in guessing the encrypted message from the ciphertext. Can we at least show that even after many such encryptions, the eavesdropper could not have learned the message with certainty?

In this work we take another look at amplification and degradation of computational hardness. We describe a rather generic setting where one can argue about amplification or degradation of computational hardness via sequential repetition of interactive protocols, and prove that in all the cases that we consider, it behaves as one would expect from the corresponding information theoretic bounds. In particular, for the example above we can prove that after committing to the same bit for n times, the receiver's advantage in guessing the encrypted bit is negligibly close to $1 - (1 - \varepsilon)^n$.

Our results for hardness amplification follow just by observing that some of the known proofs for Yao's lemmas can be easily extended also to handle interactive protocols. On the other hand, the question of hardness degradation was never considered before as far as we know, and we prove these results from scratch.

1 Introduction

This work discusses the effect of running several executions of a cryptographic protocol sequentially, on the secrecy or correctness guarantees of that protocol. An illustrating example to keep in mind is a defective bit-commitment scheme, where the sender may open the commitment in two ways with probability up to δ (binding defect) and the receiver may have probability of up to $(1 + \varepsilon)/2$ in guessing the sender's bit (secrecy defect). We ask how does sequential repetition of such a protocol effect ε and δ, in situations where the inputs to the various executions may be dependent.

This question is closely related to the issue of *robust combiners* for cryptographic protocols. Indeed, Damgård et al. considered in [2] just this kind of

R. Canetti (Ed.): TCC 2008, LNCS 4948, pp. 626–643, 2008.
© International Association for Cryptologic Research 2008

defective protocols (for both commitment and oblivious transfer), and described how a non-defective protocol can be obtained from them. Two transformations were described in [2], one running many copies of the defective protocol with the same input bit, and the other running many copies with randomly chosen inputs whose exclusive-or equals the original input bit. Damgård et al. proved that in an information-theoretic setting, if the original defects satisfy $\varepsilon + \delta < 1$ then alternating between these two transformations can reduce the secrecy and binding defects to negligible quantities. Given these results, one would like to prove the same result also in the computational setting.

To illustrate the problem with moving to the computational setting, consider using a defective bit-commitment scheme to commit twice to the same input bit. In the information theoretic setting from [2], it is clear that if the commitment scheme has secrecy defect of ε, then using it twice with the same input bit yields a secrecy defect of $1 - (1 - \varepsilon)^2 = 2\varepsilon - \varepsilon^2$. In the computational setting, however, the simple hybrid argument that is commonly used to reason about "encrypting the same message many times" can only prove a bound of 2ε on the resulting defect, which is clearly too weak of a bound. (For example, one needs to show that the resulting scheme offers some secrecy, even if the original one has secrecy defect of $\frac{2}{3}$.)

In the specific context of robust combiners for commitment and oblivious-transfer, results similar to those of Damgård et al. were recently proved in the computational setting by Wullschleger [11]. Wullschleger bypassed the problem of analyzing many executions on related inputs in the computational setting, by considering a "randomized" variant of these primitives, where the parties execute the protocol on random bits, which are considered outputs of the protocol rather than inputs to it. These variants are known to be equivalent to the standard notions of commitment and oblivious transfer, but since the parties have no inputs then the different executions are truly independent. Using results of Holenstein on hardness amplification of independent executions [5,6], Wullschleger proved that starting from a defective protocol for the randomized variants, one can obtain a non-defective protocol for the same variant.

1.1   Our Results

Although sufficient for the context of defective commitment and oblivious-transfer, Wullschleger's results do not answer the fundamental question regarding the effect of sequential repetition with related input on the secrecy and correctness guarantees of protocols. They also do not answer the question of whether the specific transformations that were described by Damgård et al. [2] work also in the computational setting. Answering these questions is the focus of the current work.

Hardness Degradation Lemmas. In Section 3 we describe a rather generic setting where one can argue about hardness amplification and degradation of interactive protocols. We formulate and prove two new lemmas, showing that the information theoretic bounds on hardness-degradation (for both secrecy and correctness) carry over also to the computational setting: Lemma 2 asserts that the secrecy degradation from "encrypting the same message t times" obeys the bound of $1 - (1 - \varepsilon)^t$. Similarly, Lemma 5 asserts that given t interactive puzzles that are δ-hard to solve, the probability of solving *at least one of them* is at most $1 - (1 - \delta)^t$. These lemmas can be thought of as mirroring Yao's XOR lemma and Yao's hardness-amplification lemma for one-way functions [12], respectively. The proofs of these hardness-degradation lemmas are similar in their high-level structure to the corresponding hardness-amplification proofs. For Lemma 2 we had to prove a new lemma (Lemma 3) that plays a role similar to the one played by Levin's "Isolation Lemma" in the proof of Yao's XOR lemma.

We complement the results for secrecy/correctness degradation with results on secrecy/correctness amplification. Specifically, we observe that some (but not all) of the known proofs for Yao's XOR lemma and Yao's hardness-amplification lemma can be used to prove amplification also for interactive protocols.[1]

Improving Defective protocols. We then consider the applicability of our hardness amplification and degradation lemmas to the analysis of the transformations from [2]. Roughly, we prove that these transformations result in a secure protocol whenever the defect parameters of the original protocol satisfy $\varepsilon + \delta \leq 1 - 1/polylog(k)$ (with k the security parameter), but our techniques cannot be applied to prove security in some cases where $\varepsilon + \delta$ is bounded away from 1 only by a polynomial fraction. In Lemma 6, we characterize exactly the range of the defect parameters (ε, δ) for which we can prove that these transformations produce a secure protocol.

2   Notations

The statistical distance between two distributions D_1, D_2 over a countable domain is the scaled sum $|\mathcal{D}_1 - \mathcal{D}_2| \stackrel{\text{def}}{=} \frac{1}{2} \sum_x |\mathcal{D}_1(x) - \mathcal{D}_2(x)|$, where the sum is taken over all the elements in the union of the support of the two distributions, and $\mathcal{D}_i(x)$ is the probability mass of x according to the distribution $\mathcal{D}_i$. We use $x \in_R S$ to denote choosing x from S uniformly at random. A positive function is *negligible* if it tends to zero faster than any polynomial, and it is *noticeable* otherwise.

An algorithm is called *efficient* if it runs in probabilistic polynomial time. A two-party protocol is a pair of algorithms, one for each party. We use the following notations to describe a two-party protocol (A, B):

- The communication transcript is denoted $\langle A(a, r_a), B(b, r_b) \rangle$.
- The event where A outputs the string x is denoted $(A(a, r_a), B(b, r_b)) \stackrel{A}{\rightarrow} x$, and similarly $(A(a, r_a), B(b, r_b)) \stackrel{B}{\rightarrow} y$ for the output of B, and $(A(a, r_a), B(b, r_b)) \rightarrow (x, y)$ for the output of both.

---

[1] Essentially, the proofs that can be extended are those where the single-instance adversary A runs the multiple-instance adversary A' on just one vector that includes the instance that A wants to solve. In the interactive case, this translates to a "non-rewinding" reduction. See more details in the proofs of Lemma 1 and Lemma 5.

In these notations, a, b are the inputs and r_a, r_b are the randomness used by the participants. We often omit the randomness (and sometimes also the input) from these notations. We use $\star$ to denote a "don't care" input or output.

3   Amplification/Degradation of Computational Hardness

In this section we prove some lemmas about amplification and degradation of computational hardness for sequential composition of protocols. (By "computational hardness" we roughly mean breaking either the secrecy or correctness of the protocol.) The amplification lemmas are straightforward extensions of Yao's XOR lemma and Yao's hardness-amplification lemma for one-way functions [12,4], but the degradation lemmas are new.

We deal with two-party protocols, where one player either tries to guess the input of the other party or tries to break the correctness of the protocol (e.g., in a commitment scheme the goal is either to learn the committed bit or to open the commitment in two different ways). We study how the computational-hardness of accomplishing these tasks is amplified or degraded when several copies of the protocol are run sequentially in various settings. We consider the following four scenarios in the setting of two parties A and B, where A has input a.

SECRECY. In this setting player B wants to learn the input of player A.

Amplification. We examine the effect of running the protocol t times, where in each invocation player A chooses a random input, subject to the condition that the XOR of the t inputs is A's original input a.

When restricted to the non-interactive case of one-way functions, this is exactly the setting for Yao's XOR lemma [12]. We note that Levin's proof [9] can be easily extended to sequential composition of interactive protocols (see also [4, Lemma 4]).

Degradation. We examine the effect of running the protocol t times, but this time player A uses the same input in every run. This "secrecy degradation" setting is dealt with in Lemma 2.

CORRECTNESS. In this setting, player A tries to break the correctness of the protocol by outputting some "forbidden value" at the end of the protocol execution (such as two different opening of the commitment).

Amplification. We consider the setting where after t runs of the protocol, player A needs to break *all the t executions*.

When restricted to the non-interactive case of one-way functions, this is exactly the setting for Yao's hardness-amplification lemma from weak to strong one-way functions [12]. Here, again, the proof of Canetti et al. [1] can be easily extended to interactive protocols.[2]

---

[2] Despite the similarities, the hardness-amplification lemma *does not follow* from the results for soundness amplification of interactive proofs. The reason is that in our case the adversary can compute the "forbidden output" at the very end, after all the executions took place. In the IP setting, on the other hand, the prover needs to "convince the verifier" after each execution and before the next one starts.

Degradation. We consider the setting where after t runs of the protocol, player A needs to break *any one of the t executions*. This "hardness degradation" setting is dealt with in Lemma 5. (The proof closely mirrors the "hardness amplification" proof from [1].)

3.1   Secrecy Amplification and Degradation

Let (A, B) be an interactive protocol where A has a single-bit input $a \in \{0, 1\}$ (and B may have no input), and let $t = t(k)$ be polynomially bounded. Denote by $(A_{=}^t, B^t)$ a t-fold sequential repetition of (A, B), where the protocol (A, B) is run t times sequentially, each time with the same input bit a. Also denote by $(A_{\oplus}^t, B^t)$ a t-fold sequential repetition of (A, B), where the input of A in each run is random and independent, subject to the condition that the XOR of the inputs in all the runs equals to the input bit of $A_{\oplus}^t$.

Definition 1 (Input Secrecy Defect). *The protocol (A, B) has an ε-bounded secrecy defect with respect to A if, for every efficient B', it holds that $\Pr[(A(a), B') \xrightarrow{B'} a] \leq (1 + \varepsilon)/2 + \mathsf{negl}(k)$, where the probability is taken over the choice of $a \in_R \{0, 1\}$ and the randomness of A and B', k is the security parameter, and negl is a negligible function.*

Lemma 1 (Yao's XOR Lemma [12] – Secrecy Amplification). *If (A, B) has an ε-bounded secrecy defect with respect to A and t is polynomially-bounded, then $(A_{\oplus}^t, B^t)$ has an ε^t-bounded secrecy defect with respect to $A_{\oplus}^t$.*

Proof (sketch): We observe that Levin's proof of Yao's XOR lemma [9] can be extended also to interactive protocols. (See a description of that proof also in [4, Lemma 4].) The reason that this particular proof extends to the interactive case (whereas the other proofs from [4] do not seem to extend) is that this proof does not need to "rewind" A:

Recall that we assume an adversary B' with advantage better than ε^t when talking to $A_{\oplus}^t$, and we want to construct an adversary B^* with advantage better than ε when talking to A. In the non-interactive case, we had a "puzzle" that came from A and we could stick that puzzle anywhere in a vector of t puzzles and let B' attempt to solve that vector. We could also stick the same puzzle in many vectors and run B' on all of them. In the interactive case, on the other hand, once we sent some messages to the real party A, we cannot "take them back" and try another interaction instead.

On a high level, the reduction following Levin's approach proceeds as follows: B^* simulates the interactions between B' and $A_{\oplus}^t$ for several runs, $i = 1, 2, \ldots$: Starting from the state that B' ended at after the $i - 1$'st run, B^* uses repeated sampling to look for a simulated execution of the i'th run after which B^* still has advantage better than ε^{t-i} in guessing the bit of $A_{\oplus}^{t-i}$ (where the probability is taken over the remaining runs). It continues in this fashion until it cannot find such an i'th run (or until it gets to the last run). Then it uses the current state of B' as a basis for a single interaction with the "real player" A. If this was the last run then it uses the output of B' as the guess of A's input bit, and otherwise it uses repeated sampling again to estimate the probability that B' outputs one (taken over the remaining runs), and compares that probability to some threshold (that it can also compute using repeated sampling).

Levin's isolation lemma then proves that if at some point B' failed to find an i'th run as above, then there is a threshold that it can set that would give it an advantage better than ε of guessing the input bit of the "real player" A.   □

Lemma 2 (Secrecy Degradation). *If (A, B) has an ε-bounded secrecy defect with respect to A and t is polynomially-bounded, then $(A_=^t, B^t)$ has an ε'-bounded secrecy defect with respect to $A_=^t$, where $\varepsilon' = 1 - (1 - \varepsilon)^t$.*

We emphasize that the simple hybrid argument that is commonly used to reason about "encrypting the same message many times" can be used in this context to prove a bound of $\varepsilon' \leq t\varepsilon$. The difficulty in the proof below is in improving the bound from $t\varepsilon$ to $1 - (1 - \varepsilon)^t$.

Proof. Let $t = t(k)$ be polynomially bounded, let $\varepsilon = \varepsilon(k)$, and denote $\varepsilon' \stackrel{\text{def}}{=} 1 - (1 - \varepsilon)^t$. We show that if there exist a randomized adversary B' of time complexity T' such that

$$\Pr_{a, r_a, r_b}[(A_=^t(a, r_a), B'(r_b)) \stackrel{B'}{\to} a] \geq \frac{1 + \varepsilon' + \rho}{2},$$

where $\rho = \rho(k)$ is noticeable, then there exists a randomized adversary B^* of time complexity $T = T' \cdot poly(kt/\varepsilon\rho)$ such that

$$\Pr_{a, r_a, r_b}[(A(a, r_a), B^*(r_b)) \stackrel{B^*}{\to} a] \geq \frac{1 + \varepsilon + \varepsilon\rho/4}{2}.$$

An alternative way to write the condition $\Pr[(A_=^t(a), B') \stackrel{B'}{\to} a] \geq \frac{1 + \varepsilon' + \rho}{2}$ is

$$\Pr[(A_=^t(1), B') \stackrel{B'}{\to} 1] - \Pr[(A_=^t(0), B') \stackrel{B'}{\to} 1] \geq \varepsilon' + \rho.$$

Below we always use this alternative formulation.

Consider breaking B' into two parts: the first part B_1' interacts with $A(a)$ only once and outputs the internal state at the end of this interaction, and the second part B_2' gets this internal state as input and then interacts with $A(a)$ for $t - 1$ more times before outputting a guess for the bit a. Denote by $\mathcal{D}_0, \mathcal{D}_1$ the probability distribution of the internal state s after B_1' interacts with $A(0)$, $A(1)$, respectively.

$$\mathcal{D}_0 \stackrel{\text{def}}{=} \left\{ s \ : \ (A(0), B_1') \stackrel{B_1'}{\to} s \right\}, \quad \text{and} \quad \mathcal{D}_1 \stackrel{\text{def}}{=} \left\{ s \ : \ (A(1), B_1') \stackrel{B_1'}{\to} s \right\}$$

(the notation $\mathcal{D}_0, \mathcal{D}_1$ is interpreted both as a probability distribution and as the corresponding support set). For any given internal state $s \in \mathcal{D}_0 \cup \mathcal{D}_1$, consider the experiment where starting from this internal state s, B_2' interacts $t - 1$ more times with A, but the input of A in all these executions is some bit a' (which may or may not be equal to the input bit a of the first execution). We denote by $p_0(s), p_1(s)$ the probabilities that B' outputs 1 in this experiment when $a' = 0$ and $a' = 1$, respectively. Namely, for every $s \in \mathcal{D}_0 \cup \mathcal{D}_1$ we denote

$$p_0(s) \stackrel{\text{def}}{=} \Pr\left[\left(A_{=}^{t-1}(0), B_2'(s)\right) \stackrel{B_2'}{\to} 1\right], \quad \text{and} \quad p_1(s) \stackrel{\text{def}}{=} \Pr\left[\left(A_{=}^{t-1}(1), B_2'(s)\right) \stackrel{B_2'}{\to} 1\right]$$

We view p_0, p_1 as random variables in $[0, 1]$, where each random variable can be chosen over either of the two probability spaces $\mathcal{D}_0$ or $\mathcal{D}_1$. Below, we use notations such as $\Pr_{\mathcal{D}_0}[p_0 > t]$ to denote the probability that we get $p_0(s) > t$ when setting $s \in_R \mathcal{D}_0$, or $E_{\mathcal{D}_1}[p_1]$ to denote the expected value of $p_1(s)$ taken over the choice $s \in_R \mathcal{D}_1$, etc.

The technical Lemma 3 below asserts roughly that either there exists some internal state s^* such that $p_1(s^*) - p_0(s^*) > 1 - (1 - \varepsilon)^{t-1}$, or there exists some probability threshold τ such that $\Pr_{\mathcal{D}_1}[p_1 > \tau] - \Pr_{\mathcal{D}_0}[p_1 > \tau] > \varepsilon$. If there exists a state s^* as in the first case, then $B_2'(s^*)$ guesses the input bit of $A_{=}^{t-1}$ with advantage better than $1 - (1 - \varepsilon)^{t-1}$ and we can continue recursively. Otherwise, we can construct B^* roughly as follows: B^* first plays the part of B_1', interacting with $A(a)$ and gets the internal state s. Then, it evaluates $p_1(s)$ (by repeated sampling), outputs 1 if $p_1(s) > \tau$ and 0 otherwise.

The actual statement of the technical lemma below is slightly more complicated, since it also includes the "slackness parameter" ρ that is needed to get the result in a uniform complexity setting. Specifically, in the first case there should be a significant probability of finding a state s^* for which $p_1(s^*) - p_0(s^*) > 1 - (1 - \varepsilon)^{t-1} + \rho$, and in the second case there should be some uniform way of finding the threshold τ.

Lemma 3. *Fix any integer t and any $\varepsilon, \rho \in [0, 1]$ such that $\rho < (1 - \varepsilon)^t$. Also let $\mathcal{D}_0, \mathcal{D}_1$ be two probability distributions and let p_0, p_1 be two random variables that are defined over both $\mathcal{D}_0$ and $\mathcal{D}_1$. If $E_{\mathcal{D}_0}[p_1] - E_{\mathcal{D}_1}[p_0] > 1 - (1 - \varepsilon)^t + \rho$, then at least one of the two conditions must hold:*

(i) Either $\Pr_{\mathcal{D}_0}\left[p_1 - p_0 > 1 - (1 - \varepsilon)^{t-1} + \rho\right] \geq \dfrac{\epsilon\rho}{2}$,

(ii) or $E_\tau\left[\Pr_{\mathcal{D}_1}[p_1 > \tau] - \Pr_{\mathcal{D}_0}[p_1 > \tau]\right] > \varepsilon(1 + \rho/2)$, where the expectation is over choosing τ uniformly at random in the interval $[1 - (1 - \varepsilon)^{t-1} + \rho, \ 1]$.

We prove Lemma 3 later in this section. Using this lemma, we now complete the proof of Lemma 2 as follows: from the assertion we have that $E_{\mathcal{D}_0}[p_1] - E_{\mathcal{D}_1}[p_0] > 1 - (1 - \varepsilon)^t + \rho$ so we can apply Lemma 3. The adversary B^* will sample $poly(k/\epsilon\rho)$ internal states $s \in_R \mathcal{D}_0$, and for each will evaluate $p_0(s)$ and $p_1(s)$ with accuracy $poly(\rho/t)$ and error $poly(\varepsilon\rho/tk)$. If it finds a state s for which $p_1 - p_0 > 1 - (1 - \varepsilon)^{t-1} + \rho(1 - 1/2t)$ then it uses $B_2'(s)$ as an adversary against the $(t - 1)$-sequential repetition $A_{=}^{t-1}(a)$ and continue by recursion.

Otherwise, B^* plays the role of B_1', interacting with $A(a)$ to produce an internal state s. Then B^* evaluates $p_1(s)$ with accuracy $poly(\varepsilon\rho)$ and error $poly(\varepsilon\rho/k)$.

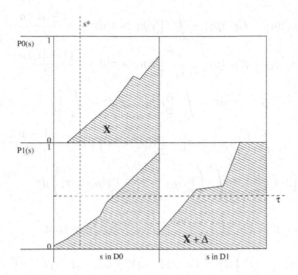

Fig. 1. An illustration of Lemma 3. We know that the gray area in the lower-right box is more than $X + (1 - (1 - \varepsilon)^t)$. We essentially prove that either there is s^* such that $p_1(s^*) - p_0(s^*) > 1 - (1 - \varepsilon)^{t-1}$ or there is τ such that $\Pr_{\mathcal{D}_1}[p_1 > \tau] - \Pr_{\mathcal{D}_0}[p_1 > \tau] > \varepsilon$.

It then chooses at random $\tau \in_R [1 - (1 - \varepsilon)^{t-1} + \rho, 1]$, and outputs 1 if $p_1(s) > \tau$ and 0 otherwise. It is not hard to see that this algorithm has expected advantage of $\varepsilon(1 + \rho/2) - poly(\varepsilon\rho/k) > \varepsilon(1 + \rho/4)$.

Proof of Lemma 3. The proof relies on the identity $E[X] \equiv \int_0^\infty \Pr[X > \tau]d\tau$, that holds for any non-negative random variable X. In our case, we have $p_0, p_1 \in [0, 1]$ so we can integrate between 0 and 1 (rather than 0 and ∞). Assume that the premise of the lemma holds but condition (i) does not, and we prove that then condition (ii) must hold. For the proof below, denote

$$\mu \overset{\text{def}}{=} 1 - (1 - \varepsilon)^{t-1} + \rho$$

If condition (i) does not hold then with all but probability $\varepsilon\rho/2$ over choosing $s \in_R \mathcal{D}_0$, we have $p_1(s) - p_0(s) \leq \mu$. This implies that, for all $\tau \in [\mu, 1]$, it holds that $\Pr_{\mathcal{D}_0}[p_1 > \tau] \leq \Pr_{\mathcal{D}_0}[p_0 > \tau - \mu] + \frac{\varepsilon\rho}{2}$, and therefore also

$$\int_\mu^1 \Pr_{\mathcal{D}_0}[p_1 > \tau]dt \leq \int_0^{1-\mu} \left(\Pr_{\mathcal{D}_0}[p_0 > \tau] + \frac{\varepsilon\rho}{2}\right)dt = \int_0^{1-\mu} \Pr_{\mathcal{D}_0}[p_0 > \tau]dt + \frac{(1 - \mu)\varepsilon\rho}{2} \quad (\star)$$

Using this inequality and the premise of the lemma, we can write:

$$1 - (1 - \varepsilon)^t + \rho \quad < \quad E_{\mathcal{D}_1}[p_1] - E_{\mathcal{D}_0}[p_0] = E_{\mathcal{D}_1}[p_1] - E_{\mathcal{D}_0}[p_1] + \int_0^\mu \Pr_{\mathcal{D}_0}[p_1 > \tau]d\tau$$

$$+ \int_\mu^1 \Pr_{\mathcal{D}_0}[p_1 > \tau]d\tau - \int_0^{1-\mu} \Pr_{\mathcal{D}_0}[p_0 > \tau]d\tau - \int_{1-\mu}^1 \Pr_{\mathcal{D}_0}[p_0 > \tau]d\tau$$

$$\overset{\text{Eq. }(\star)}{\leq} E_{\mathcal{D}_1}[p_1] - E_{\mathcal{D}_0}[p_1] + \int_0^\mu \Pr_{\mathcal{D}_0}[p_1 > \tau]d\tau + \frac{(1-\mu)\varepsilon\rho}{2} - \int_{1-\mu}^1 \Pr_{\mathcal{D}_0}[p_0 > \tau]d\tau$$

$$\leq E_{\mathcal{D}_1}[p_1] - E_{\mathcal{D}_0}[p_1] + \int_0^\mu \Pr_{\mathcal{D}_0}[p_1 > \tau]d\tau + \frac{(1-\mu)\varepsilon\rho}{2}$$

$$= \int_0^\mu \Pr_{\mathcal{D}_1}[p_1 > \tau]d\tau + \int_\mu^1 \Pr_{\mathcal{D}_1}[p_1 > \tau]d\tau - \int_\mu^1 \Pr_{\mathcal{D}_0}[p_1 > \tau]d\tau + \frac{(1-\mu)\varepsilon\rho}{2}$$

$$\leq \mu + \int_\mu^1 \left(\Pr_{\mathcal{D}_1}[p_1 > \tau] - \Pr_{\mathcal{D}_0}[p_1 > \tau] \right) d\tau + \frac{(1-\mu)\varepsilon\rho}{2}$$

$$= \mu\left(1 - \frac{\varepsilon\rho}{2}\right) + \int_\mu^1 \left(\Pr_{\mathcal{D}_1}[p_1 > \tau] - \Pr_{\mathcal{D}_0}[p_1 > \tau] \right) d\tau + \frac{\varepsilon\rho}{2}$$

Substituting back $\mu = 1 - (1-\varepsilon)^{t-1} + \rho$, we conclude that

$$\int_\mu^1 \left(\Pr_{\mathcal{D}_1}[p_1 > \tau] - \Pr_{\mathcal{D}_0}[p_1 > \tau] \right) d\tau > (1 - (1-\varepsilon)^t + \rho) - (1 - (1-\varepsilon)^{t-1} + \rho)(1 - \frac{\varepsilon\rho}{2}) - \frac{\varepsilon}{2}$$

$$= \underbrace{((1-\varepsilon)^{t-1} - \rho)}_{1-\mu}\left(\varepsilon - \frac{\varepsilon\rho}{2}\right) + \varepsilon\rho > (1-\mu)\left(\varepsilon + \frac{\varepsilon\rho}{2}\right)$$

Hence, the expected value of the difference $\Pr_{\mathcal{D}_1}[p_1 > \tau] - \Pr_{\mathcal{D}_0}[p_1 > \tau]$, taken over a uniform random choice of $\tau \in_R [\mu, 1]$, is at least $\varepsilon + \frac{\varepsilon\rho}{2}$.   □

3.2   Hardness Amplification and Degradation

Consider an interactive protocol $\mathcal{P} = (A, B)$, and let $R_{\mathcal{P}}$ be a poly-time recognizable relation that describes what it means for A to "break the protocol's correctness". Namely, after the protocol is run and B's output is some string y, a cheating A' is successful if it outputs a string x such that $(x, y) \in R_{\mathcal{P}}$. (For example, (A, B) is a commitment scheme, A is the sender, B's output is the communication transcript y, and $(x, y) \in R_{\mathcal{P}}$ if x contains two different openings that are both consistent with y.)

Let (A^t, B^t) be a t-fold sequential repetition of the protocol (A, B) with A, B having the same input (if any) but independent randomness. Define $\wedge^t(R_{\mathcal{P}})$ and $\vee^t(R_{\mathcal{P}})$ as the AND and OR of the t individual relations, namely

$$\wedge^t(R_{\mathcal{P}}) \overset{\text{def}}{=} \{(\langle x_1, \ldots, x_t \rangle, \langle y_1, \ldots, y_t \rangle) : \forall i \leq t, (x_i, y_i) \in R_{\mathcal{P}}\},$$

$$\vee^t(R_{\mathcal{P}}) \overset{\text{def}}{=} \{(x, \langle y_1, \ldots, y_t \rangle) : \exists i \leq t \text{ s.t. } (x, y_i) \in R_{\mathcal{P}}\}$$

In other words, $\wedge^t(R_{\mathcal{P}})$ represents the case that all the t copies must be broken, and $\vee^t(R_{\mathcal{P}})$ represents the case that at least one copy is broken.

Definition 2 ($R_{\mathcal{P}}$-defect). *The protocol $\mathcal{P} = (A, B)$ has a δ-bounded $R_{\mathcal{P}}$-defect with respect to B if for every efficient A' it holds that $\Pr[(A', B) \rightarrow (x, y) : (x, y) \in R_{\mathcal{P}}] \leq \delta + \mathsf{negl}(k)$, where the probability is taken over the randomness of A' and B, k is the security parameter, and negl is a negligible function.*

Lemma 4 (Hardness Amplification). *If* $\mathcal{P} = (A, B)$ *has a* δ-bounded $R_{\mathcal{P}}$-*defect with respect to* B *and* t *is polynomially bounded then* (A^t, B^t) *has a* δ^t-*bounded* $\wedge^t(R_{\mathcal{P}})$-*defect with respect to* B^t.

The proof is nearly identical to the hardness-amplification proof from [1] for the non-interactive case (and also very similar to the proof for Lemma 5 below). Again, the reason that this proof extends to the interactive case (whereas some other proofs of Yao's lemma of weak-to-strong-OWFs do not extend) is that it does not need to "rewind" the player B.   □

Lemma 5 (Hardness Degradation). *If* $\mathcal{P} = (A, B)$ *has a* δ-bounded $R_{\mathcal{P}}$-*defect with respect to* B *and* t *is polynomially bounded then* (A^t, B^t) *has a* δ'-*bounded* $\vee^t(R_{\mathcal{P}})$-*defect with respect to* B^t, *where* $\delta' = 1 - (1 - \delta)^t$.

Proof (sketch): The proof is very similar to the hardness-amplification proof from [1]. Let $t = t(k)$ be polynomially bounded, and let $\delta = \delta(k)$ be a noticeable function and $\delta' = 1 - (1 - \delta)^t$. Assume that there exists a randomized adversary A' of time complexity T' that satisfies the relation $\vee^t(R_{\mathcal{P}})$ with probability $\delta' + \rho'$ for some noticeable quantity $\rho' = \rho'(k)$. We then show that there exists a randomized adversary A^* of time complexity $T^* = T' \cdot poly(kt/\delta'\rho')$ that satisfies $R_{\mathcal{P}}$ with probability $\delta + \rho$, where ρ is the solution to $(1 - \delta - \rho)^t = (1 - \delta)^t - \rho'$. Observe that if ρ' is noticeable and t is polynomial then also ρ is noticeable. Note also that by definition, the success probability of A' is $1 - (1 - \delta - \rho)^t$.

Denote the state of A' after the i'th interaction with B by s_i (with s_0 being the initial state of A'). The adversary A^* begins by playing the role of B in the first interaction. Repeating the first interaction up to $poly(kt/\delta\rho)$ times, A^* is looking for an internal state s_1 after the first interaction such that when proceeding from this state, A' satisfies $R_{\mathcal{P}}$ for one of the last $t - 1$ runs with probability at least $1 - (1 - \delta - \rho)^{t-1}$. (Note that A' can estimate that probability by sampling.)

If A^* succeeds in finding such s_1, then it fixes that internal state and keeps looking for internal states $s_2, s_3, \ldots$ such that when proceeding from s_i, adversary A' satisfies $R_{\mathcal{P}}$ for one of the last $t - i$ runs with probability at least $1 - (1 - \delta - \rho)^{t-i}$. If A^* can find an internal state s_{t-1} from which A' satisfies $R_{\mathcal{P}}$ for the last run with probability $\geq \delta + \rho$ then we are done: A^* just uses A' from this state when interacting with the real B. Otherwise, A^* has some state s_i with $0 \leq i < t - 1$ such that A' satisfies $R_{\mathcal{P}}$ for one of the last $t - i$ runs with probability at least $1 - (1 - \delta - \rho)^{t-i}$, and yet for (almost) all continuation states s_{i+1}, A' only satisfies $R_{\mathcal{P}}$ for one of the last $t - i - 1$ runs with probability less than $1 - (1 - \delta - \rho)^{t-i-1}$.

We now consider a "matrix" M that represent the interaction of A' with B on the remaining $t - i$ runs of the protocol, when A' starts from this state s_i. (We assume that s_i includes all the randomness that A' needs for all the runs.) The columns of M are labeled by all the possibilities for the randomness of B during the $i + 1$'st run, and rows are labeled by all the possibilities for the randomness of B in runs $i + 2, \ldots, t$. Hence, each entry in the matrix corresponds to a particular interaction of A' with B on the remaining $t - i$ runs of the protocol.

Each entry in M is labeled with two bits, where the first bit is 1 if at the end of that interaction A' satisfies $R_\mathcal{P}$ for the $i+1$'st run, and the second bit is 1 if A' satisfies $R_\mathcal{P}$ for one of the last $t-i-1$ runs. By our assumption on the state s_i, we know that a random entry in this matrix is labeled with $(0,0)$ with probability at most $\gamma = (1-\delta-\rho)^{t-i}$. Denote $\alpha = (1-\delta-\rho)^{t-i-1}$ and $\beta = (1-\delta-\rho)$, so $\alpha \cdot \beta = \gamma$. Then, it must be the case that either M has (sufficiently many) columns where the fraction of entries of the form $(\star, 0)$ is no more than α, or else the conditional probability of a $(0,0)$ entry given that the entry is of the form $(\star, 0)$ is at most (only slightly more than) β.

The failure of A^* to find a continuation state s_{i+1} with sufficient residual success probability indicates that the first case does not hold, so the second case must hold. Hence, in this case A^* uses A' starting from s_i to interact with the real player B, arriving at some state s_{i+1} after this "real interaction." Then, A^* simulates many more runs of A' with B starting from this s_{i+1}. Adversary A^* looks for a run in which A' *does not satisfy* $R_\mathcal{P}$ for any of the last $t-i-1$ runs, and uses the output of A' in that run in the hope that it satisfies $R_\mathcal{P}$ for the $i+1$'st run. The conditional probability argument from above says that the odds of satisfying $R_\mathcal{P}$ for the $i+1$'st run conditioned on not satisfying it for the last $t-i-1$ runs is (only slightly less than) $1-\beta = \delta + \rho$. Indeed, a detailed argument that mirrors the proof of [1, Lemma 1] shows that this algorithm A^* has success probability noticeably larger than δ.   □

4   Fixing Defective Protocols

In [2], Damgård et al. considered defective two-party protocols such as oblivious-transfer and commitment between *a Sender* and *a Receiver*. They suggested reducing the defect by alternating between two transformations: Roughly, in a "type-R" transformation the parties run t copies of the protocol with the same input bits for the sender, and in a "type-S" transformation the sender chooses t random bits whose exclusive-or equals to its input bit and then the parties run one copy of the protocol for each of these random bits.

Below we assume that the underlying protocol has defect ε for the Sender security and defect δ for the Receiver security (such as the commitment protocol that was described in the introduction). In the information-theoretic setting that was considered in [2], it is clear that applying a type-R transformation results in a protocol with sender defect $1-(1-\varepsilon)^t$ and receiver defect δ^t, and similarly applying a type-S transformation results in a protocol with sender defect ε^t and receiver defect $1-(1-\delta)^t$. It was shown in [2] that as long as $\varepsilon + \delta < 1 - 1/poly(k)$, one can alternate between these transformations several times (with total number of copies polynomial in k) and reduce both defects to negligible quantities in k.

Our lemmas from Section 3 imply that the same bounds on the effect of type-R and type-S transformations hold also in the computational setting. One could hope, therefore, that the alternation strategy from [2] can be proven to work also in this setting. Unfortunately, this is not the case. The reason is the strategy from [2] uses a non-constant number of alternations. The proofs for hardness-amplification and degradation from Section 3 all incur a polynomial blowup in the complexity of the adversary for every alternation, and hence a non-constant number of alternations would cause a super-polynomial blowup in the adversary complexity. In Section 4.1 below we analyze the range of parameters ε, δ for which we can reduce the defect to a negligible amount using only a constant number of alternations.

Relation to Wullschleger's work. As we described in the introduction, Wullschleger recently was able to extend the results from [2] to the computational setting via a somewhat different approach. Roughly, instead of running many copies of the protocol on related inputs, he suggested to run many copies on random and independent inputs, followed by the Sender sending to the Receiver various linear combinations of these random bits and the input bit. Since the protocols are now run on independent inputs, then one can use the hardness-amplification results of Holenstein to argue about them [5,6], and these arguments still hold even in the presence of the various linear combinations that the Sender later sends to the Receiver.

Wullschleger's work yield stronger defect-reduction results than the ones that we can obtain from a direct analysis of the transformations of [2]: he is able to fix a defect of $\varepsilon + \delta < 1 - 1/poly(k)$, where we can roughly fix only when $\varepsilon + \delta < 1 - 1/polylog(k)$. However, Wullschleger's work does not shed light on what happens when a defective protocol is run several times on related inputs, and does not say what happens when the original transformations from [2] are used in the computational setting.

4.1   Iterating the Transformations

Below, we prove that repeating the transformations S and R a constant number of times results in a scheme with negligible defects as long as $\varepsilon + \delta$ is bounded away from 1 and, moreover, $\varepsilon + \delta < 1 - \min(\varepsilon, \delta)/polylog(k)$.

We begin by setting a few conventions and notations. First, we can assume without loss of generality that we always alternate between transformations S and R (since applying two successive transformations of the same type with parameters t and t' is the same as just one transformation with parameter tt'). We also assume, without loss of generality, that for $\varepsilon > \delta$ we begin with transformation S and for $\varepsilon \leq \delta$ we begin with transformation R. (Namely, we choose the first transformation to increase the larger value and decrease the smaller one.) This is without loss of generality, since we can always start with a "dummy transformation" with parameter $t = 1$.

With these two assumptions, a chain of transformations is completely characterized by the initial values ε_0, δ_0 and by the sequence of parameters $t_1, t_2, \ldots$ that indicate how many times we repeat the scheme from step i in step $i + 1$. In the analysis below we refer to this representation as a "chain".

Definition 3 (Transformation chains). *A transformation chain (or just chain) is represented by a vector* $C = \langle(\varepsilon_0, \delta_0), (t_1, t_2, \ldots, t_\ell)\rangle$ *where* $\varepsilon_0, \delta_0 \in [0, 1]$ *and* $t_i \geq 1$ *for all* i. *Given* C *as above, we can compute the values* ε_i, δ_i *for each* $i = 1, \ldots, \ell$ *as follows:*

- *If* $\varepsilon_0 \geq \delta_0$ *then for even* i *we set* $\varepsilon_{i+1} = 1 - (1 - \varepsilon_i)^{t_{i+1}}$ *and* $\delta_{i+1} = \delta_i^{t_{i+1}}$, *and for odd* i *we set* $\varepsilon_{i+1} = \varepsilon_i^{t_{i+1}}$ *and* $\delta_{i+1} = 1 - (1 - \delta_i)^{t_{i+1}}$.
- *If* $\varepsilon_0 < \delta_0$ *then we swap the even and odd rules.*

It is clear, however, that not every "chain" corresponds to a sequence of transformations that we can use. For example, it is clear that $\prod_i t_i$ must be polynomial in the security parameter k. Moreover, all the ε_i's and δ_i's must be bounded away from 1 (i.e., be at most $1 - 1/poly(k)$), since our defect definitions imply that a defect of $1 - \mathsf{negl}(k)$ is the same as a defect of 1. These conditions are captured in the following definition:

Definition 4 (Confined chains). *A chain* $C = \langle(\varepsilon_0, \delta_0), (t_1, t_2, \ldots, t_\ell)\rangle$ *is confined if there exist constants* $c, c' > 0$ *such that (a)* $\prod_{i=1}^{\ell} t_i \leq k^c$ *and (b) for all* $i \leq \ell$, *we have* $\varepsilon_i, \delta_i \leq 1 - k^{-c'}$.

Moreover, the reductions proving lemmas 1 and 2 increase the size of the adversary by a polynomial factor (even if we only use $t = 2$), so we can only apply these transformations a constant number of times. This means that, to get a scheme with negligible defect, we must find a constant-length confined chain that begins with the given $(\varepsilon_0, \delta_0)$ and ends with $\varepsilon_\ell, \delta_\ell = \mathsf{negl}(k)$. The next lemma asserts a necessary and sufficient conditions on $(\varepsilon_0, \delta_0)$ for such a chain to exist.

Lemma 6. *Fix some* $\varepsilon_0 = \varepsilon_0(k)$ *and* $\delta_0 = \delta_0(k)$ *such that* $\varepsilon_0 + \delta_0 < 1 - 1/poly(k)$. *There exist a constant-length confined chain that begins with these* $(\varepsilon_0, \delta_0)$ *and ends with* $\varepsilon_\ell, \delta_\ell = \mathsf{negl}(k)$ *if and only if* $\varepsilon_0 + \delta_0 \leq 1 - \Omega\left(\frac{\min(\varepsilon_0, \delta_0)}{polylog(k)}\right)$.

Proof. Roughly, the proof considers the quantity $a = \frac{1 - \max(\varepsilon, \delta)}{\min(\varepsilon, \delta)}$, and shows that as long as $a = 1 + o(1)$, then each iteration increases the $o(1)$ part of a by at most a factor of $O(\log k)$. Thus, we must have $a \geq 1 + \Omega(1/polylog)$ if we want a to grow beyond $1 + o(1)$ in a constant number of iterations. In the proof below we use the following facts:

1. For every $\alpha > -1$ and $x \geq 1$, $(1 + \alpha)^x \geq 1 + \alpha x$.
2. For every $0 \leq \alpha \leq \frac{1}{2}$ and $1 \leq x \leq \frac{1}{2\alpha}$, $(1 + \alpha)^x \leq 1 + 2\alpha x$.
3. For every $0 \leq \alpha \leq \frac{1}{2}$ and $1 \leq x \leq \frac{1}{\alpha}$, $(1 - \alpha)^x \leq 1 - \alpha x/2$.
4. For every $0 \leq \alpha \leq 1$, $(1 - \alpha)^{1/\alpha} < e^{-1} (\approx 0.37)$
5. For every $0 \leq \alpha < \frac{1}{2}$, $(1 - \alpha)^{1/\alpha} > 1/4$

If ($\Rightarrow$) Assume that, for some constant $c \geq 1$, it holds that $\max(\varepsilon_0, \delta_0) \leq 1 - k^{-c}$, and also $\varepsilon_0 + \delta_0 \leq 1 - \frac{\min(\varepsilon_0, \delta_0)}{\log^c(k)}$. We show a confined chain of length at most $c + 5$ such that $\varepsilon_{c+5}, \delta_{c+5} = \mathsf{negl}(k)$. Assume that $\max(\varepsilon_0, \delta_0) > k^{-c'}$ for some c' (otherwise we already have $\varepsilon_0, \delta_0 = \mathsf{negl}(k)$), and consider the following procedure for generating such a chain:

1. $H_0 := \max\{\varepsilon_0, \delta_0\}, \; L_0 := \min\{\varepsilon_0, \delta_0\}$
2. $i := 1, \; t_1 := \max\{t \in \mathbb{N} : (1 - H_0)^t > k^{-c}\}$ $// \; t_1 \le \lceil c \ln(k)/H_0 \rceil \; = \; O(k^{c'} \log k)$
3. $H_1 := 1 - (1 - H_0)^{t_1}, \; L_1 := L_0^{t_1}$ $// \; \frac{1}{2} \le H_1 < 1 - k^{-c}$

4.   **while** $(1 - H_i)/L_i < 2k$ **do** $// \; L_i > (1 - H_i)/2k > k^{-c-1}/2$
5. $t_{i+1} := \max\{t \in \mathbb{N} : (1 - L_i)^t > k^{-c}\}$ $// \; t_{i+1} \le \lceil c \ln(k)/L_i \rceil \; = \; O(k^{c+1} \log k)$
6. $H_{i+1} := 1 - (1 - L_i)^{t_{i+1}}, \; L_{i+1} := H_i^{t_{i+1}}$ $// \; \frac{1}{2} \le H_{i+1} < 1 - k^{-c}$
7. $i := i + 1$

8. $t_{i+1} := \lfloor k/(1 - H_i) \rfloor, \; H_{i+1} := 1 - (1 - L_i)^{t_{i+1}}, \; L_{i+1} := H_i^{t_{i+1}}$
9. $t_{i+2} := k, \; H_{i+2} := 1 - (1 - L_{i+1})^{t_{i+2}}, \; L_{i+2} := H_{i+1}^{t_{i+2}}$

We start by establishing some simple invariants that holds throughout all the iterations of the loop.

- For all i we have $L_i + H_i < 1$. This follows since initially we have $L_0 + H_0 < 1$, and if $x + y < 1$ then also $(1 - (1 - x)^t) + y^t < 1$ for all $t \ge 1$ so this property is preserved.
- For all i we have $L_i < \frac{1}{2} < H_i < 1 - k^{-c}$:
  - The condition $H_i < 1 - k^{-c}$ follows since the t_i's are chosen specifically to ensure it.
  - On the other hand, we always set $H_i := 1 - (1 - \alpha)^{t_i}$ for some $\alpha < 1$ and where t_i is chosen as $\max\{t : (1 - \alpha)^t > k^{-c}\}$. So either $\alpha > \frac{1}{2}$, in which case $H_i \ge \alpha > \frac{1}{2}$, or $\alpha \le \frac{1}{2}$, in which case $(1 - \alpha)^{\lceil \frac{1}{\alpha} \rceil} > \frac{1}{8} > k^{-c}$ and therefore $t_i \ge \lceil \frac{1}{\alpha} \rceil$, so $H_i > 1 - (1 - \alpha)^{\lceil \frac{1}{\alpha} \rceil} > 1 - e^{-1} > \frac{1}{2}$.
  - Finally, since $H_i > \frac{1}{2}$ and $H_i + L_i < 1$ then $L_i < \frac{1}{2}$.
- Since $L_i < \frac{1}{2}$ then $(1 - L_i)^{\frac{1}{L_i}} > 1/4$. Thus $(1 - L_i)^{\frac{c \log_2 k}{2 L_i}} > k^{-c}$, so $t_{i+1} \ge \frac{c \log_2 k}{2 L_i}$.
- Inside the loop, we always have $\frac{1 - H_i}{L_i} < 2k$ which means that $L_i > \frac{1 - H_i}{2k} > \frac{k^{-c}}{2k} = \frac{1}{2k^{c+1}}$.

We now observe that all the t_i's are polynomially bounded: Recall that $(1 - H_0)^{\lceil c \ln(k)/H_0 \rceil} < e^{-c \ln(k)} = k^{-c}$ so we must have $t_1 < \lceil c \ln(k)/H_0 \rceil < ck^{c'} \ln(k) = O(k^{c'} \log k)$ (since we assume that $H_0 \ge k^{-c'}$). Similar argument using $L_i > \frac{1}{2k^{c+1}}$ shows that in Line 5 we have $t_{i+1} = O(k^{c+1} \log k)$.

Next, we consider the quantity $a_i \stackrel{\text{def}}{=} \frac{1 - H_i}{L_i}$. First, observe that the condition $\varepsilon_0 + \delta_0 \le 1 - \frac{\min(\varepsilon_0, \delta_0)}{\log^c(k)}$ (which we can re-write as $H_0 + L_0 \le 1 - \frac{L_0}{\log^c(k)}$) implies that $\frac{1 - H_0}{L_0} - 1 = \frac{1 - (H_0 + L_0)}{L_0} \ge \frac{1}{\log^c(k)}$. Next, observe that

$$\frac{1 - H_1}{L_1} = \frac{1 - (1 - (1 - H_0)^{t_1})}{L_0^{t_1}} = \left(\frac{1 - H_0}{L_0}\right)^{t_1} > \frac{1 - H_0}{L_0} \ge 1 + \frac{1}{\log^c(k)}.$$

We now show that in each iteration of the loop, the quantity $a_i - 1$ increases by at least a factor of $\Omega(\log k)$. Denote $b_i \stackrel{\text{def}}{=} \frac{1 - L_i}{H_i}$, and note that

$$b_i - 1 = \frac{1 - L_i}{H_i} - 1 = \frac{1 - L_i - H_i}{H_i}$$

$$= \frac{L_i}{H_i} \cdot \frac{1 - L_i - H_i}{L_i} = \frac{L_i}{H_i} \cdot \left(\frac{1 - H_i}{L_i} - 1 \right) = \frac{L_i}{H_i} \cdot (a_i - 1) .$$

Observe that for each iteration of the loop, we have

$$a_{i+1} = \frac{1 - H_{i+1}}{L_{i+1}} = \frac{1 - (1 - (1 - L_i)^{t_{i+1}}))}{H_i^{t_{i+1}}} = \left(\frac{1 - L_i}{H_i} \right)^{t_{i+1}} = b_i^{t_{i+1}}$$

and therefore

$$a_{i+1} - 1 = b_i^{t_{i+1}} - 1 = (1 + (b_i - 1))^{t_{i+1}} - 1$$

$$> [1 + t_{i+1}(b_i - 1)] - 1 = t_{i+1}(b_i - 1) = t_{i+1} \frac{L_i}{H_i} \cdot (a_i - 1) > t_{i+1} \cdot L_i \cdot (a_i - 1$$

$$\geq \frac{c \log_2 k}{2 L_i} \cdot L_i \cdot (a_i - 1) = \frac{c}{2} \log_2 k \cdot (a_i - 1) = \Omega(\log k) \cdot (a_i - 1) .$$

We have seen that $a_1 - 1 > \Omega(\frac{1}{\log^c(k)})$ and that $a_{i+1} - 1 \geq \Omega(\log k) \cdot (a_i - 1)$, so after at most $c + 1$ iterations of the loop we get $a_i - 1 \geq \Omega(\log k) > 4$.

If we still do not have $a_i > 2k$ then we will do another iteration of the loop. In this iteration, we have (as usual) $t_{i+1} \geq \frac{c \log_2 k}{2 L_i}$, but now $a_i = \frac{1 - H_i}{L_i} > 5$, so $t_{i+1} \geq \frac{5c \log_2 k}{2(1 - H_i)}$. Therefore, at the end of this iteration we have

$$L_{i+1} = H_i^{t_{i+1}} = (1 - (1 - H_i))^{t_{i+1}} \leq (1 - (1 - H_i))^{\frac{5c \log_2 k}{2(1 - H_i)}}$$
$$< e^{-5c \log_2 k / 2} = e^{-5c \ln(k)/2 \ln(2)} = k^{-5c/2 \ln(2)} < k^{-3c} .$$

On the other hand, we have (as usual) $H_{i+1} \leq 1 - k^{-c}$, and therefore $a_{i+1} = \frac{1 - H_{i+1}}{L_{i+1}} > \frac{k^{-c}}{k^{-3c}} = k^{2c} > 2k$.

We conclude that the loop terminates after at most $c + 2$ iterations, so the chain is indeed of constant length. It is left to show that the chain remains confined in the last two steps, and that L_{i+2}, H_{i+2} are both negligible. Once the loop terminates, we have

$$L_{i+1} = H_i^{\left\lfloor \frac{k}{1 - H_i} \right\rfloor} < (1 - (1 - H_i))^{\frac{k}{1 - H_i} - 1} < e^{-k}/H_i < 2e^{-k} .$$

On the other hand, $\frac{1 - H_i}{L_i} > 2k$ so $L_i < \frac{1 - H_i}{2k}$ and therefore

$$H_{i+1} = 1 - (1 - L_i)^{t_{i+1}} < t_{i+1} L_i < \left\lfloor \frac{k}{1 - H_i} \right\rfloor \cdot \frac{1 - H_i}{2k} \leq \frac{1}{2} .$$

Finally, after the last step we have

$$H_{i+2} = 1 - (1 - L_{i+1})^k < k L_{i+1} < 2k e^{-k}, \quad \text{and} \quad L_{i+2} = (H_{i+1})^k < 2^{-k} .$$

This concludes the proof of the *if* direction.   □

Only if ($\Leftarrow$). Assume that $\varepsilon_0 + \delta_0 \leq 1 - poly(k)$, but $\varepsilon_0 + \delta_0 \geq 1 - o\left(\frac{\min(\varepsilon_0, \delta_0)}{polylog(k)}\right)$, and assume that $\varepsilon_0 \geq \delta_0$ (the other case is symmetric). Let $C = \langle(\varepsilon_0, \delta_0), (t_1, t_2, \ldots)\rangle$ be a confined chain with constant length.

Instead of analyzing the chain C, it will be more convenient below to analyze an "equivalent chain" C' for which $\delta_i \leq \varepsilon_i$ for all i. We get C' from C as follows: we go over the transformations one at a time, starting from the first transformation, and maintain the invariant that we always have $\delta_i \leq \varepsilon_i$. If after the next transformation we still have $\delta_{i+1} \leq \varepsilon_{i+1}$ then we leave that transformation unchanged. On the other hand, if after the next transformation (of type R with parameter t_i) we have $\delta_{i+1} \geq \varepsilon_{i+1}$ then we break it into two transformation: a type R transformation with parameter t'_i that increases δ and decreases ϵ until they are exactly equal (t'_i could be fractional), and a type S transformation with parameter $t''_i = t_i/t'_i$. In some more detail, instead of computing $\varepsilon_{i+1} = \varepsilon_i^{t_i}$ and $\delta_{i+1} = 1 - (1 - \delta_i)^{t_i}$, we do the following:

- We compute the real number $t'_i < t_i$ such that $\varepsilon_i^{t'_i} = 1 - (1 - \delta_i)^{t'_i}$,
- We set $\varepsilon'_{i+1} = \varepsilon_i^{t'_i}$ and $\delta'_{i+1} = \varepsilon'_{i+1} = 1 - (1 - \delta_i)^{t'_i}$,
- We compute $t''_i = t_i/t'_i$ and then set $\varepsilon''_{i+1} = 1 - (1 - (\varepsilon'_{i+1}))^{t''_i}$ and $\delta''_{i+1} = (\delta'_{i+1})^{t''_i}$.
- We invert the type of all the transformations until the end of the chain.

Formally, what we do is to remove t_i from the chain and replace it with t'_i, t''_i (so we get a chain which is one longer than the original one).

It is clear that the change from above only switches the roles of ε and δ (i.e., we have $\varepsilon''_{i+1} = \delta_{i+1}$ and $\delta''_{i+1} = \varepsilon_{i+1}$, and similarly for $i+2, i+3, \ldots$). It should also be noted that the resulting chain does not correspond to transformations that can be applied to the commitment scheme (since we use fractional values for the t_i's), but all the values of ε_i, δ_i are still well defined, and their sum is equal to what it was in C. Finally, the length of C' is at most twice the length of the original C, so C' still has constant length.

From now on, we therefore assume that we have a constant-length confined chain C' that starts from $\delta_0 \leq \varepsilon_0$ and maintains $\delta_i \leq \varepsilon_i$, for all i. Denote the number of transformations in C' by ℓ and assume, without loss of generality, that ℓ is even (since we can always append a last dummy transformation with $t = 1$).

Again, we consider the quantity $a_i = \frac{1-\varepsilon_i}{\delta_i}$, and the condition $\varepsilon_0 + \delta_0 \geq 1 - o\left(\frac{\delta_0}{polylog(k)}\right)$ implies that $a_0 - 1 \leq o(1/polylog(k))$. We show that the quantity $a_i - 1$ grows by at most a factor of $O(\log k)$ in every two successive transformations in the chain. It follows that $a_\ell - 1 = (a_0 - 1) \cdot O(\log^{\ell/2}(k)) = o(1/polylog(k))$, which in particular means that $\varepsilon_\ell + \delta_\ell \geq 1 - o(1) > 1/2$. In more details, we prove by induction that, for every even i, we have $a_i - 1 \leq (8c \log k)^{i/2} \cdot (a_0 - 1)$, where the constant c is the one from the "confinement" property of the chain C' (namely all the ε_i's and δ_i's are bounded by $1 - k^{-c}$).

This holds for $i = 0$ by definition, and we now proceed to the induction step. Assume that for some even $i < \ell$ it holds that $1 - a_i \leq (1 - a_0) \cdot (8c \log k)^{i/2}$.

This in particular means that $\varepsilon_i + \delta_i \geq 1 - o(1)$, and therefore (since we have $\delta_i \leq \varepsilon_i$) then $\varepsilon_i \geq \frac{1}{2} - o(1)$. We now examine how the quantity $\frac{1-\varepsilon}{\delta}$ evolves over the next two steps.

- The next (odd-numbered) transformation is of type S, so we have $\delta_{i+1} = \delta_i^{t_{i+1}}$ and $\varepsilon_{i+1} = 1 - (1 - \varepsilon_i)^{t_{i+1}}$. Since $\varepsilon_i > \frac{1}{2} - o(1)$ then $1 - \varepsilon_i < 2^{-1/2}$, and since the sequence is confined then $1 - \varepsilon_{i+1} \leq k^{-c}$. Thus we have

$$2^{-c\log_2 k} = k^{-c} \leq (1 - \varepsilon_i)^{t_{i+1}} < \sqrt{1/2}^{t_{i+1}} = 2^{-t_{i+1}/2}$$

so it follows that $t_{i+1} < 2c\log k < 1/2(a_i - 1)$ (since $1/(a_i - 1) = \omega(polylog(k))$). This means that we have

$$a_{i+1} = \frac{1 - \varepsilon_{i+1}}{\delta_{i+1}} = \left(\frac{1 - \varepsilon_i}{\delta_i}\right)^{t_{i+1}} = a_i^{t_{i+1}} = (1 + (a_i - 1))^{t_{i+1}}$$

$$\overset{\text{Fact 2}}{<} 1 + 2t_{i+1}(a_i - 1) < 1 + 2c\log k \cdot (a_i - 1)$$

Thus $a_{i+1} - 1 < 2c\log k(a_i - 1) = o(1/polylog(k))$. Let us denote $b_{i+1} \overset{\text{def}}{=} \frac{1 - \delta_{i+1}}{\varepsilon_{i+1}}$, so we have $b_{i+1} - 1 = (a_{i+1} - 1)\delta_{i+1}/\varepsilon_{i+1}$.

- The next (even-numbered) transformation is of type R, so we have $\delta_{i+2} = 1 - (1 - \delta_{i+1})^{t_{i+2}}$ and $\varepsilon_{i+2} = (\varepsilon_{i+1})^{t_{i+2}}$. Recall that we have $\delta_{i+2} \leq \varepsilon_{i+2}$ and therefore $\delta_{i+2} < 1/2 < 1 - e^{-1}$, so $(1 - \delta_{i+1})^{t_{i+2}} = 1 - \delta_{i+2} > e^{-1}$, which means that $t_{i+2} < 1/\delta_{i+1}$. Recall also that we have $\varepsilon_{i+1} \geq \varepsilon_i \geq 1/2 - o(1)$, and therefore $b_{i+1} - 1 = \frac{(a_{i+1}-1)\delta_{i+1}}{\varepsilon_{i+1}} = \frac{o(1)}{\Theta(1)} \cdot \delta_{i+1} < \delta_{i+1}/2$, so $t_{i+2} < 1/\delta_{i+1} < 1/2(b_{i+1} - 1)$. Thus we have

$$b_{i+2} = (b_{i+1})^{t_{i+2}} = (1 + (b_{i+1} - 1))^{t_{i+2}} \overset{\text{Fact 2}}{<} 1 + 2t_{i+2}(b_{i+1} - 1)$$

Hence

$$b_{i+2} - 1 < 2t_{i+2}(b_{i+1} - 1) = \frac{2t_{i+2}(a_{i+1} - 1)\delta_{i+1}}{\varepsilon_{i+1}}$$

$$< \frac{2t_{i+2} \cdot 2c\log k(a_i - 1)\,\delta_{i+1}}{\varepsilon_{i+1}} = \frac{4c\log k\,\delta_{i+1}\,t_{i+2}}{\varepsilon_{i+1}} \cdot (a_i - 1)$$

In addition, since $\delta_{i+1} < 1/2$ and $1 \leq t_{i+2} < 1/\delta_{i+1}$ then from Fact 3 above we get that

$$\delta_{i+2} = 1 - (1 - \delta_{i+1})^{t_{i+2}} > \delta_{i+1}t_{i+2}/2$$

and we also know that $\varepsilon_{i+2} \leq \varepsilon_{i+1}$. Thus, we have

$$a_{i+2} - 1 = \frac{(b_{i+2} - 1)\varepsilon_{i+2}}{\delta_{i+2}} < \frac{(b_{i+2} - 1)\varepsilon_{i+2}}{\delta_{i+1}t_{i+2}/2}$$

$$< \frac{4\,\delta_{i+1}\,t_{i+2}\,c\log k}{\varepsilon_{i+1}} \cdot (a_i - 1) \cdot \frac{2\varepsilon_{i+2}}{\delta_{i+1}t_{i+2}} = 8c\log k(a_i - 1) \cdot \frac{\varepsilon_{i+2}}{\varepsilon_{i+1}}$$

$$\leq 8c\log k \cdot (a_i - 1) < (8c\log k)^{(i+2)/2} \cdot (a_0 - 1) = o\left(\frac{1}{polylog(k)}\right)$$

This concludes the proof of the *only if* direction.

References

1. Canetti, R., Halevi, S., Steiner, M.: Hardness amplification of weakly verifiable puzzles. In: Kilian, J. (ed.) TCC 2005. LNCS, vol. 3378, pp. 17–33. Springer, Heidelberg (2005)
2. Damgård, I., Kilian, J., Salvail, L.: On the (Im)possibility of Basing Oblivious Transfer and Bit Commitment on Weakened Security Assumptions. In: Stern, J. (ed.) EUROCRYPT 1999. LNCS, vol. 1592, pp. 56–73. Springer, Heidelberg (1999)
3. Even, S., Goldreich, O., Lempel, A.: A Randomized Protocol for Signing Contracts. Communications of the ACM 28(6), 637–647 (1985)
4. Goldreich, O., Nisan, N., Wigderson, A.: On Yao's xor-lemma. Electronic Colloquium on Computational Complexity (ECCC) 2(50) (1995)
5. Holenstein, T.: Key agreement from weak bit agreement. In: STOC 2005, pp. 664–673. ACM Press, New York (2005)
6. Holenstein, T., Renner, R.: One-Way Secret-Key Agreement and Applications to Circuit Polarization and Immunization of Public-Key Encryption. In: Shoup, V. (ed.) CRYPTO 2005. LNCS, vol. 3621, pp. 478–493. Springer, Heidelberg (2005)
7. Impagliazzo, R., Luby, M.: One-way functions are essential for complexity based cryptography. In: 30th Annual Symposium on Foundations of Computer Science – FOCS 1989, pp. 230–235. IEEE Computer Society Press, Los Alamitos (1989)
8. Kilian, J.: Founding Cryptography on Oblivious Transfer. In: STOC 1988, pp. 30–31. ACM Press, New York (1988)
9. Levin, L.A.: One-way functions and pseudorandom generators. Combinatorica 7(4), 357–363 (1987)
10. Rabin, M.O.: How to exchange secrets by oblivious transfer. Technical Report TR-81, Harvard (1981)
11. Wullschleger, J.: Oblivious-Transfer Amplification. In: Naor, M. (ed.) EUROCRYPT 2007. LNCS, vol. 4515, pp. 555–572. Springer, Heidelberg (2007)
12. Yao, A.C.: Theory and applications of trapdoor functions. In: 23rd Annual Symposium on Foundations of Computer Science, November 1982, pp. 80–91. IEEE Computer Society Press, Los Alamitos (1982)

Author Index

Lecture Notes in Computer Science

Sublibrary 4: Security and Cryptology

For information about Vols. 1– 2951
please contact your bookseller or Springer

Vol. 4377: M. Abe (Ed.), Topics in Cryptology – CT-RSA 2007. XI, 403 pages. 2006.

Vol. 4356: E. Biham, A.M. Youssef (Eds.), Selected Areas in Cryptography. XI, 395 pages. 2007.

Vol. 4341: P.Q. Nguyên (Ed.), Progress in Cryptology - VIETCRYPT 2006. XI, 385 pages. 2006.

Vol. 4332: A. Bagchi, V. Atluri (Eds.), Information Systems Security. XV, 382 pages. 2006.

Vol. 4329: R. Barua, T. Lange (Eds.), Progress in Cryptology - INDOCRYPT 2006. X, 454 pages. 2006.

Vol. 4318: H. Lipmaa, M. Yung, D. Lin (Eds.), Information Security and Cryptology. XI, 305 pages. 2006.

Vol. 4307: P. Ning, S. Qing, N. Li (Eds.), Information and Communications Security. XIV, 558 pages. 2006.

Vol. 4301: D. Pointcheval, Y. Mu, K. Chen (Eds.), Cryptology and Network Security. XIII, 381 pages. 2006.

Vol. 4300: Y.Q. Shi (Ed.), Transactions on Data Hiding and Multimedia Security I. IX, 139 pages. 2006.

Vol. 4298: J.K. Lee, O. Yi, M. Yung (Eds.), Information Security Applications. XIV, 406 pages. 2007.

Vol. 4296: M.S. Rhee, B. Lee (Eds.), Information Security and Cryptology – ICISC 2006. XIII, 358 pages. 2006.

Vol. 4284: X. Lai, K. Chen (Eds.), Advances in Cryptology – ASIACRYPT 2006. XIV, 468 pages. 2006.

Vol. 4283: Y.Q. Shi, B. Jeon (Eds.), Digital Watermarking. XII, 474 pages. 2006.

Vol. 4266: H. Yoshiura, K. Sakurai, K. Rannenberg, Y. Murayama, S.-i. Kawamura (Eds.), Advances in Information and Computer Security. XIII, 438 pages. 2006.

Vol. 4258: G. Danezis, P. Golle (Eds.), Privacy Enhancing Technologies. VIII, 431 pages. 2006.

Vol. 4249: L. Goubin, M. Matsui (Eds.), Cryptographic Hardware and Embedded Systems - CHES 2006. XII, 462 pages. 2006.

Vol. 4237: H. Leitold, E.P. Markatos (Eds.), Communications and Multimedia Security. XII, 253 pages. 2006.

Vol. 4236: L. Breveglieri, I. Koren, D. Naccache, J.-P. Seifert (Eds.), Fault Diagnosis and Tolerance in Cryptography. XIII, 253 pages. 2006.

Vol. 4219: D. Zamboni, C. Krügel (Eds.), Recent Advances in Intrusion Detection. XII, 331 pages. 2006.

Vol. 4189: D. Gollmann, J. Meier, A. Sabelfeld (Eds.), Computer Security – ESORICS 2006. XI, 548 pages. 2006.

Vol. 4176: S.K. Katsikas, J. López, M. Backes, S. Gritzalis, B. Preneel (Eds.), Information Security. XIV, 548 pages. 2006.

Vol. 4117: C. Dwork (Ed.), Advances in Cryptology - CRYPTO 2006. XIII, 621 pages. 2006.

Vol. 4116: R. De Prisco, M. Yung (Eds.), Security and Cryptography for Networks. XI, 366 pages. 2006.

Vol. 4107: G. Di Crescenzo, A. Rubin (Eds.), Financial Cryptography and Data Security. XI, 327 pages. 2006.

Vol. 4083: S. Fischer-Hübner, S. Furnell, C. Lambrinoudakis (Eds.), Trust and Privacy in Digital Business. XIII, 243 pages. 2006.

Vol. 4064: R. Büschkes, P. Laskov (Eds.), Detection of Intrusions and Malware & Vulnerability Assessment. X, 195 pages. 2006.

Vol. 4058: L.M. Batten, R. Safavi-Naini (Eds.), Information Security and Privacy. XII, 446 pages. 2006.

Vol. 4047: M.J.B. Robshaw (Ed.), Fast Software Encryption. XI, 434 pages. 2006.

Vol. 4043: A.S. Atzeni, A. Lioy (Eds.), Public Key Infrastructure. XI, 261 pages. 2006.

Vol. 4004: S. Vaudenay (Ed.), Advances in Cryptology - EUROCRYPT 2006. XIV, 613 pages. 2006.

Vol. 3995: G. Müller (Ed.), Emerging Trends in Information and Communication Security. XX, 524 pages. 2006.

Vol. 3989: J. Zhou, M. Yung, F. Bao (Eds.), Applied Cryptography and Network Security. XIV, 488 pages. 2006.

Vol. 3969: Ø. Ytrehus (Ed.), Coding and Cryptography. XI, 443 pages. 2006.

Vol. 3958: M. Yung, Y. Dodis, A. Kiayias, T.G. Malkin (Eds.), Public Key Cryptography - PKC 2006. XIV, 543 pages. 2006.

Vol. 3957: B. Christianson, B. Crispo, J.A. Malcolm, M. Roe (Eds.), Security Protocols. IX, 325 pages. 2006.

Vol. 3956: G. Barthe, B. Grégoire, M. Huisman, J.-L. Lanet (Eds.), Construction and Analysis of Safe, Secure, and Interoperable Smart Devices. IX, 175 pages. 2006.

Vol. 3935: D.H. Won, S. Kim (Eds.), Information Security and Cryptology - ICISC 2005. XIV, 458 pages. 2006.

Vol. 3934: J.A. Clark, R.F. Paige, F.A.C. Polack, P.J. Brooke (Eds.), Security in Pervasive Computing. X, 243 pages. 2006.

Vol. 3928: J. Domingo-Ferrer, J. Posegga, D. Schreckling (Eds.), Smart Card Research and Advanced Applications. XI, 359 pages. 2006.

Vol. 3919: R. Safavi-Naini, M. Yung (Eds.), Digital Rights Management. XI, 357 pages. 2006.

Vol. 3903: K. Chen, R. Deng, X. Lai, J. Zhou (Eds.), Information Security Practice and Experience. XIV, 392 pages. 2006.

Vol. 3897: B. Preneel, S. Tavares (Eds.), Selected Areas in Cryptography. XI, 371 pages. 2006.

Vol. 3876: S. Halevi, T. Rabin (Eds.), Theory of Cryptography. XI, 617 pages. 2006.

Vol. 3866: T. Dimitrakos, F. Martinelli, P.Y.A. Ryan, S. Schneider (Eds.), Formal Aspects in Security and Trust. X, 259 pages. 2006.

Vol. 3860: D. Pointcheval (Ed.), Topics in Cryptology – CT-RSA 2006. XI, 365 pages. 2006.

Vol. 3858: A. Valdes, D. Zamboni (Eds.), Recent Advances in Intrusion Detection. X, 351 pages. 2006.

Vol. 3856: G. Danezis, D. Martin (Eds.), Privacy Enhancing Technologies. VIII, 273 pages. 2006.

Vol. 3786: J.-S. Song, T. Kwon, M. Yung (Eds.), Information Security Applications. XI, 378 pages. 2006.

Vol. 3108: H. Wang, J. Pieprzyk, V. Varadharajan (Eds.), Information Security and Privacy. XII, 494 pages. 2004.